IEE ELECTROMAGNETIC WAVES SERIES 28

Series Editors: Professor P. J. B. Clarricoats
Professor Y. Rahmat-Samii
Professor J. R. Wait

Handbook of MICROSTRIP ANTENNAS

Volume 1

Other volumes in this series:

Volume 1   **Geometrical theory of diffraction for electromagnetic waves** G. L. James
Volume 2   **Electromagnetic waves and curved structures** L. Lewin, D. C. Chang and E. F. Kuester
Volume 3   **Microwave homodyne systems** R. J. King
Volume 4   **Radio direction-finding** P. J. D. Gething
Volume 5   **ELF communications antennas** M. L. Burrows
Volume 6   **Waveguide tapers, transitions and couplers** F. Sporleder and H. G. Unger
Volume 7   **Reflector antenna analysis and design** P. J. Wood
Volume 8   **Effects of the troposphere on radio communications** M. P. M. Hall
Volume 9   **Schumann resonances in the earth-ionosphere cavity** P. V. Bliokh, A. P. Nikolaenko and Y. F. Flippov
Volume 10   **Aperture antennas and diffraction theory** E. V. Jull
Volume 11   **Adaptive array principles** J. E. Hudson
Volume 12   **Microstrip antenna theory and design** J. R. James, P. S. Hall and C. Wood
Volume 13   **Energy in electromagnetism** H. G. Booker
Volume 14   **Leaky feeders and subsurface radio communications** P. Delogne
Volume 15   **The handbook of antenna design, Volume 1** A. W. Rudge, K. Milne, A. D. Olver, P. Knight (Editors)
Volume 16   **The handbook of antenna design, Volume 2** A. W. Rudge, K. Milne, A. D. Olver, P. Knight (Editors)
Volume 17   **Surveillance radar performance prediction** P. Rohan
Volume 18   **Corrugated horns for microwave antennas** P. J. B. Clarricoats and A. D. Olver
Volume 19   **Microwave antenna theory and design** S. Silver (Editor)
Volume 20   **Advances in radar techniques** J. Clarke (Editor)
Volume 21   **Waveguide handbook** N. Marcuvitz
Volume 22   **Target adaptive matched illumination radar** D. T. Gjessing
Volume 23   **Ferrites at microwave frequencies** A. J. Baden Fuller
Volume 24   **Propagation of short radio waves** D. E. Kerr (Editor)
Volume 25   **Principles of microwave circuits** C. G. Montgomery, R. H. Dicke, E. M. Purcell (Editors)
Volume 26   **Spherical near-field antenna measurements** J. E. Hansen (Editor)
Volume 27   **Electromagnetic radiation from cylindrical structures** J. R. Wait
Volume 28   **Handbook of microstrip antennas** J. R. James and P. S. Hall (Editors)
Volume 29   **Satellite-to-ground radiowave propagation** J. E. Allnutt
Volume 30   **Radiowave propagation** M. P. M. Hall and L. W. Barclay (Editors)
Volume 31   **Ionospheric radio** K. Davies

Handbook of MICROSTRIP ANTENNAS

Volume 1

Edited by
J R James & P S Hall

Peter Peregrinus Ltd. on behalf of the Institution of Electrical Engineers

Published by: Peter Peregrinus Ltd., London, United Kingdom

© 1989: Peter Peregrinus Ltd.

All rights reserved. No part of this publication may be reproduced, stored in a retrieval system or transmitted in any form or by any means—electronic, mechanical, photocopying, recording or otherwise—without the prior written permission of the publisher.

While the authors and the publishers believe that the information and guidance given in this work are correct, all parties must rely upon their own skill and judgment when making use of them. Neither the authors nor the publishers assume any liability to anyone for any loss or damage caused by any error or omission in the work, whether such error or omission is the result of negligence or any other cause. Any and all such liability is disclaimed.

British Library Cataloguing in Publication Data

Handbook of Microstrip Antennas
   1. Microwave equipment: Microstrip antennas
  I. James, J. R. (James Roderick, *1933*-
  II. Hall, P. S. (Peter S) III. Institution of Electrical
Engineers IV. Series
621.381'33

ISBN 0 86341 150 9

Printed in England by Short Run Press Ltd., Exeter

Contents

Volume 1

	Foreword		xvii
	Preface		xix
	List of contributors		xxi
1	Introduction — J.R. James and P.S. Hall		1
	1.1	Historical development and future prospects	1
	1.2	Fundamental issues and design challenges	3
		1.2.1 Features of microstrip antenna technology	4
		1.2.2 Fundamental problems	7
	1.3	The handbook and advances presented	17
	1.4	Glossary of printed antenna types	24
	1.5	Summary comments	40
	1.6	References	40
2	Analysis of circular microstrip antennas — L. Shafai and A.A. Kishk		45
	2.1	Introduction	45
	2.2	Formulation of the problem	47
		2.2.1 Matrix formulation	50
		2.2.2 Excitation matrix	58
		2.2.3 Radiation fields	59
	2.3	Application 1: Circular patch antenna	63
		2.3.1 Surface fields	67
		2.3.2 Feed location	67
		2.3.3 Effect of the substrate permittivity	71
		2.3.4 Effect of the substrate thickness	74
		2.3.5 Effect of the ground-plane radius	76
		2.3.6 Effect of the ground-plane thickness	81
		2.3.7 Circular polarisation	83
		2.3.8 Effect of a central shorting pin	85
	2.4	Application 2: Wraparound microstrip antenna	85
	2.5	Application 3: Reflector antenna feeds	96
	2.6	Concluding remarks	107
	2.7	References	108

3 Characteristics of microstrip patch antennas and some methods of improving frequency agility and bandwidth — K.F. Lee and J.S. Dahele — 111

- 3.1 Introduction — 111
- 3.2 Cavity model for analysing microstrip patch antennas — 112
  - 3.2.1 Introduction — 112
  - 3.2.2 Feed modelling, resonant frequencies and internal fields — 113
  - 3.2.3 Radiation field — 115
  - 3.2.4 Losses in the cavity — 116
  - 3.2.5 Input impedance — 118
  - 3.2.6 VSWR bandwidth — 118
  - 3.2.7 Qualitative description of the results predicted by the model — 119
- 3.3 Basic characteristics of some common patches — 120
  - 3.3.1 The rectangular patch — 120
  - 3.3.2 The circular patch — 135
  - 3.3.3 The equitriangular patch — 149
  - 3.3.4 Annular-ring patch — 169
  - 3.3.5 Comparison of characteristics of rectangular, circular, equitriangular and annular-ring patches — 178
  - 3.3.6 Brief mention of other patches — 182
- 3.4 Some methods of improving the frequency agility and bandwidth of microstrip patch antennas — 187
  - 3.4.1 Introduction — 187
  - 3.4.2 Some methods of tuning MPAs — 189
  - 3.4.3 Dual-band structures — 197
  - 3.4.4 Electromagnetic-coupled patch antenna (EMCP) — 207
- 3.5 Summary — 214
- 3.6 Acknowledgments — 214
- 3.7 References — 214

4 Circular polarisation and bandwidth — M. Haneishi and Y. Suzuki — 219

- 4.1 Various types of circularly polarised antenna — 219
  - 4.1.1 Microstrip patch antennas — 220
  - 4.1.2 Other types of circularly polarised printed antennas — 222
- 4.2 Simple design techniques for singly-fed circularly polarised microstrip antennas — 224
  - 4.2.1 Rectangular type — 224
  - 4.2.2 Circular type — 232
- 4.3 More exact treatment for singly-fed circularly polarised microstrip antennas — 235
  - 4.3.1 Analysis — 236
  - 4.3.2 Conditions for circularly polarised radiation — 241
  - 4.3.3 Example — 244
- 4.4 Some considerations on mutual coupling — 249
- 4.5 Wideband techniques — 253
  - 4.5.1 Design of wideband element — 253
  - 4.5.2 Technique using parasitic element — 264
  - 4.5.3 Technique using paired element — 270
- 4.6 References — 272

5	**Microstrip dipoles — P.B. Katehi, D.R. Jackson and N.G. Alexopoulis**		**275**
	5.1	Introduction	275
	5.2	Infinitesimal dipole	276
		5.2.1 Analysis	276
		5.2.2 Substrate effects	279
		5.2.3 Superstrate effects	281
	5.3	Moment-method techniques for planar strip geometries	282
		5.3.1 Basis functions	282
		5.3.2 Reaction between basis functions	284
		5.3.3 Plane-wave-spectrum method	285
		5.3.4 Real-space integration method	286
		5.3.5 Point-dipole approximation	287
		5.3.6 Moment-method equations	287
	5.4	Centre-fed dipoles	287
		5.4.1 Single dipole	287
		5.4.2 Mutual impedance	291
	5.5	EMC dipoles	295
		5.5.1 Methods of analysis	295
		5.5.2 Single dipole	295
		5.5.3 Multiple dipoles	299
	5.6	Finite array of EMC dipoles	301
		5.6.1 Analysis	301
		5.6.2 Calculation of coefficients	302
		5.6.3 Array design	305
	5.7	Conclusions	308
	5.8	References	309
6	**Multilayer and parasitic configurations — D.H. Schaubert**		**311**
	6.1	Introduction	311
	6.2	Stacked elements for dual-frequency or dual polarisation operation	312
		6.2.1 Antennas with separate feeds for each function	313
		6.2.2 Antennas for multiple frequencies and increased bandwidth	320
	6.3	Two-sided aperture-coupled patch	330
	6.4	Parasitic elements on antenna substrate	337
	6.5	Summary	350
	6.6	References	350
7	**Wideband flat dipole and short-circuit microstrip patch elements and arrays — G. Dubost**		**353**
	7.1	Flat dipole elements and arrays	353
		7.1.1 Elementary sources	353
		7.1.2 Array designs: losses and efficiencies	367
	7.2	Short-circuit microstrip patches and arrays	374
		7.2.1 Elementary source	374
		7.2.2 Array designs	378
	7.3	References	391

8 Numerical analysis of microstrip patch antennas — J.R. Mosig, R.C. Hall and F.E. Gardiol — 393

- 8.1 Introduction — 393
  - 8.1.1 General description — 393
  - 8.1.2 The integral equation model — 394
- 8.2 Model based on the electric surface current — 395
  - 8.2.1 Geometry of the model and boundary conditions — 395
  - 8.2.2 Potentials for the diffracted fields — 397
  - 8.2.3 Green's functions — 398
  - 8.2.4 Mixed potential integral equation (MPIE) — 400
  - 8.2.5 Sketch of the proposed technique — 401
- 8.3 Horizontal electric dipole (HED) in microstrip — 403
  - 8.3.1 The vector potential — 403
  - 8.3.2 Scalar potential and the fields — 405
  - 8.3.3 Surface waves and spectral plane k — 406
  - 8.3.4 Far-field approximations — 408
  - 8.3.5 Radiation resistance and antenna efficiency — 413
- 8.4 Numerical techniques for Sommerfeld integrals — 417
  - 8.4.1 Numerical integration on the real axis — 417
  - 8.4.2 Integrating oscillating functions over unbounded intervals — 420
- 8.5 Construction of the Green's functions — 421
- 8.6 Method of moments — 423
  - 8.6.1 Rooftop (subsectional) — basis functions — 423
  - 8.6.2 Entire domain basis functions — 429
- 8.7 Excitation and loading — 431
  - 8.7.1 Several microstrip-antenna excitations — 431
  - 8.7.2 Coaxial excitation and input impedance — 432
  - 8.7.3 Multiport analysis — 434
- 8.8 Single rectangular patch antenna — 436
  - 8.8.1 Entire-domain versus subdomain basis functions — 437
  - 8.8.2 Convergence using subsectional basis functions — 440
  - 8.8.3 Surface currents — 441
- 8.9 Microstrip arrays — 443
  - 8.9.1 Array modelling — 444
  - 8.9.2 Mutual coupling — 445
  - 8.9.3 Linear array of few patches — 449
- 8.10 Acknowledgments — 452
- 8.11 References — 452

9 Multiport network approach for modelling and analysis of microstrip patch antennas and arrays — K.C. Gupta — 455

- 9.1 Introduction — 455
- 9.2 Models for microstrip antennas — 456
  - 9.2.1 Transmission-line model — 456
  - 9.2.2 Cavity model — 458
  - 9.2.3 Multiport network model — 462
- 9.3 Z-matrix characterisation of planar segments — 467
  - 9.3.1 Green's functions — 467
  - 9.3.2 Evaluation of Z-matrix from Green's functions — 468
  - 9.3.3 Z-matrices for segments of arbitrary shape — 472

			Contents ix

9.4 Edge-admittance and mutual-coupling networks 475
    9.4.1 Edge-admittance networks 475
    9.4.2 Mutual-coupling network 482
9.5 Analysis of multiport-network model 488
    9.5.1 Segmentation method 488
    9.5.2 Desegmentation method 494
9.6 Examples of microstrip antenna structures analysed by multiport-network approach 499
    9.6.1 Circularly polarised microstrip patches 499
    9.6.2 Broadband multiresonator microstrip antennas 507
    9.6.3 Multiport microstrip patches and series-fed arrays 509
9.7 CAD of microstrip patch antennas and arrays 517
9.8 Appendix: Green's functions for various planar configurations 519
9.9 Acknowledgments 522
9.10 References 522

10 Transmission-line model for rectangular microstrip antennas — A. Van de Capelle   527

10.1 Introduction 529
10.2 Simple transmission-line model 529
    10.2.1 Description of the transmission line model 529
    10.2.2 Expressions for G_s and B_s 533
    10.2.3 Expressions for the line parameters 537
10.3 Improved transmission-line model 538
    10.3.1 Description of the improved transmission-line model 538
    10.3.2 Expression for the self-susceptance B_s 541
    10.3.3 Expression for the self-conductance G_s 541
    10.3.4 Expression for the mutual conductance G_m 544
    10.3.5 Expression for the mutual susceptance B_m 548
    10.3.6 Expressions for the line parameters 551
10.4 Application of the improved transmission-line model 553
    10.4.1 Analysis and design of rectangular microstrip antennas 553
    10.4.2 Comparison with other methods 555
    10.4.3 Comparison with experimental results 556
    10.4.4 Design application 557
10.5 Transmission-line model for mutual coupling 561
    10.5.1 Description of the model 561
    10.5.2 Calculation of the model parameters 570
    10.5.3 Comparison with other methods 573
10.6 Acknowledgements 577
10.7 References 577

11 Design and technology of low-cost printed antennas — J.P. Daniel, E. Penard and C. Terret   579

11.1 Introduction 579
11.2 Analysis of simple patches and slots 580
    11.2.1 Rectangular and circular patches 580
    11.2.2 Conical antennas 601
    11.2.3 Linear and annular slots 606
11.3 Design of planar printed arrays 622
    11.3.1 Design parameters 622

		11.3.2	Cavity model analysis of mutual coupling	631
		11.3.3	Linear series array of corner-fed square patches	643
		113.4	Two-dimensional cross-fed arrays	655
	11.4	Synthesis methods for linear arrays		662
		11.4.1	Relaxation methods	663
		11.4.2	Simplex method	667
		11.4.3	Experimental results	673
	11.5	New low-cost low-loss substrate		674
		11.5.1	Substrate choice	674
		11.5.2	Fabrication procedure	678
		11.5.3	Electrical characteristics	679
		11.5.4	Environmental tests	679
		11.5.5	Examples of printed antennas on polypropylene substrate	682
	11.6	Concluding remarks		685
	11.7	References		688
12	Analysis and design considerations for printed phased-array antennas — D.M. Pozar			693
	12.1	Introduction		693
	12.2	Analysis of some canonical printed phased-array geometries		694
		12.2.1	Some preliminaries	695
		12.2.2	Infinite-planar-array solutions	706
		12.2.3	Finite-array solutions	731
	12.3	Design considerations for printed phased arrays		741
		12.3.1	Introduction	741
		12.3.2	Array architectures	745
	12.4	Conclusion		751
	12.5	Acknowledgments		751
	12.6	References		751
13	Circularly polarised antenna arrays — K. Ito, T. Teshirogi and S. Nishimura			755
	13.1	Various types of circularly polarised arrays		755
		13.1.1	Arrays of patch radiators	755
		13.1.2	Arrays of composite elements	759
		13.1.3	Travelling-wave arrays	762
		13.1.4	Other types of arrays	765
	13.2	Design of circularly polarised arrays		767
		13.2.1	Arrays of patch radiators	767
		13.2.2	Arrays of composite elements	770
		13.2.3	Design of travelling-wave arrays	777
	13.3	Practical design problems		782
		13.3.1	Mutual coupling	782
		13.3.2	Unwanted radiation	787
		13.3.3	Limitations and trade-offs	788
		13.3.4	Non-planar scanning arrays	792
	13.4	Wideband circularly polarised arrays		796
		13.4.1	Arrays of wideband elements	796
		13.4.2	Arrays of dual-frequency stacked elements	802
		13.4.3	Wideband-array techniques	804
	13.5	References		810

Volume 2

14		**Microstrip antenna feeds — R.P. Owens**	**815**
	14.1	Introduction	815
	14.2	Coupling to microstrip patches	817
		14.2.1 Co-planar coupling to a single patch	817
		14.2.2 Series-array co-planar coupling	818
		14.2.3 Probe coupling	822
		14.2.4 Aperture coupling	823
		14.2.5 Electromagnetic coupling	824
	14.3	Parallel and series feed systems	825
		14.3.1 Parallel feeds for one and two dimensions	825
		14.3.2 Series feed for one dimension	832
		14.3.3 Combined feeds	839
		14.3.4 Discontinuity arrays	843
	14.4	Direct-coupled stripline power dividers and combiners	850
		14.4.1 Simple three-port power dividers	850
		14.4.2 Isolated power dividers/combiners	852
		14.4.3 Four-port direct-coupled power dividers	854
	14.5	Other feed systems	857
		14.5.1 Alternative transmission lines	857
		14.5.2 Multiple beam-forming networks	859
	14.6	Acknowledgments	860
	14.7	References	866
15		**Advances in substrate technology — G.R. Traut**	**871**
	15.1	Considerations for substrate selection	871
		15.1.1 Impact of properties of various substrate systems on microstrip antenna performance	871
		15.1.2 Comparative list of available substrates	878
		15.1.3 Selection of metal cladding for performance	879
		15.1.4 Thermal characteristics of PTFE	880
		15.1.5 Anisotropy of relative permittivity	881
	15.2	Measurement of substrate properties	884
		15.2.1 Stripline-resonator test method	886
		15.2.2 Microstrip-resonator test method	893
		15.2.3 Full-sheet-resonance test method	897
		15.2.4 Perturbation cavity method	906
		15.2.5 Tabulated evaluation of methods for measuring relative permittivity and dissipation factor	914
	15.3	Processing laminates into antennas	916
		15.3.1 Handling incoming copper-clad laminates	917
		15.3.2 Handling prior to processing	918
		15.3.3 Safety considerations for PTFE-based substrates	919
		15.3.4 Reducing the effects of etch strain relief	919
		15.3.5 Machining of PTFE-based boards	922
		15.3.6 Bending etched antenna boards	924
		15.3.7 Bonded-board assemblies	926
		15.3.8 Plating-through holes in microstrip antenna boards	934

		15.3.9	Device attachment on microstrip antenna substrates	936

15.4 Design considerations with selected materials — 939
- 15.4.1 Environmental effects on antenna substrates — 939
- 15.4.2 Conductor losses at millimetre-wave frequencies — 942
- 15.4.3 Multilayer circuit-board technology in microstrip antennas — 944

15.5 Special features and new materials developments — 945
- 15.5.1 Substrates clad on one side with thick metal — 946
- 15.5.2 Low thermal coefficient of K' in fluoropolymer laminates — 947
- 15.5.3 Microwave laminates with a resistive layer — 947
- 15.5.4 Thermoset microwave materials — 951
- 15.5.5 Low permittivity ceramic—PTFE laminates — 953
- 15.5.6 Very-low-dielectric-constant substrates — 953

15.6 References — 954

16 Special measurement techniques for printed antennas — E. Levine — 957

- 16.1 Introduction — 957
- 16.2 Substrate properties — 958
- 16.3 Connector characterisation — 962
- 16.4 Measurements of printed lines and networks — 970
  - 16.4.1 Measurement of printed-line parameters — 971
  - 16.4.2 Measurement of printed networks — 976
- 16.5 Near-field probing — 981
- 16.6 Efficiency measurement — 991
- 16.7 Concluding remarks — 994
- 16.8 References — 995

17 Computer-aided design of microstrip and triplate circuits — J.F. Zürcher and F.E. Gardiol — 1001

- 17.1 Introduction, definition of the structure — 1001
  - 17.1.1 Outline — 1001
  - 17.1.2 Microwaves — 1001
  - 17.1.3 Transmission lines for microwaves — 1001
  - 17.1.4 Balanced stripline or triplate — 1003
  - 17.1.5 Microstrip — 1004
  - 17.1.6 Adjustments — 1005
  - 17.1.7 Multiple inhomogeneity — 1005
  - 17.1.8 Measurement problems — 1006
- 17.2 Basic relationships for uniform lines — 1006
  - 17.2.1 Uniform lines — 1006
  - 17.2.2 Conformal mapping — 1008
  - 17.2.3 Schwartz–Christoffel transform — 1011
  - 17.2.4 Zero-thickness balanced stripline — 1011
  - 17.2.5 Finite-thickness balanced stripline — 1011
  - 17.2.6 Equivalent homogeneous microstrip line — 1012
  - 17.2.7 Characteristic impedance of microstrip — 1013
  - 17.2.8 Finite-thickness homogeneous microstrip — 1014
  - 17.2.9 Microstrip-line synthesis for $b = 0$ — 1015
  - 17.2.10 Dispersion in microstrip — 1015
  - 17.2.11 Effect of an enclosure — 1015

	17.2.12	Attenuation	1016
	17.2.13	Higher-order modes and radiation	1017
17.3	Discontinuities: bends and junctions		1017
	17.3.1	Definition	1017
	17.3.2	Models	1018
	17.3.3	TEM-line models	1020
	17.3.4	Variational techniques	1020
	17.3.5	Fourier transform	1020
	17.3.6	Dielectric Green's function	1020
	17.3.7	Integral equations for inductances	1020
	17.3.8	Green's function and integral equation	1023
	17.3.9	Green's function and electrostatic-inductance computation	1023
	17.3.10	TLM (transmission-line-matrix) method	1023
	17.3.11	Waveguide model	1023
17.4	Technological realisation: Materials and manufacturing process		1023
	17.4.1	Introduction	1023
	17.4.2	Dielectric substrate	1024
	17.4.3	Comment	1024
	17.4.4	Inorganic substrates	1025
	17.4.5	Plastic substrates	1025
	17.4.6	Semiconductor substrates	1027
	17.4.7	Ferrimagnetic substrates	1027
	17.4.8	Metallisation	1027
	17.4.9	Circuit realisation	1028
	17.4.10	Etching	1028
	17.4.11	Metal deposition	1028
	17.4.12	Removal of photoresist	1029
	17.4.13	Under-etching	1030
	17.4.14	Thin and thick film	1031
17.5	Analysis and synthesis programs		1031
	17.5.1	Introduction	1031
	17.5.2	EEsof: Touchstone	1032
	17.5.3	CCC: The Supercompact Family	1033
	17.5.4	CCC: CADEC +	1033
	17.5.5	Acline	1033
	17.5.6	Thom '6: Esope	1033
	17.5.7	RCA: Midas	1034
	17.5.8	LINMIC	1034
	17.5.9	High Tech. Tournesol: Micpatch	1035
	17.5.10	Spefco Software: CiAO	1035
	17.5.11	Made-it-associates: Mama	1036
	17.5.12	Ampsa: Multimatch	1036
	17.5.13	Radar systems technology: Analop	1036
	17.5.14	Microkop/Suspend	1036
	17.5.15	Microwave software applications	1036
	17.5.16	Planim	1036
	17.5.17	DGS Associates: S/Filsyn	1037
	17.5.8	Webb Laboratories: Transcad	1037
17.6	Layouts of circuits and cutting of masks		1037
	17.6.1	Description	1037
	17.6.2	CCC: Autoart	1037
	17.6.3	EFSOF: Micad	1037

xiv Contents

		17.6.4	High Tech. Tournesol: Micros	1038
		17.6.5	British Telecom: Temcad	1039
	17.7	Insertion of components		1039
		17.7.1	Introduction	1039
		17.7.2	Discrete components	1039
		17.7.3	Mounting procedure	1042
		17.7.4	Drilling holes in the dielectric substrate	1045
		17.7.5	Deposited components	1046
	17.8	Examples		1047
		17.8.1	Design of a broadband amplifier	1047
		17.8.2	Bandpass filter design	1048
		17.8.3	Design of a miniature Doppler radar	1049
	17.9	Conclusions		1051
	17.10	Acknowledgments		1053
	17.11	References		1053

18 Resonant microstrip antenna elements and arrays for aerospace applications — A.G. Derneryd   **1057**

     18.1 Introduction   1057
     18 2 Circular antenna element   1058
     18.3 Dual-band circularly polarised antenna element   1061
     18.4 Monopulse-array antenna   1068
     18.5 Dual-polarised-array antenna   1073
     18.6 Concluding remarks   1077
     18.7 References   1078

19 Applications in mobile and satellite systems — K. Fujimoto, T. Hori, S. Nishimura and K. Hirasawa   **1079**

	19.1	Introduction		1079
	19.2	Mobile systems		1080
		19.2.1	Design considerations	1080
		19.2.2	Base stations	1085
		19.2.3	Wheeled vehicles	1087
		19.2.4	Railways	1092
		19.2.5	Pedestrian	1101
		19.2.6	Radars	1105
	19.3	Satellite system		1112
		19.3.1	Design considerations	1112
		19.3.2	Direct broadcasting reception	1113
		19.3.3	Earth stations	1124
		19.3.4	Satellite borne	1146
	19.4	References		1149

20 Conical conformal microstrip tracking antenna — P. Newham and G. Morris   **1153**

	20.1	Introduction		1153
	20.2	Single patch element		1153
		20.2.1	Choice of array element	1153
		20.2.2	Choice of substrate	1154

	20.2.3	Feeding the patch	1155
	20.2.4	Theoretical design method	1155
	20.2.5	Patch design	1158
20.3	Dual patch element		1161
	20.3.1	Choice of design	1161
	20.3.2	Location of patch phase centre	1161
	20.3.2	Design and optimisation	1162
20.4	Hybrid feeding network		1163
	20.4.1	Overview	1163
	20.4.2	Hybrid designs	1166
	20.4.3	90° bends	1168
	20.4.4	Minimum track distance	1168
	20.4.5	Feed-point terminations	1171
	20.4.6	Track lengths	1171
	20.4.7	Overall design	1172
20.5	Conical antenna array		1172
20.6	Substrate fabrication		1175
	20.6.1	Overview	1175
	20.6.2	Mask drawing and preparation	1175
	20.6.3	Etching	1176
	20.6.4	Substrate preparation	1176
	20.6.5	Triplate bonding	1177
20.7	Forming the antenna		1177
	20.7.1	Bending the substrates	1177
	20.7.2	Attachment of components	1178
	20.7.3	Final assembly	1181
20.8	Antenna performance		1181
	20.8.1	Grating-lobe suppression	1182
	20.8.2	Axial ratio	1185
	20.8.3	Antenna gain	1187
	20.8.4	Tracking slope	1188
20.9	Conclusions and future developments		1188
20.10	References		1191

21 Microstrip field diagnostics — P.G. Frayne   **1193**

21.1	Introduction	1193
21.2	Surface analytical techniques	1194
21.3	Scanning-network probe	1195
21.4	Theory of the monopole probe	1197
21.5	Resonant microstrip discs	1202
21.6	Resonant microstrip triangles	1209
21.7	Open-circuited microstriplines	1212
21.8	Antenna diagnostics	1214
	21.8.1 The rectangular patch	1215
21.9	Linear element patch array	1217
21.10	Circularly polarised patch antenna	1218
21.11	Microstrip travelling-wave antenna	1222
21.12	Acknowledgments	1224
21.13	References	1225

22		**Microstrip antennas on a cylindrical surface — E.V. Sohtell**	**1227**
	22.1	Introduction	1227
	22.2	Theoretical models for a patch on a cylinder	1227
		22.2.1 Cavity model of the patch	1228
		22.2.2 Surface-current model	1232
	22.3	Single patch application	1234
		22.3.1 Mechanical design	1234
		22.3.2 Measurements	1234
		22.3.3 Radiation-pattern comparisons	1235
	22.4	Array application	1239
		22.4.1 General	1239
		22.4.2 Theoretical treatment of finite and infinite arrays	1240
		22.4.3 Design of a phased array on C-band	1240
		22.4.4 Measured performance	1243
	22.5	Summary	1253
	22.6	References	1255
23		**Extensions and variations to the microstrip antenna concept — P.S. Hall, A. Henderson and J.R. James**	**1257**
	23.1	Introduction	1257
	23.2	Radiation pattern control	1258
		23.2.1 Reflector feeds	1258
		23.2.2 Spherical dielectric overlays	1267
	23.3	Wide-bandwidth techniques	1273
		23.3.1 Log-periodic structures	1273
		23.3.2 Dichroic dual-function apertures	1282
	23.4	Millimetre-wave hybrid antenna	1285
	23.5	Novel use of materials	1288
		23.5.1 Foam substrates for large direct-broadcast-satellite domestic receiving arrays	1288
		23.5.2 Magnetic materials and beam scanning	1292
		23.5.3 Use of very-high-permittivity substrates in hyperthermia applicators	1293
	23.6	Summary comment	1294
	23.7	References	1295

Foreword

The Handbook of Microstrip Antennas could not have been written even five years ago, for neither the technology nor the relevant analytical tools were sufficiently developed. This text arrives when the field is at a rush of activity. Fundamental mathematical tools are on hand to solve a variety of the important problems, and practical engineering results are now finding applications. Potential future capabilities and applications now look more optimistic than at any time in the history of this young technology. This new text describes vast developments in theory and practice. In two volumes, and representing the work of over thirty authors, the text is presented with such authority that it is assured a role as a key reference tool for many years.

Microstrip antennas are a new and exciting technology. Invented about twenty years ago for application as conformal antennas on missiles and aircraft, the microstrip antenna has found increasing use because it can be fabricated by lithographic techniques in monolithic circuits. Initially, microstrip patch antennas were used as individual radiators, but they soon found use in relatively large fixed beam (non scanning) arrays. More recently, they have progressed to arrays for scanning in one or two dimensions. The advantage of this technology at microwave frequencies is its compatability with large scale printed circuit fabrication. Boards are fabricated lithographically and devices mounted by robotics or automated production line techniques. Microstrip printed circuit arrays are seen as an essential key to affordable antenna technology.

At millimeter wavelengths, the benefit of microstrip arrays are enormous and so revolutionary as to create an entirely new technology; the monolithic integrated antenna array. Such an array has transmission lines, amplifiers, phase shifters and radiating elements, all on semiconductor substrates. Beyond this, these monolithic subarrays will be compatible with the integration of various solid state technologies on wafer size substrates. At these integration levels, the antenna array design and monolithic integrated circuit design cannot be separated, for the antenna architecture will need to optimise radiation, solid state device integration, board layout and thermal design. And so is born the antenna system architect!

Against this backdrop of energy and creativity, this timely and important book is the first handbook entirely dedicated to presenting a detailed overview of microstrip antenna development and theory. The vast scope of the text does justice to the broad range of research and development being undertaken throughout the world that is addressing a wide variety of microstrip elements and arrays for radiating linearly and circularly polarised waves. The text presents the work of a number of the most prominent and knowledgeable authors and so documents the state of the art at many institutions and in several countries.

This monumental handbook is a milestone in the development of microstrip antenna technology.

Robert J. Mailloux

Preface

Within two decades Microstrip Antennas have evolved as a major innovative activity within the antenna field and for both of us it has indeed been a fascinating and challenging experience to play a part in this vibrant research. In so doing the opportunity to initiate this International Handbook has arisen and this again has been a stimulating, meaningful objective that has also enriched our personal experiences through contact with numerous colleagues worldwide. It was around 1985 when it was apparent to us that the topic had raced ahead so fast that our previous IEE book "Microstrip Antenna Theory and Design" published in 1981 would soon need up-dating. Such is the vigour in Microstrip Antenna research that neither of us felt that we could do justice to the topic, at least across all its frontiers in a reasonable time scale, and it was at this point that we conferred with colleagues worldwide and this multiauthored Handbook was conceived.

As to the subject itself, it has been abundantly clear for years that it is system driven and indeed continues to be so, and that its alarming pace has promoted microstrip antennas from the ranks of a rather specialised technique to a major type of antenna technology in itself. Historically one has always associated low cost, low weight and low profile with Microstrip Antennas but this description is simplistic and inadequate in the industrial atmosphere today where many new systems owe their existence to these new radiators. In reality, the feasibility of a low profile printed radiator has inspired the system creators and there is an abundance of examples, not just in the Defence sector. For instance, we have new generations of printed paper antennas, adaptive conformal antennas sitting on the roofs of automobiles and printed antennas as true ground speed sensors in many transport scenarios.

It is indeed a stimulating topic to be associated with and we hope that the Handbook will portray this. For the in-depth researcher, however, the frontiers to push forward carry the familiar headings of bandwidth extension techniques, pattern control, minimisation of losses etc. but the scene has moved on in a decade and industry is now thirsty for significant advances, all at low cost, to meet the demand for higher performance and competitive costs. Research thus addresses critical optimisation procedures and advances are hard won. The role of substrate technology is now well appreciated and major developments have taken place to design materials that withstand a wide range of operating constraints, yet are affordable. As to the main thrust in research, it centres around the continual quest for innovative electromagnetic printed structures that satisfy the expanding system demands coupled with the ability to manufacture them and it is in the latter area where computer aided design (CAD) forms the cutting edge. Whether the manufacture of microstrip arrays can be fully automated via CAD in the immediate future is an open question that echoes throughout the Handbook and at present, further research is necessary.

In organising the Handbook we have attempted to address all these aspects giving a balanced viewpoint from both industry and research centres and the overlap between chapters is intended to be sufficient to allow meaningful comparisons between contributors to be made. The broad theme adopted is to take the reader through elements and arrays in the first volume followed by technology and applications in the second volume but as may be expected, many authors include material covering more than one aspect. Look-up charts relating items of interest to chapters and a Glossary of over one hundred different types of printed antennas form much of the Introduction to assist the reader to efficiently select those parts that are of immediate interest.

Finally, we thank all authors for their creative contributions, splendid cooperation, careful preparation of manuscripts and fellowship in the collective aim to compile a worthy international text with many years' useful life. In particular we thank Dr David Pozar and Dr Koichi Ito who helped us initially with communications in the USA and Japan respectively. We are also pleased to acknowledge the willing and professional cooperation of the publishers.

On a personal note, we have enjoyed the project and in particular the sincere experience of making new friends and acquaintances worldwide.

J. R. James
P. S. Hall

List of contributors

N. G. Alexopoulos
University of California
USA

A. R. Van de Capelle
Katholieke Universiteit Leuven
Belgium

J. S. Dahele
Royal Military College of Science
UK

J. P. Daniel
Université de Rennes I
France

A. G. Derneryd
Ericsson Radar Electronics Lab
Sweden

G. Dubost
Université de Rennes I
France

P. G. Frayne
University of London
UK

K. Fujimoto
University of Tsukuba
Japan

F. E. Gardiol
Ecole Polytechnique Fédérale de Lausanne
Switzerland

K. C. Gupta
University of Colorado
USA

P. S. Hall
Royal Military College of Science
UK

R. C. Hall
Ecole Polytechnique Fédérale de Lausanne
Switzerland

M. Haneishi
Saitama University
Japan

A. Henderson
Royal Military College of Science
UK

K. Hirasawa
University of Tsukuba
Japan

List of contributors

T. Hori
Nippon Telegraph and Telephone Corporation
Japan

K. Ito
Chiba University
Japan

D. R. Jackson
University of Houston
USA

J. R. James
Royal Military College of Science
UK

P. B. Katehi
University of Michigan
USA

A. H. Kishk
University of Mississippi
USA

K. F. Lee
University of Toledo
USA

E. Levine
Weizmann Institute of Science
Israel

G. Morris
Vega Cantley Instrument Co Ltd
UK

J. R. Mosig
Ecole Polytechnique Fédérale de Lausanne
Switzerland

P. Newham
Marconi Defence Systems
UK

S. Nishimura
University of Osaka
Japan

R. P. Owens
Thorn EMI Electronics Ltd
UK

E. Penard
Centre National D'Etudes de Télécommunications
France

D. M. Pozar
University of Massachusetts
USA

D. H. Schaubert
University of Massachusetts
USA

L. Shafai
University of Manitoba
Canada

E. V. Sohtell
Ericsson Radar Electronics Lab
Sweden

Y. Suzuki
Toshiba Corporation
Japan

C. Terret
Centre National d'Etudes de Télécommunications
France

T. Teshirogi
Radio Research Laboratories
Ministry of Posts and Telecommunications
Japan

G. R. Traut
Rogers Corporation
USA

J. F. Zurcher
Ecole Polytechnique Fédérale de Lausanne
Switzerland

Chapter 1
Introduction
J.R. James
and P.S. Hall

1.1 Historical development and future prospects

The microstrip antenna is now an established type of antenna that is confidently prescribed by designers worldwide, particularly when low-profile radiators are demanded. The microstrip, or printed, antenna has now reached an age of maturity where many well tried techniques can be relied upon and there are few mysteries about its behaviour. The fact that you are now reading an historical review is interesting in itself because all this has happened in a relatively short time span of one or two decades; such is the rate of progress in contemporary antenna technology. To imply that the topic of microstrip antennas is now static would be grossly misleading because the opposite is true with the ever increasing output of research publications and intensifying industrial R and D. The quest now is for more and more innovative designs coupled with reliable manufacturing methods. The driving force is the thirst for lower-cost, less-weight, lower-profile antennas for modern system requirements. Lower costs, however, rely on the ability of the designer to precisely control the manufacturing process, and this in turn usually demands that the prototype innovative structures can be adequately mathematically modelled and toleranced. It is in these latter respects that the challenge to the antenna expert originates, and the search for the more precise computer modelling of microstrip antennas is now the main preoccupation of designers and researchers alike, as is reflected in this handbook.

The invention of the microstrip-antenna concept has been attributed to many sources and the earliest include Greig and Englemann [1] and Deschamp [2]. At that time the emission of unwanted radiation from the then new thin stripline circuits was well appreciated and subsequently the dimensions of the substrate and conducting strip were reduced to inhibit the radiation effects, thus creating 'microstrip'. Whether the advent of the transistor influenced the rapid development of these planar printed circuits is debatable and the main interest was likely to be the development of lower-cost microwave filters etc. Lewin [3] considered the nature of the radiation from stripline but there was apparently little or no interest in making use of the radiation loss. Apart from a few references [4, 5, 6] the antenna concept lay dormant until the early 1970s [7, 8, 9] when there was an immediate need for low-profile antennas on the emerging new generation of missiles.

At this point in time, around 1970, the development of the microstrip-antenna concept started with earnest and the research publications, too numerous to itemise, started to flow. The period is perhaps most readily referenced by its workshops and major works. The most significant early workshop was held at Las Cruces, New Mexico, in 1979 [10] and its proceedings were distilled into a major *IEEE Transactions* special edition [11]. At that time two books were published by Bahl and Bhartia [12] and James, Hall and Wood [13] which remain in current use today. Another more specialised and innovative development was published as a research monograph by Dubost [14], and here the flat-plate antenna was approached from the standpoint of flat dipoles on substrates that generally only partially filled the available volume.

The early 1980s were not only a focal point in publications but also a milestone in practical realism and ultimately manufacture. Substrate manufacturers tightened their specifications and offered wider ranges of products capable of working under extreme ambient conditions. Substrate costs were, however, to remain high. It was appreciated that analytical techniques for patch elements generally fell short of predicting the fine pattern detail of practical interest and the input–impedance characteristic to sufficient accuracy. It was also appreciated that the connection of feeders to patch elements in a large array was fraught with problems and new approaches were necesary where the feeders and elements are regarded as a complete entity. More recently the term 'array architecture' has come into being as if to emphasise the importance of choice of array topology and the fact that feeders cannot necessarily be freely attached to printed elements, even if the latter are in themselves well optimised.

Recent system demands are, as previously mentioned, a dominant factor in the development of printed antennas. Communication systems spanning wider bandwidths are continually emerging and techniques for increasing the bandwidth of microstrip antennas are a growth area. Controlling the polarisation properties of printed antennas is another area of activity arising largely out of the current awareness for making greater use of the polarisation properties of waves, particularly in radar. In defence applications, systems that have an electronic, as opposed to mechanical, beam-scanning facility are attracting much research effort and the concept of 'active-array architecture' is now with us where semiconductor packages and radiating elements are integrated into planar apertures. The cost of such an array is very high and the whole concept is state-of-the-art.

This brings us to the present and how we see the immediate future of printed antennas. A seldom mentioned point is the fact that printed substrate technology is readily processed in University laboratories and continues to remain a rich source of complex electromagnetic problems; research publications will thus continue to abound, and in parallel with industrial development will most likely be dominated by two aspects:

- The search for mathematical models that will predict practical antennas more precisely and hence sharpen CAD techniques in manufacture.
- The creation of innovative antennas to match the demand for new systems.

In this latter aspect it must be emphasised that a bulky conventional microwave antenna may well out-perform its thin conformal printed counterpart. Many new systems, however, particularly in aerospace, are only made feasible with the existence of the printed antenna concept, and here lies a major driving force where new systems arise solely from innovative antenna designs.

As to the distant future, one can but extrapolate the present trends towards integrated electronically beam-scanned arrays. This leads to a vision of conformal antennas distributed over the surface of vehicles, aircraft, ships, missiles etc., thus replacing many conventional types of radiators, but the organisation and control of the radiation pattern co- and cross-polar characteristics is a complex control problem that cannot be solved by software alone and demands innovative physical concepts. Are we thus unconsciously converging on the concept of distributed sensors, so common in the insect and animal world, where information is commonly gleaned in a variety of ways to best suit a particular situation? Taking the comparison a step forward, we would therefore expect the distributed conformal apertures to require a significant back-up from signal-processing techniques, which amount to making use of temporal *a priori* information on signals and noise. Put this way these ideas are not so far-reaching because many of these adaptive concepts can be recognised in some of our new radar and communication systems, particularly for defence. In this light the printed-antenna concept would therefore appear as a gateway to system compatibility and optimal deployment of sensors, embracing the numerous facets of conformality, low costs, semiconductor integration, electronic radiation pattern control and an opportunity to exploit signal-processing techniques to the full using modern computing power. The prospects are indeed exciting and underline the importance of the microstrip-antenna concept, its continual evolution and impact on electronic systems design.

1.2 Fundamental issues and design challenges

A handbook of this type is intended as an all-embracing treatment that is both diverse and highly specialist. As such it is not possible to include comprehensive background information and we anticipate that readers wishing to recap on basic antenna theory, antenna mesurements and the rudiments of microstrip technology etc. will have no difficulty in obtaining relevant literature. It is our experience, however, that certain fundamental properties of printed antennas have been central to their evolution and limitations, and therefore embody the design challenges of the future as follows.

The microstrip antenna has many differences when compared with a conventional antenna. Most of these stem from the planar construction in which for a given substrate in the xy plane there are only two degrees of freedom, allowing the very thin printed-conductor topology to take any shape within the confines of the x and y co-ordinate directions. The first and most troublesome property is the issue of loss, principally in the thin conducting strip feeders connecting elements in large arrays. In some applications the loss in the radiating elements also creates dificulties. The radiating elements themselves have a restricted bandwidth arising from the intrinsic high-Q resonator action in the thin substrate. The generation of surface waves is equally important and cannot be avoided unless foam-type substrates are deployed allowing virtual air-spaced operation. The surface waves can corrupt radiation-pattern characteristics, particularly when low sidelobe and cross-polarisation levels are demanded. In many design specifications, problems can only be alleviated by compromising the manufacturing simplicity of the single coplanar printed assembly by employing overlaid element and feed concepts based on multilayer sandwich structures. Microstrip arrays generally require some sort of radome or weather shield, thus increasing the structure depth, but in some cases a degree of radiation-pattern enhancement is obtainable. Last but not least, mention must be made of the relatively high cost of substrates capable of providing the desired electrical and mechanical stability in operation. The substrate cost is often an inhibiting factor in what is otherwise a low-cost manufacturing process.

These above issues are of a fundamental nature and we consider it important to highlight current understanding to identify aspects which may offer particular scope for future advancement. Before addressing this we list, for completion, some of the more commonly known properties of microstrip antennas in relation to both contemporary antenna-engineering and modern electronic-systems requirements.

1.2.1 Features of microstrip antenna technology

The microstrip antenna is a newcomer to the world of antenna engineering and it is fitting to be reminded of features generally sought after when compiling an antenna specification. A typical checklist is given in Table 1.1 and it is appreciated that it is unlikely that all the performance factors are relevant or indeed critical in any given application. Equally demanding are operational and manufacturing considerations such as those listed in Table 1.2 and these are very dependent on the application in mind. The generation of thermal noise in a receiving antenna is insignificant for most conventional antennas and is clearly a new factor associated mainly with large lossy microstrip arrays. Likewise power-handling and material effects are particularly relevant for microstrip radiators, while the use of new materials such as carbon fibre necessitates careful evaluation of electrical loading, intermodulation effects etc.

Introduction 5

Table 1.1 *Antenna designers' checklist of performance factors*

Matching	Input terminals matched to source feed
Main beam	Antenna gain and beamwidth properties
Sidelobes	Constrained to desired envelope
Polarisation	Cross-polar behaviour constrained to desired envelope
Circular polarisation	Constraints on ellipticity
Efficiency	Wastage of power in antenna structure
Aperture efficiency	Relates to illumination distribution, gain and pattern characteristics
Bandwidth	Frequency range over which all above parameters satisfy specification — commonly based on input terminal impedance charactericstics
System demands	Size, weight, cost

The commonly upheld properties of microstrip antennas are listed in Table 1.3 and may be usefully compared with the general checklist of Tables 1.1 and 1.2 to ascertain the suitability of microstrip for various operational roles. However, it is important to appreciate that the interpretation of Table 1.3 is very dependent on the intended application. For instance, patch antennas on foam

Table 1.2 *Operational and manufacturing considerations*

Noise effects in receiving antennas
Power handling in transmitting antennas
Creation of hazards for personnel in near-field
Robustness to lightning strikes
Electrostatic charge effects in space applications
Effects of wind, vibration, ice, snow, rain, hail
Ambient conditions on temperature and humidity
Exposure to sunlight
Aerodynamic constraints, radomes and weather shields
Metal corrosion and creep
Mechanical and electrical stability of materials
Mechanical and electrical tolerances in manufacture
Sensitivitiy of design to manufacturing tolerances
Generation of intermodulation effects in materials

Table 1.3 *Some commonly acknowledged properties of microstrip antennas*

Advantages	Disadvantages
Thin profile	Low efficiency
Light weight	Small bandwidth
Simple to manufacture	Extraneous radiation from feeds, junctions and surface waves
Can be made conformal	Tolerance problems
Low cost	Require quality substrate and good temperature tolerance
Can be integrated with circuits	High-performance arrays require complex feed systems
Simple arrays readily created	Polarisation purity difficult to achieve

substrates may have a less desirable thick profile but good efficiency and reasonable bandwidth; in contrast a thin overlaid patch assembly with complex feed arrangements on a plastic substrate is likely to be more complicated to manufacture and not necessarily low cost. The modelling and subsequent engineering design of arrays for successful manufacture is often a factor that is originally overlooked and ultimately pushes up development costs. There are many other examples where the commonly quoted properties of Table 1.3 need qualifying, and recent experience from conferences and industrial contacts shows that academics have on occasions failed to convey a realistic impression to industry whereas industry itself has perhaps been too willing to implement the new technology without a sufficient design base that copes with the factors of Table 1.2. We have already stressed the need for advances in CAD techniques for manufacture and will specifically address this again later on, but now we return to the more general features of microstrip antennas such as the trade-offs listed in Table 1.4*a* for rectangular patch antennas. These are very approximate and can be deduced from the basic patch equations [15]. An obvious deduction which is nevertheless significant is that the use of thick low-permittivity substrates, giving essentially air spacing, gives many benefits. When the behaviour of an array of patch elements (Table 1.4*b*) is considered, feeder radiation is seen to increase for thicker lower-permittivity substrates [16, 17]. With this exception, any attempt to compact the antenna using a thin high-permittivity substrate will thus generally invoke all-round penalties in performance. These requirements are thus seen to be contrary to those for optimum operation of MICs, and this imposes restrictions on the integration of antennas and associated front-end circuitry. This perspective is valuable in emphasising the dominant characteris-

Table 1.4a *Approximate performance trade-offs for a rectangular patch*

Requirement	Substrate height	Substrate relative permittivity	Patch width
High radiation efficiency	thick	low	wide
Low dielectric loss	thin	low	—
Low conductor loss	thick	—	—
Wide (impedance) bandwidth	thick	low	wide
Low extraneous (surface wave) radiation	thin	low	—
Low cross polarisation	—	low	—
Light weight	thin	low	—
Strong	thick	high	—
Low sensitivity to tolerances	thick	low	wide

Table 1.4b *Approximate performance trade-offs for an array of circular patches*

Requirement	Substrate height	Substrate relative permittivity
High efficiency	thick	low
Low feed radiation	thin	high
Wide (impedance) bandwith	thick	low
Low extraneous surface-wave radiation	thin	low
Low mutual coupling	thick	low
Low sensitivity to tolerances	thick	low

tics of microstrip antennas and the fact that antenna volume-reduction benefits must manifest themselves as cost factors which in turn demand a high standard of engineering design to overcome.

Finally we complete our discussion of general features with a list of applications in Table 1.5 that have attracted the use of printed-antenna technology. Almost without exception the employment of microstrip technology arises because of a system demand for thin low-profile radiators. Conventional antennas are clearly disadvantaged in such applications despite their often superior performance over microstrip antennas. In some cases the system has been created around the microstrip concept as mentioned earlier on.

1.2.2 Fundamental problems

In our vision of the future we have singled out reliable CAD techniques in array manufacture and the system-led creation of innovative antennas as the major

8 Introduction

Table 1.5 *Typical applications for printed-antenna technology*

Aircraft antennas	Communication and navigation Altimeters Blind-landing systems
Missiles and telemetry	Stick-on sensors Proximity fuzes Millimetre devices
Missile guidance	Seeker monopulse arrays Integral radome arrays
Adaptive arrays	Multi-target acquisition Semiconductor integrated array
Battlefield communications and surveillance	Flush-mounted on vehicles
SATCOMS	Domestic DBS receiver Vehicle-based antenna Switched-beam arrays
Mobile radio	Pagers and hand telephones Manpack systems
Reflector feeds	Beam switching
Remote Sensing	Large lightweight apertures
Biomedical	Applicators in microwave cancer therapy
Covert antennas	Intruder alarms Personal communication

Table 1.6 *Fundamental issues that will continue to be addressed*

Bandwidth extension techniques

Control of radiation patterns involving sidelobes, beamshaping, cross-polarisation, circular polarisation, surface-wave and ground-plane effects

Reducing loss and increasing radiation efficiency

Optimal feeder systems (array architecture)

Improved lower-cost substrates and radomes

Tolerance control and operational factors

Table 1.7 *Some generic types of bandwidth-extension techniques*

Increasing antenna volume by incorporating parasitic elements, stacked substrates, use of foam dielectrics

Creation of multiple resonances in input response by addition of external passive networks and or internal resonant structures

Incorporation of dissipative loading by adding lossy material or resistors

---

Varactor and PIN diode control gives a wider effective bandwidth and is not included in the above list.

thrusts. The problem areas will however centre around the fundamental issues listed in Table 1.6. These issues are universally acknowledged and we will review some of them as follows to emphasise certain aspects which in our opinion are worthy of clarification or perhaps need various points amplified, in particular to bridge the gap between academic research and industrial implementation.

1.2.2.1 Bandwidth extension: The search for new microstrip configurations with wider bandwith has been a dominant feature of the research literature and much effort continues to be expended. No other type of antenna has been so exhaustively treated as regards its bandwidth properties, yet the literature often portrays an incomplete picture by not defining what is meant by bandwidth [18]. The many factors involved are listed in Table 1.7. A common and generally realistic assumption is that the input-impedance characteristic of a resonant patch antenna behaves as a simple tuned circuit, in which case the 3 dB bandwidth B is approximately $(100/Q)$ percent, where Q is the Q-factor of the equivalent tuned circuit. If the antenna is matched at the resonant frequency of the tuned circuit, then away from resonance the input impedance will be mismatched, creating a VSWR(>1) of S, where

$$B = \frac{(S-1)}{Q\sqrt{S}} 100\% \qquad (1.1)$$

Use of a thicker and/or lower-permittivity substrate reduces Q and hence increases B. An examination of numerous examples shows that, irrespective of whether the permittivity or substrate thickness is changed, the main effect (Table 1.7) is that B increases with the volume of the antenna, i.e. the volume of substrate between the patch and ground plane. Some examples are shown in Fig. 1.1, which also includes curves of radiation efficiency with and without allowance for the power lost to surface waves. The first point of clarification is to note that there are numerous ways of increasing the volume of a patch element by employment of thicker substrate or stacking several substrates [19] or adding

10 Introduction

parasitic elements [20], but they all belong to the same generic type of bandwidth extension technique.

A second generic technique (Table 1.7) consists of introducing multiple

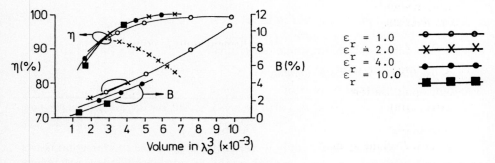

Fig. 1.1 *Patch-antenna efficiency η and bandwidth B versus resonator volume for different permittivities (Reproduced from Fig. 2 of Reference 18)*
-x-x-x- is the radiation efficiency corrected for surface-wave action ($\varepsilon_r = 2\cdot0$)

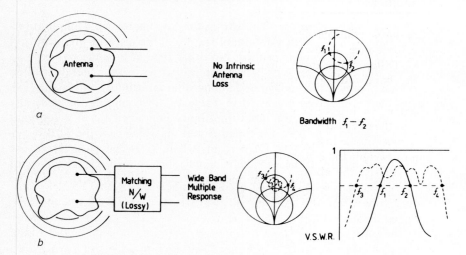

Fig. 1.2 *Patch bandwidth extension using an external passive network*
  a Antenna without network
  b Effect of matching network.

resonances in the input characteristic, as illustrated in Fig. 1.2 showing the inclusion of a passive network in the input port; the presence of the network invokes additional dissipative losses. The same bandwidth extension effects can be brought about by introducing multiple resonances within the antenna itself [18], which usually involves an increase in antenna thickness and hence volume. The important point to note is that a multiple resonance input response does not

Table 1.8 *Factors constraining the bandwidth of microstrip antenna elements and arrays*

Element	Array
Input-impedance characteristic	Surface waves
Side-lobe level	Element mutual coupling
Cross-polarisation level	Feeder radiation
Circular polarisation (axial ratio)	Corporate feed and mismatch
Pattern shape (E- and H-plane symmetry)	Scanning loss
Element gain	
Efficiency	
Feeder radition	

obey the simple relationship of eqn. 1.1, and it is difficult to relate the various multiple resonance bandwidth extension techniques that are reported in the literature. Different researchers use different VSWR or insertion-loss criteria to define the bandwidth and the insertion-loss curve shapes are likewise very different.

A third much less common technique (Table 1.7) is simply to add lossy material to the microstrip element. This technique would at first sight appear to lead to unacceptable loss, but the manufacturing simplicity has definite appeal and can outweight the other disadvantages.

We summarise the above three generic bandwidth extension techniques in Table 1.7, but emphasise that from a system designers' standpoint the definition of bandwidth based on the input-impedance characteristic is just one of many factors listed in Table 1.8 that constrain the bandwidth of an antenna element or array. For instance, the designer may decide to use a rectangular patch accompanied by several parasitic elements to achieve an impedance bandwidth specification, but then finds that the configuration fails to achieve adequate cross-polarisation levels or perhaps E- and H-plane symmetry over the band. In another instance it may be straightforward to meet all the bandwidth criteria for a selected element only to find that, when the latter is connected in an array, the bandwidth specification is not achieved because of mutual coupling or perhaps feeder-line mismatches. Research workers seldom have the opportunity to address the totality of problems in a system design, and it is a natural consequence that they focus on the optimisation of a given property in isolation from other requirements. In contrast, the industrial designer has to optimise many parameters at the same time and bandwidth is a topic area where the gulf between isolated research and system design is at its widest. The challenge facing researchers and industrial designers alike is to establish reliable designs for elements and arrays that achieve bandwidth extension under a wide selection of contraints as listed in Table 1.8. It is also highly desirable that the performance of one type of element can be quantified in relation to the performance of any other type of new element; the fact that there are in reality few generic types of bandwidth-extension techniques (Table 1.7) [18] is an important guideline.

1.2.2.2 Pattern control: There is now ample evidence to show that the radiation-pattern control of printed radiators is an order more difficult than with reflector and aperture antennas. Even for modest performance levels of sidelobes and cross-polarisation the printed-conductor topology presents many variables to optimise for a given substrate thickness and permittivity. For sidelobe and cross-polarisation levels of about $-20\,\text{dB}$ extraneous radiation due to surface waves, feeder radiation and ground-plane edge effects is not insignificant and computer models lose their precision. Surface-wave effects decrease for lower-permittivity substrates but feeder radiation is then more prominent [17]. There is evidence in the literature that much lower levels can be achieved, but generally these are pattern cuts in certain preferred planes or pertain to arrays fitted with lossy material or other special effects. A consensus of opinion is that printed antennas are at present more fitted for applications with less demanding pattern specifications. The challenge for the future thus remains the lowering of the levels of extraneous radiation in printed arrays and improved computer modelling of the overall patterns.

Some special mention needs to be made of circularly polarised elements and arrays because considerable progress has been made in this respect and it is likely to be an area for continued exploitation. It is well known that in principle a linearly polarised antenna can be converted to perfect circular polarisation by superimposing upon its radiation characteristics, those of its dual radiator having transposed E- and H-field sources. For instance, a wire dipole (electric source) would need to be combined with a wire loop (magnetic source), but in reality it is physically impossible to construct or feed such an arrangement precisely and compromises are made such as the employment of crossed-wire dipoles which yields circular polarisation in a limited region of the hemisphere and over restricted bandwidth. These and other techniques [21] are well established for conventional antennas, and the point we make here is that they are more difficult to translate to printed elements in view of the constrained planar geometry and feeder requirements. It is therefore inspiring to note the innovation that has been brought about whereby circular-polarisation characteristics have been enhanced by sequential rotation of elements [22], incorporation of finite substrate effects [23], novel feeder arrangements [24] and many more. Creating improved low-cost radiators that provide circular polarisation over wider bandwidths and larger sectors of the radiation-pattern hemisphere is a goal towards which much international effort will continue to be directed.

1.2.2.3 Efficiency and feeder architecture: The outstanding advantage of microstrip — the simplicity of the printed conductor — is also the source of one of its major disadvantages, which is the relatively high transmission-line loss. The nature of the loss is well understood and arises from the high current density at the strip edge and substrate losses. It is a fact that no worthwhile reductions in transmission loss have been achieved since the inception of microstrip, and the simplicity of the structure offers little scope for innovation in this respect. For patch elements the loss is less significant, and with an appropriate low-loss substrate and strongly radiating patch, antenna efficiencies of 95% are achievable. A conventional wire dipole antenna would have a better efficiency than the patch but the order of loss of the latter is usually very small from a systems standpoint. The main problem arises in large arrays having microstrip or other forms of printed feeder lines because feeder losses limit the gain of the aperture; in fact, beyond a certain critical aperture size the gain will actually reduce. The beamwidth will, or course, also continue to narrow. The critical size is dependent on the feeder topology, substrate etc. and a maximum gain around 35 dB is not uncommon. Fig. 1.3 shows typical computed and measured results

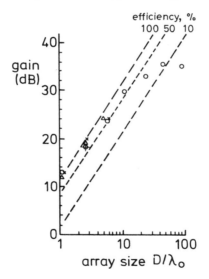

Fig. 1.3 *Patch-array gain*
   0 Calculated [17]; measured, with feed impedance + 100 Ω, × 120 Ω Δ 200 Ω

[17] and indicates that at maximum gain an efficiency of about 10% can be expected. Travelling-wave antennas show some economy of feeder loss over corporate feeds but the frequency scanning loss for large travelling-wave apertures is then the dominant limitation. Once again the simplicity of a printed feeder system gives little scope for major design changes, and more recently hybrid feeder systems are being considered incorporating more conventional

14 Introduction

cables and waveguides for the longer feeder runs. We have already emphasised in Section 1.2.2.2 the limitations on pattern control enforced by extraneous feeder radiation and any breakthrough in feeder architecture will need to address the latter. However, for some applications the radiation-pattern specifications are less critical than loss of gain and any improvements in feeder loss would be a significant advance. The future challenge is to discover new feeder architectures giving less loss, and if possible less extraneous radiation, with the knowledge that the already simplistic printed configurations offer little scope for fundamental physical change. One possible avenue for advancement lies in the integrated antenna concept whereby transistors are embedded in the feed structure to facilitate beam scanning. This may circumvent the loss problem but exacerbate extraneous radiation effects, and of course escalate costs.

1.2.2.4 Substrate technology: Substrate technology and marketing has been, and will continue to be, a key factor in the acceptance by industry of the printed-antenna concept. Earlier microstrip antennas used plastic substrates or in some cases alumina, but in recent years the use of lower-permittivity substrates is common. The substrate role thus appears to be mechanical, enabling the printed conductor to be suspended at a uniform height above the ground plane. The use of lower permittivities also reduces surface-wave effects but feeder radiation is then more difficult to suppress. Antenna designers thus require a wide range of substrates available having stable electrical and mechanical properties over the various ambient operating conditions. The major problem has been, and is likely to be in the foreseeable future, a matter of substrate cost because the world demand is relatively small compared with that of some other plastic products. This has encourage some companies to manufacture their own substrates while in other cases the substrate costs have made some large-array projects non-viable, and printed technology is then seen to be costly in contradiction to the commonly upheld properties of Table 1.3. It is also noted that many microstrip antennas will require some sort of weather shield or perhaps a radome, which again is a cost factor. Substrate technology thus offers a challenge to material manufacturers to create lower-cost high-performance stable substrates. Clearly this is a somewhat circular problem which appears to demand a higher-volume market to initiate an immediate advance; conversely such an advance would open up a larger-volume market. Such a situation is not uncommon, and with the considerable manufacturing interest in substrates that is building up (Table 1.9) antenna designers should be optimisitic about the way substrate technology is likely to develop in the next few years.

1.2.2.5 Manufacture and computer-aided design (CAD): The microstrip antenna has been widely mathematically modelled for many years and yet from the manufacturers' standpoint there is a dearth of ready-to-use design equations, and hence reliable CAD packages. This situation has arisen partly owing to the mathematical difficulties associated with practical geometries and partly

Table 1.9 *Representative substrate list*

ε_r	Material	Supplier
1·0	Aeroweb (honeycomb)	Ciba Geigy, Bonded Structures Div., Duxford, Cambridge, CB2 4QD
1·06	Eccofoam PP-4 (flexible low-loss plastic foam sheet)	Emerson & Cumming Inc, Canton, Massachusetts, USA (Colville Road, Acton, London. W3 8BU, UK)
1·4	Thermoset microwave foam material	Rogers Corp., Bo 700, Chandler, AZ 85224, USA. (Mektron Circuit Systems Ltd., 119 Kingston Road, Leatherhead, Surrey, UK)
2·1	RT Duroid 5880 (microfiber Teflon glass laminate)	Rogers Corp.
2·32	Polyguide 165 (polyolefin)	Electronized Chemical Corp., Burlington, MA 01803, USA
2·52	Fluorglas 600/1 (PTFE impregnated glass cloth)	Atlantic Laminates, Oak Materials Group, 174 N. Main St., Franklin, MH 0323, USA. (Walmore Defence Components, Laser House, 132/140 Goswell Road, London, EC1V 7LE)
2·62	Rexolite 200 (cross-linked styrene copolymer)	Atlantic Laminates
3·20	Schaefer Dielectric Material, PT (polystyrene with titania filler)	Marconi Electronic Devices Ltd., Radford Crescent, Billericay, Essex, CM12 0DN, UK
3·5	Kapton film (copper clad)	Dupont (Fortin Laminating Ltd., Unit 3, Brookfield Industrial Estate, Glossop, Derbyshire, UK)
3·75	Quartz (fuzed silica)	A & D Lee Co. Ltd., Unit 19, Marlissa Drive, Midland Oak Trading Estate, Lythalls Lane, Coventry, UK

Introduction

6·0	RT Duroid 6006 (ceramic-loaded PTFE)	Rogers Corp.,
9·9	Alumina	Omni Spectra Inc, 24600 Hallwood Ct. Farmington, Michigan, 48024, USA (Omni Spectra, 50 Milford Road, Reading, Berks, RG1 8LJ, UK)
10·2	RT Duroid 6010 (ceramic-loaded PTFE)	Rogers Corp.
11	Sapphire	Tyco Saphikin (A & D Lee Co Ltd., Unit 19, Marlissa Drive, Midland Oak Trading Estate, Lythalls Lane, Coventry, UK)

The brief details in the Table are intended to give readers an insight into the range and types of materials available. Mention of any particular product does not imply our endorsement. Likewise exclusion of a material does not imply adverse comment and we presume that some excellent products have been omitted.

to the many varieties of patch antennas and the fact that designs must conform to the vagaries of system requirements. Horn, wire and other conventional metal antennas can be modelled to a high accuracy with well established formulas and this is also true of electrically larger apertures such as the reflector antenna. In contrast to these homogeneous electromagnetic systems the modelling complexity arises largely from the presence of a finite-sized dielectric slab that gives rise to the factors noted in Table 1.8 and elsewhere. This complexity is compounded when the number of elements in an array increases and when the fine detail of patch feeding or complex feed networks is required. Mutual coupling, surface-wave effects and feed radiation manifest themselves as relatively small effects in a small array but they quickly take charge of sidelobe and cross-polarisation levels in the region of -20 dB as the number of elements increase. When viewed in the light of increasingly tight requirements, modelling accuracy is seen to be the key parameter to successful implementation. As an example, the use of inaccurate CAD may well be more expensive for the manufacturer than design by hand in the long run. Likewise the range of applicability of the package needs to be understood if inherently good models are not to be brought to bear on the wrong problems. However, despite the unlikely possiblity of extending present-day numerical analysis to arrays of patches in the immediate future, the problem is not unsurmountable provided that a close liaison exists between the CAD and antenna designer. Indeed there is a trend for manufacturers to evolve their own CAD packages based on a mixture of simple closed-form expressions for the radiation mechanism backed up by empirical results for the particular array in question. Some degree of iteration is commonly included but with the decision

Introduction 17

making and convergence under the designers control. Such an approach has many merits since tolerance and operational effects can be added in gradually to create a reliable manufacturing tool. A disadvantage of the approach is that it must be re-established together with empirical data when a change is made in the design, and furthermore the experience is confined to the particular manufacturer. As we have mentioned already, some manufacturers have been surprised by the need to underpin printed-array manufacture with positive modelling in view of the commonly acknowledge property (Table 1.3) that the antennas are 'simple and low cost'. There is ample evidence, however, showing that the simplicity and low-cost properties are realisable once modelling has been accomplished, and the latter is a one-off development cost and perhaps no more than a few months of a printed-antenna specialist's time. The challenge for the immediate future thus lies in the evolution of reliable interactive CAD packages for printed-array manufacture that are capable of wider usage and of gaining universal acceptance. In the long term one might expect some advances in the rigorous analysis of microstrip-antenna geometries embodying practical features which in turn will translate into more precise manufacturing techniques.

1.3 The handboook and advances presented

Many, if not most, of the international community of printed-antenna specialists have contributed to this handbook, which necessarily portrays the state-of-the-art at the time of going to press. The contributions reflect the authors' specialisation which in some cases is fairly wide ranging. This has meant that it has not always been possible for the editors to maintain a full thematic flow throughout the handbook. However, the chapters have been generally ordered in the following way

- element analysis and design
- array aspects
- microstrip technology
- applications

To assist the reader we have already listed in Table 1.5 the general application areas for microstrip antennas. In Table 1.10 the content of the Handbook is resolved in more detail to identify with the various topic areas within the subject of printed antennas. It can be seen that patch theory and design still concerns many researchers, with current emphasis being on basic characterisation and innovation for controlling in particular bandwidth and polarisation purity. The same applies to arrays, with additional topics such as mutual coupling in scanning arrays gaining more attention. Technology is addressed by several authors with contributions on substrates, connectors, radomes and computer-aided design and manufacturing. However, little on environmental factors has

18 Introduction

Table 1.10 *Topic areas in printed-antenna technology addressed in the handbook*

Introduction	Chapter
	1 2 3 4 5 6 7 8 9 10 11 12 13 14 15 16 17 18 19 20 21 22 23
General review	● ●
Glossary of types	●

Fundamental concepts	
	1 2 3 4 5 6 7 8 9 10 11 12 13 14 15 16 17 18 19 20 21 22 23
Microstrip lines	● ● ●
Microstrip discontinuities	● ● ● ● ● ● ● ●
Measurements	● ● ● ● ● ●

Patches	
	1 2 3 4 5 6 7 8 9 10 11 12 13 14 15 16 17 18 19 20 21 22 23
Theory	● ● ● ● ● ● ● ● ● ● ● ● ● ● ● ●
Design	● ● ● ● ● ● ● ● ● ● ● ● ● ●
Feeds	● ● ● ● ● ● ● ● ● ● ●
Circular polarisation	● ● ● ●
Dual polarisation	● ● ● ● ●
Multilayer/parasitic	● ● ● ● ● ● ● ●
Wideband	● ● ● ●
Dual/agile frequency	● ● ●
Slot/dipole radiators	● ● ● ● ●
Bandwidth	● ● ● ● ● ● ●
Efficiency	● ● ● ● ● ● ● ● ● ●
Cross-polarisation	● ● ● ● ●

Introduction   19

Arrays

	1	2	3	4	5	6	7	8	9	10	11	12	13	14	15	16	17	18	19	20	21	22	23
Linear — series-fed				●							●	●	●					●			●		●
Linear — parallel-fed			●									●	●	●				●			●	●	●
Two-dimensional						●						●	●	●				●			●	●	
Array feeds						●	●				●	●	●					●				●	
Phased arrays					●				●			●							●		●		●
Wideband/multioctave																		●					●
Hybrid dielectric																							●
Slot arrays							●					●							●				●
Omni arrays											●			●									
Circular polarisation					●								●						●	●			
Cross polarisation							●						●						●				
Feed effects								●				●	●			●					●		
Mutual coupling								●	●			●	●								●		●
Surface waves			●					●					●									●	●
Sidelobes		●										●	●								●	●	●

Technology

	1	2	3	4	5	6	7	8	9	10	11	12	13	14	15	16	17	18	19	20	21	22	23
Substrates	●																						
Connectors			●												●	●	●	●		●	●	●	
Radomes			●				●								●	●	●		●	●			
Manufacture/tolerance			●		●						●	●		●	●		●	●		●			
Environment			●								●			●	●			●		●			
Power handling									●					●	●								
CAD/CAM	●																●						

Applications

	1	2	3	4	5	6	7	8	9	10	11	12	13	14	15	16	17	18	19	20	21	22	23
Satellite antennas																						●	●
Satellite reception																			●				
Reflector feeds			●	●																			
Conformal																						●	
Mobiles						●														●			
Phased arrays								●	●										●				
Dual frequency										●									●			●	
Monopulse trackers																			●				
Millimetric antennas																		●	●	●			●
Hyperthermia applicators																							●

20 *Introduction*

Introduction 21

Table 1.11 *Summary of handbook chapters*

Element analysis and design

2 Shafai and Kishk: Analysis of circular microstrip antennas: An analysis is presented based on the equivalence principle involving both conducting and dielectric boundaries. This allows substrate edges to be accounted for. The method is used to optimise a circular disc on a finite-sized circular ground plane for low cross-polarisation and also a wrap around antenna.

3 Lee and Dahele: Characterisation of microstrip patch antennas and some methods of improving frequency agility and bandwidth: The basic characteristics of patches are reviewed here and conclusions on comparative performance are made. Bandwidth is identified as being crucial and methods of overcoming the limitations by making the patches frequency agile are presented.

4 Haneishi and Suzuki: Circular polarisation and bandwidth: The methods of obtaining circular polarisation from patches are described in this chapter together with design techniques. Again, as in the previous chapter, bandidth-extension methods are noted and, in particular, element pairing which is described more fully in Chapter 13.

5 Katehi, Jackson and Alexopoulos: Microstrip dipoles: The analysis and design of narrow-strip microstrip dipoles is presented here. For electromagnetically coupled dipoles an improvement in feed radiation is noted together with methods for offsetting mutual coupling in arrays. Bandwidth and superstrate effects are also discussed.

6 Schaubert: Multilayer and parasitic configurations: This chapter exhaustively reviews multilayer configurations and emphasises advances for wide bandwidth, multiple frequency and dual polarisation. Such structures increase the antenna thickness, and, as a contrast, antennas with coplanar parasitics are also described.

7 Dubost: Wideband flat dipole and short-circuit microstrip patch elements and arrays: Elements and arrays developed from the flat dipole concept are described in addition to short-circuited quarter-wavelength patch elements and arrays. The dipole work can perhaps be viewed as a parallel development with microstrip and has produced antennas whose performance is highly competitive.

8 Mosig, Hall and Gardiol: Numerical analysis of microstrip patch antennas: An integral-equation formulation is solved by the moments method to give solutions for arbitrary shaped wide patches, including input impedance, radiation patterns and surface-wave effects.

9 Gupta: Multiport network approach for modelling and analysis of microstrip patch antennas and arrays: Here patches possessing separable geometries in whole or part are analysed using a planar model involving impedance

22 Introduction

Table 1.11 *(Cont)*

matrices. Radiation loading is included by means of edge admittances. The author presents several illustrative examples and discusses the extension of the work to CAD methods. Further progress in the application of analysis such as this and others in this handbook is expected in the near future.

10 Van de Capelle: Transmission-line model for rectangular microstrip antennas: The application of the transmission-line model to patch analysis is described. Various improvements to the basic model are noted, such as connections for mutual coupling between the radiating edges, that enable good agreement with measurements to be obtained. However, the attraction here is the method's simplicity and easy adaption to CAD, an example of which is given.

Array Aspects

11 Daniel, Penard and Terret: Design and technology of low-cost printed antennas: The design of elements and arrays with the emphasis on low-cost technology is important for successful application in many areas. Design and construction including array sythesis is described here. In addition, some technology and substrate innovations are included which can be compared to materials detailed in Chapter 15.

12 Pozar: Analysis and design considerations for printed phased-array antennas: The effects of scanning of printed phased arrays are derived using moments methods for both infinite and finite arrays of patches. Blind spots due to surface-wave effects are noted to be particularly severe where high-dielectric-constant substrates are used for millimetric integrated arrays. Some alternative integration technologies are discussed that mitigate these and other problems.

13 Ito, Teshirogi and Nishimura: Circularly-polarised-array antennas; Various types of circularly polarised arrays are reviewed together with the possible feeding arrangements. Some practical problems are considered including pattern control and bandwidth. Some examples of practical arrays are also highlighted.

Microstrip Technology

14 Owens: Microstrip antenna feeds: Printed antenna feeds are sometimes given insufficient consideration at the outset of an array design, thus degrading the array performance. Feed design is extensively reviewed here and comparative examples drawn from the literature are used to give engineering direction. Although considerable work has been done, further progress is expected as the importance of good feed design for printed antennas is more widely appreciated.

Introduction 23

15 Traut: Advances in substrate technology: Microwave substrates are one of the important 'enabling technologies' in printed antennas. Advances in this area are presented which given the reader some insight into manufacturing and environmental factors that affect the antenna's progress from conception to use. Progress here is determined to some extent by the volume of production, and it is hoped that as the applications proliferate substrate technology will continue to improve.

16 Levine: Special measurement techniques for printed antennas: Measurement characterisation of connectors, lines and discontinuities, together with analysis, form an important foundation to good antenna design. Such measurement characteristics are described here with particular emphasis on accuracy and applicability to design. Near-field probing can also form a useful diagnostic tool, and examples are given together with a novel method for efficiency measurement.

17 Zurcher and Gardiol: CAD of microstrip and triplate circuits: As noted above, computer-aided design is likely to be an increasingly important factor in printed antenna design. This Chapter is dedicated to CAD of microstrip and triplate systems and highlights some of the important aspects such as characterisation of components, materials, manufacturing, analysis and synthesis.

Applications

18 Derneryd: Resonant microstrip antenna elements and arrays for aerospace applications: Four examples of resonant microstrip antennas and arrays are presented here for various requirements including dual frequency, monopulse radiation pattern and dual polarisation. Important design freatures are noted together with some environmental aspects.

19 Fujimoto, Hori, Nishimura and Hirasawa: Applications in mobile and satellite systems: Mobile and satellite systems are an important area for low-cost low-profile printed antennas. Various examples of such antennas are given in this Chapter. It is likely that the explosive increase in information systems in the future will accelerate the development of antennas to meet these diverse needs.

20 Newham and Morris: Conical conformal microstrip tracking antenna: A conical conformal microstrip tracking antenna is a severe requirement that involves difficult manufacturing and fundamental electromagnetic problems to be solved. Here the authors have described progress to date in this very challenging area that is likely to require much more research and innovation for some time to come.

21 Frayne: Microstrip field diagnostics: The near-field probing technique noted in Chapter 16 is described here in some detail together with extensive results both for microstrip patches and patch arrays with feed networks.

24 Introduction

22 Sohtell: Microstrip antennas on a cylindrical surface: Patch arrays on cylindrical bodies is likely to be an important future application of conformal concepts. Modifications to the basic patch-design expressions due to the curvature are presented here together with design and performance of a representative cylindrical array.

23 Hall, Henderson and James: Extensions and variations to the microstrip-antenna concept: There are a wide variety of specialist applications that spawn innovative concepts in the use of microstrip antennas. This final Chapter highlights some of these, including applications where microstrip is combined with other radiating or transmission structures to form hybrid antennas. Operation over multi-octave bandwidths or at millimetric frequencies are also design challenges that are covered, together with applications involving very high and very low dielectric-constant materials. Such specialised requirements are likely to continue to lead antenna designers to further innovative progress in the microstrip field.

appeared in the literature and Chapter 15 contains one of the few appraisals of this area of printed technology. A diverse range of applications is also noted that mirror the list given in Table 1.5. In terms of printed-antenna techniques microstrip is the predominant one discussed in the handbook although some contributions on printed dipoles and ground-plane slots are presented. Although both these elements are likely to have similar electromagnetic properties to microstrip radiators, the increased complexity involved in manufacture seems to have been the overriding factor that has influenced designers away from them. This is particularly true for slot antennas in triplate stripline where shorting pins or holes are needed to prevent parallel-plate mode excitation. It is thus clear that the conceptual simplicity of microstrip remains one of its most attractive features.

As a final aid to the reader Table 1.11 highlights the advances described in each chapter with particular reference to the fundmental issues and challenges identified in Section 1.2.

1.4 Glossary of printed antenna types

The various forms of printed elements and arrays are very numerous and it is useful for designers to have a check list at hand. With this in mind we have composed the glossary giving an outline sketch, some key references and a few supporting comments and an indication of which chapter in the handbook deals with each type. Although the glossary is by no means exhaustive, the 70 entries on radiating elements and the 37 on arrays reflect the wide range of flexibility and scope for innovation that microstrip offers.

Introduction 25

Relevant Chapter numbers are given together with References in brackets.

(a) Patches

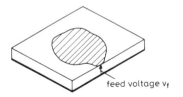

The *generic microstrip patch* is an area of metallisation supported above a ground plane and fed against the ground at an appropriate point or points

Principal shapes

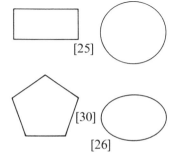

The freedom in the xy plane gives rise to the possibility of a multiplicity of possible shapes. Only a few have been seriously examined such as the *rectangular* or *square patch* and *disc* [25], *ellipse* [26], *equilateral*[27], or right-angled isoceles [28] *triangle*, *annular ring* [29] and *pentagon* [30].

2, 3,
4, 8,
9, 10,
11, 18,
21, 22

Characteristics of these principal shapes are generally similar [Chapter 3] with fundamental modes having broadside beam. Bandwidth and physical area vary between shapes. The annular ring gives increased bandwidth, gain and sidelobe levels for higher-order modes but becomes physically large.

Patches can be short-circuited along a null voltage plane to form the *shorted patch* or *hybrid microstrip antenna* [31]. Impedance and resonant frequency remain the same as for a full-size patch but for low dielectric constant the bandwidth is increased.

7, 11,
20

26 Introduction

Variants on principal shapes

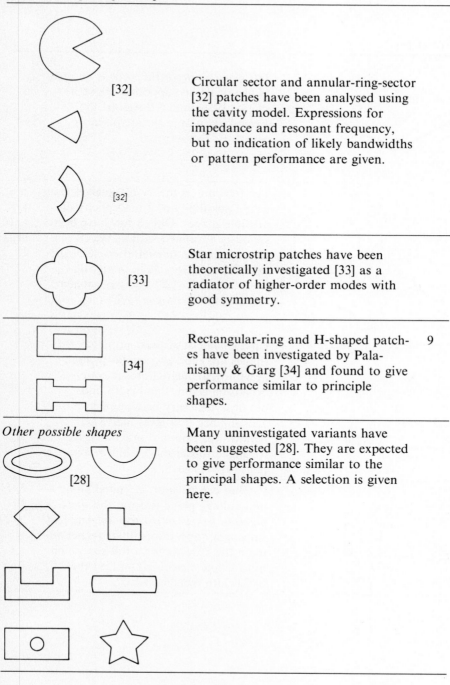

[32] Circular sector and annular-ring-sector [32] patches have been analysed using the cavity model. Expressions for impedance and resonant frequency, but no indication of likely bandwidths or pattern performance are given.

[33] Star microstrip patches have been theoretically investigated [33] as a radiator of higher-order modes with good symmetry.

[34] Rectangular-ring and H-shaped patches have been investigated by Palanisamy & Garg [34] and found to give performance similar to principle shapes.

Other possible shapes

[28] Many uninvestigated variants have been suggested [28]. They are expected to give performance similar to the principal shapes. A selection is given here.

Circularly polarised patches

Single point feeds

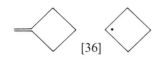
[36]

[37]

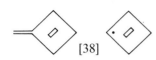
[38]

[39]

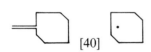
[40]

[26]

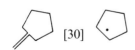
[30]

Single-point feeding gives circular polarisation with constructional simplicity. The feed excites two orthogonal degenerate modes [35]. The 90° excitation phase difference is obtained by detuning the two modes by a variety of geometrical distortions giving the following types:
rectangular patch [36]
notched square patch [37]
slotted square patch [38]
notched disc [39]
truncated corner square [40]
ellipse [26]
pentagon [30]

4, 9

The input VSWR bandwidth is wider than that of an isolated mode but the axial-ratio bandwidth is much narrower with axial ratio rising to about 6 dB at the edge of the 2:1 VSWR bandwidth. The above technique is shown for the fundamental mode. Use of internal slots in discs for higher-order modes has also been shown [41].

Multiple-point feed

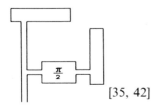
[35, 42]

Circular polarisation can be produced using multiple-point feeding by:

Two offset rectangular patches [35], which is also used in arrays [42]. Here the offset phase centre leads to a more rapid degradation of axial ratio off boresight than the following

4, 11, 13, 18, 19, 21

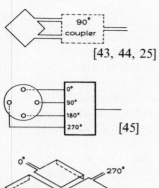

[43, 44, 25]

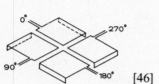

[45]

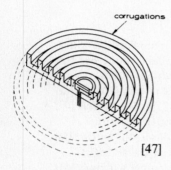

[46]

[47]

Single patches fed at two points [43, 44, 25]

Four-point feeding [45], which suppresses cross-polarisation generated by higher-order modes within the patch and this improves axial ratio

Short-circuit patches arranged to produce a *crossed slot* [46] which improves cross-polarisation at low angles

A patch in a corrugated ground plane [47], which also improves low-angle performance

Wideband patches			
Thick patches		Use of *low-dielectric-constant* ($\varepsilon_r \approx 1\cdot0$), *thick* ($h/\lambda_0 > 0\cdot1$) *substrate* results in bandwidths $> 10\%$ [48]. Alternatively patches on thinner substrates can be broadbanded (up to 30%) by *external matching circuits* [49]. Use of thick substrates leads to impedance-matching problems that can be overcome by use of *matching gaps* in the probe [50] or patch [51].	2, 4, 6, 7, 11, 18
	[48]		
	[49]		
	[50]		
	[51]		

Shaped patches [52] [53]	Patch shaping such as *steps* [52] or *conical depression* [53] also yields wide bandwidths	
Parasitics [54] [55]	*Multiple stacked patches* with the upper patches acting as electromagnetically coupled parasitics can be designed for extended bandwidths. Examples using *coaxial-probe feeding* [54] and *microstrip-line feeding* [55] have been made with the latter designed for an alumina ($\varepsilon_r = 10$) base substrate.	3
[56] [57, 58] [59]	Parasitics can also be mounted coplanar as *thin resonators* [56], as *additional patches* either gap or line coupled to square [57] or triangular patches [58]. Alternatively many *thin parasitics* [59] can be gap coupled to form a wideband patch. Some of these configurations exhibit variations in the radiation pattern with frequency.	3
[60] [61]	*Short-circuited quarter-wavelength parasitics* have also been applied to square [60] or circular [61] patches. A band-width increase of 2 is obtained in the square-patch case. In the circular case cross-polarisation is substantially reduced.	

30 Introduction

Other wideband forms

[62]

The *microstrip spiral* [62] gives about 40% bandwidth with limited efficiency. The spiral is limited to less than one turn, as further turns give rise to radiation-pattern degradation.

Dual-frequency patches

Multiple layers

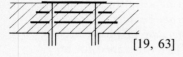
[19, 63]

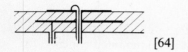
[64]

[65]

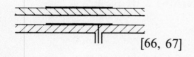
[66, 67]

Multiple-layer patches having two- [19] and three- [63] frequency operation use direct probe connection to the top patch and gap coupling to the lower ones. Direct connections to both patches in *two-frequency designs* [64, 65] have also been made. Tuning of the two frequencies is also possible by an *adjustable-height upper patch* using discs [66] and annular rings [67].

2, 3, 4, 5, 6, 18, 19

Single layers

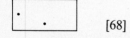
[68]

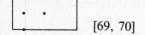
[69, 70]

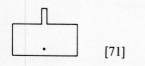
[71]

Single-layer two-frequency patches with *orthogonal polarisations* [68] using two feed points. Use of shorting pins [69, 70] allows operation with the same polarisation. The use of *tabs* in *rectangular* [71] or *circular* [72] patches also permits dual-frequency operation.

3, 4, 6

[72]

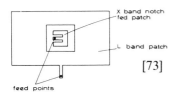
[73]

High-frequency patches have been located within low-frequency patches to give *orthogonal polarisation* [73] or same *circular polarisation* [74] using frequency-sensitive coupling stubs

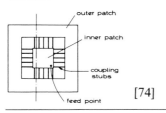
[74]

Other patch variants

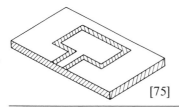
[75]

In the *coplanar stripline* patch [75] the input line is fed against the upper ground plane. The overall performance is similar to conventional patches with reduced cross-polarisation and mutual coupling.

[76]

[77]

The *electromagnetically coupled patch* [76] allows reduction in feed radiation by locating it closer to the ground plane than the patch. The effect of dielectric covers (*superstrates*) is noted [77], where, in addition to element protection, enhanced gain can be obtained.

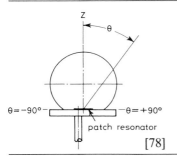
[78]

The use of *superimposed dielectric spheres* [78] on patches result in improved gain and reduced cross-polarisation.

23

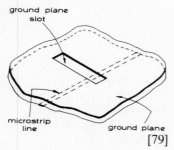

[79]

The *ground-plane slot* [79] fed by a microstrip line can be used as a bi-directional element or as a unidirectional one by the addition of a reflector.

11, 13

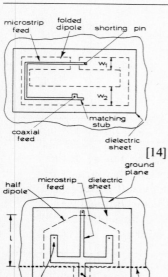

[14]

Folded dipoles can be operated close to a ground plane by means of appropriate matching circuits [14]. Many variants are possible giving wide bandwidth and low cross-polarisation. Construction is more complex than the basic microstrip patch although, owing to the use of matching circuits, wider impedance bandwidths may be possible.

7

[81]

Microstrip patches and arrays can be combined with the reflector concept. Their use as *feeds* [80, 81] allows integration with microwave integrated circuits, but with lower bandwidth and cross-polarisation compared to conventional feeds.

2, 23

Introduction   33

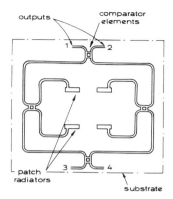

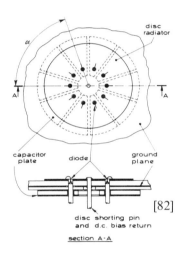

Patches have also been used in the *reflect array* configuration [82] where beam scanning is achieved by varying the phase of the reflected wave by pin diodes. A single element is shown here. Phase shift in the circularly polarised system is obtained by varying the angle of the short-circuit plane.

[82]

Conformal antennas

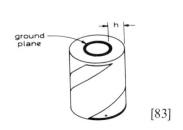

[83]

The flexibility of the microstrip concept allows use in conformal applications. Examples are:

— *Spiral slot* [83]

2, 7, 19, 20, 22

34 Introduction

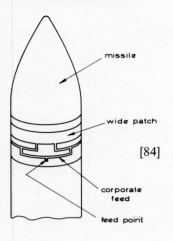

[84]

— *Wrap-around antenna* [84]: a single wide-quarter wavelength patch wrapped around a cylindrical body

— *Cylindrical* [85, 86] or *spherical* [87] patch arrays

[85]

Active patch

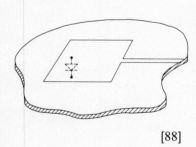

[88]

Active devices can be integrated into patches. An example with a Gunn diode [88] demonstrated the principle, but had high cross-polarisation and low patch efficiency.

(b) Arrays

Feed structures

Patch connection [89] [90] [91] [92] [76]	Patch elements for arrays can be connected by *through the substrate* pin connections (through hole plating or via holes), to one or more layers [89] of feed circuits located behind the ground plane in microstrip or triplate. Mechanical simplification can be achieved by *aperture coupling* to a *parallel* [90] or a *perpendicular* [91] microstripline. Feeding can also be from a *coplanar microstrip circuit* [92] which involves pattern perturbation due to feed radiation. This can be reduced by electromagnetic coupling to *overlaid patches* [76]. Connector effects give rise to fundamental limits to array action [93] due to radiation from the discontinuity in the guiding structure. Pin connections to patches give rise to higher-order modes [94] that perturb the radiation pattern and increase cross-polarisation levels.	5, 6, 11, 12, 14, 16, 18, 19, 20, 22
Feed circuits [95] [95]	Feed structures for many elements take various forms. *Corporate feeds* [95, 96] for either one- or two-dimensional arrays give wideband action, whilst series-fed arrays give narrow bandwidth with broadside beam when resonant or wide bandwidth with a scanning beam when travelling wave.	4, 5, 7, 11, 12, 13, 14, 18, 19, 20, 22

36 Introduction

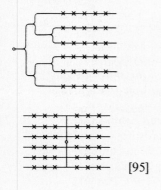

[95]

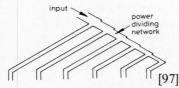

[97]

Wideband squintless operation is obtained with the *series-compensated* feed [97]

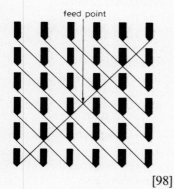

[98]

The *cross-fed* arrangement [98] gives narrowband action with a tapered distribution and hence low sidelobes with equal-width feed lines

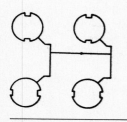

[22]

Sequentially rotated feeding [22] gives wide-bandwidth axial ratio and input VSWR for circularly polarised patch arrays

---

Array structures

Two-dimensional arrays

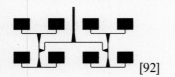

[92]

Microstrip patch arrays can be fed by any of the feed structures noted above. A typical *linearly polarised patch array* [92] fed by a coplanar microstrip corporate feed is shown here.

7, 11, 12, 13, 14, 18, 19, 20, 22

Introduction 37

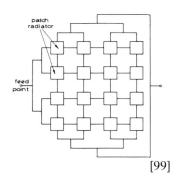
[99]

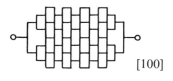
[100]

Dual polarisation is obtained by dual series interconnection of patches [99]. *Chain type* structures for linear polarisation having rectangular [100] and triangular or honeycomb shapes [101]. Both the series array of patches and chain arrays are resonant and thus have a narrow bandwidth.

One-dimensional arrays: linear polarisation

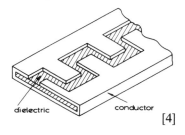
[4]

[102]

[103]

[103]

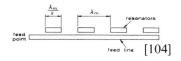
[104]

Early microstrip antennas were *series fed* one-dimensional arrays of Gutton and Bassinot [4] and Dumanchin [102] in 1955 and 1959, respectively. Since then many forms of series-fed array have been developed.

Arrays can be formed using resonant elements or meandering microstrip lines, in which the radiation is determined by radius of curvature or line width.

Examples of arrays using resonant elements are:

    comb line [103]

    parasitically coupled patch array [104]

    series-connected patches [105]

9, 10, 11, 13, 14, 18, 19

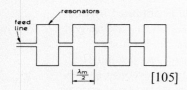

[105]

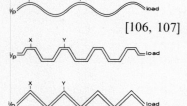

[106, 107]

Examples of arrays using meandering microstrip lines are:

    serpent line [106, 107]
    triangle or trapezoidal line [106]
    rampart line [108]
    chain line [109]
    Franklin line [110]

[106]

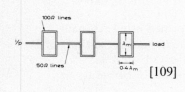

[108]

[109]

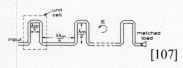

[110]

Circular polarisation

Circular polarisation is obtained from the *rampart line* [108], *chain line* [111] and *herringbone line* [112].

13, 14, 18

[107]

[111]

Introduction 39

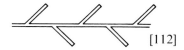
[112]

Other array forms

Various other forms of series fed linear arrays exist:

13, 23

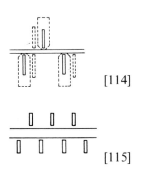
[113]

A $\lambda_m/2$ wide line can be made to radiate by feeding with an *asymmetric step* [113]. Alternatively *angled slots* can also force the line to radiate.

A combination of strip dipoles and slots can be used to form a *circularly polarised linear array* [114].

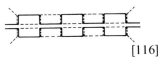
[114]

The high losses in long microstrip series arrays can be reduced by replacing the line by a *dielectric rod* [115]; radiation occurs by coupling the line energy to microstrip patches.

[115]

An *omni-directional array* has been made by forming alternating resonators in the line and ground plane [116].

[116]

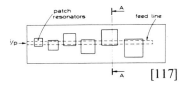
[117]

Multi-octave bandwidth operation can be obtained by a series-fed *log-periodic* arrangement of patches [117].

1.5 Summary comments

The historical development and future prospects of the microstrip antenna are reviewed to portray an invention that is now reaching maturity while its supporting research and development continues to expand unabated, driven by system demands for conformal low-cost radiators. Future activity will be dominated by both the creation of innovative designs to match system demands and the search for improved CAD techniques in array manufacture. The concept of distributed conformal sensors with integral signal processing is one projection for the distant future. For completeness the more common features of microstrip antennas, their applications and typical antenna design criteria are listed and briefly described prior to critically reviewing the outstanding design problems that are fundamental to microstrip antennas. The viewpoint of both researcher and antenna manufacturer is usefully taken to identify knowledge gaps and challenging issues that are vital to the advancement of the printed-antenna concept. The importance of the contributions in the present handbook in advancing the state-of-the-art is emphasied and each chapter briefly highlighted. Finally a glossary of microstrip antenna types is presented as an initial guideline for the designer.

1.6 References

1 GREIG, D. D. and ENGLEMAN, H. F.: 'Microstrip — a new transmission technique for the kilomegacycle range', *Proc. IRE*, 1952, **40**, pp. 1644–1650
2 DESCHAMPS, G. A.: 'Microstrip microwave antennas'. 3rd USAF Symposium on Antennas, 1953
3 LEWIN, L.: 'Radiation from discontinuities in stripline', *Proc. IEE*, 1960, **107C**, pp. 163–170
4 GUTTON, H. and BAISSINOT, G.: 'Flat aerial for ultra high frequencies'. French Patent No. 703113, 1955
5 FUBINI, E. G. *et al.*: 'Stripline radiators'. IRE Nat. Con. Rec., 3, 1955, pp. 51–55
6 McDONOUGH, J. A.: 'Recent developments in the study of printed antennas'. IRE Nat. Conv. Rec., 5, Pt. 1, 1957, pp. 173–176
7 DENLINGER, E. J.: 'Radiation from microstrip resonators', *IEEE Trans*, 1969, **MTT-17**, pp. 235–236
8 HOWELL, J. Q.: 'Microstrip antennas'. IEEE AP-S. Int. Symp. Digest, 1972, pp. 177–180
9 MUNSON, R. E.: 'Conformal microstrip antennas and microstrip phased arrays', *IEEE Trans.*, 1974, **AP-22**, pp. 74–78
10 Proc. of Workshop on Printed Circuit Antenna Technology, 17–19, Oct. 1979, New Mexico State Univ., Las Cruces, New Mexico
11 *IEEE Trans*, 1981, **AP-29**, (1)
12 BAHL, I. J. and BHARTIA, P.: 'Microstrip antennas' (Artech House, Dedham, Mass, 1980)
13 JAMES, J. R., HALL, P. S. and WOOD, C.: 'Microstrip antenna theory and design' (IEE, Peter Peregrinus, 1981)
14 DUBOST, G.: 'Flat radiating dipoles and applications to arrays' (Research Studies Press, Antenna Series No 1, 1980)
15 JAMES, J. R., HENDERSON, A., and HALL, P. S.:'Microstrip antenna performance is determined by substrate constraints', *Microwave Syst. News*, 1982, **12**, (8), pp. 73–84

16 HALL, P. S., and JAMES, J. R.' 'Cross polarisation behaviour of series fed microstrip linear arrays', *IEE Proc.*, 1984, **131H**, pp. 247–257
17 HALL, P. S., and HALL, C. M.: 'Coplanar corporate feed design effects in microstrip patch array design', *IEE Proc.*, 1988 **135, H.**
18 HENDERSON, A., JAMES, J. R., and HALL, C. M.: 'Bandwidth extension techniques in printed conformal antennas'. Military Microwaves, MM 86, Brighton, June 1986, pp. 329–334
19 LONG, S. A., and WALTON, M. D.: 'Dual frequency stacked circular disc antenna', *IEEE Trans.*, 1979, **AP-27**, pp. 270–273
20 KUMAR, G., and GUPTA, K. C.: 'Non radiating edge and four edges gap coupled multiple resonator broad band microstrip antennas,' *IEEE Trans.*, 1985, **AP-33**, pp. 173–177
21 RUDGE, A. W., MILNE, K., OLVER, A. D., and KNIGHT, P.: 'Handbook of antenna design' (IEE, Peter Peregrinus, 1982) pp. 24–28
22 TESHIROGI, T., TANAKA, M., and CHUJO, W.: 'Wideband circularly polarised array with sequential rotation', Proc ISAP, Tokyo, Japan, Aug 1985, pp. 117–120
23 KISHK, A. A., and SHAFAI, L.: 'Effect of various parameters of circular microstrip antennas on their radiation efficiency and the mode excitation', *IEEE Trans.*, 1986, **AP-34**, pp. 969–977
24 HORI, T., TERADA, N., and KAGOSHIMA, K.: 'Electronically steerable spherical array antenna for mobile earth station'. IEE Conf. on Ant. and Prop., ICAP 87, York, pp. 55–58
25 HOWELL, J.Q.: 'Microstrip Antennas', *IEEE Trans*, 1975, **AP-23**, pp. 90–93
26 SHEN, L. C.: 'The elliptical microstrip antenna with circular polarisation', *IEEE Trans*, 1981, **AP-29**, pp. 90–94
27 LUK, K. M., LEE, K. F., and DAHELE, J. S.: 'Theory and experiment on equilateral triangular microstrip antenna'. Proc 16th European Microwave Conference, 1986
28 Reference 12, pp. 139–153
29 CHEW, W. C.: 'Broadband annular ring microstrip antenna', *IEEE Trans*, 1982, **AP-30**, pp. 918–922
30 WEINSCHEL, H. D.: 'Cylindrical array of circularly polarised microstrip antennas'. IEEE AP-S Int. Symp. Dig., 1975, pp. 177–180
31 PENARD, E., and DANIEL, J. P.: 'Open and hybrid microstrip antennas', *IEE Proc.*, 1984, **131, H,** (1)
32 RICHARDS, W. F., OU, J. D., and LONG, S.A.: 'Theoretical and experimental investigation of annular, annular sector and circular sector microstrip antennas', *IEEE Trans*, 1984, **AP-12**, pp. 864–866
33 PARASNIS, K., SHAFAI, L., and KUMAR, G.: 'Performance of star microstrip as a linearly and circularly polarised TM_{21} mode radiator', *Electron. Lett.*, 1986, **22**, pp. 463–464
34 PALANISAMY, V and GARG, R.: 'Rectangular ring and H-shaped microstrip antennas: Alternatives to rectangular patch antenna', *Electron Lett.*, 1985, **21**, pp. 874–876
35 Reference 13, pp. 194–224
36 SANFORD, G. G., and MUNSON, R. E.: 'Conformal VHF antenna for the Apollo-Soyuz test project'. IEE Int. Conf. on Antennas for Aircraft and Spacecraft, London, pp. 130–135
37 OSTWALD, L. T., and GARVIN, C. W.: 'Microstrip command and telemetry antennas for communications and technology satellites'. IEE Int. Conf. on antennas for Aircraft and Spacecraft, London, pp. 217–222
38 KERR, J.: 'Microstrip antenna developments'. Workshop on Printed Antenna Technology, New Mexico State University, 1979, pp. 3.1–3.20
39 HANEISHI, M., *et al.*: 'Broadband microstrip array composed of single feed type circularly polarised microstrip element'. IEEE AP-S Int. Symp. Dig, May 1982, pp. 160–163
40 SHARMA, P. C., and GUPTA, K. C.: 'Analysis and optimised design of single feed circularly polarised microstrip antennas', *IEEE Trans.* 1983, **AP-31**, (6)
41 MARTIN PASCUAL, C., FONTECHA, J. L., VASSAL'LO, J., and BARBERO, J.: 'Land mobile antennas for satellite communications'. IEE Conf. on Mobile Satellite Systems, 1984, pp. 145–149

42 HUANG, J.: 'Technique for an array to generate circular polarisation with linearly polarised elements'. *IEEE Trans*, 1986, **AP2-34**, (9)
43 BRAIN, D. J., and MARK, J. R.: 'The disc antenna — A possible L band aircraft antenna'. IEE Conf. Publ. 95, Satellite Systems for Mobile Communications and Surveillance, 1973, pp. 14–16
44 MUNSON, R. E.: 'Conformal microstrip antennas and microstrip phased arrays,' *IEEE Trans*, 1974, **AP-22**, pp. 74–78
45 CHIBA, T., SUZUKI, Y., MIYANO, N., MIURA, S., and OHMORI, S.: 'Phased array antenna using microstrip antennas'. 12th European Microwave Conference, Finland, 1982, pp. 475–477
46 SANFORD, G. G., and KLEIN, L.: 'Recent developments in the design of conformal microstrip phased arrays'. IEE Conf. Publ. 160, Maritime and Aeronautical Satellites for Communication and Navigation, London, 1978, pp. 105–108
47 BAILEY, M.C.: 'A broad beam circularly polarised antenna'. IEEE AP–S Symp. Stanford, USA, 1977
48 CHANG, E., LONG, S. A., and RICHARDS, W. F.: 'Experimental investigation of electrically thick rectangular microstrip antennas', *IEEE Trans*, 1986, **AP-34**, (6)
49 GRIFFIN, J. M., and FORREST, J. R.: 'Broadband circular disc microstrip antenna', *Electron. Lett.*, 1982, **18**, pp. 26–269
50 FONG, K. S., PUES, H. F., and WITHERS, M. J.: 'Wideband multilayer coaxial fed microstrip antenna element', *Electron Lett.* 1985, **21**, pp. 497–499
51 HALL, P. S.: 'Probe compensation in thick microstrip patches', *Electron. Lett.*, 1987, **23**, pp. 606–607
52 PODDAR, D. R., CHATTERJEE, J. S., and CHOWDHURY, S. K.: 'On some broad band microstrip resonators', *IEEE Trans*, 1983, **AP-31**, (1)
53 CHATTERJEE, J. S.: 'Conically depressed microstrip patch antenna', *IEE Proc*, 1983, **130, H**, pp. 193–196
54 SABBAN, A.: 'New broadband stacked two layer microstrip antenna'. IEEE AP–S Symp., Houston, 1983, pp. 63–66
55 HALL, P. S., WOOD, C., and GARRETT, C.: 'Wide bandwidth microstrip antennas for circuit integration', *Electron Lett.* 1979, **15**, pp. 458–460
56 SCHAUBERT, D. H., and FARRAR, F. G.: 'Some conformal printed circuit antenna designs'. Proc Workshop of Printed Antenna Technology, New Mexico State University, 1979, pp. 5.1–5.21
57 KUMAR, G., and GUPTA, K. C.: 'Non-radiating edge and four edges gap coupled multiple resonator broad band microstrip antennas', *IEEE Trans.*, 1985, **AP-33**, pp. 173–177
58 BHATNAGAR, P. S., DANIEL, J. P., MAHDJOUBI, K., and TERRET, C.: 'Hybrid edge, gap and directly coupled triangular microstrip antenna', *Electron Lett.*, 1986, **22**, pp. 853–855
59 AANANDAN, C. K., and NAIR, K. G.: 'A compact broad band microstrip antenna', *Electron Lett.*, 1986, **22**, pp. 1064–1065
60 WOOD, C.: 'Improved bandwidth of microstrip antennas using parasitic elements', *IEE Proc.*, 1980, **127H**, pp. 231–234
61 PRIOR, C., and HALL, P. S.: 'Microstrip disc antenna with short circuit annular ring, *Electron Lett.*, 1985, **21**, pp. 719–721
62 WOOD, C.: 'Curved microstrip lines as compact wideband circularly polarised antennas', *IEE J. MOA*, 1979, **3**, pp. 5–13
63 MONTGOMERY, N. W.: 'Triple frequency stacked microstrip element', IEEE AP–S, Boston, MA, June 1984, pp. 255–258
64 SENSIPER, S., WILLIAMS, D., and MCKONE, J. P.: 'An integrated global positioning satellite antenna low noise amplifier system'. ICAP 87, York, IEE Conf. Publ. 274, 1987, pp. 51–54
65 JONES, H. S., SCHAUBERT, D. H., and FARRAR, F. G.: 'Dual frequency piggyback antenna', US Patent No 4 162 499, 24 July 1979

Introduction 43

66 DAHELE, J. S., and LEE, K. F.: 'Dual frequency stacked microstrip antenna'. IEEE AP-S Int. Symp. Dig. 1982, pp. 308–311
67 DAHELE, J. S., LEE, K. F., and WONG, D. P.: 'Dual frequency stacked annular ring microstrip antenna', *IEEE Trans*, 1987, **AP-35**,
68 GRONAU, G., MOSCHURING, H., and WOLFF, I.: 'Input impedance of a rectangular microstrip resonator fed by a microstrip network on the backside of the substrate'. 14th European Microvwave Conference, Liege, Sept. 1984, pp. 625–630
69 ZHONG, S. S., and LO, Y. T.: 'Single-element rectangular microstrip antenna for dual frequency operation', *Electron Lett.*, 1983, **19**, pp. 298–300
70 KUBOYAMA, H., HIRASAWA, K., and FUJIMOTO, K.: 'Post loaded microstrip antenna for pocket size equipment at UHF'. Int. Symp. on Ant. & Prop., Japan, 1985, pp. 433–436
71 RICHARDS, W. F., DAVIDSON, S. E., and LONG, S. A.: 'Dual band reactively loaded microstrip antenna'. *IEEE Trans*, 1985, **AP-33**, pp. 556–561
72 McILVENNA, J., and KERNWEIS, N.: 'Modified circular microstrip antenna elements', *Electron Lett.*, 1979, **15**, pp. 207–208
73 KERR, J.: 'Other microstrip antenna applications', Proc 1977 Antenna Applications symposium, Illinois, USA
74 SANFORD, G. G., and MUNSON, R. E.: 'Conformal VHF Antenna for the Apollo-Soyuz Test Project'. IEE Conf. Publ. 128, Antennas for Aircraft and Spacecraft, 1975, pp. 130–135
75 GREISER, J. W.: 'Coplanar stripline antenna', *Microwave J.*, Oct. 1976, pp. 47–49
76 OLTMAN, H. G.: 'Electromagnetically coupled microstrip dipole antenna elements'. 8th European Microwave Conference, Paris, Sept. 1978, pp. 281–285
77 ALEXOPOULUS, N. G., and JACKSON, D. R.: 'Fundamental superstrate (cover) effects on printed circuit antennas', *IEEE Trans*, 1984, **AP-32**, pp. 807–816
78 JAMES, J. R., HALL, C. M., and ANDRASIC, G.: 'Microstrip elements and arrays with dielectric overlays' *IEE Proc.*, 1986, **133**, H, (6)
79 YOSHIMURA, Y.: 'A microstrip slot antenna', *IEEE Trans*, 1972. **MTT 22**, pp. 760–762
80 HALL, P. S., and PRIOR, C. J.: 'Microstrip feeds for prime focus red reflector antennas', *IEE Proc.* 1987, **134**, H, pp. 185–193
81 OLTMAN, H. G., WEEMS, D. M., LINDGREN, G. M., and WALTON, F. D.: 'Microstrip components for low cost millimeter wave missile seekers', AGARD Conf. Proc. 245, Millimeter and Submillimeter Wave Propagation and Circuits, Munich, 1978, pp. 27.1–27.9
82 MONTGOMMERY, J. P.: 'Microstrip reflect array antenna element'. Proc 1978 Antenna Applications Symposium, Illinois, USA
83 SINDORIS, A. R., SCHAUBERT, D. H., and FARRAR, F. G.: 'The spiral slot — A unique microstrip antenna'. IEE Conf. Proc. Ant. & Prop., London, 1978, pp. 150–154
84 MUNSON, R. E.: 'Conformal microstrip antennas and microstrip phased arrays', *IEEE Trans*, 1974, **AP-22**, pp. 74–78
85 SOHTELL, E. V., and STARSKI, J. P.: 'Cylindrical microstrip patch phased array antenna–Chalscan C'. Military Microwaves Conf., Brighton, June 1986, pp. 317–322
86 LUK, K. M., LEE, K. F., and DAHELE, J. S.: 'Input impedance and Q factors and cylindrical–rectangular microstrip patch antennas', ICAP 87, York, IEE Conf. Publ. 274, 1987, pp. 95–99
87 SEEHAUSEN, G: 'Polarisation control of conformal arrays consisting of linerly polarised elements', ICAP 83, Norwich, IEE Int. Conf. on Ant. & Prop., 1983, pp. 154–157
88 THOMAS, H. J., FUDGE, D. L., and MORRIS, G.: 'Active patch antenna', Proc. Military Microwave Conf., June 1984, pp. 246–249
89 OWENS, R. P., and SMITH, A. C.: 'Dual band, dual polarisation microstrip antenna for X band satellite communications', Military Microwaves Conf., Brighton, June 1986, pp. 323–328
90 POZAR, D. M.: 'Microstrip antenna aperture coupled to a microstrip line', *Electron Lett.*, 1985, **21**, pp. 49–50
91 BUCK, A.C., and POZAR, D. M.: 'Aperture coupled microstrip antenna with a perpendicular feed, *Electron Lett.*, 1986, **22**, pp. 125–126

92 HALL, P. S., and PRIOR, C. J.: 'Radiation control in corporately fed microstrip patch arrays'. JINA 86, Journeees Internationales de Nice sur les Antennes, 1986, pp. 271–275
93 HENDERSON, A., and JAMES, J. R.: 'Design of microstrip antenna feeds — Pt 1: Estimation of radiation loss and design implications', *IEE Proc*, 1981, **128H**, (1), pp 19–25
94 LO, Y. T., SOLOMON, D., and RICHARDS, W. F.: 'Theory and experiment on microstrip antennas', *IEEE Trans*, 1979, **AP-27**, pp. 137–145
95 Reference 13, pp. 116 and 161
96 HALL, P. S., and JAMES, J. R.: 'Design of microstrip antenna feeds — Pt 2: Design and performance limitations of triplate corporate feeds', *IEE Proc.*, 1981, **128H**, pp. 26–34
97 ROGERS, A.: 'Wideband squintless linear arrays', *Marconi Rev.*, 1972, **187**, pp. 221–243
98 WILLIAMS, J. C.: 'Cross fed printed aerials'. Proc 7th European Microwave Conference, Copenhagen, Sept 1977, pp. 292–296
99 DERNERYD, A. G.: 'Linearly polarised microstrip antennas', *IEEE Trans.*, 1976, **AP-24**, pp. 846–851
100 TIURI, M., HENRIKSSON, J., and TALLQUIST, S.: 'Printed circuit radio link antenna', 6th European Microwave Conference, Rome, Sept 1976, pp. 280–282
101 HILL, R.: 'Printed planar resonant arrays', ICAP 87, York, IEE Int. Conf. on Ant. & Prop., 1987, pp. 473–476
102 DUMANCHIN, R.: 'Microstrip aerials'. French Patent Application 855234, 1959
103 JAMES, J. R., and HALL, P. S.: 'Microstrip antennas and arrays — Pt. 2: New design technique'. IEE J. MOA 1977, **1**, pp. 175–181
104 CASHEN, E. R., FROST, R., and YOUNG, D. E.: 'Improvements relating to aerial arrangements'. British Provisional Patent (EMI Ltd) Specification 1294024.
105 METZLER, T.: 'Microstrip series arrays', *IEEE Trans.*, 1981, **AP-29**, pp. 174–178
106 TRENTINI, VON G,: 'Flachantenna mit periodisch gebogenem leiter', *Frequenz*, 1960 **14**, pp. 230–243
107 SKIDMORE, D. J., and MORRIS, G.: 'Design and performance of covered microstrip serpent antennas'. ICAP 83, Norwich, IEE Int. Conf. on Ant. and prop., 1983, pp. 295–300
108 HALL, P. S.: 'Microstrip liner array with polarisation control'. *IEE Proc.*, 1983, **130H**, pp. 215–224
109 TIURI, M., TALLQUIST, S., and URPO, S.: 'The chain antenna', IEEE SP-S Int. Symp., Atlanta, USA, pp. 274–277
110 NISHIMURA, S., NAKANO, K., and MAKIMOTO, T.: 'Franklin-type microstrip line antenna', IEEE AP-S, Int. Symp., Seattle, pp. 134–137
111 HENRIKSSON, J., MARKUS, K., and TIURI, M.,: 'Circularly polarised travelling wave chain antenna'. Proc 9th European Microwave Conf., Brighton, 1979
112 JAMES, J. R., and WILSON, G. J.: UK Patent Specification No 1529361, 18 Oct. 1978
113 MENZEL, W.: 'New travelling wave antenna in microstrip', Proc. 8th European Microwave Conference, Paris, 1978, pp. 302–206
114 ITO, K., ITOH, K., and KOGO, H.: 'Improved design of series fed circularly polarised printed linear arrays', *IEE Proc.*, 1986, **133**, H, pp. 462466
115 JAMES, J. R., JOHN, G., and HALL, C.M.: 'Millimetric-wave dielectric-microstrip antenna array', *IEE Proc.*, 1984, **131**, H, pp. 341–350
116 HILL, R.: 'Twin line omni-directional aerial configuration', Proc. 8th European Microwave Conference, Sept, 1978, pp. 307–311
117 HALL, P. S.: 'Multi-octave bandwidth log periodic microstrip antenna array', *IEE Proc.* 1986, **133**, H, pp. 127–137

Chapter 2

Analysis of circular microstrip antennas

L. Shafai and A. A. Kishk

2.1 Introduction

Microstrip antennas are finding increasing popularity owing to their advantages in size, cost, conformity to the supporting structure and ease of fabrication [1] [2], [3]. To analyse their impedance and radiation properties many elaborate analytical techniques are proposed and used. Numerical methods are also developed and have received increasing attention in recent years, these being primarily based on Sommerfeld-type integral equations. All these methods, which are discussed in following Chapters of this handbook have one important assumption in common: they assume that the dielectric substrate and the supporting ground plane are infinite in extent. The solutions are therefore valid for infinite geometries, or when the substrate and ground-plane dimensions are relatively large. The assumption does not introduce a severe difficulty in impedance calculations since microstrip geometries are inherently resonant structures and their impedance characteristic is primarily controlled by the printed elements. However, difficulty arises in predicting the radiation patterns, where, for small antenna dimensions, diffraction effects alter the side and back radiations. Consequently, the Geometrical Theory of Diffraction is occasionally used in conjunction with other methods to improve the radiation-pattern predictions [4, 5].

Accurate formulation of the electromagnetic problem of microstrip antennas is feasible. But, for finite substrate and ground-plane sizes the formulation must be solved numerically. In this Chapter we present a general formulation which is based on the concept of equivalence principle, and provides integral equations for the field distribution on the surfaces of the conductors and dielectric substrate. The formulation is exact and satisfies all boundary conditions. However, since it involves the field distributions on the substrate and ground plane, the numerical solution of the resulting integral equations is efficient only for small antenna dimensions. The problem is considerably simpler for axisymmetric geometries, where the surface distributions can be expanded in terms of the azimuthal modes representing the physical modes of the structure. Consequently, since microstrip antennas support only a limited number of modes, the numerical solutions for accurate field representations are readily obtainable. In addition, the modal expansion of the fields reduces the problem to the solution of matrix equations for each individual mode and simplifies computation considerably. For this reason, all computed results in this Chapter are presented for circularly symmetric configurations. The formulations can, however, be used with additional labour for the investigation of other microstrip configurations as well.

For instance, when the microstrip geometry is non-circular, or even arbitrary in shape, one can use a surface-patch segmentation over the conducting and dielectric surfaces. The current and surface distributions can then be represented by appropriate basis functions over these patches to convert the integral equations to a matrix equation using a moment method. The solution of the resulting matrix equation gives the surface distributions over the conducting and dielectric surfaces. However, since the segmentation is over the entire surface the matrix size is normally large. In addition, the method gives numerical results for the surface distributions and fails to provide information on the modal excitation. This difficulty can be overcome by expanding the current distributions in terms of patch eigen functions, in which case the procedure becomes similar to the case of axisymmetric configurations discussed previously.

Since the existing solutions in the literature predict microstrip-antenna impedance properties accurately, no attempt is made here to investigate the antenna impedances. Instead, emphasis is put on predicting the radiation patterns and investigating the effects of microstrip dimensions on them. Consequently, to simplify the analysis, excitation sources are replaced by simple electric dipoles. No significant effect is anticipated by this source simplification, since the resonant nature of microstrip antennas controls their mode excitation, and thus radiation patterns.

The generated equations are used to investigate the radiation properties of three different antennas; namely, a circular microstrip patch antenna, a wrap-around antenna and the reflector feeds. They are fundamentally different antennas and selected to provide complementary analysis and design information. For instance, the circular patch is one of the basic microstrip antennas. Its radiation characteristics and mode excitations are studied in length and the effect of the ground-plane size and other dimensional or material parameters on its radiation patterns are investigated. The results, although computed for a circular patch geometry, provide information for precise understanding of the radiation properties of resonant patch antennas. The wrap-around antenna is selected to show that the formulation can be used to investigate any axisymmetric antenna configuration. It is also shown that multiple source excitation can be used to control mode excitation, and consequently the radiation patterns. The last example, i.e. the reflector feed, is included to indicate the usefulness of the method for design of precision antennas. The reflector feeds should not only provide an efficient illumination function, but their cross-polarisation and the

Analysis of circular microstrip antennas 47

phase-centre location must also be controlled precisely. The numerical method provided in this Chapter enables accurate and efficient generation of pattern data, which, when coupled with optimisation algorithms, gives antenna-design parameters to meet stringent performance requirements.

2.2 Formulation of the problem

The electromagnetic problem involving microstrip antennas deals with determination of the field components in the presence of conductors and dielectrics. The boundary conditions to be satisfield are therefore of mixed type. This requires vanishing of the tangential electric-field components on the conductors and continuity of the tangential electric and magnetic components over the dielectrics. Because practical geometries are finite in size, an exact analytical solution cannot be found to satisfy all boundary conditions. A numerical solution must therefore be utilised. In this regard, two formulation types can be developed. One involves volume integral equations for the polarisation currents in the dielectrics and the induced surface currents on the conductors. This type of formulation is not convenient to work with, but is general enough to handle inhomogeneous dielectrics. For homogeneous dielectrics a convenient formulation can be developed in terms of the tangential field components over the boundary surfaces. The resulting integral equation includes all boundary conditions and the formulation is therefore an exact one. Thus the solution accuracy will depend on the management of the problem thereafter and on the numerical algorithms used to determine the unknown surface distributions. In the following Sections we shall provide integral-equation formulations only for the surface distributions, and present numerically generated data for several known antenna configurations.

The formulations may be derived from the use of the equivalence principle [6, 7]. To proceed we select a general electromagnetic problem shown in Fig. 2.1a, where a homogeneous dielectric material is sandwiched between two conducting layers. The surfaces S_{ce}, S_{cd} and S_{de} refer, respectively, to the boundaries between the conductors and the exterior region, the conductors and the dielectric, and the exterior region with the dielectric. Similarly, $\bar{E}^d$, $\bar{H}^d$ and $\bar{E}^e$, $\bar{H}^e$ refer to the field vectors within the dielectric and exterior regions, respectively. The dielectric region has a volume V_d, bounded by surfaces S_{cd} and S_{de}, and its material parameters are ε_d and μ_d. The exterior region has a volume V_e and its permittivity and permeability are defined by ε_e and μ_e, respectively. The excitation sources are provided by impressed electric and magnetic currents $\bar{J}^{id}$ and $\bar{M}^{id}$, within the dielectric.

We now may invoke the equivalence principle to reduce the complex original problem to two simpler ones [8], involving the exterior and interior regions. Fig. 2.1b shows the external equivalence. The combined volume of the conductors and the dielectric is bounded by S_{ce} and S_{de} and supports equivalent currents $\bar{J}_{ce}$,

48 Analysis of circular microstrip antennas

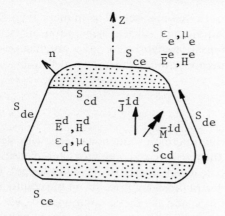

Fig. 2.1a *Original problem*

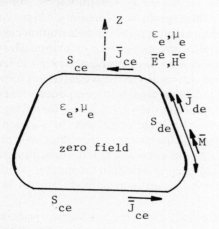

Fig. 2.1b *External equivalence*

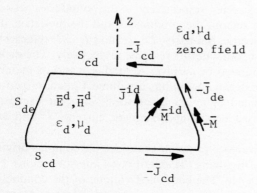

Fig. 2.1c *Internal equivalence*

$\bar{J}_{de}$ and $\bar{M}$. These currents radiate in a homogeneous medium (ε_e, μ_e) and produce ($\bar{E}^e$, $\bar{H}^e$) in V_e and zero field within the bounded region. Here, $\bar{J}_{ce}$ is the electric current on S_{ce} and $\bar{J}_{de}$ and $\bar{M}$ are the electric and magnetic currents on S_{de}. The internal equivalence is shown in Fig. 2.1c, where the volume V_d is enclosed by S_{cd} and S_{de}. The equivalent currents are $-\bar{J}_{cd}$, $-\bar{J}_{de}$ and $-\bar{M}$ and together with $\bar{J}^{id}$ and $\bar{M}^{id}$ radiate in a homogeneous medium (ε_d, μ_d) to produce ($\bar{E}^d$, $\bar{H}^d$) in V_d and zero field elsewhere. Again, $-\bar{J}_{cd}$ is the electric current on S_{cd}, and $-\bar{J}_{de}$ and $-\bar{M}$ are the electric and magnetic currents on S_{de}. Since, in the original problem of Fig. 2.1a the surfaces S_{ce} and S_{cd} are perfectly conducting, they support only equivalent electric currents in Figs. 2.1b and 2.1c. The negative-sign relationship between the aperature currents of Figs. 2.1b and 2.1c is dictated by the zero-field stipulations and the continuity of the tangential electric and magnetic field components across the aperature surface S_{de} of Fig. 2.1a. However, the selection of the negative sign for $-\bar{J}_{cd}$ in Fig. 2.1c is not mandatory and is made to match the negative sign of the aperature currents.

In the above example, the application of the equivalence principle reduced a complex multi-region problem to two simpler ones involving homogeneous regions. The field components in each region can therefore be found readily from the equivalent currents. However, these equivalent currents are still unknown and must be determined. This can be achieved by enforcing the boundary conditions on the field vectors of the original problem in Fig. 2.1a. The boundary conditions to be satisfied are:

$$\begin{aligned}\hat{n} \times \bar{E}^d &= 0 & \text{on } S_{cd} \\ \hat{n} \times \bar{E}^e &= 0 & \text{on } S_{ce} \\ \hat{n} \times \bar{E}^d &= \hat{n} \times \bar{E}^e & \text{on } S_{de} \\ \hat{n} \times \bar{H}^d &= \hat{n} \times \bar{H}^e & \text{on } S_{de}\end{aligned} \quad (2.1)$$

and the surface equivalent currents are

$$\begin{aligned}\bar{J}_{cd} &= \hat{n} \times \bar{H}^d & \text{on } S_{cd} \\ \bar{J}_{ce} &= \hat{n} \times \bar{H}^e & \text{on } S_{ce} \\ \bar{J}_{de} &= \hat{n} \times \bar{H}^e \\ \bar{M} &= -\hat{n} \times \bar{E}^e \end{aligned}\Bigg\} \text{on } S_{de} \quad (2.2)$$

Again, the currents $\bar{J}_{cd}$, $\bar{J}_{ce}$ and $\bar{J}_{de}$ are the equivalent electric currents on each respective surface and $\bar{M}$ is the magnetic current on the interface surface between the dielectric and the exterior region. The field components in eqns. 2.1 can be determined from these equivalent currents and provide the following field relationships:

$$-\bar{E}_{tan}^d(\bar{J}_{cd} + \bar{J}_{de}, \bar{M}) = -\bar{E}_{tan}^d(\bar{J}^{id}, 0) \quad \text{on } S_{cd} \quad (2.3)$$

$$-\bar{E}_{tan}^e(\bar{J}_{ce} + \bar{J}_{de}, \bar{M}) = 0 \quad \text{on } S_{ce} \quad (2.4)$$

$$-\bar{E}_{tan}^e(\bar{J}_{ce} + \bar{J}_{de}, \bar{M}) - \bar{E}_{tan}^d(\bar{J}_{cd} + \bar{J}_{de}, \bar{M})$$
$$= -\bar{E}_{tan}^d(\bar{J}^{id}, 0) \quad \text{on } S_{de} \quad (2.5)$$

$$-\bar{H}^e_{tan}(J_{ce} + J_{de}, \bar{M}) - \bar{H}^d_{tan}(J_{cd} + J_{de}, \bar{M})$$
$$= -\bar{H}^d_{tan}(J^{id}, 0) \quad \text{on } S_{de} \qquad (2.6)$$

where $\bar{E}^e(J, \bar{M})$ and $\bar{E}^d(J, \bar{M})$ represent the electric fields due to the currents J and $\bar{M}$, radiating in media characterised by ε_e, μ_e, and ε_d, μ_d, respectively. $\bar{H}^e(J, \bar{M})$ and $\bar{H}^d(J, \bar{M})$ are the associated magnetic fields. Note that, since the equivalent currents are still unknown, the field eqns. 2.3 – 2.6 represent integral equations for these currents. These integral equations can be generated using appropriate vector potentials, in terms of which the field vectors are given by

$$\bar{E}^q(J, \bar{M}) = -j\omega \bar{A}^q - \nabla \Phi^q - \frac{1}{\varepsilon_q} \nabla \times \bar{F}^q \qquad (2.7)$$

$$\bar{H}^q(J, \bar{M}) = -j\omega \bar{F}^q - \nabla \Psi^q + \frac{1}{\mu_q} \nabla \times \bar{A}^q \qquad (2.8)$$

where

$$\bar{A}^q = \mu_q \int_s J\, G^q\, ds \qquad (2.9)$$

$$\bar{F}^q = \varepsilon_q \int_s \bar{M}\, G^q\, ds \qquad (2.10)$$

$$\phi^q = \frac{1}{\varepsilon_q} \int_s \sigma\, G^q\, ds \qquad (2.11)$$

$$\Psi^q = \frac{1}{\mu_q} \int_s m\, G^q\, ds \qquad (2.12)$$

and

$$\sigma = \frac{-1}{j\omega} \nabla_s \cdot J \qquad (2.13)$$

$$m = \frac{-1}{j\omega} \nabla_s \cdot \bar{M} \qquad (2.14)$$

The function G^q is the scalar Green's function and is given by

$$G^q = \frac{\exp(-jk_q R)}{4\pi R} \qquad (2.15)$$

where $R = |r - r'|$ is the distance between the field point r and the source point r' on the surface, $k_q = \omega(\varepsilon_q \mu_q)^{1/2}$ is the propagation constant of the region and q represents e or d.

2.2.1 Matrix formulation

The above formulations provide integral equations valid for any combination of dielectric and conducting bodies of arbitrary shape. They can be solved for the unknown currents by a non-linear optimisation routine or after linearisation of the relationships by an application of a moment method. However, since the

Analysis of circular microstrip antennas 51

formulations involve the surface distributions on the dielectric, their numerical solution for arbitrarily shaped geometries may require excessive computer time and storage. The problem is considerably simpler for axisymmetric geometries,

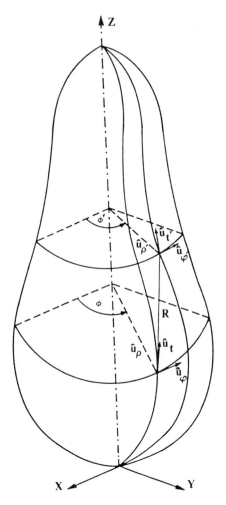

Fig. 2.2 *Geometry of the body of revolution*

where the field vectors can be expressed in Fourier series of the azimuthal co-ordinate. A solution can therefore be generated separately for each Fourier component, resulting in reduced computation time and storage. This is particularly important in microstrip antennas, which are highly resonant and often support only one of the azimuthal modes. For this reason, we shall restrict the remaining material of this Chapter to the development of solutions for axisymmetric geometries.

Fig. 2.2 shows a simple representation of an axisymmetric object, generally known as a body of revolution. The surface tangents can be defined along the generating curve t and the azimuthal co-ordinate ϕ. They are shown in Fig. 2.2 and form an orthogonal curvi-linear co-ordinate system on the surface of the body. Because the geometry is rotationally symmetric the surface co-ordinates can be represented conveniently in terms of ϱ, ϕ and z co-ordinates in a cylindrical system, with the origin on the axis of the body. We define the orthogonal surface tangents by their unit vectors $\hat{u}_t$ and $\hat{u}_\phi$ and the outward normal by the direction of its unit vector $\hat{n}$ given by

$$\hat{n} = \hat{u}_\phi \times \hat{u}_t \qquad (2.16)$$

On the surface of the body we define a field point by its co-ordinates (t, ϕ) or (ϱ, ϕ, z) and a source point by (t', ϕ') or (ϱ', ϕ', z'). Their respective unit tangent vectors are $(\hat{u}_t, \hat{u}_\phi)$ and $(\hat{u}'_t, \hat{u}'_\phi)$. The unit vector $\hat{u}_\phi$ is orthogonal to the z-axis, but $\hat{u}_t$ and $\hat{z}$ are at an angle v. This angle is assumed to be positive if $\hat{u}_t$ points away from the z-axis. Similarly, v' is the angle between $\hat{u}'_t$ and the z-axis at (t', ϕ'). The relationships among these unit vectors can be determined by an inspection of Fig. 2.2, and are given by

$$\hat{u}_t = \sin v\, \hat{u}_\varrho + \cos v\, \hat{u}_z \qquad (2.17)$$

$$\hat{u}_{t'} = \sin v' \cos(\phi' - \phi)\hat{u}_\varrho + \sin v' \sin(\phi' - \phi)\hat{u}_\phi + \cos v'\hat{u}_z \qquad (2.18)$$

$$\hat{u}_{\phi'} = -\sin(\phi' - \phi)\hat{u}_\varrho + \cos(\phi' - \phi)\hat{u}_\phi \qquad (2.19)$$

In addition, if the positional vectors of points (t, ϕ) and (t', ϕ') are $\bar{r}$ and $\bar{r}'$, respectively, then

$$\bar{r} = \varrho\, \hat{u}_\varrho + z\, \hat{u}_z \qquad (2.20)$$

and

$$\bar{r}' = \varrho' \cos(\phi' - \phi)\, \hat{u}_\varrho + \varrho' \sin(\phi' - \phi)\hat{u}_\phi + z'\hat{u}_z \qquad (2.21)$$

where

$$|\bar{r} - \bar{r}'| = [(\varrho - \varrho')^2 + (z - z')^2 + 4\varrho\varrho' \sin^2 \frac{\phi - \phi'}{2}]^{1/2} \qquad (2.22)$$

Also, by definition, the surface gradient of a scalar function Φ on the body of revolution is given by

$$\nabla_s \Phi = \hat{u}_t \frac{\partial \Phi}{\partial t} + \hat{u}_\phi \frac{\partial \Phi}{\varrho \partial \phi} \qquad (2.23)$$

and the surface divergence of a vector function $\bar{W}$ is defined as

$$\nabla_s \cdot \bar{W} = \frac{1}{\varrho} \frac{\partial}{\partial t}(\varrho\, \bar{W} \cdot \hat{u}_t) + \frac{1}{\varrho} \frac{\partial}{\partial \phi}(\bar{W} \cdot \hat{u}_\phi) \qquad (2.24)$$

The unknown currents $\bar{J}$ and $\bar{M}$ can now be decomposed into two components

Analysis of circular microstrip antennas 53

along the unit vectors $\hat{u}_t$ and $\hat{u}_{\phi'}$ and expressed in the form [9 – 11]

$$\bar{J}(\bar{r}') = \hat{u}_{t'} J^t(t', \phi') + \hat{u}_{\phi'} J^{\phi}(t', \phi') \qquad (2.25)$$

$$\bar{M}(\bar{r}') = \hat{u}_{t'} M^t(t', \phi') + \hat{u}_{\phi'} M^{\phi}(t', \phi') \qquad (2.26)$$

where J^t, J^{ϕ} and M^t, M^{ϕ} are the current components along $\hat{u}_{t'}$ and $\hat{u}_{\phi'}$, respectively.

The electric current $\bar{J}$ exists on both conducting and dielectric surfaces, but $\bar{M}$ exists only on the dielectric. If the electric and magnetic surface currents are expanded into N_c and N_d expansion functions, respectively, the surface currents can be represented by

$$\bar{J}(\bar{r}') = \sum_{n=-M_o}^{M_o} \sum_{j=1}^{N_c} I_{nj}^t J_{nj}^t(t', \phi') \hat{u}_{t'} + I_{nj}^{\phi} J_{nj}^{\phi}(t', \phi') \hat{u}_{\phi'} \qquad (2.27)$$

$$\bar{M}(\bar{r}') = \eta_e \sum_{n=-M_o}^{M_o} \sum_{j=N_c+1}^{N_c+N_d} M_{nj}^t K_{nj}^t(t', \phi') \hat{u}_{t'}$$
$$+ M_{nj}^{\phi} K_{nj}^{\phi}(t', \phi') \hat{u}_{\phi'} \qquad (2.28)$$

where J_{nj}^t, J_{nj}^{ϕ}, K_{nj}^t, K_{nj}^{ϕ} are expansion functions defined by

$$J_{nj}^t = J_{nj}^{\phi} = K_{nj}^t = K_{nj}^{\phi} = f_j(t') e^{jn\phi'} \qquad (2.29)$$

The range $-M_o$ to $+M_o$ gives the total number of azimuthal modes. The coefficients I_{nj}^t, I_{nj}^{ϕ}, M_{nj}^t, M_{nj}^{ϕ} are the current coefficients to be determined by solving the matrix equation which results when eqns. 2.27 and 2.28 are substituted via eqns. 2.25 and 2.26 into the integral eqns. 2.3 – 2.6. The procedure involves taking the inner products of the field equations with certain testing functions and integrating them over the surface. The testing functions are defined by

$$\bar{W}_{li}^t = \hat{u}_t f_i(t) e^{-jl\phi} \qquad (2.30)$$

$$\bar{W}_{li}^{\phi} = \hat{u}_{\phi} f_i(t) e^{-jl\phi} \qquad (2.31)$$

The inner product of two vectors $\bar{P}$ and $\bar{Q}$ is defined by their scalar product and integrated over the surface of the body; that is

$$<\bar{P}, \bar{Q}> = \int_s \bar{P} \cdot \bar{Q} \, ds' \qquad (2.32)$$

The expansion and testing functions, i.e. $[\bar{J}_{nj}, \bar{K}_{nj}]$ and $[\bar{W}_{li}]$, as defined by eqns. 2.29, 2.30 and 2.31 are orthogonal over the period 0 to 2π in ϕ. This means that the inner products of $\bar{W}_{li}^p$ and $\bar{J}_{nj}^p$ ($p = t$ or ϕ) vanish for $l \neq n$ and the contribution of different azimuthal mode separates. The resulting equations therefore involve only a particular mode of index n. This is the major simplification that is introduced by the mode orthogonality in axisymmetric objects. Accordingly we obtain a separate matrix equation for each mode.

For explicit evaluation of the matrix elements, one must choose $f_i(t)$. It is known that subsectional expansions, using flat pulse or triangle pulse functions, give rise to well conditioned matrices. Flat-pulse current expansions with point matching were first used to solve the scattering problems of conducting spheres. However, with such expansions the moment-method solutions of the surface currents do not converge rapidly to the exact solutions. The triangular-function expansions, on the other hand, converge satisfactorily and provide accurate solutions. For this reason triangular pulse functions are used here to represent both current expansions and the testing functions [9]. A minor deviation from the literature is to use $\varrho f_i(t)$ instead of $f_i(t)$, which is defined by four impulses given by

$$\varrho f_i(t) = \sum_{p=1}^{4} T_{p+4i-4} \, \delta(t - t_{p+2i-2}) \qquad (2.33)$$

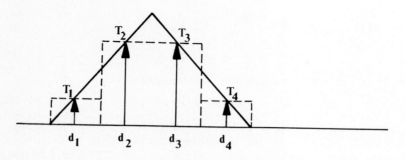

Fig. 2.3 *Triangle function approximation*

where $\delta(t)$ is the unit impulse function and its coefficients T are defined in Fig. 2.3, which for $i = 1$ are given by

$$T_1 = \frac{d_1^2}{2(d_1 + d_2)}$$

$$T_2 = \frac{(d_1 + 0\cdot 5 d_2)d_2}{d_1 + d_2}$$

$$T_3 = \frac{(d_4 + 0\cdot 5 d_3)d_3}{d_3 + d_4} \qquad (2.34)$$

$$T_4 = \frac{d_4^2}{2(d_3 + d_4)}$$

Similarly, the derivative of $\varrho f_i(t)$ is approximated by four impulses as

$$\frac{d}{dt}[\varrho f_i(t)] = \sum_{p=1}^{4} T'_{p+4i-4} \delta(t - t_{p+2i-2}) \qquad (2.35)$$

and are shown in Fig. 2.4, where the coefficient T', for $i = 1$, are given by

$$T'_1 = \frac{d_1}{d_1 + d_2}$$

$$T'_2 = \frac{d_2}{d_1 + d_2} \qquad (2.36)$$

$$T'_3 = \frac{-d_3}{d_3 + d_4}$$

$$T'_4 = \frac{-d_4}{d_3 + d_4}$$

Now, all information needed to proceed with transferring the integral equations to a system of linear matrix equations is known. Following procedures well known in the application of moment methods the matrix equation for the nth Fourier component of currents can be written as

$$[\bar{T}_n][\bar{I}_n] = [\bar{V}_n], \qquad n = 0, \pm 1, \pm 2, \ldots, \pm Mo \qquad (2.37)$$

where $\bar{T}_n$ is a square matrix representing the impedance and the admittance sub-matrices, $\bar{I}_n$ is a column matrix for the unknown expansion coefficients of $\bar{J}$ and $\bar{M}$, and $\bar{V}_n$ is the excitation column matrix. Each mode has a matrix equation of the form [12]

$$\begin{bmatrix} Z^e_{ce,ce} & 0 & Z^e_{ce,de} & Y^e_{ce,de} \\ 0 & \eta_r Z^d_{cd,cd} & \eta_r Z^d_{cd,de} & Y^d_{cd,de} \\ Z^e_{de,ce} & \eta_r Z^d_{de,cd} & Z^e_{de,de} + \eta_r Z^d_{de,de} & Y^e_{de,de} + Y^d_{de,de} \\ Y^e_{de,ce} & Y^d_{de,cd} & Y^e_{de,de} + Y^d_{de,de} & -Z^e_{de,de} - \frac{1}{\eta_r} Z^d_{de,de} \end{bmatrix}$$

$$\times \begin{bmatrix} I_{ce,n} \\ I_{cd,n} \\ I_{de,n} \\ M_n \end{bmatrix} = \begin{bmatrix} 0 \\ -V^d_{cd,n} \\ -V^d_{de,n} \\ -I^d_{de,n} \end{bmatrix} \qquad (2.38)$$

where $\eta_r = \eta_d/\eta_e$; and V^d_n is the excitation sub-matrix due to the electric-field sources in the dielectric and I^d_n is the excitation due to the magnetic-field sources in the dielectric region, respectively. The sub-matrices Z and Y with superscripts e and d denote the impedance and admittance matrices for the exterior and interior media, respectively. The first pair of suffixes identify field surface and the second pair of suffixes identify the source surface where the Fourier mode n is implied. $I_{ce,n}$, $I_{cd,n}$, $I_{de,n}$ and M_n are the unknown expansion coefficients of the electric and magnetic currents on S_{ce}, S_{cd} and S_{de}, respectively. In the above equations, each sub-matrix Y_n^q or Z_n^q consists of four submatrices. They are given by

$$[Y_n^q] = \begin{bmatrix} Y_n^{q,tt} & Y_n^{q,t\phi} \\ Y_n^{q,\phi t} & Y_n^{q,\phi\phi} \end{bmatrix} \tag{2.39}$$

and

$$[Z_n^q] = \begin{bmatrix} Z_n^{q,tt} & Z_n^{q,t\phi} \\ Z_n^{q,\phi t} & Z_n^{q,\phi\phi} \end{bmatrix} \tag{2.40}$$

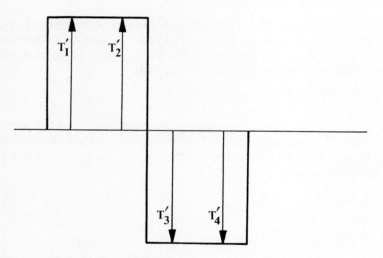

Fig. 2.4 *Derivative of the triangle function approximation*

and their elements have the form

$$\begin{bmatrix} (Y_n^{q,tt})_{ij} \\ (Y_n^{q,\phi t})_{ij} \\ (Y_n^{q,t\phi})_{ij} \\ (Y_n^{q,\phi\phi})_{ij} \end{bmatrix} = \sum_{p=1}^{4} T_{p'} \sum_{l=1}^{4} T_{l'} \begin{bmatrix} (Y_1)_{ij} \\ (Y_2)_{ij} \\ (Y_3)_{ij} \\ (Y_4)_{ij} \end{bmatrix} \tag{2.41}$$

$(Y_1)_{ij} =$

$$\begin{vmatrix} k_q[(\varrho_j - \varrho_i)\cos v_j - (z_j - z_i)\sin v_j]G_2 - k_q\varrho_i G_1 \cos v_j, & i \neq j \\ \dfrac{\pi}{k_q^2 d_i \varrho_i} - k_q \varrho_i G_I \cos v_i, & i = j \end{vmatrix} \tag{2.42}$$

$(Y_2)_{ij} = jk_q(\varrho_j \sin v_i \cos v_j - \varrho_i \sin v_j \cos v_i - (z_j - z_i)\sin v_i \sin v_j)G_3 \tag{2.43}$

$$(Y_3)_{ij} = jk_q(z_j - z_i)G_3 \tag{2.44}$$

$$(Y_4)_{ij} =$$
$$\begin{vmatrix} k_q([\varrho_j - \varrho_i)\cos v_i - (z_j - z_i)\sin v_i]G_2 k_q\varrho_i G_1 \cos v_i, & i \neq j \\ \dfrac{\pi}{k_q^2 d_i \varrho_i} + k_q \varrho_i G_I \cos v_i, & i = j \end{vmatrix} \tag{2.45}$$

$$(Z_n^{q,tt})_{ij} = j \sum_{p=1}^{4} \sum_{l=1}^{4} T_{p'} T_{l'} (G_5 \sin v_{i'} \sin v_{j'} + G_4 \cos v_i \cos v_{j'}')$$
$$- T'_{p'} T'_{l'} G_4 \tag{2.46}$$

$$(Z_n^{q,\phi t})_{ij} = -\sum_{p=1}^{4}\sum_{l}^{4} T_{p'} T_{l'} G_6 \sin v_{j'} + \frac{n}{k\varrho_i'} T_{\varrho'} T'_{l'} G_4 \tag{2.47}$$

$$(Z_n^{q,t\phi})_{ij} = \sum_{p=1}^{4}\sum_{l}^{4} T_{p'} T_{l'} G_6 \sin v_{i'} + \frac{n}{k\varrho_j'} T'_{p'} T_{l'} G_4 \tag{2.48}$$

$$(Z_n^{q,\phi\phi})_{ij} = j \sum_{p=1}^{4}\sum_{l}^{4} T_{p'} T_{l'} (G_5 - \frac{n^2}{k^2 \varrho_i' \varrho_j'} G_4) \tag{2.49}$$

and

$$\begin{aligned} p' &= p + 4i - 4 \\ l' &= l + 4j - 4 \\ i' &= p + 2i - 2 \\ j' &= l + 2i - 2 \end{aligned} \tag{2.50}$$

where

$$G_1 = 2\int_0^\pi d\phi' G'_q \sin^2\left(\frac{\phi'}{2}\right)\cos(n\phi') \tag{2.51}$$

$$G_2 = \int_0^\pi d\phi' \, G'_q \cos\phi' \cos(n\phi') \tag{2.52}$$

$$G_3 = \int_0^\pi d\phi' \, G'_q \sin\phi' \sin(n\phi') \tag{2.53}$$

$$G_4 = \int_0^\pi d\phi' \, G'_q \cos(n\phi') \tag{2.54}$$

$$G_5 = \int_0^\pi d\phi' \, G'_q \cos\phi' \cos(n\phi') \tag{2.55}$$

$$G_6 = \int_0^\pi d\phi' \, G'_q \sin\phi' \sin(n\phi') \tag{2.56}$$

with

$$G_q = \frac{e^{-jk_q R}}{k_q R} \tag{2.57}$$

$$G'_q = -\frac{1 + jk_q R}{k_q^3 R^3} e^{-jk_q R} \tag{2.58}$$

$$R = \left[(\varrho - \varrho')^2 + (z - z')^2 + 4\varrho\varrho' \sin^2\left(\frac{\phi'}{2}\right)\right]^{1/2} \tag{2.59}$$

at $t = t'$, R is approximated by

$$R_e = \left[\left(\frac{d_{i'}}{4}\right)^2 + 4\varrho_{i'}{}^2 \sin^2\left(\frac{\phi'}{2}\right)\right]^{1/2} \tag{2.60}$$

2.2.2 Excitation matrix

Microstrip antennas are normally excited by a transmission line or a coaxial probe. To solve the problem numerically one must model the exciting source, from which the elements of the excitation matrix can be determined. However, a precise modelling of either source, i.e. the junction between the transmission line or the coaxial probe with the microstrip patch, although feasible is a difficult task. On the other hand, microstrip antennas are highly resonant structures and within their operating frequency band one of the Fourier components, i.e. the modes, dominates. This means, one can represent the excitation source with a simple elementary source, such as an electric dipole, without affecting the solution accuracy. The representation of the source by a single electric dipole is quite adequate if the substrate thickness is small, or the width of the transmission-line feed is not excessively large. Otherwise, multiple dipole sources must be used. For instance, when the substrate thickness is so large that the current distribution along the coaxial probe is not constant, a linear array of electric dipoles along the probe length may be used. In such a case, the excitation of dipoles must correspond to the current distribution of the probe. Similarly, for simulating wide transmission-line junctions, multiple dipole excitations may be used, where their excitation must be weighted by the field distribution under the line. In the following analysis we consider only the case of thin substrates or transmission lines, and represent the excitation source by a single electric dipole. This simple form of source representation simplifies the excitation matrix considerably.

Simulating the excitation by an electric dipole, its electric and magnetic fields can be computed from

$$\vec{E}^{inc,q} = -j\omega \vec{A}^q - \nabla \Phi^q \tag{2.61}$$

$$\vec{H}^{inc,q} = \frac{1}{\mu_q} \nabla \times \vec{A}^q \tag{2.62}$$

where

$$\vec{A}^q = \frac{-jk_q \mu_q}{4\pi r} \vec{I}l \, h_0^{(2)}(k_q|\vec{r} - \vec{r}'|) \tag{2.63}$$

$$\Phi^q = \frac{\eta_q}{4\pi r} \vec{I}l \cdot \nabla h_0^{(2)}(k_q|\vec{r} - \vec{r}'|) \tag{2.64}$$

and $h_0^{(2)}$ is the spherical Hankel function of the second kind and zero order and $\bar{I}l$ is the dipole moment in the z-direction. If the Hankel function is represented by

$$h_0^{(2)}(k_q|\bar{r} - \bar{r}'|) = \sum_{-\infty}^{\infty} G_m e^{jm(\phi - \phi')} \quad (2.65)$$

with

$$G_m = G_{-m} =
\begin{cases}
\sum_{n=m}^{\infty} a_n h_n^{(2)}(k_q r') j_n(k_q r) P_n^m(\cos\theta) P_n^m(\cos\theta'), & r' > r \\
\sum_{n=m}^{\infty} a_n h_n^{(2)}(k_q r) j_n(k_q r') P_n^m(\cos\theta) P_n^m(\cos\theta'), & r' < r
\end{cases} \quad (2.66)$$

where $a_n = (2n + 1)(n - m)!/(n + m)!$; then using eqn. 2.61 for $\bar{E}^{inc}$, its excitation matrix elements can be calculated from

$$(V_m^q)_i^t = \frac{\eta_q |Il|}{2}\left[-k_q^2 \int_0^{t_u} dt \varrho f_i(t) \cos\nu\, G_m \right.$$

$$\left. - \int_0^{t_u} dt \frac{d}{dt}[\varrho f_i(t)] \frac{1}{\varrho} \frac{\partial G_m}{\partial \theta}\right] \quad (2.67)$$

$$(V_m^q)_i^t = \frac{\eta_q |Il|}{2}\left[jm \int_0^{t_u} dt \frac{f_i(t)}{\varrho} \frac{\partial G_m}{\partial \theta}\right] \quad (2.68)$$

where t_u is the t co-ordinate of the upper end of the generating curve. Similarly, using eqn. 2.62 for $\bar{H}^{inc}$, its matrix element can be calculated from

$$(I_m^q)_i^t = \frac{-jk_q|Il|}{4}\left[\int_0^{t_u} dt\, f_i(t) \sin\nu\, G_m\right] \quad (2.69)$$

$$(I_m^q)_i^\phi = \frac{-jk_q|Il|}{2}\left[\int_0^{t_u} dt\, \varrho f_i(t) \frac{\partial G_m}{\partial r}\right] \quad (2.70)$$

2.2.3 Radiation fields

Once the induced currents $\bar{J}$ and $\bar{M}$ on the surface are determined after the solution of the matrix equation, the field components E_θ and E_ϕ at a far-field point (r_0, θ_0, ϕ_0) can be determined [12] as

$$E_\theta = \frac{-j\omega\mu_e}{4\pi r_0} e^{-jkr_0} F_1(\theta_0, \phi_0) \quad (2.71)$$

and

$$E_\phi = \frac{-j\omega\mu_e}{4\pi r_0} e^{-jkr_0} F_2(\theta_0, \phi_0) \quad (2.72)$$

with

$$F_1(\theta_0, \phi_0) = \int_s (\bar{J} \cdot \hat{i}_\theta + \frac{1}{\eta_e} \bar{M} \cdot \hat{i}_\phi) e^{-jk\hat{r}_0 \cdot \bar{r}'} ds \qquad (2.73)$$

$$F_2(\theta_0, \phi_0) = \int_s (\bar{J} \cdot \hat{i}_\phi - \frac{1}{\eta_e} \bar{M} \cdot \hat{i}_\theta) e^{-jk\hat{r}_0 \cdot \bar{r}'} ds \qquad (2.74)$$

where S is the exterior antenna surface, $\bar{r}_0$ is a unit vector in the direction from the origin of the co-ordinates to the field point, $\bar{r}'$ is the positional vector of the source point (x', y', z') on the antenna, and $\hat{i}_\theta$ and $\hat{i}_\phi$ are unit vectors in the direction of increasing θ and ϕ, respectively. Referring to the field point, these vectors can be written as

$$\hat{i}_\theta = \cos\theta_0 \cos\phi_0 \hat{a}_x + \cos\theta_0 \sin\phi_0 \hat{a}_y - \sin\theta_0 \hat{a}_z \qquad (2.75)$$

$$\hat{i}_\phi = -\sin\phi_0 \hat{a}_x + \cos\phi_0 \hat{a}_y \qquad (2.76)$$

and the dot product in the exponential term of eqns. 2.73 and 2.74 can be shown to be

$$\hat{r}_0 \cdot \bar{r}' = (x'\cos\phi_0 + y'\sin\theta_0)\sin\theta_0 + z'\cos\theta_0$$
$$= \varrho'\cos(\phi' - \phi_0)\sin\theta_0 + z'\cos\theta_0 \qquad (2.77)$$

where (ϱ', ϕ', z') are the cylindrical coordinates of the point (x', y', z'). If we now substitute for $\bar{J}$ and $\bar{M}$ in eqns 2.73 and 2.74, and evaluate the dot products $(\hat{u}_{t'} \cdot \hat{i}_\theta)$, $(\hat{u}_t \cdot \hat{i}_\phi)$ as

$$\hat{u}_{t'} \cdot \hat{i}_\theta = \sin v' \cos\theta_0 \cos(\phi' - \phi_0) - \cos v' \sin\theta_0 \qquad (2.78)$$

$$\hat{u}_{t'} \cdot \hat{i}_\phi = \sin v' \sin(\phi' - \phi_0) \qquad (2.79)$$

and

$$\hat{u}_\phi \cdot \hat{i}_\theta = -\sin\theta_0 \sin(\phi' - \phi_0) \qquad (2.80)$$

$$\hat{u}_\phi \cdot \hat{i}_\phi = \sin\phi' \sin\phi_0 + \cos\phi' \cos\phi_0$$
$$= \cos(\phi' - \phi_0) \qquad (2.81)$$

then $F_1(\theta_0, \phi_0)$ and $F_2(\theta_0, \phi_0)$ take the form

$$F_1(\theta_0, \phi_0) = \int_0^{t_u} \int_0^{2\pi} \sum_{m=-M}^{M} \sum_{j=1}^{N} f_j(t') e^{jm\phi'} e^{jk[\varrho'\cos(\phi' - \phi_0)\sin\theta_0 + z'\cos\theta_0]}$$

$$\cdot \left[I_{mj}^t [\sin v' \cos\theta_0 \cos(\phi' - \phi_0) - \cos v' \sin\theta_0] \right.$$

$$- I_{mj}^\phi \cos\theta_0 \sin(\phi' - \phi_0)$$

$$+ \frac{1}{\eta} K_{mj}^t \sin v' \sin(\phi' - \phi_0)$$

$$\left. + \frac{1}{\eta} k_{mj}^\phi \cos(\phi' - \phi_0) \right] \varrho' d\phi' dt' \qquad (2.82)$$

and

$$F_2(\theta_0, \phi_0) = \int_0^{t_u} \int_0^{2\pi} \sum_{m=-M}^{M} \sum_{j=1}^{N} f_j(t') e^{jm\phi'} e^{jk[\varrho'\cos(\phi' - \phi_0)\sin\theta_0 + z'\cos\theta_0]}$$

$$\cdot \left[I_{mj}^t \sin v' \sin(\phi' - \phi_0) + I_{mj}^{\phi} \cos(\phi' - \phi_0) \right.$$

$$+ \frac{1}{\eta} K_{mj}^t [\sin v' \cos\theta_0 \cos(\phi' - \phi_0) - \cos v \sin\theta_0]$$

$$\left. + \frac{1}{\eta} k_{mj}^{\phi} \cos\theta_0 \sin(\phi' - \phi_0) \right] \varrho' d \phi' dt' \tag{2.83}$$

Changing the order of summations and integrations and using the integral representation for the Bessel function of the first kind, namely

$$J_n(\varrho) = \frac{j^n}{2\pi} \int_0^{2\pi} e^{-j\varrho\cos\alpha} e^{-jn\alpha} d\alpha \tag{2.84}$$

the integration in eqns. 2.82 and 2.83 involving the azimuthal co-ordinate ϕ' can be evaluated as

$$\int_0^{2\pi} e^{jm\phi'} e^{jk\varrho'\cos(\phi' - \phi_0)\cos\theta_0} d\phi' = 2\pi j^m e^{jm\phi_0} J_m(k\varrho' \sin\theta_0) \tag{2.85}$$

$$\int_0^{2\pi} \begin{Bmatrix} \cos(\phi' - \phi_0) \\ \sin(\phi' - \phi_0) \end{Bmatrix} e^{jm\phi'} e^{jk\varrho'\cos(\phi'-\phi_0)\sin\theta_0} d\phi'$$

$$= \pi j^m e^{jm\phi_0} \begin{Bmatrix} j[J_{m+1}(k\varrho'\sin\theta_0) - J_{m-1}(k\varrho'\sin\theta_0)] \\ [J_{m+1}(k\varrho'\sin\theta_0) + J_{m-1}(k\varrho'\sin\theta_0)] \end{Bmatrix} \tag{2.86}$$

The far-field functions can then be represented by

$$F_1(\theta_0, \phi_0) = \sum_{m=-M_o}^{M_o} \sum_{j=N_c+1}^{N_c+N_d} [(R_m^{t\theta,J})_j I_{mj}^t + (R_m^{t\theta,M})_j k_{mj}^t + (R_m^{\phi\theta,J})_j I_{mj}^{\phi}$$

$$+ (R_{mj}^{\phi\theta,M}) k_{mj}^{\phi}] \tag{2.87}$$

$$F_2(\theta_0, \phi_0) = \sum_{m=-M_o}^{M_o} \sum_{j=N_c+1}^{N_c+N_d} [(R_m^{t\phi,J})_j I_{mj}^t + (R_m^{t\phi,J})_j k_{mj}^t + (R_m^{\phi\phi,J})_j I_{mj}^{\phi}$$

$$+ (R_{mj}^{\phi\phi,M}) k_{mj}^{\phi}] \tag{2.88}$$

where

$$(R_m^{t\theta,J})_j = \sum_{q=1}^{4} C [j(J_{m+1} - J_{m-1})\sin v' \cos\theta_0 - 2\cos v' \sin\theta_0 J_m] \tag{2.89}$$

$$(R_m^{\phi\theta,J})_j = -\sum_{q=1}^{4} C [(J_{m+1} + J_{m-1})\cos\theta_0 \tag{2.90}$$

$$(R_m^{t\phi,J})_j = \sum_{q=1}^{4} C [(J_{m+1} + J_{m-1})\sin v' \tag{2.91}$$

$$(R_m^{\phi\phi,J})_j = \sum_{q=1}^{4} C\,[j(J_{m+1} - J_{m-1}) \tag{2.92}$$

with

$$C = \pi j^m e^{jm\phi_0} e^{jkz'\cos\theta_0} T_{q'} \tag{2.93}$$

and

$$(R_m^{t\theta,M})_j = \frac{1}{\eta}(R_m^{t\phi,J})_j \tag{2.94}$$

$$(R_m^{t\theta,M})_j = \frac{1}{\eta}(R_m^{t\phi,J})_j \tag{2.95}$$

$$(R_m^{t\phi,M})_j = -\frac{1}{\eta}(R_m^{t\theta,J})_j \tag{2.96}$$

$$(R_m^{\phi,\phi,M})_j = \frac{1}{\eta}(R_m^{\phi\theta,J})_j \tag{2.97}$$

The far-field component can finally be put in a compact form as

$$\begin{bmatrix} E_\theta \\ \\ E_\phi \end{bmatrix} = \frac{-j\omega\mu_e e^{-jkr_0}}{4\pi r_0} \sum_{m=-M}^{m=M} \sum_j \begin{bmatrix} (R_m^{t\theta,J})_j\ (R_m^{t\theta,M})_j\ (R_m^{\phi\theta,J})_j\ (R_m^{\phi\theta,M})_j \\ \\ (R_m^{t\phi,J})_j\ (R_m^{t\phi,M})_j\ (R_m^{\phi\phi,J})_j\ (R_m^{\phi\phi,M})_j \end{bmatrix} \begin{bmatrix} I_{m,j}^t \\ M_{m,j}^t \\ I_{m,j}^\phi \\ M_{m,j}^\phi \end{bmatrix} \tag{2.98}$$

This completes the formulation of the problem and determination of the radiated fields from the computed equivalent currents. The external field near the antenna or the field within the dielectric substrate can also be determined, but are omitted here for brevity.

Two different properties of the Y and Z submatrices are used to reduce the computation time and cost. One is the matrix symmetry. This property is evident from eqns. 2.42 – 2.49 and gives

$$\begin{aligned}
(Z_m^{tt})_{ij} &= (Z_m^{tt})_{ji} \\
(Z_m^{t\phi})_{ij} &= -(Z_m^{t\phi})_{ji} \\
(Z_m^{\phi\phi})_{ij} &= (Z_m^{\phi\phi})_{ji} \\
(Y_m^{tt})_{ij} &= -(Y_m^{\phi\phi})_{ji} \\
(Y_m^{\phi t})_{ij} &= -(Y_m^{\phi t})_{ji} \\
(Y_m^{t\phi})_{ij} &= -(Y_m^{t\phi})_{ji}
\end{aligned} \tag{2.99}$$

This means that only one half of the matrix needs to be created. The other half can be generated from the above symmetric property. The second property is the mode symmetry. This enables one to generate the Y and Z submatrices of the modes with negative indices from the positive ones, as

$$\begin{aligned} Z^{tt}_{-m} &= Z^{tt}_{m} \\ Z^{\phi t}_{-m} &= -Z^{\phi t}_{m} \\ Z^{t\phi}_{-m} &= -Z^{t\phi}_{m} \\ Z^{\phi\phi}_{-m} &= Z^{\phi\phi}_{m} \end{aligned} \quad (2.100)$$

The same relations are valid for the Y submatrices. Similarly, the elements of the excitation for an electric dipole, generated using eqns. 2.67 – 2.70, satisfy the following relationships:

$$\begin{aligned} (V_{-m})^t &= V^t_m \\ (V_{-m})^\phi &= -V^\phi_m \\ (I_{-m})^t &= -I^t_m \\ (I_{-m})^\phi &= I^\phi_m \end{aligned} \quad (2.101)$$

Thus the solution of the matrix equations, given by eqns. 2.38, gives the following symmetric relationships satisfield by the current coefficients:

$$\begin{aligned} I^t_{-n} &= I^t_n \\ M^t_{-n} &= -M^t_n \\ I^\phi_{-n} &= -I^\phi_n \\ M^\phi_{-n} &= M^\phi_n \end{aligned} \quad (2.102)$$

Accordingly, one only needs to compute the coefficients of the positive modes, i.e. one half of the mode coefficients. This results in a major reduction of computation time and cost. The mode symmetry can also be used in calculating the far-field components from eqn. 2.98. The needed relationships are

$$\begin{aligned} R^{t\theta,J}_{-m} &= R^{t\theta,J}_{m} \\ R^{\phi\theta,J}_{-m} &= -R^{\phi\theta,J}_{m} \\ R^{t\phi,J}_{-m} &= -R^{t\phi,J}_{m} \\ R^{\phi\phi,J}_{-m} &= R^{\phi\phi,J}_{m} \end{aligned} \quad (2.103)$$

which can reduce the summation over the modes to one from $m = 0$ to $+M$.

2.3 Application 1: Circular patch antenna

The circular patch antenna is one of the fundamental microstrip geometries and its impedance and radiation characteristics have been investigated extensively.

64 Analysis of circular microstrip antennas

However, the methods used so far assumed an infinite size for the ground plane and substrate. The solutions are therefore approximate and lack the influence of the finite substrate and ground-plane dimensions. Their accuracy therefore depends on the type of application. For instance, because microstrip patch geometries are highly resonant, their impednace characteristics are dominated by the patch dimensions. The ground plane size, provided it is reasonably larger than the patch, has a negligible effect. Similarly, the radiation near the broadside direction is determined primarily by the patch itself. Th finite size of the

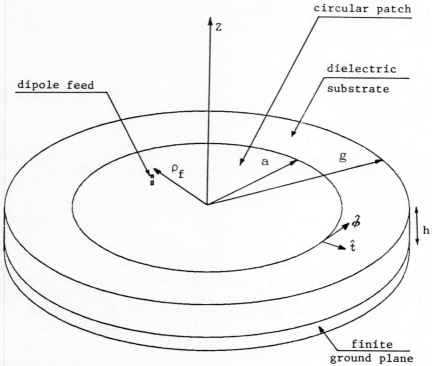

Fig. 2.5 *Microstrip antenna geometry*

substrate or ground plane influences the radiation at wide angles, and particularly behind the antenna. Thus, when radiation patterns along the broadside direction are necessary or their approximate form is adequate, the analytic solutions can provide sufficient information. The numerical method presented in this Chapter enables one to determine the radiation characteristics in all space. In addition, the accuracy of the generated results can be very high. Thus it is a useful method for generating solutions for high-precision work, such as in a reflector-antenna feed design. In this Section we present a few representative results for a circular patch antenna.

The geometry of a circular patch microstrip antenna is shown in Fig. 2.5, where the excitation is simulated by an electric dipole immersed in the dielectric substrate under the conducting patch. The radius is selected as [2]

$$a_e = a[1 + \frac{2h}{\pi a \varepsilon_r}(\ln\frac{\pi a}{2h} + 1{\cdot}7726]^{1/2} \qquad (2.104]$$

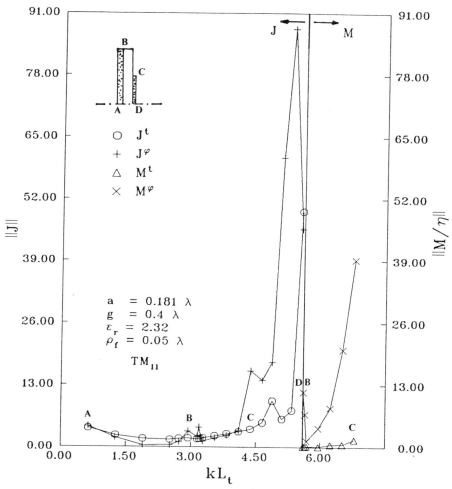

Fig. 2.6 *The computed electric and magnetic surface currents of the TM_{11} mode on the outside boundary (Reproduced from Reference 14 © 1986 IEEE)*

where a is the radius of the conducting patch, a_e is the effective radius due to the spread of the fringing field from the patch edge to the ground plane, h is the dielectric thickness and ε_r is the relative permittivity of the dielectric substrate.

66 *Analysis of circular microstrip antennas*

The effective radius is calculated from

$$a_e = \frac{K_{nm}}{2\pi\sqrt{\varepsilon_r}} \tag{2.105}$$

where K_{nm} is the mth zero of the derivative of the Bessel function of order n. The effective patch radius is therefore a function of substrate height, the dielectric permittivity and the order of the excited mode. The effects of these parameters, as well as the ground-plane size, on the radiation characteristics of the patch are investigated in the following Sections [14].

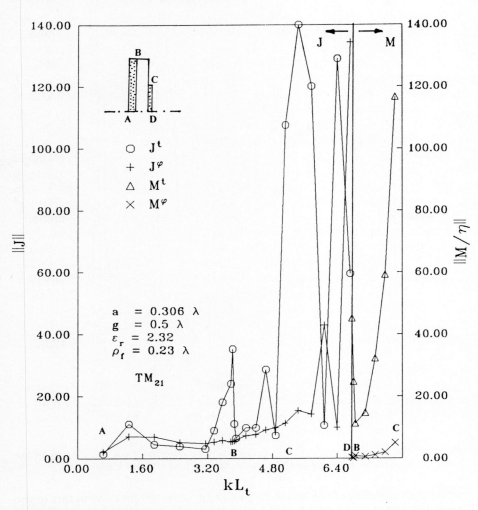

Fig. 2.7 *The computed electric and magnetic surface currents of the TM_{21} mode on the outside boundary (Reproduced from Reference 14 © 1986 IEEE)*

2.3.1 Surface fields

In microstrip antennas the radiation is normally from the periphery of the patch, where the fringing field is maximum. However, since the exciting dipole launches guided modes of the parallel-plate region under the patch, it is desirable to compute the surface-field distributions on the conducting and dielectric surfaces of the antenna to understand their behaviour. These distributions are given by the equivalent currents $\bar{J}$ and $\bar{M}$. For two selected modes, i.e. the dominant TM_{11} and the higher TM_{21} modes, the computed results are shown in Figs. 2.6 and 2.7. In Fig. 2.6 the patch radius is $a = 0.181\lambda$, where λ is the wavelength in free space and the patch resonates in the dominant TM_{11} mode. The horizontal axis shows the length of the contour along the generating curve. Since the geometry is rotationally symmetric only one half of the surface contour is shown. The external surface currents are plotted with respect to their locations on the surface, where points A to B correspond to the ground plane, points B to C represent the dielectric substrate which supports both electric and magnetic currents and points C to D correspond to the patch surface. An examination of this Figure reveals that the electric current is the strongest on the patch surface and has a negligible value on the ground plane. Its equivalent distribution on the dielectric, i.e. the tangential magnetic field on the dielectric, is also small. However, it shows some slight increase near B, on the substrate termination, which is an indication of surface-wave excitation. The distribution of the magnetic current $\bar{M}$, i.e. the tangential electric field on the substrate, is shown on the right side of the Figure. It increases progressively from B to C, indicating strong fringing field near C. The contributions to the antenna radiation are therefore mainly from J^ϕ on the patch and M^ϕ on the substrate.

The surface distributions for the TM_{21} excitation are shown in Fig. 2.7. Again, the currents on the ground plane are small, but J^t shows stronger values on the substrate termination near B. Here, both J^t and J^ϕ are strong on the patch and have rapid variations. The magnetic current $\bar{M}$ again increases rapidly from B to C, near the patch edge. The main radiation zones are similar to the TM_{11} mode case, being the upper patch surface, the dielectric surface near the patch and its truncated end near the ground plane.

2.3.2 Feed location

For coaxial feeds, the location is usually selected to provide a good impedance match. Since, we simulate the excitation by an electric dipole we ignore the impdenace of the feed and investigate the effect of its location on the excitation efficiency of various modes. Also, different modes have different radiation patterns and affect the overall antenna pattern at different angular regions. For this reason, rather than computing the magnitude of the Fourier coefficients of the currents we compute the peak intensity of their radiation fields. Fig. 2.8 shows the effect of the feed position ϱ_f on the excitation of the first three modes, when the patch is resonant at the TM_{11} mode. The dominant mode has the strongest excitation efficiency of the other modes, i.e. TM_{01} and TM_{21} modes, increase progressively as the feed moves to the patch edge, but their peak radiation level is always below −15 dB. These modes radiate conical beams and their effect will manifest at an angular range near 45° off the z-axis. However, the TM_{11} mode has a broad beam and its pattern roll-off is about 4 dB near the 45° angle. Thus, other modes will not affect significantly the radiation of the TM_{11} mode for $0 \leqslant \theta \leqslant 45°$, and the co-polar patterns will be decided primarily by the dominant mode. Their contributions will be significant for $\theta > 45°$ and, in particular, for determining the cross-polarisation which, from Fig. 2.8, shows a peak at about −25 dB range. Here the cross-polarisation is computed in $\phi = 45°$ plane, in which it maximizes. The results also show that increasing the substrate height generally increases the excitation efficiency of the other modes.

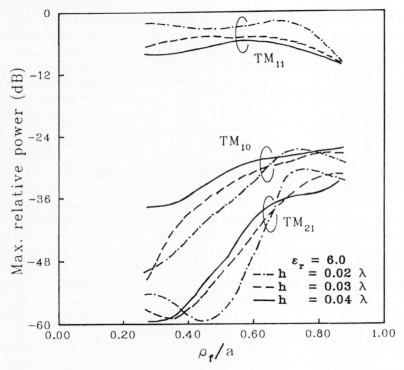

Fig. 2.8 *The effect of the feed position on the excitation efficiency of TM_{11} mode (Reproduced from Reference 14 © 1986 IEEE)*

The excitation efficiencies for a patch dominant at the TM_{21} mode are shown in Fig. 2.9. The results are plotted for two different substrate permittivities, and show similar excitations. Again the dominant mode has the strongest excitation, but its peak radiation increases for $\varrho_f > 0.68a$ and decreases thereafter. The peak radiations of the other modes have more complex behaviour and minimise at $\varrho_f = 0.75a$. Again, their contributions are below the $-25\,\text{dB}$ range. However, since the TM_{21} mode has a null along the z-axis, the contribution of the TM_{11} mode will cause a minor peak at this location. The results of Figs. 2.8 and 2.9 indicate that the resonance nature of a microstrip patch controls the excitation of the azimuthal modes, and the resonant modes can easily be excited significantly above the adjacent modes simply by selecting an appropriate location for the feed. With this type of excitation the contributions of the adjacent modes manifest themselves mainly in the cross-polarisation. They may be ignored if the antenna cross-polarisation is not the main concern. Also, the substrate permittivity seems to have a small effect on the mode excitation.

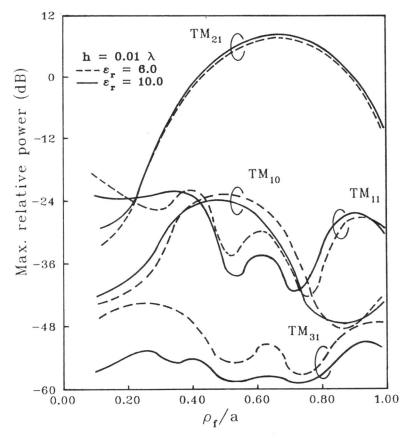

Fig. 2.9 *The effect of the position on the excitation efficiency of TM_{21} mode (Reproduced from Reference 14 © 1986 IEEE)*

We now present a few results for the radiation patterns. Fig. 2.10 shows the computed patterns for the TM_{11} mode and Fig. 2.11 for the TM_{21} mode. In both cases the feed location is selected to maximise the excitation of the dominant mode. In Fig. 2.10, the TM_{11} mode is dominant and the radiation pattern is computed by including four modes, i.e. the TM_{11} mode along with adjacent TM_{01}, TM_{21} and TM_{31} modes. The radiation peak is in the broadside with a significant radiation level behind the ground plane, owing to its finite size. In Fig. 2.11 the TM_{21} mode is resonant and the radiation patterns are generated by including the first five modes, i.e. TM_{01} to TM_{41} modes. To examine the accuracy of the computed results, sample calculations are also compared with experi-

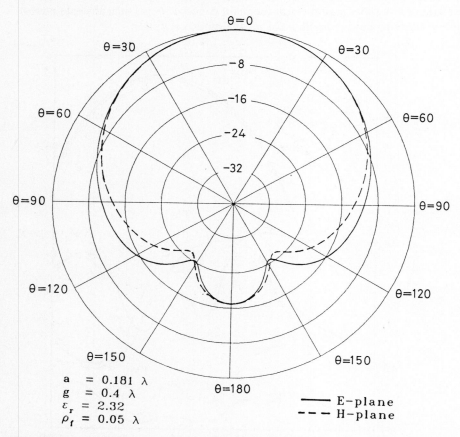

Fig. 2.10 *The radiation patterns of a circular patch for the dominant mode excitation*
$t = 0.02 \lambda$
Ground plane thickness = 0.01λ

ment. Figs. 2.12 and 2.13 show the comparison for the TM_{11} mode. The computed and experimental patterns are identical in the upper half plane and deviate slightly thereafter, owing to the coupling between the antenna and its support structure.

2.3.3 Effect of the substrate permittivity

Increasing the substrate permittivity reduces the patch size and consequently the size of the radiation zone. One therefore expects to see a broadening of the radiation pattern. This is shown in Figs. 2.14 and 2.15 for a TM_{11} mode patch and in both E- and H-planes. Since the ground-plane sizes are all the same, the antennas have equal sizes. The results show that the broadening is taking place only in the E-plane. The H-plane patterns are independent of the substrate permittivity, but show an increase in the level of the back radiation, which is also evident in the E-plane patterns. Note that, for the selected antenna dimensions,

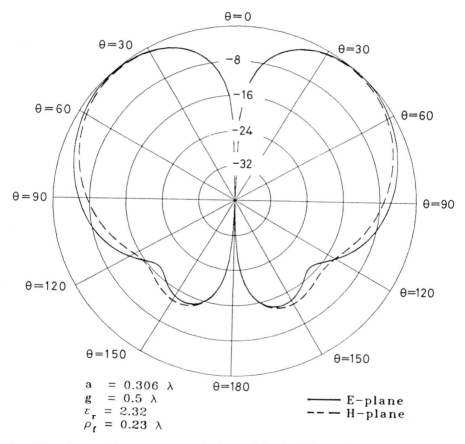

Fig. 2.11 *The radiation patterns of a circular patch for the TM_{21} mode excitation: data same as Fig. 2.10*

the small permittivity of $\varepsilon_r = 2.32$ gives nearly symmetric radiation patterns with small cross-polarisation. Since increasing ε_r broadens only the E-plane pattern, the pattern symmetry deteriorates by increasing the substrate permit-

72 Analysis of circular microstrip antennas

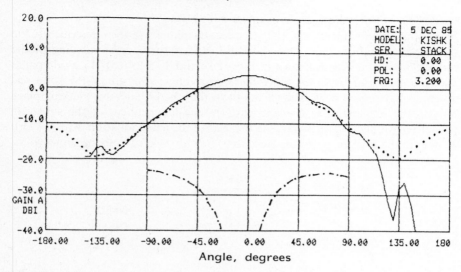

Fig. 2.12 Measured and computed data in H-plane of a circular patch excited with a coaxial probe (Reproduced from Reference 14 © 1986 IEEE)
$\varepsilon_r = 2.54$, $g = 4.5$ cm, $h = 0.159$ cm, $f = 3.2$ GH$_z$, feed at edge
——— measured
· · · · computed

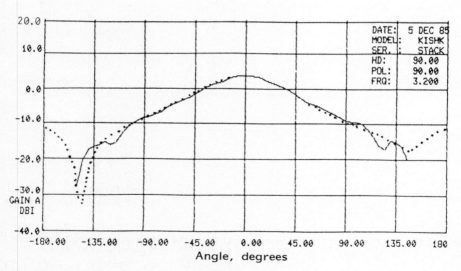

Fig. 2.13 Measured and computed data in E-plane of the case in Fig. 2.12 (Reproduced from Reference 14 © 1986 IEEE)
——— measured, data same as Fig. 2.12
· · · · computed

Analysis of circular microstrip antennas 73

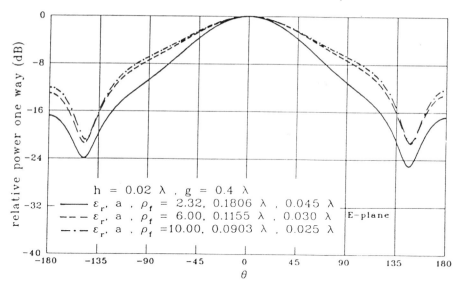

Fig. 2.14 *E-plane radiation patterns of a circular patch with different substrate permittivities (Reproduced from Reference 14 © 1986 IEEE)*

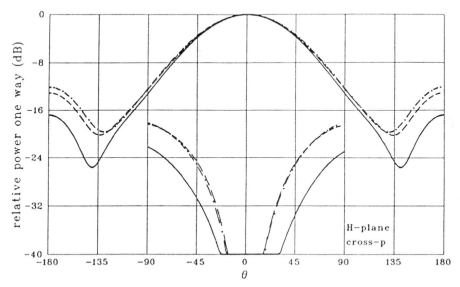

Fig. 2.15 *H-plane and cross-polarisation patterns of Fig 2.14 (Reproduced from Reference 14 © 1986 IEEE)*

tivity. This means the antenna cross-polarisation will increase, which is evident from the results presented in Fig. 2.15. Here, the cross-polarisations are computed in the $\phi = 45°$ plane, where it has the maximum magnitude.

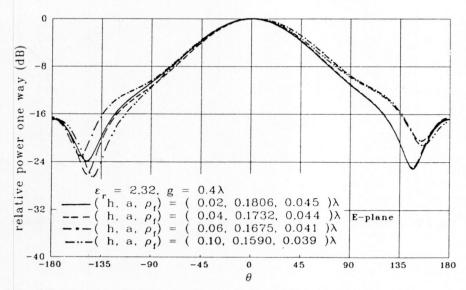

Fig. 2.16 E-plane radiation patterns of a circular patch with different substrate heights (Reproduced from Reference 14 © 1986 IEEE)

2.3.4 Effect of the substrate thickness

The bandwidth of microstrip antennas increases by increasing the substrate height. It is therefore desirable to study its effect on the radiation patterns. For the TM_{11} mode patch, representative results are shown in Figs. 2.16 and 2.17. For $h < 0.06\lambda$ the beamwidth of the H-plane patterns decreases slightly by increasing h, but increases to some degree in the E-plane. The relationship reverses for $h > 0.06\lambda$. Consequently, the cross-polarisation increases initially with h, but tends to decrease afterwards. Also, it is interesting to note that the effect of the substrate may resemble a thinner one with a higher substrate permittivity, which, from Fig. 2.14 may affect the E-plane patterns significantly. However, the results of Fig. 2.16 show otherwise, where E-plane patterns are relatively independent of h. This can be understood by considering the effect of these two parameters. A higher permittivity reduces the patch size and the extent of the fringing fields. Consequently, the radiation is due to a narrow magnetic current ring around the patch periphery, which normally gives asymmetric radiation patterns. A thicker substrate, on the other hand, does not reduce the patch size significantly, but extends the zone of the fringing fields, thus resulting in a broad radiation ring.

Analysis of circular microstrip antennas

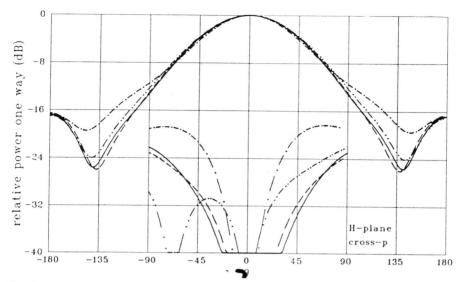

Fig. 2.17 *H-plane and cross-polarisation patterns of Fig. 2.16 (Reproduced from Reference 14 © 1986 IEEE)*

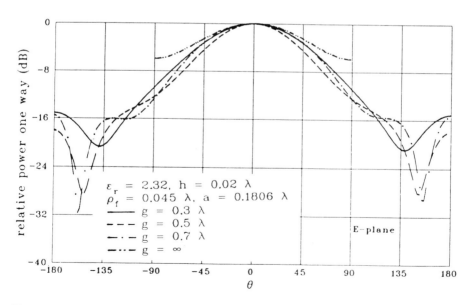

Fig. 2.18 *E-plane radiation patterns of a circular patch with different ground plane diameter for the dominant TM_{11} mode (Reproduced from Reference 14 © 1986 IEEE)*

2.3.5 Effect of the ground-plane radius

Since the ground plane controls the back radiation, its size has a pronounced effect on the patch radiation pattern. For the TM_{11} mode case, sample computed patterns are shown in Figs. 2.18 and 2.19. Again, cross-polarisation are computed in the $\phi = 45°$ plane. In the H-plane, shown in Fig. 2.19, pattern beamwidth decreases by increasing the ground-plane size. Consequently, the infinite ground plane has the most rapid pattern roll-off. In the E-plane, on the other hand, the beamwidth decreases initially by increasing the ground-plane radius g, but increases for $g > 0.7\lambda$. The infinite ground plane gives the broadest beam, which approaches -6 dB at the horizontal plane. The cross-polarisation therefore increases rapidly by increasing the ground-plane size from its optimum radius. These results show that the assumption of an infinite ground plane in approximate analysis of microstrip antennas will have a serious effect on the correct prediction of the radiation patterns, particularly for angular ranges beyond 45° off the main beam. The prediction of the cross-polarisation will, in fact, be an impossible task.

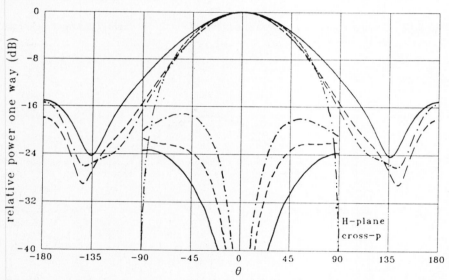

Fig. 2.19 *H-plane and cross-polarisation patterns of Fig. 2.18 (Reproduced from Reference 14 © 1986 IEEE)*

For a TM_{21} mode excitation the corresponding computed results are shown in Figs. 2.20 and 2.21. The effect of the ground-plane size on the radiation patterns is similar to the TM_{11} mode case. The beamwidth in the H-plane decreases progressively with the ground-plane size, and for the infinite ground plane the pattern roll-off is the largest. The angle for the peak radiation, which is around 45° off the z-axis, is, however, almost independent of the ground plane. The E-plane patterns also show similar behaviour to those of the TM_{11}

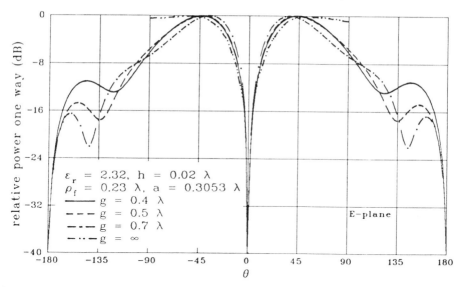

Fig. 2.20 *E-plane radiation patterns of a circular patch with different ground-plane diameter for the dominant TM_{21} mode (Reproduced from Reference 14 © 1986 IEEE)*

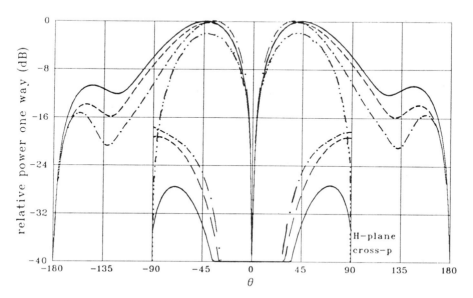

Fig. 2.21 *H-plane and cross-polarisation radiation patterns of Fig. 2.20 (Reproduced from Reference 14 © 1986 IEEE)*

mode, and their beamwidth initially decreases by increasing g, but increases for larger ground planes. For an infinite ground plane the pattern remains relatively constant beyond the peak of the pattern at about 45°. It is therefore evident that the assumption of an infinite ground plane will not provide a meaningful pattern shape for the TM_{21} mode, where the main feature of the patterns manifest itself beyond the 45°.

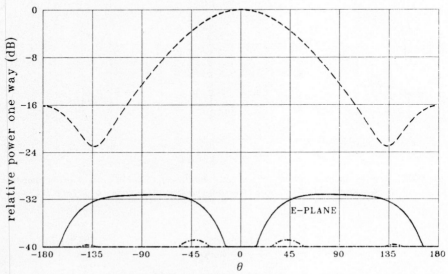

Fig. 2.22 *E-plane radiation patterns of TM_{01}, TM_{11} and TM_{21} modes of a circular patch $a = 0.1806\,\lambda$, $g = 0.3\,\lambda$, $\varrho_f = 0.05\,\lambda$, $t = 0.02\,\lambda$, $\varepsilon_r = 2.32$, and the ground plane thickness is zero.*
——— TM_{01}
– – – TM_{11}
–·–·– TM_{21}.

The above results indicate that, the radiation characteristics of various modes can easily be controlled by the ground-plane size. So far, the total patterns are shown. It may be desirable to examine the effect of the ground-plane radius on the mode-excitation efficiencies. To investigate this, the case of the TM_{11} mode patch is selected and the mode patterns are computed for two ground-plane radii of $0.3\,\lambda$ and $0.4\,\lambda$. This range of ground-plane radius gives the most symmetric co-polar patterns, with minimum cross-polarisations. The computed patterns for the $0.3\,\lambda$ antenna are shown in Figs. 2.22 – 2.24. The E-plane patterns in Fig. 2.22 are all in the $\phi = 0$ plane. The H-plane patterns are, however, in the H-plane of each mode, being $\phi = 90°$ and $\phi = 45°$ for the TM_{11} and TM_{21} modes, respectively. The corresponding results for the $0.4\,\lambda$ ground plane are shown in Figs. 2.25 – 2.27. These results indicate that the small ground plane, with $g = 0.3\,\lambda$, all non-resonant modes are well below the

Analysis of circular microstrip antennas 79

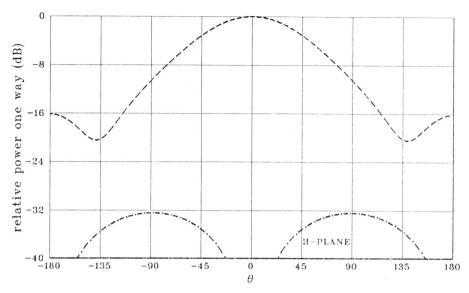

Fig. 2.23 *H-plane radiation patterns of the cases in Fig. 2.22*

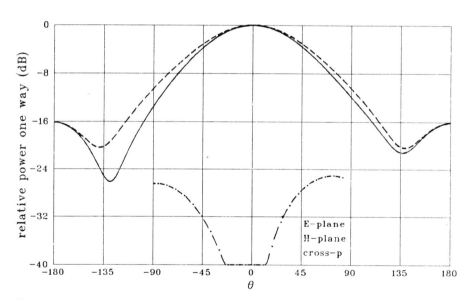

Fig. 2.24 *Total radiation patterns of the case in Fig. 2.22*
——— *E*-plane
– – – *H*-plane
–·–·– cross-polarisation.

80 Analysis of circular microstrip antennas

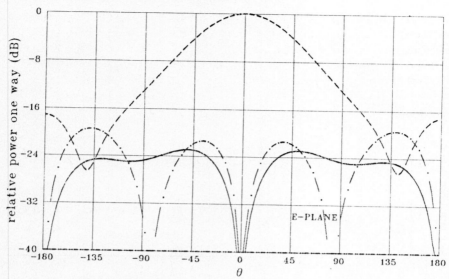

Fig. 2.25 *E-plane radiation patterns of TM_{01}, TM_{11}, TM_{21} and TM_{31} modes of a circular patch $a = 0.1806\, \lambda$, $g = 0.4\, \lambda$, $p_f = 0.05\, \lambda$, $t = 0.02\, \lambda$, $\varepsilon_r = 2.32$, and the ground plane thickness is zero*
————— TM_{01}
– – – TM_{11}
—·—·— TM_{21}
—··—··— TM_{31}

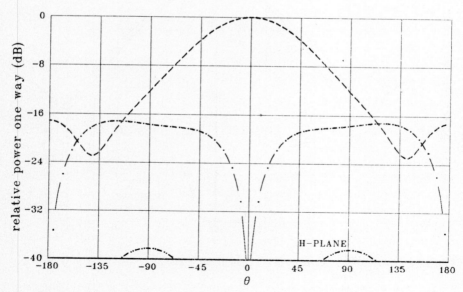

Fig. 2.26 *H-plane radiation patterns of the cases in Fig. 2.25*

dominant TM_{11} mode and their peak amplitudes are less than -30 dB. Increasing the ground-plane radius to 0.4λ increases the excitation of both TM_{01} and TM_{21} modes, and their peak amplitudes approach -17 dB range. The generated cross-polarisation is therefore larger in magnitude and increases from the -25 dB level of 0.3λ ground plane to more than -20 dB for the 0.4λ case.

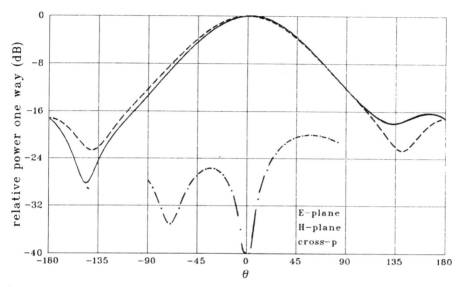

Fig. 2.27 *Total radiation patterns of the case in Fig. 2.25*
———— *E*-plane
— — — *H*-plane
—·—·— Cross-polarisation

2.3.6 Effect of the ground-plane thickness
The results of the previous Section indicate that the size of the ground plane affects the excitation efficiency of non-resonant modes. In some applications such as reflector feeds, the ground plane may not be infinitesimally thin but may have a finite thickness. Since the thickness of the ground plane affects the reflection coefficients of various modes, at its terminal edge, it may also affect their excitation efficiency. This is investigated here for the TM_{11} excitation of the patch and a ground-plane radius of 0.4λ. The results are shown in Figs. 2.28 – 2.30. Again, each of the modes of the patterns are generated in their respective *E*- and *H*-planes. The overall patterns for the co-polarisation are in the $\phi = 0°$ and $90°$ planes, the principal planes of the TM_{11} mode and the cross-polarisation are in the $45°$ plane. Comparing these results with those in Figs. 2.25 – 2.27 for zero ground-plane thickness, one notes that increasing the thickness of the ground plane to 0.05λ has reduced the excitation efficiencies of the higher TM_{21} and TM_{31} modes. The excitation of the TM_{01} mode, on the other hand, has

82 Analysis of circular microstrip antennas

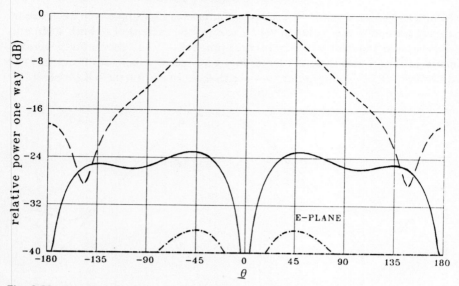

Fig. 2.28 *Same as Fig. 2.25 with the ground-plane thickness of 0·05 λ*

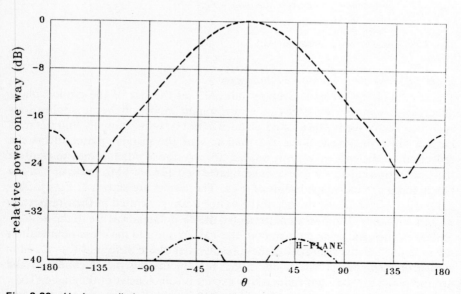

Fig. 2.29 *H-plane radiation patterns of the cases in Fig. 2.28*

remained the same. One therefore concludes that the thickness of the ground plane can also be used to modify the mode excitation. In other words, in designing microstrip antennas with small ground plane and for a high degree of mode purity, one must optimise not only the resonant patch size but also the size and thickness of the ground plane. This is, of course, valid for a given location of the excitation source, which in practice is determined by the impedance requirements. However, as shown earlier, the feed location has its own effect on the mode excitation and can be used as a parameter if warranted.

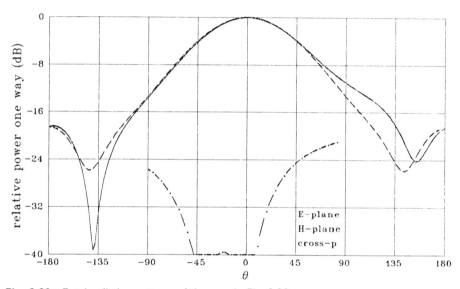

Fig. 2.30 Total radiation patterns of the case in Fig. 2.28
――――― E-plane
― ― ― H-plane
―·―·― Cross-polarisation

2.3.7 Circular polarisation

Many techniques have been proposed in the literature to generate circular polarisation with a microstrip patch antenna. In most methods a geometrical deformation is used to generate both symmetric and asymmetric modes to cause a circularly polarised radiation. These methods are convenient to generate circular polarisation when only one sense of polarisation is needed, and can be implemented by a single feed. However, when a polarisation diversity is required, one must use two separate feed arrangements. In such a case, a symmetric patch with two separate feed points and an appropriate phase switch will be sufficient. Also, to generate circularly polarised radiations with a low axial ratio, one needs an antenna with a nearly symmetric radiation pattern. The results of previous anlaysis indicated that the pattern symmetry can be controlled by

84   Analysis of circular microstrip antennas

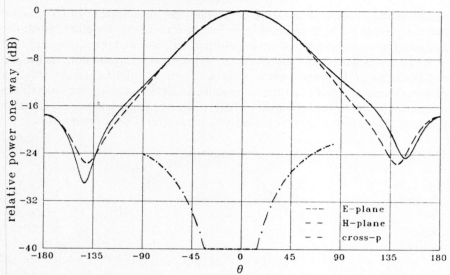

Fig. 2.31   Radiation patterns of a circular patch fed by two dipoles for circular polarisation $a = 0.1806\lambda$, $g = 0.4\lambda$, $\varrho_f = 0.05\lambda$, $t = 0.02\lambda$, $\varepsilon_r = 2.32$, and the ground plane thickness of 0.1λ
——— E-plane
– – – H-plane
–·–·– Cross-polarisation.

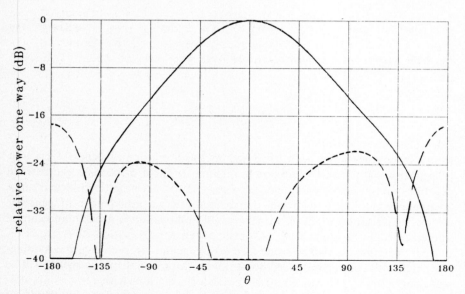

Fig. 2.32   Co-polar and cross-polar circularly polarised radiation patterns of the example of Fig. 2.31

modifying the ground-plane size and thickness. To investigate the quality of the circularly polarised radiation we present a few computed results for circular patches, fed at two locations, with a 90° phase difference.

The antenna selected is resonant at the TM_{11} mode and has a ground-plane radius of 0.4λ. The circularly polarised E_θ and E_ϕ components, and their difference as the cross-polarisation, are shown in Fig. 2.31. These principal-plane electric vectors represent the envelopes of measurement data with a rotating linearly polarised test antenna. The actual right- and left-handed circular polarisation vectors are shown in Fig. 2.32. The peak of the left-hand vector, which is the cross-polarisation, is below -20 dB level in the upper half plane, which is the main region of concern. Its pattern shape is identical to the difference pattern of Fig. 2.31 and indicates a fairly good circular polarisation. The computed data for the case of 0.3λ ground plane of zero thickness or for the 0.4λ ground plane of 0.05λ thickness are not shown here. However, their results can easily be determined from the co-polar and cross-polar patterns already provided. Since they have lower non-resonant mode excitations with reduced levels of cross-polarisation, their generated circular polarisations will also be more superior. Thus, the foregoing results indicate that, by a proper selection of the feed-point location or the size or thickness of the ground plane, circularly polarised patch antennas with extremely low axial ratio can be designed. One only needs to optimise the antenna dimensions properly.

2.3.8 Effects of a central shorting pin

So far the computed results were presented for a standard circular patch geometry. The patch dimension is therefore selected to resonate at a particular mode. However, it is reported in the literature that, by using a central pin to short the upper patch to the ground plane, one may improve the purity of the resonant mode. The previous results show that this may not be the case. Since, with a finite ground-plane size and thickness, the mode excitation can easily be controlled by modifying their dimensions. An addition of a shorting pin acts as an extra parameter to control the mode excitation. For a given antenna dimension, one can readily find a pin radius that minimised the non-resonant mode excitations. This can be done easily with the current method, provided that the pin does not destroy the rotational symmetry of the configuration. Since the introduction of the pin increases the resonance size of the patch, perhaps the most important property of the pin is to control the antenna gain by increasing the patch size. This may be a useful parameter to use in the design of higher-gain patch antennas.

2.4 Application 2: Wrap-around microstrip antenna

The wrap-around antenna refers to a microstrip-ring conformal antenna that is embedded in a missile or cylinder body. Its various configurations are considered in literature and investigated [15], [3]. Analytic solutions, using cylindrical Green's functions, can be obtained [16] by assuming that the cylinder length is infinite, so that boundary conditions on its surface can be satisfied. In practice, however, the missile shape and the location of the antenna on the body influence

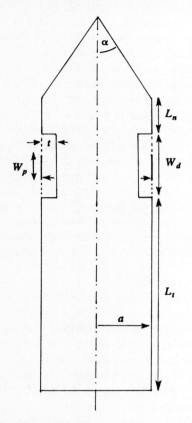

Fig. 2.33 *Cross-section geometry of wrap around antenna for missile geometry*

the antenna radiation patterns. The accurate determination of the radiation patterns must again be determined numerically. Since the configuration is rotationally symmetric, it can be analysed readily by the current method. In this section, the radiation patterns for some of useful geometrical shapes are computed. We have selected this antenna because of its complex shape. Although it is a microstrip antenna, the conducting patch and the ground plane, i.e. the cylinder, are not planar and the mode configuration are different. Here, the modes under the patch form the azimuthal modes of the cylindrical zone and their excitation is dependent on the cylinder radius. In practice, for a single exciting source all modes are present, but their magnitudes decrease with the mode number. Thus, when a particular mode must be excited, the exciting source must eliminate others, which can be achieved by the axial symmetry of multiple source excitations. For each excited mode, the radiation patterns then depend on the shape of the cylindrical surface. In most practical applications the zero-order mode is used and the present study will provide the computed data for its radiation patterns.

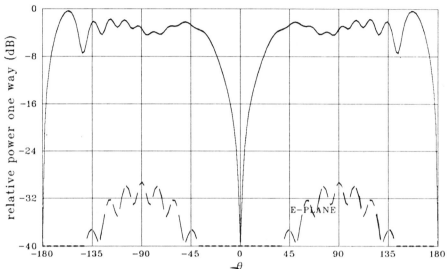

Fig. 2.34 *Radiation patterns of wrap-around antennas for the zero and 4th order modes.*
——— Zero mode
– – – 4th mode
$W_p = \lambda_d/4$, $L_t = 4 \cdot 1 \lambda$, $W_d = 0 \cdot 6 \lambda$, $L_n = 0 \cdot 1 \lambda$, $\alpha = 30°$, $\varepsilon_r = 2 \cdot 32$, $a = 0 \cdot 257 \lambda$, $t = 0 \cdot 02 \lambda$

Fig. 2.33 shows the cross-section of the antenna geometry that is investigated. It consists of a conducting ring conformal to the cylinder surface and is supported by a dielectric substrate, which is embedded in the cylinder. The radius of the cylinder is selected to be $a = 0 \cdot 257 \lambda$, which represents that of a typical small rocket. The excitation is due to four dipoles at the lower edge of the ring, which are angularly separated by 90°. Since the cylinder radius is small, the selection of four excitations ensures that the azimuthal pattern is omnidirectional. The azimuthal symmetry of the excitation means that only $4K\pi$ modes are allowed to be excited, where K is an integer. All intermediate modes cancel out. To investigate the mode excitation we select a geometry and compute the radiation patterns of the first two modes, i.e. $K = 0$ and 4. The results are shown in Fig. 2.34, where the dominant mode is for $K = 0$, the zero-order mode. The next mode for $K = 4$ is weakly excited and its contribution is below -30 dB range. The next higher mode for $K = 8$ is far too weak to be shown on the plot. These

88 Analysis of circular microstrip antennas

results indicate that, for the selected radius of the cylinder, only the zero-order mode has a significant value and all other modes can be neglected. Also, since E_ϕ is zero for the zero-order mode, all radiation patterns in this section will show the plots of the E_θ component only. Here, we present the dependance of E_θ on the antenna parameters. Fig. 2.35 shows the computed patterns when the radiating ring is located at the base of the cone, i.e. $L_n = 0$. The computed patterns show the effect of the cone angle on the radiation patterns, where α is the half cone angle and $\alpha = 90°$ refers to the geometry of a finite cylinder. Since the patterns are all similar, they are progressively shifted down by 4 dB to improve the clarity. The results show that, although the antenna is located at the upper end of the cylinder, the main beam is in the backward direction and the radiation towards the cone tip is small. Also, for the selected cone angles, the effect of the cone is not significant. In this example, the substrate permittivity is 2·32 and the width of the ring is $0.5 \lambda_d$; i.e. one half wavelength in the substrate. Other dimensional parameters are shown on the Figure.

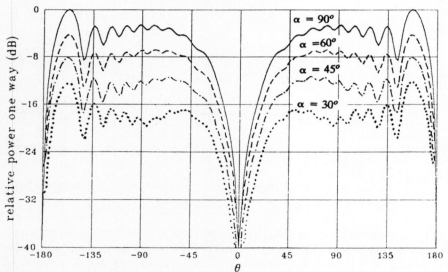

Fig. 2.35 *Radiation patterns of wrap-around antennas with different nose angle α*
───── $\alpha = 90°$
─ ─ ─ $\alpha = 60°$
─·─·─ $\alpha = 45°$
· · · · $\alpha = 30°$
$L_t = 4.1 \lambda$, $W_d = 0.6 \lambda$, $L_n = 0.0$, $W_p = \lambda_d/2$, $\varepsilon_r = 2.32$, $a = 0.257 \lambda$

For $\alpha = 30°$, Fig. 2.36 shows the effects of moving the antenna away from the cone tip and changing the cylinder length, by retaining all other parameters constant. It is evident that increasing the separation from the cone base improves the forward radiation. Also reducing the antenna separation from the cylinder base reduces the back-lobe level. Otherwise, the pattern shape stays the same. This means that the separation distances of the antenna from the cylinder ends can be used to control the radiation intensities in the forward and backward directions. In the above examples the width of the ring was selected to be one half wavelength in the substrate. The effects of ring width on the radiation patterns are shown in Fig. 2.37, where the ring widths are $0.5\lambda_d$, $0.4\lambda_d$ and $0.25\lambda_d$, respectively. Reducing the ring size reduces the broadside radiation.

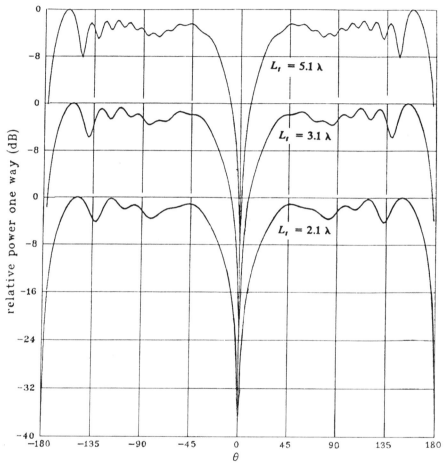

Fig. 2.36 *Radiation patterns of wrap-around antennas with different tail lengths*
From up to down: $L_t = 5.1\lambda$, $L_t = 3.1\lambda$ and $L_t = 2.1\lambda$, respectively. $W_d = 0.6\lambda$, $L_n = 0.1\lambda$, $W_p = \lambda_d/2$, $\varepsilon_r = 2.32$, $a = 0.257\lambda$, $\alpha = 30°$

Here, patterns are normalised by the main beam peaks. The quarter-wavelenth ring is an end-fire antenna and radiates mainly in the back direction. The effect of the substrate permittivity is shown in Fig. 2.38, where the width of the ring is again $0.5\lambda_d$. Increasing the permittivity rapidly reduces the broadside radiation. Finally, Fig. 2.39 shows the effects of moving the antenna towards the cylinder base. It indicates that, by increasing the antenna separation from the cone, the forward radiation increases.

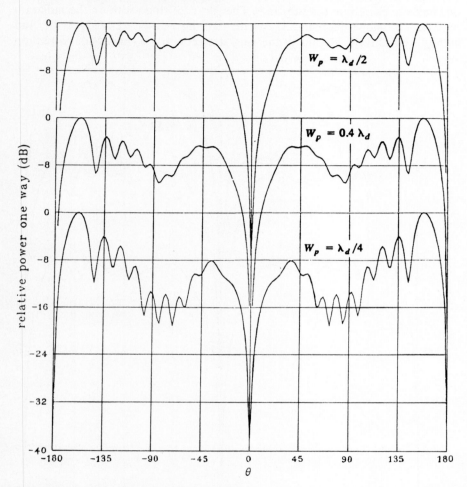

Fig. 2.37 *Radiation patterns of wrap-around antennas with different patch widths*
——— $W_p = \lambda_d/2$
– – – $W_p = 0.4\lambda_d$
–·–·– $W_p = \lambda_d/4$
$L_t = 4.1\lambda$, $\alpha = 30°$, $\varepsilon_r = 2.32$, $a = 0.257\lambda$, $t = 0.02\lambda$

To complete this study, the radiation characteristics of the ring antenna on a dielectric-coated cylinder with the antenna located symmetrically from two ends. Fig. 2.41 shows the effect of the cylinder radius on the patterns. The excitation is again due to four dipoles. For a small cylinder radius the pattern is symmetric and radiates mainly near the broadside direction, but the beam peak is around 70°. Increasing the cylinder radius moves the beam peak initially towards 90° and then towards the back direction. The effect of the substrate thickness is also studied and shown in Fig. 2.42. Decreasing the substrate thickness moves the beam peak towards the broadside and improves its symmetry. The effect of the substrate permittivity is shown in Fig. 2.43. Larger permittivities broaden the radiation pattern, which is partly due to the reduction of the effective ring width.

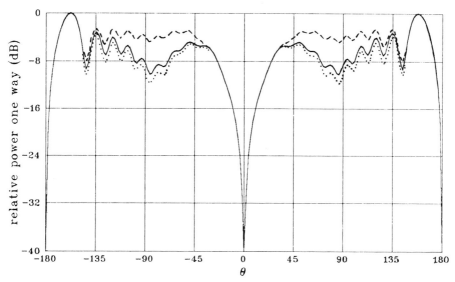

Fig. 2.38 *Radiation patterns of wrap-around antennas with different dielectric constants*
———— $\varepsilon_r = 4$
– – – $\varepsilon_r = 2\cdot 32$
· · · $\varepsilon_r = -10$

Since the analytic solution of wrap-around antennas on infinite cylinders is known, it is useful to generate their numerical solutions as well for comparison. To handle the problem, the image theory of infinite cylinders is applied to modify the original geometry. Fig. 2.44 shows the original and equivalent problems. In the original problem a ring antenna, excited by four dipoles, is supported on a dielectric-coated infinite cylinder. Applying the image theory one can determine the image of dipoles and the conducting ring inside the cylinder. The equivalent problem, in the cross-section of the cylinder, thus has eight dipoles and additional central ring representing the image of the original ring inside the cylinder. The two rings are separated by the dielectric substrate and excited by eight dipoles. The numerical solution is obtained for the finite geometry, i.e. a truncated cylinder of height $0\cdot 6\,\lambda$, and the corresponding radiation patterns are shown in Fig. 2.45 for different dielectric permittivities. The computed results agree well with the published analytical data [16]. The radiation is in the broadside direction, for the selected ring width of $0.5\lambda_d$, and the pattern broadens by increasing the permittivity. These results show that the

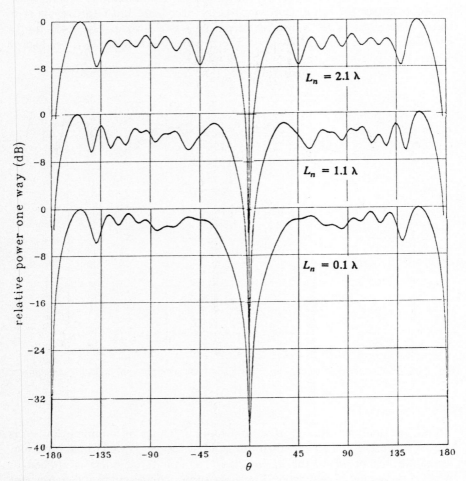

Fig. 2.39 *Radiation patterns of wrap-around antennas with different L_t and L_n*
From up to down ($L_n = 2.1 \lambda$, $L_t = 3.2 \lambda$), ($L_n = 1.1 \lambda$, $L_t = 4.1 \lambda$) and ($L_n = 0.1 \lambda$, $L_t = 5.1 \lambda$), respectively
$W_d = 0.6 \lambda$, $W_p = \lambda_d/2$, $\varepsilon_r = 2.32$, $a = 0.257 \lambda$, $\alpha = 30°$

numerical method presented here can also be used to investigate infinite structures and the accuracy of the generated results is satisfactory. The usefulness of the method, however, is in handling finite geometries where analytic methods fail.

Analysis of circular microstrip antennas 93

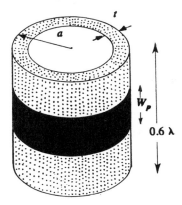

Fig. 2.40 *Geometry of a wrap-around antenna on a finite dielectric-coated cylinder*

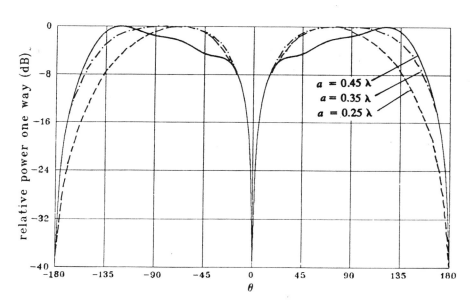

Fig. 2.41 *Effect of the radius on the radiation patterns*
——— $a = 0{\cdot}45\,\lambda$
—·—·— $a = 0{\cdot}35\,\lambda$
— — — $a = 0{\cdot}25\,\lambda$
$W_p = \lambda_d/2,\ t = 0{\cdot}02\,\lambda$

94   Analysis of circular microstrip antennas

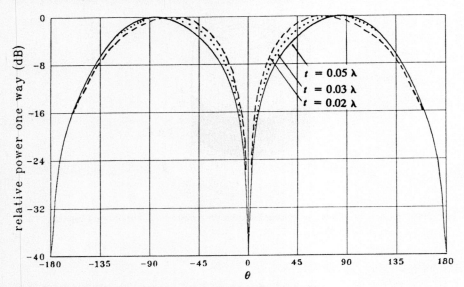

Fig. 2.42 *Effect of coating thickness on the radiation patterns*
——— $t = 0.05 \lambda$
· · · · · $t = 0.03 \lambda$
— — — $t = 0.02 \lambda$

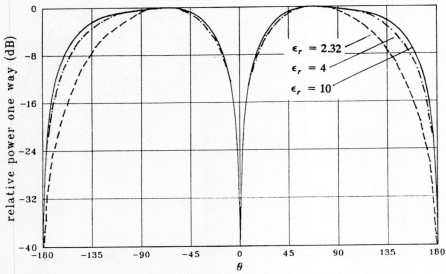

Fig. 2.43 *Effect of dielectric constant on the radiation patterns*
——— $\varepsilon_r = 10$
— — — $\varepsilon_r = 4$
—··—··— $\varepsilon_r = 2.32$
$W_p = \lambda_d/2$
$a = 0.25 \lambda, \ t = 0.02 \lambda$

Analysis of circular microstrip antennas 95

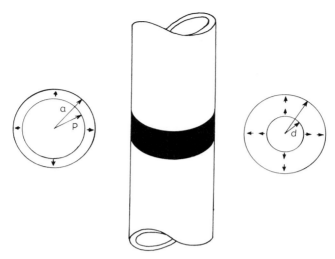

Fig. 2.44 *Geometry of wrap-around antenna of an infinite cylinder* Left cross-section indicates the original problem with four dipoles for excitation; right is the cross-section using the image theory.

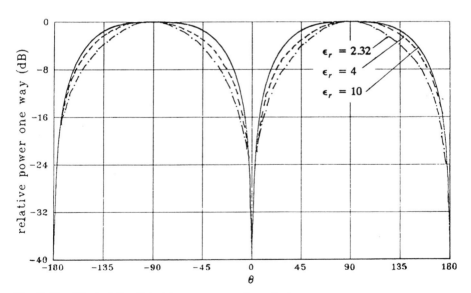

Fig. 2.45 *Effect of the dielectric constant on the radiation patterns*
—·—·— $\varepsilon_r = 2\cdot32$
— — — $\varepsilon_r = 4\cdot0$
———— $\varepsilon_r = 10$
$a = 0\cdot257\,\lambda$, $t = 0\cdot02\,\lambda$

2.5 Application 3: Reflector antenna feeds

In high-gain applications microstrip antennas may be used as feeds for reflector antennas [17 – 20]. The merits of microstrip feeds, however, depend on the type of application. In symmetric prime focus systems, the feed nomrally blocks the central region of the aperture and causes reductions in aperture efficiency and gain factor, raises the sidelobe levels and causes undesirable diffraction effects. The rise of the antenna sidelobes also increases the antenna noise temperature. Because the size of the feed depends on the operating frequency and the reflector f/D, where f and D are the reflector focal length and diameter, the aperature blockage is most severe in small paraboloid reflectors. A larger feed blocks a larger portion of the reflector central region and also requires heavier support structures. The latter further blocks the aperture, reducing the reflector performance and limiting the cross-polarisation performance. A microstrip feed is normally smaller and reduces the central blockage and its subsequent degrading effects. Furthermore, it is low cost and light weight, which reduces the complexity of the supporting structure, and can be integrated readily with its associated electronics.

The simplicity of microstrip elements also offers additional features with other reflector configurations. In offset paraboloids and dual reflector systems a small array can be used to control the reflector illumination and provides a limited scan capability with reduced sidelobes and coma lobe difficulties. Such arrays can also be used in non-paraboloidal reflectors, such as spherical reflectors, to improve the aperture efficiency and reduce the abberation. Their main advantage, however, is in the reduction of the system complexity. Microstrip feed arrays can be integrated readily with their associated circuitry and electronics, such as the power dividers, phase shifters and amplifiers. Here, we will only address the design approach and determine the performance levels for wide-angle feeds that are used with symmetric paraboloids. The array designs and their associated problems are beyond the scope of this Chapter and are discussed in subsequent Chapters.

In symmetric paraboloid reflectors the system performance is controlled primarily by the feed [21, 22]. A desirable feed must illuminate the reflector efficiently and cause small spillover. This means that the feed pattern must be broad within the cone of the reflector and roll off rapidly thereafter. It should also have negligible back radiation. The shape of the feed pattern controls the reflector efficiency, but with a symmetric system does not affect the reflector cross-polarisation. Thus, for low cross-polarisation the feed must also have a good polarisation property. From Ludwig's third definition, for minimum cross-polarisation, the feed pattern must be symmetric and have a unique phase centre [23].

A circular patch antenna is a good candidate as a feed for a symmetric reflector. Its pattern shape can be controlled readily by the size and thickness of the ground plane. Fortunately, symmetric patterns are achievable with small ground planes, which reduces the blockage. There are, however, a few problems to be overcome. The back radiation of a microstrip antenna with a small ground plane is high and its bandwidth is normally narrow. The level of the back radiation can be reduced by incorporating peripheral chokes. Generally, adding a single quarter-wavelength choke on the periphery of a waveguide feed reduces its back radiation by about 10 dB [24]. Such a reduction of the back radiation in microstrip antennas is also expected. Additional chokes can further reduce the back-lobe level, but at the expense of increased aperture blockage. In microstrip feeds one should select one or perhaps two chockes, since a large number of chokes will increase the feed size. Microstrip antennas are small in size and peripheral chokes will increase their relative size considerably, thus eliminating one of their main advantages. The limitation in the microstrip antenna band-width can also be overcome by using any of the many methods which are avaiable in literature. However, broadening the bandwidth should not affect the pattern symmetry and shape.

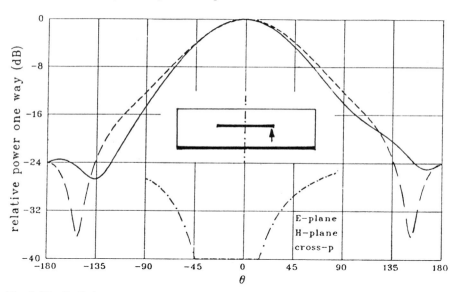

Fig. 2.46 *Radiation patterns of a circular microstrip patch, covered by a dielectric thickness of 0.1 λ*
 $a = 0.17 \lambda$, $g = 0.4 \lambda$, $\varrho_f = 0.17 \lambda$ and $\varepsilon_r = 2.32$
 ——— *E*-plane
 – – – *H*-plane
 –·–·– Cross-polar

Here, we present a design example. The previous results for a circular microstrip patch indicated that the ground-plane size and thickness can be used as parameters to equalize the *E*- and *H*-plane patterns. It was also shown that, for a ground plane radius around 0.4 λ, the pattern symmetry is satisfactory. This

98 Analysis of circular microstrip antennas

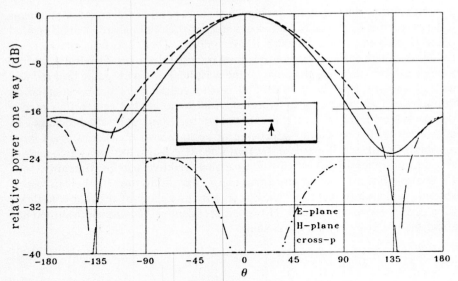

Fig. 2.47 Radiation patterns of a circular patch covered by a dielectric
$g = 0.3\lambda$, other data same as Fig. 2.46
——— E-plane
– – – H-plane
–·–·– Cross-polar

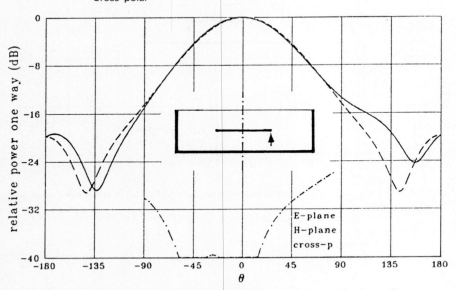

Fig. 2.48 Radiation patterns of a covered circular patch with a conducting collar
Data same as Fig. 2.46
——— E-plane
– – – H-plane
–·–·– Cross-polar

Analysis of circular microstrip antennas

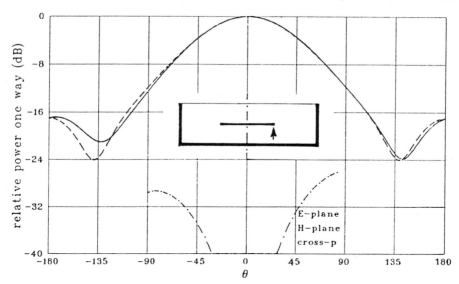

Fig. 2.49 *Radiation patterns of a covered circular patch with a conducting collar*
$g = 0.3\lambda$; other data same as Fig. 2.48
——— E-plane
— — — H-plane
—·—·— Cross-polar

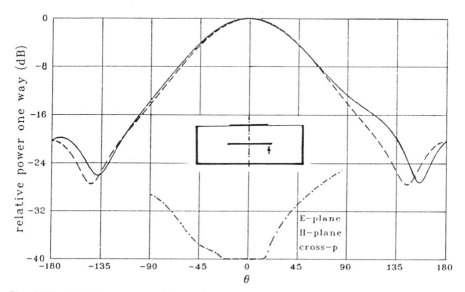

Fig. 2.50 *Radiation patterns of a two-layer stacked microstrip (Reproduced from Reference 20 © 1986 IEEE* Diameters: 0.32λ (up); 0.34λ (bottom; $g = 0.4\lambda$)

results in a feed of diameter less than one wavelenth, which is considerably smaller than commonly used waveguide feeds with chokes. To retain the geometrical symmetry and to increase the bandwidth one may use a stacked patch configuration [25]. This means the resonant patch will be covered by another dielectric-substrate which will alter its resonance frequency and the radiation pattern. To investigate the latter we have shown the radiation patterns of the new structure for ground-plane radii of 0.4 and 0.3 λ in Figs. 2.46 and 2.47. The symmetry of the patterns is satisfactory, but not perfect. To improve the geometrical rigidity we then incorporate a peripheral collar around the substrate and compute the new radiation patterns. They are shown in Figs. 2.48 and 2.49 for the previous configurations. The addition of the collar limits the radiation from the substrate termination and considerably improves the pattern symmetry. The cross-polarisation is thus improved. Finally, we add the upper patch to the configuration. The radiation patterns of the final design are shown in Figs. 2.50 and 2.51, respectively for 0.4 λ and 0.3 λ ground planes. It is interest-

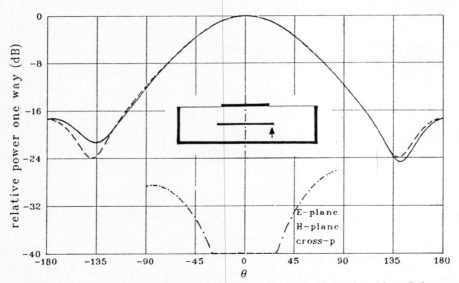

Fig. 2.51 *Radiation patterns of a two-layer stacked microstrip (Reproduced from Reference 20 © 1986 IEEE)*
 $g = 0.3 \lambda$; other data same as Fig. 2.50
 ———— *E*-plane
 ———— *H*-plane
 —·—·— Cross-polar

ing to note that the pattern characteristics remain unchanged and excellent pattern symmetries are found for both antenna geometries. The cross-polarisations for both cases are below -24 dB, but the back radiations are high. The latter will be reduced later by incorporating chokes. For the present, we investigate the sources of the cross-polarisation. We select the case of Fig. 2.48 where the ground-plane radius is 0.4λ and compute the cross-polarisation of different modes. Fig. 2.52 shows the contribution of the first four modes to the cross-polarisation. As expected, the TM_{01} and TM_{11} modes have the main contributions. However, since they have different azimuthal dependenmces their combined cross-polarisation is asymmetric. Note that the feed cross-polarisation is maximum in the $\phi = 45°$ plane and all presented data are in this plane. Fig. 2.52 also shows that adding the contribution of the higher-order modes reduces the cross-polarisation of the TM_{01} and TM_{11} modes. The overall cross-polarisation is high at about -24 dB, but within the small angular region of $\pm 45°$ is below the -30 dB range.

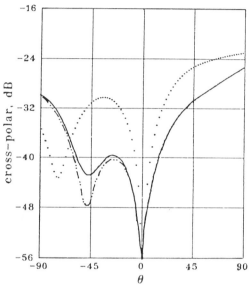

Fig. 2.52 *Effect of different modes on the cross-polarisation of antenna in Fig. 2.48. (Reproduced from Reference 20 © 1986 IEEE)*
····· $TM_{01} + TM_{11}$
···—·· $TM_{01} + TM_{11} + TM_{21}$
——— $TM_{01} + TM_{11} \ TM_{21} + TM_{31}$.

To reduce the back radiation we have used two different choke configurations. In Figs. 2.53 and 2.54 the antennas of Figs. 2.50 and 2.51 are incorporated with a choke behind the ground plane. Their pattern characteristics in the forward directions remain unchanged, but the back radiation decreases to around -24 dB. This type of choke geometry is not as efficient as the peripheral chokes in reducing the back radiation, but does not increase the feed diameter. The results with a peripheral choke are shown in Fig. 2.55, where the back radiation decreases to about -30 dB range. Adding a second choke behind the ground plane reduces the back lobe by an additional 2 dB. A second peripheral

102 Analysis of circular microstrip antennas

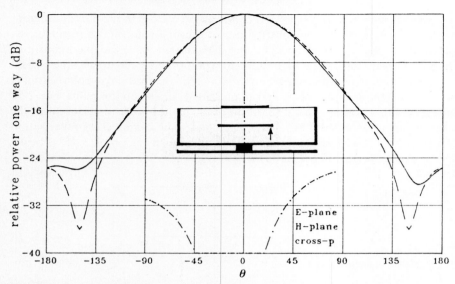

Fig. 2.53 *Radiation patterns of antenna in Fig. 2.51 with a $\lambda/4$ back choke (Reproduced for Reference 20 © 1986 IEEE)*
———— E-plane
– – – H-plane
–·–·– Cross-polar

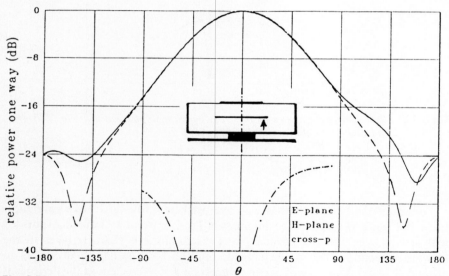

Fig. 2.54 *Radiation patterns of antenna in Fig. 2.50 with a $\lambda/4$ back choke (Reproduced for Reference 20 © 1986 IEEE)*
———— *E*-plane
– – – *H*-plane
–·–·– Cross-polar

Analysis of circular microstrip antennas 103

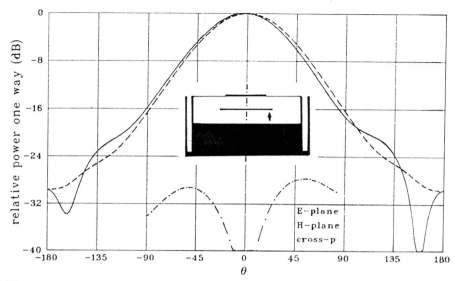

Fig. 2.55 Radiation patterns of antenna in Fig. 2.50 with a $\lambda/4$ side choke (Reproduced from Reference 20 © 1986 IEEE)
——— E-plane
– – – H-plane
–·–·– Cross-polar

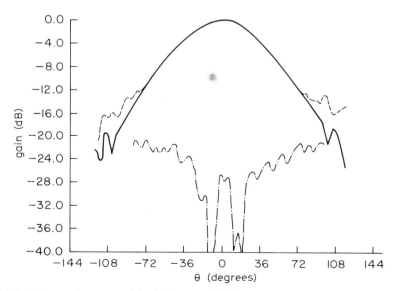

Fig. 2.56 Measured patterns of feed shown in Fig. 2.55
——— E-plane
– – – H-plane
–·–·– Cross-polarisation at 45° plane

choke can also be incorporated, but may not be necessary since the back lobe is already low and a new choke will enlarge the feed size. The feed with the peripheral choke was also fabricated and tested. Its measured E- and H-plane patterns, as well as the cross-polarisation in the $\phi = 45°$ plane shown in Fig. 2.56, which are at the band centre frequency of 4.6 GHz. The principal plane patterns agree with the computed data, but the cross-polarisation is higher. The measured peak cross-polarisation is about -21 dB, which is about 7 dB higher than the computed one. It also shows a central peak at the boresight, which indicates the misalignment of the test set up. Also, within a bandwidth of 500 MHz, i.e. 11%, the co-polar patterns remained nearly symmetric. We expect that, by improving the fabrication tolerances and a proper alignment of the test range the measured cross-polarisation should approach the computed ones. The return loss of the feed was also measured and is shown in Table 2.1. Further improvement of these return losses can be achieved by modifying the feed-point location.

Table 2.1 *Measured return losses of feeds*

Feed 1, Fig. 2.51		Feed 2, Fig. 2.55	
Frequency, GHz	Return loss, dB	Frequency, GHz	Return loss, dB
4.10	9.5	4.30	10.0
4.15	11.0	4.35	11.0
4.20	13.0	4.40	11.5
4.25	14.0	4.45	12.0
4.30	17.0	4.50	12.5
4.35	18.0	4.55	12.5
4.40	16.0	4.60	12.5
4.45	15.5	4.65	12.5
4.50	14.0	4.70	12.0
4.55	11.0	4.75	11.5
4.60	9.0	4.80	10.0

Table 2.2 summarises the performance of the above antennas, where the beamwidths at 3 dB and 10 dB levels, as well as the peak cross-polarisation, are provided. To evaluate the performance of these feeds on a reflector antenna the data on the gain factor, spill-over efficiency and the corresponding aperture angles must be known. They are calculated and shown in Table 2.3. It shows that the aperture angle varies from 60° to 71° and the gain factor rises from 72.5% to 74.24%. The feed performance is therefore reasonable. The computed gain factors are somewhat smaller than those of waveguide feeds with a corrugated flange. However, their aperture blockage is small owing to their small size. Thus, when used on small reflectors, they should provide comparable performance. These microstrip feeds are, on the other hand, light weight and easy to fabricate and can readily be integrated with receiving electronics.

Table 2.2a Feed characteristics at the resonant frequency f_0

Case of Fig.	Peak cross-pol. $0 \leqslant \theta \leqslant 90°$ (dB)	Gain (dB)	Beamwidths, deg			
			3 dB		10 dB	
			E-plane	H-plane	E-plane	H-plane
2.46	−25·2	7·98	74·4	78·7	143·4	156·5
2.47	−23·9	7·60	77·3	86·0	145·3	169·5
2.48	−25·2	8·37	72·3	72·5	141·8	141·9
2.49	−25·9	7·45	82·6	82·4	160·3	162·5
2.50	−24·7	8·19	74·6	73·4	148·1	143·3
2.51	−26·0	7·43	83·1	82·7	161·6	162·5
2.54	−26·1	8·36	74·2	73·4	142·9	142·2
2.55	−27·7	9·08	65·9	66·9	128·4	130·4

Table 2.2b Data of Table 2.2 at $f = 1 \cdot 05 f_0$

Case of Fig.	Peak cross-pol. $0 \leqslant \theta \leqslant 90°$ (dB)	Gain (dB)	Beamwidths, deg			
			3 dB		10 dB	
			E-plane	H-plane	E-plane	H-plane
2.46	−24·9	7·5	75·4	76·1	145·1	154·2
2.47	−24·7	7·9	76·82	82·96	140·6	165·7
2.48	−28·7	7·13	80·29	85·94	170·1	171·1
2.49	−26·0	7·42	81·67	88·66	159·1	177·7
2.54	−25·9	7.35	82.65	76.95	163.5	148.2
2.55	−25·3	9·05	66·10	67·21	129·5	131·1

Table 2.2c Data of Table 2.2 at $f = 0 \cdot 95 f_0$

Case of Fig.	Peak cross-pol. $0 \leqslant \theta \leqslant 90°$ (dB)	Gain (dB)	Beamwidths, deg			
			3 dB		10 dB	
			E-plane	H-plane	E-plane	H-plane
2.46	−26·1	7·68	73·1	78·1	144·3	158·1
2.47	−26·94	7·35	81·3	87·31	145·40	171·32
2.48	−26·50	7·50	73·4	81·56	145·70	161·38
2.49	−26·5	7·32	80·3	88·52	153·36	172·84
2.54	−25·1	8·24	72·7	74·4	141·1	143·8
2.55	−23·1	8·61	68·1	69·0	139·7	134·6

106 Analysis of circular microstrip antennas

Table 2.3 also shows the location of the phase centre of each antenna calculated over its aperture angle, given in column 2 [26]. Their location is measured from the ground plane, i.e. $z = 0$, and are all positive, indicating that the phase centres are above the ground plane. However, it is interesting to compare

Table 2.3a *Reflector aperture angles, gain factors, spill-over efficiencies and phase-centre locations above the ground plane for various feeds,* $f = f_0$

Case of Fig.	Aperture angle, deg	Gain factor, %	Spill-over efficiency, %	phase centre, λ
2.46	68	72·93	85·43	0·075
2.47	71	73·85	86·70	0·0675
2.48	66	73·14	85·23	0·1127
2.49	71	73·37	85·50	0·1167
2.50	66	72·86	84·16	0·1202
2.51	71	73·47	85·48	0·1212
2.54	66	74·29	86·24	0·1186
2.55	60	73·83	84·67	0·1827

Table 2.3b *Data of Table 2.3 at* $f = 1·05 f_0$

Case of Fig.	Aperture angle, deg	Gain factor, %	Spill-over efficiency, %	phase centre, λ
2.46	67	72·1	84·9	0·08
2.47	71	73·9	88·1	0·09
2.48	67	72·3	86·6	0·144
2.49	71	73·50	85·5	0·121
2.54	71	73·0	87·0	0·142
2.55	60	73·83	84·67	0·174

Table 2.3c *Data of Table 2.3 at* $f = 0·95 f_0$

Case of Fig.	Aperture angle, deg	Gain factor, %	Spill-over efficiency, %	phase centre, λ
2.46	69	73·2	85·8	0·065
2.47	71	74·1	86·2	0·068
2.48	68	73·34	87·1	0·085
2.49	71	73·53	85·6	0·071
2.54	66	74·34	86·8	0·114
2.55	66	71·2	83·9	0·416

the cases of Figs. 2.46 and 2.47 with the remaining ones, which have the peripheral conducting collar. In the former cases the phase-centre location is just above the ground plane. Since the total thickness of the dielectric is 0·1 λ, the phase centres are inside the dielectric and under the lower patch. In the remaining cases, all phase centres are outside the dielectric. From these results the following important conclusion can be drawn: In microstrip antennas, in Figs. 2.46 and 2.47, the radiation is mostly from the aperture between the patch and the ground plane. Incorporating the side collar raises the radiation zone to the periphery around the upper patch.

The performance of the above antennas listed in Tables 2.2a and 2.3a is also studied as a function of frequency. Within $\pm 5\%$ frequency variation the computed results are shown in Tables 2.2b, 2.2c and 2.3b, 2.3c. An examination of these results reveals that the feed performance remains relatively constant within the band, in the magnitude of the peak cross-polarisation and the reflector gain factor. The feed gain, however, decreases to some degree, regardless of increasing or decreasing the frequency. This is primarily due to the increased excitation of the modes adjacent to the TM_{11} mode.

2.6 Concluding remarks

In this Chapter a general numerical method has been presented that enables one to solve antenna problems involving conductors and dielectrics. While the formulation is applicable to arbitrary antenna shapes, the matrix formulation was provided only for axisymmetric configurations. The method was then used to investigate the radiation properties of three distinctly different antenna types. The circular microstrip patch antenna was selected to study the radiation mechanism of a typical microstrip antenna element. The wraparound antenna was chosen to show that the method can be used to design or analyse complex antenna candidates. The last example, i.e. the reflector feed, was included to show the design steps involving high-precision antennas, with stringent amplitude and phase-pattern requirements.

The circular patch antenna was studied in some detail to show the effect of various material and dimensional parameters on its radiation patterns. For instance, the results showed that the ground-plane size has a significant influence on the radiation patterns beyond 45° off the symmetric axis. In addition, it was shown that, by selecting an appropriate ground-plane size, nearly symmetric radiation pattern with very low cross-polarisation can be obtained. On the other hand, the feed-point location was shown to influence the excitation of non-resonant modes, which also contribute to the cross-polarisation. The information provided in this Chapter is intended to help the reader in understanding the radiation mechanism of microstrip antennas and use of various parameters to control them. While the results are computed for circular patch antennas, they can also be used for square-patch configurations, and with judicious qualifications, for other patch geometries as well. Also, the results are valid only for single, i.e. isolated microstrip antennas. When antenna elements in a practical array environment are considered, their radiation characteristics will be affected by their mutual coupling and the element location within the array. The main effect of the mutual coupling will manifest itself in the mode excitation, which is considered in a later chapter. The element location within the array affects its ground-plane size, and thus its radiation patterns. For large arrays the ground plane is large and its effect can be neglected. For small arrays the peripheral elements will 'see' a smaller ground plane than the central ones and their radiation patterns will be affected accordingly. However, the ground-plane effect in array applications becomes significant mainly in phased arrays, where the beam must be scanned for low elevation angles.

2.7 References

1 JAMES, J.R., HALL, P.S., and WOOD, C.: 'Microstrip antenna, theory and design' (Peter Peregrinus, 1981)
2 BAHL, I.J., and BHARTIA, P.: 'Microstrip antennas' (Artech House, 1980, Dedham, Mass.)
3 JOHNSON, R.C., and JASIK, H. (Eds.): 'Antenna engineering handbook' (McGraw-Hill, NY, 1984) 2nd edn., chap. 7
4 HUANG, J.: 'Finite ground plane effect on microstrip antenna radiation patterns,' *IEEE Trans.*, 1983, **AP-31,** 649-653
5 LIER, E.: 'Rectangular microstrip patch antennas.' Ph.D. Dissertation, University of Trondheim, Norway, June 1982
6 MAUTZ, J.R., and Harrington, R.F.: 'Boundary formulation for aperture coupling problem,' *Archiv fur Elekronik & Ubertrangungstechnik,* 1980, **34,** pp. 377-384
7 MEDGYESI-MITSCHANG, L.N., and PUTNAM, J.M.: 'Electromagnetic scattering from axially inhomogeneous bodies of revolution,' *IEEE Trans.*, 1984, **AP-32,** pp. 797-806
8 HARRINGTON, R.F.: 'Time harmonic electromagnetic fields' (McGraw-Hill, NY, 1961) Sec. 3-5
9 MAUTZ, J.R., and HARRINGTON, R.F.: 'H-field, E-field and combined field solutions for conducting bodies of revolution,' *Archiv fur Elektronik & ubertragungstechnik,* 1978, **32,** pp. 175-164
10 MAUTZ, J.R., and HARRINGTON, R.F.: 'Electromagnetic scattering from a homogeneous material body of revolution, *Archiv fur Elektronik & Ubertragungstechnik,* 1979, **33,** pp. 71-80
11 ISKANDER, K.A., SHAFAI, L., FRADSEN, A., and HANSEN, J.E.: 'Application of impedance boundary conditions to numerical solution of corrugated circular horns,' *IEEE trans.*, 1982, **AP-30,** pp. 366-372
12 KISHK, A.A.: 'Different integral equations for numerical solution of problems involving conducting or dielectric objects and their combination.' Ph.D., Dissertation, University of Manitoba, Winnipeg, Canada, 1986
13 YAGHJIAN, A.D.: 'Augmented electric and magnetic-field integral equations,' *Radio Science,* 1981, **16,** pp. 987-1001
14 KISHK, A.A., and SHAFAI, L.: 'The effect of various parameters of circular microstrip antennas on their radiation efficiency and the mode excitation,' *IEEE Trans.*, 1986 **AP-34,** pp. 969-977
15 MUNSON, R.E.: 'Conformal microstrip antennas and microstrip phase arrays,' *IEEE Trans.*, 1974, **AP-22,** pp. 74-78
16 FONSECA, S.B.A., and GIAROLA, A.J.: 'Pattern coverage of microstrip wraparound antennas.' Int. Conf. on Antennas and Propag., ICAP 83, Norwich, England, P. 1, 1983, pp. 300-304
17 KERR, J.L.: 'Microstrip antenna developments,' Proc. Workshop on Printed Circuit Antenna Technology, New Mexico State University, USA, Oct. 1979, pp. 3.1-3.20

18 HALL, P.S., and PRIOR, C.J.: 'Wide bandwidth microstrip reflector feed element.' 15th European Microwave Conference, Paris, 1985, pp. 1029–1044
19 PRIOR, C.J., and HALL, P.S.: 'Microstrip disc antenna with short circuit annular ring,' *Electron. Lett.* 1985, **21**, pp. 719–721
20 KISHK, A.A., and SHAFAI, L.: 'Radiation characteristics of a circular microstrip feed,' Conference on Antennas and Comm., Montech 86, Montreal, Canada, 1986, pp. 89–92
21 CLARRICOATS, P.J.B., and OLVER, A.D.: 'Corrugated horns for microwave antennas.' IEE Electromagnetic Wave Series 18 (Peter Peregrinus, 1984)
22 RUDGE, A.W., MILNE, K., OLVER, A.D., and KNIGHT, P., (Eds.): 'The hand-book of antenna design' Vol. 1. IEE Electromagnetic Wave Series 15 (Peter Peregrinus, 1982)
23 LUDWIG, A.C.: 'The definition of cross-polarisation,' *IEEE Trans.*, 1973, **AP-21**, pp. 116–119
24 SHAFAI, L., and KISHK, A.A.: 'Coaxial waveguides as primary feeds for reflector antennas and their comparison with circular waveguides,' *Archiv Fur Elektronik & Ubertragungstchnik*, 1985, **39**, pp. 8–15
25 OLTMAN, H.G.: 'Electromagnetically coupled microstrip dipole antenna,' *IEEE Trans.*, 1986, **AP-34**, pp. 467-50
26 SHAFAI, L., and KISHK, A.A.: 'Phase centre of small primary feeds and its effect on the feed performance,' *IEE Proc.*, 1985, **132**, pp. 207–214

Chapter 3
Characteristics of microstrip patch antennas and some methods of improving frequency agility and bandwidth

K.F. Lee and J.S. Dahele

3.1 Introduction

The development of microstrip antennas arose from the idea of utilising printed-circuit technology not only for the circuit components and transmission lines but also for the radiating elements of an electronic system. The basic geometry of a microstrip patch antenna (MPA) is shown in Fig. 3.1. A conducting patch is printed on the top of a grounded substrate. The shape of the patch can in principle be arbitrary. In practice, the rectangular, the circular, the equitriangular and the annular ring are common shapes. The feed can be either a coaxial cable (Fig. 3.1a) or a strip line (Fig. 3.1b), which guides the electromagnetic energy from the source to the region under the patch. Some of this energy crosses the boundary of the patch and radiates into space. The MPA is a relatively new form of radiator. In addition to compatibility with integrated-circuit technology, it offers other advantages such as thin profile, light weight, low cost and conformability to a shaped surface. The main disadvantage is its inherent narrow bandwidth (typically a few percent) arising from the fact that the region under the patch is basically a resonant cavity with a high quality factor.

The MPA was first proposed by Deschamps in 1953 [1]. However, it was only in the past 15 years or so that extensive research was devoted to this type of antennas. This was motivated by the advantages mentioned above, which make the microstrip antenna an attractive candidate for use in high-speed moving vehicles such as aircraft, missiles, rockets and communication satellites. By 1981, two textbooks [2, 3] and a special journal issue [4] containing two review articles [5, 6] were devoted to the subject. A wealth of information is now available about the microstrip patch antenna as a radiating element, primarily for the case when the substrate thickness is much smaller than a wavelength. This Chapter attempts to present some of this information, including some developments since 1981.

The plan of the Chapter is as follows. In Section 3.2, the cavity model method of analysing MPAs is described. The basic characteristics of common patch shapes are presented in Section 3.3. Some methods of improving frequency agility and bandwidth are discussed in Section 3.4. Section 3.5 contains concluding remarks.

3.2 Cavity model for analysing microstrip patch antennas

3.2.1 Introduction

Let us consider the basic geometry of a microstrip patch antenna shown in Fig. 3.1, where the z-axis is perpendicular to the plane of the patch. Electromagnetic waves are first guided along the coaxial or stripline and then spread out under the patch. When they reach the boundary of the patch, some are reflected and some radiate into open space. There are two lines of approach to deduce the radiation fields. One is to find the current distributions along the antenna structure and then obtain the radiation fields from these current sources. The other is to find the fields at the exit region. These fields act as equivalent sources, from which the radiation fields are obtained.

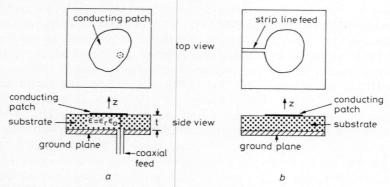

Fig. 3.1 *Microstrip patch antenna with (a) coaxial feed and (b) stripline feed*

Under the two approaches mentioned above, a number of methods of analysis have been developed. The main ones are the transmission-line model, the cavity model and the integral equation method. The transmission-line model in its original form is limited to rectangular or square patches; however, extension to other shapes is possible. The integral-equation method is perhaps the most general: it can treat arbitrary patch shapes as well as thick substrates. However, it requires considerable computational effort and provides little physical insight. Both the transmission-line model and the integral-equation method are treated elsewhere in this Handbook.

In this section, we shall introduce the cavity model. Most of the results obtained using this model are for electrically thin substrates. For this case, the cavity model offers both simplicity and physical insight. It also appears to yield results accurate enough for many engineering purposes. Our discussion will be restricted to the thin-substrate case. At the time of writing, the extension of the model to thick substrates is still in the early stage of exploration.

3.2.2 Feed modelling, resonant frequencies and internal fields
The simplicity of the cavity model can be traced to the assumption that the thickness of the substrate is much less than a wavelength, i.e. $t \ll \lambda$. The following observations then follow from this assumption:

(i) The electric field E has only the z component and the magnetic field H has only the transverse components in the region bounded by the conducting patch and the ground plane.
(ii) The fields in the aforementioned region do not vary with z.
(iii) Since the electric current in the microstrip must not have a component normal to the edge, it follows from Maxwell's equations that the tangential component of H along the edge is negligible.

As a result of (i)–(iii), the region between the patch and the ground plane can be considered as a cavity bounded by electric walls on the top and bottom, and by a magnetic wall on the side. The fact that assumption (i) does not hold near the edge because of the existence of fringing fields is taken into account by extending the edge slightly. This model has long been used in the analysis of microstrip resonators. However, the application to microstrip patch antenna appears to be due to Lo *et al.* [7], Richards *et al.* [8] and Derneryd [9, 10].

Writing Maxwell's equations for the region under the patch, we have

$$\nabla \times E = -j\omega\mu_0 H \qquad (3.1)$$

$$\nabla \times H = j\omega \varepsilon E + J \qquad (3.2)$$

$$\nabla \cdot E = \varrho/\varepsilon \qquad (3.3)$$

$$\nabla \cdot H = 0 \qquad (3.4)$$

ε in eqns. 3.2 and 3.3 is the permittivity of the substrate, the permeability of which is assumed to be μ_0. The current density J in eqn. 3.2 is due to the feed, which is usually in the form of a coaxial cable or a stripline. The advantages of the coaxial feed are that the desired impedance characteristic can be obtained by proper location of the inner conductor (see Section 3.3) and that the cable can be placed under the ground plane to minimise coupling between the feed and the antenna patch. The disadvantage is that the structure is not completely monolithic and becomes more difficult to produce. This advantage is avoided in a stripline feed, which, however, introduces some radiation of its own and offers less flexibility in obtaining the proper impedance. Usually, a quarter-wave line with a proper characteristic impedance is necessary to transform the antenna impedance to that of the stripline.

It is appropriate at this point to discuss the modelling of a feed current which has been used in the development of the cavity model. Consider first a coaxial-line feed. It can be represented by a cylindrical band of electric current flowing from the ground plane to the patch, plus an annular ring of magnetic current at the coaxial opening in the ground plane [11]. The latter can be neglected with little error, and the former can be idealised by assuming that it is equivalent to a uniform current of some effective angular width $2w$, centered on the feed axis. For example, for a circular patch fed at a distance d from the centre, it is illustrated in Fig. 3.2 and described by

$$\boldsymbol{J} = \hat{z}J(\psi)\delta(\varrho - d)/d \tag{3.5}$$

where

$$J(\psi) = \begin{cases} J, & \pi - w < \psi < \pi + w \\ 0 & \text{elsewhere.} \end{cases} \tag{3.6}$$

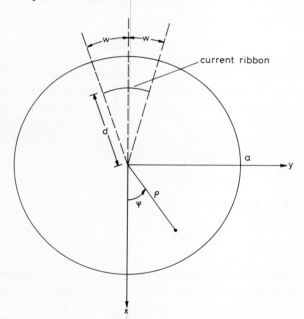

Fig. 3.2 *Modelling of a coaxial feed by a current ribbon for a circular patch*

The effective angular width $2w$ is a parameter chosen so that good agreement between the theoretical and experimental impedances is obtained. Usually, the arc length $2wd$ is several times the physical dimension of the inner conductor.

If the feed is a stripline, it can be replaced by an equivalent current source obtainable from the transverse fields in the plane where the stripline connects the patch [12]. From uniqueness concepts, only the tangential magnetic field $\boldsymbol{H}$ backed by a perfect magnetic conductor is needed. Hence the stripline feed can be modelled by a z-directed equivalent current source of some effective width $2w$. For the circular patch, it is of the same form as eqn. 3.5 except that d is replaced by the patch radius a.

In both the coaxial and the stripline feed, the z-directed current is assumed to be independent of z on account of the thinness of the dielectric region. Hence $\nabla \cdot \boldsymbol{J} = -j\omega\varrho = 0$ and eqn. 3.3 reduces to

$$\nabla \cdot \boldsymbol{E} = 0 \tag{3.7}$$

From eqns. 3.1, 3.2, 3.7 and 3.4, we obtain

$$(\nabla^2 + k_d^2)E_z = j\omega\mu_0 J_z \tag{3.8}$$

where $k_d = \omega\sqrt{\mu_0 \varepsilon}$ is the wavenumber in the dielectric. The electric-wall condition is automatically satisfied since $\boldsymbol{E} = E_z \hat{z}$ while the magnetic-wall condition implies that

$$\frac{\partial E_z}{\partial n} = 0 \tag{3.9}$$

on the sides of the cavity.

To solve eqn. 3.8 subject to the boundary conditions, we first find the eigen functions of the homogeneous wave equation

$$(\nabla^2 + k_d^2)E_z = 0 \tag{3.10}$$

subject to the same boundary conditions. Let the eigen functions of eqn. 3.10 be ψ_{mn} and the eigen values of k_d be k_{mn}.

Assuming the eigen functions to be orthogonal, the solution to eqn. 3.8 is

$$E_z = j\omega\mu_0 \sum_m \sum_n \frac{1}{k_d^2 - k_{mn}^2} \frac{\langle J_z \psi_{mn}^* \rangle}{\langle \psi_{mn} \psi_{mn}^* \rangle} \psi_{mn} \tag{3.11}$$

where $*$ denotes complex conjugate and

$$\langle J_z \psi_{mn}^* \rangle = \iiint J_z \psi_{mn}^* \, dv \tag{3.12}$$

$$\langle \psi_{mn} \psi_{mn}^* \rangle = \iiint \psi_{mn} \psi_{mn}^* \, dv \tag{3.13}$$

In eqns. 3.12 and 3.13, integration is over the domain of the patch.

The resonant frequencies are obtained from setting $k_d^2 - k_{mn}^2 = 0$ and are given by

$$f_{mn} = k_{mn}/2\pi\sqrt{\mu_0 \varepsilon}. \tag{3.14}$$

3.2.3 Radiation field

To calculate the radiation field, consider a closed surface S shown in Fig. 3.3. The top face of S lies just outside the patch and the bottom face lies just outside the ground plane. The vertical face of S coincides with the magnetic wall of the cavity. The fields exterior to S can be calculated from the equivalent sources on S and their images; the latter is necessary to account for the ground plane, which is assumed to be infinite in extent for the purpose of analysis. Since the tangential electric fields on the top and bottom faces, as well as the tangential magnetic field on the vertical surface, are zero, the only contribution to the equivalent sources are the tangential electric field E_t on the vertical surface of the cavity. Together with its image, the total equivalent magnetic current is

$$M = 2E_t \times \hat{n} \tag{3.15}$$

where $\hat{n}$ is the unit outward normal.

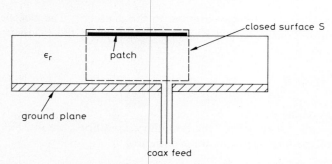

Fig. 3.3 *Application of the equivalence principle to calculate the radiation from a microstrip patch antenna*

If the substrate thickness t is much less than the wavelength λ, its effect on the radiation field is small and M can be assumed to radiate in free space. Using the free-space Green's function, the electric potential F at a point r is given by

$$F(r) = \frac{\varepsilon_0 t}{4\pi} \int M(r') \frac{e^{-jk_0|r-r'|}}{|r-r'|} dl' \tag{3.16}$$

where integration is over the perimeter of the patch.

The fields in the far-zone are given by

$$H(r) = -j\omega F(r) \tag{3.17}$$

$$E_\theta(r) = \zeta_0 H_\phi(r) \tag{3.18}$$

$$E_\phi(r) = -\zeta_0 H_\theta(r) \tag{3.19}$$

where $\zeta_0 = \sqrt{\mu_0/\varepsilon_0}$.

3.2.4 Losses in the cavity

The losses in the cavity under the patch comprise dielectric loss P_d, conductor loss P_c, radiation loss P_r and surface-wave loss P_{sw}. James *et al.* [13] estimated that surface-wave excitation is not important if $t/\lambda_0 < 0.09$ for $\varepsilon_r \simeq 2.3$ and $t/\lambda_0 < 0.03$ for $\varepsilon_r \simeq 10$, where λ_0 is the free-space wavelength. The criterion given by Wood [14] is more quantitative: $t/\lambda_0 < 0{\cdot}07$ for $\varepsilon_r = 2{\cdot}3$ and $t/\lambda_0 < 0{\cdot}023$ for $\varepsilon_r = 10$ if the antenna is to launch no more than 25% of the total radiated power as surface waves. More recent work by Fonseca et al. [15] showed that the size of the patch is also a parameter. For simplicity, we shall use Wood's criterion and assume that it is satisfied in subsequent discussions.

The dielectric loss P_d and the conductor loss P_c are calculated from the electric field under the cavity, while the radiation loss P_r is calculated from the far-zone electromagnetic field. They are given by

$$P_d = \frac{\omega \varepsilon \delta}{2} \iiint |E_z|^2 \, dv \tag{3.20}$$

$$P_c = R_s \iint |H|^2 \, ds \tag{3.21}$$

$$P_r = \frac{1}{4\zeta_0} \int_0^{2\pi} \int_0^{\pi} |E|^2 \, r^2 \sin\theta d\theta \, d\phi \tag{3.22}$$

The quantity δ in eqn. 3.20 is the loss tangent of the dielectric and R_s in eqn. 3.21 is the surface resistivity of the conductors.

The radiation or antenna efficiency is the ratio of radiated power to input power:

$$e\% = \frac{P_r}{P_r + P_d + P_c} \times 100\% \tag{3.23}$$

In calculating the losses, it is usual to make use of the resonance approximation [8]. This arises from the observation that, if the frequency is close to the resonant frequency of a particular mode, the factor $1/(k_d^2 - k_{mn}^2)$ in eqn. 3.11 is very large and the contribution to E_z, and hence to the radiation field E, is due mainly to the resonant-mode term. The electric energy stored at resonance is

$$W_e = \frac{\varepsilon}{4} \iiint |E_z|^2 \, dv = P_d/2\omega\delta \tag{3.24}$$

and the total stored energy at resonance is

$$W_T = 2W_e = \frac{\varepsilon}{2} \iiint |E_z|^2 \, dv = P_d/\omega\delta. \tag{3.25}$$

The effective loss tangent of the cavity, taking into account the three losses P_d, P_c and P_r, is given by

$$\delta_{eff} = P_T/(\omega W_T) \tag{3.26}$$

where

$$P_T = P_d + P_c + P_r. \tag{3.27}$$

A number of quality Q factors are defined as follows:

Dielectric Q: $Q_d = \omega W_T/P_d = 1/\delta$ \hfill (3.28)

118 Characteristics of microstrip patch antennas

$$\text{Conductor } Q: Q_c = \omega W_T/P_c = \sqrt{\pi f \mu_0 \sigma}\, t \qquad (3.29)$$

$$\text{Radiation } Q: Q_r = \omega W_T/P_r \qquad (3.30)$$

$$\text{Total } Q: Q_T = \omega W_T/P_T = 1/\delta_{eff} \qquad (3.31)$$

In eqn. 3.29, σ is the conductivity of the patch and the ground plane.

3.2.5 Input impedance

The input impedance at the feed of the antenna is given by

$$Z = R + jX = V/I = E_{av}t/I \qquad (3.32)$$

where E_{av} is the average value of the electric field at the feed point and I is the total current. For example, if the feed is modelled by eqn. 3.5, we have

$$E_{av} = \frac{1}{2w} \int_{\pi-w}^{\pi+w} E_z(d, \phi)\, d\phi \qquad (3.33)$$

and

$$I = -J(2wd) \qquad (3.34)$$

Unlike the calculations of δ_{eff}, it was found that non-resonant modes must be included in the calculation of input impedance if good agreement between theory and experiment was to be obtained. The appropriate equation for E_z is therefore eqn. 3.11, which contains the factor $1/(k_d^2 - k_{mn}^2)$. To keep this term finite at resonance, the permittivity of the dielectric must be considered complex. If only the dielectric loss is considered, we have

$$\varepsilon = \varepsilon_0 \varepsilon_r (1 - j\delta) \qquad (3.35)$$

$$k_d^2 = \omega^2 \mu_0 \varepsilon = k_0^2 \varepsilon_r (1 - j\delta), \qquad (3.36)$$

However, Richards et al. [16] found that better agreement with experiment was obtained if, instead of the loss tangent of the dielectric, the effective loss tangent is used. Thus, in calculating the input impedance, eqn. 3.11 is modified to read

$$E_z = j\omega \mu_0 \sum_m \sum_n \frac{1}{k_{eff}^2 - k_{mn}^2} \frac{\langle J_z \psi_{mn}^* \rangle}{\langle \psi_{mn} \psi_{mn}^* \rangle} \psi_{mn} \qquad (3.37)$$

where

$$k_{eff}^2 = k_0^2 \varepsilon_r (1 - j\delta_{eff}) \qquad (3.38)$$

A typical impedance versus frequency curve is illustrated in Fig. 3.4. There is usually some reactance at the resonant frequency of a mode due to the contributions from the non-resonant modes.

3.2.6 VSWR bandwidth

The bandwidth of an antenna is the range of frequencies within which the performance of the antenna, with respect to some characteristic, conforms to a specific standard. In the case of the microstrip patch antenna which is basically a strongly resonant device, it is usually the variation of impedance, rather than pattern, which limits the standard of performance. If the antenna impedance is matched to the transmission line at resonance, the mismatch off resonance is related to the VSWR. The value of VSWR which can be tolerated then defines the bandwidth of the antenna. If this value is to be less than S, the usable bandwidth of the antenna is related to the total Q-factor by [11]

$$\text{Bandwidth } (BW) = \frac{100(S-1)}{Q_T \sqrt{S}} \% \ (S \geqslant 1) \tag{3.39}$$

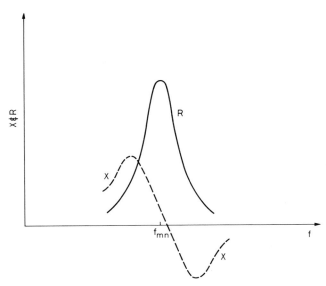

Fig. 3.4 *Typical impedance characteristics around the resonant frequency of a mode*

For $S = 2$, which is a common standard, the above equation reduces to

$$BW = \frac{100}{\sqrt{2}\, Q_T} \% \tag{3.40}$$

While eqn. 3.39 is the most commonly used definition for bandwidth and is the one we use in this Chapter, it should be pointed out that this is not a universal definition. For example, some authors define the bandwidth as $1/Q_T$.

3.2.7 Qualitative description of the results predicted by the model

In Section 3.3, the equations presented above will be used to obtain the specific results for a number of microstrip patch antennas. It is perhaps instructive to describe here the qualitative features which are common to MPAs. These features follow naturally from viewing the MPA as a leaky cavity.

(i) There are an infinite number of resonant modes, each characterised by a resonant frequency.

(ii) Because of fringing fields at the edge of the patch, the patch behaves as if it has a slightly larger dimension. Semi-empirical factors are usually introduced to obtain these effective dimensions. These factors vary from patch to patch.

(iii) Each resonant mode has its own characteristic radiation pattern. The lowest mode usually radiates strongest in the broadside direction. The pattern of this mode is broad, with half-power beamwidths of the order of 100°.

(iv) For coaxial-fed antennas, the input impedance is dependent on the feed position. The variation of input resistance at resonance with feed position essentially follows that of the cavity field. For the lowest mode, it is usually large when the feed is near the edge of the patch and decreases as the feed moves inside the patch. Its magnitude can vary from tens to hundreds of ohms. By choosing the feed position properly, an effective match between the antenna and the transmission line can be obtained.

(v) Since the cavity under the patch is basically a resonator, the total Q and the impedance bandwidth are dependent on the thickness of the substrate t and its permittivity ε. For low values of ε_r, the bandwidth generally increases with increasing t and decreases with increasing ε_r. This is presumably due to the fact that the stored energy W_T decreases with t and increases with ε_r while the total loss P_T is insensitive to these changes. However, detailed analysis (Section 3.3) shows that the bandwidth and Q are complicated functions of frequency, substrate thickness and the permittivity.

(vi) For thin substrates, the impedance bandwidth varies from less than one to several percent.

In the next Section, the results obtained by applying the formulas of this Section to rectangular, circular, equitriangular and annular-ring patches will be presented.

3.3 Basic characteristics of some common patches

A number of canonical patch shapes can be analysed by straightforward application of the cavity model. Of these, the rectangular, the circular, the equitriangular and the annular ring are the common shapes used in practice. They will be considered in detail in this Section. An example comparing the characteristics of these patches will be given in Section 3.4, while other patch shapes will be briefly mentioned in Section 3.5.

3.3.1 The rectangular patch

3.3.1.1 Introduction: The rectangular patch (Fig. 3.5) is probably the most commonly used microstrip antenna. It is characterised by the length a and the width b. The electric field of a resonant mode in the cavity under the patch is given by

$$E_z = E_0 \cos(m\pi x/a) \cos(n\pi y/b) \qquad (3.41)$$

where $m, n = 0, 1, 2\ldots$
The resonant frequency is

$$f_{mn} = k_{mn} c/(2\pi\sqrt{\varepsilon_r}) \qquad (3.42)$$

where

$$k_{mn}^2 = (m\pi/a)^2 + (n\pi/b)^2 \qquad (3.43)$$

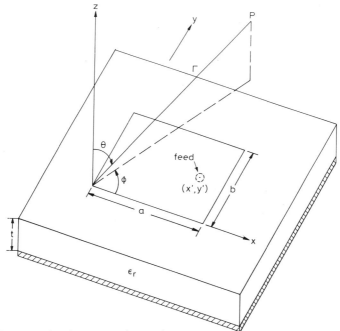

Fig. 3.5 *Geometry for the rectangular patch*

Eqn. 3.42 is based on the assumption of a perfect magnetic wall. To account for the fringing fields at the perimeter of the patch, the following empirical formula can be used for the effective dimensions [7]:

$$a_e = a + t/2 \qquad (3.44)$$
$$b_e = b + t/2 \qquad (3.45)$$

A more accurate but lengthy formula, suggested by James *et al.* [3] is

$$f_{r1} = f_{r0} \frac{\varepsilon_r}{\sqrt{\varepsilon_e(a)\varepsilon_e(b)}} \frac{1}{(1+\Delta)} \qquad (3.46)$$

where

$$\Delta = \frac{t}{a}\left[0{\cdot}882 + \frac{0{\cdot}164(\varepsilon_r - 1)}{\varepsilon_r^2} + \frac{(\varepsilon_r + 1)}{\pi\varepsilon_r}\right.$$
$$\left. \times \left\{0{\cdot}758 + \ln\left(\frac{a}{t} + 1{\cdot}88\right)\right\}\right] \quad (3.47)$$

$$\varepsilon_e(u) = \frac{\varepsilon_r + 1}{2} + \frac{\varepsilon_r - 1}{2}\left[1 + \frac{10t}{u}\right]^{-1/2} \quad (3.48)$$

where f_{r0} is the resonant frequency given by eqn. 3.42.

Eqn. 3.46 is found to yield resonant frequencies which are within 3% of experimental values, while the perfect magnetic-wall model gives errors up to 20%.

The far-field, losses and Q, and input impedance can be calculated by applying the equations of Section 3.2. Since for the rectangular patch they are well documented [3, 7, 8], they will not be reproduced here. In what follows, we shall present numerical results based on these equations to illustrate the basic characteristics of the rectangular patch. Experimental results will also be given.

3.3.1.2 Illustrative results: We present in this Section numerical, and in some cases experimental, results illustrating the basic characteristics of the rectangular patch antenna.

(a) Magnetic-current distribution
The electric-field and magnetic-surface-current distributions on the side wall for TM_{10}, TM_{01} and TM_{20} modes are illustrated in Fig. 3.6. For the TM_{10} mode, the magnetic currents along b are constant and in phase while those along a vary sinusoidally and are out of phase. For this reason, the b edge is known as the radiating edge since it contributes predominantly to the radiation. The a edge is known as the non-radiating edge. Similarly, for the TM_{01} mode, the magnetic currents are constant and in phase along a and are out of phase and vary sinusoidally along b. The a edge is thus the radiating edge for the TM_{01} mode.

(b) Radiation patterns
The modes of the greatest interest are TM_{10} and TM_{01}. However, the TM_{03} mode has also received some attention. These three modes all have broadside radiation patterns. The computed patterns for $a = 1{\cdot}5b$ and two values of ε_r are shown in Figs. 3.7a–f. In the principal planes, the TM_{01} and TM_{03} modes have similar polarisations while that of the TM_{10} mode is orthogonal to the other two. As will be discussed in Section 3.4.3.2, the TM_{01} and TM_{03} modes can be utilised to operate the rectangular patch as a dual-frequency antenna. The patterns do not appear to be sensitive to a/b or t. However, they change appreciably with ε_r.

Characteristics of microstrip patch antennas 123

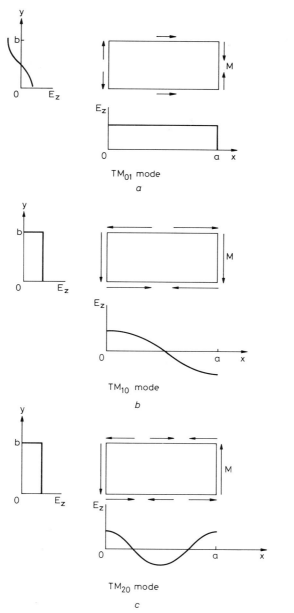

Fig. 3.6 *Electric field and magnetic-surface-current distributions in walls for several modes of a rectangular microstrip patch antenna*
 a TM_{01} mode
 b TM_{10} mode
 c TM_{20} mode

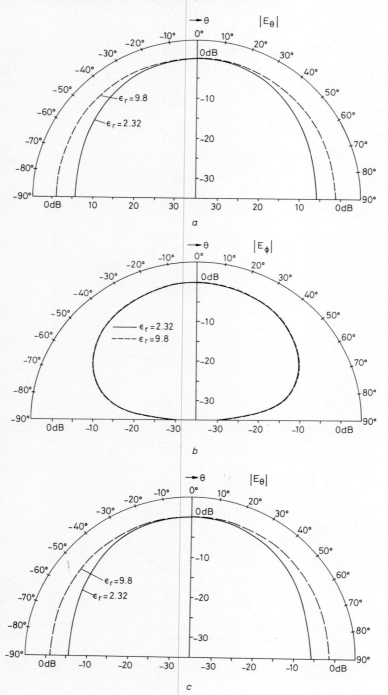

Characteristics of microstrip patch antennas 125

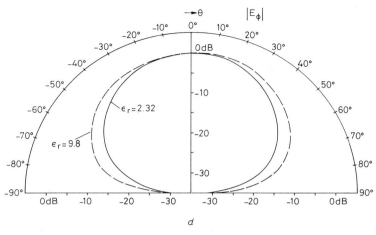

d

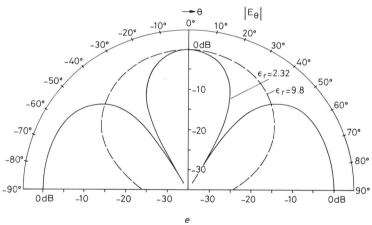

e

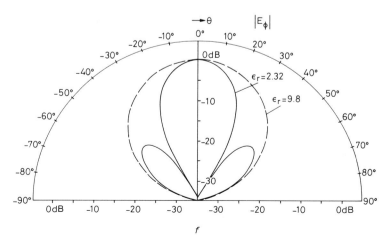

f

126 Characteristics of microstrip patch antennas

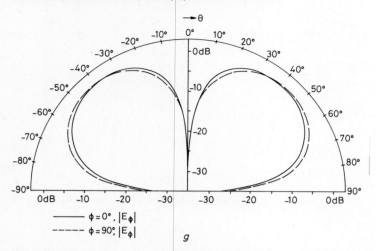

Fig. 3.7 *Relative field patterns for a rectangular patch with a/b = 1·5, f_{mn} = 1 GHz, and (i) ε_r = 2·32, t = 0·318, 0·159, 0·0795 cm; (ii) ε_r = 9·8, t = 0·127, 0·0635, 0·0254 cm*
(a) TM_{10}, $\phi = 0°$
(b) TM_{10}, $\phi = 90°$
(c) TM_{01}, $\phi = 90°$
(d) TM_{01}, $\phi = 0°$
(e) TM_{03}, $\phi = 90°$
(f) TM_{03}, $\phi = 0°$
(g) TM_{11}, $\varepsilon_r = 2·32$

The patterns of most of the other modes have maxima off broadside. For example, those of the TM_{11} mode are illustrated in Fig. 7g.

Fig. 3.8 shows the computed and measured radiation patterns of the TM_{10} and TM_{01} modes obtained by Lo et al. [7] for a rectangular patch with $a = 11·43$ cm, $b = 7·6$ cm, $\varepsilon_r = 2·62$ and $t = 0·159$ cm. Both E_θ and E_ϕ were measured in each of the two cuts, $\phi = 0°$ and $\phi = 90°$. It was found that one component of polarisation was negligible when compared to the other in each case and is not shown.

(c) Radiation efficiency
Let us first obtain some idea of the relative magnitudes of the power dissipated in the metal, the power dissipated in the dielectric, and the antenna radiation efficiency. These are described by the quantities P_c/P_T, P_d/P_T and P_r/P_T, respectively. The cases of (i) $a = 1·5b$, $\varepsilon_r = 2·32$, $t = 0·159$ cm and (ii) $a = 1·5b$, $\varepsilon_r = 9·8$, $t = 0·0635$ cm are illustrated in Fig. 3.9 for the TM_{10} mode. It is seen that, for both $\varepsilon_r = 9·8$ and $\varepsilon_r = 2·32$, the loss due to the conductor is larger than the loss due to the dielectric. The ratio P_c/P_T decreases rapidly as frequency increases.

The radiation efficiency $e = P_r/P_T$ of the TM_{10}, TM_{01} and TM_{03} modes as a function of resonant frequent is shown in Figs. 3.10a–c for a patch with

$a = 1.5b$. For $\varepsilon_r = 2.32$ and $\varepsilon_r = 9.8$, the results for three thicknesses are given. In general, the efficiency increases with the thickness of the substrate and decreases with increasing ε_r.

In using these curves, the criterion given by Wood for the avoidance of excessive surface-wave excitation mentioned in Section 3.2.4 should be kept in mind. For $\varepsilon_r = 2.32$ and $\varepsilon_r = 9.8$, the cut-off frequencies (below which the surface wave is less than 25% of total radiated power) are $21/t$ and $6.9/t$ GHz, respectively, where t is in millimetres. These correspond to 6·60 GHz for $\varepsilon_r =$

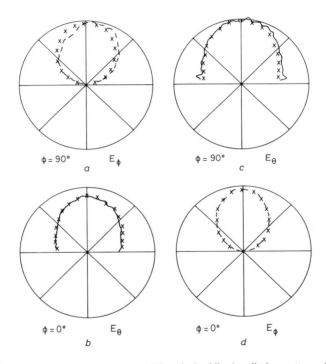

Fig. 3.8 *Theoretical (x) and measured (solid or dashed line) radiation patterns in $\phi = 0°$ and $\phi = 90°$ planes of a rectangular patch antenna with $a = 11.43$ cm, $b = 7.6$ cm, $\varepsilon_r = 2.62$, $t = 0.159$ cm. (Reproduced from Reference 7 p. 140 © 1979 IEEE)*
(a) and (b) at resonant frequency 804 MHz of (1, 0) mode
(c) and (d) at resonant frequency 1187 MHz of (0, 1) mode

2·32, $t = 0.318$ cm and 5·43 GHz for $\varepsilon_r = 9.8$, $t = 0.127$ cm. For the other cases, the cut-off frequencies occur beyond 10 GHz.

(d) Directivity and gain
The directivity D of an antenna is defined as the ratio of power density in the main beam to the average power density while the gain $G = eD$. For a rectangular patch with $a = 1.5b$, the directivities as a function of resonant frequency

for the three broadside modes are illustrated in Fig. 3.11. The directivity of the TM_{03} mode is largest and that of the TM_{10} mode the smallest. It is not sensitive to substrate thickness and resonant frequency. The gain, on the other hand, increases with resonant frequency in the manner shown in Figs. 3.12a–c.

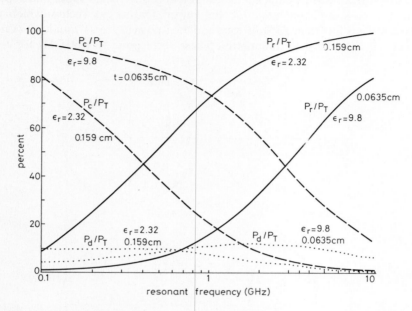

Fig. 3.9 *Metallic (P_c), dielectric (P_d) and radiation (P_r) losses for the TM_{10} mode as a function of resonant frequency for (a) a = 1·5b, ε_r = 2·32, t = 0·159 cm and (b) a = 1·5b, ε_r = 9·8, t = 0·0635 cm*

(e) *Total Q and bandwidth*

For the three broadside modes, the variation of total Q with resonant frequency is shown in Figs. 3.13a–c for the case $a = 1\cdot5b$. The bandwidth, as defined by eqn. 3.43, is shown in Figs. 3.14a–c. It is seen that the TM_{01} mode has the lowest Q and therefore the largest bandwidth compared to the other two modes. For $\varepsilon_r = 2\cdot32$, the bandwidth for a given mode increases with substrate thickness except for frequencies below about 0·7 GHz. The behaviour is more complicated for $\varepsilon_r = 9\cdot8$. For this case, there appears to be a range of frequencies for which a thinner substrate actually yields a larger bandwidth.

(f) *Input impedance*

Richards et al. [8] have reported calculated and measured values of the input impedance of a coaxial-fed rectangular patch with $\varepsilon_r = 2\cdot62$ and $t = 0\cdot159$ cm. The Smith chart plot for the TM_{01} mode is shown in Fig. 3.15 for three feed positions. The variation of the input resistances at resonance of the TM_{10} and TM_{01} modes is shown in Fig. 3.16. It is seen that the input resistance is largest

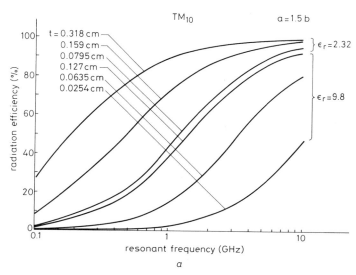

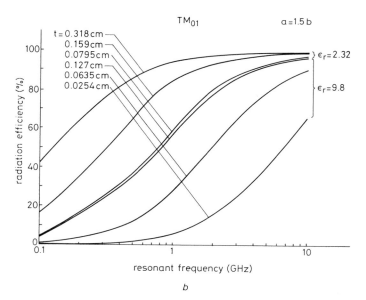

Fig. 3.10 *Radiation efficiency as a function of resonant frequency for a rectangular patch with $\sigma = 5{\cdot}8 \times 10^7\,S/m$, $\delta = 0{\cdot}0005$, $a = 1{\cdot}5b$ and (i) $\varepsilon_r = 2{\cdot}32$, $t = 0{\cdot}318, 0{\cdot}159, 0{\cdot}0795\,cm$; (ii) $\varepsilon_r = 9{\cdot}8$, $t = 0{\cdot}127, 0{\cdot}0635, 0{\cdot}0254\,cm$*
(a) TM_{10}
(b) TM_{01}
(c) TM_{03}

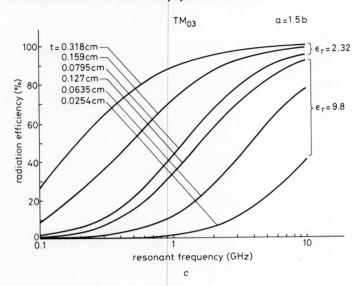

c

when fed at the edge of the patch, but it can attain the convenient value of 50 Ω when the feed position is chosen properly.

More detailed theoretical results for the input resistance at resonance are shown in Figs. 3.17a–c for the TM_{10}, TM_{01} and TM_{03} modes for $\varepsilon_r = 2\cdot32$, $t = 0\cdot159$ cm and $a/b = 1\cdot5$. The resistances are plotted as a function of feed position parametric in the resonant frequencies. It is seen that the values vary somewhat with the resonant frequency. It should also be noted that, for the TM_{03} mode, the variation with feed position is not a monotonically decreasing function, which is the case for the TM_{10} and TM_{01} modes.

It is clear from these illustrations that, for a coaxial feed, matching the antenna impedance to the transmission-line impedance can be accomplished simply by putting the feed at the proper location. There is less flexibility in the case of a stripline feed. In this case, a quarter-wave transformer may be added to effect matching. Alternatively, an insert into the patch can be made.

To conclude this Section, we point out that a rectangular patch with a single feed produces linearly polarised radiation. If circular polarisation is desired, the most direct approach is to use two feeds located geometrically 90° apart and with a relative phase shift of 90°. This arrangement excites two orthogonal modes, each providing a linearly polarised wave at right angles to each other and at phase quadrature.

Circular polarisation can also be produced by a nearly square patch, where one pair of sides resonates at a slightly higher frequency than the other pair. If the phase difference at the centre frequency between the pairs of sides is $\pi/2$, circular polarisation results. Lo and Richards [17, 18] have shown that the sides

Characteristics of microstrip patch antennas

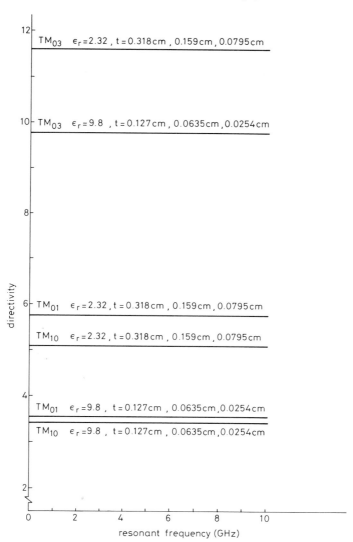

Fig. 3.11 *Directivities (absolute value) of the TM_{10}, TM_{01} and TM_{03} modes as a function of resonant frequency for a rectangular patch with $a = 1·5b$ and (ii) $\varepsilon_r = 2·32$, $t = 0·318$, $0·159$, $0·0795$ cm; (ii) $\varepsilon_r = 9·8$, $t = 0·127$, $0·0635$, $0·0254$ cm*

a and b must satisfy $a/b = 1 + 1/Q$ and the feed must be located along the line $y = \pm bx/a$. The plus and minus signs produce left-hand and right-hand circular polarisations, respectively, in the direction normal to the patch.

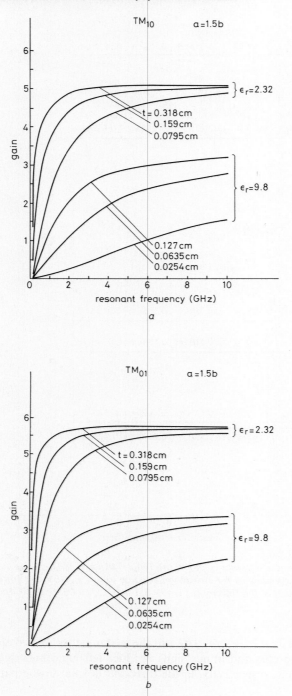

Characteristics of microstrip patch antennas

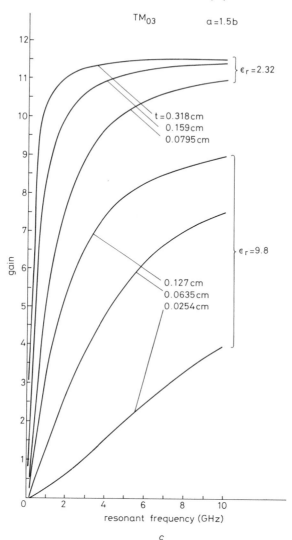

Fig. 3.12 *Gain (absolute value) as a function of resonant frequency for a rectangular patch with $\sigma = 5\cdot 8 \times 10^7\, S/m$, $\delta = 0\cdot 0005$, $a = 1\cdot 5b$ and (i) $\varepsilon_r = 2\cdot 32$, $t = 0\cdot 318, 0\cdot 159, 0\cdot 0795\,cm$; (ii) $\varepsilon_r = 9\cdot 8$, $t = 0\cdot 127, 0\cdot 0635, 0\cdot 0254\,cm$*
  (a) TM_{10}
  (b) TM_{01}
  (c) TM_{03}

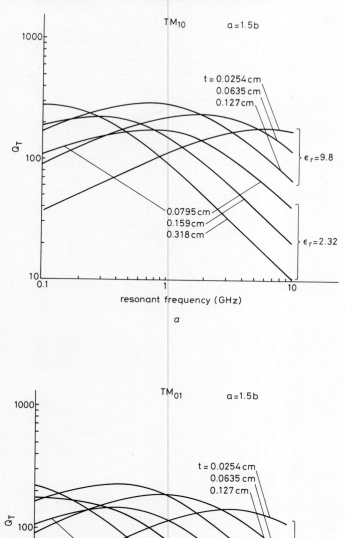

Characteristics of microstrip patch antennas 135

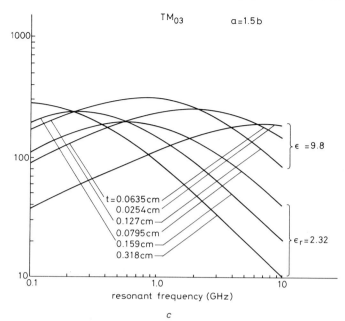

c

Fig. 3.13 *Total Q factor as a function of resonant frequency for a rectangular patch with $\sigma = 5.8 \times 10^7$ S/m, $\delta = 0.0005$, $a = 1.5b$ and (i) $\varepsilon_r = 2.32$, $t = 0.318$, 0.159, 0.0795 cm; (ii) $\varepsilon_r = 9.8$, $t = 0.117$, 0.0635, 0.0254 cm*
(a) TM_{10}
(b) TM_{01}
(c) TM_{03}

3.3.2 The circular patch

3.3.2.1 Introduction: The geometry of the circular patch or disc (Fig. 3.18) is characterised by a single parameter, namely, its radius a. In this respect, it is perhaps the simplest geometry since other shapes require more than one parameter to describe them. The mathematical analysis, however, involves Bessel functions.

The electric field of a resonant TM_{nm} mode in the cavity under the circular patch is given by

$$E_z = E_0 J_n(k_{nm}\varrho) \cos n\psi \qquad (3.49)$$

where ϱ and ψ are the radial and azimuthal co-ordinates, respectively. E_0 is an arbitrary constant, J_n is the Bessel function of the first kind of order n and

$$k_{nm} = X_{nm}/a \qquad (3.50)$$

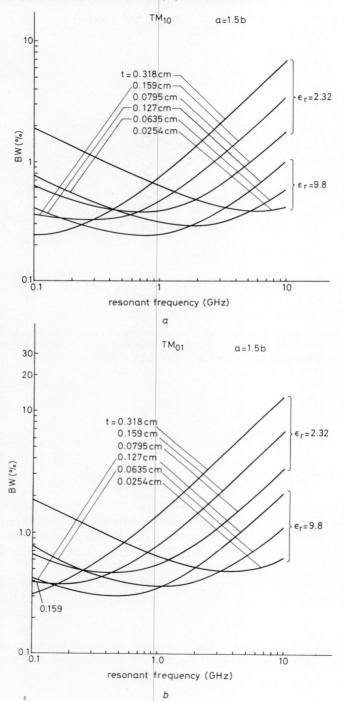

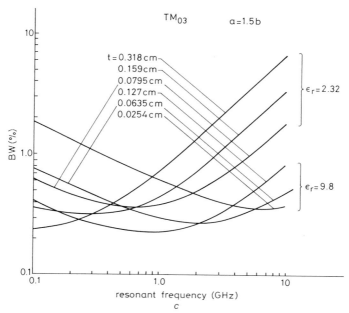

Fig. 3.14 *Bandwidth as a function of resonant frequency for a rectangular patch with $\sigma = 5.8 \times 10^7 \, S/m$, $\delta = 0.0005$, $a = 1.5b$ and (i) $\varepsilon_r = 2.32$, $t = 0.318$, 0.159, $0.0795 \, cm$; (ii) $\varepsilon_r = 9.8$, $t = 0.117$, 0.0635, $0.0254 \, cm$*
 (a) TM_{10}
 (b) TM_{01}
 (c) TM_{03}

In eqn. 3.50, X_{nm} are the roots of the equation

$$J'_n(x) = 0 \tag{3.51}$$

where differentiation is with respect to x.

The first five non-zero roots of eqn. 3.51 are shown in Table 3.1. The resonant frequency of a TM_{nm} mode is given by

$$f_{nm} = \frac{X_{nm}}{2\pi a \sqrt{\mu_0 \varepsilon}} = \frac{X_{nm} c}{2\pi a \sqrt{\varepsilon_r}} \tag{3.52}$$

where c is the velocity of light in free space.

Eqn. 3.52 is based on the assumption of a perfect magnetic wall and neglects the fringing fields at the open-end edge of the microstrip patch. To account for these fringing fields, an effective radius a_e, which is slightly larger than the physical radius a, is introduced [19]:

$$a_e = a \left[1 + \frac{2t}{\pi a \varepsilon_r} \left(\ln \frac{\pi a}{2t} + 1.7726 \right) \right]^{1/2} \tag{3.53}$$

$$f_{nm} = \frac{X_{nm} c}{2\pi a_e \sqrt{\varepsilon_r}} \tag{3.54}$$

Eqn. 3.53 is obtained by considering the radius of an ideal circular parallel-plate capacitor which would yield the same static capacitance after fringing is taken into account. Although the result is borrowed from the static case, it appears to yield theoretical resonant frequencies which are within 2.5% of measured values.

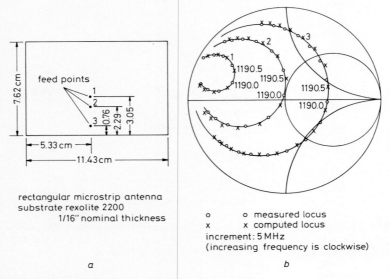

Fig. 3.15 *Impedance of the TM_{10} mode of a rectangular patch antenna of a = 11·43 cm, b = 7·6 cm, ε_r = 2·62, t = 0·159 cm. (Reproduced from Reference 8 p. 39 © 1981 IEEE)*
a Feed placement for impedance measurements
b Comparison of measured (0) and computed (x) impedance loci

Table 3.1 *The first five non-zero roots of $J_n(x) = 0$*

(n, m)	X
(1, 1)	1·841
(2, 1)	3·054
(0, 2)	3·832
(3, 1)	4·201
(1, 2)	5·331

As with the rectangular patch, the expressions for the far field, losses and Q, input impedance etc. are well documented [2] for the circular patch and will not be reproduced here. The characteristics obtained from these equations will be illustrated in the next Section.

3.3.2.2 Illustrative results: In this Section, we present graphical illustrations of the circular microstrip patch antenna. They include the magnetic-current distribution, radiation patterns, efficiency, directivity and gain, bandwidth and Q, and input impedance.

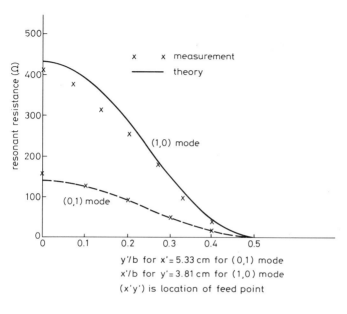

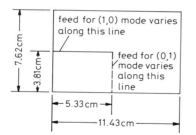

Fig. 3.16 *Variation of resonant resistance with feed position of the TM_{10} and TM_{01} mode in a rectangular patch antenna with $a = 11·43\,cm$, $b = 7·62\,cm$, $\varepsilon_r = 2·62$, $t = 0·159\,cm$. (Reproduced from Reference 8 p. 42 © 1981 IEEE)*

(i) *Magnetic current distribution*
The magnetic-current distribution around the edge of the disc for the *nm*th mode is proportional to $\cos n(\psi - \pi)$. This is illustrated in Fig. 3.19 for $n = 0$, 1, 2 and 3. It is independent of ψ for modes with $n = 0$ and undergoes three sinusoidal periods for modes with $n = 3$.

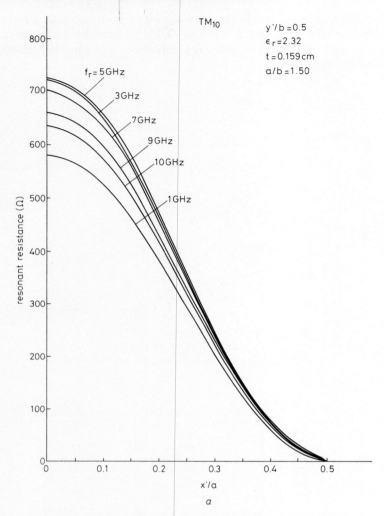

Fig. 3.17 *Variation of resonant resistance with resonant frequency for a rectangular patch antenna with $a/b = 1.5$, $\varepsilon_r = 2.32$, $t = 0.159$ cm*
(a) TM_{10}
(b) TM_{01}
(c) TM_{03}

(ii) *Radiation patterns*

Lo and co-workers [7] were among the first to obtain theoretical and experimental radiation patterns of the circular disc. Some of their results are shown below. In their experiment, a disc with a radius of 6·7 cm and a dielectric thickness of 1·5 mm was used. The relative permittivity was 2·62. For this disc, the first resonance, i.e. that of the TM_{11} mode, was at 794 MHz. The second

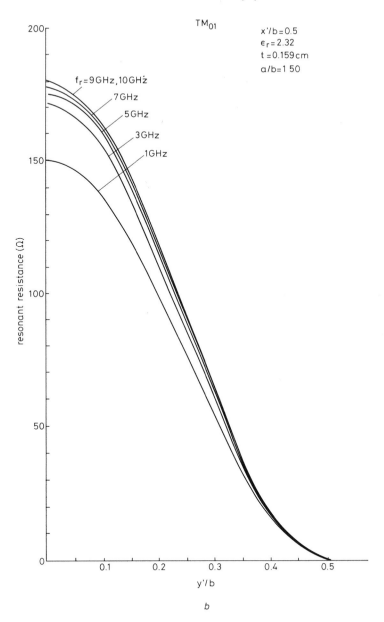

b

resonance, i.e. that of the TM_{21} mode, was at 1324 MHz. The calculated and measured radiation patterns in the $\phi = 0°$ and $\phi = 90°$ planes are shown in Fig. 3.20. Qualitatively, the calculated and measured results showed reasonable agreement. No attempt was made to take into account the effects of a finite

142 Characteristics of microstrip patch antennas

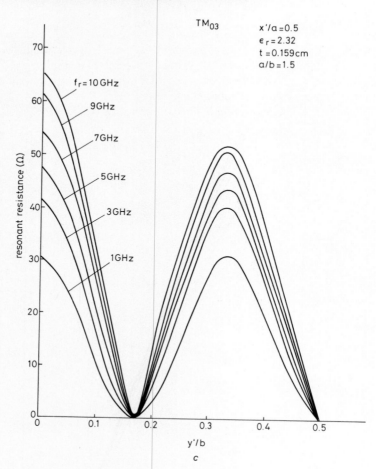

c

ground plane in the theory, which was about 12 wavelengths on a side for frequencies near 794 MHz.

The radiation patterns of the higher-order modes (0, 2) and (3, 0) also exhibit nulls in the broadside direction. Since the higher-order modes are seldom used in practice, only the characteristics of the lowest mode will be illustrated in subsequent sections.

(iii) *Efficiency*

The radiation efficiency for the lowest mode TM_{11} as a function of resonant frequency for various dielectric substrate is shown in Fig. 3.21. It is seen that the efficiency increases with increasing substrate thickness and decreasing dielectric constant.

(iv) Directivity and gain

The directivity versus resonant frequency is plotted in Fig. 3.22. A circular patch on an alumina substrate has a directivity of about 3·5, which is almost independent of substrate thickness and resonant frequency. If the substrate is Duroid, it has a maximum directivity of about 5·3, which decreases with increasing resonant frequency and dielectric thickness.

The gain of the lowest mode is illustrated in Fig. 3.23.

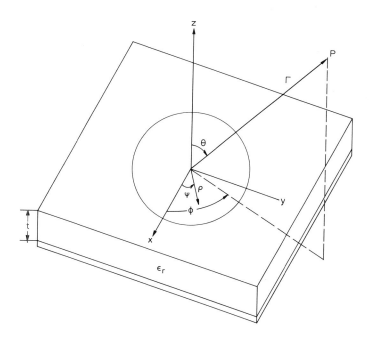

Fig. 3.18 *Geometry of the circular patch antenna*

(v) Total Q and bandwidth

The variation of total Q with resonant frequency for the lowest mode is shown in Fig. 3.24. The bandwidth, as defined by eqn. 3.40, is shown in Fig. 3.25. Its dependences on substrate thickness and ε_r are similar to the rectangular patch.

(vi) Input impedance

Richards *et al.* have reported calculated and measured values of the input impedance of a coaxially fed circular patch as a function of radial feed position. This is shown in Fig. 3.26 for the TM_{11} mode. It is seen that the input resistance is largest when fed at the edge of the patch, but it can attain the convenient value of 50 Ω when the feed location is chosen properly.

144 Characteristics of microstrip patch antennas

Dahele and Lee [20] have studied experimentally the effect of substrate thickness on the input impedance of a coaxially fed circular patch. Their results are shown in Fig. 3.27.

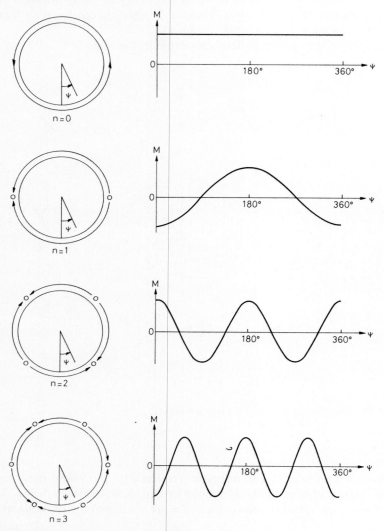

Fig. 3.19 *Surface magnetic-current distribution of the nmth mode in the circular patch antenna*

To conclude this Section, we point out that, as in the case of the rectangular patch, feeding the circular patch at a single point results in linearly polarised radiation. Circular polarisation on boresight can be obtained using two feeds at

Characteristics of microstrip patch antennas 145

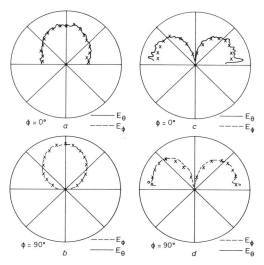

Fig. 3.20 *Theoretical (x) and measured radiation patterns in the $\phi = 0°$ and $\phi = 90°$ planes of a circular patch with radius a = 6·75 cm, ε_r = 2·32 (Reproduced from Reference 7 p. 141 © 1979 IEEE)*
(a) and (b) At 794 MHz of mode (1, 1)
(c) and (d) At 1324 MHz if mode (2, 1)

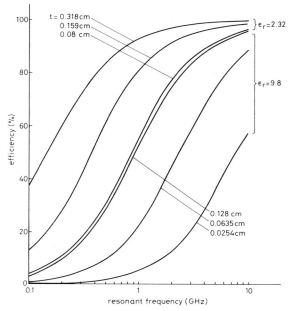

Fig. 3.21 *Radiation efficiency versus resonant frequency for the TM_{11} mode of the circular patch with $\sigma = 5·8 \times 10^7 S/m$, $\delta = 0·0005$ and (i) ε_r = 2·32, t = 0·318, 0·159, 0·0795 cm; (ii) ε_r = 9·8, t = 0·127, 0·0635, 0·0254 cm*

146 Characteristics of microstrip patch antennas

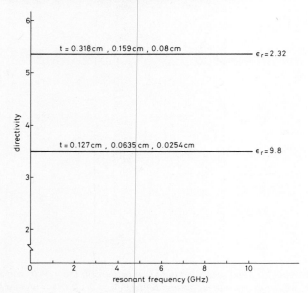

Fig. 3.22   Directivity versus resonant frequency for the TM_{11} mode of the circular patch: (i) $\varepsilon_r = 2\cdot32$, $t = 0\cdot318, 0\cdot159, 0\cdot0795$ cm; (ii) $\varepsilon_r = 9\cdot8$, $t = 0\cdot127, 0\cdot0635, 0\cdot0254$ cm

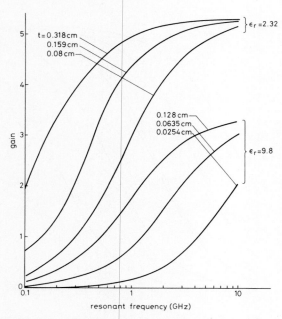

Fig. 3.23   Gain versus resonant frequency for the TM_{11} mode of the circular patch with $\sigma = 5\cdot8 \times 10^7$ S/m, $\delta = 0\cdot0005$ and (i) $\varepsilon_r = 2\cdot32$, $t = 0\cdot318, 0\cdot159, 0\cdot0795$ cm; (ii) $\varepsilon_r = 9\cdot8$, $t = 0\cdot127, 0\cdot0635, 0\cdot0254$ cm

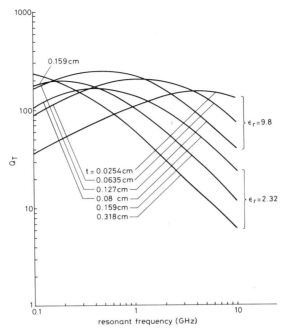

Fig. 3.24 Total Q versus resonant frequency for the TM_{11} mode of the circular patch with $\sigma = 5\cdot 8 \times 10^7\,S/m$, $\delta = 0\cdot 0005$ and (i) $\varepsilon_r = 2\cdot 32$, $t = 0\cdot 318, 0\cdot 159, 0\cdot 0795\,cm$; (ii) $\varepsilon_r = 9\cdot 8$, $t = 0\cdot 127, 0\cdot 0635, 0\cdot 0254\,cm$

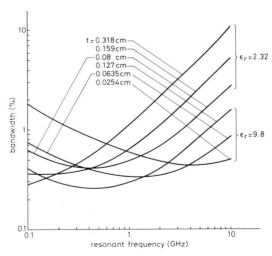

Fig. 3.25 Bandwidth versus resonant frequency for the TM_{11} mode of the circular patch with $\sigma = 5\cdot 8 \times 10^7\,S/m$, $\delta = 0\cdot 0005$ and (ii) $\varepsilon_r = 2\cdot 32$, $t = 0\cdot 318, 0\cdot 159, 0\cdot 0795\,cm$; (ii) $\varepsilon_r = 9\cdot 8$, $t = 0\cdot 127, 0\cdot 0635, 0\cdot 0254\,cm$

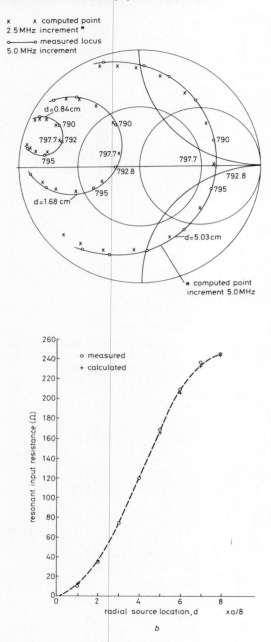

Fig. 3.26 *(a) Calculated and measured TM_{11} mode input impedance loci for several radial feed locations. (b) Variation of TM_{11} resonant input resistance with radial feed position for a circular patch with a = 6·7 cm on Rexolite 2200 substrate (ε_r = 2·62) of thickness 0·159 cm (Reproduced from Reference 8 p. 41, © 1981, IEEE)*

$\psi = \psi_1$ and $\psi_1 + 90°$ and excited in phase quadrature. Alternatively, a slightly elliptical patch with the right amount of ellipticity and fed at the appropriate location can produce circularly polarised waves. This will be discussed further in Section 3.3.6.

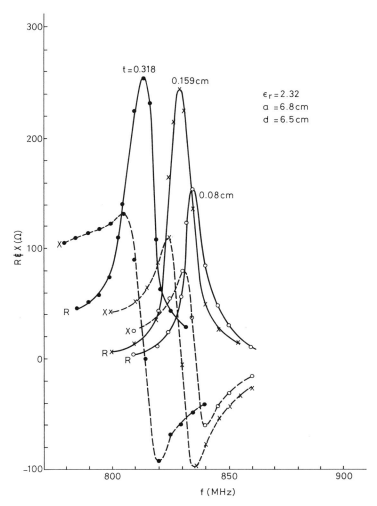

Fig. 3.27 *Measured real part (R) and imaginary part (X) of input impedance as function of frequency for the TM_{11} mode of a circular patch with a = 6·8 cm, ε_r = 2·32 and three dielectric thicknesses. The feed is at d = 6·5 cm (Reproduced from Reference 20 p. 359 © 1983 IEEE)*

3.3.3 The Equitriangular patch

Several triangular patch shapes are amendable to analysis by the cavity model. These include the 45°–45°–90°, 30°–60°–90°, and the 60°–60°–60° equitriangular (equilateral triangular) patches. However, unlike the rectangular and circular patches which have been studied extensively, there are only a handful of investigation on the triangular patches [2, 21–24]. In this Section, the equitriangular patch is treated in detail. The geometry, for the case of a coaxial feed, is shown in Fig. 3.28. The presentation follows closely that of Luk et al. [24].

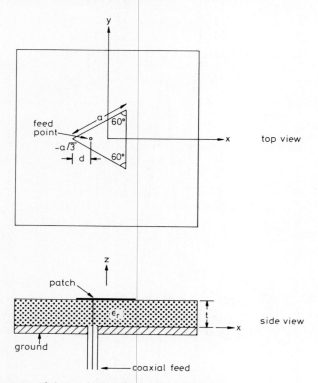

Fig. 3.28 *Geometry of the equitriangular patch antenna*

3.3.3.1 Formulas based on the cavity model: The solutions for the fields in an equitriangular waveguide with perfect electric walls have been described by Schelkunoff [25]. It follows from the duality principle of electromagnetism that the TM modes with perfect magnetic walls are the same as those of TE modes with perfect electric walls. Starting with the solutions given in Schelkunoff, we obtain the following results for the equitriangular patch antenna.

(i) *Resonant frequencies*
The formula for the resonant frequencies of z-independent TM modes satisfying the perfect magnetic wall boundary condition is

$$f_{mn} = \frac{2c}{3a\sqrt{\varepsilon_r}} (m^2 + mn + n^2)^{1/2} \qquad (3.55)$$

For the TM_{10} and TM_{21} modes, radiation is strongest in the broadside. Although there is a slight dip at the broadside for the TM_{20} mode, the radiation is still very strong in this direction. For convenience, we shall refer all three modes as broadside modes. Reference to the Figures shows that the polarisations of the three modes are the same at $\theta = 0°$. This suggests that the equitrian-

Table 3.2 *Theoretical resonant frequencies of an equitriangular patch with sidelength a = 10 cm*

	f_{mn}, GHz	
(m, n)	$\varepsilon_r = 2\cdot32, t = 0\cdot159\,\text{cm}$	$\varepsilon_r = 9\cdot8, t = 0\cdot0635\,\text{cm}$
(1, 0)	1·3	0·64
(1, 1)	1·84 (2·25)	0·90 (1·10)
(2, 0)	2·6	1·28
(2, 1)	3·44	1·66
(3, 0)	3·9	1·91

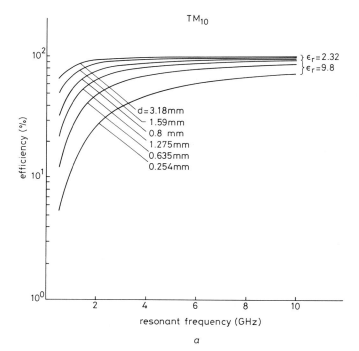

Fig. 3.30 *Radiation efficiency versus resonant frequency for an equitriangular patch with $\sigma = 5\cdot8 \times 10^7\,S/m$, $\delta = 0\cdot0005$ and (i) $\varepsilon_r = 2\cdot32$, $t = 0\cdot318, 0\cdot159, 0\cdot0795\,cm$; (ii) $\varepsilon_r = 9\cdot8$, $t = 0\cdot127, 0\cdot0635, 0\cdot0254\,cm$*
  (a) TM_{10}
  (b) TM_{20}
  (c) TM_{21}

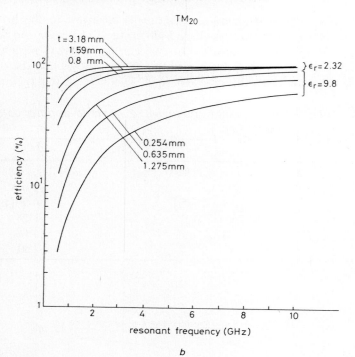

b

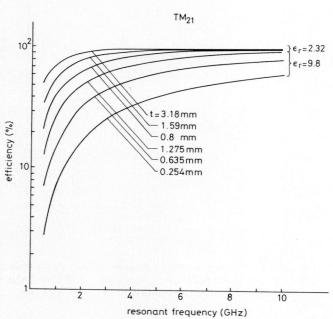

c

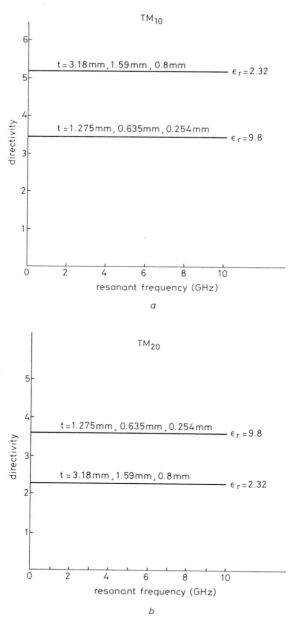

Fig. 3.31 *Directivity versus resonant frequency for an equitriangular patch with: (i) $\varepsilon_r = 2.32$, $t = 0.318, 0.159, 0.0795$ cm; (ii) $\varepsilon_r = 9.8$, $t = 0.127, 0.0635, 0.0254$ cm*
 (a) TM_{10}
 (b) TM_{20}
 (c) TM_{21}

162 Characteristics of microstrip patch antennas

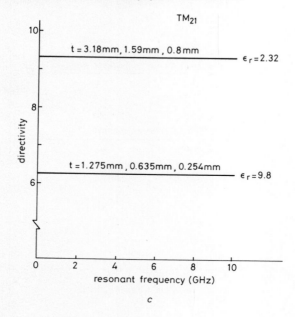

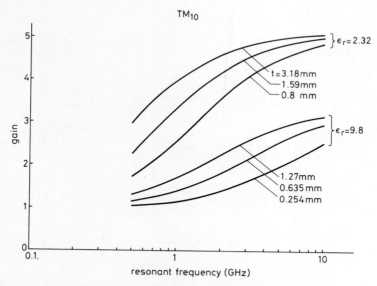

Fig. 3.32 *Gain versus resonant frequency for an equitriangular patch with $\sigma = 5\cdot 8 \times 10^7$ S/m, $\delta = 0\cdot 0005$ and (i) $\varepsilon_r = 2\cdot 32$, $t = 0\cdot 318$, $0\cdot 159$, $0\cdot 0795$ cm; (ii) $\varepsilon_r = 9\cdot 8$, $t = 0\cdot 127$, $0\cdot 0635$, $0\cdot 0254$ cm*
  (a) TM_{10}
  (b) TM_{20}
  (c) TM_{21}

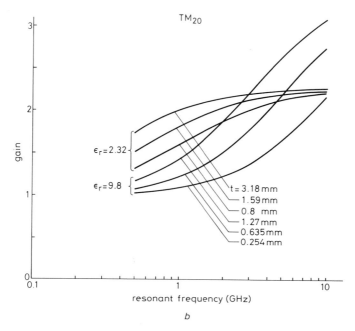

b

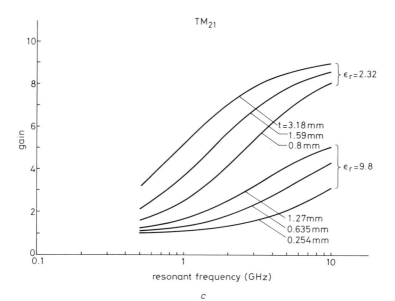

c

gular patch can be operated at the resonant frequencies of the three broadside modes with similar pattern and polarisation characteristics. We shall further show in the next Section that it is possible to find a position for a coaxial feed such that the input impedance of all three modes are in the range of 50–100 Ω.

(ii) *Radiation efficiency, directivity, gain, total Q and bandwidth*
The radiation efficiency, directivity, gain, total Q and bandwidth as a function of resonant frequency for the three lowest broadside modes, i.e. TM_{10}, TM_{20}, and TM_{21} are shown in Figs. 3.30–3.34. The results of e, Q_T and BW are similar. However, the directivities and gains are significantly different.

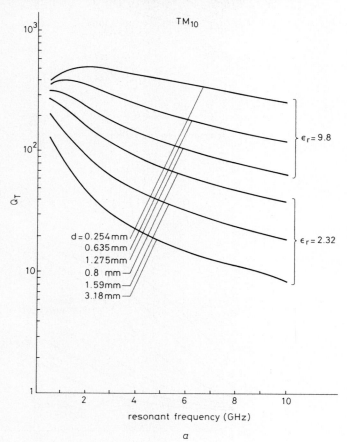

Fig. 3.33 *Total Q factor versus resonant frequency for an equitriangular patch with σ = $5 \cdot 8 \times 10^7$ S/m, δ = 0·0005 and (i) ε_r = 2·32, t = 0·318, 0·159, 0·0795 cm; (ii) ε_r = 9·8, t = 0·127, 0·0635, 0·0254 cm*
(a) TM_{10}
(b) TM_{20}
(c) TM_{21}

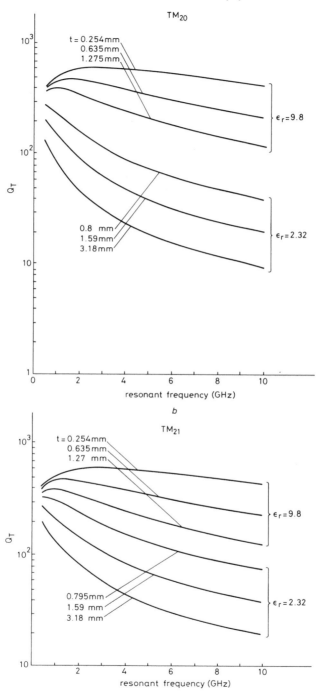

166   Characteristics of microstrip patch antennas

Compared to the rectangular and the circular patches, major differences are found in the Q_T and BW curves. While these parameters depend on ε_r, t and f in a complicated manner for the rectangular and circular patches, their behaviour for the equitriangular patch is simple: Q_T decreases (BW increases) with decreasing ε_r and increasing t irrespective of frequency.

(iii) *Input impedances and their variations with feed position*

The input impedances and their variations with feed positions are important characteristics of patch antennas. In the cavity-model theory for the equitriangular patch, the coaxial is modelled by a current ribbon of effective width $2w$ along the x-axis. Usually this is several times the physical diameter of the coaxial inner conductor and the input impedance is not a sensitive function of $2w$. For a patch with $a = 10$ cm, $\varepsilon_r = 2 \cdot 32$ and $t = 0 \cdot 159$ cm, the input impedances of the three broadside modes, TM_{10}, TM_{20} and TM_{21} are shown in Fig. 3.35. The value of $2w$ used in the computation is 6 mm.

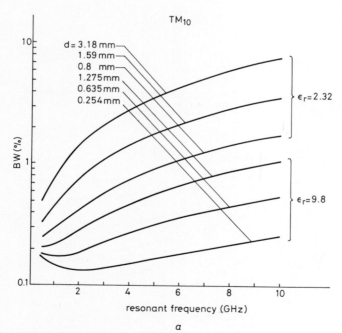

a

Fig. 3.34   *Bandwidth versus resonant frequency for an equitriangular patch with $\sigma = 5 \cdot 8 \times 10^7$ S/m, $\delta = 0 \cdot 0005$ and (i) $\varepsilon_r = 2 \cdot 32$, $t = 0 \cdot 318, 0 \cdot 159, 0 \cdot 0795$ cm; (ii) $\varepsilon_r = 9 \cdot 8$, $t = 0 \cdot 127, 0 \cdot 0635, 0 \cdot 0254$ cm*
  (a) TM_{10}
  (b) TM_{20}
  (c) TM_{21}

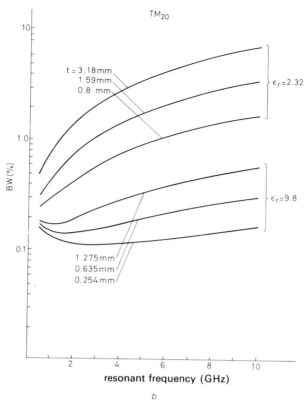

b

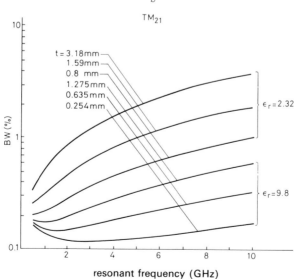

c

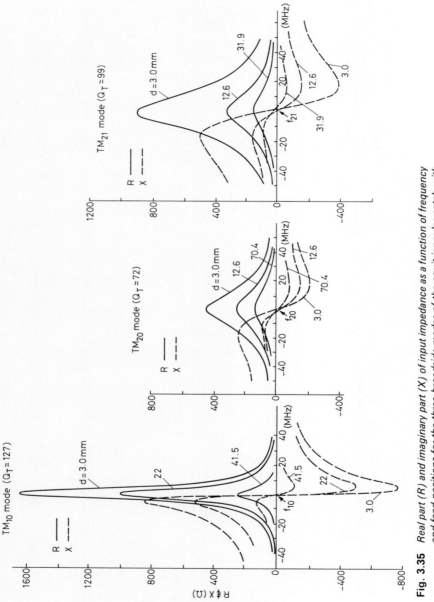

Fig. 3.35 *Real part (R) and imaginary part (X) of input impedance as a function of frequency and feed positions for the three broadside modes of the equitriangular patch with sidelength $a = 10\,cm$, $\varepsilon_r = 2.32$ and $t = 0.159\,cm$*

The variations of R and X at resonance with feed position d are shown in Fig. 3.36. It is seen that R decreases with increasing d only up to a certain value of d. The distance d can be chosen so that, for a particular mode, it assumes the value necessary to match the characteristic impedance of the feed. If the antenna is to be used for more than one mode, it is desirable that the input impedances for these modes are not too different, in addition to having the same polarisation and similar radiation patterns. In this connection, it is interesting to point out that, by placing the feed at an appropriate location, the input resistances at

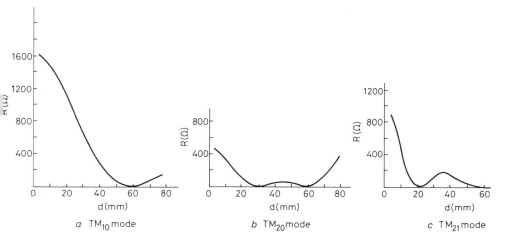

Fig. 3.36 *Resonant resistance versus feed position of the three broadside modes of the equitriangular patch with sidelength a = 10 cm, ε_r = 2·32, t = 0·159 cm*

resonance of the three broadside modes can be made to fall in the range of 50–100 Ω. For example, if d is equal to 4·7 cm, we obtain R = 100, 50 and 60 Ω for modes TM_{10}, TM_{20} and TM_{21}, respectively.

Comparison of theory and experiment for the equilateral triangular patch is relatively scarce in the literature. In Reference 24, a comparison was made on resonant frequencies and input impedances, and reasonable agreements were obtained.

3.3.4 Annular-ring patch

3.3.4.1 Introductory remarks: While the rectangular and the circular patches are probably the most extensively studied patch shapes, the annular ring has also received considerable attention [26–34]. There are several interesting features associated with this patch. First, for a given frequency, the size is substantially smaller than that of the circular patch when both are operated in the lowest mode (see example in Section 3.3.5). In application to arrays, this allows the elements to be more densely situated, thereby reducing the grating-lobe problem. Secondly, it is possible to combine the annular ring with a second microstrip element, such as a circular disc within its aperture, to form a compact dual-band antenna system [26]. Thirdly, the separation of the modes can be controlled by the ratio of outer to inner radii. Finally, it has been found that, by operating in one of the higher-order broadside modes, i.e. TM_{12}, the impedance bandwidth is several times larger than is achievable in other patches of comparable dielectric thickness.

The annular ring has been analysed using the cavity model [2, 26, 30], the spectral-domain technique in Fourier–Hankel transform domain [28] and the

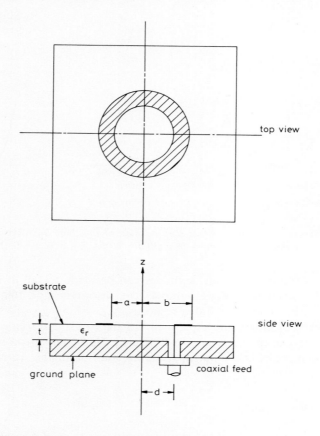

Fig. 3.37 *Geometry of the annular-ring patch antenna*

use of the method of matched asymptotic expansion [27]. In what follows, the results obtained using the cavity model will be presented, together with some comparisons with experiment.

3.3.4.2 Cavity-model theory
(i) *Resonant frequencies, internal and radiation fields*
Consider an annular ring patch with outer radius b and inner radius a, as shown in Fig. 3.37. Assuming that only TM modes exist, the resonant frequencies are determined by

$$f_{nm} = \frac{k_{nm} c}{2\pi \sqrt{\varepsilon_r}} \qquad (3.76)$$

where k_{nm} are the roots of the characteristic equation

$$J'_n(kb) Y'_n(ka) - J'_n(ka) Y'_n(kb) = 0 \qquad (3.77)$$

In eqn. 3.77, $J_n(x)$ and $Y_n(x)$ are Bessel functions of the first and second kind, order n, respectively, and the prime denotes derivatives with respect to x. Letting $C = b/a$, eqn. 3.77 takes the form

$$J'_n(CX_{nm}) Y'_n(X_{nm}) - J'_n(X_{nm}) Y'(CX_{nm}) = 0 \qquad (3.78)$$

where

$$X_{nm} = k_{nm} a \qquad (3.79)$$

For the case $C = 2$, the roots of eqn. 3.78 are given in Table 3.3.

Table 3.3 *Roots of the characteristic equation $J'_n(X_{nm}C) Y'_n(X_{nm}) - J'_n(X_{nm}) Y'_n(X_{nm}C) = 0$, where $C = b/a = 2$*

n \ m	1	2	3	4	5
0	–	3·1966	6·3123	9·4445	12·5812
1	0·6773	3·2825	6·3532	9·4713	12·6012
2	1·3406	3·5313	6·4747	9·5516	12·6612
3	1·9789	3·9201	6·6738	9·6842	12·7607
4	2·5876	4·4182	6·9461	9·8677	12·8989
5	3·1694	4·9929	7·2868	10·1000	13·0750

The cases when $n = 1$ and $n = 2$ are also of particular interest. The roots are shown in Figs. 3.38 and 3.39. Note that the spacings between the roots are dependent on a/b. This parameter can therefore be used to control the frequency separation of the modes. For the general equation 3.77, solutions presented in the form of a mode chart are given in Reference 33.

To account for the fact that a small fraction of the field exists outside the dielectric, it is customary to use an effective permittivity ε_e in place of ε_r in eqn.

172 *Characteristics of microstrip patch antennas*

3.76. The formula for ε_e as given by Schneider [31] is

$$\varepsilon_e = \frac{1}{2}(\varepsilon_r + 1) + \frac{1}{2}(\varepsilon_r - 1)\left(1 + \frac{10t}{W}\right)^{-1/2} \tag{3.80}$$

where

$$W = (b - a) \tag{3.81}$$

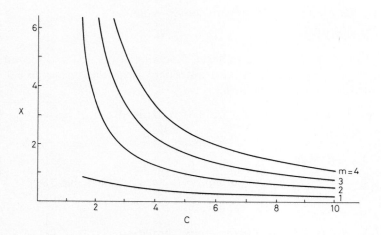

Fig. 3.38 *The roots X, of eqn. 3.78, for n = 1, as a function of C = b/a*

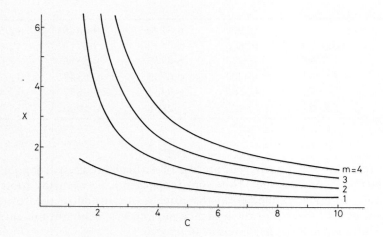

Fig. 3.39 *The roots, X, of eqn. 3.78, for n = 2, as a function of C = b/a*

To account for the fringing fields along the curved edges of the ring, it has been suggested that the outer and inner radii be modified according to

$$b_e = b + \frac{1}{2}(W_e(f) - W) \qquad (3.82)$$

$$a_e = a - \frac{1}{2}(W_e(f) - W) \qquad (3.83)$$

where

$$W_e(f) = W + \frac{W_e(0) - W}{1 + (f/f_p)^2} \qquad (3.84)$$

$$W_e(0) = 120\pi t/z_0\sqrt{\varepsilon_e} \qquad (3.85)$$

$$f_p = z_0/2\mu_0 t \qquad (3.86)$$

μ_0 is the permeability and z_0 is the quasi-static characteristic impedance of a microstrip line of width W.

A pair of empirical formulas for the modified radii, sufficient for many engineering purposes, are given by [33]

$$a_e = a - (3/4)t \qquad (3.87)$$

$$b_e = b + (3/4)t \qquad (3.88)$$

For the given values of a and b, a_e and b_e are calculated. Then the characteristic equation is solved by replacing a and b by a_e and b_e. After solving the characteristic equation for k_{nm}, the resonant frequencies are determined from

$$f_{nm} = \frac{ck_{nm}}{2\pi\sqrt{\varepsilon_e}} \qquad (3.89)$$

It should be pointed out that the correction to the resonant-frequency formula involves both the effective permittivity and the effective radii. This is somewhat different from the cases of the circular, the rectangular, and the equitriangular patches, which involve the effective dimensions only. Lee and Dahele [30] had shown that, in the case of the annular ring, good agreement between theory and experiment can be obtained only if both effective quantities are used.

The electric field under the patch is given by

$$E_z = E_0[J_n(k_{nm}\varrho)Y'_n(k_{nm}a) - J'_n(k_{nm}a)Y_n(k_{nm}\varrho)]\cos n\psi \qquad (3.90)$$

The far-zone electric field is

174 Characteristics of microstrip patch antennas

$$E_\theta = \frac{j^n 2tk_0 E_0}{\pi k_{nm}} \frac{e^{-jk_0 r}}{r} \cos n\phi \left[J'_n(k_0 a \sin\theta) - J'_n(k_0 b \sin\phi) \right.$$

$$\left. \times \frac{J'_n(k_{nm}a)}{J'_n(k_{nm}b)} \right] \tag{3.91}$$

$$E_\phi = \frac{-j^n 2nt E_0}{\pi k_{nm}} \frac{e^{-jk_0 r}}{r} \frac{\cos\theta \sin n\phi}{\sin\theta} \left[\frac{J_n(k_0 a \sin\theta)}{a} - \frac{J_n(k_0 b \sin\theta)}{b} \right.$$

$$\left. \times \frac{J'_n(k_{nm}a)}{J'_n(k_{nm}b)} \right] \tag{3.92}$$

Using eqns. 3.91 and 3.92, the relative radiation patterns for the various modes can be plotted.

For the case $b/a = 2$ and $\varepsilon_r = 2\cdot 32$, the patterns of the TM_{11}, TM_{12}, TM_{21} and TM_{22} modes are illustrated in Figs. 3.40 and 3.41. It is seen that, for the TM_{11} and TM_{12} modes, the strongest radiation occurs in the broadside direction ($\theta = 0°$). On the other hand, radiation patterns for the TM_{21} and TM_{22} modes have nulls in the broadside direction, with the strongest radiation occurring at oblique angles. The radiation patterns for the TM_{11} and TM_{12} modes have been verified by experiment [34].

(ii) *Losses and Q*
The dielectric, copper and radiation losses, as well as the total energy stored at resonance, are given by the following expressions:

$$P_d = \frac{\omega E_0^2 \varepsilon t \delta}{\pi k_{nm}^2} \left[\frac{J'^2_n(k_{nm}a)}{J'^2_n(k_{nm}b)} \left(1 - \frac{n^2}{k_{nm}^2 b^2}\right) - \left(1 - \frac{n^2}{k_{nm}^2 a^2}\right) \right] \tag{3.93}$$

$$P_c = \frac{2}{\pi} R_s \frac{E_0^2}{\omega^2 \mu_0^2} \left[\frac{J'^2_n(k_{nm}a)}{J'^2_n(k_{nm}b)} \left(1 - \frac{n^2}{k_{nm}^2 b^2}\right) - \left(1 - \frac{n^2}{k_{nm}^2 a^2}\right) \right] \tag{3.94}$$

$$P_r = \frac{2\omega k_0^2 t E_0^2}{\zeta_0 k_{nm}^2} I_1 \tag{3.95}$$

$$W_T = \frac{\varepsilon t E_0^2}{4\pi k_{nm}^2} \left[\frac{J'^2_n(k_{nm}a)}{J'^2_n(k_{nm}b)} \left(1 - \frac{n^2}{k_{nm}^2 b^2}\right) - \left(1 - \frac{n^2}{k_{nm}^2 a^2}\right) \right] \tag{3.96}$$

Characteristics of microstrip patch antennas 175

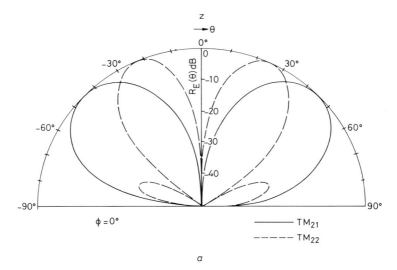

Fig. 3.40 *Sketches of the radiation patterns of the TM_{11} and TM_{12} modes of the annular-ring patch antenna with $b/a = 2$*
(a) $\phi = 0°$
(b) $\phi = 90°$

176   Characteristics of microstrip patch antennas

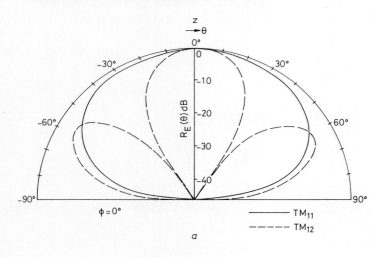

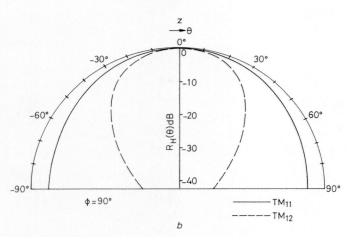

Fig. 3.41 *Sketches of the radiation patterns of the TM_{11} and TM_{12} modes of the annular-ring patch antenna with $b/a = 2$*
(a) $\phi = 0°$
(b) $\phi = 90°$

where the quantity I_1 is the integral:

$$I_1 = \int_0^{\pi/2} \frac{n^2 \cos^2\theta}{k_0^2 \sin\theta} \left\{ \frac{J_n(k_0 a \sin\theta)}{a} - \frac{J_n(k_0 b \sin\theta)}{b} \frac{J'_n(k_{nm}a)}{J'_n(k_{nm}b)} \right\}^2 \\ + \sin\theta \left\{ J'_n(k_0 a \sin\theta) - J'_n(k_0 b \sin\theta) \frac{J'_n(k_{nm}a)}{J'_n(k_{nm}b)} \right\}^2 \right] d\theta \quad (3.97)$$

The effective loss tangent, comprising the three kinds of losses, is given by

$$\delta_{eff} = \delta + \frac{1}{t\sqrt{\sigma\mu_0 \pi f}}$$

$$+ \frac{2\omega\mu_0 t I_1}{\zeta_0 \varepsilon_r \left[\frac{J_n'^2(k_{nm}a)}{J_n'^2(k_{nm}b)}\left(1 - \frac{n^2}{k_{nm}^2 b^2}\right) - \left(1 - \frac{n^2}{k_{nm}^2 a^2}\right)\right]} \quad (3.98)$$

The total Q factor is the inverse of the effective loss tangent.

(iii) *Coaxial-fed annular ring*
For an annular ring fed by a coaxial line at a distance d from the centre, which we modelled by a uniform current ribbon of effective width $2w$ (eqn. 3.5), the expression for E_z in the cavity is

$$E_z = j\omega\mu_0 J \left[\sum_{m=1}^{\infty}\sum_{n=0}^{\infty} R_{nm}(J_n(k_{nm}\varrho) Y_n'(k_{nm}a) - J_n'(k_{nm}a)\right.$$

$$\left. \times Y_n(k_{nm}\varrho)\right] \cos n\psi \quad (3.99)$$

where

$$R_{nm} = \frac{k^2 \pi \sin(2nw)\cos n\pi [J_n(k_{nm}d)Y_n'(k_{nm}a) - J_n'(k_{nm}a)Y_n(k_{nm}d)]}{\varepsilon_{0n} n(k_d^2 - k_{nm}^2)\left[\frac{J_n'^2(k_{nm}a)}{J_n'^2(k_{nm}b)}\left(1 - \frac{n^2}{k_{nm}^2 b^2}\right) - \left(1 - \frac{n^2}{k_{nm}^2 a^2}\right)\right]}$$

(3.100)

and $\varepsilon_{on} = \begin{cases} 1 \text{ for } n \neq 0 \\ 2 \text{ for } n = 0 \end{cases}$

The θ and ϕ components of the far-zone electric field are

$$E_\theta = j^{n+1} \frac{2tk_0\omega\mu_0 J}{\pi} \frac{e^{-jk_0 r}}{r} \sum_n \sum_m \frac{R_{nm}}{k_{nm}} \cos n\phi$$

$$\times \left[J_n'(k_0 a \sin\theta) - J_n'(k_0 b \sin\theta)\frac{J_n'(k_{nm}a)}{J_n'(k_{nm}b)}\right] \quad (3.101)$$

$$E_\phi = -j^{n+1} \frac{2t\omega\mu_0 J}{\pi} \frac{e^{-jk_0 r}}{r} \frac{\cos\theta}{\sin\theta} \sum_n \sum_m \frac{nR_{nm}}{k_{nm}} \sin n\phi$$

$$\times \left[\frac{J_n(k_0 a \sin\theta)}{a} - \frac{J_n(k_0 b \sin\theta)}{b}\frac{J_n'(k_{nm}a)}{J_n'(k_{nm}b)}\right] \quad (3.102)$$

178 Characteristics of microstrip patch antennas

The input impedance is

$$Z = j\omega\mu_0 t \left\{ \sum_m \sum_n \right.$$

$$\times \left. \frac{\pi k_{nm}^2 \left(\dfrac{\sin 2nw}{2nw}\right)^2 [Y'_n(k_{nm}a)J_n(k_{nm}d) - J'_n(k_{nm}a)Y_n(k_{nm}d)]^2}{2\varepsilon_{on}(k_{eff}^2 - k_{nm}^2)\left[\dfrac{J'^2_n(k_{nm}a)}{J'^2_n(k_{nm}b)}\left(1 - \dfrac{n^2}{k_{nm}^2 b^2}\right) - \left(1 - \dfrac{n^2}{k_{nm}^2 a^2}\right)\right]} \right\}$$

(3.103)

where

$$k_{eff} = k_0 \sqrt{\varepsilon_e(1 - j\delta_{eff})} \tag{3.104}$$

3.3.4.3 Broadside modes TM_{11} and TM_{12}: The most interesting finding for the annual ring patch is perhaps the relatively wide-band property of the TM_{12} mode. This was first predicted theoretically by Chew [27] and by Ali et al. [28] using the matched asymptotic expansion technique and the vector Hankel transform, respectively. Experimental verification of the theoretical prediction was first reported by Dahele and Lee [29]. Lee and Dahele [30] also obtain this theoretically within the framework of the cavity model, i.e. using the formulas of Section 3.3.4.2. For an annular ring patch with $b/a = 2$, $a = 3.5$ cm, $\varepsilon_r = 2.32$, $t = 0.159$ cm, the variation of input impedance with frequency of the TM_{11} and TM_{12} modes is illustrated in Figs. 3.42 and 3.43 for two feed positions, one near the inner edge ($d/a = 1.05$) and the other near the outer edge ($d/a) = 1.95$. It is seen that, for the TM_{11} mode, the impedance is not sensitive to the feed position and the impedance bandwidth is very narrow ($< 1\%$). On the other hand, the input impedance of the TM_{12} mode is very sensitive to the feed position. With the feed near the outer edge, the value at resonance is only about 20Ω. With the feed near the inner edge, it attains the convenient value of about 60Ω at resonance. The bandwidth is about 4%, which is several times that of the TM_{11} mode. This is also larger than the bandwidth achievable with the rectangular, the circular, or the equitriangular patches with the same dielectric constant and thickness, as reference to the corresponding Figures in Sections 3.3.1–3.3.3 shows. The theoretical results agree with the conclusion of Chew [27], and Ali et al. [28] who analysed the problem using considerably more complicated methods. Experimentally, the above predictions had been verified by Dahele and Lee [29] and Lee and Dahele [30].

Detailed theoretical results based on the cavity model for the characteristics of the TM_{11} and TM_{12} modes are shown in Figs. 3.44–3.48. Note that, in addition to larger bandwidth, the TM_{12} mode also has a larger directivity. Similar results were obtained in Reference 35.

The input impedance and bandwidth of the annular ring patch antenna have also been determined by modelling the antenna as a section of radial line loaded with wall admittances [36]. The results are in qualitative agreement with the other methods mentioned above.

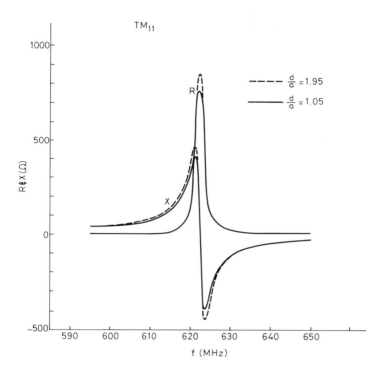

Fig. 3.42 *Theoretical input impedance of the TM_{11} mode of an annular-ring patch with b = 7·0 cm, a = 3·5 cm, ε_r = 2·32, t = 0·159 cm, fed at two radial locations*

3.3.5 Comparison of characteristics of the rectangular, circular, equitriangular and annular ring patches

It is instructive at this point to present an example comparing the characteristics of the rectangular, the circular, the equitriangular and the annular ring patches. Let us take the operating frequency to be 2 GHz and fabricate the patches on a substrate material of thickness t = 1·59 mm and ε_r = 2·32. If the patches are designed to operate in the lowest mode, a rectangular patch with an aspect ratio 1·5 has dimensions b = 3·28 cm, a = 4·92 cm; a circular patch has radius 4·92 cm; an equitriangular patch has side length 6·57 cm; and an annular ring

180 *Characteristics of microstrip patch antennas*

with $b/a = 2$ has $b = 1.84$ cm, $a = 0.92$ cm. The characteristics of the lowest mode for the four patches are shown in Table 3.4. It is seen that all are broadside modes. The circular patch has the smallest beamwidth in both planes. The annular ring patch has the smallest physical area. The circular patch has the largest physical area but it also has the largest bandwidth, efficiency and gain.

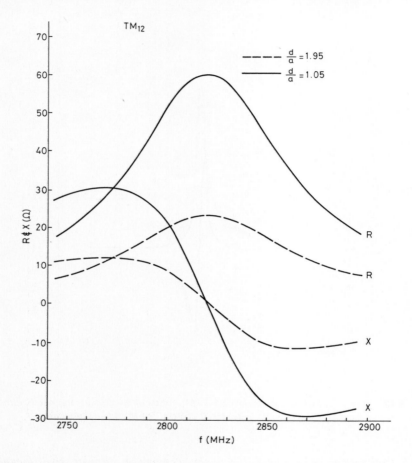

Fig. 3.43 *Theoretical input impedance of the TM_{12} mode of an annular-ring patch with $b = 7.0$ cm, $a = 3.5$ cm, $\varepsilon_r = 2.32$, $t = 0.159$ cm, fed at two radial locations*

The picture changes somewhat if the annular ring with $b = 2a$ is designed to operate in the TM_{12} mode. This is shown in the last column in Table 3.4. The beamwidth is much narrower while both the gain and the bandwidth are considerably larger. However, these improvements are achieved at the expense of increasing the size of the patch. For the TM_{12} mode to resonate at 2 GHz,

Table 3.4 Comparison of characteristics of rectangular, circular, equitriangular and annular-ring patch antennas ($\varepsilon_r = 2.32$, $t = 1.59$ mm, $f = 2$ GHz)

Characteristics	Rectangular ($a = 1.5b$)	Circular disk	Equitriangular	Annular ring ($b = 2a$)	
				TM_{11}	TM_{12}
1. *Radiation*					
Beam position	TM_{10}	TM_{11}	TM_{10}		
	Broadside	Broadside	Broadside	Broadside	Broadside
3 db beamwidth:					
E-plane	102°	100°	100°	103°	30°
H-plane	85°	80°	88°	81°	47°
1st sidelobe level:					
E-plane	—	—	—	—	−6 dB
H-plane	—	—	—	—	—
Directivity	7·0 dB	7·1 dB	7·1 dB	7·1 dB	10·9 dB
Efficiency	87%	94%	87%	86%	97%
Gain	6·1 dB	6·8 dB	6·2 dB	6·1 dB	10·6 dB
2. *Bandwidth*					
2:1 VSWR	0·7%	1·1%	0·78%	0·70%	3·8%
3. *Physical dimensions*	$a = 4.92$ cm	$a = 4.92$ cm	$a = 6.57$ cm	$b = 1.84$ cm	$b = 8.9$ cm
	$b = 3.28$ cm			$a = 0.92$ cm	$a = 4.45$ cm
Area	16·1 cm²	24·3 cm²	18·1 cm²	10·6 cm²	249 cm²

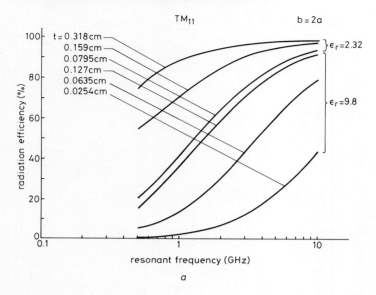

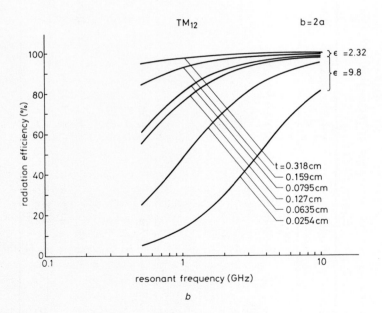

Fig. 3.44 *Radiation efficiency versus resonant frequency for the annular-ring patch with $b = 2a$: (i) $\varepsilon_r = 2.32$, $t = 0.318\,cm$, $0.159\,cm$, $0.0795\,cm$ (ii) $\varepsilon_r = 9.8$, $t = 0.127\,cm$, $0.0635\,cm$, $0.0254\,cm$*
(a) TM_{11}
(b) TM_{12}

$b = 8.9$ cm and $a = 4.45$ cm, yielding an area of 249 cm^2. This turns out to be a specific example of a general principle: increase in bandwidth can only be achieved at the expense of increasing the volume of the resonator.

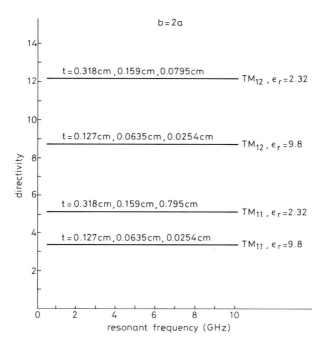

Fig. 3.45 *Directivities of the TM_{11} and TM_{12} modes for the annular-ring patch with $b = 2a$: (i) $\varepsilon_r = 2.32$, $t = 0.318$ cm, 0.159 cm, 0.0795 cm, (ii) $\varepsilon_r = 9.8$, $t = 0.127$ cm, 0.0635 cm, 0.0254 cm*

3.3.6 Brief mention of other patches

Besides the rectangular, circular, equitriangular and the annular ring, a number of other patch shapes have been studied in the literature. They include the right-angled isosceles triangular patch [2, 7, 37], the annular sector [7, 37], the circular sector [37], the rectangular ring [38], the H-shaped patch [38] and the elliptical patch [38–41]. The analysis of the rectangular ring and the H-shaped patch requires the segmentation method, while the other patches mentioned above can be analysed using the simple cavity model. However, except for the elliptical patch which offers the possibility of generating circularly polarised waves using a single feed, the other shapes do not appear to contain any features which are not obtainable from the rectangular, circular, equitriangular or annular ring patches. For this reason, only the elliptical patch will be briefly discussed below.

The geometry of the elliptical patch is shown in Fig. 3.49. Experimental study

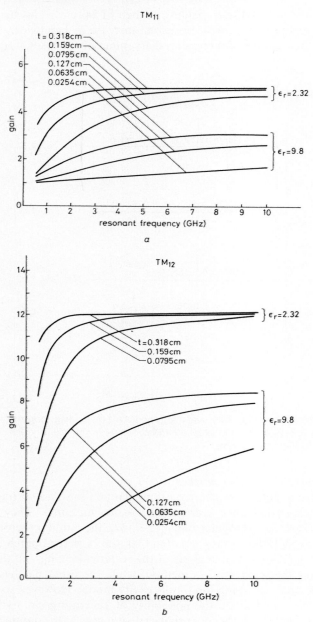

Fig. 3.46 *Gain versus resonant frequency for the annular-ring patch with b = 2a, σ = 5·8 × 10⁷ S/m, δ = 0·0005, and (i) ε_r = 2·32, t = 0·318 cm, 0·159 cm, 0·0795 cm, (ii) ε_r = 9·8, t = 0·127 cm, 0·0635 cm, 0·0254 cm*
(a) TM_{11}
(b) TM_{12}

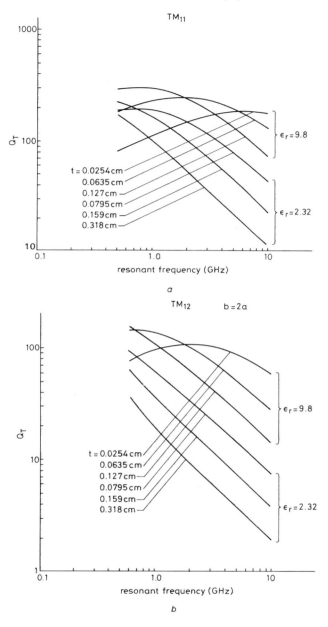

Fig. 3.47 Total Q factor versus resonant frequency for the annular-ring patch with $b = 2a$, $\sigma = 5.8 \times 10^7$ S/m, $\delta = 0.0005$ and (i) $\varepsilon_r = 2.32$, $t = 0.318$ cm, 0.159 cm, 0.0795 cm, (ii) $\varepsilon_r = 9.8$, $t = 0.127$ cm, 0.0635 cm, 0.0254 cm
(a) TM_{11}
(b) TM_{12}

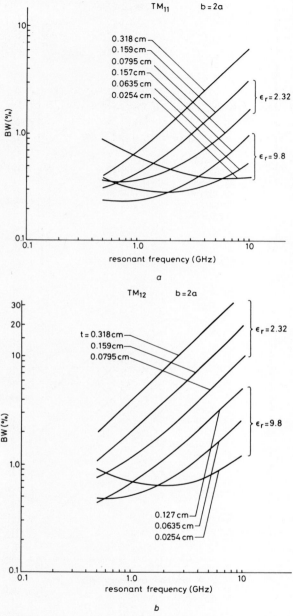

Fig. 3.48 *Bandwidth versus resonant frequency for the annular-ring patch with $b = 2a$, $\sigma = 5\cdot 8 \times 10^7 \, S/m$, $\delta = 0\cdot 0005$ and (i) $\varepsilon_r = 2\cdot 32$, $t = 0\cdot 318\,cm$, $0\cdot 159\,cm$, $0\cdot 0795\,cm$, (ii) $\varepsilon_r = 9\cdot 8$, $t = 0\cdot 127\,cm$, $0\cdot 0635\,cm$, $0\cdot 0254\,cm$*
  (a) TM_{11}
  (b) TM_{12}

of this antenna was reported by Yu [39] and later by Long *et al.* [41]. Theoretical studies were carried out by Shen [40] using the cavity model, by Lo and Richards [18] using a perturbation method, and by Despande and Bailey [42] using moment method. The main conclusions of these studies are summarised as follows:

(*a*) The radiation in the direction perpendicular to the patch is in general elliptically polarised. However, with proper selection of both the feed position and the eccentricity of the ellipse, circular polarisation can be obtained.
(*b*) The desired circular polarisation is best achieved by limiting the eccentricity of the ellipse to a range of 10–20%. This corresponds to *a* (semi-major axis) and *b* (semi-minor axis) differing by only a few percent. The perturbation method of Lo *et al.* [18] yields the formulas

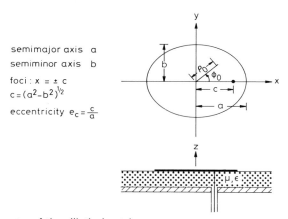

Fig. 3.49 *Geometry of the elliptical patch*

$$c/a = 1.0887/Q \tag{3.105}$$

or

$$a/b = 1 + 1.0887/Q \tag{3.106}$$

where the quality factor Q can be assumed to be that of the circular patch of radius a or b. For example, if $Q = 46.35$, $b/a = 0.976$.
(*c*) The feed point should be on a radial line making 45° relative to the semi-major axis, i.e. $\phi_0 = \pm 45°$. The positive sign yields left-hand while the negative sign yields right-hand circular polarisation.
(*d*) To achieve an operating frequency f, the semi-major axis should be chosen to be

$$a = \frac{p}{f\sqrt{\mu_0 \varepsilon}} \tag{3.107}$$

where p is a constant ranging from 0.27 to 0.29.

188 Characteristics of microstrip patch antennas

(e) To achieve an impedance of 50 Ω, the feed point should be at a distance ϱ_0 from the centre on the $\phi_0 = \pm 45°$ line, where $\varrho_0 \simeq 0.28a$.

3.4 Some methods for improving the frequency agility and bandwidth of microstrip patch antennas

3.4.1 Introduction

As mentioned in Section 3.1, the microstrip patch antenna (MPA), being basically a leaky cavity, is inherently narrow band. The pattern bandwidth is usually many times larger than the impedance bandwidth, which therefore is the parameter controlling the frequency response of the antenna. For this reason, our subsequent discussion on bandwidth will refer to impedance. For a single patch operating at the lowest mode, typical bandwidth is from less than 1% to several percent for thin substrates satisfying the criteria $t/\lambda_0 < 0.07$ for $\varepsilon_r \simeq 2.3$ and $t/\lambda_0 < 0.023$ for $\varepsilon_r \simeq 10$. When these inequalities are satisfied, the effect of surface wave is assumed to be unimportant. For comparison purposes, a half-wave dipole with a radius/length ratio equal to 0.01 has a bandwidth of about 16%, while a medium-length helix operating in the axial mode has a bandwidth of about 70%.

One way of obtaining a relatively wide bandwidth is to use an annular ring patch and operate it in the TM_{12} mode, as the results of Section 3.3.4 indicate. The price one has to pay is that the size of the patch is considerably larger than that of the rectangular, circular, equitriangular or the annular ring operated in the lowest mode. In this Section, we discuss a number of other methods which have been developed for overcoming the bandwidth problem.

Let us illustrate our discussion of bandwidth with a series of frequency response characteristics depicted in Fig. 3.50. Let the response in (a) represent the ideal characteristics, in which the input resistance is constant over a wide range of frequencies. A typical MPA response, however, is that shown in (b). Since this is not satisfactory for most purposes, a great deal of attention has been devoted in recent years to improving the bandwidth characteristics of MPAs. One line of attack is to widen the absolute bandwidth of the antenna as much as possible, as illustrated in (c). This in principle can be achieved simply by increasing the thickness of the substrate. However, this introduces several problems. First, a thick substrate supports surface waves, which will produce undesirable effects on the radiation pattern as well as reducing the radiation efficiency of the antenna. Secondly, as the thickness of the substrate increases, problems associated with the feeding of the antenna arise. Thirdly, higher-order cavity modes with fields depending on z may develop, introducing further distortions in the pattern and impedance characteristics. It is therefore of interest to develop more sophisticated methods of improving the absolute bandwidth of MPAs, and a great deal of research has been devoted to this effort.

There has also been a substantial amount of effort devoted to increasing the frequency agility of MPAs. One line of approach is to consider methods whereby the operating frequency of the antenna can be tuned over a range of values so that the same antenna can be used for several adjacent channels. This is the single-band tunable case illustrated in Fig. 3.50d. In another scheme, dual-frequency antennas with resonant frequencies separated by a certain range have

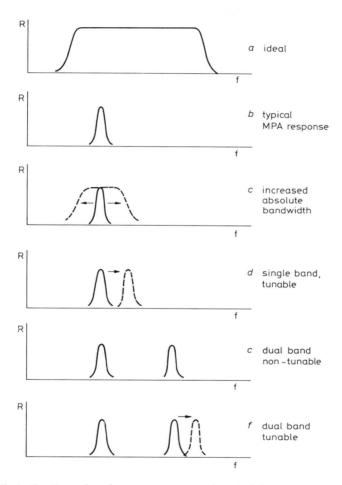

Fig. 3.50 *Illustrating the various frequency-response characteristics*

been developed (Fig. 3.50e). These dual-frequency structures are useful in situations where the antenna is required to operate in two distinct frequencies which may be too far apart for a single antenna to perform efficiently at both frequencies. Related to this is the dual-band tunable configuration, in which one or both of the resonances are tunable. The case for which only the upper resonance is tunable is illustrated in Fig. 3.50f.

190   Characteristics of microstrip patch antennas

In the next two Sections, some of the methods that have been developed to provide the characteristics illustrated in Fig. 3.50 will be described.

3.4.2 Some methods of tuning MPAs

We shall describe four methods of tuning the resonant frequencies of MPAs. These utilise (i) varactor diodes, (ii) shorting pins, (iii) optically controlled pin diodes and (iv) adjustable air gap. The advantages and disadvantages of these methods will be discussed.

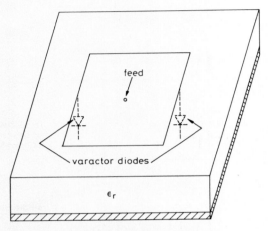

Fig. 3.51   *Illustrating the use of varactor diodes for tuning*

3.4.2.1 Varactor diodes: For a given set of patch dimensions, the resonant frequency is primarily governed by the value of the relative permittivity ε_r of the substrate. If some means is available to alter ε_r, the resonant frequency will change. One method of achieving this is to introduce varactor diodes between the patch and the ground plane, as shown in Fig. 3.51. The diodes are provided with a bias voltage, which controls the varactor capacitance and hence the effective permittivity of the substrate. Bhartia and Bahl [43] performed an experiment on this method and the results are shown in Fig. 3.52. The resonant frequency f_r of the lowest mode of the rectangular patch increases with the bias voltage, owing to the increase of the diode capacitance. It is seen that, in this experiment, a tuning range of some 20% was achieved with a 10 V bias. The range increased to about 30% with a 30 V bias. Note that the curve of f_r versus bias voltage is not a linear one.

Since the paper by Bhartia and Bahl [43], there appeared to be no further reports in the literature on this method, either experimentally or theoretically.

3.4.2.2 Tuning using shorting posts (pins): The value of ε_r can also be changed by introducing shorting posts (pins) at various points between the patch and the ground plane. These shorting posts present an inductance, and therefore alter the effective permittivity of the substrate. In the context of microstrip antennas, the method was first introduced by Schaubert *et al.* [44] in 1981. It is illustrated in Fig. 3.53. Using two posts, the experimental results obtained are shown in Fig. 3.54. It is seen that the resonant frequency is dependent on the separation of the two posts and a tuning range of some 18% is obtained as the separation varies between 0 and the whole width of the patch.

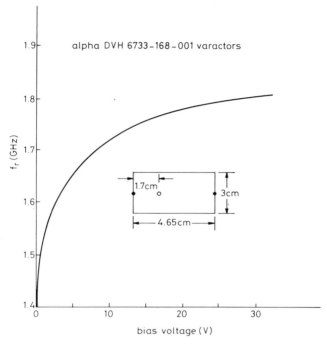

Fig. 3.52 *Resonant frequency versus bias voltage for a varactor-loaded rectangular patch antenna (Reproduced from Reference 43 p. 306 © IEEE 1982)*

Schaubert *et al.* [44] developed a theory of shorting pins based on the transmission-line model, and the predictions (shown in Fig. 3.54) agree reasonably well with experimental data. However, because the transmission line model is not capable of predicting the variations in the inductive component of a load as its position is varied within the element, it fails to predict certain trends in the resonant frequency of a short-loaded patch as the shorting pin approaches the patch edge. This model also cannot predict the impedance of the element very accurately because the field distribution between the ground plane and the patch of a loaded element is much too complicated to be adequately represented by a single-mode transmission-line model. It should be noted, however, that the transmission-line model has been further developed for rectangular as well as for circular patches with shorting pins by Sengupta and co-workers [45, 46].

192　Characteristics of microstrip patch antennas

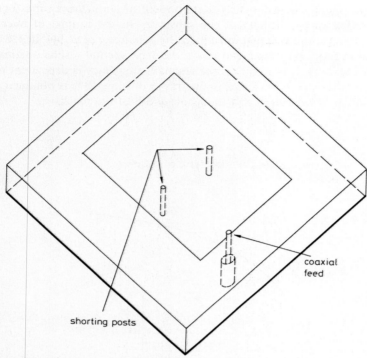

Fig. 3.53 *Illustrating the use of shorting posts for tuning the resonant frequency of a patch antenna*

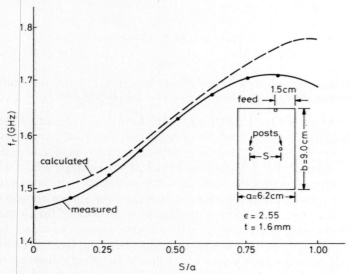

Fig. 3.54 *Resonant frequency versus separation of posts for a 6·2 × 9·0 cm rectangular patch antenna with $\varepsilon_r = 2·55$, $t = 1·6$ mm (Reproduced from Reference 44 p. 119 © IEEE 1981)*

Lo and coworkers [47] have developed the cavity model for MPAs with lumped linear loads in general and shorting pins in particular. They have also applied the shorting-pin method to design dual-frequency structures. This will be discussed in Section 3.4.3.2.

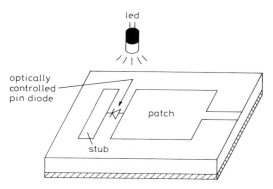

Fig. 3.55 *Tuning using optically controlled pin diode*

3.4.2.3 Optically tuned patch antenna: A method of tuning the resonant frequencies of patch antennas utilising optically controlled pin diodes was recently reported [48]. The scheme is illustrated in Fig. 3.55. A stub is connected to the patch by means of an optically controlled pin diode. When the diode is reversed biased, it acts as an open circuit and the patch resonates at the frequency for which it is designed, say f_r. When the diode is forward biased, it acts as a short circuit and the resonant frequency becomes that of the patch and the stub, i.e. $f_r - \Delta f$. In the experiment, f_r was 10·285 GHz and $f_r - \Delta f$ was 10·207 GHz. These are the limits in the range of tuning. By illuminating the diode with light, the diode impedance can be varied from a high value to a low value. As a result, the resonant frequency is optically tuned. It was found in their experiment that an illumination of 1 W/cm^2 resulted in a 15 MHz downward shift in the frequency. This method clearly needs further development as the range of tuning reported was extremely limited.

Discussion: The three methods described so far suffer from the following disadvantages:

(i) The design of the patches is complicated by the added components such as varactor diodes, optically controlled pin diodes and their associated biasing circuit. In the case of shorting pins, their precise positions are also important.
(ii) For high frequencies (say > 10 GHz), the patch sizes are small and it is difficult to accommodate the diodes and shorting posts underneath each patch.
(iii) The added complications in design multiply for an array consisting of a large number of elements.

194 Characteristics of microstrip patch antennas

The potential advantage of the three methods is the possibility of electronic tuning. For example, there were suggestions that the shorting pins could take the form of switching diodes so that the frequency can be changed by electronically switching the diodes on and off. However, to the authors' knowledge, a real demonstration of such electronic switching applied to MPAs has yet to be reported in the literature.

In Section 3.4.2.4, we describe a somewhat different method of tuning the resonant frequency of an MPA, i.e. utilising an adjustable air gap between the substrate and the ground plane.

3.4.2.4 Tuning using an adjustable air gap

Introduction: By introducing an air gap between the substrate and the ground plane in a microstrip patch antenna the effective permittivity of the cavity will change. This can be used to tune the resonant frequency of a microstrip patch antenna as discussed below.

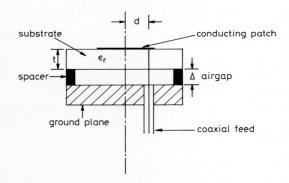

Fig. 3.56   *Geometry of a microstrip patch antenna with air gap*

The geometry of a microstrip antenna with an airgap is shown in Fig. 3.56. Consider the cavity under the conducting patch. It is made of two layers: a substrate of thickness t and an air region of thickness Δ. Compared to the case with no air gap the effective permittivity of the cavity is evidently smaller. As a result the resonant frequencies of the various modes will increase. Since the effective permittivity decreases as Δ increases, tending towards the free-space value ε_0 as $\Delta \to \infty$, it follows that the resonant frequencies can be tuned by adjusting the air-gap width Δ. As a by product the bandwidth will also increase partly due to the increase in the height of the dielectric medium and partly because the effective permittivity is smaller.

Based on the above idea, Dahele and Lee [49–52] have carried out a series of experimental and theoretical studies on the microstrip antenna with air gaps. Some of their results are presented below.

Experimental results: The first configuration studied by Dahele and Lee was the circular patch. The radius of the patch was 5 cm fabricated on Duroid material of thickness 0·159 cm and relative permittivity 2·32. The width of the air gap is controlled by using spacers between the substrate and the ground plane. In the experiment, spacers of 0·5 mm and 1·0 mm were used. The antenna was provided with a coaxial feed near the edge of the disc at a distance $d = 4·75$ cm from the centre. This feed position is chosen as it is well known that it yields a larger resistance at resonance compared to a feed which is closer to the centre. The

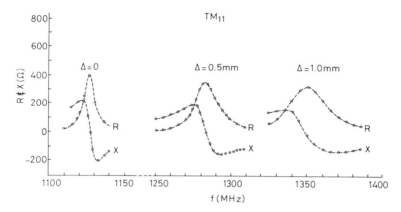

Fig. 3.57 *Measured input impedance of the TM_{11} mode of a 5 cm circular-disc microstrip antenna for three values of air-gap width Δ. $\varepsilon_r = 2·32$, $t = 0·159$ cm (After Reference 51 pp. 455–460)*

Table 3.5 *Measured resonant frequencies and impedance bandwidths of the first few modes of a 5 cm-radius circular-disc microstrip antenna for three values of the air-gap width*

	$\Delta = 0$		$\Delta = 0·5$ mm		$\Delta = 1·0$ mm	
	f_{nm}	% BW	f_{nm}	% BW	f_{nm}	% BW
TM_{11}	1128 MHz	0·89	1286 MHz	1·48	1350 MHz	2·07
TM_{21}	1879 MHz	0·85	2136 MHz	2·15	2256 MHz	2·61
TM_{31}	2596 MHz	0·77	2951 MHz	1·63	3106 MHz	2·02

$\varepsilon_r = 2·32$, $t = 0·159$ cm; fed at 4·75 cm from the centre

measured resonant frequencies are shown in Table 3.5. For the lowest mode TM_{11}, there is a tuning range of about 20% in frequency and a more than twofold increase in the bandwidth as Δ goes from 0 to 1·0 mm. Similar behaviour is recorded for the other modes. The measured input impedances of the TM_{11} mode as a function of frequency are shown in Fig. 3.57. The upward shift in the resonant frequency and the widening of the bandwidth are clearly seen.

As for the radiation pattern it was found that the air gap did not have a significant effect on the pattern.

Another antenna studied was the annular-ring patch. The effect of an air gap on the two broadside modes TM_{11} and TM_{12} are shown in Table 3.6 for an annular ring of outer radius 7·0 cm and inner radius 3·5 cm fabricated on Duroid material of thickness 0·159 cm and $\varepsilon_r = 2\cdot32$. As in the circular patch there is an upward shift in the resonant frequencies and a widening of the bandwidths. It is significant that, for the TM_{12} mode, the bandwidth attains a value of 8·6% when Δ is equal to 1·0 mm.

Table 3.6 *Measured resonant frequencies and impedance bandwidths of the TM_{11} and TM_{12} modes of an annular-ring microstrip antenna for three values of the air-gap width Δ*

	$\Delta = 0$		$\Delta = 0\cdot5$ mm		$\Delta = 1\cdot0$ mm	
	f_{nm}	% BW	f_{nm}	% BW	f_{nm}	% BW
TM_{11}	626 MHz	0·6	720 MHz	0·7	778 MHz	0·8
TM_{12}	2757 MHz	4·0	3040 MHz	8·0	3240 MHz	8·6

Inner radius $a = 3\cdot5$ cm, outer radius $b = 7\cdot0$ cm, $\varepsilon_r = 2\cdot32$, $t = 0\cdot159$ cm. The feed is placed at $d/a = 1\cdot05$ where d is the distance from the centre

Theory: Lee and Dahele has developed the theory of the two-layered microstrip antenna using the cavity model. The original assumptions of the model are modified to account for the two layers as follows:

(i) Owing to the close proximity between the conducting patch and the ground plane only transverse magnetic (TM) modes are assumed to exist. The z-component of the electric field, however, is a function of z since the cavity is two-layered.
(ii) The cavity is assumed to be bounded by perfect electric walls on the top and on the bottom and by a perfect magnetic wall along the edge.
(iii) Across the dielectric–air interface the tangential electric field and the normal electric flux density are continuous.

Based on the above assumptions detailed analysis for the circular and annular-ring patch were carried out and good agreement between theory and experiment was obtained. In the interest of brevity, except for the resonant frequencies, the theoretical formulas will not be included here. The formula for the resonant frequency, however, is a very simple one and is given by

$$f_{nm}(\Delta) = f(0) \sqrt{\frac{\varepsilon}{\varepsilon_{eff}}} \tag{3.108}$$

where $f_{nm}(0)$ is the resonant frequency when there is no air gap and ε_{eff} is the effective permittivity of the two-layered medium:

$$\varepsilon_{eff} = \frac{\varepsilon(t + \Delta)}{(t + \Delta\varepsilon_r)} \tag{3.109}$$

Eqns. 3.108 and 3.109 are valid for any patch shape. Note that, as the air gap width Δ increases, ε_{eff} decreases and the resonant frequency increases. The dependence of $f_{nm}(\Delta)$ on Δ, however, is not a linear one.

Discussion: As in the other methods the adjustable air gap as a means of tuning the resonant frequencies has advantages and disadvantages. The advantages are:

(i) No costly components are added.
(ii) It can be applied to patches of any shape. There is no need to know the details of the fields in the cavity.
(iii) The method is particularly attractive for an array made up of a great number of elements as illustrated in Fig. 3.58. If the elements are fed by striplines, the resonant frequencies of all the elements, and therefore of the array, can be tuned by a single adjustment of the air gap width Δ.

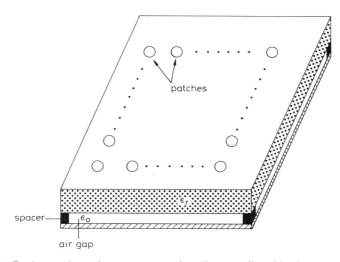

Fig. 3.58 *Tuning a microstrip antenna array by using an adjustable airgap*
Stripline feeds are assumed

The disadvantages are:

(i) The width of the air gap has to be changed mechanically. Electronic tuning appears to be difficult.
(ii) The antenna is slightly thicker. This however is compensated for by an increase of the bandwidth.

198 Characteristics of microstrip patch antennas

To end this Section we point out that it is possible to alter the resonant frequency by inserting a piece of dielectric in the air region, as illustrated in Fig. 3.59. The relative permittivity of the inserted dielectric can be either the same as that of the substrate or different. Both the thickness and the permittivity of the inserted dielectric will determine the resultant resonant frequency.

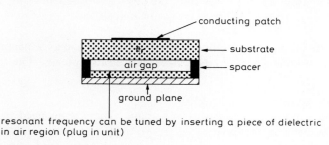

resonant frequency can be tuned by inserting a piece of dielectric in air region (plug in unit)

Fig. 3.59 *Altering the resonant frequency by inserting a piece of dielectric in the air gap*

3.4.3 Dual-band structures

There has been considerable interest in the development of dual-frequency microstrip antennas. The characteristics of this class of antennas is illustrated in Fig. 3.50e. They are useful when the antenna is required to operate in two distinct frequencies which are too far apart for a single antenna to perform efficiently at both frequencies while the behaviour of the antenna in the range of intermediate frequencies is of little or no concern. Several methods of obtaining the dual-frequency characteristics have been developed. We begin with the method of simply stacking two patches together.

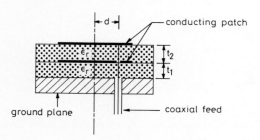

Fig. 3.60 *Non-tunable dual-frequency stacked microstrip antenna*

3.4.3.1 Stacked circular-disc antenna: The first experimental report on a dual-frequency structure using two stacked circular patches was that of Long and Walton [53]. The geometry is shown in Fig. 3.60. The discs were photo-etched on separate substrates and aligned so that their centres were along the same

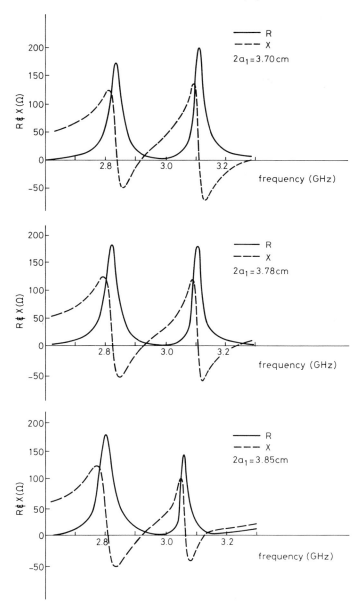

Fig. 3.61 *Real and imaginary parts of impedance of stacked circular patches etched on a dielectric with $\varepsilon_r = 2\cdot47$*
(a) $2a_1 = 3\cdot70$ cm
(b) $2a_1 = 3\cdot78$ cm
(c) $2a_1 = 3\cdot85$ cm
(Reproduced from Reference 53 p. 271 © IEEE 1979)

line. The sizes of the two discs and their spacings were varied and the resultant behaviour of the antenna characteristics measured. The antenna was fed by means of a coaxial line. The centre conductor passed through a clearance hole in the lower disc and is connected electrically to the upper disc. If one considers the two regions under the patch as two resonant cavities it is clear that the system behaves as a pair of coupled cavities. Since the fringing fields are different for the upper and lower cavities, two resonant frequencies are expected even if the diameters of the two discs are the same. While the qualitative explanation is relatively simple the quantitative theory for this structure is still lacking. In what follows the experimental results of Long and Walton are described.

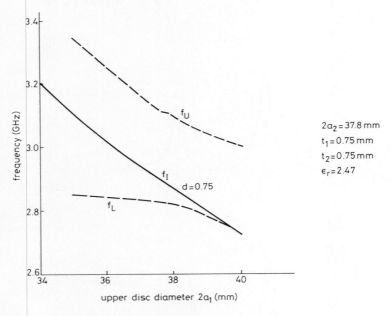

Fig. 3.62 *Resonant frequencies versus upper disc diameter of stacked circular patches etched on a dielectric with $\varepsilon_r = 2.47$, $2a_2 = 3.78$ cm, $t_1 = t_2 = 0.75$ mm. (Reproduced from Reference 53 p. 271 © IEEE 1979)*

Fig. 3.61 shows the real and imaginary parts of the input impedance for $2a_2 = 3.78$ cm, $t_1 = t_2 = 0.075$ cm and three values of $2a_1$. The resonant frequencies as a function of the upper disc diameter are shown in Fig. 3.62. Also shown is the resonant frequency of the lowest mode for a single disc of diameter $2a$ and substrate thickness $t = 0.075$ cm, taking into account the fringing field through the effective diameter. It is seen that the lower resonant frequency is relatively constant, remaining near the value of a single disc with $2a = 3.78$ cm and $d = 0.075$ cm. The upper resonance, on the other hand, is highly dependent on the size of the upper disc. Radiation patterns were also taken by Long and

Walton which showed that they were similar to the radiation pattern of the lowest mode for the single circular patch.

While the results of Long and Walton [53] showed that it is possible to design for the separation of the resonant frequencies by choosing the diameters of the upper and lower discs, this is not very convenient in practice because of the lack of formulas to predict the frequencies. Also once they are designed and etched, it is not possible to alter or tune the separation of the two resonant frequencies. The configuration as presented by Long and Walton was therefore a dual-band non-tunable antenna of the type illustrated in Fig. 3.50e.

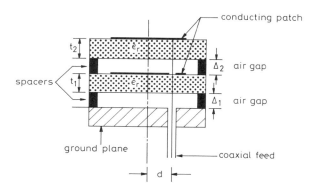

Fig. 3.63 *Tunable dual-frequency stacked microstrip antenna utilising the air-gap idea*

Dahele and Lee [54] have applied the air gap idea to study dual-frequency stacked discs. The geometry is shown in Fig. 3.63 in which air gaps between the lower substrate and the ground plane and/or between the two substrates are introduced. Either of the air gap widths can be set to zero. Their experimental results performed with two stacked discs of 7 cm radii, each etched on substrates with $\varepsilon_r = 2.32$ and thickness 0.159 cm are shown in Fig. 3.64 and 3.65. In Fig. 3.64, with the lower air gap set to zero the upper air gap is seen to increase the resonant frequency of the upper resonance. In Fig. 3.65 the upper air gap is set to zero and the effect of the lower air gap is studied. It is seen that the effect is more complicated since it shifts not only the lower but also the upper resonance. In both cases the bandwidth of the lower resonance is substantially broadened.

Dahele and Lee [55] have also studied a structure consisting of two stacked annular ring patches as shown in Fig. 3.66. This structure was also found to exhibit dual-frequency behaviour. As in the case of circular discs an upper air gap was found to be a convenient method of altering the separation of the frequency bands.

3.4.3.2 Single-element dual-frequency microstrip antenna: It is possible for a single-element microstrip antenna to operate at many frequencies corresponding to the various resonant modes pertaining to the structure. However, for most applications it is required that the radiation pattern, the polarization and the impedance be similar if not identical in all the frequency bands of operation.

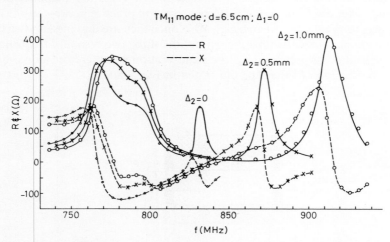

Fig. 3.64 *Measured input impedances of the TM_{11} mode of a pair of stacked circular discs of 7 cm radius for three values of the upper air gap: $d = 6.5$ cm, $\Delta_1 = 0$, $\varepsilon_r = 2.32$, $t = 0.159$ cm (After Reference 51 pp. 455–460)*

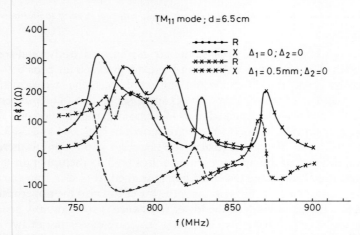

Fig. 3.65 *Measured input impedances of the TM_{11} mode of the stacked circular discs of Fig. 3.63 for two values of the lower air-gap. $d = 6.5$ cm, $\Delta_2 = 0$, $\varepsilon_r = 2.32$, $t = 0.159$ cm (After Reference 51 pp. 455–460)*

This immediately rules out many modes. Furthermore, for a given geometry all the resonant frequencies are related in fixed ratios, providing no flexibility for the designer.

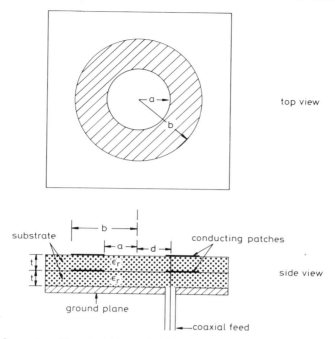

Fig. 3.66 *Geometry of the stacked annular-ring antenna*

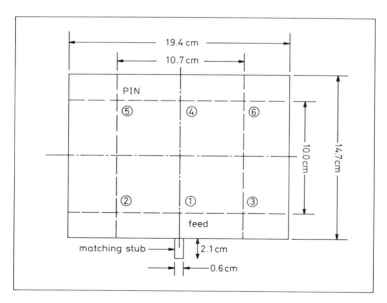

Fig. 3.67 *Geometry of a rectangular patch antenna with six possible shorting pins and a short matching stub. (After Reference 56 pp. 298–300)*
All dimensions in centimetres

If for a particular patch shape two modes can be found which produce similar radiation patterns with the same polarisation, dual frequency is possible with a single patch. For the rectangular patch the two modes (0, 1) and (0, 3) satisfy this requirement. However, their resonant frequencies are related by a fixed ratio of approximately 3, the exact value being dependent on the edge effect. Suppose now shorting pins are placed on the nodal lines of the (0, 3) model field, there will be little effect on the (0, 3) mode but a strong effect on the (0, 1) mode. This offers a way of altering the separation of the two frequency bands. The insertion of pins at proper locations can also be used to tune the input impedance for the (0, 1) mode while the feed location is chosen first for the desired impedance for the (0, 3) mode.

The above idea has been successfully demonstrated experimentally by Zhong and Lo [56]. A multiport-cavity-model theory has also been developed by Lo and coworkers which appears to predict the effects of shorting pins on frequency and impedance well. We limit here to a summary of the experimental results of Zhong and Lo.

Table 3.7 *Resonant frequencies for (0, 1) and (0, 3) modes against shorting pins used (After Reference 56)*

Number of pins	Pin position	f_{01}	f_{03}	f_{03}/f_{01}
		MHz	MHz	
0	–	613	1861	3·04
1	(1)	664	1874	2·82
2	(1) (2)	706	1865	2·64
3	(1) (2) (3)	792	1865	2·36
4	(1) (2) (3) (6)	813	1865	2·29
5	(1) (2) (3) (5) (6)	846	1865	2·20
6	(1) to (6)	891	1865	2·09

The geometry of the rectangular patch in their experiment is shown in Fig. 3.67. It is made of 1/8 in copper-cladded Rexolite 2200 with six shorting-pin positions. The effects of successively adding more and more pins (each approximately 0·05 cm in diameter) at the positions indicated in Fig. 3.67 are shown in Table 3.7. It is seen that the ratio of the two operating frequencies f_{03}/f_{01} can be varied approximately from 3 to 2. Since all these pins are located on the (0, 3) nodal lines f_{03} remain constant at approximately 1865 MHz while f_{01} is varied from 613 to 891 MHz. In order for the impedances of the two bands to be close to 50 Ω at resonance it is necessary to attach a short capacitive stub of 0·6 cm × 2·1 cm. With the stub added, the bandwidth with reference to 3:1 VSWR is about 2% for the low band and almost 8% for the high band. Typical low- and high-band patterns in both E and H planes are shown in Figs. 3.68. It is seen that, while the two modes radiate strongest in the broadside, the directivities of the two modes are quite different. By increasing the number of pins the two frequencies can be brought to a ratio of about 1·8. If a smaller ratio is desired it is found that it can be achieved by introducing slots in the patch. This, however, makes the fabrication of the patch somewhat complicated.

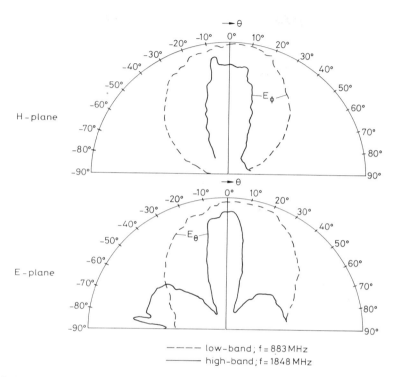

Fig. 3.68 *Typical radiation patterns in H- and E-planes for antennas shown in Fig. 3.67 with six pins inserted (After Reference 56 pp. 298–300)*

The rectangular patch is not the only geometry capable of providing dual-frequency operation. In Section 3.3.3, it was shown that the TM_{10}, TM_{20} and TM_{21} modes of the equitriangular patch are all broadside modes with similar polarisations in the broadside direction. Moreover, by choosing the location of the feed properly, the impedances of these modes do not vary greatly. It thus appears that it is possible to utilise the equitriangular patch for dual- or even triple-frequency operation.

3.4.3.3 Dual-band microstrip antennas with reactive loading: A dual-frequency microstrip antenna can also be obtained simply by loading it with a reactive load. If the reactive load takes the form of a short-circuited length of microstrip transmission line the low-profile characteristic of a microstrip patch antenna is retained. Such a structure was suggested by Davidson et al. [57] and was demonstrated to work experimentally.

Fig. 3.69 shows the dual-band rectangular microstrip patch antenna with a monolithic load studied in the experiment of Davidson et al. The patch was of

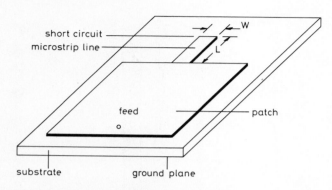

Fig. 3.69 *Dual-frequency rectangular patch antenna with monolithic reactive loading (After Reference 57 p. 936–937)*

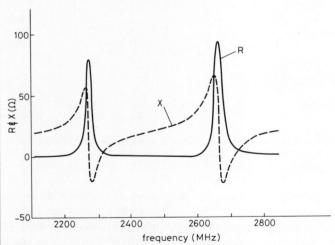

Fig. 3.70 *Impedance of edge-loaded, 4 × 6 cm patch antenna with L = 4·0 cm, W = 0·33 cm, ε_r = 2·17, t = 0·079 cm; coaxially fed near the edge and at the centre of the 6 cm side (After Reference 57 pp. 936–937)*

dimension 6 cm × 4 cm etched on a substrate with $\varepsilon_r = 2\cdot17$ and thickness 0·079 cm. It is coaxially fed near the edge and at the centre of the 6 cm side. For a line length of $L = 4\cdot0$ cm and width $w = 0\cdot33$ cm, the impedance characteristics is shown in Fig. 3.70. Good pattern performance was observed at each of the resonant frequencies (2·275 GHz and 2·666 GHz, respectively).

The separation of the resonances can be varied by (i) changing the length of the microstrip line and (ii) introducing an inset dimension S with an accompanied gap spacing G between the line and the radiator, as shown in Fig. 3.71. The results for the resonant frequencies are shown in Table 3.8.

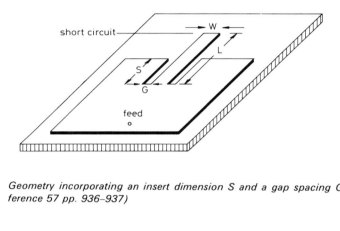

Fig. 3.71 *Geometry incorporating an insert dimension S and a gap spacing G (After Reference 57 pp. 936–937)*

Table 3.8 *Resonant frequencies of monolithic microstrip elements (after Reference 57)*

W	L	G	S	f_L	f_u
cm	cm	cm	cm	GHz	GHz
0·33	4·0	1·0	1·5	2·356	2·494
0·33	4·0	0	0	2·275	2·666
0·33	8·4	0	0	2·339	2·628
0·33	4·0	0·7	1·5	2·437	2·494
0·33	4·0	0·3	1·5	2·471	2·514

Discussion: In summary three methods of obtaining dual-frequency characteristics for microstrip patch antennas have been described. The method using shorting pins use two different modes. As such the radiation patterns, while similar in the broad sense, do vary in detail as well as in directivity. The separation of the resonances can be controlled by the number of pins, but it is difficult to have them close together unless additional features such as slots are introduced in the patch. These additional design features appear to be difficult to accommodate at high frequencies where the patch size is small.

The advantage of using shorting pins to realise dual-frequency characteristics and to control the frequency separation is that it is a single-patch geometry, thereby retaining the low profile characteristic of microstrip antennas. This advantage is also shared by the monolithic reactive-loading method. In the reactive-loading method the two frequencies are separated by 10–20%. The separation can be controlled by several parameters associated with the reactive load. However, once a design is etched it is not possible to tune the antenna.

For the case when the separation is in the range of 10–20% it appears that the stacked geometry discussed in Section 3.4.3.1 offers the advantages of operating in the same mode and the flexibility of tuning the separation by means of an air gap. This structure, however, is thicker than the single patch and the low-profile characteristics of the microstrip antenna is slightly compromised.

3.4.4 Electromagnetic-coupled patch antenna (EMCP)

3.4.4.1 Introduction: As mentioned in Section 3.4.1, a great deal of research has been devoted to increasing the absolute bandwidth of MPAs. The methods fall into three categories: electromagnetic-coupled patches (EMCP), use of parasitic elements and log-periodic arrangement of an array of patches. We shall discuss in this Section only the EMCP since it is related to the tunable stacked geometry of Section 3.4.3.1. The use of parasitic elements and log-periodic arrangement are covered in other Chapters of the Handbook.

As mentioned in Section 3.4.1 it is possible to increase the absolute bandwidth of MPAs by simply using thicker substrates. This, however, introduces several problems. The first is the excitation of surface waves, which distorts the normal radiation pattern and introduces additional loss; the second is the excitation of higher-order modes with z dependence, which introduces further distortions on the pattern and impedance characteristics. The third is that the application of common feeding techniques, i.e. direct feeding by either a coplanar microstrip line or a perpendicular coaxial line, becomes increasingly difficult for the following reasons. Consider first a coaxial feed. Since the probe (extension of inner conductor of the coaxial line) introduces a series reactance almost proportional to the substrate thickness, the lead inductance will become significant with respect to the antenna radiation resistance for thick substrates and will therefore prevent proper matching. Consider next a patch which is edge-fed by a coplanar microstrip line. For a fixed impedance level the line width is almost proportional to the dielectric thickness. Since the patch dimensions for a fixed resonant frequency are only weakly dependent on the dielectric thickness (through the fringing field) the width of the feed line will become non-negligible as the substrate reaches a certain thickness. As a result the radiation pattern of the antenna will be disturbed partly due to the covering of the radiating patch edge by the line and partly due to increased radiation from the feed line.

In view of the above problems, electromagnetic coupling (instead of direct coupling) has been studied as a possible feed technique for electrically thick MPAs. In particular, promising results have been obtained for the stacked dual-patch geometry which we now discuss.

3.4.4.2 Stacked dual-patch geometry: The basic geometry of the stacked dual-patch electromagnetic-coupled microstrip antenna is shown in Fig. 3.72. Each conducting patch is fabricated on an electrically thin substrate and separated by a region of air or foam with $\varepsilon_r \simeq 1$. The structure looks similar to the tunable dual-frequency antenna of Section 3.4.3.1, but is different in two aspects. First, the thickness of the air region is several times the substrate thickness, while in the tunable version discussed earlier, it is a fraction of the substrate thickness. Secondly, rather than being fed directly by a transmission line, the top element is excited via electromagnetic coupling from the lower element, which is located closer to the ground plane and is connected directly to a feed line. The top and bottom patches are referred to as the radiating and the feeding patches, respectively.

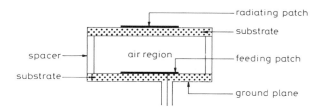

Fig. 3.72 *Electromagnetic-coupled patch antenna*

When the air region is small two resonances are expected, as in the case discussed in Section 3.4.3.1. Experimental studies showed that, as the air region exceeds a certain thickness, the lower resonance disappears and only one resonance remains. The single resonance condition can also be obtained by designing the diameter of the radiating element to be larger than the feeding element.

Electromagnetic-coupled patches appeared to be first discussed by Sabban [60]. Circular, annular-ring, rectangular and square patches in the S band (2-4 GHz), etched with substrates about 0·01 λ thick, were reported to yield bandwidths ranging from 9% to 15%. His paper contained very little information on the air-gap width and the relative sizes of the elements, other than the statement that the radiating element was larger than the feeding element and that the antennas exhibited a single resonance rather than dual resonances.

A more detailed experimental study was carried by Bhatnagar *et al.* [61]. The elements were triangular patches operating in the S band, and foam material with $\varepsilon_r = 1$ was introduced as the air gap between the dielectric layers (Fig. 3.73). The side lengths of both the top and bottom equitriangular patches were 37 mm fabricated on a substrate of thickness 1·6 mm and relative permittivity 2·55. The lower patch was provided with a coaxial feed at a distance

$F_d = 13\cdot5$ mm. The width Δ of the air gap is controlled by using foam material of uniform thickness.

The functional behaviour of the impedance characteristics is given in Table 3.9. The results for $\Delta = 3$ mm showed an increase in the bandwidth at lower resonance and a sizable radiation resistance at the second resonance, so that the structure may be operated as a dual-frequency antenna. Separation of the resonances was about 12% at $\Delta = 0$ and 18% at $\Delta = 1\cdot5$ mm.

For $\Delta > 3$ mm the first resonance disappears. The second resonant frequency increased and the real and imaginary parts of the impedance increased with Δ. At $t = 5$ mm the bandwidth was 595 MHz, which was about 17·5% at the centre frequency of 3·407 GHz.

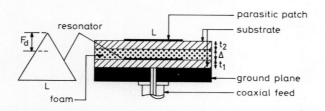

Fig. 3.73 *Geometry of electromagnetic-coupled triangular patch antenna (After Reference 61 pp. 864–865)*

Table 3.9 *Characteristics of the stacked triangular patch antenna (after Reference 61)*

	First resonance			Second resonance		
Δ	f	% BW	Maximum resistance	f	% BW	Maximum resistance
mm	GHz		Ω	GHz		Ω
0	3·1	3·5	150	3·54	2·5	90
1·5	3·12	4·8	150	3·81	3·1	55
3	3·135	6·4	75	3·77	3·2	48
4	–	–	–	3·72	10·5	52
5	–	–	–	3·61	17·46	55
6	–	–	–	3·56	14·8	62·5
9	–	–	–	3·45	8·6	105

The radiation patterns in both the E-plane and the H-plane at various frequencies within the impedance bandwidth are shown in Fig. 3.74 and Fig. 3.75, respectively. The beamwidth varied from 75° to 85° in the H-plane and 55° to 65° in the E-plane. The cross-polar level was better than 16 dB in the H-plane and 20 dB in the E-plane. It is interesting to note that the directivity of the antenna was larger than that of an ordinary microstrip patch, the beamwidth of which was greater than 85°–90°.

The configurations of EMCP studied by Sabban and Bhatnagar *et al.* can be

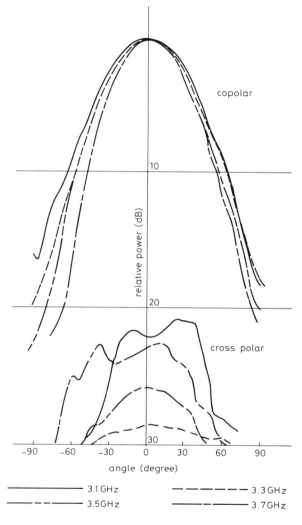

Fig. 3.74 *E-plane radiation patterns of the antenna of Fig. 3.73 with L = 37 mm, $t_1 = t_2 = 1·6$ mm, $\Delta = 5$ mm and $F_d = 13·5$ mm (After References 61 pp. 864–865)*
——— 3·1 GHz
— — — 3·3 GHz
—··— 3·5 GHz
—·—·— 3·7 GHz

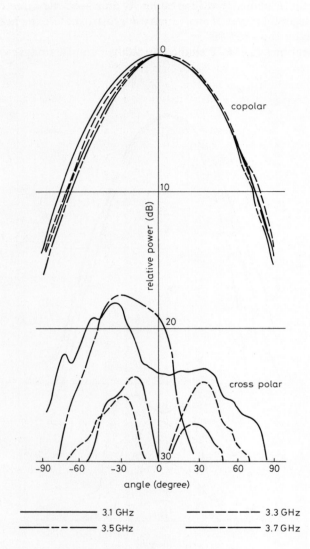

Fig. 3.75 H-plane radiation patterns of the antenna of Fig. 3.73 with L = 37 mm, $t_1 = t_2 = 1.6$ mm, $\Delta = 5$ mm and $F_d = 13.5$ mm (After Reference 61 pp. 864–865)
———— 3·1 GHz
– – – 3·3 GHz
–···– 3·5 GHz
–·–·– 3·7 GHz

described as the 'normal' type. If the upper patch is fabricated on the underside of the substrate an 'inverted' configuration is cobtained. Fig. 3.76 illustrates these two types of configurations. The advantage of the inverted type is that there is a protective dielectric cover for the upper conducting patch. It has been studied by Chen *et al.* [62] and by Dahele *et al.* [63].

Further studies of the EMCP antenna were carried out by Lee *et al.* [64], using rectangular patches etched on Cuflon substrates ($\varepsilon_r = 2 \cdot 17$). Exciting the TM_{01} mode at about 10 GHz, they recorded the variation of pattern shape, 3 dB beamwidth and bandwidth with the separation Δ, for Δ between 0 and $0 \cdot 37 \lambda_0$. This is beyond the range studied by previous authors. It was found that, depending on Δ, the characteristics of the antenna can be separated into three

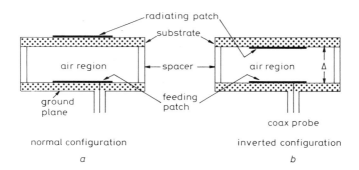

Fig. 3.76 *Normal and inverted configurations of the electromagnetic-coupled patch antenna*

regions. Region 1 is associated with bandwidths exceeding 10%; region 2 has abnormal radiation characteristics and region 3 is associated with narrow beamwidth and high gain. The value of Δ separating these regions depends on the dielectric material between the two layers. The gain in region 3 is 9–10 dB, compared to 5·3 dB for the single patch. It begins at $\Delta = 0 \cdot 31 \lambda_0$ for air dielectric and at $\Delta = 0 \cdot 21 \lambda_0$ for Teflon. The bandwidth in region 3, however, is only about 1·3% for air and 0·85% for Teflon.

It is evident from the experimental results of the authors cited above that, by operating in region 1, the EMCP offers a promising way of achieving bandwidths in excess of 10%, while reducing the problems encountered by simply increasing the substrate thickness. If high gain rather than large bandwidth is desired, the EMCP can be operated in region 3. Although analytical methods are available [65], little work has been done to apply them to this interesting antenna. As such, the experimental results described above have not been quantitatively explained, nor are there formulas available which would aid the design in terms of resonant frequency, impedance, bandwidth and gain. Such theoretical research is urgently needed.

3.5 Summary

This Chapter begins with introducing the simple cavity model for analysing microstrip patch antennas with thin substrates. The formulas obtained from this model for rectangular, circular, equitriangular and annular-ring patches are then presented. The radiation pattern, efficiency, directivity, gain, quality factor and impedance of these antennas are illustrated, mainly for the broadside modes. Experimental results are included or referenced where available.

After a brief mention of some other geometries, notably the elliptical patch, the Chapter proceeds to discuss some methods of improving the frequency agility of microstrip patch antennas. They include the use of varactor diodes, shorting pins, optically controlled diodes, adjustable air gap, stacked geometries and reactive loading. The advantages and disadvantages of these methods are discussed. Finally, the method of increasing the absolute bandwidth by electromagnetic coupling of a fed and a parasitic patch in a stacked geometry is discussed.

3.6 Acknowledgments

The authors wish to acknowledge the assistances of Dr. K.M. Luk and Mr. T. Huynh. Dr. Luk provided the numerical data for the majority of illustrations for the rectangular, equitriangular and annular-ring patches, while T. Huynh contributed to some of the computations for the rectangular and circular patches.

3.7 References

1. DESCHAMPS, G. A.: 'Microstrip microwave antennas'. Presented at the 3rd USAR Symposium on Antennas, 1953
2. BAHL, I. J., and BHARTIA, P.: 'Microstrip antennas' (Artech House, Mass., 1980)
3. JAMES, J. R., HALL, P. S., and WOOD, C.: 'Microstrip antenna theory and design' (Peter Peregrinus, 1981)
4. *IEEE Trans.*, Jan. 1981, **AP-29**
5. CARVER, K. R., and MINK, J. W.: 'Microstrip antenna technology', *IEEE Trans.*, 1981, **AP-29**, pp. 2–24
6. MAILLOUX, R. J., McILEVENNA, J. F., and KERNWEIS, N. P.: 'Microstrip array technology', *IEEE Trans.*, 1981, **AP-29**, pp. 25–39
7. LO, Y. T., SOLOMON, D., and RICHARDS, W. F.: 'Theory and experiment on microstrip antennas', *IEEE Trans.*, 1979, **AP-27**, pp. 137–145
8. RICHARDS, W. F., LO, Y. T., and HARRISON, D. D.: 'An improved theory for microstrip antennas and applications', *IEEE Trans.*, 1981, **AP-29**, pp. 38–46
9. DERNERYD, A. G.: 'Analysis of the microstrip disk antenna element', *IEEE Trans.*, 1979, **AP-27**, pp. 660–664
10. DERNERYD, A. G.: 'Extended analysis of rectangular microstrip resonator antenna', *IEEE Trans.*, 1979, **AP-27**, pp. 846–849

11 DERNERYD, A. G.: 'Microstrip array antenna'. Proc. 6th European Microwave Conference, 1976, pp. 339–343
12 RICHARDS, W. F., LO, Y. T., and SOLOMON, D.: 'Theory and application for microstrip antennas'. Proc. Workshop on Printed Circuit Antenna Technology', New Mexico University, Las Cruces, 1979, pp. 8.1–8.23
13 JAMES, J. R., and HENDERSON, A.: 'High-frequency behaviour of microstrip open-circuit terminations', *IEE J Microwaves, Optics & Acoustics*, 1979, **3**, pp. 205–218
14 WOOD, C.: 'Analysis of microstrip circular patch antennas', *IEE Proc.*, 1981 **128H**, pp. 69–76
15 FONSECA, S. B. A., and GIAROLA, A. J.: 'Microstrip disk antennas. Pt. I: Efficiency of space wave launching', *IEEE Trans.*, 1984, **AP-32**, pp. 561–567
16 RICHARDS, W. F., LO, Y. T., and HARRISON, D. D.: 'Improved theory of microstrip antennas', *Electron. Lett.*, 1979, **15**, pp. 42–44
17 RICHARDS, W. F., LO, Y. T., and SIMON, P.: 'Design and theory of circularly polarized microstrip antennas'. IEEE AP-S International Symposium Digest, June 1979, pp. 117–120
18 LO, Y. T., and RICHARDS, W. F.: 'Perturbation approach to design of circularly polarized microstrip antennas'. *Electron. Lett.*, 1981, **17**, pp. 383–385
19 SHEN, L. C., LONG, S. A., ALLERDING, M. R., and WALTON, M. D.: 'Resonant frequency of a circular disc, printed-circuit antenna', *IEEE Trans.*, 1977, **AP-25**, pp. 595–596
20 DAHELE, J. S., and LEE, K. F.: 'Effect of substrate thickness on the performance of a circular-disk microstrip antenna', *IEEE Trans.*, 1983, **AP-31**, pp. 358–360
21 HELSZAJN, J., and JAMES, D. S.: 'Planar triangular resonators with magnetic walls', *IEEE Trans.*, 1978, **MTT-26**, pp. 95–100
22 KEUSTER, E. F., and CHANG, D. C.: 'A geometrical theory for the resonant frequencies and Q factors of some triangular microstrip patch antennas', *IEEE Trans.*, 1983, **AP-31**, 27–34
23 DAHELE, J. S., and LEE, K. F.: 'Experimental study of the triangular microstrip antenna'. IEEE AP-S International Symposium Digest, 1984, pp. 283–286
24 LUK, K. M., LEE, K. F., and DAHELE, J. S.: 'Theory and experiment on the equilateral triangular microstrip antenna'. Proc. 16th European Microwave Conference, 1986, pp. 661–666
25 SCHELKUNOFF, S. A.: 'Electromagnetic waves' (Van Nostrand, New York, 1943) Chap. 10
26 MINK, J. W., 'Circular ring microstrip antenna elements'. IEEE AP-S Int. Symp. Digest, June 1980, pp. 605–608
27 CHEW, W. C.: 'A broad-band annular-ring microstrip antenna', *IEEE Trans.*, 1982, **AP-30**, pp. 918–922
28 ALI, S. M., CHEW, W. C., and KONG, J. A.: 'Vector Hankel transform analysis of annular-ring microstrip antenna'. *IEEE Trans.*, 1982, **AP-30**, pp. 637–644
29 DAHELE, J. S., and LEE, K. F.: 'Characteristics of annular-ring microstrip antenna', *Electron. Lett.*, 1982, **28**, pp. 1051–1052
30 LEE, K. F., and DAHELE, J. S.: 'Theory and experiment on the annular-ring microstrip antenna', *Ann. des Telecomm.*, 1985, **40**, pp. 508–515
31 SCHNEIDER, M. V.: 'Microstrip lines for microwave integrated circuits', *Bell Syst. Tech. J.*, 1969, **48**, pp. 1421–1444
32 OWENS, R. P.: 'Curvature effect in microstrip ring resonators', *Electron. Lett.*, 1976, **12**, pp. 356–357
33 WU, Y. S., and ROSENBAUM, F. J.: 'Mode chart for microstrip ring resonators', *IEEE Trans.*, 1973, **MTT-21**, pp. 487–489
34 DAS, A., DAS, S. K., and MATHUR, S. P.: 'Radiation characteristics of higher-order modes in microstrip ring antenna', *IEE Proc.*, 1984, **131**, pp. 102–106
35 EL-KHAMY, S. E., EL-AWADI, R. M., and EL-SHARRAWY, E-B. A.: 'Simple analysis and design of annular ring microstrip antennas', *IEE Proc.*, 1986, **133H**, pp. 198–202
36 BHATTACHARYYA, A. K., and GARG, R.: 'Input impedance of annular ring microstrip antenna using circuit theory approach', *IEEE Trans.*, 1985, **AP-33**, pp. 369–374
37 RICHARDS, W. F., OU, J. D., and LONG, S. A.: 'A theoretical and experimental investiga-

tion of annular, annular sector, and circular sector microstrip antennas', *IEEE Trans.*, 1984, **AP-12**, pp. 864–866
38 PALANISAMY, V., and GARG, R.: 'Rectangular ring and H-shaped microstrip antennas – Alternatives to rectangular patch antenna', *Electron. Lett.*, 1985, **21**, pp. 874–876
39 YU, I. P.: 'Low profile circularly polarized antenna', NASA Report N78-15332, 1978
40 SHEN, L. C.: 'The elliptical microstrip antenna with circular polarization', *IEEE Trans.*, 1981, **AP-29**, pp. 90–94
41 LONG, S. A., SHEN, L. C., SCHAUBERT, D. H., and FARRAR, F. G.: 'An experimental study of the circular-polarized elliptical printed circuit antenna', *IEEE Trans.*, 1981, **AP-29**, pp. 95–99
42 BAILEY, M. C., and DESHPANDE, M. D.: 'Analysis of elliptical and circular microstrip antennas using moment method', *IEEE Trans.*, 1985, **AP-33**, pp. 954–959
43 BHARTIA, P., and BAHL, I.: 'A frequency agile microstrip antenna', IEEE AP-S Int. Symp. Digest, 1982, pp. 304–307
44 SCHAUBERT, D. H., FARRAR, F. G., SINDORIS, A. R., and HAYES, S. T.: 'Microstrip antennas with frequency agility and polarization diversity', *IEEE Trans.*, 1981, **AP-29**, pp. 118–123
45 SENGUPTA, D. L.: 'Resonant frequency of a tunable rectangular patch antenna', *Electron. Lett.*, 1984, **20**, pp. 614–615
46 LAN, G. L., and SENGUPTA, D. L.: 'Tunable circular patch antennas', *Electron. Lett.*, 1985, **21**, pp. 1022–1023
47 LO, Y. T., and RICHARDS, W. F.: 'Theoretical and experimental investigations of a microstrip radiator with multiple linear lumped loads', *Electromagnetics*, 1983, **3**, pp. 371–385
48 DARYOUSH, A. S., BONTZOS, K., and HERCSFELD, P. R.: 'Optically tuned patch antenna for phased array applications'. IEEE AP-S Int. Sym. Digest, 1986, pp. 361–364
49 DAHELE, J. S., LEE, K. F., and HO, K. Y.: 'Mode characteristics of annular-ring and circular disc microstrip antennas with and without airgaps'. IEEE AP-S Int. Sym. Digest, 1983, pp. 55–58
50 LEE, K. F., HO, K. Y., and DAHELE, J. S.: 'Circular-disk microstrip antenna with an air gap', *IEEE Trans.*, 1984, **AP-32**, pp. 880–884
51 DAHELE, J. S., and LEE, K. F.: 'Theory and experiment on microstrip antennas with airgaps', *IEE Proc.*, 1985, **132**H, pp. 455–460
52 LEE, K. F., and DAHELE, J. S.: 'The two-layered annular ring microstrip antenna', *Int. J. Electronics*, 1986, **61**, pp. 207–217
53 LONG, S. A., and WALTON, W. D.: 'A dual frequency stacked circular disc antenna', *IEEE Trans.*, 1979, **AP-27**, pp. 270–273
54 DAHELE, J. S., and LEE, K. F.: 'A dual-frequency stacked microstrip antenna'. IEEE AP-S Int. Sym. Digest, 1982, pp. 308–311
55 DAHELE, J. S., LEE, K. F., and WONG, D. P.: 'Dual-frequency stacked annular-ring microstrip antenna', *IEEE Trans.*, 1987, **AP-35**
56 ZHONG, S. S., and LO, Y. T.: 'Single-element rectangular microstrip antenna for dual-frequency operation', *Electron. Lett.*, 1983, **19**, pp. 298–300
57 DAVIDSON, S. E., LONG, S. A., and RICHARDS, W. F.: 'Dual-band microstrip antennas with monolithic reactive loading', *Electron. Lett.*, 1985, **21**, pp. 936–937
58 DAHELE, J. S., and LEE, K. F.: 'Top-loaded single and coupled microstrip monopoles'. IEEE AP-S Int. Sym. Digest, 1983, pp. 47–50
59 McILVENNA, J., and KERNWEIS, N.: 'Modified circular microstrip antenna elements', *Electron. Lett.*, 1979, **15**, pp. 207–208
60 SABBAN, A.: 'A new broadband stacked two-layer microstrip antenna'. IEEE AP-S Int. Sym. Digest, 1983, pp. 63–66
61 BHATNAGAR, P. S., DANIEL, J.-P., MAHDJOUBI, K., and TERRET, C.: 'Experimental study on stacked triangular microstrip antennas', *Electron. Lett.*, 1986, **22**, pp. 864–865

62 CHEN, C. H., TULINTSEFF, A., and SORBELLO, R. M.: 'Broadband two-layer microstrip antenna'. IEEE AP-S Int. Symp. Digest, 1984, pp. 251–254
63 DAHELE, J. S., TUNG, S. H., and LEE, K. F.: 'Normal and inverted configurations of the broadband electromagnetic coupled microstrip antenna'. IEEE AP-S Int. Sym. Digest, 1986, pp. 841–844
64 LEE, R. Q., LEE, K. F., and BOBINCHAK, J.: 'Characteristics of a two-layer electromagnetically coupled rectangular patch antenna', *Electron. Lett.*, 1987, **23**, pp. 1070–1072; also IEEE AP-S Int. Sym. Digest, 1988, pp. 948–951
65 RIVERA, J., and ITOH, T.: 'Analysis of an electromagnetically coupled patch antenna'. IEEE AP-S Int. Sym. Digest, 1983, pp. 170–173

Chapter 4

Circular polarisation and bandwidth

M. Haneishi and Y. Suzuki

Microstrip antennas are widely used as an efficient radiator in many communication systems [1]. One of the most interesting applications is their use for transmitting or receiving systems required for circular polarisation [2–5]. A circularly polarised microstrip antenna can be classified into two categories, e.g. single- or dual-fed types. The classification of an antenna is based upon the number of feeding points required for circularly polarised waves. The singly-fed antenna is useful, because it can excite circular polarisation without using an external polariser. Therefore, it is important to understand the radiation mechanism of the antenna. However, one of the most serious problem in such an antenna is the considerable narrowness of the bandwidth compared to ordinary microwave antennas. This is a serious problem for the practical application of this antenna. For this reason, it is also important to study some wideband techniques.

In this Chapter, the various types of circularly polarised antennas are first briefly introduced in Section 4.1. In Section 4.2, a simple design method for a singly-fed antenna is described together with some useful design data. This method is useful in understanding its radiation mechanism and to roughly design it. However, if the general radiation mechanism and an accurate design method are required, then the more exact treatment, developed in Section 4.3, is necessary. In Section 4.4, some considerations of mutual coupling are described. Finally, three kinds of wideband techniques are introduced in Section 4.5.

4.1 Various types of circularly polarised antennas

There are many types of circularly polarised (CP) printed antennas, which are widely used as efficient radiators in many communication systems. Fig. 4.1 shows basic arrangements for various types of CP-wave printed antennas. In this Section, we describe briefly techniques for designing such CP printed antennas.

4.1.1 Microstrip patch antennas

A microstrip antenna is one of the most effective radiators for exciting circular polarisation. A circularly polarised microstrip antenna is categorised into two types by its feeding systems: one is a dual-feed CP antenna with an external polariser such as 3 dB hybrid, and the other is a singly-fed one without a polariser. The classification of antennas is based upon the number of feeding point required for CP excitation.

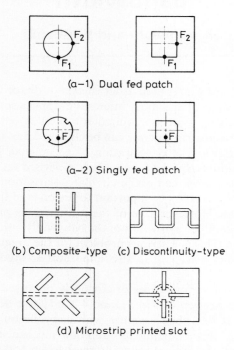

Fig. 4.1 *Various types of circularly polarised printed antennas*

(a) *Dual-fed CP patch antenna:* The fundamental configurations of a dual-fed CP patch antenna are shown in both Fig. 4.1 (a–1) and Fig. 4.2 (a). The patches are fed with equal amplitude and 90° out of phase by using an external polariser. As shown in the Figure, these antennas are also divided into two categories by the shape of an external polariser: one is the 3 dB hybrid type and the other is an offset-feeding one.

As is well known, a 3 dB hybrid such as a branch-line coupler produces fields of equal amplitude but 90° out of phase at its centre frequency. Therefore, setting the outputs of such a hybrid to the edges of the patch, the antenna acts as a CP radiator. It is necessary to note that each input terminal of a hybrid, however, gives an opposite sense of circular polarisation. In the present case, both the input VSWR and ellipticity bandwidth are broad, since a 3 dB hybrid, in general, has a broadband nature.

The other category is the offset-feeding CP antenna. Here, offset feeding lines, with one quarter wavelength longer than the other, are set at the edges of the patch, as shown in Fig. 4.2a. One of the most serious disadvantages of this type of antenna is the narrow bandwidth, since the frequency dependency of an offset-feeding line is greater than that of the usual hybrid.

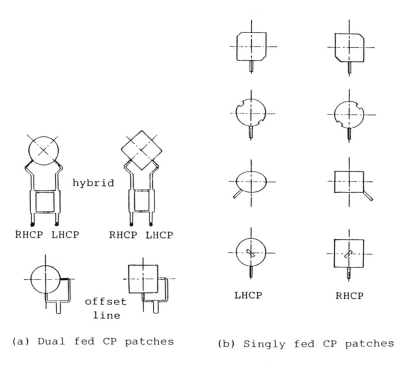

Fig. 4.2 *Typical arrangements for circularly polarised microstrip antennas*
LHCP: Left-hand circular polarisation
RHCP: Right-hand circular polarisation

(*b*) *Singly-fed CP patch antenna:* A singly-fed CP antenna may be regarded as one of the simplest radiators for exciting circular polarisation. The typical configurations of this antenna are shown in Fig. 4.2b. It is important to note that the generated mode in this case is usually excited in an electrically thin cavity region of the microstrip antenna. Accordingly, the operational principle of this antenna is based on the fact that the generated mode can be separated into two orthogonal modes by the effect of a perturbation segment such as a slot or other truncated segment [6–7,10–11]. Consequently, by setting the perturbation segment to the edge of the patch, the generated mode is separated into two orthogonal modes 1 and 2. The typical amplitude and phase diagrams after perturbation are shown in Fig. 4.3, together with typical samples of antennas.

The radiated fields excited by these two modes are, in general, perpendicular to each other, and orthogonally polarised in the boresight direction. When the amount of perturbation segment is adjusted to the optimum value, modes 1 and 2 are excited in equal amplitude and 90° out of phase at the centre frequency, as shown in the Figure. This enables the antenna to act as a CP radiator in spite of single feeding. This antenna has several advantages compared to dual-fed ones and can excite CP radiation without using an external polariser. Design techniques will be described in detail in Sections 4.2 and 4.3.

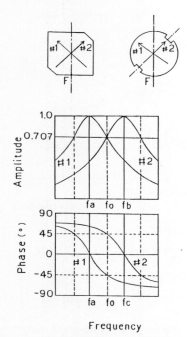

Fig. 4.3 *Amplitude and phase diagrams for singly-fed circularly polarised microstrip antennas*

4.1.2 Other types of circularly polarised printed antennas

In this section, we describe briefly the fundamental design procedures for others types of CP antennas.

(*a*) *Composite type of CP printed antenna:* Fig. 4.4a shows the fundamental configuration of a composite-type CP antenna [8]. The antenna is composed of the combination of a half-wavelength-long strip conductor and a slot in the ground plane. If the microstrip feeding line is short-circuited at $l = 0$, a standing-wave voltage V_l and current I_l occurs along the microstrip feeding line. In the present case, on setting the radiating elements such as the strip and the slot to the maximum positions of V_l and I_l, these radiating elements radiate transverse and longitudinal electric fields E_{st} and E_{sl}, respectively, in the boresight direction. The fields E_{st} and E_{sl} can be excited in equal amplitude and at 90° out

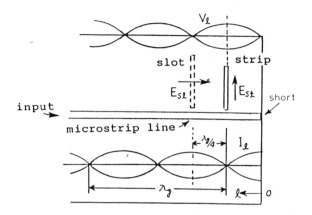

(a) Radiating element for composite-type circularly polarised printed antenna[8]

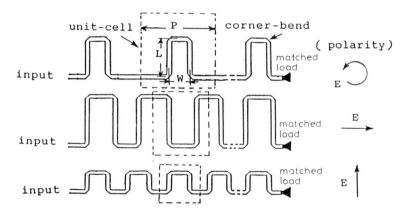

(b) Various arrangements of rampart line antennas[6]

Fig. 4.4 *Typical arrangements for travelling-wave-type printed antennas*

of phase, if the strip and the slot are spaced one-quarter wavelength apart and the coupling between the radiating element and feeder is controlled to be identical in value. Therefore, this type of antenna acts as CP radiator without using any external polariser. Details of design techniques for this antenna will be discussed in Chapter 13.

(b) *Discontinuity type of CP printed antenna:* A rampart-line antenna is a typical radiator for the discontinuity type [6,9]. Fig. 4.4b shows typical rampart line antennas that act as CP and LP radiators. Each radiator consists of a microstrip meander line having a series of corner bends. The antenna also has a matched load at the open end of the meander line. In this system, radiation occurs mainly from the discontinuity section of the meander line such as a corner bend. Therefore, both CP and LP rampart line antennas can be easily fabricated by controlling the length L, width W and period P of the meander line. If L, W and P are adjusted to $\lambda_g/2$, $\lambda_g/4$ and λ_g, respectively, for a unit cell of the meander line (λ_g is the wavelength of the travelling wave along the meander line), the antenna acts as a CP radiator. When $L = 2\lambda_g/3$, $W = \lambda_g/3$ and $P = 2\lambda_g/3$, the antenna radiates horizontal polarisation, while $L = W = \lambda_g/4$, $P = \lambda_g/2$ excites vertical polarisation, as shown in Fig. 4.4b. Details of the design procedure will be discussed in Chapter 13.

4.2 Simple design techniques for singly-fed circularly polarised microstrip antennas

This Section gives a brief description of design techniques for singly-fed radiators together with some useful experimental results. The approach is based on the variational method, and is useful for understanding the mechanism of CP radiation from such singly-fed radiators.

4.2.1 Rectangular type

In general, microstrip antennas are divided into two types by the shape of radiating element: rectangular type and circular type. However, since the rectangular patch antenna is considered to be a fundamental device for exciting CP radiation, the design techniques for this type are discussed first.

(a) *Fundamental configuration of rectangular CP-wave antenna:* The fundamental configurations of the antenna and its co-ordinate system are shown in Fig. 4.5. In type A, the feeding point F is placed on the x- or y-axis, whereas in type B, F is placed on the diagonal axis. In both cases, the perturbation segment Δs is set at an appropriate location in the patch element to excite CP radiation. Here, we describe briefly the sense of direction for the CP-wave. Right-hand or left-hand CP radiation can be achieved by setting feeding points at appropriate locations such as $F(\pm \varrho_0, 0)$ and $F(0, \pm \varrho_0)$, as shown in Fig. 4.6.

(b) *Effect of perturbation segment:* The effect of perturbation segment Δs for the type-A antenna is described first, since this type of radiator is a basic device for exciting CP radiation.

The eigen functions ϕ_a and ϕ_b, which are excited in an electrically thin cavity region of the square patch, are generally given mathematically by the following equations, while a perfect magnetic wall is assumed as a boundary condition at the antenna peripheries ($x = \pm a/2, y = \pm b/2$);

$$\begin{aligned} \phi_a &= V_0 \sin k\, x \\ \phi_b &= V_0 \sin k\, y \end{aligned} \right\} \quad (4.1)$$

where $V_0 = \sqrt{2}/a$ and $k = \pi/a$.

The eigen function ϕ_a is concerned with the field distribution of TM$_{100}$ mode, and ϕ_b with that of TM$_{010}$ mode. By setting the perturbation segment Δs at an appropriate position of the antenna, as shown in Fig. 4.5, two orthogonally polarised modes are excited in a cavity region of the antenna.

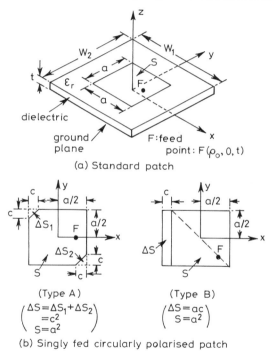

Fig. 4.5 *Fundamental configurations of singly-fed rectangular patches (From Reference 11)*

The new eigen function ϕ' and the new eigen value k', after perturbation by the segment, are determined by the following equations [11, 12, 15]:

$$\left. \begin{aligned} \phi' &= P\phi_a + Q\phi_b \\ k'^2 &= \frac{\iint_{s + \Delta s_1 + \Delta s_2} (P\nabla\phi_a + Q\nabla\phi_b)^2\, dS}{\iint_{s + \Delta s_1 + \Delta s_2} (P\phi_a + Q\phi_b)^2\, dS} \end{aligned} \right\} \quad (4.2)$$

where P and Q are unknown expansion coefficients of the new eigen function ϕ'.

The new eigen value k' of the antenna can be derived by employing the following matrix, since eqn. 4.2 is a variational-expression form:

$$\det \begin{vmatrix} k^2 + q_1 - k'^2(1 + p_1) & q_{12} - k'^2 p_{12} \\ q_{12} - k'^2 p_{12} & k^2 + q_2 - k'^2(1 + p_2) \end{vmatrix} = 0 \quad (4.3)$$

In case of the type-A antenna, the parameters in eqn 4.3, such as $p_1, p_2, q_1, q_2, p_{12}$ and q_{12}, are expressed by the following equations [11]:

$$\left. \begin{array}{l} q_1 = q_2 = q_{12} = 0 \\ p_1 = p_2 = 2(\Delta s/S) \\ p_{12} = -2(\Delta s/S) \end{array} \right\} \quad (4.4)$$

Substituting eqn. 4.4 into eqn. 4.3, the new eigen values k_a' and k_b' for type A are given as

$$\left. \begin{array}{l} k_a'^2 = k^2(1 - 4\Delta s/S) \\ k_b'^2 = k^2 \end{array} \right\} \quad (4.5)$$

where k_a' and k_b' correspond to the eigenvalues of the new orthogonal eigen functions, ϕ_a' and ϕ_b', respectively.

Using eqn. 4.5, new sets of resonant frequencies for the ϕ_a' and ϕ_b' modes are easily obtained as follows:

$$\left. \begin{array}{l} f_a = f_{0r} + \Delta f_a' = f_{0r}(1 - 2\Delta s/S) \\ f_b = f_{0r} + \Delta f_b' = f_{0r} \end{array} \right\}$$

where f_{0r} is the resonant frequency for a normal square patch before perturbation, and $\Delta f_a'$ and $\Delta f_b'$ are the shifts of resonant frequencies for the ϕ_a' and ϕ_b' modes after perturbation.

Normalising the new eigen functions for the ϕ_a' and ϕ_b' modes, the unknown expansion coefficients P and Q are determined as follows [11, 12]:

For ϕ_a' mode,

$$\left. \begin{array}{l} P_a = (1/\sqrt{2})(1 - 2\Delta s/S) \simeq (1/\sqrt{2}) \\ Q_a = (-1/\sqrt{2})(1 - 2\Delta s/S) \simeq (-1/\sqrt{2}) \end{array} \right\}$$

For ϕ_b' mode

$$P_b = Q_b = (1/\sqrt{2})$$

Finally, using eqns. 4.1, 4.2 and the expansion coefficient, the new eigen functions ϕ_a' and ϕ_b' are given in a closed form by

$$\left. \begin{array}{l} \phi_a' \simeq (\phi_a - \phi_b)/\sqrt{2} = V_0(\sin kx - \sin ky)/\sqrt{2} \\ \phi_b' \simeq (\phi_a + \phi_b)/\sqrt{2} = V_0(\sin kx + \sin ky)/\sqrt{2} \end{array} \right\} \quad (4.6)$$

Circular polarisation and bandwidth

The eigen values used in eqn. 4.6 are assumed to be $k_a' = k_b' = k$ by means of first-order approximation.

Furthermore, the turn ratios N_a' and N_b', which correspond to the energy distribution ratios for both the ϕ_a' and ϕ_b' modes after perturbation, are defined as [11]

$$\begin{aligned} N_a' &= (\sqrt{S}/a)(\sin kx - \sin ky) \\ N_b' &= (\sqrt{S}/a)(\sin kx + \sin ky) \end{aligned} \quad (4.7)$$

In the case of the type-B antenna shown in Fig. 4.5, the eigen functions ϕ_a', ϕ_b' and other parameters can also be derived by similar calculations employed for type A. The equations obtained by these calculations are as follows:

$$\left. \begin{aligned} p_1 &= p_2 = (3/2)(\Delta s/S) \\ p_{12} &= -(1/2)(\Delta s/S) \\ q_1 &= q_2 = (1/2)(\Delta s/S)k^2 \\ q_{12} &= +(1/2)(\Delta s/S)k^2 \\ k_a'^2 &= k^2(1 - 2\Delta s/S) \\ k_b'^2 &= k^2 \\ f_a &= f_{0r} + \Delta f_a' = f_{0r}(1 - \Delta s/S) \\ f_b &= f_{0r} + \Delta f_b' = f_{0r} \\ \phi_a' &= (\sqrt{2}/a)(1 - \Delta s/S)\sin kx \\ \phi_b' &= (\sqrt{2}/a)(1 - \Delta s/2S)\sin ky \\ N_a' &= \sqrt{S}\,\phi_a' \simeq \sqrt{2}\sin kx \\ N_b' &= \sqrt{S}\,\phi_b' \simeq \sqrt{2}\sin ky \end{aligned} \right\} \quad (4.8)$$

where $V_{00} = 1/a$ and $k = \pi/a$.

Using eqns. 4.1–4.7, we can derive the equivalent circuit for the type-A antenna. Furthermore, the equivalent circuit of the type-B antenna after perturbation can also be derived using the relations given in eqns. 4.8. The circuit for both the types of antennas is shown in Fig. 4.7. In this circuit, T_A' and T_B' represent ideal transformers having turn ratios N_a' and N_b', and V_f is input voltage applied to the 1–1' terminal.

(c) Condition required for CP-wave radiation: In this Section, conditions for exciting CP-wave radiation are determined by use of the preceding equivalent circuit. As is well known, the equivalent conductances G_a' and G_b' in the circuit are expressed as the sum of the radiation, dielectric and copper losses. However, in normal patches having adequate radiation efficiency above 90%, radiation loss is dominant compared with the other losses.

Consequently, the equivalent conductances G'_a and G'_b are mainly caused by the radiated fields resulting from the patch antenna. In other words, the induced voltages $\dot{V}_a$ and $\dot{V}_b$ generated on G'_a and G'_b can be assumed to correspond to the radiated fields caused by the orthogonal ϕ'_a and ϕ'_b modes.

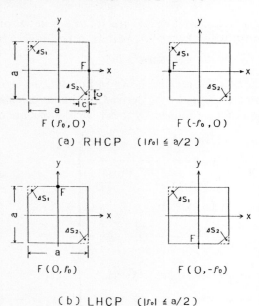

Fig. 4.6 *Feeding locations required for circular polarisation*
RHCP: Right-hand circular polarisation
LHCP: Left-hand circular polarisation
ϱ_0 = feed location

Applying network analysis to the equivalent circuit, the complex amplitude ratio $\dot{V}_a/\dot{V}_b$ in the two orthogonal modes is given as follows:

$$(\dot{V}_b/\dot{V}_a) = (N'_b/N'_a) * (\dot{Y}_a/\dot{Y}_b)$$

$$= \left(\frac{N'_b}{N'_a}\right) \frac{\left\{\frac{f_a}{Q_0} + j\left(f - \frac{f_a^2}{f}\right)\right\}}{\left\{\frac{f_b}{Q_0} + j\left(f - \frac{f_b^2}{f}\right)\right\}} \quad (4.9)$$

where $\dot{Y}_a$ and $\dot{Y}_b$ are input admittances for the orthogonally polarised ϕ'_a and ϕ'_b modes, respectively. In addition, the unloaded Q factors in the above equation are expressed as $Q_{0a} = Q_{0b} = Q_0$ to first-order approximation, where Q_{0a} and Q_{0b} are the unloaded Q factors of the ϕ'_a and ϕ'_b modes.

From eqn. 4.9, radiation of CP waves by these radiators may be expected if

$(\dot{V}_b/\dot{V}_a) = \pm j$ is satisfied. Accordingly, these antennas act as a CP radiator by setting the relative amplitude and phase between the two orthogonal modes at $|\dot{V}_b/\dot{V}_a| = 1$ and $\arg(\dot{V}_b/\dot{V}_a) = \pm 90°$, respectively.

Applying the above conditions to eqn. 4.9, turns ratios are required to satisfy the relation $|N'_b/N'_a| = 1$. In addition, when this restriction is applied to the type-A antenna, it is necessary to place the feeding point F on the x-axis for $(N'_b/N'_a) = 1$. Contrariwise, the feeding point F is required to be placed on the y-axis by another restriction $(N'_b/N'_a) = -1$.

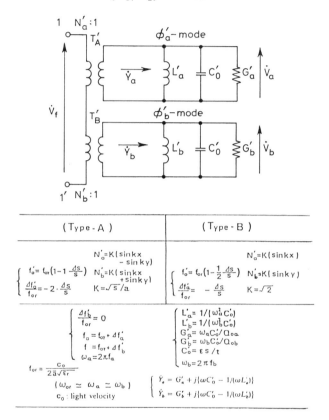

Fig. 4.7 *Equivalent circuit for rectangular circularly polarised patch antennas*

Setting the feeding point at each location, the expression for the complex amplitude ratio is shown as follows:

$$(\dot{V}_b/\dot{V}_a) = \pm \frac{\left(\dfrac{f_a}{Q_0}\right) + j\left(f - \dfrac{f_a^2}{f}\right)}{\left(\dfrac{f_b}{Q_0}\right) + j\left(f - \dfrac{f_b^2}{f}\right)}$$

By application of the CP conditions satisfying $|\dot{V}_b/\dot{V}_a| = 1$ and $\arg(\dot{V}_b/\dot{V}_a) = \pm 90°$ to the above equation, an important relation between Q_0 and $(\Delta s/S)$ is obtained as follows:

$$\frac{(Q_0^2 - 1)Q_0^2}{2Q_0^2 - 1}(M^2 + N^2) = MN\left\{1 + \frac{(2Q_0^2 - 1)MN}{M^2 + N^2}\right\} \tag{4.10}$$

where $M = (1 + m\Delta s/S)$, $N = (1 + n\Delta s/S)$, and m and n are the constants in $f_a = f_{0r}(1 + m\Delta s/S)$ and $f_b = f_{0r}(1 + n\Delta s/S)$.

In case of the type-A radiator, constants m and n are shown as $m = -2$ and $n = 0$, as mentioned previously. Substituting these values into eqn. 4.10, and carrying out some modifications of the equation, the most important expression for designing purposes is easily obtained as follows:

$$|\Delta s/S|Q_0 = 1/2 \tag{4.11}$$

This expression is simple in form but very useful for actual design of the type-A radiator.

Furthermore, in case of the type-B antenna, a basic equation for design can also be derived by the use of similar techniques, and the expression is as follows:

$$|\Delta s/S|Q_0 = 1$$

The relations for both basic expressions are illustrated in Fig. 4.8b by solid lines. They help to provide important design parameters such as the amount of perturbation $(\Delta s/S)$ required for CP radiation.

(d) Basic design procedures for CP-wave antennas: In designing CP-wave antennas, it is necessary to estimate the value of unloaded $Q(Q_0)$ as a function of substrate thickness t for an antenna. Therefore, theoretical values of Q_0 were calculated for a typical sample ($a = 9.14$ mm, $t = 0.6$ mm, $\varepsilon_r = 2.55$) employing a commonly used technique [6, 10]. The theoretical values agreed well with the experimental ones, as shown in Fig. 4.8a. After determining the value of Q_0, the design of an antenna can be achieved by the following procedures:

(i) Using the relations shown in Fig. 4.8a, the unloaded $Q(Q_0)$ of the square patch is chosen so as to ensure that the radiation efficiency η of the patch will exceed 90%.
(ii) The amount of perturbation $(\Delta s/S)$ required for CP-wave excitation is determined using the relation between Q_0 and $(\Delta s/S)$ shown in Fig. 4.8b.
(iii) Finally, the input impedance of the test antenna is matched to that of the feed network by the offset loading technique of coaxial probe or using a quarter-wavelength transformer.

The approaches described above are performed without considering the effect due to fringing fields. However, when the fringing effect is taken into consideration [6], the procedures help to provide more accurate design parameters required for CP radiation.

(e) Radiation characteristics of CP antennas: In order to verify the validity of the above design procedures, typical samples of CP antennas were fabricated and tested at X-band. These antennas were fed with a coaxial probe to avoid the influence of unwanted radiation from the feeding networks. The radiation

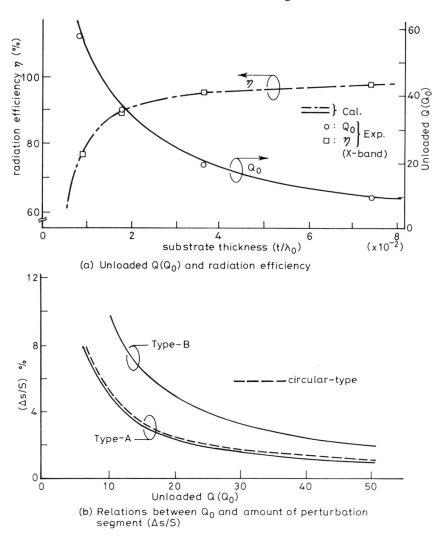

Fig. 4.8 *Fundamental design chart for singly-fed circularly polarised patch antennas*

patterns measured by a spinning dipole are shown in Fig. 4.9. As seen from the Figure, the ellipticity of the test antenna is less than 0·5 dB in the boresight direction. Furthermore, the ellipticity is within 1·5 dB in the desired angular region of 45°.

232 Circular polarisation and bandwidth

Fig. 4.10 shows the measured impedance characteristics of the typical CP-wave antenna. From these results, it is found that loop 1 in the impedance-plot locus depends on the degree of mode separation; namely, loop 1 becomes larger in area with an increase in mode separation, and converged to a point when the mode separation is reduced. In any case, however, the best ellipticity can be obtained at or near the peak of loop 1 in the impedance locus.

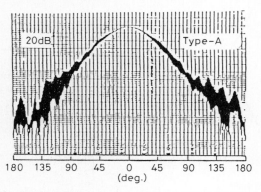

(a)   Pattern of type-A circularly polarised patch

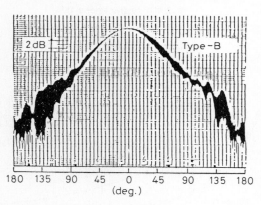

(b)   Pattern of type-B circularly polarised patch

Fig. 4.9 *Typical radiation patterns of rectangular singly-fed circularly polarised patches, (From Reference 11)*
$[t/\lambda_0 = 0.018$, $\varepsilon_r = 2.55$, $\tan\delta = 0.0018$, $(\varrho_0/(a/2)) = 0.3$, $W_1 = W_2 = \lambda_0$ (Fig. 4.5a), and X-band]

4.2.2 Circular type

This Section gives a brief description of a design technique for a circular antenna. The geometry and feed system of the antenna are shown in Fig. 4.11. In this antenna, the perturbation segment Δs is also located at a specific location. The degenerate mode is also separated from the dominant mode (TM_{110}) in the antenna into two orthogonal modes by the effect of the perturbation segment.

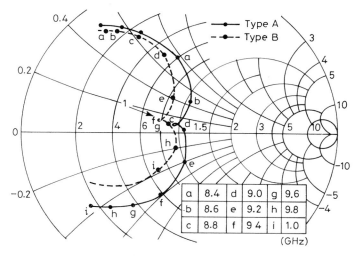

Fig. 4.10 *Typical measured impedance characteristics for rectangular singly-fed circularly polarised patch antennas (From Reference 11)*
$t/\lambda_0 = 0{\cdot}018$, $\varepsilon_r = 2{\cdot}55$, $\tan\delta = 0{\cdot}0018$ and $(\varrho_0/(a/2)) = 0{\cdot}3$

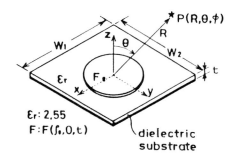

(a) Standard patch

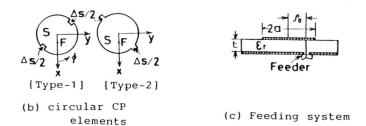

(b) circular CP elements

(c) Feeding system

Fig. 4.11 *Fundamental configurations for circular singly-fed circularly polarised patch antennas*

234 Circular polarisation and bandwidth

The equivalent circuit after perturbation is useful for network analysis of the radiator. Fortunately, the equivalent circuit for the antenna can easily be obtained by employing the same procedures as in the previous Section. The equivalent circuit for the antenna after perturbation is shown in Fig. 4.12 [7].

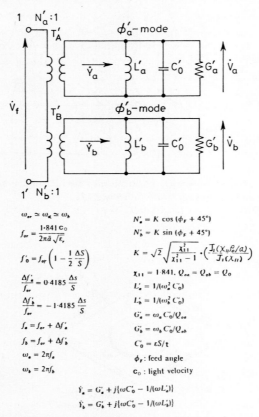

Fig. 4.12 *Equivalent circuit for circular singly-fed circularly polarised patch antenna (From Reference 7)*

The CP-radiation condition for the circular patch can be determined using the above equivalent circuit. Namely, by application of the design procedures employed for the rectangular patches, the CP-wave condition for the circular one is obtained as

$$Q_0 |\Delta s/S| = 1/\chi_{11}$$

where $(\Delta s/S)$, Q_0 and χ_{11} correspond to the amount of perturbation, the unloaded Q and the eigen value of the dominant TM_{110} mode, respectively.

Using the above equation, the relation that gives a CP-radiation condition for the antenna is indicated by the dotted-line in Fig. 4.8b. This Figure helps to provide important design parameters such as the amount of perturbation required for CP-wave radiation. In actual design, however, it is important to note that the Q_0 of the circular patch becomes equivalent to that of the rectangular one, if both the patches are designed to have the same resonant frequencies.

For this reason, the circular CP antennas can be designed as easily as rectangular antennas by employing the relations shown in Fig. 4.8. However, the fringing effect is disregarded in above approaches. If the effect of fringing fields is taken into consideration [6], the experimental results agree well with theory, as mentioned below.

In order to verify the validity of the design procedure, some circular CP patches were fabricated and tested at X-band. These samples were fabricated using a substrate consisting of copper-clad 0·6 mm-thick Teflon glass fibre with a dielectric constant of 2·55 and a loss tangent of approximately 0·0018. The boresight ellipticity of the test antenna was about 0·5 dB or less, and the radiation patterns revealed as a high a level of performance as those of the rectangular patch. In addition, the ellipticity bandwidth within 3 dB was about 1% with a substrate thickness of $(t/\lambda_0) = 0.019$, and about 2% with a thickness of $(t/\lambda_0) = 0.037$. These results indicate that, as the substrate thickness increases, the ellipticity bandwidth also increases.

Furthermore, the trend of the impedance locus plot of the antenna coincided with that of the rectangular one shown in Fig. 4.10.

4.3 More exact treatment of singly-fed circularly polarised microstrip antennas

The patch radiator can easily be modified from a circular, square or rectangular shape so as to excite circularly polarised waves with a single feed as mentioned previously. In addition to these shapes [4, 5], a specially shaped pentagonal [2], triangular [13] or elliptical radiator [3] can also radiate circular polarisation. Furthermore, it has been known that the polarisation and resonant frequency can be conveniently controlled by inserting posts at suitable locations within the patch boundary [14]. However, it is not generally easy to analyse such antennas accurately, so designers are often forced to use cut-and-try methods to realise the desired characteristics.

In this Section, an analysis, based on variational method [15] and modal expansion technique [16], is briefly summarised for an arbitrarily shaped microstrip antenna with multi-terminals before starting the discussion concerning a singly-fed circularly polarised antenna. Using the results of analysis the conditions for producing circularly polarised waves are derived, and then it is shown that a microstrip antenna, in general, can radiate circularly polarised waves at two kinds of frequencies with a single feed. Finally, one design example is given in order to confirm experimentally the several theoretical predictions concerning the feed points and the operating frequencies to radiate them.

4.3.1 Analysis [17]

The present method is based on the variational method applied to arbitrarily shaped microstrip planar circuits with multi-terminals [15] and the modal expansion technique [16]. The following approach is more suitable and useful in the analysis, and the design of a singly-fed circularly polarised microstrip antenna than that based on the moment method [19]. In the present method, the eigen values and orthonormalised eigen functions are derived from the Rayleigh–Ritz method under the Neumann boundary conditions [20]. The formula for the mutual impedance is derived using the relations between the terminal voltages, stored energies and radiated and dissipated powers. Also the equivalent circuit applicable to the microstrip antenna with multi-terminals is obtained.

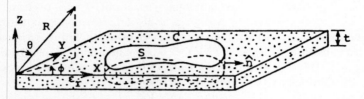

Fig. 4.13 *Structure of analytical model and co-ordinate system*

(a) Green function: The geometry of an analytical model and the co-ordinate system employed are shown in Fig. 4.13. The arbitrarily shaped patch is located on the surface of the grounded dielectric substrate with thickness t and the dielectric constant ε_r. Usually, the patch is fed either by microstrip feed lines or coaxial probes. However, microstrip feed lines lead to problems of coupling with the patch radiator and problems of radiation, though there is an advantage in that they can be etched together with the patch radiator. Accordingly, the following discussion is restricted to the case of coaxial-probe use, because we want to separate the problem of the antenna itself from that of the feed network. In this Figure, C denotes a boundary line for the patch radiator, S is its area and $\hat{n}$ is a unit vector normal and outward to the boundary C. In many practical applications, the substrate is electrically thin, so that only a Z component of the electric field and the X and Y components of the magnetic field exist in the region bounded by the patch radiator and ground plane. Assuming $e^{j\omega t}$ time variation, the electric field E_z associated with a current source J_z located at (x_0, y_0) must satisfy

$$(\nabla_T^2 + k^2)E_z = -j\omega\mu_0 J_z(x_0, y_0) \tag{4.12}$$

where ∇_T is the transverse part of the del operator with respect to the Z-axis, ω signifies angular frequency, and $k^2 = \varepsilon_r k_0^2$, with k_0 being a free-space wave number. If the perfect magnetic wall is assumed on the boundary, the solution

Circular polarisation and bandwidth

to this problem may be given by

$$E_z(x, y) = \iint G(x, y/x_0, y_0) J_z(x_0, y_0) dx_0 dy_0 \quad (4.13)$$

where $G(x, y/x_0, y_0)$ is a Green function generally expressed using the eigen values and eigen functions as

$$G(x, y/x_0, y_0) = j\omega\mu_0 \sum_{l=1}^{N} \frac{\varphi^{(l)}(x, y)\varphi^{(l)*}(x_0, y_0)}{k^{(l)2} - k^2} \quad (4.14)$$

In this equation, $k^{(l)}$ and $\varphi^{(l)}$ are the eigen value and eigen function for the l th mode, respectively, and can be derived by employing the Rayleigh–Ritz method [20] for an arbitrarily shaped microstrip antenna.

(b) Mutual impedance and equivalent circuit: In a multi-terminal microstrip antenna, if the power is supplied only to the q th terminal and the other terminals are all open, the electric field $E_{zq}(x, y)$ associated with the terminal current I_q located at (x_q, y_q) can be expanded in terms of series of eigen functions as

$$E_{zq}(x, y) = I_q \sum_{l=1}^{N} F(l, x_q, y_q) \varphi^{(l)}(x, y) \quad (4.15)$$

where the unknowns $F(l, x_q, y_q)$ are functions of the mode number l and the terminal location (x_q, y_q). Eqn. 4.15 implies that a mutual impedance can be expressed as a superposition of that for each mode as follows:

$$Z_{p,q} = E_{zq}(x_p, y_p)/I_q = \sum_{l=1}^{N} Z_{p,q}^{(l)} \quad (4.16)$$

where $Z_{pq}^{(l)}$ is the mutual impedance between the p th and q th terminals for the l th mode and can be expressed by

$$Z_{p,q}^{(l)} = \frac{V_p^{(l)} V_q^{(l)*}}{j2\omega(W_{eq}^{(l)} - W_{mq}^{(l)}) + P_{rq}^{(l)} + P_{cq}^{(l)} + P_{dq}^{(l)}} \quad (4.17)$$

In the above equation, $V_p^{(l)}$ and $V_q^{(l)}$ are the terminal voltages, $W_{eq}^{(l)}$ and $W_{mq}^{(l)}$ are the time-averaged electric and magnetic stored energies, $P_{rq}^{(l)}$ is the radiated power, and $P_{cq}^{(l)}$ and $P_{dq}^{(l)}$ are the powers dissipated in the conductor walls and the dielectric, respectively. These parameters can be derived from the fields within the patch boundary according to the perturbation theory. As a result, the mutual impedance can be expressed as

$$Z_{p,q} = \sum_{l=1}^{N} \frac{M_p^{(l)} \cdot M_q^{(l)*}}{j\omega C + \frac{1}{j\omega L^{(l)}} + g^{(l)}} \quad (4.18)$$

where

$$M_p^{(l)} = \sqrt{S} \varphi^{(l)}(x_p, y_p) \quad (4.19)$$

$$M_q^{(l)} = \sqrt{S}\varphi^{(l)}(x_q, y_q) \tag{4.20}$$

$$C = \frac{\varepsilon_r \varepsilon_0}{t} S \tag{4.21}$$

$$\omega^{(l)} = \frac{k^{(l)}}{\sqrt{\varepsilon_r \varepsilon_0 \mu_0}} \,(= 2\pi f^{(l)}) \tag{4.22}$$

$$L^{(l)} = \frac{1}{(\omega^{(l)})^2 C} \tag{4.23}$$

$$g^{(l)} = g_c^{(l)} + g_d^{(l)} + g_r^{(l)} \tag{4.24}$$

$$g_c^{(l)} = \frac{2R_s}{t\mu_0} \left(\frac{\omega^{(l)}}{\omega}\right)^2 C \tag{4.25}$$

$$g_d^{(l)} = \omega C \tan \delta \tag{4.26}$$

$$g_r^{(l)} = \frac{2S}{t^2} P_{r0}^{(l)} \tag{4.27}$$

with R_s being the real part of the surface impedance of the conductor walls and tan δ being the loss tangent of the dielectric substrate. Also, $P_{r0}^{(l)}$ is written as

$$P_{r0}^{(l)} = \frac{1}{2} \operatorname{Re} \left\{ \iint (E_0^{(l)*} \times H_0^{(l)}) \cdot \hat{R} \sin\theta \, d\theta \, d\phi \right\} \tag{4.28}$$

where

$$H_0^{(l)} = \frac{-j\omega\varepsilon_0}{4\pi} t \left\{ 2\cos\left(\frac{k_0 t}{2}\cos\theta\right) \right\} \oint_c (\hat{n} \times \hat{z})\varphi^{(l)}$$

$$\exp\{jk_0(x\cos\phi + y\sin\phi)\sin\theta\}\, dl \quad \left(0 \leqslant \theta \leqslant \frac{\pi}{2}\right) \tag{4.29}$$

$$E_0^{(l)} = 120\pi H_0^{(l)} \times \hat{R}, \quad (0 \leqslant \theta \leqslant \pi/2) \tag{4.30}$$

and $\hat{R}$ is unit vector of R-axis direction in polar co-ordinate, the asterisk means complex conjugation and Re{·} means real part in the brace. Eqn. 4.18 implies that an equivalent circuit for a multi-terminal microstrip antenna can be represented by the network model shown in Fig. 4.14, where the first resonant circuit, for $l = 1$, has $\omega^{(l)} = 0$ and thus corresponds to the mode resonating at zero frequency. This equivalent circuit is useful for network analysis of the microstrip antennas.

(c) Input impedance: When the microstrip antenna has multi-terminals, the coupling among the terminals in the cavity must be considered in order to derive the accurate input impedance. Such input impedance is called an 'active input

impedance'. When viewed from each feed point, it is generally defined as

$$Z_{in}^{(q)} = \frac{V_q^{(in)}}{I_q} - Z_{0q}, \quad q = 1, 2, 3, \ldots M \qquad (4.31)$$

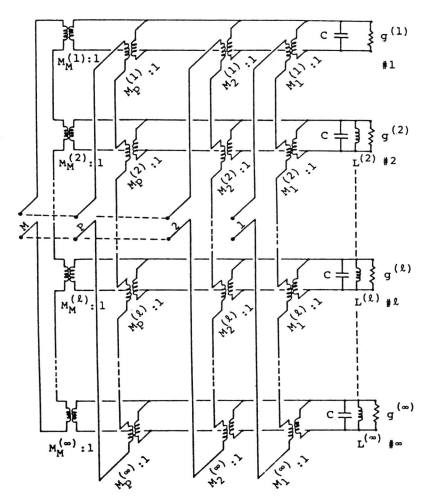

Fig. 4.14 *Equivalent circuit for multi-terminal microstrip antenna (From Reference 17)*

where $V_q^{(in)}$ is the input voltage to the q th terminal and Z_{0q} is the characteristic impedance of the q th terminal. I_q is the current flowing in the q th terminal and found as a solution of the following matrix equation:

$$[I] = [Z']^{-1}[V^{(in)}] \qquad (4.32)$$

where $[V^{(in)}]$ is the input voltage vector whose typical term is $V_q^{(in)}$ and $[Z']$ is the impedance matrix whose typical term is given by

$$Z'_{p,q} = \begin{cases} Z_{p,q} + Z_{0q}, & p = q \\ Z_{p,q}, & p \neq q \end{cases} \quad (4.33)$$

In practice, when using coaxial probes as feed lines, the correction is approximated by adding the following terms [21] to the self-impedance terms:

$$jX_L = j\sqrt{\frac{\mu_0}{\varepsilon_r \varepsilon_0}} \tan(k_0 t) \quad (4.34)$$

However, when using striplines, such a correction is not needed as the striplines are regarded as a part of the patch radiator, and the eigen values and eigen functions are derived for the patch boundary including the striplines.

(d) Radiation field, directive gain and radiation efficiency: The total radiation field can be calculated as a superposition of that for each mode. The radiation field for the l th mode can be represented by

$$E^{(l)} = \sum_{q=1}^{M} E_q^{(l)} = \sum_{q=1}^{M} \Omega_q^{(l)} E_0^{(l)} \quad (4.35)$$

So the total radiation field is obtained from

$$E_\theta = \sum_{l=1}^{N} \left\{ \sum_{q=1}^{M} \Omega_q^{(l)} \right\} E_0^{(l)} \cdot \hat{\theta} \quad (4.36)$$

$$E_\phi = \sum_{l=1}^{N} \left\{ \sum_{q=1}^{M} \Omega_q^{(l)} \right\} E_0^{(l)} \cdot \hat{\phi} \quad (4.37)$$

where $\Omega_q^{(l)}$ is an unknown coefficient, and can be determined for each mode from the boundary condition with respect to the voltage at the q th terminal as follows:

$$\Omega_q^{(l)} = \frac{\sqrt{2S}}{t} \frac{M_q^{(l)*}}{j\omega C + \frac{1}{j\omega L^{(l)}} + g^{(l)}} I_q \quad (4.38)$$

Next, the directive gain U at $\theta = 0$ and the radiation efficiency η are defined by

$$U = \frac{|E_\theta(\theta = 0)|^2 + |E_\phi(\theta = 0)|^2}{60 P_r} \quad (4.39)$$

$$\eta = \frac{P_r}{P_c + P_d + P_r} \quad (4.40)$$

where P_r, P_c and P_d are the total radiated power and the dissipated powers due

Circular polarisation and bandwidth

to the conductor and dielectric losses, and can be written as

$$P_c = \frac{R_s}{(120\pi)^2} \sum_{l=1}^{N} \left(\frac{k^{(l)}}{k_0}\right)^2 \left|\sum_{q=1}^{M} \Omega_q^{(l)}\right|^2 \quad (4.41)$$

$$P_d = \frac{t\omega}{2} \varepsilon_r \varepsilon_0 \tan\delta \sum_{l=1}^{N} \left|\sum_{q=1}^{M} \Omega_q^{(l)}\right|^2 \quad (4.42)$$

$$P_r = \sum_{l=1}^{N} P_{r0}^{(l)} \left|\sum_{q=1}^{M} \Omega_q^{(l)}\right|^2 \quad (4.43)$$

4.3.2 Conditions for circularly polarised radiation [18]

In general, ideal circularly polarised waves are obtained when the ratio of the two orthogonally polarised radiation-field components is equal to $\pm j$. Solving this relationship with respect to the frequency, two kinds of frequencies at which the circularly polarised waves are radiated can be derived through an iterative process. Also, all the corresponding optimum feed locations can be determined numerically.

(a) Radiation field from singly-fed circularly polarised microstrip antenna: If the contributions from the non-resonant modes are very weak and may be ignored, except those for the two strongest orthogonal modes necessary to radiate the circularly polarised waves, the total radiation field may be written as

$$\boldsymbol{E}(\theta) = \{\Omega^{(v)}(x_0, y_0)\boldsymbol{E}_0^{(v)}(\theta, \omega) + \Omega^{(v+1)}(x_0, y_0)\boldsymbol{E}_0^{(v+1)}(\theta, \omega)\} \quad (4.44)$$

where (x_0, y_0) is a feed point. In the above equation, the vth and $(v + 1)$th modes are chosen as the two wanted orthogonal modes. If a co-ordinate system can be fixed for convenience so as to align the X-axis with the direction of the vth field vector $\boldsymbol{E}_0^{(v)}(\theta, \omega)$ and the Y-axis with the direction of the $(v + 1)$th field vector $\boldsymbol{E}_0^{(v+1)}(\theta, \omega)$ on the boresight, then the far field given by eqn. 4.44 can be expressed on the boresight as

$$\boldsymbol{E}(0) = E_x \hat{\boldsymbol{x}} + E_y \hat{\boldsymbol{y}} \quad \text{at } \theta = 0, \quad (4.45)$$

where

$$E_x = \boldsymbol{E}(0) \cdot \hat{\boldsymbol{x}} = \Omega^{(v)}(x_0, y_0)\boldsymbol{E}_0^{(v)}(0, \omega) \cdot \hat{\boldsymbol{x}} \quad (4.46)$$

$$E_y = \boldsymbol{E}(0) \cdot \hat{\boldsymbol{y}} = \Omega^{(v+1)}(x_0, y_0)\boldsymbol{E}_0^{(v+1)}(0, \omega) \cdot \hat{\boldsymbol{y}}. \quad (4.47)$$

Eqn. 4.45 can also be modified as

$$\boldsymbol{E}(0) = E_L(\hat{\boldsymbol{x}} + j\hat{\boldsymbol{y}}) + E_R(\hat{\boldsymbol{x}} - j\hat{\boldsymbol{y}}) \quad (4.48)$$

where E_R and E_L denote the right-hand and left-hand circularly polarised

242 Circular polarisation and bandwidth

components, respectively, and they are written as

$$E_L = \frac{1}{2j}(jE_x + E_y) = \frac{1}{2}\{\Omega^{(v)}(x_0, y_0)\boldsymbol{E}_0^{(v)}(0, \omega) \cdot \hat{\boldsymbol{x}}$$
$$- j\Omega^{(v+1)}(x_0, y_0)\boldsymbol{E}_0^{(v+1)}(0, \omega) \cdot \hat{\boldsymbol{y}}\} \quad (4.49)$$

$$E_R = \frac{1}{2j}(jE_x - E_y) = \frac{1}{2}\{\Omega^{(v)}(x_0, y_0)\boldsymbol{E}_0^{(v)}(0, \omega) \cdot \hat{\boldsymbol{x}}$$
$$+ j\Omega^{(v+1)}(x_0, y_0)\boldsymbol{E}_0^{(v+1)}(0, \omega) \cdot \hat{\boldsymbol{y}}\} \quad (4.50)$$

Therefore, from $E_L = 0$ or $E_R = 0$, the following equation can be obtained:

$$\frac{\Omega^{(v+1)}(x_0, y_0)\boldsymbol{E}_0^{(v+1)}(0, \omega) \cdot \hat{\boldsymbol{y}}}{\Omega^{(v)}(x_0, y_0)\boldsymbol{E}_0^{(v)}(0, \omega) \cdot \hat{\boldsymbol{x}}} = \mp j \begin{cases} \text{for } E_L = 0 \\ \text{for } E_R = 0 \end{cases} \quad (4.51)$$

When eqn. 4.51 is satisfied, the resultant fields become the circular polarisations and are expressed as

$$\boldsymbol{E}(0) = \Omega^{(v)}(x_0, y_0)[\boldsymbol{E}_0^{(v)}(0, \omega) \cdot \hat{\boldsymbol{x}}](\hat{\boldsymbol{x}} \mp j\hat{\boldsymbol{y}}) \begin{cases} \text{RHCP} \\ \text{LHCP} \end{cases} \quad (4.52)$$

where RHCP and LHCP mean right-hand and left-hand circular polarisation, respectively.

(b) CP operating frequency and optimum feed location: A microstrip antenna may become singly-fed circularly polarised antenna when its dimensions are adjusted to suitable values as mentioned previously. In addition, when the operating frequency and feed point are chosen correctly, good circularly polarised waves can be radiated. The frequency at which the ideal circularly polarised waves are excited is called the CP operating frequency. This Section indicates how the CP operating frequency and the corresponding optimum feed location are derived.

Substituting eqns. 4.20 and 4.38 into eqn. 4.51, the following expression can be derived for the CP operating conditions:

$$\frac{\varphi^{(v+1)}(x_0, y_0)}{\varphi^{(v)}(x_0, y_0)} \frac{j\left(\omega C - \frac{1}{\omega L^{(v)}}\right) + g^{(v)}(\omega)}{j\left(\omega C - \frac{1}{\omega L^{(v+1)}}\right) + g^{(v+1)}(\omega)} =$$

$$\mp j \frac{\boldsymbol{E}_0^{(v)}(0, \omega) \cdot \hat{\boldsymbol{x}}}{\boldsymbol{E}_0^{(v+1)}(0, \omega) \cdot \hat{\boldsymbol{y}}} \begin{cases} \text{for RHCP} \\ \text{for LHCP} \end{cases} \quad (4.53)$$

Through comparison between the coefficients of the real parts and the imaginary parts on both sides of the above complex equation, the following simultaneous equations can be derived:

$$g^{(v)}(\omega) = \pm \frac{B(\omega)}{\beta(x_0, y_0)} \left\{ \omega C - \frac{1}{\omega L^{(v+1)}} \right\} \quad \begin{cases} \text{for RHCP} \\ \text{for LHCP} \end{cases} \quad (4.54)$$

$$\left\{ \omega C - \frac{1}{\omega L^{(v)}} \right\} = \mp \frac{B(\omega)}{\beta(x_0, y_0)} g^{(v+1)}(\omega) \quad \begin{cases} \text{for RHCP} \\ \text{for LHCP} \end{cases} \quad (4.55)$$

where

$$\beta(x_0, y_0) = \frac{\varphi^{(v+1)}(x_0, y_0)}{\varphi^{(v)}(x_0, y_0)} \quad (4.56)$$

$$B(\omega) = \frac{E_0^{(v)}(0, \omega) \cdot \hat{x}}{E_0^{(v+1)}(0, \omega) \cdot \hat{y}} \quad (4.57)$$

By eliminating $\pm B/\beta$ from eqns. 4.54 and 4.55, a biquadratic equation with respect to the CP operating angular frequency can be obtained as follows:

$$\omega^4 + \omega^2 [U^2(\omega) - \{(\omega^{(v+1)})^2 + (\omega^{(v)})^2\}] + \{\omega^{(v)} \cdot \omega^{(v+1)}\}^2 = 0 \quad (4.58)$$

where

$$U(\omega) = \frac{\sqrt{g^{(v)}(\omega) g^{(v+1)}(\omega)}}{C} \quad (4.59)$$

In this case the significant roots of eqn. 4.58 are given by

$$\omega = \sqrt{\frac{[\{(\omega^{(v)})^2 + (\omega^{(v+1)})^2\} - U^2(\omega)] \pm \sqrt{D(\omega)}}{2}} \quad (\geq 0) \quad (4.60)$$

where

$$D(\omega) = [\{\omega^{(v)} - \omega^{(v+1)}\}^2 - U^2(\omega)] \cdot [\{\omega^{(v)} + \omega^{(v+1)}\}^2 - U^2(\omega)] \geq 0 \quad (4.61)$$

Eqn. 4.60 shows that eqn. 4.58 has two significant roots, provided that the following CP operating condition, derived from an inequality of eqn. 4.61, is satisfied:

$$|\omega^{(v)} - \omega^{(v+1)}| \geq U(\omega) \quad (4.62)$$

Physically, this implies that the microstrip antennas can produce circular polarisation at two kinds of frequencies with a single feed. But, if the inequality of eqn. 4.62 is not satisfied, good circularly polarised waves cannot be produced from such antennas. However, eqn. 4.60 is not a closed-form expression, because the conductance components are, in general, a function of operating frequency. So it is difficult analytically to find out the CP operating frequencies. Accordingly, they are determined approximately through an iterative process;

namely, the $(\mu + 1)$th iterative solution is approximated by

$$\omega_{\mu+1} = \sqrt{\frac{[\{(\omega^{(v)})^2 + (\omega^{(v+1)})^2\} - U^2(\omega_\mu)] \pm \sqrt{D(\omega_\mu)}}{2}} \tag{4.63}$$

using the μth iterative solution ω_μ. Fortunately, satisfactory convergence for eqn. 4.63 is usually obtained by about three iterations. Correct choice of the feed point however, is, also very important for good circularly polarised radiation. All the feed-location loci, consisting of optimum feed points, are determined numerically by substituting the convergence results of eqn. 4.63 into eqn. 4.54. Next, let us consider the case when the equality in eqn. 4.62 is satisfied. Then instead of eqn. 4.60, only one CP operating frequency is given by

$$\omega = \sqrt{\omega^{(v)} \omega^{(v+1)}} \tag{4.64}$$

The theory developed here is quite adequate for obtaining initial design data on the CP operating frequencies and the corresponding feed-location loci. However, if one wants to realise a near-perfect singly-fed circularly polarised antenna, then experimental trimming may be necessary in the final design stages to revise the errors due to the material-tolerances effects of the substrate used. Examples of calculations based on this theory are presented in the following Section.

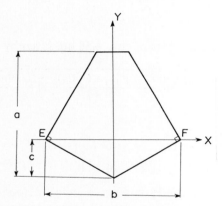

Fig. 4.15 *Plan view of pentagonal microstrip antenna*

4.3.3 Example [13, 18]

In this Section, an example is given of a singly-fed circularly polarised microstrip antenna designed on the basis of the theory developed in the previous Section. Several measured results are also presented for comparison with the calculated ones. The antenna used in the experiments was made of copper-clad 3·2 mm-thick Teflon glass fibre with a nominal dielectric constant $\varepsilon_r = 2·55$ and a loss tangent of approximately 0·0018; it was fed by a coaxial probe to avoid the degradation of ellipticity by unwanted radiation from the feed network.

Now, let us consider the patch radiator shown in Fig. 4.15. This Figure shows a plan view of the patch radiator whose angles $\angle E$ and $\angle F$ are right angles. The shape of such a pentagon can be prescribed completely by two parameters c/a and b/a; i.e. the pentagon becomes a rectangle when $c/a = 0$, an isosceles triangle when $c/a = 1$, and it becomes a special pentagon proposed by Weinschel [2] when $b/a = 1\cdot0603$ and $c/a = 0\cdot3061$. So it is interesting to investigate the variation of CP operating frequency with respect to the aspect ratio c/a in the case of $b/a = 1\cdot0603$. This can be derived iteratively from eqn. 4·63; the solid lines in Fig. 4.16 show the theoretical relations. In this Figure, the two chain-dotted lines denote the resonant frequencies for the two orthogonal modes contributing to the circular polarisation and the pair of dots indicate the experimental results for the pentagon proposed by Weinschel. These results show that antennas having dimensions satisfying the condition of eqn. 4.62 can

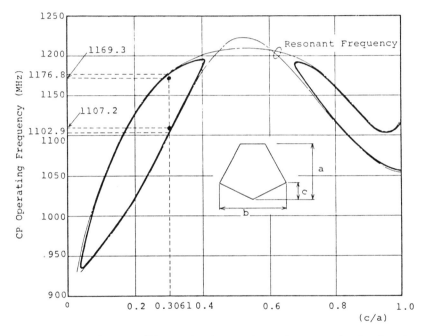

Fig. 4.16 *Relations between CP operating frequencies and aspect ratio c/a of pentagonal microstrip antenna with b/a = 1·0603, a = 100 mm, t = 3·2 mm, ε_r = 2·55, and tan δ = 0·0018 (From Reference 18)*
—— *calculated,* ● *measured*

always radiate circularly polarised waves at two frequencies. However, when eqn. 4.62 is not satisfied, i.e. c/a approaches 0 or 0·5 in this example, such antennas cannot radiate any pure circularly polarised waves. In addition to the above condition concerning patch dimensions and CP operating frequencies, in order practically to obtain good circular polarisation, the antenna must be fed

246 *Circular polarisation and bandwidth*

at a location so that the two orthogonal radiation fields, contributing to the circular polarisation, are excited with equal amplitude. Such a feed location can be determined numerically by substituting the corresponding CP operating frequency into eqn. 4.54. The numerical results for several aspect ratios are shown as feed-location loci in Figs. 4.17a–f [37]. These Figures show the various pairs of feed-location loci when the aspect ratio c/a is varied with $b/a = 1\cdot0603$ and $a = 100$ mm. In these Figures, Γ_1 and Γ_2 show the loci when the CP operating frequency is f_{c1} and Γ_3 and Γ_4 show the loci when it is f_{c2}, noting that Γ_1 and Γ_4, indicated by the solid line, correspond to the LHCP and Γ_2 and Γ_3, indicated by the broken line, correspond to the RHCP. It is found from these results that the triangular microstrip antenna can also radiate circularly polarised waves at two frequencies. The triangular microstrip antenna has the attraction that the area necessary for the patch radiator can be small; namely one half to three quarters that of the nearly square one.

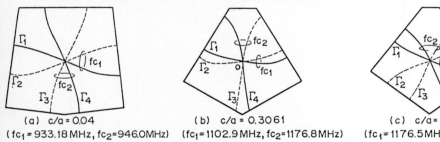
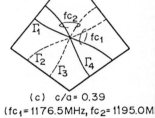

(a) $c/a = 0.04$
($fc_1 = 933.18$ MHz, $fc_2 = 946.0$ MHz)

(b) $c/a = 0.3061$
($fc_1 = 1102.9$ MHz, $fc_2 = 1176.8$ MHz)

(c) $c/a = 0.39$
($fc_1 = 1176.5$ MHz, $fc_2 = 1195.0$ M

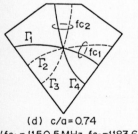
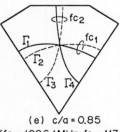
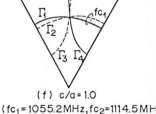

(d) $c/a = 0.74$
($fc_1 = 1150.5$ MHz, $fc_2 = 1183.6$ MHz)

(e) $c/a = 0.85$
($fc_1 = 1096.1$ MHz, $fc_2 = 1138.4$ MHz)

(f) $c/a = 1.0$
($fc_1 = 1055.2$ MHz, $fc_2 = 1114.5$ MH

Fig. 4.17 *Variation of feed-location loci with respect to aspect ratio c/a for pentagonal microstrip antenna with $b/a = 1\cdot0603$, $a = 100$ mm, $t = 3\cdot2$ mm, $\varepsilon_r = 2\cdot55$, and $\tan\delta = 0\cdot0018$ (From Reference 37)*
- - - - RHCP, ——— LHCP

Let us therefore consider the triangular microstrip antenna in detail. Fig. 4.18 shows a plan view of the isosceles-triangular patch radiator with the loci Γ_1 to Γ_4 of the theoretical feed location for each CP operating frequency, where $a = 76$ mm and $b/a = 0\cdot98$. In this Figure, Γ_1 and Γ_2 indicate the loci when the

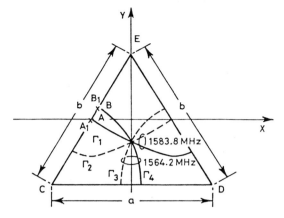

Fig. 4.18 *Plan view of isosceles triangular microstrip antenna and feed-location loci in case of b/a = 0·98, a = 76mm, t = 3·2mm, ε_r = 2·55, and tan δ = 0·0018 (From Reference 13)*
——— RHCP, - - - - LHCP

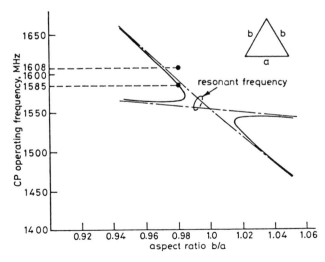

Fig. 4.19 *Relations between CP operating frequencies and aspect ratio for isosceles triangular microstrip antenna with a = 76mm, t = 3·2mm, ε_r = 2·55 and tan δ = 0·0018 (From Reference 13)*
——— calculated, ● measured

Circular polarisation and bandwidth

CP operating frequency is 1583·8 MHz and Γ_3 and Γ_4 indicate the loci when it is 1564·2 MHz. In general, the shape of the isosceles-triangular patch shown in Fig. 4.18 can be prescribed completely by introducing the aspect ratio b/a as a parameter. The solid lines in Fig. 4.19 show the variation of CP operating frequency when b/a is varied, with $a = 76$ mm; the pair of dots represent the measured results when the aspect ratio is 0·98. From these results it can be noted that the circularly polarised waves are always excited at two different frequencies when b/a is smaller than about 0·985 or greater than about 1·015 in this case. Next, let us consider the axial-ratio characteristics. Fig. 4.20 shows the boresight axial ratio with respect to frequency, when $b/a = 0·98$ and $a = 76$ mm. The

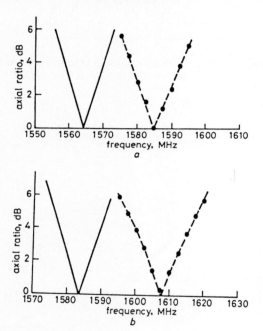

Fig. 4.20 *Axial-ratio characteristics for isosceles triangular microstrip antenna with $b/a = 0·98$, $a = 76$ mm, $t = 3·2$ mm, $\varepsilon_r = 2·55$, and $\tan \delta = 0·0018$ (From Reference 13)*

a Calculated results for point B-fed case in Fig. 4.18 and measured results for point B_1-fed case

b Calculated results for point A-fed case in Fig. 4.18 and measured results for point A_1-fed case

solid line in Fig. 4.20a represents the calculated results when selecting point B in Fig. 4.18 as a feed point, and the broken line the measured results with B_1 as a feed point. In Fig. 4.20b, the solid line represents the calculated results when selecting point A in Fig. 4.18 as a feed point and the broken line the measured results with A_1 as a feed point. From these Figures it can be appreciated that good circularly polarised waves are excited at two different frequencies and that their bandwidths for 3 dB axial ratio are about 0·5–0·6%. Although there are frequency differences of about 20–25 MHz and slight different of feed point between theory and experiment, both agrees well and also yield the excellent circular polarisation.

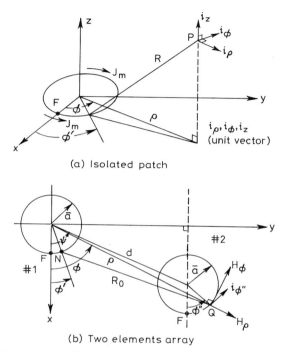

Fig. 4.21 *Co-ordinate systems for circular patch and its array*

4.4 Some considerations on mutual coupling

In the design of an array, it is important to estimate the mutual coupling between microstrip patch antennas [22–23]. In this Section, we present a simple method for calculating the mutual coupling of patch antennas. The technique based on the EMF method is simple and very effective for estimating the mutual coupling and the mutual admittance of antennas [22]. The geometry of the analytical model and the co-ordinate system employed here are illustrated in Fig. 4.21. Two patches are located on the same surface of a grounded dielectric substrate having thickness t and the dielectric constant ε_r. Using this co-ordinate system, and considering the field distribution of the dominant mode (TM_{110}) excited in a cavity region of the patch, the magnetic current J_m due to the dominant mode is given by

$$J_m(\phi') = 2J_m \cos(\phi')\delta(\bar{a})\delta(z)\boldsymbol{i}_{\phi'} \qquad (4.65)$$

250 Circular polarisation and bandwidth

where J_m corresponds to the z-component of the electric field at the periphery of the antenna, and J_m is the maximum amplitude of an equivalent magnetic current due to the dominant mode. Also, $\bar{a}$ is an effective radius that contains a fringing effect [6], $\delta(z)$ is the delta function and $i_{\phi'}$ is the ϕ'-directed unit vector at the point $N(\bar{a}, \phi', 0)$ in the spherical co-ordinates.

In order to simplify the following estimations, the equivalent magnetic current J_m is assumed to exist only in the xy-plane, as shown in Fig. 4.21. Then the vector potential A_m and the magnetic field H at $P(\varrho, \phi, z)$ generated by this magnetic current J_m are given mathematically by

$$A_m = \frac{\varepsilon_0 \bar{a}}{4\pi} \left\{ i_\phi \int_{-\pi}^{\pi} J_m(\phi') \cos(\phi' - \phi) \frac{e^{-jk_0 R}}{R} d\phi' \right.$$

$$\left. - i_\varrho \int_{-\pi}^{\pi} J_m(\phi') \sin(\phi' - \phi) \frac{e^{-jk_0 R}}{R} d\phi' \right\} \quad (4.66)$$

$$H = \{(\nabla\nabla \cdot A_m)/j\omega\varepsilon_0\mu_0\} - j\omega A_m$$

where $J_m(\phi') = 2J_m \cos(\phi')$, ω is the angular frequency, $k_0 (= \omega\sqrt{\varepsilon_0\mu_0})$ is the free-space propagation constant and ε_0 and μ_0 are the free-space permittivity and permeability, respectively. R is the distance between the magnetic current J_m and the observation point P, and is given by

$$R = \{\bar{a}^2 + \varrho^2 - 2\bar{a}\varrho\cos(\phi' - \phi) + z^2\}^{1/2}$$

Using eqn. 4.66 and the co-ordinate system shown in Fig. 4.21b, the magnetic field H^q at the point Q can be derived analytically by the following equation [22], while it corresponds to the magnetic field at the periphery of patch antenna 2:

$$H_\varrho^q = (K_m/\omega\mu_0) \int_{-\pi}^{\pi} \frac{\sin(\phi')}{R_0} \left\{ \varrho - \frac{\bar{a}}{\varrho} m_2(d, \phi', \phi'', \psi) \right\} \left(jk_0 + \frac{1}{R_0} \right)$$

$$\cdot \frac{e^{-jk_0 R_0}}{R_0} d\phi' - K_m \omega \varepsilon_0 \bar{a} \int_{-\pi}^{\pi} \cos(\phi') \frac{m_1(d, \phi', \phi'', \psi)}{\varrho}$$

$$\cdot \frac{e^{-jk_0 R_0}}{R_0} d\phi'$$

$$H_\phi^q = (-K_m\bar{a}/\omega\mu_0) \int_{-\pi}^{\pi} \frac{\sin(\phi') m_1(d, \phi', \phi'', \psi)}{R_0 \varrho} \left(jk_0 + \frac{1}{R_0} \right)$$

$$\cdot \frac{e^{-jk_0 R_0}}{R_0} d\phi' + K_m \omega \varepsilon_0 \bar{a} \int_{-\pi}^{\pi} \cos(\phi') \frac{m_2(d, \phi', \phi'', \psi)}{\varrho}$$

$$\cdot \frac{e^{-jk_0 R_0}}{R_0} d\phi' \quad (4.67)$$

Circular polarisation and bandwidth 251

where

$$m_1(d, \phi', \phi'', \psi) = d\sin(\phi' - \psi) + \bar{a}\sin(\phi' - \phi'')$$
$$m_2(d, \phi', \phi'', \psi) = d\cos(\phi' - \psi) + \bar{a}\cos(\phi' - \phi'')$$
$$R_0 = R|_{z=0}, \quad K_m = J_m/(j4\pi)$$
$$\varrho = \{d^2 + \bar{a}^2 + 2\bar{a}d\cos(\psi - \phi'')\}^{1/2}$$
$$\cos(\phi) = \{d\cos(\psi) + \bar{a}\cos(\phi'')\}/\varrho$$
$$\sin(\phi) = \{d\sin(\psi) + \bar{a}\sin(\phi'')\}/\varrho$$

and H_ϱ^q and H_ϕ^q are the components of the magnetic field $\boldsymbol{H}^q$ at the point Q.

By application of the EMF method to eqn. 4.67, the mutual admittance Y_{12} between the two patches can be easily obtained as follows:

$$\begin{aligned} Y_{12} &= \frac{-2\bar{a}}{J_{m1}(0)J_{m2}^*(0)} \int_{-\pi}^{\pi} (i_\varrho H_\varrho^q + i_\phi H_\phi^q) \cdot \{J_{m2}^*(\phi'')i_{\phi''}\}d\phi'' \\ &= \frac{-2\bar{a}}{J_{m1}(0)J_{m2}^*(0)} \int_{-\pi}^{\pi} \left\{ H_\varrho^q \frac{d\sin(\psi - \phi'')}{\varrho} \right. \\ &\quad \left. + H_\phi^q \frac{d\cos(\psi - \phi'') + \bar{a}}{\varrho} \right\} \{J_{m2}^*(0)\cos(\phi'')\}d\phi'' \end{aligned} \quad (4.68)$$

where $J_{m2}^*(\phi'')$ is the complex conjugate of $J_{m2}(\phi'')$ and $i_{\phi''}$ is the ϕ'' directed unit vector at the point Q.

Substituting eqn. 4.67 into eqn. 4.68, and carrying out numerical integration for eqn. 4.68, the values of mutual admittance $\tilde{Y}_{12} (= G_{12} + B_{12})$ can be determined numerically, as shown by the solid lines in Fig. 4.22. It is important to note that the results shown in Fig. 4.22 are expressed in term of the normalised admittance $\tilde{Y}_{12} (= Y_{12}/G_{11})$ between the two patch antennas. Here, G_{11} denotes the self conductance for an isolated patch antenna, and is shown as $G_{11} = G_{12}|_{d=0}$. In order to verify the estimates of $\tilde{Y}_{12}$, experimental work was carried out at S-band using typical samples. The theoretical values based on eqn. 4.68 agree well with the experimental ones within the desired range, as shown in the Figure.

In the design of an array, the mutual coupling $|S_{12}|$ for a patch antenna is also an important factor. The mutual coupling for an antenna is therefore described here, together with the experimental results. Using the normalised admittance $\tilde{Y}_{12}$ and the scattering S-matrix, we can express the mutual coupling by the following equation:

$$|S_{12}| = \left| \frac{-2\tilde{Y}_{12}}{(1 + \tilde{Y}_0)^2 - \tilde{Y}_{12}^2} \right| \quad (4.69)$$

where $\tilde{Y}_0$ and $\tilde{Y}_{12}$ correspond to the normalised self and mutual admittances, respectively.

252 *Circular polarisation and bandwidth*

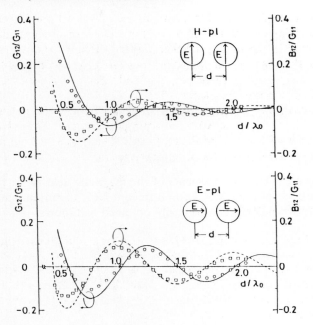

Fig. 4.22 *Normalised mutual admittance $\tilde{Y}_{12}$ between two microstrip antennas, where $t/\lambda_0 = 0.018$ (From Reference 22)*

——— G_{12}/G_{11} } theory, $\circ$ } experimental
- - - - B_{12}/G_{11} } $\square$

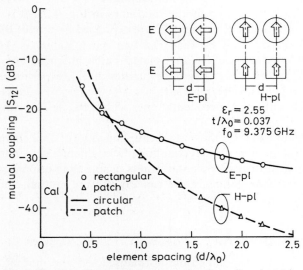

Fig. 4.23 *Comparison of $|S_{12}|$ between circular and rectangular microstrip antennas*

Circular polarisation and bandwidth 253

The mutual coupling $|S_{12}|$ for a circular patch was estimated here using eqn. 4.69. The calculated values of coupling for a circular patch are shown in Fig. 4.23. As shown in the figure, the theoretical coupling for a rectangular patch obtained by the same calculation coincides fairly well with that for a circular patch. The samples of both rectangular and circular patches are designed to have the same resonant frequency.

In this Section, we have presented a simple method for estimating the mutual coupling of patch antennas. After estimating the coupling, the calculated values were compared with the experimental ones. The calculated values agree well with the experimental results in spite of neglecting the effects of dielectric substrate.

4.5 Wideband techniques

As is well known, the bandwidth of microstrip antennas can basically be increased by increasing the thickness of the substrate and decreasing its dielectric constant [24, 27]. Also, it is well known that it is quite effective to mount a parasitic element on the original patch radiator [25]. On the other hand, in the case of an array, the bandwidth can also be increased collectively by arranging the antenna elements in a certain way [26].

In the following Sections, the wideband techniques for the first two methods are described fully, and finally the last method is also described.

4.5.1 Design of wideband elements

The bandwidth of a microstrip antenna depends on the patch shape, the resonant frequency, and the dielectric constant and thickness of the substrate. In this Section, the relations between these parameters are derived and a design method is described for the wideband microstrip antenna. The wider bandwidth is usually obtained by employing a thick substrate with low dielectric constant [24, 27]. However, such an antenna, in general, has two major problems which should be considered. One concerns the surface-wave radiation and the other concerns the unwanted mode generation. The former may impose a limiting factor on the maximum usable thickness of any substrate, because a practical microstrip antenna is usually designed so as not to radiate any surface waves. However, the latter problem cannot be ignored for microstrip antennas having a bandwidth greater than about 6%. Although a wideband microstrip antenna has these problems, it has the advantage that no balun (or balanced-to-unbalanced transformer) is required, so that the bandwidth can be accurately and analytically estimated. This advantage makes it possible to design the microstrip antenna taking into account not only the resonant frequency but also the bandwidth.

In this Section, the relations between the design parameters are briefly summarised first. Next, in order to make it easy to understand the design procedure, a design example is given for a circular microstrip antenna whose bandwidth for a VSWR less than 2·0 is 8·75% [28]. Finally, two methods of cancelling out the higher-order modes caused by lowering the antenna Q-factor is briefly described.

(a) Relations between parameters necessary for design [29]: The antenna bandwidth is, in general, represented as a function of the unloaded Q factor and the input VSWR [24]. Accordingly, if the requirements of the input VSWR and the bandwidth are specified, the desired value for the unloaded Q factor can be determined. In this Section, the relation between them is first derived and it is also shown that the product of the bandwidth and unloaded Q takes the maximum value for the special characteristic impedance of feeder used. Generally, the input admittance of the microstrip antenna may be approximated, using the relative bandwidth denoted by B_r and the unloaded Q denoted by Q_0, as follows:

$$Y_{in} \doteqdot g^{(l)'}\{1 + jQ_0 B_r\} \tag{4.70}$$

where

$$g^{(l)'} = g^{(l)}/M^{(l)}$$

and $g^{(l)}$ is the conductance component given by eqn. 4.24. $M^{(l)}$ is the turns ratio for the lth resonant circuit in Fig. 4.14, and is given by eqn. 4.19 or 4.20 for any feed point. If the transmission line with characteristic admittance of G_0 is connected to this antenna, the input VSWR is given by

$$\varrho = \frac{|G_0 + Y_{in}| + |G_0 - Y_{in}|}{|G_0 + Y_{in}| - |G_0 - Y_{in}|} \tag{4.71}$$

Substituting eqn. 4.70 in the above, the equation giving the relation between the bandwidth, unloaded Q, and input VSWR can be obtained by

$$Q_0 B_r = \sqrt{(\beta\varrho - 1)(1 - \beta/\varrho)} \tag{4.72}$$

where β is the coupling coefficient and is defined by

$$\beta = G_0/g^{(l)'} \tag{4.73}$$

Eqn. 4.72 implies that, if an input VSWR of $< \varrho$ is required, the product of the Q factor and bandwidth necessary for the antenna is related only to the coupling coefficient. As a special case, let us consider the case of $\beta = 1$, i.e. $G_0 = g^{(l)}$; then eqn. 4.72 results in [24]

$$Q_0 B_r = (\varrho - 1)/\sqrt{\varrho} \tag{4.74}$$

However, it should be noted that eqn. 4.74 does not usually give the maximum value for the product of the bandwidth and Q factor. It can be obtained from

$$Q_0 B_r = \sqrt{(\varrho^2 - 1)(1 - 1/\varrho^2)}/2 \tag{4.75}$$

Circular polarisation and bandwidth

when the following condition is satisfied for the coupling coefficient:

$$\beta = \beta_0 = (\varrho + 1/\varrho)/2 \qquad (4.76)$$

The results of eqns. 4.74 and 4.75 are illustrated by the broken and solid lines in Fig. 4.24, respectively. Using this Figure, the unloaded Q necessary for the design of the antenna can be determined graphically if the requirements for the bandwidth and input VSWR are specified. Also, the chain-dotted line in Fig. 4.24 indicates the result obtained from eqn. 4.76. This relation is useful in determining the position of the feed point and the characteristic impedance necessary for the feeder used.

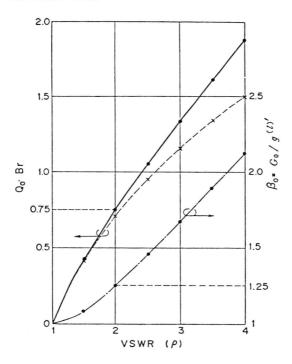

Fig. 4.24 *Relationship between Q_0, B_r, β_0 and input VSWR (From Reference 29)*

Next, the relation between the resonant frequency, the dimension of the patch radiator, and the thickness and dielectric constant of the substrate used is derived. Let S be the physical area of the patch radiator and t be the substrate thickness. Now, if $\sqrt{S}/t$ is introduced as a parameter, the product of the resonant frequency $f^{(1)}$ and $\sqrt{S}$ can generally be expressed as a function of $\sqrt{S}/t$ and the substrate dielectric constant ε_r. Fig. 4.25 shows the relations between $f^{(2)}\sqrt{S}$ and $\sqrt{S}/t$ with ε_r as a parameter for a circular microstrip antenna. The dots in the Figure show the measured results.

On the other hand, if the radiation conductance can be regarded as a dominant factor compared with the other conductance components, the unloaded Q denoted by Q_0 can also be approximated as a function of ε_r and $\sqrt{S}/t$. Fig. 4.26 shows the relation between Q_0 and $\sqrt{S}/t$ with ε_r as a parameter, for the circular microstrip antenna. In this Figure the dots show the measured results.

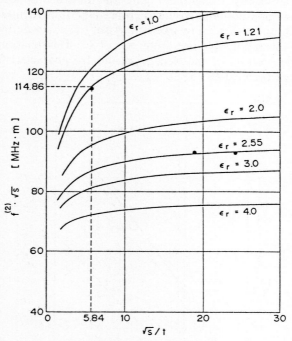

Fig. 4.25 *Relations between $f^{(2)}\sqrt{S}$ and $\sqrt{S}/t$ with ε_r as a parameter, for circular microstrip antenna (From Reference 29)*

Although only one example of a circular microstrip antenna is given here, the relationships shown in Figs. 4.25 and 4.26 are typical, and the above approach is applicable to any shape of antenna including rectangular and triangular microstrip.

(b) Design example [29]: In this Section, the specific design procedure is given for the example of a wideband circular microstrip antenna. Let us assume the following requirements for the frequency range and input VSWR:

Frequency range: 1530–1670 MHz

Input VSWR: less than 2·0

In this case, the centre or resonant frequency and the relative bandwidth are

$$f^{(2)} = 1600 \text{ MHz} \tag{4.77}$$

$$B_r = 0.0875 (= 8.75\%) \tag{4.78}$$

In eqn. 4.77, the second-order mode (TM_{110}) is chosen as a dominant mode. From Fig. 4.24, the product of the maximum bandwidth and Q_0, when $\varrho = 2 \cdot 0$, is

$$Q_0 B_r = 0.75 \qquad (4.79)$$

From the above equation and eqn. 4.78, it is found that Q_0 necessary for this antenna is

$$Q_0 \doteq 8.6 \qquad (4.80)$$

It follows from Fig. 4.26 that when a substrate of

$$\varepsilon_r = 1.21 \qquad (4.81)$$

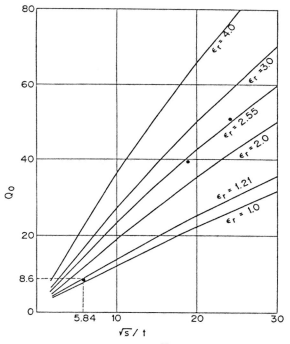

Fig. 4.26 *Relations between unloaded Q and $\sqrt{S}/t$ with ε_r as a parameter for circular microstrip antenna (From Reference 29)*

is used, the $\sqrt{S}/t$ value necessary to get the Q_0 value of eqn. 4.80 is

$$\sqrt{S}/t = 5.84 \qquad (4.82)$$

Accordingly, in the case of a circular microstrip antenna whose $\sqrt{S}/t$ value is equal to the above, it can be seen from Fig. 4.25 that

$$\sqrt{S} f^{(2)} = 114.86 \qquad (4.83)$$

258 Circular polarisation and bandwidth

Eqns. 4.77 and 4.83 show that

$$\sqrt{S} \doteqdot 71 \cdot 79 \, \text{mm} \tag{4.84}$$

and the radius of the patch radiator is 40·5 mm. From the above result and eqn. 4.82, the thickness of the substrate to be used is calculated as

$$t = 12 \cdot 3 \, \text{mm} \tag{4.85}$$

In summary, the circular microstrip antenna to meet the proposed requirements has the following specifications:

    Dielectric constant of substrate: 1·21

    Thickness of substrate: 12·3 mm

    Patch radius: 40·5 mm

Such a microstrip antenna can be made using paper honeycomb materials as a substrate. Fig. 4.27 shows a circular microstrip antenna manufactured according to the above specification. Also, Fig. 4.28 shows the return-loss characteristics for this antenna, where the calculated and measured results are indicated by the solid and broken lines, respectively. Both results show the wideband performance of about 8·75% for VSWR $\leq$ 2·0, which agrees well with the requirement in eqn. 4.78.

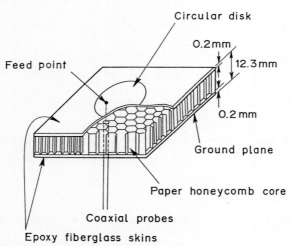

Fig. 4.27 *Structure of circular microstrip antenna consisting of epoxy Fiberglass skins and paper honeycomb core (From Reference 29)*

(c) Influences of unwanted modes and countermeasures against them [28]: The bandwidth of a microstrip antenna can be increased by employing a thick substrate with low dielectric constant, as shown in Fig. 4.28. However, it is expected that some unwanted modes will be generated as well as the wanted

mode, and their influence may become a serious problem, when the bandwidth is expanded without careful consideration.

In this Section, the influence of lowering the quality factor is described. For example, the antenna shown in Fig. 4.27 is a fairly wideband antenna whose relative bandwidth is 8·75%, and the influence of the unwanted modes can no longer be ignored. The mode closest to the wanted one is the TM_{210} mode in this case. The influence of the TM_{210} mode may be most prominent when the antenna is used as a circularly polarised one with dual feeds. In that case, the TM_{210} mode gives rise to some coupling between two terminals. The measured results for this coupling are shown, together with the calculated ones, in Fig. 4.29, where the solid line shows the measured results and the broken line shows the calculated ones. Also the chain-dotted line shows as a reference measured results for an antenna having a fairly high quality factor. In this Figure, the coupling is less

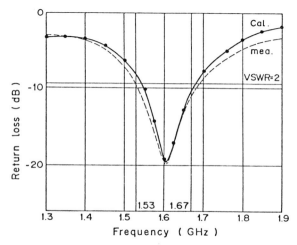

Fig. 4.28 *Return-loss characteristics for circular microstrip antenna shown in Fig. 4.27 (From Reference 29)*
$a = 40·5$ mm, $t = 12·3$ mm, and $\varepsilon_r = 1·21$

than -50 dB at the resonant frequency, which is small enough for the antenna with a high quality factor, and the coupling increases to about -28 dB for an antenna with a low quality factor. These results suggest that the axial ratio may be degraded owing to the influence of this coupling when the latter antenna is used as a dual-fed circularly polarised antenna, although it is fed by a perfect 90° hybrid. Fig. 4.30 illustrates the axial-ratio characteristics for such an antenna, where the solid line shows the calculated results and the dots show the measured ones. This Figure shows that the best axial ratio is only of the order of 1·3 dB.

In order to improve the axial ratio it is necessary to investigate the mechanism of the degradation. When a microstrip antenna has a low quality factor, the

260 Circular polarisation and bandwidth

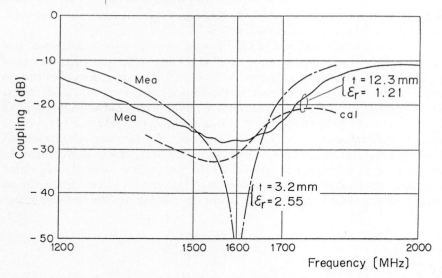

Fig. 4.29 *Coupling characteristics between orthogonal ports for circular microstrip antenna with dual feeds (From Reference 28)*

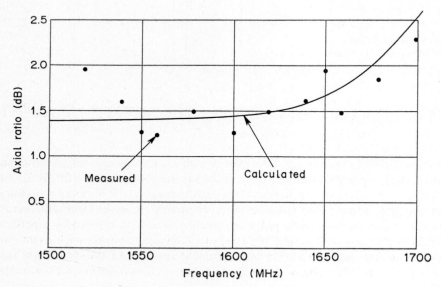

Fig. 4.30 *Axial-ratio characteristics for dual-fed circularly polarised circular microstrip antenna*
  $a = 40.5$ mm, $t = 12.3$ mm, and $\varepsilon_r = 1.21$

Circular polarisation and bandwidth 261

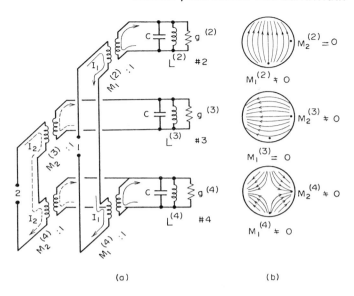

Fig. 4.31 *Equivalent circuit for dual-fed circular microstrip antenna having low-quality factor and the corresponding inner-surface current flows on the patch radiator*
a Equivalent circuit
b Current flows (● signs denote the feed point)

equivalent circuit can be approximated by Fig. 4.31a. This Figure implies that the coupling may arise between two orthogonal terminals through the 4th resonant circuit, in which the current due to the unwanted mode flows. Therefore, the current flowing in the 4th resonant circuit must somehow be suppressed. Fig. 4.32 shows one method of cancelling out such a current by adding more two terminals to the original two, and feeding from four terminals 90° out of phase with equal amplitude. Fig. 4.33 shows the axial-ratio characteristics with such a method of feeding, where the solid line represents the calculated results and the broken line the measured ones. As expected, the axial ratio is improved remarkably compared with Fig. 4.30 for the case of the antenna fed from only two terminals.

The above is particularly useful when the antenna is used only as a single element. When it is used as an element comprising an array antenna, it is also possible to cancel out the cross-polarised component radiated from each element at any observation point in free space. For convenience, let us consider a two-element array radiating elliptically polarised waves from each element, as shown in Fig. 4.34. If they are both RHCP and the excitation ratio is $1:\gamma e^{j\Delta}$, the total electric field radiated from them can be determined from Fig. 4.34 as

$$E = \{\alpha_1 + \gamma e^{j\Delta}(\beta_1 \cos \delta + j\beta_2 \sin \delta)\}\hat{x} \\ + j\{-\alpha_2 - j\gamma e^{j\Delta}(\beta_1 \sin \delta - j\beta_2 \cos \delta)\}\hat{y} \quad (4.86)$$

262 *Circular polarisation and bandwidth*

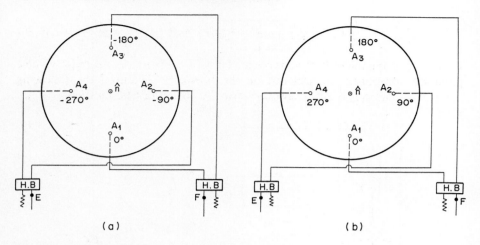

Fig. 4.32 *Actual feeding methods for circularly polarised circular microstrip antenna with four feed point (From Reference 28)*
  a Right-hand circular polarisation
  b Left-hand circular polarisation

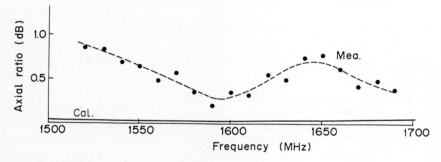

Fig. 4.33 *Axial-ratio characteristics improved by feeding from four terminals as shown in Fig. 4.32a (From Reference 28)*
  $a = 40 \cdot 5$ mm, $t = 12 \cdot 3$ mm, and $\varepsilon_r = 1 \cdot 21$

on the boresight direction, where δ indicates the physical rotation angle of the polarisation ellipse of the no. 2 element against that of the no. 1 element on an XY plane as shown in Fig. 4.34. Dividing the above field into the two components of co-polarisation and cross-polarisation, it can be expressed as

$$E = E_{co}(\hat{x} - j\hat{y}) + E_{cross}(\hat{x} + j\hat{y}) \tag{4.87}$$

where

$$E_{co} = E_R = \frac{1}{2}[(\alpha_1 + \alpha_2) + \gamma e^{j\Delta} e^{j\delta}(\beta_1 + \beta_2)] \tag{4.88}$$

$$E_{cross} = \frac{1}{2}[(\alpha_1 - \alpha_2) + \gamma e^{j\Delta} e^{-j\delta}(\beta_1 - \beta_2)] \qquad (4.89)$$

with E_R denoting the RHCP component. Therefore, in order to make the reverse polarised components cancel out on the boresight direction, the following complex excitation condition between two elements must be satisfied:

$$\gamma e^{j\Delta} = -\frac{\alpha_1 - \alpha_2}{\beta_1 - \beta_2} e^{j\delta} \qquad (4.90)$$

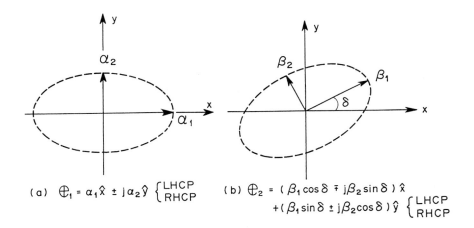

Fig. 4.34 Two kinds of elliptically polarised waves radiated from two-element array
  a Polarisation ellipse of no. 1 element
  b Polarisation ellipse of no. 2 element
  LHCP: left-hand circular polarisation
  RHCP: right-hand circular polarisation

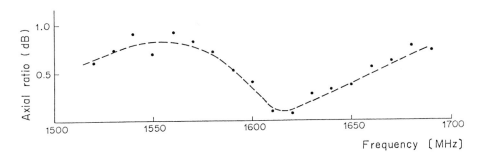

Fig. 4.35 Axial-ratio characteristics improved by employing paired elements (From Reference 28)
  a = 40·5 mm, t = 12·3 mm, and ε_r = 1·21

264 Circular polarisation and bandwidth

Since an ordinary array antenna consists of the same elements, the above relationship can be reduced to

$$\gamma e^{j\Delta} = -e^{j\delta} = e^{j(\delta+\pi)} \tag{4.91}$$

because it can be assumed that $\alpha_1 = \beta_1$ and $\alpha_2 = \beta_2$. The radiation field for the RHCP, being a co-polarised component in this case, can be expressed as

$$E_{co} = E_R = -j(\alpha_1 + \alpha_2)e^{j\delta}\sin\delta \tag{4.92}$$

The above equation implies that the co-polarised component takes the maximum value

$$|E_{co}| = |E_R| = |\alpha_1 + \alpha_2| \tag{4.93}$$

when the following condition is satisfied for the rotation angle δ:

$$\delta = 90° \tag{4.94}$$

On the other hand, the excitation condition for LHCP, instead of RHCP as above, can be similarly derived and is given by

$$\gamma e^{j\Delta} = -e^{-j\delta} = e^{j(\pi-\delta)} \tag{4.95}$$

when $\alpha_1 = \beta_1$ and $\alpha_2 = \beta_2$. The resultant LHCP component is represented as a function of δ, as in the case of RHCP. So the same maximum value as in RHCP case can be obtained, when the same condition (eqn. 4.94) is satisfied for δ. It is concluded from the above discussion that the axial ratio may be improved, in case of an array antenna, by arranging for the paired elements to have a rotation angle $\delta = 90°$. Fig. 4.35 shows the measured results for the axial-ratio characteristics of such paired elements. As expected, the resultant axial ratio is improved remarkably compared with the results of Fig. 4.30, and is of the same order as that obtained by the previous four terminal-fed cases.

4.5.2 Technique using parasitic element [34]

The bandwidth of a microstrip antenna can also be increased by employing a parasitic element [25, 30–32]. In this Section, the type shown in Fig. 4.36a is analysed using the Hankel-transformed domain-analysis method [33], and it is shown from theory and experiment to achieve an increase in the bandwidth. In this Figure, the two substrates are stacked so that they are parallel and the two circular etched disc conductors are concentric. The upper one is used as a parasitic element.

(a) Characteristic equation in the Hankel-transformed domain: In general, Green's function in the real domain is a very complicated convolutional integral or summation form, as shown in eqn. 4.14. However, it is known that Green's function becomes of simple algebraic form if it is expressed in the Hankel-transformed domain. In this domain, it can be deduced from Reference 33 that the electric fields on the boundary are related to the corresponding current distribu-

Circular polarisation and bandwidth 265

tions, using an impedance matrix known as a matrix Green's function, as follows:

$$\begin{bmatrix} [\tilde{E}(\alpha)]_{(+)} \\ [\tilde{E}(\alpha)]_{(-)} \end{bmatrix} = \begin{bmatrix} [Z(\alpha)]_{++} & [Z(\alpha)]_{+-} \\ [Z(\alpha)]_{-+} & [Z(\alpha)]_{--} \end{bmatrix} \begin{bmatrix} [\tilde{I}(\alpha)]_{(+)} \\ [\tilde{I}(\alpha)]_{(-)} \end{bmatrix} \quad (4.96)$$

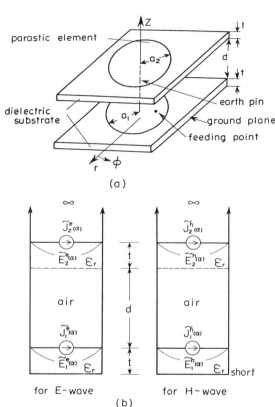

Fig. 4.36 *Circular microstrip antenna with parasitic element and its spectral-domain equivalent circuit (From Reference 34)*
  a Structure of analytical model and co-ordinate system
  b Equivalent circuit for E- and H-waves

where the sub-vectors $[\tilde{E}(\alpha)]_{(\pm)}$ and $[\tilde{I}(\alpha)]_{(\pm)}$ consist of two elements, respectively, and are

$$[\tilde{E}(\alpha)]_{(\pm)} = \begin{bmatrix} \tilde{E}_1^{(\pm)}(\alpha) \\ \tilde{E}_2^{(\pm)}(\alpha) \end{bmatrix} = \frac{1}{2} \begin{bmatrix} \tilde{E}_1^e(\alpha) \pm \tilde{E}_1^h(\alpha) \\ \tilde{E}_2^e(\alpha) \pm \tilde{E}_2^h(\alpha) \end{bmatrix} \quad (4.97a)$$

$$[\tilde{I}(\alpha)]_{(\pm)} = \begin{bmatrix} \tilde{I}_1^{(\pm)}(\alpha) \\ \tilde{I}_2^{(\pm)}(\alpha) \end{bmatrix} = \frac{1}{2} \begin{bmatrix} \tilde{J}_1^e(\alpha) \pm \tilde{J}_1^e(\alpha) \\ \tilde{J}_2^e(\alpha) \pm \tilde{J}_2^h(\alpha) \end{bmatrix} \quad (4.97b)$$

The tilde means the quantities in the Hankel-transformed domain, and each element is defined from its tangential components, which consists of both E-wave and H-wave, as

$$\tilde{F}^{(\pm)}(\alpha) = \int_0^\infty \{F_r(r) \pm jF_\phi(r)\} J_{n\pm 1}(\alpha r) r \, dr \tag{4.98}$$

with $F_r(r)$ and $F_\phi(r)$ being the tangential components and $J_{n\pm 1}(x)$ being the Bessel function of the first kind with $(n + 1)$ or $(n - 1)$th order. Also the subscripts 1 and 2, being the order of elements in each sub-matrix of eqn. 4.97, are referred to the lower and upper conductors. The sub-matrices in eqn. 4.96 can be represented by

$$[Z(\alpha)]_{++} = [Z(\alpha)]_{--} = \frac{1}{2} \{[Z(\alpha)]_e + [Z(\alpha)]_h\} \tag{4.99a}$$

$$[Z(\alpha)]_{+-} = [Z(\alpha)]_{-+} = \frac{1}{2} \{[Z(\alpha)]_e - [Z(\alpha)]_h\} \tag{4.99b}$$

where the matrices $[Z(\alpha)]_e$ and $[Z(\alpha)]_h$ denote the impedance matrices for E-wave and H-wave, respectively, and are written as

$$[Z(\alpha)]_e = \begin{bmatrix} Z_{11}^e(\alpha) & Z_{12}^e(\alpha) \\ Z_{21}^e(\alpha) & Z_{22}^e(\alpha) \end{bmatrix} \tag{4.100a}$$

$$[Z(\alpha)]_h = \begin{bmatrix} Z_{11}^h(\alpha) & Z_{12}^h(\alpha) \\ Z_{21}^h(\alpha) & Z_{22}^h(\alpha) \end{bmatrix} \tag{4.100b}$$

They can be easily determined from the corresponding admittance matrices, which consist of the following elements:

$$Y_{11}^l(\alpha) = jY' \cot(\beta' t) - \frac{Y_{12}^l(\alpha)}{Y'} [Y' \cos(\beta d) \cos(\beta' t)$$

$$- Y \sin(\beta d) \sin(\beta' t)] \tag{4.101a}$$

$$Y_{12}^l(\alpha) = Y_{21}^l(\alpha) = \frac{-jYY'}{Y \cos(\beta d) \sin(\beta' t) + Y' \cos(\beta' t) \sin(\beta d)} \tag{4.101b}$$

$$Y_{22}^l(\alpha) = -Y - \frac{Y_{12}^l(\alpha)}{Y} [Y \cos(\beta d) \cos(\beta' t)$$

$$- Y' \sin(\beta d) \sin(\beta' t)] \tag{4.101c}$$

with $l = e$ or h and

$$\beta = \sqrt{\omega^2 \mu_0 \varepsilon_0 - \alpha^2} \tag{4.102a}$$

$$\beta' = \sqrt{\omega^2 \mu_0 \varepsilon_0 \varepsilon_r - \alpha^2} \tag{4.102b}$$

Circular polarisation and bandwidth

$$Y = \begin{cases} \dfrac{\omega\varepsilon_0}{\beta} & \text{for } l = e \text{ (}E\text{-wave)} \\ \dfrac{\beta}{\omega\mu_0} & \text{for } l = h \text{ (}H\text{-wave)} \end{cases} \quad (4.102c)$$

$$Y' = \begin{cases} \dfrac{\omega\varepsilon_0\varepsilon_r}{\beta'} & \text{for } l = e \text{ (}E\text{-wave)} \\ \dfrac{\beta'}{\omega\mu_0} & \text{for } l = h \text{ (}H\text{-wave)} \end{cases} \quad (4.102d)$$

These elements can be derived from the spectral-domain equivalent circuit shown in Fig. 4.36b. Next, the unknown current distributions are expanded on each circular conductor in the real domain as follows:

$$I_{ri}(r) = \sum_v A_{iv} f_{riv}(r) \qquad (4.103a)$$

$$I_{\phi i}(r) = \sum_v B_{iv} f_{\phi iv}(r) \qquad (4.103b)$$

where A_{iv} and B_{iv} are expansion coefficients with $i = 1$ and 2 being referred to the lower and upper conductors, while $f_{riv}(r)$ and $f_{\phi iv}(r)$ are basis functions. In this case, the Hankel-transformed current distribution $\tilde{I}_i^{(\pm)}(\alpha)$ in eqn. 4.97b is expressed as

$$\begin{aligned}\tilde{I}_i^{(\pm)}(\alpha) &= \int_0^\infty \{I_{ri}(r) \pm jI_{\phi i}(r)\} J_{n\pm 1}(\alpha r) r\, dr \\ &= [A_i][\hat{f}_{ri}(\alpha)]_{(\pm)} \pm j[B_i][\hat{f}_{\phi i}(\alpha)]_{(\pm)} \end{aligned} \quad (4.104)$$

where $[A_i]$ and $[B_i]$ are unknown vectors with A_{iv} and B_{iv}, and $[\hat{f}_{ri}(\alpha)]_{(\pm)}$ and $[\hat{f}_{\phi i}(\alpha)]_{(\pm)}$ are the transformed basis function vectors with $\hat{f}_{riv}^{(\pm)}(\alpha)$ and $\hat{f}_{\phi iv}^{(\pm)}(\alpha)$. Substituting eqn. 4.104 into eqn. 4.97b and then substituting the resulting equation and eqn. 4.97a into eqn. 4.96, the following matrix equation can be obtained:

$$\begin{bmatrix} \tilde{E}_1^{(+)}(\alpha) \\ \tilde{E}_2^{(+)}(\alpha) \\ \tilde{E}_1^{(-)}(\alpha) \\ \tilde{E}_2^{(-)}(\alpha) \end{bmatrix} = \begin{bmatrix} [Z(\alpha)]_{++} & [Z(\alpha)]_{+-} \\ \hline [Z(\alpha)]_{-+} & [Z(\alpha)]_{--} \end{bmatrix} \begin{bmatrix} [A_1][\hat{f}_{r1}(\alpha)]_{(+)} + j[B_1][\hat{f}_{\phi 1}(\alpha)]_{(+)} \\ [A_2][\hat{f}_{r2}(\alpha)]_{(+)} + j[B_2][\hat{f}_{\phi 2}(\alpha)]_{(+)} \\ [A_1][\hat{f}_{r1}(\alpha)]_{(-)} - j[B_1][\hat{f}_{\phi 1}(\alpha)]_{(-)} \\ [A_2][\hat{f}_{r2}(\alpha)]_{(-)} - j[B_2][\hat{f}_{\phi 2}(\alpha)]_{(-)} \end{bmatrix}$$

(4.105)

Taking inner products of eqn. 4.105 with all the transformed basis functions $\hat{f}_{ri\mu}^{(\pm)}(\alpha)$ and $\hat{f}_{\phi i\mu}^{(\pm)}(\alpha)$ according to Galerkin's method, the left-hand sides of all the resulting equations vanish owing to the boundary conditions. Thus it is possible to choose the following combinations in order to avoid divergence of integrals appearing in the characteristic equation:

$$\langle \hat{f}_{r1\mu}^{(+)}(\alpha)\tilde{E}_1^{(+)}(\alpha)\rangle + \langle \hat{f}_{r1\mu}^{(-)}(\alpha)\tilde{E}_1^{(-)}(\alpha)\rangle = 0 \quad (4.106a)$$

$$\langle \hat{f}_{\phi1\mu}^{(+)}(\alpha)\tilde{E}_1^{(+)}(\alpha)\rangle - \langle \hat{f}_{\phi1\mu}^{(-)}(\alpha)\tilde{E}_1^{(-)}(\alpha)\rangle = 0 \quad (4.106b)$$

$$\langle \hat{f}_{r2\mu}^{(+)}(\alpha)\tilde{E}_2^{(+)}(\alpha)\rangle + \langle \hat{f}_{r2\mu}^{(-)}(\alpha)\tilde{E}_2^{(-)}(\alpha)\rangle = 0 \quad (4.106c)$$

$$\langle \hat{f}_{\phi2\mu}^{(+)}(\alpha)\tilde{E}_2^{(+)}(\alpha)\rangle - \langle \hat{f}_{\phi2\mu}^{(-)}(\alpha)\tilde{E}_2^{(-)}(\alpha)\rangle = 0 \quad (4.106d)$$

where a bracket $\langle \cdot \rangle$ in the above equations is employed for the following infinite integral:

$$\langle \tilde{F}(\alpha)\rangle \equiv \int_0^\infty \tilde{F}(\alpha)\, \alpha\, d\alpha \quad (4.107)$$

Finally, eqn. 4.106 can be rewritten in a matrix form and is given by

$$[P]\begin{bmatrix}[A_1]\\ [B_1]\\ [A_2]\\ [B_2]\end{bmatrix} = 0 \quad (4.108)$$

where the matrix $[P]$ is defined as follows:

$$[P] = \begin{bmatrix} [P_{rr}^{11}] & [P_{r\phi}^{11}] & [P_{rr}^{12}] & [P_{r\phi}^{12}] \\ [P_{\phi r}^{11}] & [P_{\phi\phi}^{11}] & [P_{\phi r}^{12}] & [P_{\phi\phi}^{12}] \\ [P_{rr}^{21}] & [P_{r\phi}^{21}] & [P_{rr}^{22}] & [P_{r\phi}^{22}] \\ [P_{\phi r}^{21}] & [P_{\phi\phi}^{21}] & [P_{\phi r}^{22}] & [P_{\phi\phi}^{22}] \end{bmatrix} \quad (4.109)$$

and the elements of the sub-matrices $[P_{rr}^{ij}]$ through $[P_{\phi\phi}^{ij}]$ with $i = 1$ or 2 and $j = 1$ or 2 are given by

$$P_{rr\mu\nu}^{ij} = \langle Z_{ij}^e(\alpha)\hat{f}_{ri\mu}^e(\alpha)\hat{f}_{rj\nu}^e(\alpha) + Z_{ij}^h(\alpha)\hat{f}_{ri\mu}^h(\alpha)\hat{f}_{rj\nu}^h(\alpha)\rangle \quad (4.110a)$$

$$P_{r\phi\mu\nu}^{ij} = \langle Z_{ij}^e(\alpha)\hat{f}_{ri\mu}^e(\alpha)\hat{f}_{\phi j\nu}^h(\alpha) + Z_{ij}^h(\alpha)\hat{f}_{ri\mu}^h(\alpha)\hat{f}_{\phi j\nu}^e(\alpha)\rangle \quad (4.110b)$$

$$P_{\phi r\mu\nu}^{ij} = \langle Z_{ij}^e(\alpha)\hat{f}_{\phi i\mu}^h(\alpha)\hat{f}_{rj\nu}^e(\alpha) + Z_{ij}^h(\alpha)\hat{f}_{\phi i\mu}^e(\alpha)\hat{f}_{rj\nu}^h(\alpha)\rangle \quad (4.110c)$$

$$P_{\phi\phi\mu\nu}^{ij} = \langle Z_{ij}^e(\alpha)\hat{f}_{\phi i\mu}^h(\alpha)\hat{f}_{\phi j\nu}^h(\alpha) + Z_{ij}^h(\alpha)\hat{f}_{\phi i\mu}^e(\alpha)\hat{f}_{\phi j\nu}^e(\alpha)\rangle \quad (4.110d)$$

with

$$[\hat{f}_{ri}(\alpha)]_{\substack{e\\h}} = [\hat{f}_{ri}(\alpha)]_{(+)} \pm [\hat{f}_{ri}(\alpha)]_{(-)} \quad (4.111a)$$

$$[\hat{f}_{\phi i}(\alpha)]_{\substack{e\\h}} = [\hat{f}_{\phi i}(\alpha)]_{(+)} \pm [\hat{f}_{\phi i}(\alpha)]_{(-)} \quad (4.111b)$$

From eqn. 4.108, the determinant of $[P]$ must vanish for a non-trivial solution to exist. Therefore, the characteristic equation for this problem can be written as

$$\Omega(\omega) \equiv \det[P] = 0 \qquad (4.112)$$

where ω is the complex resonant angular frequency whose real and imaginary parts correspond to the resonant angular frequencies and the damping factors, respectively.

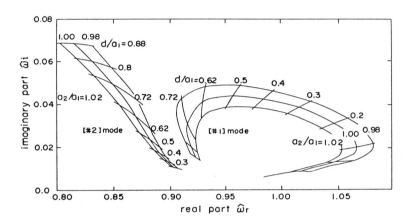

Fig. 4.37 *Contour of complex resonant frequencies of circular microstrip antenna with parasitic element (From Reference 34)*
$t = 1.6$ mm and $\varepsilon_r = 2.55$, $\hat{\omega} = \omega/\omega_0 = \hat{\omega}_r + j\hat{\omega}_i$

(b) Electrical characteristics: In this case, two basis functions for each current component provide satisfactory accuracy, so that the characteristic equation results in a form of determinant of size 8×8. Fig. 4.37 shows numerical results for the complex resonant frequencies solved from eqn. 4.112, where the thickness and dielectric constant for the substrate used are $t = 1.6$ mm and $\varepsilon_r = 2.55$, respectively. The Figure shows the contour map for the real and imaginary parts of normalised complex resonant frequencies with a_2/a_1 and d/a_1 as parameters. From this Figure, it is found that two dominant resonant modes, which exhibit a double-tuned characteristic, exist in this antenna. So the input VSWR characteristics can be calculated by considering a double-tuned performance. In this case, two resonant resistances R_1 and R_2 can be determined uniquely as follows:

$$R_1 = X_2 \frac{1 + X_1^2}{X_2 - X_1} \qquad (4.113a)$$

$$R_2 = X_1 \frac{1 + X_2^2}{X_1 - X_2} \qquad (4.113b)$$

because

$$X_1 = Q_1 \left\{ \frac{\omega}{\omega_{01}} - \frac{\omega_{01}}{\omega} \right\} \qquad (4.114a)$$

$$X_2 = Q_2 \left\{ \frac{\omega}{\omega_{02}} - \frac{\omega_{02}}{\omega} \right\} \qquad (4.114b)$$

$$\omega_0 = \frac{1\cdot 841 c}{a_1 \sqrt{\varepsilon_r}} \text{ in Fig. 4.37}$$

where ω_{01} and ω_{02} are resonant angular frequencies for two dominant modes and c is the velocity of light in vacuum. Fig. 4.38 shows the calculated and measured input VSWR characteristics for the antenna with $d = 10$ mm, $a_1 = 20\cdot 8$ mm, $a_2 = 21\cdot 0$ mm, $t = 1\cdot 6$ mm, and $\varepsilon_r = 2\cdot 55$. In this Figure, the results for the antenna without a parasitic element are also shown as a reference, and it is seen that the effect of the parasitic element is considerable.

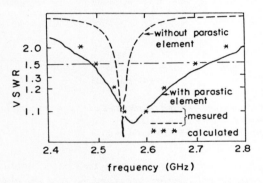

Fig. 4.38 *Calculated and measured VSWR characteristics for circular microstrip antenna with parasitic element (From Reference 34)*
$d = 10$ mm, $a_1 = 20\cdot 8$ mm, $a_2 = 21.0$ mm, $t = 1\cdot 6$ mm, and $\varepsilon_r = 2\cdot 55$

4.5.3 Technique using paired element

Circularly polarised microstrip antennas including single-fed patches are widely used as effective radiators in many communication systems [6, 10].

However, the most serious problem with such antennas is the narrowness of the ellipticity and impedance bandwidth compared with ordinary microwave antennas. Several techniques for the expansion of bandwidth have been reported in the literature [29–34]. However, most of these broadband techniques, including double-stacked CP patches, are applicable to isolated CP patch antennas. Hence, other wideband techniques using sequential arrangements of antenna elements have been developed in recent years [26, 35]. The simplest device for such a sequential array is a paired element, and it is used as an effective radiator

Circular polarisation and bandwidth 271

in some special applications [36]. In this Section, we briefly describe design procedure of such a paired element.

Fig. 4.39a shows the fundamental arrangement of a microstrip paired-element unit. The patch elements are rotated orthogonally on the coplanar plane and are fed in uniform amplitude but 90° out of phase through the sequentially rotated feeding points F_1 and F_2.

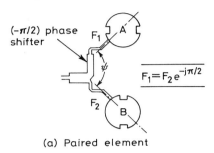

(a) Paired element

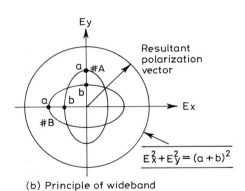

(b) Principle of wideband

Fig. 4.39 *Paired element and its polarisation pattern.*

A sub-array composed of such paired elements demonstrates the broadband nature in spite of using narrow-band patch elements [26]. In order to evaluate the performance, the polarisation pattern of an antenna is briefly described.

In general, individual elements of a pair show elliptical patterns of polarisation, as shown in Fig. 4.39b. The polarisation ellipses marked A and B correspond to those of each element of a pair, while the E_x- and E_y-axes correspond to the horizontal and vertical components of the radiated electric field.

The polarisation ellipses for an individual CP patch vary rapidly with change of frequency. However, if the CP patch of a pair is arranged orthogonally and is fed uniformly in amplitude but 90° out of phase, the resultant polarisation pattern due to the pair can be shown as a trace of perfectly circular polarisation over a wide frequency range, as shown in the figure.

In order to verify the performance, a 2 × 2 element sub-array unit having two pairs was constructed and tested at X-band. With regards the feeding system, the input impedance for each element of the pairs was matched to that of the main feeder M_1 by means of $\lambda_g/4$ impedance transformers, T_a, T_b, T_c and T_m, where λ_g is the wavelength in stripline. Fig. 4.40b shows the typical measured ellipticity bandwidth for the sub-array unit. The ellipticity bandwidth for an isolated CP-patch antenna is also shown for comparison in the Figure. It is shown that the sub-array unit using the paired element contributes enormously to the improvement of the ellipticity bandwidth compared with a single CP-patch element. The 3 dB ellipticity bandwidth obtained by this sub-array unit is about five times the value obtained by an ordinary CP-patch element, as shown in the figure.

A more detailed description of the sequential array is given in Chapter 13.

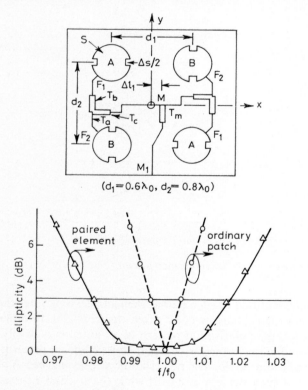

Fig. 4.40 *Sub-array unit and its ellipticity bandwidth*

4.6 References

1   MAILLOUX, R. J., McILVENNA, J. F., and KERNWEIS, N. P.: 'Microstrip array technology', *IEEE Trans.*, 1981, **AP-29**, pp. 25–37

2 WEINSCHEL, H. D.: 'A cylindrical array of circularly polarized microstrip antennas', *Int. Symp. Dig. Antennas Propagat. Soc.*, June 1975, pp. 177–180
3 SHEN, L. C.: 'The elliptical microstrip antenna with circular polarization', *IEEE Trans.*, 1981, **AP-29**, pp. 90–94
4 RICHARDS, W. F., LO, Y. T., and HARRISON, D. D.: 'An improved theory for microstrip antennas and applications', *IEEE Trans.*, 1981, **AP-29**, pp. 38–46
5 KERR, J. L.: 'Microstrip polarization techniques'. Proc. Antenna Applications Symp., Allerton Park, IL, Sept. 1978
6 JAMES, J. R., HALL, P. S., and WOOD, C.: 'Microstrip antenna'. (Peter Peregrinus, 1981) chap. 7
7 HANEISHI, M., NAMBARA, T., and YOSHIDA, S.: 'Study on elliptical properties of singly-fed circularly polarised microstrip antennas', *Electron Lett.*, 1982, **18**, pp. 191–193
8 ITOH, K.: 'Circularly polarised printed array composed of strip dipole and slot', *Microwave J.*, April 1987, pp. 143–153
9 JAMES, J. R., HALL, P. S., WOOD, C., and HENDERSON, A.: 'Some recent developments in microstrip antenna design', *IEEE Trans.*, 1981, **AP-29**, pp. 124–128
10 CARVER, K. R., and MINK, J. R.: 'Microstrip antenna technology', *IEEE Trans., Antennas & Propagt.*, 1981, **AP-29**, pp. 2–24
11 HANEISHI, M., and YOSHIDA, S.: 'A design method of circularly polarised rectangular microstrip antenna by one-point feed', *Electron & Commun in Japan*, 1981, **54**, pp. 46–54
12 OKOSHI, T., and MIYOSHI, T.: 'Planar circuits', (Ohm Publishing (in Japanese), 1973)
13 SUZUKI, Y., MIYANO, N., and CHIBA, Y.: 'Circularly polarised radiation from singly-fed equilateral-triangular microstrip antenna', *IEE Proc.* 1987, **134**, pp. 194–198
14 SCHAUBERT, D. H., FARRAR, F. G., SINDORIS, A., and HAYES, S. T.: 'Microstrip antennas with frequency agility and polarization diversity', *IEEE Trans.*, 1981, **AP-29**, pp. 118–123
15 OKOSHI, T., and MIYOSHI, T.: 'The planar circuits – An approach to microwave integrated circuitry', *IEEE Trans.*, 1972, **MTT-20**, pp. 245–252
16 CARVER, K. R.: 'A modal expansion theory for the microstrip antenna', *Int. Symp. Dig. Antennas Propagat. Soc.*, June 1979, pp. 101–104
17 SUZUKI, Y., and CHIBA, T.: 'Computer analysis method for arbitrarily shaped microstrip antenna with multi-terminals', *IEEE Trans.*, 1984, **AP-32**, pp. 585–590
18 SUZUKI, Y., and CHIBA, T.: 'Improved theory for a singly-fed circularly polarized microstrip antenna', *Trans. IECE Japan*, 1985, **E68**, pp. 76–82
19 For example, NEWMAN, E. H., and TULYATHAN, P.: 'Analysis of microstrip antennas using moment methods', *IEEE Trans.*, 1981, **AP-29**, pp. 47–53
20 MORSE, P. M., and FESHBACH, H.: 'Methods of theoretical physics: Pt. II. (McGraw-Hill, NY, 1953), pp. 1112–1119
21 CARVER, K. R.: 'Input impedance to probe fed microstrip antennas', *Int. Symp. Dig. Antennas Propagat. Soc.*, June 1980, pp. 617–620
22 HANEISHI, M., YOSHIDA, S., and TABETA, M.: 'A design of back-feed type circularly polarised microstrip antenna having symmetrical perturbation segment', *Electron & Commun. in Japan*, 1981, **2**, pp. 52–60
23 POZAR, D. M.: 'Input impedance and mutual coupling of rectangular microstrip antennas', *IEEE Trans.*, 1982, **AP-30**, pp. 1191–1197
24 DERNERYD, A. G., and LIND, A. G.: 'Extended analysis of rectangular microstrip resonator antennas', *IEEE Trans.*, 1979, **AP-27**, pp. 846–849
25 ITAMI, H., and HORI, T.: 'Broad band circular polarized microstrip antenna'. *Int. Conv. Rec. IECE* (in Japanese), 1982, p. 642
26 HANEISHI, M., YOSHIDA, S., and GOTO, N.: 'A broadland microstrip array composed of single-feed type circularly polarized microstrip antennas', in *Int. Symp. Dig. Antennas Propagat. Soc.*, May 1982, pp. 160–163
27 MURPHY, L.: 'SEASAT and SIR-A microstrip antennas', Proc. Workshop on Printed Circuit Antenna Technology, Oct. 1979, paper 18

28 CHIBA, T., SUZUKI, Y., MIYANO, N., MIURA, S., and OHMORI, S.: 'A phased array antenna using microstrip patch antennas', 12th European Microwave Conference, Sept. 1982, pp. 472–477
29 SUZUKI, Y., and CHIBA, T.: 'Designing method of microstrip antenna considering the bandwidth', *Trans. IECE Japan*, 1984, **E67**, pp. 488–493
30 WOOD, C.: 'Improved bandwidth of microstrip antennas using parasitic elements', *IEE Proc.*, 1980, **127**, pp. 231–234
31 LONG, S. A., and WALTON, M. D.: 'A dual-frequency stacked circular disc antenna', *Int. Symp. Dig. Antennas Propagat. Soc.*, June 1978, pp. 260–263
32 SANFORD, G. G.: 'Multiple resonance radio frequency microstrip antenna structure', US Patent 4070676, Jan. 1978
33 ARAKI, K., and ITOH, T.: 'Hankel transform domain analysis of open circular microstrip radiating structures', *IEEE Trans.*, 1981 **AP-29**, pp. 84–89
34 ARAKI, K., UEDA, H., and TAKAHASHI, M.: 'Numerical analysis of circular disk microstrip antennas with parasitic elements', *IEEE Trans.*, 1986, **AP-34**, pp. 1390–1394
35 TESHIROGI, T., TANAKA, M., and CHUJO, W.: 'Wideband circularly polarised array antennas with sequential rotations and phase shift of elements'. *Proc. Int. Symp. on Antennas & Propagt., Japan*, Vol. 1, Aug. 1985, pp. 117–120
36 HANEISHI, M., HAKURA, Y., SAITO, S., and HASEGAWA, T.: 'A low-profile antenna for DBS reception'. *Int. Symp. Dig. Antennas Propagat. Soc.*, June 1987, pp. 914–917
37 SUZUKI, Y.: 'Analysis of microstrip antennas based on the planar circuit theory and its applictions (in Japanese)'. Doctoral dissertation, Tokyo Inst. Technol., Nov. 1984

Chapter 5

Microstrip dipoles

P.B. Katehi, D.R. Jackson and N.G. Alexopoulos

5.1 Introduction

Microstrip dipoles have been studied extensively during the last 20 years. They are planar elements which consist of a pair of collinear thin-strip conductors printed on the surface of a dielectric slab (Fig. 5.1). They resemble the free-space cylindrical dipoles in the sense that radiation results from a harmonically varying dipole moment. Microstrip dipoles are attractive elements owing to their desirable properties such as simplicity, small size and linear polarisation.

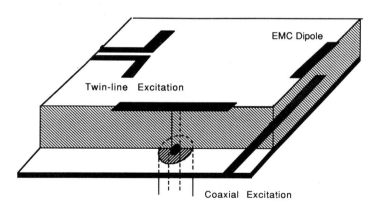

Fig. 5.1 *Excitation mechanisms for a microstrip dipole*

They are well suited for higher frequencies in particular, where the substrate may be electrically thick. In this case the bandwidth of the dipoles may be quite significant. For thicker substrates it is also possible to alter the radiation properties by the use of a superstrate layer, making dipoles a possible candidate in a substrate–superstrate geometry.

When designing microstrip dipoles, the choice of feed mechanism is very

276 Microstrip dipoles

important and should be made taking into consideration the following two factors: theoretical modelling and practical implementation. Fig. 5.1 shows the most commonly used mechanisms: the coaxial feed, the twin-line feed and the coupled-line feed (EMC dipole). In the twin-line feed a voltage is applied directly to the arms of the dipole. In the coaxial feed the two dipole arms are shorted together, with the dipole becoming essentially a narrow patch antenna with a probe feed. The EMC dipole excitation is realised through electromagnetic coupling to the feed line, with no direct contact. Because of its simplicity, the EMC feed represents the most desirable way to feed a dipole from a microstrip line.

Even if practical excitation mechanisms are employed in the design of microstrip dipoles, more ideal ones may be considered for their analysis. The reason lies in the fact that most of their radiation properties are independent of the excitation (i.e. bandwidth, efficiency, radiation pattern etc.). Throughout this Chapter the microstrip dipole is studied extensively as a single radiator as well as an array element. Furthermore, infinitesimally small, centre-fed and EMC dipoles are presented separately, and the dependence of their properties on the electric characteristics of the dielectric layers is discussed. The study of microstrip dipoles is concluded by presenting a design technique for an array of EMC dipoles which accounts for the mutual interactions between dipole elements.

5.2 Infinitesimal dipole

The simplest dipole structure which can be studied is the infinitesimal dipole. An analysis of the infinitesimal dipole is important because all the radiation characteristics of full-size dipoles may be obtained simply from this solution. Only for near-field (impedance) calculations is it essential to analyse the full-size dipole with a moment-method technique.

5.2.1 Analysis
A horizontal electric dipole (HED) is shown in Fig. 5.2. In general, the dipole may be embedded within an arbitrary number of layers, although two layers are sufficient to cover most cases of practical interest, including microstrip dipoles with a protective top (superstrate) layer, or EMC dipoles with a transmission line at the interface ($z = b$).

In the classical Sommerfeld solution, components of the magnetic vector potential at x, y, z due to a source at x', y', z' are written as

$$\Pi_x = \int_0^\infty f(\lambda, z, z') J_0(\lambda \varrho) d\lambda \tag{5.1}$$

$$\Pi_z = \cos\phi \int_0^\infty g(\lambda, z, z') J_1(\lambda \varrho) d\lambda \tag{5.2}$$

with ϱ, ϕ describing cylindrical coordinates [1]. The functions f and g are complicated functions of λ, z, z', and are given in Reference 2. The time dependence is $e^{+j\omega t}$, and is suppressed here. The functions f and g are of the form

$$f(\lambda) = \frac{a(\lambda, z, z')}{D_e(\lambda)} \tag{5.3}$$

$$g(\lambda) = \frac{b(\lambda, z, z')}{D_e(\lambda)D_m(\lambda)} \tag{5.4}$$

where a and b are analytic functions of λ except for the branch-type singularity due to the wavenumber

$$k_{z0} = \left(k_0^2 - \lambda^2\right)^{1/2}$$

which appears in the expressions [2]. The functions $D_e(\lambda)$ and $D_m(\lambda)$ have zeros on the real axis at λ_p, producing poles in the integrand. The zeros of $D_e(\lambda)$ are

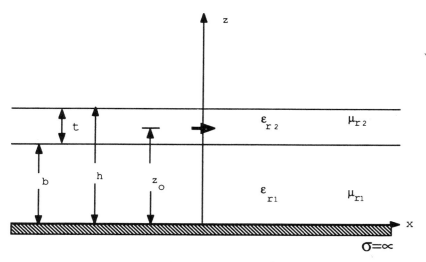

Fig. 5.2 *Substrate–superstrate geometry with horizontal electric dipole (HED) embedded*

the TE-mode surface-wave propagation constants, while those of $D_m(\lambda)$ correspond to the TM-mode surface waves. These poles are in the region $k_0 < \lambda_p < k_{max}$, where $k_{max} = \max(k_1, k_2)$. The path of integration goes around the poles, as shown by contour C in Fig. 5.3.

By using symmetry properties [1] the integrations may be extended to $(-\infty, +\infty)$ and the path deformed to an integral around the branch cut (contour C_b) plus integrals around the poles in the right-half plane. The potentials may then be written as

$$\Pi_x = \tfrac{1}{2} \int_{C_b} f(\lambda, z, z') H_0^{(2)}(\lambda\varrho) d\lambda$$

$$-\pi j \sum_{\lambda_{TE}} H_0^{(2)}(\lambda_p \varrho) \operatorname{Res}\{f(\lambda_p)\} \quad (5.5)$$

$$\Pi_z = \tfrac{1}{2} \int_{c_b} g(\lambda, z, z') H_0^{(2)}(\lambda \varrho) d\lambda$$

$$-\pi j \sum_{\lambda_{TE}+\lambda_{TM}} H_0^{(2)}(\lambda_p \varrho) \operatorname{Res}\{g(\lambda_p)\}. \quad (5.6)$$

The branch-cut integral contribution may be identified as the radiation field.

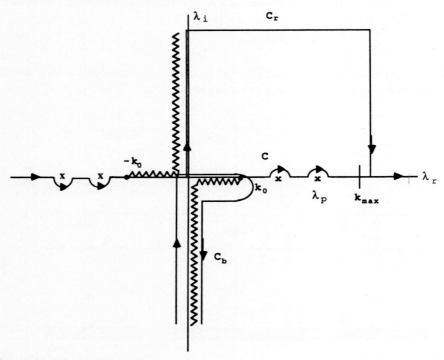

Fig. 5.3 *Contours of integration*

The residue contributions at the poles give the surface-wave fields. A steepest-descent method may be used to find the far-zone radiation field, although a reciprocity method is simpler [2]. The Poynting vector from the far-zone radiation field may be integrated over a hemisphere to find the radiated power. This reduces to a one-dimensional numerical integration [2].

The surface-wave Poynting vector may be integrated over a large cylinder to find a closed-form expression for the power in a surface wave. The surface waves are orthogonal with each other and with the radiation field in the lossless case [3], so the total power is simply the sum of all the powers. The radiation efficiency is defined in the lossless case as

$$e_r = \frac{P_{rad}}{P_{rad} + P_{sw}} \qquad (5.7)$$

with P_{rad} and P_{sw} the radiated and total surface-wave powers, respectively.

5.2.2 Substrate effects

The effect of substrate thickness on the efficiency of an infinitesimal dipole on top of a single layer is shown plotted against the electrical substrate thickness b/λ_d (where $\lambda_d = \lambda_0/\sqrt{\varepsilon_r}$) in Fig. 5.4 for various substrates. Note that the efficiency decreases for increasing substrate thickness, and is lower for higher ε_r.

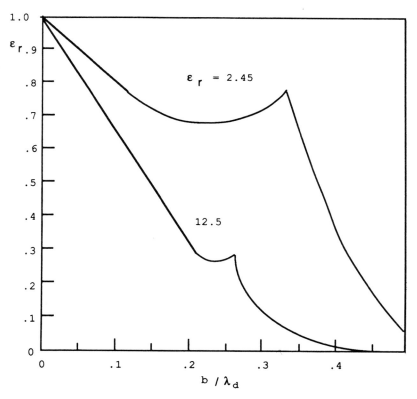

Fig. 5.4 *Efficiency of HED versus substrate thickness for dipole on top of single substrate layer*

The efficiency approaches 1·0 for thin substrates, but the radiated power of the dipole then becomes very low, as seen from Fig. 5.5. This points toward one of the practical limitations of using resonant-length dipoles on thin substrates, namely low input resistance. A patch antenna does not have this disadvantage since the resonant resistance is fairly independent of substrate thickness [4].

280 Microstrip dipoles

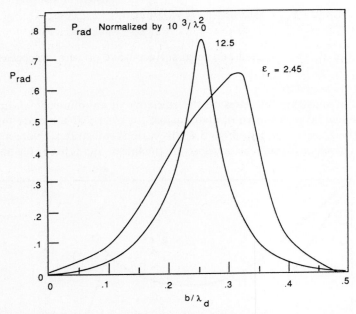

Fig. 5.5 *Radiated power of HED (Watts) versus substrate thickness for unit-strength dipole on top of single substrate layer*

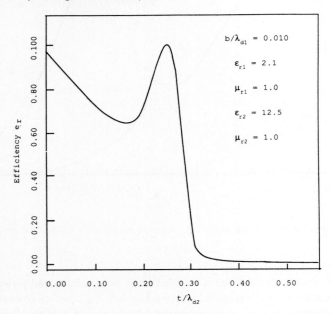

Fig. 5.6 *Efficiency of HED versus superstrate thickness, showing 100% efficiency at $t/\lambda_{d2} = 0.233$ (Reproduced from Reference, 2, © 1984 IEEE)*
The dipole is at the interface ($z_0 = b$)

Another limitation of resonant-length dipoles on thin substrates is a very small bandwidth – much lower than that of a patch (this is discussed in Section 5.4.2). Because of these limitations, dipoles find best application for thicker substrates where bandwidth and resistance are no longer serious limitations, and where the patch antenna becomes non-resonant [5].

5.2.3 Superstrate effects

It is interesting to note that a superstrate (cover) layer on top of a microstrip dipole may significantly influence the radiation properties. For example, if the substrate is thin enough, a superstrate layer may be used to eliminate surface-wave excitation, resulting in an efficiency of 100% [2]. An example of this is shown in Fig. 5.6 for a GaAs superstrate over a Teflon substrate. In a different application, a substrate–superstrate geometry may be used to produce 'radiation into the horizon', a phenomenon in which the far-zone radiation field

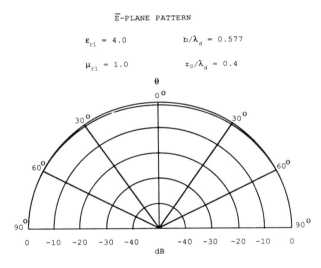

Fig. 5.7 *Radiation pattern of HED demonstrating the radiation into the horizon effect (Reproduced from Reference 6, © 1985 IEEE)*
The dipole is embedded within a single substrate layer here

remains non-zero down to the layer surface. One result is the possibility of producing very nearly omnidirectional patterns [6]. An example of this is shown in Fig. 5.7 for a dipole embedded within a single substrate layer of thickness b (or, equivalently, the superstrate material is the same as the substrate).

A third application of a superstrate layer is in the production of high-gain patterns about any desired angle θ_p in space. By using one or more superstrate layers of the proper thickness, narrow-beam patterns may be produced as the superstrate ε_{r2} becomes large [7]. An example of this for $\theta_p = 45°$ is shown in

282 Microstrip dipoles

Fig. 5.8 using a superstrate with $\varepsilon_{r2} = 100$. This narrow-beam effect is produced by weakly attenuated leaky waves which exist on the structure [8].

All the above effects pertain to any type of microstrip element in a substrate–superstrate geometry. However, except for the first effect (increased efficiency), these methods all require electrically thick layers, making dipoles an attractive candidate.

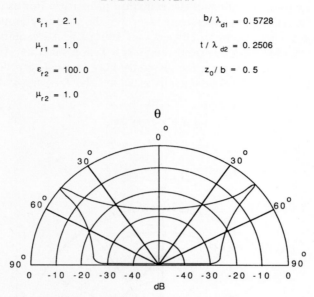

Fig. 5.8 *Radiation pattern of HED demonstrating the high-gain effect (Reproduced from Reference 7, © 1985 IEEE)*
The dipole is embedded within the substrate with a superstrate layer on top

5.3 Moment-method techniques for planar strip geometries

The analysis of geometries comprising dipoles and transmission lines involves strips which are usually narrow compared to a wavelength ($w \ll \lambda_0$). The geometry of a strip is shown in Fig. 5.9. Because the strip is narrow, current may be assumed in the $\hat{x}$ direction only. The narrow-strip assumption also allows for certain techniques to improve the computational efficiency of the analysis. In this Section methods for analysing strip structures are discussed.

5.3.1 Basis Functions
For the moment-method solution of strip geometries, including dipoles and transmission lines, the current on the strips may be represented using sub-

domain piecewise-sinusoidal basis functions of the form

$$B(x,y) = J_x = \eta(y)\,\xi(x) \qquad |y| < w/2 \qquad (5.8)$$
$$|x| < d$$

where

$$\xi(x) = \frac{\sin k_e(d - |x|)}{\sin k_e d}.$$

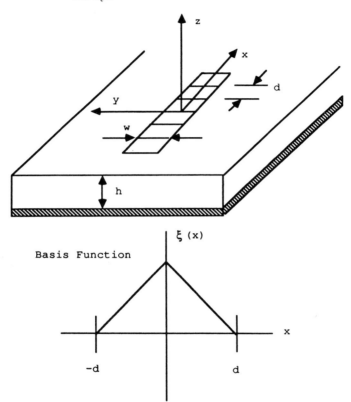

Fig. 5.9 *Planar strip in a layered geometry, divided into subsections. Also shown is the basis function variation in x*

Three useful choices for the transverse variation are the pulse function, the modified Maxwell function [9] and the Maxwell function:

$$\eta(y) = \frac{1}{w} \qquad |y| < w/2 \qquad (5.9)$$

$$\eta(y) = \frac{4}{5w}\left[1 - \left|\frac{2y}{w}\right|^3\right] \qquad |y| < w/2 \qquad (5.10)$$

284   Microstrip dipoles

$$\eta(y) = \frac{2}{\pi w}\left[1 - \left|\frac{2y}{w}\right|^2\right]^{-1/2} \quad |y| < w/2. \tag{5.11}$$

The normalisation constants here are chosen to give a unit current at $x = 0$. The Maxwell function more closely represents the true current on a narrow strip. The modified Maxwell function is similar to the Maxwell function except near the edges. Because it lacks the singular behaviour at the edges, the Fourier transform of the modified Maxwell function decays faster for large values of the transform variable, which makes it computationally more efficient for spectral-domain analysis techniques.

5.3.2 Reaction between basis functions

The fundamental step in a Galerkin moment-method solution of strip problems is the computation of reaction between two basis functions:

$$Z_{mn} = -\langle J_{xm}, J_{xn} \rangle = -\int_{S_m} E_{xn}(x, y) J_{xm}(x, y)\, dx dy \tag{5.12}$$

with

$$E_{xn}(x, y) = \int_{S_n} J_{xn}(x', y') G_{xx}(x, y; x', y)\, dx' dy' \tag{5.13}$$

with S_m and S_n the basis-function surfaces. G_{xx} is the $\hat{x}\hat{x}$ component of the dyadic Green's function for $E_x(x, y, z_m)$ due to a source at x', y', z_n. This is given by

$$G_{xx} = k^2 \Pi_x + \frac{\partial^2 \Pi_x}{\partial x^2} + \frac{\partial^2 \Pi_z}{\partial x \partial z}. \tag{5.14}$$

If more than one strip is involved in the problem under consideration, it is convenient to choose the x-axis offset between strips as an integral multiple of the subsection length d, so that the impedance submatrices will all be Toeplitz. Depending on the particular problem, d is typically in the range $0.01\lambda_0 < d < 0.05\lambda_0$. In this case $\xi(x)$ is close to a piecewise linear function, and the value of k_e in eqn. 5.8 is not critical. The choice of transverse distribution $\eta(y)$ has some influence in the reaction values obtained, especially for the self-term (no offset between basis functions). Experience has shown, however, that the current amplitudes, obtained from the moment-method matrix solution, are fairly independent of the choice of $\eta(y)$.

For the solution of the strip problems, the starting point is the calculation of arrays $Z_i(m)$ where the mth element is reaction between basis functions separated by $(m-1)d$ between centres in the $\hat{x}$-direction. The index i refers to the particular submatrix in the Galerkin impedance matrix, corresponding to basis functions on different strips. The y-directed offset may be different for each submatrix if the strips have transverse offset. z_m and z_n may be different for each submatrix as well.

5.3.3 Plane-wave spectrum method

Two distinct methods have been used in the literature for the calculation of reaction: the Fourier transform (plane-wave spectrum) method and the real-space integration technique. In the spectral method, eqns. 5.1 and 5.2 are first transformed into rectangular (Fourier) transform form, and substituted into eqns. 5.12 – 5.14. The resulting integrations in x, y, x', y' result in Fourier transforms of the current [10,11]. The result is

$$Z_{mn} = \frac{2}{\pi} \int_0^{\pi/2} \int_0^{\infty} \left[\left(k^2 - \lambda_x^2 \right) f(\lambda, z_m, z_n) + \lambda_x^2 \, h(\lambda, z_m, z_n) \right]$$
$$\cdot \tilde{J}_x^2 (\lambda_x, \lambda_y) \cos(\lambda_x \Delta_x) \cos(\lambda_y \delta_y) \, d\lambda \, d\bar{\phi} \qquad (5.15)$$

where

$$\tilde{J}_x = \tilde{\xi}(\lambda_x) \, \tilde{\eta}(\lambda_y)$$

with

$$\tilde{f}(\alpha) = \int_{-\infty}^{+\infty} f(x) \, e^{-j\alpha x} \, dx$$

and

$$h(\lambda, z_m, z_n) = \frac{1}{\lambda} \frac{\partial g}{\partial z}$$

$$\lambda_x = \lambda \cos \bar{\phi}$$

$$\lambda_y = \lambda \sin \bar{\phi}.$$

The separation between basis-function centres is denoted as Δx, Δy here. For the basis function choices (eqns. 5.9 – 5.11) the transforms may be evaluated in closed form [10]. The integral on $(0, \infty)$ in eqn. 5.15 is along the Sommerfeld contour C in Fig. 5.3. A pole-extraction technique may be used to account for the poles on the real axis [11]. A simpler way is just to deform the contour to go around the poles as shown by contour C_r in Fig. 5.3 [12].

An advantage of the spectral approach is that self-term problems are avoided. However, the convergence of the Sommerfeld integral in eqn. 5.15 becomes worse as the separations Δx, Δy become large compared to the respective basis function dimensions, for the case $z_m = z_n$. This is because the functions f and h tend to constants as $\lambda \to \infty$ in this case, resulting in a rapidly oscillating, slowly converging integral. For $z_m \neq z_z$, as is possible for the reaction between currents on different strips, the terms f and h decay exponentially in λ, and there is relatively little trouble for most values of Δx, Δy. To speed up the computation for the case $z_m = z_n = z$, several numerical techniques may be employed. The first is the use of a Filon integration method to account for the oscillatory cosine terms [13]. A second technique is to subtract from the integrand a limiting-behaviour term with constants $f(\infty, z, z)$, $h(\infty, z, z)$, so that the integral converges much faster. The integral of the extracted term may be evaluated by identifying it as the reaction between currents in a grounded homogeneous half-space of effective wave number k_e [14]. This reaction may be evaluated using a free-space Green's function with a real-space integration. In this case the pulse choice of $\eta(y)$ (eqn. 5.9) is most convenient, since the resulting real-space reaction integral reduces to a one-dimensional integral [14].

Another type of extraction may be employed for the case $z = z_m = z_n$ when z is sufficiently far from a layer boundary, as for a strip embedded within a layer. In this case, a term corresponding to the incident field in a grounded homogeneous space of wave number k_i is extracted [15], where k_i is the wave number of the layer. The resulting integrand then decays exponentially. The reaction in the grounded homogeneous half-space is computed as before. This extraction fails for a strip at the interface of different media, and is therefore of more limited use than the first type of extraction.

Another method for improving the convergence of eqn. 5.15 is to asymptotically approximate the transform $\tilde{J}_x$ for large λ in a sufficiently simple form so that the tail integral over (A, ∞) may be performed analytically, for some large number A. This is the most straightforward technique, but the tail integral must be reformulated for each specific choice of $\eta(y)$.

Finally, as an alternative to trying to accelerate the convergence of eqn. 5.15 for large separation between basis functions, the reaction may be computed by a different technique, as discussed in Sections 5.3.4 and 5.3.5.

5.3.4 Real-space integration method

The reaction Z_{mn} may also be computed by integrating the electric field directly in the spatial variables. To avoid a prohibitive amount of calculation, a δ-function testing procedure is used at the strip centre instead of a Galerkin method, so that impedance elements are defined as

$$Z_{mn} = -\int_x E_{nx}(x, 0) J_{mx}(x, 0) \, dx. \qquad (5.16)$$

A technique due to Katehi and Alexopoulos [16,17] computes this impedance term by directly applying the electric-field operator (eqn. 5.14) to the Sommerfeld form of potentials (eqns. 5.1 and 5.2). An integration by parts is used to shift the derivatives to the current function $J_{nx}(x', y')$, and various algebraic manipulations are employed, including an analytical tail integral evaluation. The resulting expression involves a single Sommerfeld-type integral of a rapidly converging series. This formulation does not suffer from convergence difficulties to the same degree that eqn. 5.15 does.

Although a comprehensive comparison of computational efficiency has not been performed, it is felt that the real-space method is somewhat comparable to the spectral method when one of the accelerating techniques mentioned previously is used. Both methods have been used to generate the results of this Chapter.

5.3.5 Point-dipole approximation

The reaction between widely separated basis functions may often need to be computed, especially for mutual-impedance problems. In this case, the most efficient technique is to approximate the currents as point dipoles located at the basis-function centres. Computing reaction is then equivalent to finding the E_x field of an $\hat{x}$-directed point source, which may be obtained efficiently. One way to perform this calculation is by directly applying the electric-field operator (eqn. 5.14) to eqns. 5.1 and 5.2. It is well known that the resulting Sommerfeld integrals are nonconvergent when $z = z'$, however, and therefore cannot be evaluated directly. One technique for overcoming this difficulty is to extract terms from the integrand to give convergence [18]. Another way is to extend the integration contour to $(-\infty, +\infty)$ and deform around the branch cut, as described in Section 5.2.1. The numerical integration around the branch cut converges very rapidly for large radial separation ϱ between dipoles, owing to the exponential decay of the Hankel functions in eqns. 5.5 and 5.6 along the imaginary axis. For separations larger than $0.25\lambda_0$ this is usually the most efficient of the two methods.

5.3.6 Moment-method equations

Consider an arbitrary set of x-directed strips having a 1V δ-gap voltage source at some point on one of the strips. Let the basis functions be numbered $1, \ldots N$ with the source at the centre of basis-function number s, at $x = x_s$. The current representation is then

$$J_x(x, y) = \sum_{n=1}^{N} I_n B_n(x, y) \qquad (5.17)$$

with $B_n(x, y) = B(x - x_n, y - y_n)$. Because $\eta(y)$ in eqns. 5.9 – 5.11 is normalised, I_n represents the current in amperes at $x = x_n$. Enforcing $E_x = \delta(x - x_s)$ on the strips by applying Galerkin's method using eqn. 5.17 results in the set of equations

$$[Z_{mn}][I_n] = [\delta_{ns}] \qquad (5.18)$$

$$\delta_{ns} = \begin{cases} 1 & n = s \\ 0 & n \neq s \end{cases}$$

which is then solved to find the currents on the strips.

5.4 Centre-fed dipoles

5.4.1 Single dipole

The analysis described in Section 5.3.6 may be applied to find the current distribution for a centre-fed dipole, shown in Fig. 5.10. A variational expression for the input impedance is [19]

$$Z_{in} = -\frac{1}{I_{in}^2} \int_s \int_{s'} J_x(\vec{r}) G_{xx}(\vec{r},\vec{r}') J_x(\vec{r}') ds ds'. \tag{5.19}$$

Using eqn. 5.17 this reduces to

$$Z_{in} = \frac{1}{I_{in}^2} \sum_{m,n} Z_{mn} I_m I_n \tag{5.20}$$

which, in view of eqn. 5.18, reduces further to simply

$$Z_{in} = \frac{1}{I_{in}}. \tag{5.21}$$

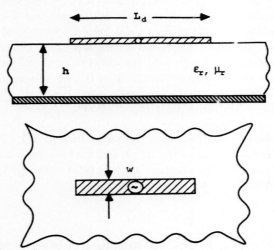

Fig. 5.10 *Centre-fed strip dipole*

This simple formula for input impedance is thus accurate to second order owing to the relation between Galerkin's method and the variational method [20]. From input impedance, the resonant length and bandwidth may be determined.

Approximate formulas for resonant dipoles may also be used [10]. A dipole on a substrate layer has a resonant length

$$\frac{L_r}{\lambda_0} \approx \frac{1}{\sqrt{2(\varepsilon_r + 1)}} \tag{5.22}$$

provided $h \gg w$ and $w \ll \lambda_0$. At resonance the resistance is

$$R_r \approx 120 \left(\frac{L_r}{\lambda_0}\right)^2 \frac{P_T}{15\pi^2/\lambda_0^2} \text{ ohms} \tag{5.23}$$

where P_T is the total power (watts) produced by a unit-strength point dipole on the substrate. Owing to the behavior of P_T (see Fig. 5.5) R_r is very small for thin substrates.

Microstrip dipoles 289

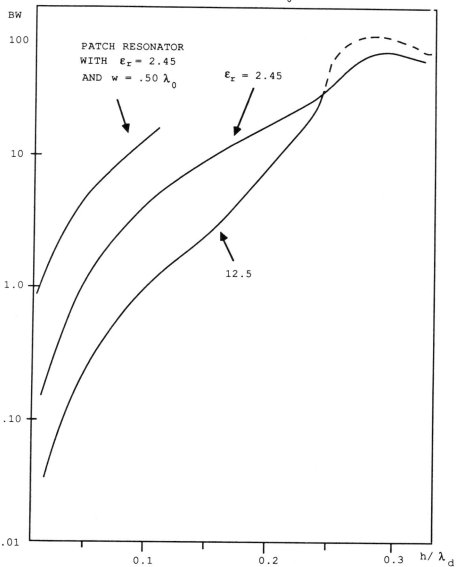

Fig. 5.11 *Bandwidth (%) of centre-fed dipole versus substrate thickness for two different substrates.*
The dashed lines indicate that no frequency is found for which $X_{in} = 0$. In this case f_0 was chosen to minimise $|X_{in}|$, with $X_{in}(f_0)$ then subtracted from all values. A patch resonator is shown for comparison for $h/\lambda_d < 0.1$

290 *Microstrip dipoles*

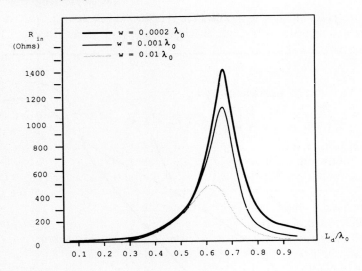

Fig. 5.12a *Real part of the input impedance of a centre-fed strip dipole with $\varepsilon_r = 2\cdot 45$ and $h = 0\cdot 2\lambda_0$*

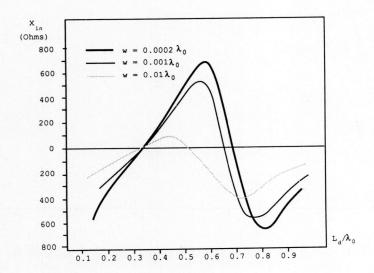

Fig. 5.12b *Imaginary part of the input impedance of a centre-fed strip dipole with $e_r = 2\cdot 45$ and $h = 0\cdot 2\lambda_0$*

The substrate thickness has a dominant effect on bandwidth. The bandwidth for different substrates with a dipole width of $0.05\lambda_0$ is shown in Fig. 5.11. Here the bandwidth is defined as

$$BW(\%) = 100\left(\frac{f_2 - f_1}{f_0}\right) \qquad (5.24)$$

with f_0 the resonant frequency ($X_{in} = 0$) and f_1, f_2 the frequencies at which SWR = 2.0 on a feed line having a match at f_0. The maximum bandwidth occurs for $h/\lambda_d \approx 0.30$, and increases for larger ε_r. For substrates thinner than this, the bandwidth is relatively independent of ε_r for a given physical thickness h/λ_0. For comparison, the bandwidth of a microstrip patch resonator is shown for the $\varepsilon_r = 2.45$ case. The patch has a much higher bandwidth for thin substrates, a conclusion reached previously by Pozar [5].

The effect of dipole width is seen in Figs. 5.12a, and b for a substrate with $\varepsilon_r = 2.45$. The resonant input resistance is insensitive to width, as is the resonant length. The slope dX_{in}/dL at f_0 is lower for wider dipoles, indicating a greater bandwidth.

However, the slope is not extremely sensitive to width. Only when the substrate becomes thin does the width have a dramatic effect on bandwidth, owing to the cavity-resonator effect.

5.4.2 Mutual impedance

Mutual impedance between centre-fed dipoles may be calculated in different ways. In the moment method, dipole no. 1 is excited with no. 2 short-circuited. After solving eqn. 5.18 for the currents, the formula [10]

$$Z_{12} = \frac{-1}{I_1(0)}\left(\frac{p}{1-p^2}\right)$$

with

$$p = I_2(0)/I_1(0) \qquad (5.25)$$

may be employed. Alternatively, the classical reciprocity formula [21]

$$Z_{12} = \frac{-1}{I_1(0)I_2(0)}\int_{s_2} E_{x1}\, J_{x2}\, ds \qquad (5.26)$$

may be used, with piecewise-sinusoidal currents assumed on the dipoles. The reaction may be evaluated using eqn. 5.15. For narrow strips, the dipoles may be assumed filamentary and the integral (eqn. 5.26) evaluated directly using the real-space technique, or by using the point-dipole approximation to find E_x (Section 5.3.5). Results for filamentary dipoles obtained in this way are shown in Figs. 5.13a,b for broadside and endfire dipoles on a substrate with $\varepsilon_r = 10$. The slow decay of Z_{12} in the endfire case is due to the dominant TM_0 mode, which is strongest at $\phi = 0$ [22].

292 Microstrip dipoles

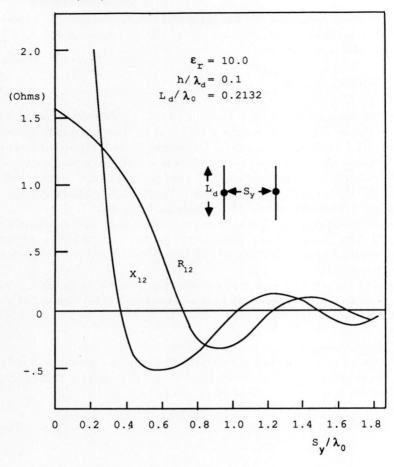

Fig. 5.13a *Mutual impedance versus separation for resonant-length filamentary dipoles in broadside configuration with $\varepsilon_r = 10\cdot 0$ (Reproduced from Reference 10, © 1986 IEEE)*

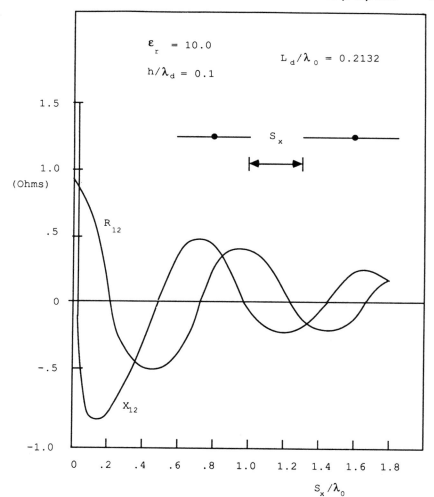

Fig. 5.13b *Mutual impedance versus separation for resonant-length filamentay dipoles in endfire configuration with $\varepsilon_r = 10 \cdot 0$ (Reproduced from Reference 10, © 1986 IEEE)*

294 Microstrip dipoles

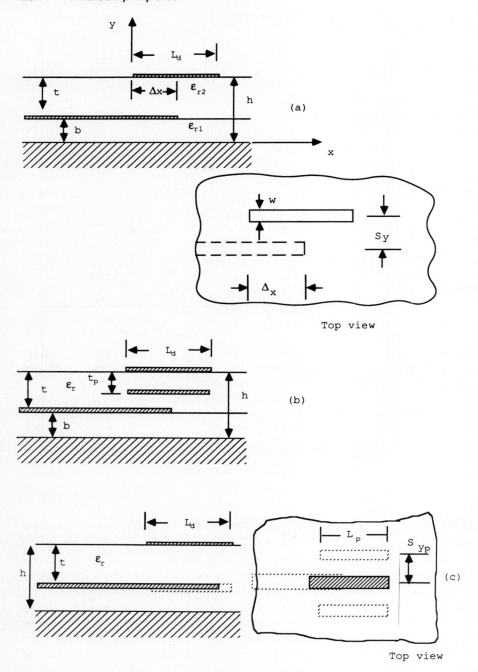

Fig. 5.14 *Various configurations of dipoles electromagnetically coupled to a microstrip line*

5.5 EMC Dipoles

5.5.1 Methods of analysis

Fig. 5.14 shows several ways in which one or more dipoles may be electromagnetically coupled to a transmission line. In each case the analysis involves a moment-method solution together with simple transmission-line theory [23]. The transmission line is assumed close to the ground plane, so that a TEM-like field propagates on the line. A δ-gap source is taken near the end of the line farthest from the coupled end. The exact location is not critical. The source sets up a standing-wave current on the line which is essentially a sinusoidal current, except within a small region near the coupled end where the current is perturbed by the dipoles. Away from this coupling region, the line may be regarded as terminated by a equivalent self-impedance $Z_s = R_s + j X_s$ at some arbitrary value of x, where

$$Z_s = \frac{1 + \Gamma(x)}{1 - \Gamma(x)} \tag{5.27}$$

with

$$\Gamma(x) = \frac{\text{SWR} - 1}{\text{SWR} + 1} e^{+j2\beta(x - x_{min})}. \tag{5.28}$$

In eqn. 5.28 x_{min} is the position of a minimum and β is the propagation constant on the line, which may be determined from the distance between current minima on an isolated line or by separate analysis [9,10].

In this formulation only the reaction between piecewise-sinusoidal-basis functions is required. An alternative formulation using traveling-wave-basis functions on the line may also be developed [24]. An advantage of this latter formulation is the use of fewer unknowns for the line current, although it requires different types of basis functions. For the results of this Chapter, only piecewise-sinusoidal basis functions were used.

5.5.2 Single dipole

A single EMC dipole can be either overcoupled, matched, or undercoupled according to the amount of coupling to the feed line (Fig. 5.15). Of all the possible parameters which affect the coupling, the most critical is the distance t between the line and dipole. If t is too large, then no choice of dipole length or offset will yield an input match, and the dipole is said to be undercoupled. If the line is sufficiently close to the dipole, an input match may be achieved by varying the dipole length and either the longitudinal (x-axis) or transverse (y-axis) offset, or both. The dipole is then said to be overcoupled. In this case the locus of points for the centre of the matched dipole is somewhat elliptical in shape, roughly centered about the end of the line [25]. For a given substrate, this implies a maximum distance t_{max} for which an input match may be achieved,

296 Microstrip dipoles

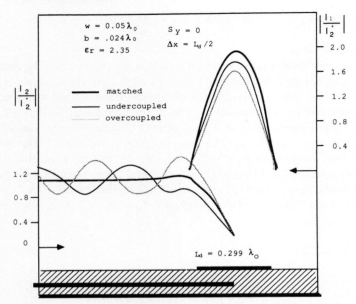

Fig. 5.15 *Current amplitude on the strip dipole (I_1) and microstrip line (I_2) (Reproduced from Reference 23 © 1984 IEEE)*
I_2^+ is the incident current amplitude. The three cases correspond to $h = 0.079\lambda_0$ (matched), $h = 0.084\lambda_0$ (undercoupled), and $h = 0.070\lambda_0$ (overcoupled)

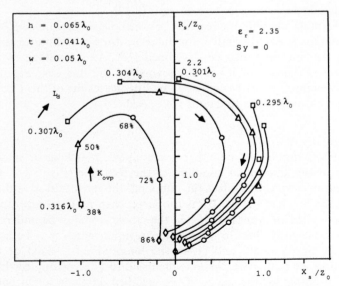

Fig. 5.16 *Z_s/Z_0 as a function of dipole length L_d and longitudinal offset (Reproduced from Reference 23 © 1984 IEEE)*
Offset is measured by percent overlap as $K_{ovp} = (\Delta x/L_d) \times 100$. The impedance reference plane is at the position of a current maximum on the line

where the locus collapses to a single point [25]. In this case the dipole may be said to be critically coupled. Figs. 5.16 and 5.17 show how an input match may be achieved for the overcoupled case by varying either the longitudinal or transverse offset, respectively, for $\varepsilon_r = \varepsilon_{r1} = \varepsilon_{r2} = 2\cdot35$.

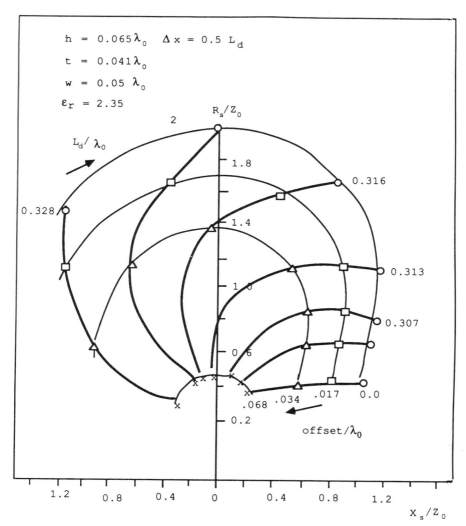

Fig. 5.17 Z_s/Z_0 as a function of dipole length L_d and transverse offset (Reproduced from Reference 23, © 1984 IEEE
The longitudinal offset is 50% of the dipole length. The impedance reference plane is at the position of a current maximum on the line

298 *Microstrip dipoles*

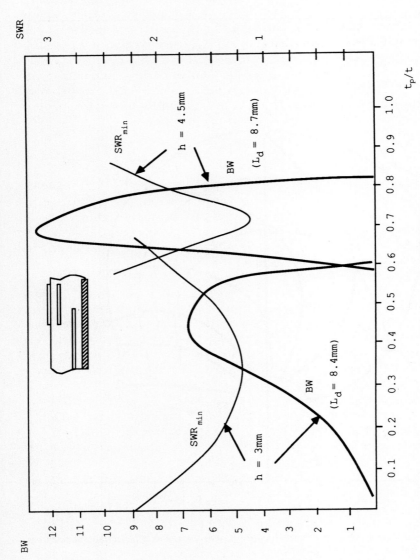

Fig. 5.18 *Bandwidth (%) and minimum achieved SWR as functions of the ratio t_p/t for two values of the substrate thickness h. (Reproduced from Reference 26, © 1987 IEEE)*
$\varepsilon_r = 2.35$, $w = 1.5$ mm and $b = 0.72$ mm. The two dipoles are of the same size, with a 50% longitudinal overlap. The minimum SWR was obtained by varying the frequency

The bandwidth of an EMC dipole is almost identical to that of a centre-fed dipole on the same substrate, provided the EMC dipole is matched. Hence, it is desirable to keep the dipole significantly above the ground plane to obtain a reasonable bandwidth. On the other hand, it is also desirable to keep the line as close as possible to the ground plane, to minimise line radiation. However, for a given substrate thickness h/λ_0 corresponding to a prescribed bandwidth, the line height should be chosen so that $b > h - t_{max}$ to avoid being undercoupled, which will reduce bandwidth. Hence, the height $b = h - t_{max}$ is a good trade-off between bandwidth and line radiation. An improved coupling may be achieved by using a top layer with $\varepsilon_2 > \varepsilon_1$ between the line and dipole. This improves coupling by increasing the capacitance in between [10]. However, a more pronounced improvement is possible by using multiple dipoles.

5.5.3 Multiple dipoles

The bandwidth/line-radiation trade-off may be improved by using multiple dipoles. A dominant theme in these schemes is the introduction of additional capacitance between the line and the main radiating dipole. This is usually accomplished by placing one or two coupling dipoles (parasitics) either in a stacked fashion between the line and top dipole (Fig. 5.14b) or coplanar to, and near the end of, the line (Fig. 5.14c) [26].

For the stacked configuration, the bandwidth for SWR < 2·0 is shown plotted against t_p/t in Fig. 5.18 for two different values of h (3 mm and 4·5 mm) with corresponding dipole lengths of 8·4 mm and 8·7 mm, respectively, at a frequency of 10 GHz. These lengths are close to the input-match values for a single EMC dipole. The transmission-line height b is constant at 0·72 mm. Also shown is the minimum achieved SWR as the frequency is varied for each value of t_p. From this Figure it can be seen that the thinner substrate has a lower optimum bandwidth. Also, as the substrate thickness changes from 3 mm to 4·5 mm the range of t_p for SWR < 2·0 becomes smaller and shifts toward higher values. Therefore, like the case of a single dipole, there is a maximum value h_{max} such that, for $h > h_{max}$, the SWR is always larger than 2·0 for this value of b. The addition of another dipole between the top dipole and line may further reduce the SWR and increase the bandwidth in this case.

When the parasitics are on the same level with the line and of comparable length to the radiating dipole, the bandwidth may be improved as shown in Fig. 5.19. Appropriate positioning of the parasitics may improve the bandwidth even more. An advantage of this configuration is possibly fewer alignment difficulties during fabrication. However, the improvement in bandwidth/line-radiation is much more dramatic for the case of stacked dipoles.

300 Microstrip dipoles

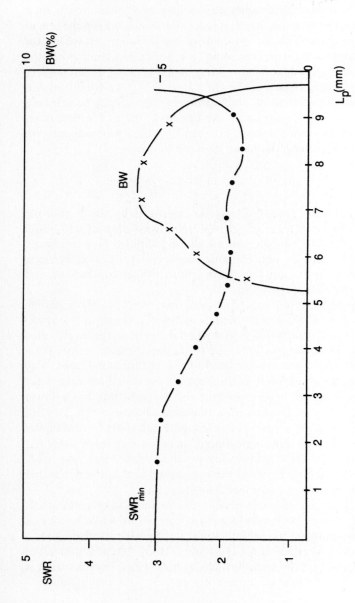

Fig. 5.19 *Minimum achieved SWR and bandwidth (%) as functions of the parasitic dipole length L_p for the case of Fig. 5.14c (Reproduced from Reference 26, © 1987 IEEE)*

$\varepsilon_r = 2\cdot35$, $h = 3\cdot0$ mm, $t = 2\cdot28$ mm, $w = 1\cdot5$ mm, $L_d = 8\cdot72$ mm, and $S_{yp} = 1\cdot6$ mm, where S_{yp} is the transverse offset for the parasitic dipoles

All dipoles have a 50% longitudinal overlap. The minimum SWR was obtained by varying the frequency

5.6 Finite array of EMC dipoles

5.6.1 Analysis

A finite array of EMC dipoles is shown in Fig. 5.20. Away from the coupled ends, the current on the mth line may be written as the sum of incident and backward waves, of the form

$$J_m^i(x, y) = I_m^i \eta(y) e^{-j\beta x}$$
$$J_m^b(x, y) = I_m^b \eta(y) e^{+j\beta x} \tag{5.29}$$

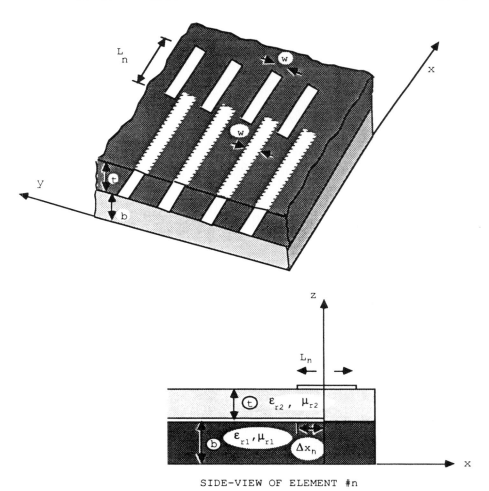

Fig. 5.20 *Finite array of EMC dipoles*

with $x = 0$ at the line end. The current on the mth dipole may be assumed sinusoidal as

$$J_m^d(x, y) = I_m^d \eta(y) \xi(x) \tag{5.30}$$

Here $\xi(x)$ is given by eqn. 5.8 with d replaced by $L_m/2$. Because of linearity, the line and dipole current amplitudes may be written as

$$I_m^b = -\Gamma_m I_m^i + \sum_{n \neq m} I_n^d B_{mn} \tag{5.31}$$

$$I_m^d = E_m I_m^i + \sum_{n \neq m} I_n^d D_{mn} \tag{5.32}$$

where B_{mn}, D_{mn} are back-scattering and dipole coefficient accounting for mutual interaction between dipoles. Γ_m is the voltage reflection coefficients at the end of line m when all dipoles except number m are absent. E_m is similarly an excitation coefficient in the isolated case. These equations may be written in matrix form as

$$[I^b] = -[\Gamma][I^i] + [B][I^d] \tag{5.33}$$

$$[I^d] = [E][I^i] + [D][I^d] \tag{5.34}$$

where $[\Gamma]$, $[E]$ are diagonal matrices and $[B]$, $[D]$ are zero on the diagonal. $[I^b]$ and $[I^d]$ are column vectors.

Eqn. 5.34 yields

$$[I^i] = [E]^{-1}([U] - [D])[I^d] \tag{5.35}$$

with $[U]$ the identity matrix.

Substituting into eqn. 5.33 yields

$$[I^b] = \left(-[\Gamma][E]^{-1}([U] - [D]) + [B]\right)[I^d]. \tag{5.36}$$

Eqn. 5.35 gives the incident currents required to produce the desired set $[I^d]$ (which determines the radiation pattern). Eqn. 5.36 gives the resultant back-scattered currents. From this the scattering matrix is found to be

$$[S] = [\Gamma] - [B]([U] - [D])^{-1}[E] \tag{5.37}$$

for the case of identical lines.

5.6.2 Calculation of coefficients

The transmission-line analysis used to calculate Γ in Section 5.5.1 may be extended to calculate E, as well as B_{mn}, D_{mn} between two line–dipole pairs. The formula for E is

$$E = \frac{I^d}{I_p}(1 + |\Gamma|)e^{-j\beta x_{max}} \tag{5.38}$$

with I_p the line current at a maximum ($x = x_{max}$). The line may be excited as discussed in Section 5.5.1.

Microstrip dipoles

To calculate B_{mn} and D_{mn}, dipole n may be excited with a δ-gap source (line n may be absent), with line–dipole pair m passive. Then

$$D_{mn} = \frac{I_m^d}{I_n^d} - E_m \frac{I_0}{I_n^d} \tag{5.39}$$

$$B_{mn} = -\frac{I_0}{I_n^d} e^{j2\beta l} + \Gamma_m \frac{I_0}{I_n^d} \tag{5.40}$$

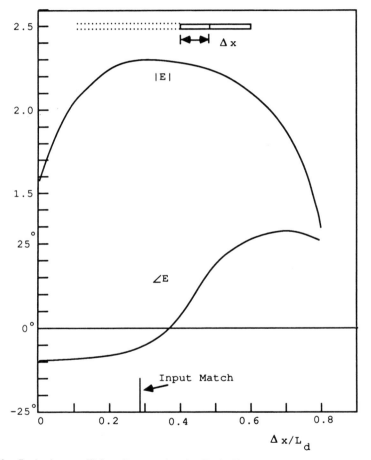

Fig. 5.21 *Excitation coefficient E versus longitudinal offset for a single EMC dipole*
$\varepsilon_{r1} = 2\cdot 2$, $\varepsilon_{r2} = 2\cdot 35$, $b = 0\cdot 03$ in, $t = 0\cdot 06$ in, $L_d = 0\cdot 367$ in, and $w = 0\cdot 059$ in with $f = 10\,\text{GHz}$

where $I_0 = \frac{1}{2} j I_p e^{-j\beta l}$ is the incident-current amplitude, with l the line length. As for the calculation of Γ, a formulation using traveling-wave-basis functions could also be used here.

304 Microstrip dipoles

Results were calculated for the case $\varepsilon_{r1} = 2\cdot2$, $\varepsilon_{r2} = 2\cdot35$, $b = 0\cdot03$ in., $t = 0\cdot06$ in, $w = 0\cdot059$ in at a frequency of 10 GHz. The dipole lengths are taken as $0\cdot367$ in, the value required for an input match in the isolated case with only longitudinal offset. In Fig. 5.21 the excitation coefficient E is shown plotted

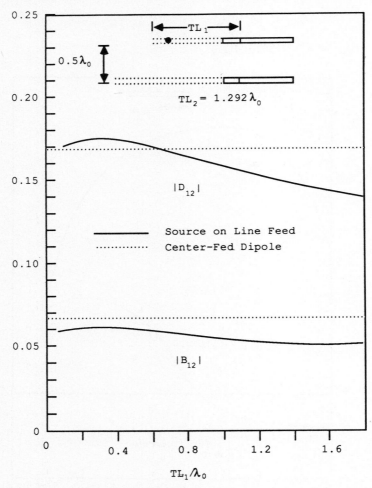

Fig. 5.22 *Mutual coupling coefficients D_{12}, B_{12} for two line–dipole pairs*
Coefficients are calculated with dipole no. 1 excited in two ways: with a centre feed (no line) and by coupling to a line of length TL_1. Dimensions are the same as in Fig. 5.21 with dipole offsets chosen to give a match in the isolated case

against offset for a single line–dipole pair. The magnitude of E is a maximum close to the input-match point ($\Delta_x = 0\cdot114$ in). In Fig. 5.22 results for B_{12}, D_{12} between two pairs are shown. Dipole no. 1 is excited as a source in two different ways: first, by a δ-gap feed at the centre (line no. 1 absent), and secondly, when used as an EMC dipole excited by line no. 1. The results for the EMC case depend on the feed-line length (TL_1) to some extent owing to spurious coupling between line no. 1 and dipole no. 2. The agreement is fairly good, however. In Fig. 5.23, S_{12} results [27] are shown for the case measured by Stern and Elliott

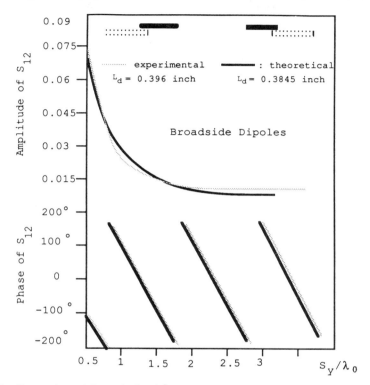

Fig. 5.23 *Comparison of theoretical and measured S_{12} for two line–dipole pairs. (Reproduced from Reference 27, © 1987 IEEE)*
The two dipoles are broadside with separation S_y between centres. Measured values taken from Reference 29: $\varepsilon_r = 2\cdot35$, $b = 0\cdot037$ in, $t = 0\cdot0485$ in, $w = 0\cdot059$ in with $f = 10\,\text{GHz}$. The reference planes have been placed at the first current maximum from the end of the lines

[28,29]. The dipoles were taken to be slightly shorter than the experimental lengths since rounded dipoles were used in the measurement.

5.6.3 Array design

There are two possibilities for an array design: to have an input match on each line in the active state, or to relax the match condition and simply put matching transformers or stubs on each line. The second case is simpler because all dipole lengths and offsets may be chosen the same ($L_n = L_d$). Eqns. 5.35 and 5.36 give the line currents, with the active reflection coefficient for line m given by

306 Microstrip dipoles

$$\Gamma_m^a = -I_m^b/I_m^i. \tag{5.41}$$

From this the matching network may be designed. In particular, knowing Γ_m^a determines the distance from the end of each line at which the impedance is purely real. At this point quarter-wave transformers may be placed to impedance match to the desired feed-line impedance. The input power on line m is $P_m = \frac{1}{2} Z_{0m} |I_m^i|^2 (1 - |\Gamma_m^a|^2)$ where Z_{0m} is the impedance of line m, which couples to the mth dipole. The phase of I_m^i determines the necessary phase delay, and hence line length, of line m.

In the first case, an iterative procedure may be used to solve for each dipole length and offset to give $\Gamma_m^a = 0$. This could be achieved by starting with an initial length and offset to give an input match in the isolated case, and then using eqn. 5.35 to find I_m^i. The new length and offset are chosen to satisfy

$$\Gamma_m = \frac{1}{I_m^i} \sum_{n \neq m} I_n^d B_{mn}.$$

The process then repeats. This is similar to the iteration scheme used by Elliott [28] to design an array of EMC dipoles, which was based on an experimental evaluation of coupling.

By utilising one of the design techniques described above, a complete array of EMC dipoles may be designed, which will have prescribed dipole currents in the presence of mutual coupling. The dipole currents directly determine the array pattern, neglecting line radiation. The design equations 5.33, 5.35 and 5.41 permit the direct design of the array feed network using standard corporate-feed power dividers once the coupling coefficients have been obtained.

Table 5.1 *Summary of coefficient values and results*

$$\Gamma = 0$$
$$E = 2{\cdot}26 \angle -4{\cdot}1°$$

$$B_{12} = 0{\cdot}0714 \angle 143{\cdot}9°$$
$$B_{13} = 0{\cdot}0192 \angle -25{\cdot}2°$$
$$B_{14} = 0{\cdot}00855 \angle 155{\cdot}6°$$

$$D_{12} = 0{\cdot}157 \angle 148{\cdot}0°$$
$$D_{13} = 0{\cdot}0425 \angle -24{\cdot}2°$$
$$D_{14} = 0{\cdot}0191 \angle 155{\cdot}7°$$

$$I_1^i = I_4^i = 0{\cdot}331 \angle -1{\cdot}3°$$
$$I_2^i = I_3^i = 0{\cdot}531 \angle -1{\cdot}9°$$

$$I_1^b = I_4^b = 0{\cdot}058 \angle -144{\cdot}0°$$
$$I_2^b = I_3^b = 0{\cdot}105 \angle -146{\cdot}4°$$

$$\Gamma_1^a = \Gamma_4^a = 0{\cdot}175 \angle -37{\cdot}3°$$
$$\Gamma_2^a = \Gamma_3^a = 0{\cdot}198 \angle -35{\cdot}5°$$

As an example, a 4-element linear array with a $0.5\lambda_0$ element spacing was designed to give a broadside beam with a dipole excitation ratio of $0.65:1.0:1.0:0.65$. The board materials and line-dipole widths were the same as those in Figs. 5.21 – 5.22. The dipoles were chosen to be of resonant length at the design frequency of 10.0 GHz with an offset chosen to give a match in the isolated case. A summary of the coefficient values and results obtained from the equations in Sections 5.6.1 and 5.6.2 is given in Table 5.1.

Based on these results the ratio of powers into the feed lines is calculated as $P_2/P_1 = P_3/P_4 = 2.55$. Because $\angle\Gamma_1^a \doteq \angle\Gamma_2^a$ in this particular design, the

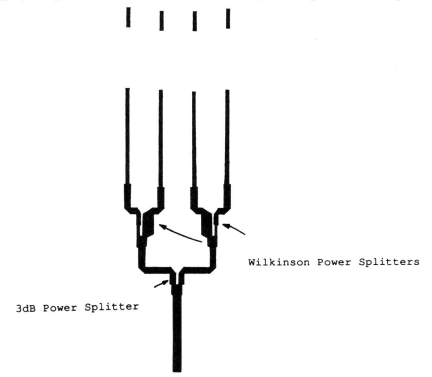

Fig. 5.24 *Diagram of the feed network for a four-element array (not actual size)*
The dipoles are shown displaced from the feed network for clarity

impedance is purely real and a minimum at approximately the same distance from the end of the line on each feed line, at a distance $s = 4.43$ cm (λ_g on the feed lines is 2.02 cm). At this location the active impedance is approximately $40\,\Omega$ (from simple transmission line calculation). Hence, a feed network which gives the required power split into impedances of $40\,\Omega$ is required. Such a feed network was designed using standard Wilkinson power splitters [30], and is shown in Fig. 5.24 [31]. After a 1:1 power split, each Wilkinson splitter further

308 Microstrip dipoles

splits the power to a 2·55:1·0 ratio. The 40 Ω lines coming out of each Wilkinson splitter meet the 62 Ω lines of width $0·05\,\lambda_0$ at a distance s from the ends, allowing each Wilkinson splitter to feed into a matched load. The inner two lines were made slightly shorter than the outer lines to account for a small unbalanced phase shift through the Wilkinson splitters, which was observed experimentally.

The theoretical and measured radiation patterns for this array are shown in Fig. 5.25. The theoretical pattern is found from a reciprocity method [2]. The

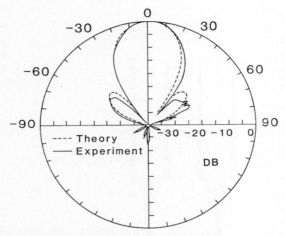

Fig. 5.25 *Theoretical (dashed line) and measured (solid line) patterns for the four-element array*
The theoretical pattern is at 10·0 GHz while the measured pattern is at 9·9 GHz.
$\varepsilon_{r1} = 2·2, \varepsilon_{r2} = 2·35$, $b = 0·03$ in, $t = 0·06$ in, $L_d = 0·367$ in and $w = 0·059$ in. The offsets are $\Delta_x = 0·114$ in

measurements were performed at a frequency of 9·9 GHz since this was found experimentally to be the optimum frequency for the array. At this frequency the SWR on the 50 Ω input feed line was approximately 1·4.

Finally, it should be mentioned that in large arrays of EMC dipoles, accurate results may be obtained by using an infinite array analysis, involving a summation of Floquet modes [32]. Results for an infinite EMC dipole array have recently been obtained [33].

5.7 Conclusions

Microstrip dipoles are generally characterized as being narrower than patch antenna elements, and are not usually probe-fed like the patch. These features allow the microstrip dipole to be a useful radiating element in many applications. In particular, due to the small size, the dipole may be useful when space limitations are important. Dipoles have low radiation resistance and narrow bandwidth for thin substrates, in comparison with the patch antenna. However, the dipole may be used as a resonant element for thicker layers, for which the bandwidth may be quite considerable, and the input resistance no longer a limitation. Dipoles thus find best application for thicker substrates.

In this chapter a general method has been presented for analyzing strip configurations, which include the microstrip dipole as well as the feeding microstrip lines. The analysis technique discussed is flexible, allowing for a wide variety of different configurations, including the single dipole, the mutual coupling between two dipoles, or the electro magnetically coupled (EMC) dipole.

One of the most practical methods for feeding a microstrip dipole is by electromagnetic coupling to a microstrip line. If the dipole is sufficiently close to the line, an input match can always be achieved by varying the dipole length and the offset from the line, in either the longitudinal or transverse directions. In the EMC dipole it is usually desired to minimize the line radiation as much as possible, while maintaining an input match. This implies a line height above the ground plane for which the dipole is critically coupled. To improve the bandwidth-line radiation trade-off, multiple dipoles may be coupled to the line, either in a stacked configuration, or coplanar with the line. The stacked configuration gives the best improvement in bandwidth, for a given line height.

A design procedure for a finite array of EMC dipoles may be developed using a matrix description of the line and dipole currents, together with a moment-method solution for the necessary mutual coupling coefficients. Two design procedures were discussed. One is an iterative procedure which yields an input match on each line, but requires each dipole length and offset to be different. The other design procedure allows each dipole length and offset to be the same, but requires matching transformers on each line. This procedure is simpler, not requiring any iterations. In both cases, the design equations allow for the direct determination of the necessary feed network.

5.8 References

1 SOMMERFELD, A.: 'Partial differential equations' (Academic Press, 1962)
2 ALEXOPOULOS, N. G., and JACKSON, D. R.: 'Fundamental superstrate (cover) effects on printed circuit antennas', *IEEE Trans.*, 1984, **AP-32**, pp. 807–816
3 COLLIN, R. E.: 'Field theory of guided waves' (McGraw-Hill, 1960)
4 CARVER, K. R, and MINK, J. W.: 'Microstrip antenna technology', *IEEE Trans.*, 1981, **AP-29**, pp. 2–24
5 POZAR, D. M.: 'Considerations for millimeter wave printed antennas', *IEEE Trans.*, 1983, **AP-31**, pp. 740–747
6 ALEXOPOULOS, N. G., JACKSON, D. R., and KATEHI, P. B.: 'Criteria for nearly omnidirectional radiation patterns for printed antennas', *IEEE Trans.*, 1985, **AP-33**, pp. 195–205

7 JACKSON, D. R. and ALEXOPOULOS, N. G.: 'Gain enhancement methods for printed circuit antennas', *IEEE Trans.*, 1985, **AP-33,** pp. 976–987
8 JACKSON, D. R. and OLINER, A. A.: 'A leaky-wave analysis of the high-gain printed antenna configuration', *IEEE Trans.*, 1988, **AP-36,** pp. 905–910
9 DENLINGER, E. J.: 'A frequency dependent solution for microstrip transmission lines', *IEEE Trans.*, 1971, **MTT-19,** pp. 30–39
10 JACKSON, D. R., and ALEXOPOULOS, N. G.: 'Analysis of planar strip geometries in a substrate-superstrate configuration', *IEEE Trans.*, 1986, **AP-34,** pp. 1430–1438
11 UZUNOGLU, N. K., ALEXOPOULOS, N. G., and FIKIORIS, J. G.: 'Radiation properties of microstrip dipoles', *IEEE Trans.*, 1979, **AP-27,** pp. 853–858,
12 NEWMAN, E. H., and FORRAI, D.: 'Scattering from a microstrip patch', *IEEE Trans.*, 1987, **AP-35,** pp. 245–251
13 FILON, L. N. G.: 'On a quadrature formula for trigonometric integrals', *Proc. Roy. Soc. Edin.*, 1928, **49,** pp. 38–47
14 POZAR, D. M.: 'Improved computational efficiency for the moment method solution of printed dipoles and patches', *Electromagnetics,* 1983, **3**(3–4), pp. 299–309.
15 YANG, H. Y.: 'Frequency dependant modelling of passive integrated circuit components' Ph. D. dissertation, University of California, Los Angeles, 1988
16 KATEHI, P. B., and ALEXOPOULOS, N. G.: 'Real axis integration of Sommerfeld integrals with applications to printed circuit antennas', *J. Math. Phys,* **24,** (3), pp. 527–533
17 KATEHI, P. B.: 'A generalized solution to a class of printed circuit antennas' Ph.D. Dissertation, University of California, Los Angeles, 1984
18 JACKSON, D. R. and ALEXOPOULOS, N. G.: 'An asymptotic extraction technique for evaluating Sommerfeld-type integrals,' *IEEE Trans.*, 1986, **AP-34,** pp. 1467–1470
19 RUMSEY, V. H.: 'Reaction concept in electromagnetic theory', *Phys. Rev.*, 1954, **94,** pp. 1483–1491
20 JONES, D. S.: 'A critique of the variational method in scattering problems', *IRE Trans.*, 1956, **AP-4,** pp. 297–301
21 HARRINGTON, R. F.: 'Time harmonic electromagnetic fields' (McGraw-Hill, 1961)
22 ALEXOPOULOS, N. G. and RANA, I. E.: 'Mutual impedance computation between printed dipoles', *IEEE Trans.*, 1981, **AP-29,** pp. 106–111
23 KATEHI, P. B., and ALEXOPOULOS, N. G.: 'On the modeling of electromagnetically coupled microstrip antennas – The printed strip dipole', *IEEE Trans.*, 1984, **AP-32,** pp. 1179–1186
24 JACKSON, R. W. and POZAR, D. M.: 'Full-wave analysis of microstrip open-end and gap discontinuities,' *IEEE Trans.*, 1985, **MTT-33,** pp. 1036–1042
25 OLTMAN, H. G. and HUEBNER, D. A.: 'Electromagnetically coupled microstrip dipoles', *IEEE Trans.*, 1981, **AP-29,** pp. 151–157
26 KATEHI, P. B., ALEXOPOULOS, N. G. and HSIA, I. Y.: 'A bandwidth enhancement method for microstrip antennas', *IEEE Trans.*, 1987, **AP-35,** pp. 5–12
27 KATEHI, P. B.: 'A generalized method for the evaluation of mutual coupling in microstrip arrays', *IEEE Trans.*, 1987, **AP-35,** pp. 125–133
28 ELLIOTT, R. S. and STERN, G. J.: 'The design of microstrip dipole arrays including mutual coupling Part I: Theory', *IEEE Trans.*, 1981, **AP-29,** pp. 757–760
29 STERN, G. J. and ELLIOTT, R. S.: 'The design of microstrip dipole arrays including mutual coupling. Part II: Experiment', *IEEE Trans.*, 1981, **AP-29,** pp. 761–765
30 HOWE, H.: 'Stripline circuit design' (Artech House, 1974)
31 DINBERGS, A.: 'Analysis and design of an array of electromagnetically coupled microstrip dipoles'. Masters thesis, University of Houston, 1988
32 POZAR, D. M., and SCHAUBERT, D. H.: 'Scan blindness in infinite phased arrays of printed dipoles', *IEEE Trans.*, 1984, **AP-32,** pp. 602–610
33 CASTANEDA, J. and ALEXOPOULOS, N. G.: 'Infinite arrays of microstrip dipoles with a superstrate (cover) layer', *IEEE Intl. Symp. Digest,* 1985, Vol 2, pp. 713–718

Chapter 6

Multilayer and parasitic configurations

D.H. Schaubert

6.1 Introduction

A standard configuration for a microstrip antenna is a single patch of conductor supported above a ground plane by a simple dielectric substrate and directly contacting an appropriate transmission line in order to couple power between the resonant patch antenna and the transmitter or receiver circuit. This is a simple configuration that is rugged and relatively easy to fabricate, but it is limited in its functional capabilities. The focus of this Chapter is antennas that consist of two or more metallic patches supported by one or more dielectric layers, or that consist of one metallic patch that is not directly contacting the transmission line that feeds it. These configurations are more complicated to design and fabricate, but they offer performance features that are not usually obtainable from the single-patch, single-dielectric configuration. These features include increased bandwidth, multiple frequency operation, dual polarisation, lower sidelobe levels, and ease of integration.

The examples presented here do not represent all of the configurations that have been successfully demonstrated, but they do represent many of the fundamental methods that have been successful. One method often employed is to stack patch radiators one above the other with intervening dielectric layers. This allows two or more resonant patches to share a common aperture area. The patches may be fed individually from microstriplines or coaxial probes, or only one or two may be fed directly while the others are coupled parasitically. Several examples are presented in Section 6.2. Another method that is employed utilises a single resonant patch that is coupled electromagnetically to a microstrip feed line. This form of parasitic coupling usually involves two layers of substrate, which may be on the same or opposite sides of the ground plane. If the feedline and patch are on opposite sides of the ground plane, a small aperture can be used to efficiently couple power through the ground plane. This configuration is presented in Section 6.3.

The final method presented here consists of one or more patches on the surface of a single substrate layer. These patches may be coupled parasitically to each other, and possibly to the feed line. These configurations are generally easier to fabricate than the multilayer configurations, but the performance often is not as desirable. Some examples are presented in Section 6.4.

6.2 Stacked elements for dual-frequency and dual-polarisation operation

Stacked elements have the advantage of providing two or more metallic patches within the same aperture area. This allows the antenna designer to obtain multiple frequencies with or without dual polarisation. Three typical configurations are depicted in Fig. 6.1, where case *a* represents a triple-frequency dual-polarised antenna [1], case *b* represents two linearly polarised elements operating at different frequencies [2], and case *c* represents two circularly polarised elements operating at different frequencies. These examples are representative of stacked configurations, which may use a single feed for multiple frequencies and different feeds for each polarisation (case *a*) or separate feeds for each frequency and polarisation (cases *b* and *c*). The dielectric substrates may differ in thickness or permittivity in order to control the bandwidths and sizes of the metallic radiator.

This Section is organised into two Subsections. The first describes antennas that utilise separate feeds for each frequency and polarisation. The second describes antennas that utilise a single feed to obtain multiple frequencies. Before presenting details of these stacked antennas, it is desirable to list some of their general advantages and disadvantages.

Advantages	*Disadvantages*
Multiple functions share common aperture.	Stacked substrates must be aligned and bonded.
Stagger tuning increases bandwidth.	Increased thickness and weight of the antenna structure.
Separately tuned radiators benefit from frequency and/or polarisation isolation	Fabrication of feed can be difficult, particularly when upper feed must attach to lower antenna.
Many configurations are possible to meet a variety of needs	Increasing total substrate thickness increases excitation of surface waves, resulting in lowered efficiency.
Different substrates may be selected for upper and lower antennas.	

Most of these advantages relate to increases in performance, whereas most of the disadvantages relate to fabrication and mechanical concerns. One performance parameter that has not received sufficient attention in the literature is the efficiency of antennas designed for increased bandwidth. Decreased efficiency has been reported for some configurations, but available experimental and analytical data are not sufficient to quantify the relationship between bandwidth and efficiency for the many configurations that have been demonstrated.

6.2.1 Antennas with separate feeds for each function

Stacked patches with separate feeds can take a variety of forms. Two of these are shown in Fig. 6.1*b* and *c*. Another form is depicted in Fig. 6.2. In all these configurations, the outer conductor of the upper feed connects the lower patch to the ground plane (Fig. 6.1*d*). This short-circuit connection, which actually presents a small inductive load to the antenna, can often be placed to have minimal effect on the antenna performance. However, it also can be placed to achieve desirable tuning effects, which will be described below.

The use of one-half-wavelength and one-quarter wavelength elements in the stacked configurations offers the designer considerable flexibility in selecting the operating frequencies of the antenna. Of course, the use of one-quarter-wavelength elements restricts the radiated field from that element to be linearly polarised, and requires the fabrication of via connections to form a short circuit along one side of the antenna. Nevertheless, this configuration is desirable for many applications and the designer can stack a one-quarter-wavelength patch and a one-half-wavelength patch of comparable sizes in order to obtain operating frequencies that are separated by approximately one octave. Independent control of ε_{lower} and ε_{upper} (Fig. 6.1*d*) provides another means of adjusting the two operating frequencies.

The piggy-back antenna depicted in Fig. 6.1*b* and described below is one example of an antenna that uses a one-half-wavelength lower element and a smaller one-quarter-wavelength upper element that is tuned to operate at a frequency moderately close to that of the lower element. This configuration leads to different beamwidths at the two closely spaced operating frequencies. (Care must be exercised if extremely close operating frequencies of the same polarisation are required because the stacked elements can exhibit strong mutual coupling, as described in Section 6.2.2.) As noted above, one of the principal disadvantages of the one-quarter-wavelength antennas is the need to fabricate a short-circuit boundary by plated-through holes, rivets or soldered pins.

The shapes of the patches that are stacked is somewhat arbitrary, although most designs that have been reported use similar shapes for the upper and lower patches. Square (or rectangular) and circular patches are the most commonly used shapes, but the wedge shape depicted in Fig. 6.1*b* is useful for conformal antennas on small conical bodies.

Two-frequency antennas: The piggy-back antenna in Fig. 6.1*b*, or any similar configuration, has been found to work well for radiating or receiving two independent, linearly polarised signals at different frequencies from a common aperture. The input impedance and radiation patterns of the piggy-back antenna

314 Multilayer and parasitic configurations

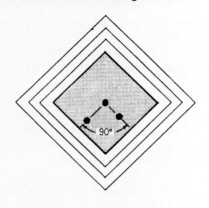

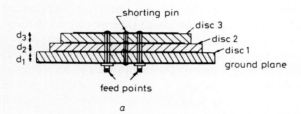

a

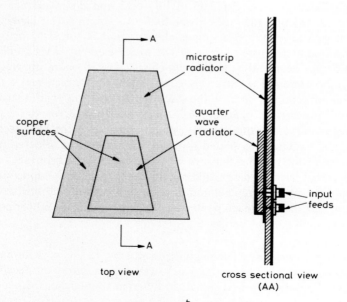

b

are typical of those obtained from ordinary patch antennas, although any asymmetries of the stacked geometry may lead to slight asymmetries in the patterns. Some asymmetries are evident in the E-plane patterns of Fig. 6.3. The ground plane for the antenna is modest in size, so that some diffraction effects

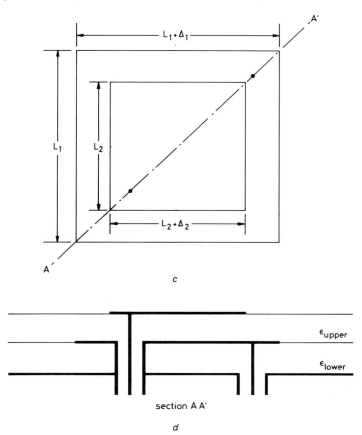

Fig. 6.1 *Typical stacked-patch configurations*
  a Triple-frequency, dual-polarised antenna [After Reference 1]
  b Piggy-back antenna for two linearly polarised frequencies
  c Two-frequency, circularly polarised antenna
  d Cross-section of stacked patches that utilise upper feed as inductive post in lower patch

are also evident. The operating frequencies and impedance bandwidths of each element are approximately the same as they would be in the absence of the other, except that the lower element must be considered as a post-tuned antenna [3]. The feed for the upper element may not, in general, pass through the lower element at a point where the electric field between the lower patch and the ground plane is zero. Then, the effect is similar to an inductive post in a

316 Multilayer and parasitic configurations

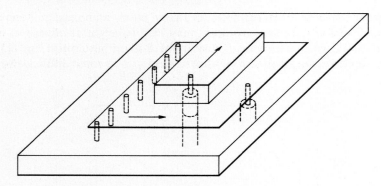

Fig. 6.2 *Orthogonally polarised version of piggy-back antenna with low-frequency, one-quarter-wavelength patch on bottom*

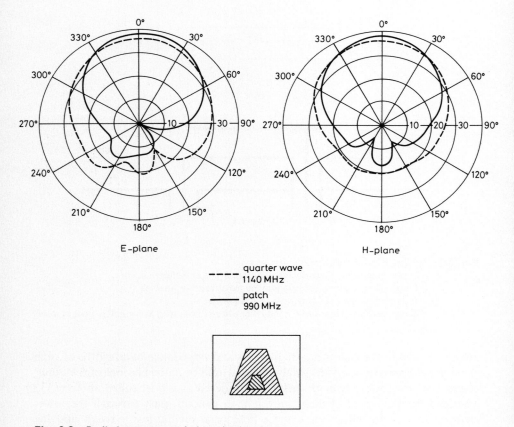

Fig. 6.3 *Radiation patterns of piggy-back antenna*

rectangular waveguide cavity. Several papers have analysed the post-tuned antenna [4–6] and have succeeded in predicting quite accurately the resonant frequency and input impedance. However, a first-order approximation is easily obtained by using the transmission-line model [7] for a rectangular antenna as

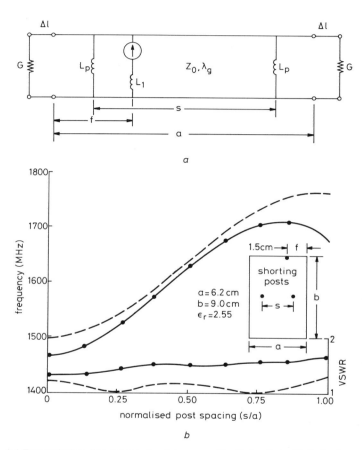

Fig. 6.4 *(a) Transmission-line model of patch symmetrically loaded with inductive posts. (b) Operating frequency (upper curves) and VSWR (lower curves) of post-tuned antenna on 1·6 mm Teflon fibreglass substrate*
- - - *calculated,*
— *measured [taken from Reference 3]*

is done in Reference 3. Typical results taken from Reference 3 are shown in Fig. 6.4. The values used in the calculations for the post reactances were obtained from

$$X_p = \frac{377}{\sqrt{\varepsilon_r}} \tan \frac{2\pi t}{\lambda} \quad (t = \text{substrate thickness}) \tag{6.1}$$

which is Carver's estimate [8] that the post's reactance should be the same as the input impedance to a piece of short-circuited transmission line. The use of only one inductive post yields similar results, but the amount of frequency increase caused by the post is approximately one-half that caused by two posts (Fig. 6.5).

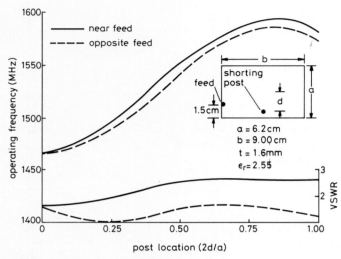

Fig. 6.5 *Measured operating frequency and VSWR of antenna with one post near the feed or one post opposite the feed*

Two-Polarisation Antennas: A simple rectangular patch antenna with two feeds located along perpendicular centre lines of the antenna is an easy choice for radiating or receiving two independent, orthogonal, linearly polarised signals. However, this antenna has the disadvantage that the width of the patch for the horizontal mode is fixed by the desired operating frequency of the vertical mode, which determines L_v (Fig. 6.6). Therefore, the H-plane beamwidth of the horizontal mode is fixed by the vertical frequency, and vice versa. This limitation can often be overcome by using a stacked configuration such as the one in Fig. 6.7. The upper feed of the model shown passes through the lower element at a voltage null so that the primary effect on the lower antenna is some loading due to the upper substrate, but this effect is easily compensated in the design. Thus, the patch widths and H-plane beamwidths can be controlled as long as the performance requirements allow the designer to fit the upper vertically polarised antenna within the boundaries of the lower, horizontally polarised antenna. If the use of differing permittivities for the two substrates is permissible, a higher-permittivity upper substrate helps to reduce the size of the upper patch.

The radiation patterns and input impedances of the individual antennas are similar to those of comparable antennas operating in an isolated environment. Some typical radiation patterns are shown in Fig. 6.8. The impedances of the

Multilayer and parasitic configurations 319

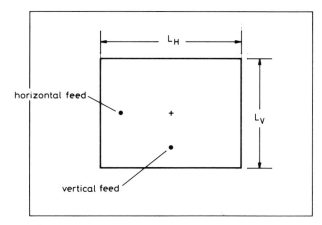

Fig. 6.6 *Rectangular patch antenna with feeds for horizontal and vertical polarisation at different frequencies*

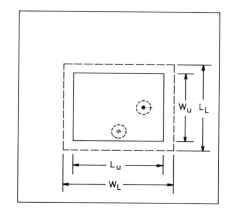

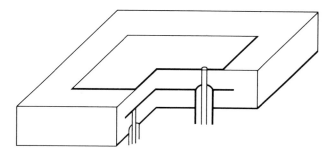

Fig. 6.7 *Stacked patches for orthogonal linear polarisation at two independent frequencies*

320 Multilayer and parasitic configurations

two elements, when both are tuned to operate at 3.465 GHz, are shown in Fig. 6.9. The isolation between the feeds for the two orthogonally polarised elements is greater than 18 dB over the operating band.

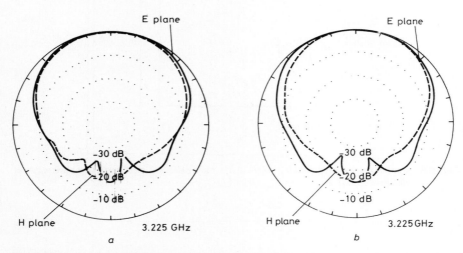

Fig. 6.8 Typical radiation patterns of stacked orthogonal patches on a small ground plane
  a Upper patch
  b Lower patch

6.2.2 Antennas for multiple frequencies and increased bandwidth

One form of the stacked patch antenna for multiple frequencies is depicted in Fig. 6.10a. This antenna is similar to the one in Fig. 6.1a [1], where three patches permit three separate operating frequencies to be obtained at each of the orthogonally polarised feed ports. This form is typical of that used for multiple-frequency antennas where the feed probe passes through a clearance hole in the lower patch and connects to the upper patch. This method of feeding couples strongly to the resonance of each patch, even though the resonant frequencies may be far part. This strong coupling is probably the result of currents on the feed probe directly exciting the cavity of each patch antenna through which it passes.

A second form of the stacked patch antenna is depicted in Fig. 6.10b, where only the lower patch is fed directly and the upper patch is coupled parasitically. This form of the antenna is widely used for increased bandwidth at a single operating frequency [9–11], and it can be used for dual polarisation by inserting a second, orthogonally polarised feed similarly to the multiple-frequency antenna in Fig. 6.1a.

The characteristics of both forms of the stacked, single-feed patches have been investigated and compared to a single patch. For the data presented below, the patches were circular and fabricated on separate sheets of Duroid 5870

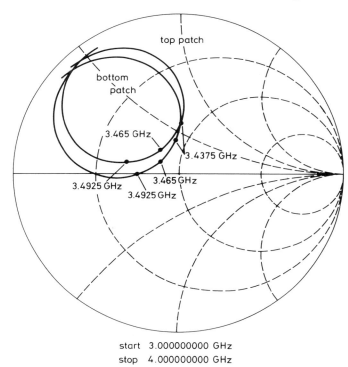

Fig. 6.9 *Input impedance of stacked orthogonal patches*
Each substrate is 1·6 mm thick, $\varepsilon_r = 2·2$; upper patch is 29 × 23 mm and is fed 11·6 mm from edge; lower patch is 27 × 36 mm and is fed 8·0 mm from edge

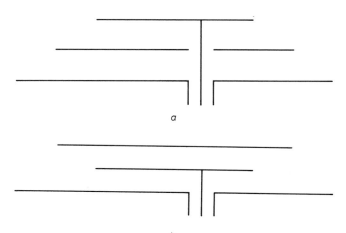

Fig. 6.10 *Single feed configurations for multiple frequencies or increased bandwidth*
    a Top-feed model
    b Bottom-feed model

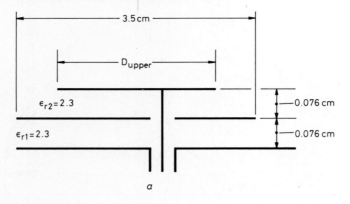

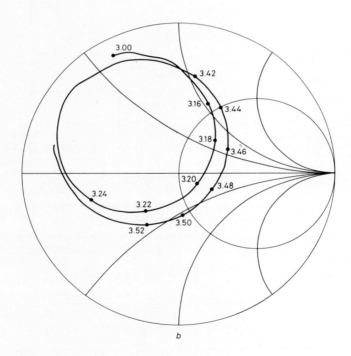

Fig. 6.11 *Input impedances of dual-frequency, top-fed circular discs*
  a Dimensions of the model
  b D_{upper} = 3·3 cm.
  c D_{upper} = 3·45 cm
  d D_{upper} = 3·55 cm

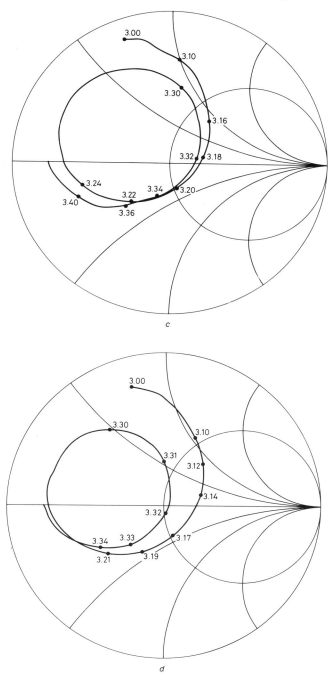

Fig. 6.11 *Cont.*

324 Multilayer and parasitic configurations

($\varepsilon_r = 2\cdot3$) substrate and then carefully aligned in the stacked configuration. The input impedance of the multiple-frequency form is shown in Fig. 6.11 for three different diameters of the upper patch. These and additional data are summarised in Table 6.1, which tabulates the maximum value of input resistance and the associated frequency and Q. As noted by Long and Walton [12], the resonant frequency and resistance of the lower patch are relatively unaffected by changes in the diameter of the upper patch. However, the resonant frequency and resistance of the upper patch both decrease as the patch diameter increases. In fact, increasing the upper patch diameter to 3·6cm results in an upper resonance at 3·277 GHz with a resistance of 16 ohms. It appears that the upper patch in this configuration should always be smaller (or only slightly larger) than the lower patch.

Table 6.1 *Impedance characteristics of multiple-frequency stacked patches*

Upper patch diameter cm	R_{max}, ohms	f_r, GHz	Q
3·3	75	3·185	61
	93	3·465	60
3·4	75	3·190	58
	83	3·351	80
3·45	78	3·175	40
	72	3·316	92
3·5	78	3·180	50
	62	3·331	115
3·55	80	3·137	30
	54	3.317	144

Lower patch diameter: 3·5 cm
$\varepsilon^{lower} = \varepsilon^{upper} = 2\cdot32$
$d^{lower} = d^{upper} = 0\cdot079$ cm $= 0\cdot030$ in
Feed is 0.6 cm from centre

An equivalent circuit has been used successfully to model the input impedance of the stacked, dual-frequency antenna. This circuit, depicted in Fig. 6.12, consists of two coupled parallel resonant circuits and a series inductor to model the feed inductance that is usually observed in probe-fed patches. The values shown in Fig. 6.12 are for the 3·45 cm-diameter upper patch. These values were obtained by optimising the fit between the impedance of the measured data and the model. The mutual inductance represents electromagnetic coupling between the two discs.

The wide-bandwidth form of stacked patches (Fig. 6.10b) has also been studied and three typical results are shown in Fig. 6.13. The best results for increased bandwidth are obtained when the two patches are nearly the same

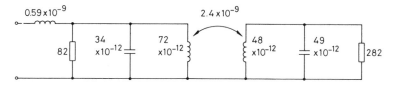

Fig. 6.12 *Equivalent circuit to model input impedance of Fig. 6.11c*

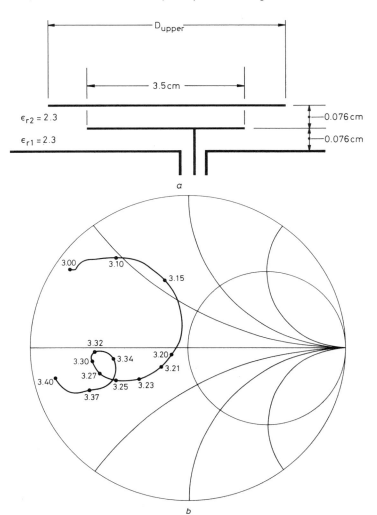

Fig. 6.13 *Input impedance of wide-bandwidth bottom-feed circular discs*
  a Dimensions of the model
  b D_{upper} = 3·45 cm
  c D_{upper} = 3·50 cm
  d D_{upper} = 3·55 cm

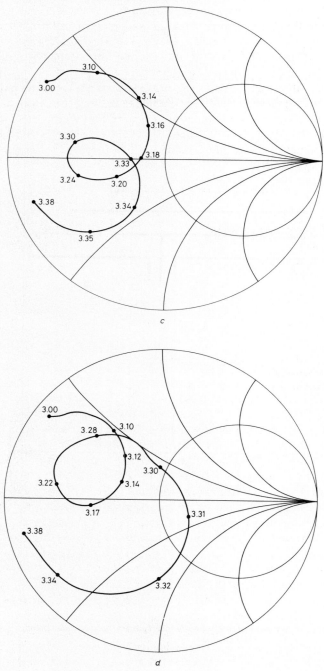

Fig. 6.13 Cont.

size. Since the feed probe does not pass through the cavity of the upper patch, the coupling to that patch is very weak unless it is comparable in size, or larger than the lower patch. Thus, the preferred configuration for multiple frequencies is to feed the top disc and use a larger bottom disc, while the preferred configuration for increased bandwidth is to feed the lower disc and use an equal or slightly larger top disc. By properly adjusting the feed distance from the patch centre, the impedance loop in Fig. 6.13c can be made to encircle the centre of the Smith chart. This has produced a model fabricated from two sheets of 0·076 cm-thick (0·030 in) Duroid 5870 ($\varepsilon_r = 2·3$) having a VSWR < 2 bandwidth of 5% at 3·3 GHz. This compares with approximately 3% bandwidth for a single patch on 0·15 cm-thick substrate.

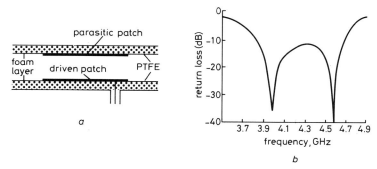

Fig. 6.14 *Stacked discs for increased bandwidth [After Reference 10]*
  a Antenna structure
  b Return loss

Chen *et al.* [10] have provided data on the operation of antennas in the form of Fig. 6.10b for either wide bandwidth or dual frequencies. By using a relatively thick, low-permittivity foam substrate between the patches, they have achieved a 20% bandwidth for VSWR < 2 and a 10% bandwidth for VSWR < 1·22. A typical result is shown in Fig. 6.14. They measured a gain of 7·9 dB at 4·25 GHz for the 10% bandwidth antenna. They also noted that stacked patches with the lower patch on a relatively thin substrate exhibit lower cross-polarised radiation than a single patch fabricated on a thick substrate to achieve comparable bandwidth. This result is consistent with observations that direct radiation from feed probes can be significant for antennas on thick substrates [13]. For dual-frequency operation, Chen *et al.* present a graph showing the relationship between the patch diameters, separations and operating frequencies. However, they do not indicate if the results are limited to their particular choices of dielectric substrates.

An analysis of stacked, circular patch antennas has been conducted by Araki *et al.* by using the spectral domain Green's function [14]. They solve the eigenvalue problem to find the complex resonant frequencies of the structure, and they calculate the input impedance. Their results are compared to measurements in Fig. 6.15, which shows good agreement for a wide-bandwidth example. No results are presented for a dual-frequency example, but their analysis should be valid for this case, as well.

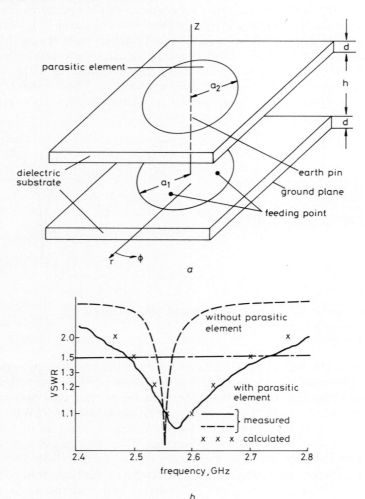

Fig. 6.15 *(a) Geometry of stacked circular patches for spectral domain computations (b) VSWR characteristics for $h = 10.0$, $a_1 = 20.8$ mm, $a_2 = 21.0$ mm, $d = 1.6$ mm, $\varepsilon_r = 2.55$ [After Reference 14]*

Another antenna configuration that resembles those of Fig. 6.10 has been proposed by Paschen [15]. However, this antenna actually functions more like that of Fong et al. [16], as is apparent from alternative configurations shown by Paschen. Three antenna configurations (from Paschen [15], Hall [17], and Fong et al. [16]) are shown in Fig. 6.16. The objective of these configurations is to create a series capacitance, which adds a degree of freedom that can be used in conjunction with the inductance of the feed probe and the resonant patch impedance to obtain increased bandwidth in the manner of Griffin and Forrest [18]. Alternative configurations suggested by Paschen are shown in Fig. 6.17.

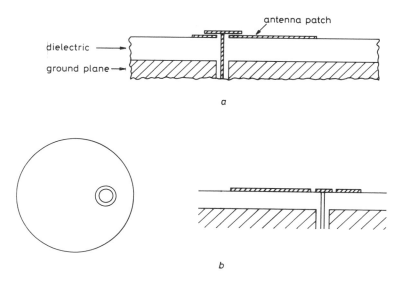

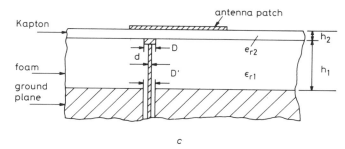

Fig. 6.16 *Proximity coupling by means of series capacitance between patch and feed probe*
  a Paschen's design [After Reference 15]
  b Hall's design [After Reference 17]
  c Fong et al design [After Reference 16]

The cylindrical form of the capacitor has been used to fabricate an L-band antenna covering the global-positioning satellite frequencies of 1227 MHz and 1575 MHz as well as the NDS frequency of 1381 MHz. Air dielectric was used and a quadrature feed network for circular polarisation was implemented on a thin circuit at the bottom of the antenna cavity (Fig. 6.18). By adjusting the

330   Multilayer and parasitic configurations

probe diameter to control probe inductance and also adjusting the cylindrical capacitor, the antenna was tuned to provide VSWR less than 2 for a bandwidth of 25%. A typical radiation pattern is also shown.

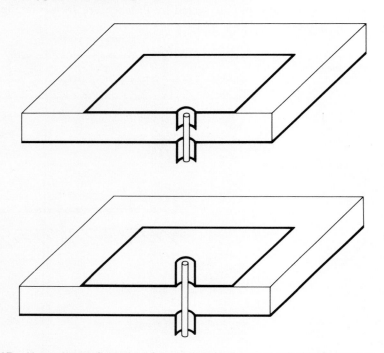

Fig. 6.17 *Alternative configurations for coupling by means of cylindrical capacitors*

6.3 Two-sided aperture-coupled patch

Microstrip antennas have been popular elements for planar and conformal arrays. A traditional means of fabricating these arrays that takes maximum advantage of the economies of printed-circuit fabrication involves the layout of radiating elements and a feed network on a single surface of a grounded substrate. This minimises the size and weight of the total array and requires the installation of only one coaxial connector to feed the array, thus reducing the cost. However, the microstrip feed network radiates small amounts of power that can degrade the sidelobe and polarisation characteristics of the array. Also, the radiators and feed lines occupy much of the available area, leaving little space for the phase shifters that are required for beam steering. Furthermore, for monolithic phase shifters, the substrate must be GaAs or another appropriate material, which is not a desirable substrate for the radiators [19].

Most of these problems can be alleviated by using a two-layer structure that has the radiating elements and their substrate on one side of a ground plane and

Multilayer and parasitic configurations 331

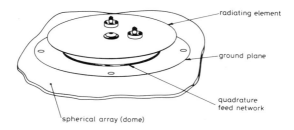

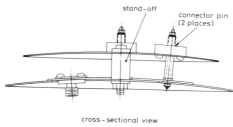

a

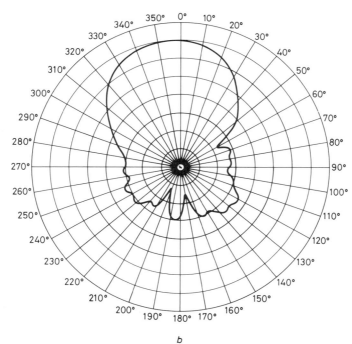

b

Fig. 6.18 *Circularly-polarised L-band antenna utilising air dielectric and cylindrical capacitor coupling [After Reference 15]*
 a Antenna for use on spherical surface
 b Typical radiation pattern at 1381 MHz

the feed network and its substrate on the other. In the past, arrays fabricated in this fashion have utilised a via connection probe at each element in order to transfer power from the feed network to the radiators (Fig. 6.19). However, these via connections are increasingly difficult to fabricate as the frequency increases and excess probe inductance makes the antenna difficult to match. Also, the use of GaAs as a feed substrate complicates the fabrication because it is more difficult to drill the via hole.

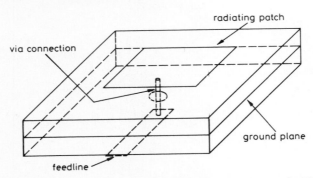

Fig. 6.19 *Two-sided design with via connection between feed line and radiating patch*

A useful alternative to the via connection probe is aperture coupling [20, 21]. In this configuration, power is coupled from the microstripline feed to the radiating patch through an electrically small aperture in the ground plane (see Fig. 6.20). No electrical connection is required and the performance is relatively insensitive to small errors in the alignment of the two circuits. Single elements have performed well at frequencies as high as K_u-band [22].

Several additional advantages are obtained by the use of a two-sided configuration. These include isolation of the feed network from the radiating aperture of an array, which eliminates the spurious-feed-network radiation that can degrade polarisation and sidelobe levels [13]. Also, the two-sided configuration provides two distinct microstripline media so that the antenna substrate can be chosen to optimise the performance of the radiating patches (e.g. low ε_r to improve radiation, increase bandwidth, and move scan blindness further from broadside), and the feed substrate can be chosen independently to optimise feed performance (e.g. high ε_r to reduce circuit size or the use of GaAs for active integrated feeds). Furthermore, the feed substrate may be composed of many diced wafers without introducing substrate discontinuities into the radiating side of the array.

The aperture-coupled patch antenna resembles a traditional microstrip-array element with a microstrip patch antenna on a substrate with relative permittivity ε_r^a and a feed network on a substrate with relative permittivity ε_r^f. These are separated by a common ground plane. In the traditional configuration, a via connection carries power from the feed network to the radiating patch. In the new design, the via connection is replaced by an electrically small aperture (Fig. 6.20) that couples power efficiently to the patch and is easy to manufacture. An open-circuited length of the microstripline extending beyond the aperture provides an additional degree of freedom to be used for impedance matching and bandwidth enhancement. This stub, together with the aperture length, can be used to control the input impedance over a wide range of values, as illustrated in Fig. 6.21. The calculated results were obtained by using the moment-method technique of Reference 21, and have been found to be in good agreement with measured results.

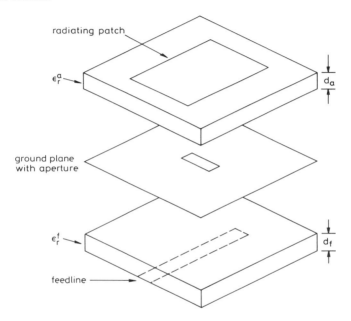

Fig. 6.20 *Microstrip radiator electromagnetically coupled to microstripline feed in two-sided configuration [After Reference 21]*

The performance illustrated in Fig. 6.21 can be understood in terms of the equivalent circuit [23] in Fig. 6.22, which is a two-port representation of the microstrip feed line as it passes the aperture. The effect of the aperture and patch is approximately a lumped series load on the microstripline of characteristic impedance Z_c. The input impedance of the patch, Z_{in}, can be obtained by evaluating the series combination of the aperture/patch circuit and a short stub terminated in an open circuit. The series inductance represents the effect of the electrically small (below resonance) aperture, and the impedance Z_{oc} represents the open-end effects of the microstrip stub. As the stub increases in length, the input impedance at a fixed frequency approximately follows a constant-resistance circle in Fig. 6.21a, with the reactance increasing according to the reactance of the open-circuited stub. The effect of increasing the aperture size is

334 Multilayer and parasitic configurations

similar to that of increasing the size of a hole coupling power from a waveguide to a resonant cavity. When the aperture is small, the patch is undercoupled and the resonant resistance is less than the characteristic impedance of the feed line. As the aperture size increases, the coupling and the resonant resistance increase. A wide range of resistance and reactance values can be achieved by adjusting the aperture length and the stub length (It has been found that narrow slot apertures offer the most effective coupling in this configuration.)

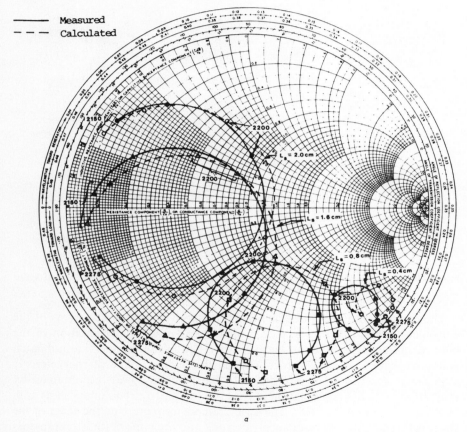

a

Fig. 6.21 *Impedance of aperture coupled patch $\varepsilon_r^f = \varepsilon_r^a = 2\cdot 54$, $d_a = d_f = 0\cdot 16$ cm; patch length = 4 cm; patch width = 3 cm, aperture width = 0·16 cm*
a Impedance for various stub lengths with aperture length = 1·12 cm
b Impedance for various aperture lengths with stub length = 2·0 cm
[Reproduced from Reference 21]

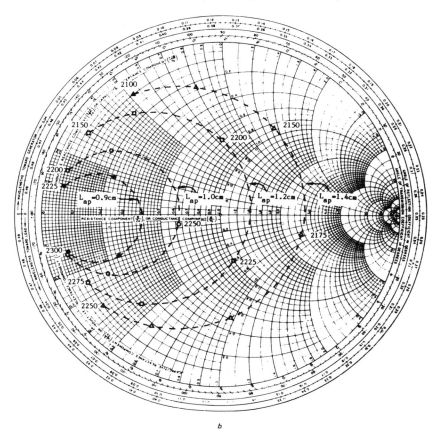

Fig. 6.21 *Cont.*

The bandwidth and radiation patterns of the aperture-coupled patch antenna are essentially the same as those of a probe-fed antenna on the same substrate. The peak radiation from the aperture on the feed side of the ground plane has been computed and measured to be at least 20 dB below the peak of the patch radiation for the antennas that have been tested. This is an important characteristic of the antenna that makes it useful in planar arrays, as compared to a simple microstripline-fed slot that has a bidirectional radiation pattern.

Arrays of aperture-coupled patches can be built by using series feeding or corporate feeding. Examples of E-plane arrays are illustrated in Fig. 6.23. Both types of arrays have been built at C-band. An eight-element corporate-fed array with patches on 0·159 cm $\varepsilon_r^a = 2·2$ and feed lines on 0·064 cm $\varepsilon_r^f = 10·2$ has performed favourably, demonstrating the feasibility of using a substrate like GaAs for the feed network and a lower-permittivity substrate for the radiating elements. The use of a low-ε_r antenna substrate will increase the angle at which scan blindness occurs due to surface waves on the antenna substrate. However, a blindness also will occur due to surface waves on the feed substrate, and methods may be needed to control this phenomenon at very short wavelengths where the substrate thickness may exceed $0.02-0.03\lambda_0$.

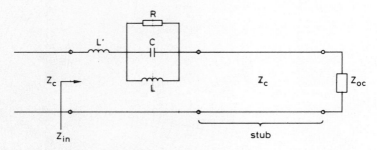

Fig. 6.22 *Equivalent two-port network for aperture backed by a patch antenna and fed by a microstripline*

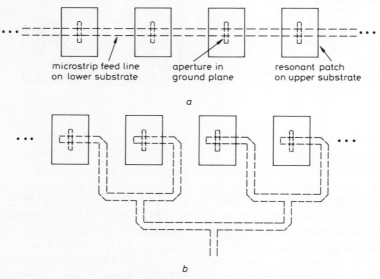

Fig. 6.23 *Aperture-coupled microstrip antenna arrays*
  a Series feed
  b Corporate feed

The array dimensions in Table 6.2 could be scaled to yield operation at 25 GHz, which would lead to a feed substrate thickness of 0.005 in. The use of 0.005-in GaAs ($\varepsilon_r = 12.8$) for the feed substrate would require only a slight modification in the dimensions and would allow monolithic phase shifters to be integrated with the feed network. Mutual coupling levels in the array are given by the S parameters in Table 6.3, and the input impedance of a typical interior element is shown in Fig. 6.24. Radiation patterns of the array, fed by an external coaxial power divider and appropriate lengths of line to steer the beam, are shown in Fig. 6.25. These patterns agree well with expected results and the radiation behind the ground plane is 20 dB below the main beam, despite the relatively small ground plane (approximately $2\lambda_0 \times 4\lambda_0$).

Table 6.2 *Eight-element, E-plane array*

Patch length = 1·78 cm	Element spacing = 3·0 cm
Patch width = 2·54 cm	Stub length = 0·42 cm
Slot length = 0·83 cm	Slot width = 0·056 cm
$\varepsilon_r^{ant} = 2 \cdot 22$	$\varepsilon_r^{feed} = 10 \cdot 2$
$d^{ant} = 0 \cdot 159\, cm$	$d^{feed} = 0 \cdot 064\, cm$

Table 6.3 *Mutual coupling in array with $\lambda_0/2$ spacing*

	$|S_{ij}|$, dB	$\angle S_{ij}$, deg
S_{41}	−30·7	−79
S_{42}	−24·8	102
S_{43}	−21·6	−46
S_{45}	−22·1	−52
S_{46}	−27·0	103
S_{47}	−30·3	−110
S_{48}	−31·8	70

A variation of the aperture-coupled patch is shown in Fig. 6.26 [24]. This configuration permits the feed network to occupy as much space as is needed in the depth dimension. This use of depth has been essential in most microwave phased arrays that have been fabricated. Other variations with perpendicular feed substrates have been built with non-contacting feeds [25], which offer significant advantages in fabrication.

6.4 Parasitic elements on antenna substrate

The microstrip antenna design is so appealing that engineers are inclined to capitalise on its conformability, manufacturability, and ruggedness for applications that are contrary to its inherent electrical characteristics. In particular, considerable effort has been devoted to increasing the operating bandwidth of the microstrip antennas. Several of these efforts have been described in previous

338 Multilayer and parasitic configurations

Sections of this Chapter. However, a recent study [26] has shown that the bandwidth obtainable from a microstrip antenna is approximately proportional to its volume measured in λ_0^3. This phenomenon is consistent with accepted antenna theory [27]. Therefore, the bandwidth of a simple microstrip antenna

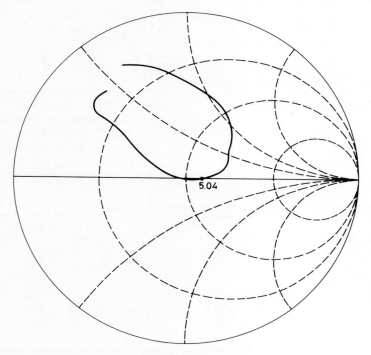

Fig. 6.24 *Input impedance of element 6 in corporately fed E-plane array having dimensions in Table 6.2. All other elements terminated in 50 Ω*

can be increased by increasing its length, width, or substrate thickness. However, the length of a fundamental-mode antenna must be approximately one-half wavelength in the dielectric substrate, so the antenna can be lengthened only if the substrate permittivity is lowered. Unfortunately, feed lines and probes radiate more on low-ε_r substrates, so this technique must be used with care when cross-polarisation and sidelobe levels are important. Increasing the antennas's width is fairly straightforward, but higher modes of the antenna can be excited if the width is increased to one or two wavelengths. Also, elements larger than approximately $\lambda_0/2$ cannot be used in scanning arrays owing to undesirable grating lobes.

In this Section, some alternatives to multiple-layer structures and to increasing the substrate thickness, which can lead to increased feed radiation and surface wave excitation, are presented. The fundamental approach here is to

Multilayer and parasitic configurations 339

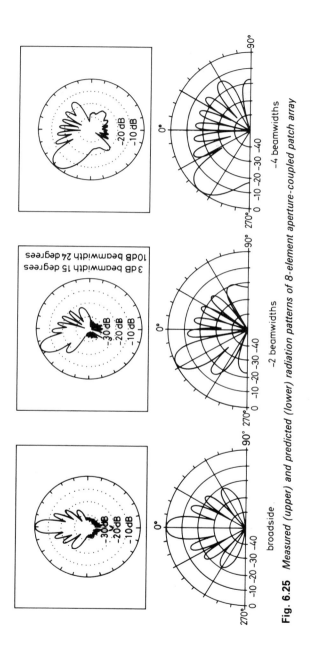

Fig. 6.25 Measured (upper) and predicted (lower) radiation patterns of 8-element aperture-coupled patch array

340 Multilayer and parasitic configurations

create a double-tuned resonance by adding parasitic resonators on the same substrate surface as the primary microstrip antenna. Many of the elements described here have two disadvantages that must be weighed against the benefits of increased bandwidth: (i) the physical area of the element is increased, and (ii) the radiation pattern exhibits asymmetries that may change with frequency.

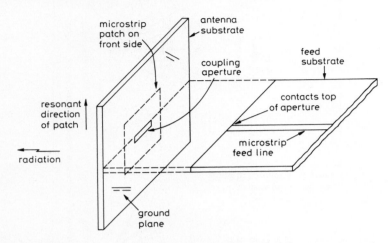

Fig. 6.26 *Aperture-coupled patch with perpendicular feed network. (Reproduced from Reference 24)*

The first example (Fig. 6.27) utilises narrow conducting strips adjacent to the driven radiator [28] in order to alter its impedance and radiation properties. These strips couple to the non-radiating edges of the antenna and significantly modify its impedance. A square, edge-fed microstrip antenna with VSWR = 4 can be matched to VSWR = 1·2 by using the parasitic strips. The antenna works best when the parasitic strips are slightly longer than the patch. The performance of the antenna is strongly affected by the separation between the strips and the patch, and the best performance has been obtained when the separation is 2·5 to 3 times the substrate thickness.

The interaction between the strips and the patch changes the resonant frequency of the patch by a few percent. However, it is possible to broaden the impedance bandwidth of the antenna by stagger-tuning the strips. Fig. 6.27b shows the VSWR of the antenna with and without the strips. Radiation patterns of the parasitic-tuned antenna are shown in Fig. 6.28. The *H*-plane radiation pattern is slightly skewed by the asymmetry of the stagger-tuned strips.

The coupled-resonator approach has been extended to include up to four parasitic elements [29] and to provide for direct as well as electromagnetic coupling to the parasitic elements [30]. Parasitic elements coupled to the radiating edge of the antenna are described in Reference 29, where the segmentation method [32] has been used to accurately predict the input VSWR. Examples of these antennas are depicted in Fig. 6.29. All of the structures described in References 29–31 exhibit multiple tuning, and an element like the one in Fig. 6.29a has been measured to have 25% impedance bandwidth (VSWR < 2) at

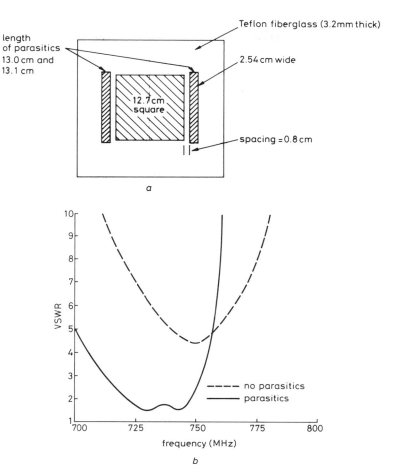

Fig. 6.27 *Parasitically tuned antenna with narrow strips adjacent to nonradiating edges [After Reference 28]*
  a Dimensions for UHF model
  b VSWR with and without parasitics

3·16 GHz on 0·125-in substrate, $\varepsilon_r = 2·55$. In order to achieve this bandwidth, the total element area is approximately one wavelength square. Also, the radiation pattern changes with frequency within the band of operation as the segments contribute with differing amplitudes and phases. This is illustrated in the

E-plane patterns of Fig. 6.30, which shows the reported performance of the antenna in Fig. 6.31. The VSWR < 2 bandwidth is 24% (about seven times that of a single patch on the same substrate). However, the radiation pattern changes appreciably at the four frequencies shown in these plots. No data on cross-

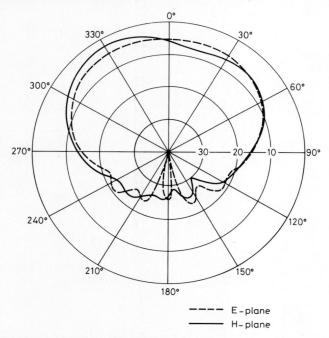

Fig. 6.28 *Radiation patterns of parasitically tuned antenna [After Reference 28]*

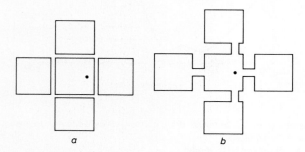

Fig. 6.29 *Antennas with four parasitic elements*
  a Electromagnetically coupled parasitics [After Reference 29]
  b Directly coupled parasitics [After Reference 30]

polarised radiation are presented. The usefulness of this, or the other parasitically coupled antennas, will depend on the designer's ability to accept increased element area and pattern asymmetries and variations across the operating band.

Multilayer and parasitic configurations 343

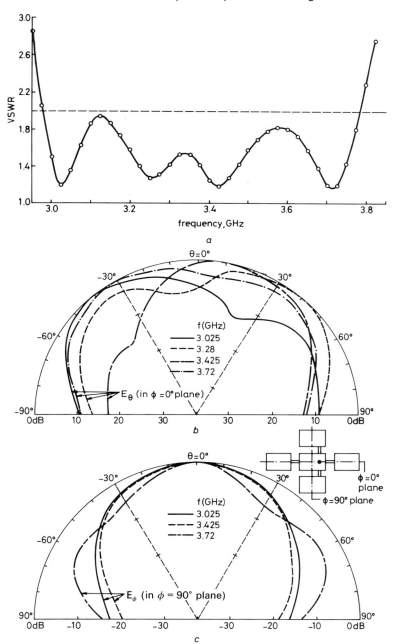

Fig. 6.30 *Measured performance of antenna in Fig. 6.31* [After Reference 30]
  a VSWR
  b E_θ in $\phi = 0°$ plane
  c E_ϕ in $\phi = 90°$ plane

344 Multilayer and parasitic configurations

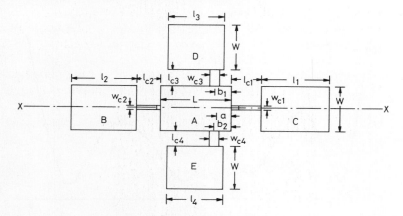

Fig. 6.31 *Dimensions of antenna used to obtain results of Fig. 6.30*
$L = 3.0$ cm $W_{c1} = W_{c2} = 0.025$ cm
$W = 2.0$ cm $W_{c3} = W_{c4} = 0.44$ cm
$l_1 = 2.85$ cm $b_1 = b_2 = 0.71$ cm
$l_2 = 2.635$ cm $a = 0.48$ cm
$l_3 = l_4 = 2.35$ cm $\varepsilon_r = 2.55$ cm
  $h = 0.318$ cm

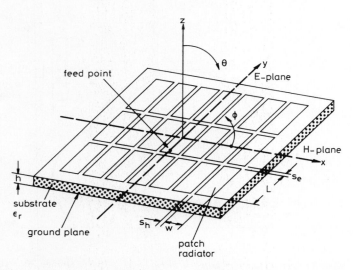

Fig. 6.32 *Planar array of parasitically coupled microstrip elements [Reproduced from Reference 33]*

An extension of parasitically coupled elements leads to a parasitically coupled array [33]. In this configuration, a single driven patch is coupled to closely spaced adjacent patches, which are coupled to additional patches to form a linear or planar array (Fig. 6.32). Arrays of this type have been developed experimentally to produce linear or circular polarisation. The distance between patches is reported to be 0·1 to 2 times the substrate thickness and can be

Fig. 6.33 *Three parasitically coupled arrays [Reproduced from Reference 33]*

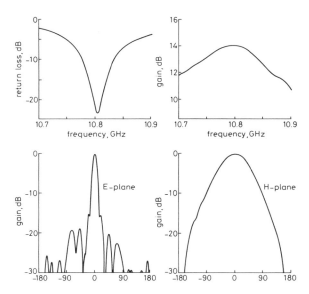

Fig. 6.34 *Measured performance of 7-element, E-plane linear array of parasitically coupled patches [Reproduced from Reference 33]*

adjusted to control the power distribution in rows and columns for sidelobe minimisation. This distance also affects the input impedance, which is approximately equal to the impedance of a single patch divided by the number of patches.

346 Multilayer and parasitic configurations

The performances of the three antennas in Fig. 6.33 are reported in Reference 33. The linear array operates at 10·8 GHz and has the characteristics shown in Fig. 6.34. The 5 × 3 array is reported to have 9 dB gain at 8·55 GHz and −26 dB sidelobes. The 2 × 2 array is designed to produce circular polarisation by feeding the driven patch at a corner. The ellipticity was measured to be less than 2 dB over a bandwidth of 130 MHz at 5·83 GHz.

Another antenna that consists of a driven element and additional metallisation on the substrate surface is the microstrip disc antenna with a short-circuited annular ring [34]. The antenna (Fig. 6.35) can be considered as a microstrip disc

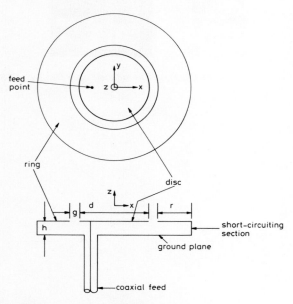

Fig. 6.35 *Microstrip disc antenna with annular ring*
Centre frequency = 5·21 GHz, h = 3·18 mm, d = 19·5 mm, g = 2·25 mm, r = 9 mm, ε_r 2·5 [Reproduced from Reference 34]

with a parasitic annular ring, or as a cylindrical cavity with an annular slot. The dimensions of the cavity produce a resonance at approximately the same frequency as the resonance of the microstrip disc. The element produces a circularly symmetric radiation pattern with 10 dB beamwidth equal to 160°, which is appropriate for illuminating a reflector with $F/D = 0\cdot 3$. By adjusting the slot width g, VSWR < 2 bandwidths of 10% have been obtained, but significant gain reduction is experienced at the upper portions of the operating band. It is speculated in Reference 34 that the losses are due to a second resonance that is contributing to the increased bandwidth. The radiation patterns in Fig. 6.36

(taken at band centre) illustrate the wide beamwidth and low cross-polarisation levels obtainable with the antenna. Maximum cross-polarisation within the 10 dB beamwidth is reported to be −21 dB over the band 5·00–5·44 GHz.

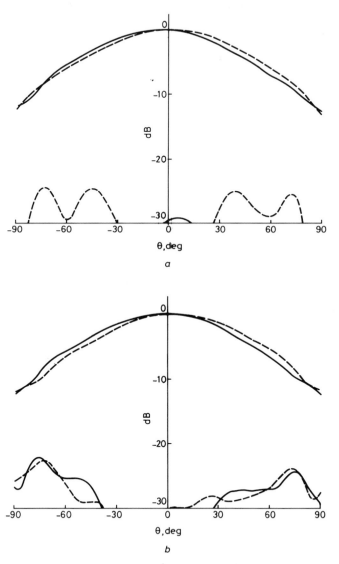

Fig. 6.36 *Radiation patterns of disc antenna with short-circuited annular ring [Reproduced from Reference 34]*
 a Principal planes
 b Diagonal planes

Table 6.4 Antenna characteristics

Antenna type	Section	Performance characteristics	Fabrication requirements
Piggy-back	6.2.1	Independent antennas share aperture Multiple frequency and/or polarisation Upper feeds form tuning posts in lower antennas	Align and bond multiple layers Coaxial feed lines soldered to ground plane and to lower patches
Stacked patches with upper patch fed	6.2.2	Multiple frequencies Can be dual polarised	Align and bond multiple layers Upper patches smaller
Stacked patches with lower patch fed	6.2.2	Increased bandwidth Can be dual polarisation Reduces cross-polarisation May reduce efficiency	Align and bond multiple layers Upper patch larger or same size
Capacitive feed tuning	6.2.2	Increased bandwidth May reduce efficiency No effect on pattern	Precise control of capacitor gap
Aperture coupling	6.3	Feed network isolated from radiation More parameters to control impedance Can be used in most cases where probe could be used	No via connection Microstrip stub occupies space on substrate Independent choice of substrates Alignment not critical, but bonding is required

		Patch radiation and bandwidth not affected Some radiation from stub and aperture	
Parasitic beside driven patch	6.4	Increased bandwidth Usually some pattern asymmetry Pattern may change with frequency Large bandwidth requires greater than $\lambda_0/2$ size	Single-layer fabrication Direct or electromagnetic coupling to parasitics

6.5 Summary

Microstrip antennas composed of multiple conducting patches or feed lines electromagnetically coupled to the resonant, radiating patch offer several advantages over the traditional single patch connected to a feed line or feed probe. These advantages include increased bandwidth or multiple frequency operation, dual polarisation, and control of input impedance. However, the antennas are often more complicated to fabricate or they may require more surface area, so that the designer may be required to sacrifice one desirable feature in order to obtain another one. Table 6.4 contains a summary of several key features of the antennas described in this Chapter, and can be used as a guide in choosing the best configuration for a particular purpose.

Most of the structures in this Chapter have been modelled by using either the cavity model or moment methods, and the segmentation method has also been applied to some of the configurations. In general, the analysis techniques provide a good qualitative model for the antennas, thus providing the designer with the insights needed to develop a functional antenna. Many of the existing analyses also have shown good quantitative agreement for the cases appearing in the literature, but the ranges of validity are not generally known, and the computed results may not be sufficiently accurate for many practical values of substrate permittivity and thickness or for variations in patch geometry. Nonetheless, multiple-layer microstrip antennas and antennas utilising parasitic coupling to the feed or to other resonant patches offer distinct advantages for many systems applications, and are likely to be used in antenna systems that require the specialised features available from these antennas.

6.6 References

1 MONTGOMERY, N. W.: 'Triple-frequency stacked microstrip element.' IEEE Ant. and Prop. Intl. Sym., 1984, Boston, MA, pp 255–258
2 JONES, H. S. Jr., SCHAUBERT, D. H. and FARRAR, F. G.: 'Dual-frequency piggyback antenna.' US Patent 4 162 499, 24 July 1979
3 SCHAUBERT, D. H., FARRAR, F. G., SINDORIS, A. R., and HAYES, S. T.: 'Microstrip antennas with frequency agility and polarization diversity.' *IEEE Trans.* 1981, **AP-29**, pp. 118–123
4 RICHARDS, W. F., and LO, Y. T.: 'A wide-band, multiport theory for thin microstrip antennas.' IEEE Ant. and Prop. Intl Sym., 1981, Los Angeles, CA, pp 7–10
5 RICHARDS, W. F., and LO, Y. T.: ;Theoretical and experimental investigation of a microstrip radiator with multiple lumped linear loads.'*Electromagnetics*, 1983, **3**, pp. 371–385
6 SENGUPTA, D. L.: 'Transmission line model analysis of rectangular patch antennas,' *Electromagnetics*, 1984 **4**, pp. 355–376
7 MUNSON, R. E.: 'Conformal microstrip antennas and microstrip phased arrays,' *IEEE Trans.*, 1974, **AP-22**, pp. 74–78
8 CARVER, K. R.: 'Input impedance to probe-fed microstrip antennas.' IEEE Ant. and Prop. Intl. Sym., 1980, Quebec, Canada, pp. 617–620
9 SABBAN, A.: 'A new broadband stacked two-layer microstrip antenna.' IEEE Ant. and Prop. Intl. Sym., 1983, Houston, TX, pp 63–66

Multilayer and parasitic configurations   351

10 CHEN, C. H., TULINTSEFF, A., and SORBELLO, R. M.: 'Broadband two-layer microstrip antenna.' IEEE AP-S Sym. Digest, 1984, Boston, MA, pp. 251–254
11 HOLZHEIMER, T., and MILES, T. O.: 'Thick, multilayer elements widen antenna bandwidths,' *Microwaves & RF*, Feb. 1985, pp. 93–95
12 LONG, S. A., and WALTON, M. D.: 'A dual-frequency stacked circular-disc antenna,' *IEEE Trans.*, 1979, **AP-27**, pp. 270–273
13 HALL, P. S., and PRIOR, C. J.: 'Radiation control in corporately fed microstrip patch arrays.' Digest of 1986 Journees Internationales de Nice sur les Antennes (JINA '86), Nice, France, pp. 271–175
14 ARAKI, K, UEDA H., and TAKAHASHI, M.: 'Hankel transform domain analysis of complex resonant frequencies of double-tuned circular disc microstrip resonators/radiators,' *Electron. Lett.*, 1985, **21**, pp. 277–279
15 PASCHEN, D. A.: 'Practical examples of integral broadband matching of microstrip antenna elements,' Antenna Applications Sym., Univ. of Illinois, Urbana, IL, 1986
16 FONG, K. S., PUES, H. F., and WITHERS, M. J.: 'Wideband multilayer coaxial-fed microstrip antenna element,: *Electron Lett.*, 1985, **21**, pp. 497–499
17 HALL, P. S.: 'Probe compensation in thick microstrip patches,' *Electron lett.*, 1987, **23**, pp. 606–607
18 GRIFFIN, J. M., and FORREST, J. R.: 'Broadside circular disc microstrip antenna,' *Electron. Lett.*, 1982, **18**, pp. 266–269
19 POZAR, D. M., and SCHAUBERT, D. H.: 'Comparison of architectures for monolithic phased arrays,' *Microwave J.* 1988, **29**, pp. 93–104
20 POZAR, D. M.: 'A microstrip antenna aperture coupled to a microstripline,' *Electron Lett.*, 1985, **21**, pp. 49–50
21 SULLIVAN, P. L., and SCHAUBERT, D. H.: 'Analysis of an aperture coupled microstrip antenna,' *IEEE Trans.*, 1986, **AP-34**, pp. 977–984
22 SCHAUBERT, D. H., JACKSON, R. W., and POZAR, D. M.,' 'Antenna elements for integrated phased arrays,' Antenna Applications Sym., Univ. of Illinois, Urbana, IL, 1985
23 SULLIVAN, P. L., and SCHAUBERT, D. H.: 'Analysis of an aperture coupled microstrip antenna,' RADC-TR-85-274, Rome Air Development Center, Feb. 1986
24 BUCK, A. C., and POZAR, D. M.: 'Aperture coupled microstrip antenna with a perpendicular feed,' *Electron Lett.*, 1986, **22**, pp. 125–126
25 POZAR, D. M., and JACKSON, R. W.: 'An aperture coupled microstrip antenna with a proximity feed on a perpendicular substrate, *IEEE Trans.*, 1987, **AP-35**, pp. 728–731
26 HENDERSON, A., JAMES, J. R. and HALL, C. M.: 'Bandwidth extension techniques in printed conformal antennas,' Proc. Military Microwaves '86, Brighton, England, 1986, pp. 329–334
27 HARRINGTON, R. F.: 'Time harmonic electromagnetic fields' (McGraw-Hill, NY 1961) pp. 307–311
28 SCHAUBERT, D. H. and FARRAR, F. G.: 'Some conformal printed circuit antenna designs.' Proc. Printed Circuit Ant. Tech. Workshop, New Mexico State Univ., Las Cruces, 1979
29 KUMAR G. and GUPTA, K. C.: 'Nonradiating edges and four edges gap-coupled multiple resonator broad-band microstrip antennas,' *IEEE Trans.*, 1985, **AP-33**, pp. 173–178
30 KUMAR G. and GUPTA, K. C.: 'Directly coupled multiple resonator wide-band microstrip antennas,' *IEEE Trans.,* 1985., **AP-33**, pp. 588–593
31 KUMAR G. and GUPTA, K. C.: 'Broadband microstrip antennas using additional resonators gap-coupled to the radiating edges,' *IEEE Trans.*, 1984, **AP-32**, pp. 1375–1379
32 GUPTA, K. C. and SHARMA, P. C.: 'Segmentation and desegmentation techniques for analysis of planar microstrip antennas,' *IEEE Intl.* Sym. on Ant. and Prop., June 1981, pp. 19–22
33 ENTSCHLADEN, H. and NAGEL, U.: Microstrip patch array antenna,' *Electron Lett.,* 1984, **20**, pp. 931–933
34 PRIOR, C. J. and HALL, P. S.: 'Microstrip disc antenna with short-circuited annular ring,' *Electron. Lett.*, 1985, **21**, pp. 719–721

Chapter 7
Wideband flat dipole and short-circuit microstrip patch elements and arrays

G. Dubost

This chapter comprises of two sections. Section 7.1 discusses a flat dipole which is a wide- bandwidth hybrid radiating source, originated and developed in France, and used in flat arrays having several hundred or more of such elements. Its low radiation resistance is advantageously matched to the characteristic resistance of the stripline used to feed it. In each array, spurious radiation is avoided because the feed network is completely shielded. Section 7.2 describes the short-circuited patch acting at quarter-wave resonance. Its large beamwidth in the E-plane and weak coupling in the H-plane are characteristics particularly suitable for use in microstrip-phased arrays with beam steering over a large angular sector. Because of their small thickness compared with wavelength they can be used with advantage in flat arrays having omnidirectional radiation or a directional deflected beam. These two radiating dipoles are studied theoretically by means of models which are equivalent to several lossy coupled transmission lines operating in the quasi-transverse electromagnetic mode.

7.1 Flat dipole elements and arrays

Fig. 7.1 shows several models of the usual wide-bandwidth flat dipoles. The specific properties of each model are given in Table 7.1.

7.1.1 Elementary sources

The flat dipole is used as a folded slot dipole symmetrically fed across a gap. When isolated in Fig. 7.1 (2) or used in a large array in Fig. 7.1 (3), it is fed by means of a Lecher line or a stripline network, respectively. It has already been described in References 2 – 5. Fig. 7.2 shows the flat dipole which is linearly polarized, parallel to, and at a distance H from, a reflector. A dielectric sheet is placed between the radiating structure and the reflector, so that it can be fabricated by means of printed circuits. The analysis and synthesis are presented in sections 7.1.1.1 and 7.1.1.2. The first model Fig. 7.1 (1), which was asymmetrical, has been described previously [1] and is used in arrays [3]. A broadband

Table 7.1 Properties of the usual wide-bandwidth flat dipoles

Model number (Fig. 7.1)	References	Definition and specific properties	Advantages	Defaults
(1)	1	Asymmetrical and isolated flat folded dipole: LP.	Small size	High CPL
(2)	2, 5, 16, 19, 25	Symmetrical and isolated flat folded dipole: LP.	Wideband; especially Refs. 19, 25: low CPL	
(3)	5, 6, 20, 21, 22, 23, 24	Symmetrical flat folded dipole used in arrays: LP	Wideband: low CPL.	
(4)	3, 12	Cross flat dipoles CP	Wideband: low CPL.	Large size
(5)	9, 10, 15	Flat dipole used in dual polarised arrays	Wideband especially with directors [15]: Low CPL	Large size $\phi = 0.6\lambda_o$

LP.: Linearly polarised
DP.: Dual polarised

CP.: Circularly polarised
CPL.: Cross-polarisatiuon level

Wideband flat dipole and short-circuit microstrip 355

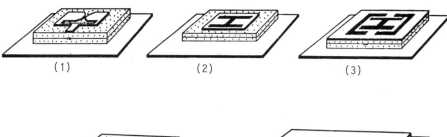

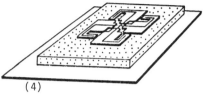

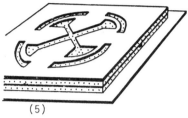

Fig. 7.1 *Usual wide-bandwidth flat dipoles*

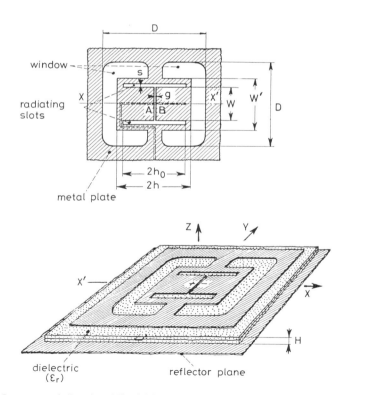

Fig. 7.2 *Geometry of directional flat folded dipole*

circularly polarised flat radiating source is shown in Fig. 7.1 (4) and described in Reference 3. A dual polarised model in Fig. 7.1 (5) is described next.

We present a flat radiating source which is able simultaneously to receive or transmit two frequencies of orthogonal linear polarisations, and subsequently a circularly polarised wave of either the right-hand or left-hand sense. It is a new flat dipole arrangement [(5) in Fig. 7.1], [9, 10]. Circular polarisation can be obtained from a trivial patch antenna when using two orthogonally phased feeds, but it has a narrow circularly polarised bandwidth [11] when non-isolating power splitters are used. To broaden the bandwidth, we can make use of a thick substrate of low dielectric constant, with a thickness of approximately one-tenth of a wavelength, and isolating power splitters (Wilkinson) as in Reference 18. The higher modes must be suppressed [12, 3, 13, 14] when symmetrical feed structures are used.

The radiating source (Fig. 7.3) is an enlargement of the linear wide-bandwidth flat symmetrical folded dipole [4, 5]. The radiating structure is composed of two symmetrical crossed and overlapped flat folded dipoles with orthogonal electric moments. Each dipole is composed of two metallic plates which are fed in opposite phase. The two dipoles, nos. 7, 8 and nos. 9, 10, are etched onto one of the two metallic faces of the first printed circuit (1). The radiation in one half-space is possible when the whole metallic face (11) of an identical second printed circuit (2) is used. The other metallic faces of the two printed circuits, (1) and (2), support the two stripline central conductors, (5) and (6), which are insulated with a thin dielectric sheet (3).

Each dipole is coupled by its feeding line through a gap by means of an open quarter-wave stripline section, so that the two edges are fed in opposite phase. The model operates between 3·45 and 3·85 GHz [15] with a VSWR of less than 2, when directors are used to broaden the bandwidth.

7.1.1.1 Analysis in quasi-TEM approximation: Self and mutal impedances, and far fields are next studied. We have shown previously that the symmetrical flat dipole operates with the widest bandwidth at its third resonance [4], when a model equivalent to several lossy and uncoupled transmission lines, operating in a quasi-transverse electromagnetic mode, is taken into account.

Radiation impedance and bandwidth: Fig. 7.2 shows the flat dipole parallel to, and at a distance H from, a reflector plane. A dielectric sheet is placed between the radiating source and the reflector, so that it can be realised by means of printed circuits. The stripline feeding network is shown. The coupling between the radiating element and the feeding stripline is realised by means of a quarter-wavelength open stripline. The two large plates are fed in opposite phase from a gap AB. Thus the symmetrical plane perpendicular to the xx' axis and to the reflector plane is at zero potential. It is then possible, but not necessary, to short-circuit middle points of the folded arm at the reflector plane without disturbance to the electromagnetics properties.

The input impedance is due to the transformation of two impedances; namely the radiation impedances of the two radiating slots. This transformation is achieved along a non-radiating slot line. Each half slot can be considered as two

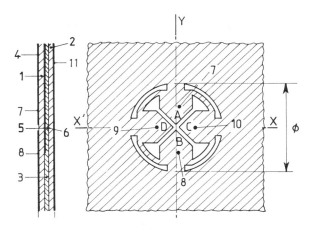

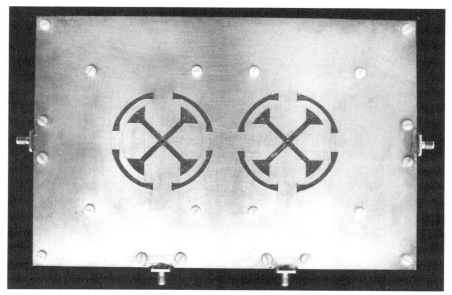

Fig. 7.3 *Wide-bandwidth dual polarised microstrip antenna*

transmission lines. To explain theoretically the wider bandwidth, in conformity with experimental work, it was necessary to introduce coupling between the two transmission lines [5]; ie., across the two radiating slots (Fig. 7.2). The input impedance Z_r relative to the middle AB of the gap is due to the transformation of two impedances $2Z'_r$ which are the radiation impedances of the two radiating slots. This transformation is carried out along a non-radiating slot line of total length W and characteristic resistance R_c. Each half slot, having a radiation impedance equal to Z'_r, can be considered as two coupled transmission lines of length l with characteristic impedances Z and Z', and C_{12} being the coupling capacitance per unit of length. The length l is given by the expression

$$l = h_0 + \Delta h_0 + s/2 \tag{7.1}$$

with

$$\Delta h_0 = 0.41 H \frac{(\varepsilon_e + 0.3)(s/H + 0.26)}{(\varepsilon_e - 0.26)(s/H + 0.81)}$$

where s is the width of the slot of total length $2h_0$, Δh_0 the increase in length due to an end effect. The characteristic resistance Z and Z' are those of microstrip-lines of equal H and thickness and widths, respectively, of $W/2$ and $(W'-W)/2 - s$. The conducting currents are strictly located on the edges of every slot. Radiation is taken into account by introducing, for equivalent coupled lines, attenuation constants per unit length α and α' given by the following equations:

$$\tanh(\alpha l) = \frac{8\pi^3}{5} \frac{1}{Z} \mu_0^{5/2} \varepsilon_0^{3/2} f^4 (Hl)^2 \tag{7.2}$$

$$\tanh[\alpha'(l + H)] = \frac{16\pi^3}{3} \frac{1}{Z'} \mu_0^{5/2} \varepsilon_0^{3/2} f^4 (Hl)^2 \tag{7.3}$$

where f is the frequency and tanh is the hyperbolic tangent.

The two coupled transmission lines are divided into N equal four-port sections. When the boundary conditions are applied, the potential and the electric-current distributions, and the input impedance Z_r are deduced.

Fig. 7.4 shows the theoretical input admittance $1/Z_r$, obtained with $N = 250$, and experimental points in a wide frequency band with the following parameters (in millimetres):

$$2h = 11.4, \quad W' = 8.0, \quad W = 3.8, \quad 2h_0 = 8.7$$
$$s = 0.9, \quad g = 0.4, \quad H = 3.2, \quad \varepsilon_r = 2.2$$

We deduced:

$$Z = 114\,\Omega, \quad Z' = 125\,\Omega, \quad l = 5.9\,\text{mm},$$
$$R_c = 170\,\Omega, \quad C_{12} = 18\,\text{pF/m}$$

Owing to the coupling capacitance across the two slots, a fourth resonance (11·2 GHz) was observed not far from the third one (10·3 GHz), which explains the increased bandwidth. From another model [6] a bandwidth of 16% has been measured between 11·25 and 13·2 GHz for a VSWR lower than 2. The radiation resistance is always located at about $100\,\Omega$ whatever the model parameters.

Mutual impedance and coupling: Fig. 7.5 shows the configuration of two coupled flat dipoles in a parallel position (or in the *H*-plane) separated by a distance *D* and located above a perfectly conducting ground plane at a height *H*. The mutual coupling may, in principle, be due to either guided waves, or

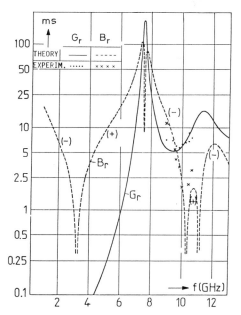

Fig. 7.4 *Theoretical input admittance $Y_r = 1/Z_r$ and experimental points of a directional flat folded dipole ($Y_r = G_r + jB_r$)*

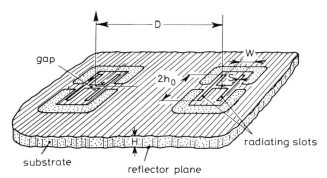

Fig. 7.5 *Configuration of two flat folded dipoles in H-plane*

space waves, or both. In our theory, guided waves will be neglected. Then the mutual impedance Z_{12} and the coupling coefficient $C(\text{dB})$ are given by the following expressions [7,8]:

$$Z_{12} = -\frac{4}{\pi}\sqrt{\frac{\mu_0}{\varepsilon_0}}\, tg^2\left(\frac{kl}{2}\right)\left(\frac{H}{D}\right)^2 \frac{1}{\varepsilon_e}\exp(-jk_0 D) \tag{7.4}$$

$$C = 20\log_{10}\left|\frac{2Z_{12}R_N}{(Z_r + R_N)^2 - Z_{12}^2}\right| \tag{7.5}$$

where R_N is the normalisation resistance and ε_e is the equivalent relative dielectric constant with $k_0 = 2\pi/\lambda_0$ and $K = k_0\sqrt{\varepsilon_e}$.

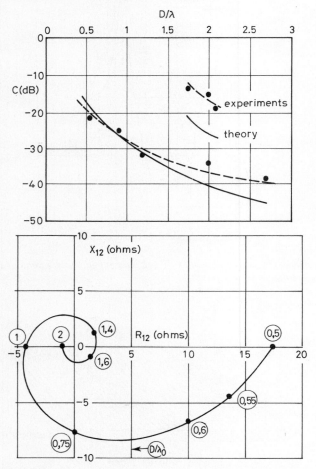

Fig. 7.6 Mutual impedance $Z_{12} = R_{12} + jX_{12}$ and coupling factor in terms of D/λ_0 at $f = 9.5\,GHz$ in H-plane

The validity condition of eqns. 7.4 and 7.5 is:

$$\pi\frac{W}{\lambda_0}\sqrt{\frac{\varepsilon_0}{\mu_0}}|Z_r|(\varepsilon_r - 1) \ll 1 \tag{7.6}$$

Fig. 7.6 shows, for $h_0 = 4{\cdot}35\,\text{mm}$, $H = 3{\cdot}2\,\text{mm}$, $W = 3{\cdot}8\,\text{mm}$, $s = 0{\cdot}9\,\text{mm}$, $\varepsilon_r = 2{\cdot}2$ and $f = 9{\cdot}5\,\text{GHz}$, the theoretical mutual impedance and coupling factor in terms of their distance D, together with some experimental points. It was seen that, for microstrip patches, the E-plane coupling always exceeds that of the H-plane coupling. It is the opposite for the flat dipole. Nevertheless the coupling levels encountered, which are more critical for the H-plane (parallel position as in Fig. 7.5), should not involve any difficulty of array design. The main coupling effect is via space waves [7, 8]. In effect, for the arrays which are described in Section 7.1.2, the ratio of the mean distance between two adjacent flat dipoles to the wavelength in free space is greater than $0{\cdot}8$. Then the coupling factor is always lower than $-25\,\text{dB}$.

Radiated fields and polarisation: We assume the flat-dipole reflector plane to be infinite. The flat-dipole moment vector is parallel to the xx'-axis (Fig. 7.2). By using electric $\boldsymbol{I}$ and polarization $\boldsymbol{I}_p$ current distributions and applying the volume equivalence theorem we can calculate the far field radiated by the antenna in the E-plane from the following expressions [5]:

$$\boldsymbol{I}(x) = \frac{2V_e}{Z_r}\left[\cos kx - \frac{\sin k|x|}{\text{tg}kl}\right]\hat{x} \tag{7.7}$$

$$\boldsymbol{I}_p(x) = j\omega(\varepsilon - \varepsilon_0)W(\cos kx + \text{tg}kl\,\sin k|x|)V_e\,\hat{z}\frac{x}{|x|} \tag{7.8}$$

$2V_e$ is the potential which is applied between the two gap edges.

In the E-plane and after integration the co-polar pattern is deduced from eqn. 7.7:

$$E_e = \cos\phi\,\sin(k_0 H\cos\phi)\left\{\frac{\sin k_0 l(\sqrt{\varepsilon_e} - \sin\phi)}{k_0 l(\sqrt{\varepsilon_e} - \sin\phi)} + \frac{\sin k_0 l(\sqrt{\varepsilon_e} + \sin\phi)}{k_0 l(\sqrt{\varepsilon_e} + \sin\phi)}\right.$$

$$\left. - \cot g(k_0 l\sqrt{\varepsilon_e})\left[\frac{\sin^2 \frac{k_0 l}{2}(\sqrt{\varepsilon_e} - \sin\phi)}{\frac{k_0 l}{2}(\sqrt{\varepsilon_e} - \sin\phi)} + \frac{\sin^2 \frac{k_0 l}{2}(\sqrt{\varepsilon_e} + \sin\phi)}{\frac{k_0 l}{2}(\sqrt{\varepsilon_e} + \sin\phi)}\right]\right\} \tag{7.9}$$

In the H-plane, the polarisation current distribution due to the substrate has no effect, and the normalised far field radiated is given by the equation:

$$E_H = \frac{\sin(k_0 H\cos\theta)}{\sin k_0 H}\cos\left[k_0\frac{W + s}{2}\sin\theta\right] \tag{7.10}$$

In eqns 7.9 and 7.10 no effect of the mutal coupling are taken into account, since the equations apply to an isolated source and an infinite reflector plane.

7.1.1.2 Synthesis in quasi-TEM approximation: A theoretical model equivalent to several lossy coupled transmission lines was used in section 7.1.1.1 to explain the bandwidth and the radiation resistance of the flat dipole. Conversely it is useful to find the geometrical parameters of such an antenna and its radiation resistance when the frequency band is known. The number of parameters is reduced, and to obtain an attractive bandwidth their values are bounded. From Fig. 7.2 we take [16]: $W'/\lambda_0 = 2(W/_0 + s/\lambda_0)$, $g/\lambda_0 = 0.014$,

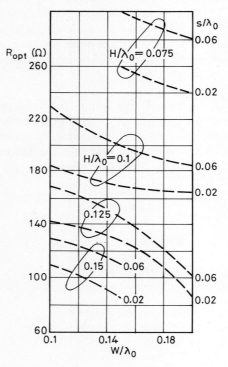

Fig. 7.7 *Optimal resistance R_{opt}*

$\varepsilon_r = 2.2$. As $2h_0 \leq 2l \leq 2h$ we take $2l = 2h$, with l given by eqn. 7.1. Then the variable parameters are H/λ_0, W/λ_0 and s/λ_0, and using Figures. 7.7 – 7.9, we can define the size of the whole antenna and its optimal resistance R_{opt} when the frequency band is given. In Fig. 7.8 the optimised bandwidth ($B\%$) is due to the input impedance Z_r, related to R_{opt} for a VSWR below 2; an example is given. The frequency band is equal to $11.3 - 13.2$ GHz ($B = 15.5\%$). With a normalised thickness $H = 3.2$ mm ($\varepsilon_r = 2.2$), we obtain, for the mean frequency ($\lambda_0 = 24.5$ mm), $H/\lambda_0 = 0.13$. In Figs 7.8 and 7.9 we deduce, for $s/\lambda_0 = 0.02$, that $W/\lambda_0 = 0.1$ (then $W'/\lambda_0 = 0.24$) and $l/\lambda_0 = 0.204$. From eqn. 7.1 we deduce that $\Delta h_o/\lambda_0 = 0.029$ and $2h_0/\lambda_0 = 0.33$. The final antenna dimensions are: $W = 2.45$ mm, $W' = 5.9$ mm, $2h = 10$ mm, $s = 0.5$ mm, $2h_0 = 8$ mm, $H = 3.2$ mm, $g = 0.34$ mm. For these parameters and from Fig. 7.7 we obtain $R_{opt} = 140\,\Omega$. Synthesis resulting (Fig. 7.7 and 7.8) indicate that the

Wideband flat dipole and short-circuit microstrip

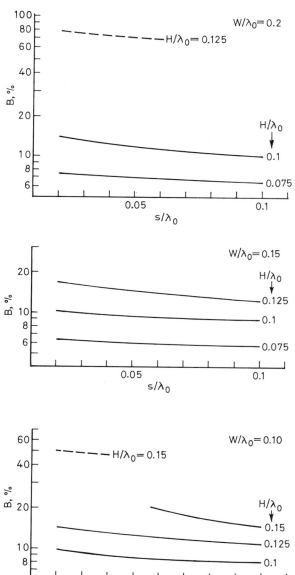

Fig. 7.8 *Optimised bandwidth B(%) of input impedance Z_r related to R_{opt} (Fig. 7.7) for a VSWR ≤ 2*
$W'/\lambda_0 = 2(W/\lambda_0 + s/\lambda_0)$, $g/\lambda_0 = 0.014$ $\varepsilon_r = 2.2$]
--- flat dipole is acting at the 3rd and 4th resonances
—— flat dipole is acting at the 3rd resonance

antenna bandwidth increases and radiation resistance decreases as its transverse area (in relation to the square wavelength) and coupling between the two equivalent radiating lines of each slot increase. Elsewhere, it was observed that the flat dipole performs best when a substrate of low dielectric constant is used [17], this is a general property [18]. For these reasons we obtained, using $\varepsilon_r = 1$, $H = 5$ mm, $W' = 9 \cdot 2$ mm, $W = 7$ mm, $g = 1 \cdot 2$ mm, $2h = 9 \cdot 6$ mm, $2h_0 = 7$ mm, $s = 0 \cdot 6$ mm, the correct behaviour for an experimental model acting within one actave, i.e. between 8 and 16 GHz [19, 25].

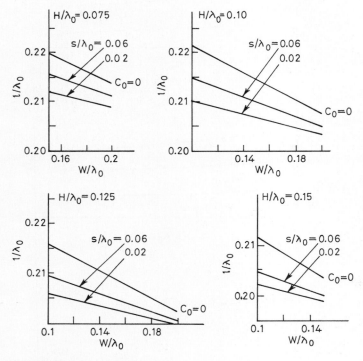

Fig. 7.9 *Half-slot effective length l/λ_0*

7.1.1.3 Moments-method analysis: Recently we have studied a wide-bandwidth flat dipole, fed through the gap by means of a Lecher line which is located on a thin printed circuit parallel to a reflector plane and isolated by a sheet of air (Fig. 7.10). The general scattering problem of an arbitrary shaped tri-dimensional antenna solved by the moments method and the finite-difference approach applied to integral equations has explained the antenna behaviour. The wire-grid model is suitable for expressing the boundary conditions along the various edges of the antenna, which are, in fact, the boundary conditions on the thin wire surfaces. The antenna is equivalent to an array of conducting square meshes. Each mesh is composed of four segments which are crossed cylindrical conducting wires. The influence of the conducting-wire diameter on the radiation impedance and bandwidth is considerable. The diameter chosen is equal to one quarter of the mesh dimension. The scattering problem is reduced to a flat-shaped bi-dimensional structure. In Fig. 7.11 we present the complete

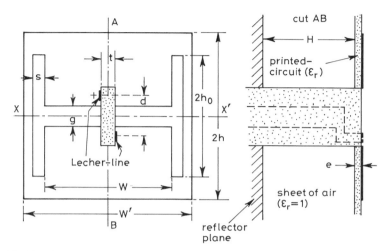

Fig. 7.10 *Very large bandwidth flat folded dipole configuration*

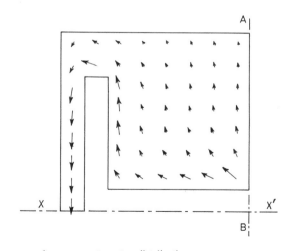

Fig. 7.11 *Average surface-current vector distribution*

distribution, which corresponds to the average surface current vector for each mesh. The length vector is proportional to the current amplitude. In Fig. 7.12 we present the theoretical and experimental input impedance related to the

Table 7.2 *Parameters used in Fig. 7.12*

Parameters,	W''	W	$2h$	$2h_0$	g	s	H	d	e	t	ε_r
Experiments	9·2	7·0	9·6	7·0	1·2	0·6	5·0	2·4	0·4	0·8	2·17
Theory	9·6	7·2	9·6	7·2	1·2	0·6	5·0	1·2	0	0	1

middle of the gap in terms of frequency (GHz). The impedance is normalised to 100 Ω, which is the Lecher-line characteristic resistance.: Table 7.2 gives the parameters used in Fig. 7.12 (in mm)..

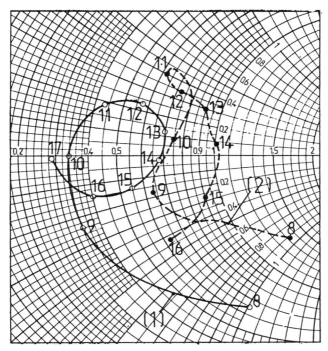

Fig. 7.12 *Theoretical (1) and experimental (2) input impedance of the large bandwidth flat folded dipole in terms of frequency (GHz)*
1 = 100 Ω

7.1.2 Array designs. Losses and efficiencies
A number of flat arrays with several hundred or more flat dipoles have been designed and constructed. In every array each source radiates through a window cut in one of the two metallic shields of the stripline feeding network (Fig. 7.2).

7.1.2.1 Large gain: Fig. 7.13 shows an array of 1024 radiating sources operating between 11·7 and 12·4 GHz with a measured isotropic linear maximum gain of 37 dB [20]. This high-gain array has been designed to receive radio-broadcasting signals sent out by geostationary satellites. It is composed of 1024 (32 × 32) flat dipoles (section 7.1.1.1) fed by a stripline network (Fig. 7.14). Like the elementary source, the array is constructed using two large compressed printed-circuit sheets with no direct connection between the feed network and the radiating sources. The symmetrical feed network comprises unequal stripline power splitters joined by stripline transmission lines, the characteristic impedance of which is 75 Ω. The distance between two adjacent sources is $0.89\lambda_0$, where λ_0 is the wavelength in air at the mean frequency (12·1 GHz). The square array of a $0.5\,m^2$ area was manufactured using two printed-circuit sheets ($\varepsilon_r = 2.17$) 1·6 mm thick. Circular polarisation is produced by a polariser embedded in a radome. Fig. 7.16 shows an example of the measured patterns in one diagonal plane at 12·1 GHz. Between 11·7 and 12·5 GHz the experimental maximum linear isotropic gain was equal to 36.9 ± 0.3 dB, and the efficiency is better than 48% between 11·7 and 12·4 GHz.

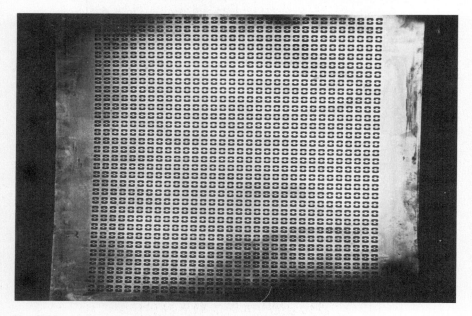

Fig. 7.13 *High-gain array (11·7 – 12·4 GHz) of flat folded dipole.*

The feeding arrangement is shown in Fig. 7.14 and the feeder dielectric and metallic losses are equal to 2·5 dB. The cross-polarisation level on the axis is lower than −30 dB, and, together with the polariser, the measured array ellipticity ratio along the principal axis is better than 1·5 dB over the frequency range. Fig. 7.15 shows the radiation patterns in E- and H-planes measured at 12·1 GHz.

7.1.2.2 Low sidelobe level: Two passive arrays with flat dipoles showed a sidelobe level lower than −30 dB. The first array [21, 22] operates between 5·25 and 5·45 GHz in a system used to detect natural resources, and is carried on an aircraft. It is composed of 128 symmetrical flat dipoles which are fed by lines and splitters realised using stripline technique by means of two printed-circuit sheets, as has been shown for the high-gain array (Fig. 7.14); the dimensions are

0·85 × 0·40 × 0·05 m. The linear isotropic maximum measured gain at 5·25 GHz is 26·5 dB, the efficiency is about 40% and the measured sidelobe level is lower than −29 dB for a theoretical level of −39 dB. Fig. 7.17 shows an example of radiation patterns in E- and H-planes at 5.25 GHz for an array covered with a radome. The second array [23] (Fig. 7.18) is incorporated in a

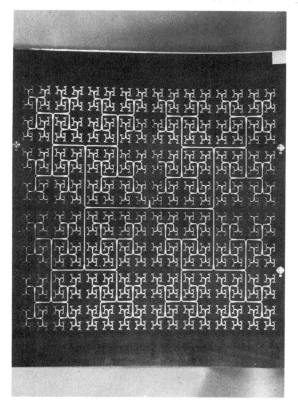

Fig. 7.14 *High-gain array (11·7 – 12·4 GHz): feeding arrangements*

Table 7.3 *Measured values for the array in Fig. 7.18*

f, GHz	9·1	9·3	9·6	9·9	10·1
Max. gain, db	30·9	26·9	31·9	31·3	31·4
3 dB beamwidth, deg (H-plane)	2°40′	2°35′	2°40′	2°30′	2°30′
First sidelobe, dB (H-plane)	< −28	< −28	< −36	< −30	< −29

The efficiency at 9·6 GHz is equal to 33·3% (or −4·78 dB). The cross-polarisation level is lower than −35 dB.

system used on ground-surveillance radars. It is composed of 512 flat dipoles, it uses stripline technology and features low sidelobe radiation: it operates between 9·4 and 10·1 GHz. The distance between two adjacent sources equals

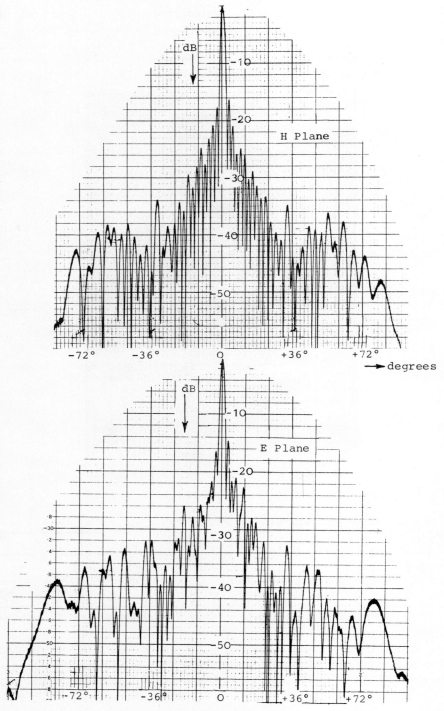

Fig. 7.15 *Patterns at 12·1 GHz. Large gain array (see Figs. 7.13 and 7.14) (From Reference 20)*

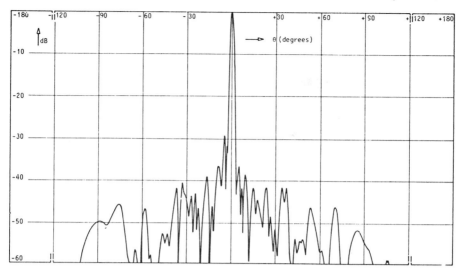

Fig. 7.16 *Large-gain array (see Figs. 7.13 and 7.14) (Reference 20)*
Pattern in a diagonal plane at 12·1 GHz

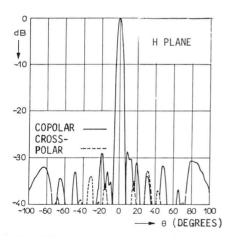
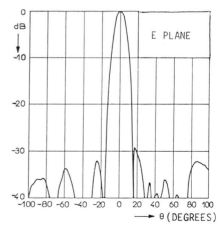

Fig. 7.17 *Low-side lobe-level array.*
Patterns in *E*- and *H*-planes at 5·25 GHz

$0.9 \lambda_0$ in the H-plane and $0.8 \lambda_0$ in the E-plane. Some measured values are given in Table 7.3.

7.1.2.3 Variable directivities: A flat array with four fixed beams of different directivities and gains is considered [24]. This array supports 256 flat dipoles separated by a distance of $0.85 \lambda_0$ together with 60 electronic switches, integrated in a stripline structure (Fig. 7.19). The global efficiency is about 50%. For each of the four states we can form sum ($\sum$) and difference (Δ) patterns in the two

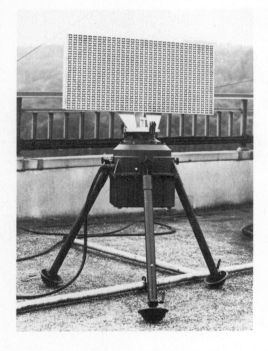

Fig. 7.18 *Low-side lobe-level array of flat folded dipoles (9·4 – 10·1 GHz)*

orthogonal principal planes of the array, which acts between 11·7 and 12·4 GHz. Fig. 7.20 shows the ratio $\Delta/\sum$ for each of the four states in terms of the deviation angle θ. The various beamwidths obtained at the mean frequency are 4°3′, 7°7′, 12°6′ and 27°, and the cross-polarisation level is always lower than -25 dB. The efficiency is given by the formula $\eta\% = 100 \lambda^2 G_M / 4\pi s$, where s is the geometrical area (0.12 m^2) and G_M is the maximum isotropic linear gain. With a measured value $G_M = 31$ dB, we obtained $\eta = 50\%$. For each of the four states the mean switching losses are about 1·6 dB.

7.1.2.4 Very large bandwidth: This section discusses broad-angular-coverage and wide-bandwidth antennas, which are increasingly used, for example in airborne communication systems, electronic warfare systems or ECM. We show [25] cylindrically shaped radiating elements working in octave bandwidth (Fig. 7.21). In azimuth, the theoretical angular-sector coverage can be equal to 360°. In the different meridian planes or site planes, the beamwidth and sidelobe level may be of any proportion. The antennna consists of radiating elements arranged in vertical arrays photo-etched on a printed circuit which is wrapped around a cylindrical dielectric lens of $2a$ diameter. Each radiating element is a flat dipole, used as a folded slot dipole symmetrically fed. When the substrate has air as the

Fig. 7.19   *Multiple-beam flat array (11·7 – 12·4 GHz)*

medium and if the thickness is increased, the flat dipole acts over wide bandwidth (sction 7.1.1.3). Thus, when it is fed symmetrically with a Lecher line, the operating range covers one octave. Each array, which is aligned along a generating line of the cylindrical lens, can be considered as a separate channel. In a plane perpendicular to the axis the antenna acts as a Luneberg lens. In effect, the n constant-index-of-refraction surfaces are cylinders of radius r such that $n(r) = [2 - (r/a)^2]^{1/2}$. The dielectric lens was machined from a cylindrical Teflon rod. It has longitudinal grooves which are uniformly distributed around the lens axis and which are small compared with the vacuum wavelength [26].

In the meridian section of the lens, which contains one array the optical path is deduced, by means of integration, from the continuous refraction law. Applying an asymptotic development of the Kottler formula, the radiated far field is calculated from the electromagnetic field distribution along the lens outside surface, as has been done previously, but for a small bandwidth using short-circuited flat dipoles at quarter-wave resonance [27, 28].

The antenna is composed of 64 folded flat dipoles arranged in eight vertical

374 Wideband flat dipole and short-circuit microstrip

arrays, and it acts in the 8 – 16 GHz range. The diameter of the cylindrical lens is equal to 86 mm. The beamwidth in the frequency range lies between 24° and 10° in the azimuth plane, and between 15° and 7° in the site planes. Whatever the azimuth direction, the minimum isotropic linear gain which can be expected over an angular coverage of 110°, and in the 8 – 16 GHz frequency range, is 18 dB. The maximum isotropic linear measured gain G_M(dB) is compared with the theoretical maximum directivity D_M(dB) in Fig. 7.22. The impedance matching of each array is given in Fig. 7.23. Fig. 7.24 shows the measured ratio Δ/Σ between difference (Δ) and sum (Σ) patterns expressed on a linear scale.

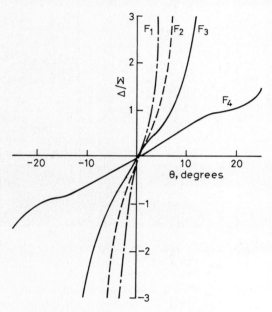

Fig. 7.20 Ratio Δ/Σ of sum (Σ) and difference (Δ) patterns for the four states (F_1 to F_4) at 12·1 GHz (see Fig. 7.19)

7.2 Short-circuit microstrip patches and arrays

7.2.1 Elementary source

Self impedance and bandwidth: Fig. 7.25 shows the half short-circuited patch configuration. It looks like a half short-circuited flat dipole acting at a quarter-wave resonance. The relation between radiation conductance G_r and G_0 can be determined from [29]

$$\frac{G_r}{G_0} = \frac{16\pi^3}{3} \left(\frac{W}{\lambda_0}\right)^2 \left(\frac{1}{4} - \frac{H}{\lambda}\right)^2 \tag{7.11}$$

Wideband flat dipole and short-circuit microstrip 375

Fig. 7.21 *Broad-angular-coverage and wide-bandwidth antenna*

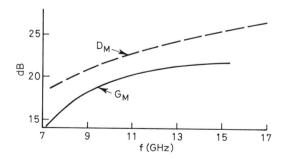

Fig. 7.22 *Maximum measured gain G_M and maximum directivity D_M*

Eqn. 7.11 expresses the development of the input impedance of a lossy short-circuited line whose length is assumed equal to $H' = H + h$. The losses comprise the radiation resistance of the patch. The main problem lies in determining the line attenuation constant. In order to do this, the radiation resistance of the half short-circuited patch, parallel to the perfect reflector plane, is identified with that of a loop constituted by the half short-circuited dipole and its electrical image in relation to the reflector plane [3].

When finite conductivity σ and lossy dielectric substrate are taken into account, as has been done for the opened flat dipole acting at half-wave resonance (Patch) [30], the input conductance G_t is given as follows:

$$\frac{G_t}{G_0} = \frac{16\pi^3}{3}\left(\frac{W}{\lambda_0}\right)^2\left(\frac{1}{4} - \frac{H}{\lambda}\right)^2 + \frac{\pi}{4}\frac{W}{H}\sqrt{\varepsilon_e}\left(\text{tg}\delta + \frac{d_s}{H}\right) \quad (7.12)$$

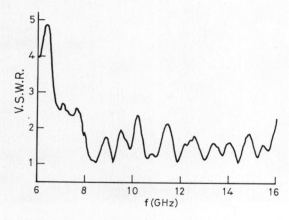

Fig. 7.23 Broad-angular-coverage and wide-bandwidth antenna
Impedance matching of each array

where d_s $(\pi f \mu_0 \sigma)^{-1/2}$ is the skin depth, δ is the dielectric loss angle, $\lambda_0 = \lambda\sqrt{\varepsilon_e}$ and $G_0 = (\varepsilon_0/\mu_0)^{1/2}$

The bandwidth $B\%$ related to VSWR is given by

$$B(\%) = \frac{100(\text{VSWR}-1)}{\sqrt{\text{VSWR}}}\left[\frac{64\pi^2}{3}\frac{1}{\sqrt{\varepsilon_e}}\frac{WH}{\lambda_0^2}\left(\frac{1}{4}-\frac{H}{\lambda}\right)^2 + \text{tg}\delta + \frac{d_s}{H}\right] \quad (7.13)$$

Lastly, the radiating source efficiency at resonance is given by

$$\eta(\text{dB}) = -10\log_{10}\left[1 + \frac{3}{64\pi^2}\sqrt{\varepsilon_e}\frac{\lambda_0^2}{WH}\left(\frac{1}{4}-\frac{H}{\lambda}\right)^{-2}\left(\text{tg}\delta + \frac{d_s}{H}\right)\right] \quad (7.14)$$

The quarter-wave resonance condiction is $h + H \simeq \lambda/4$.

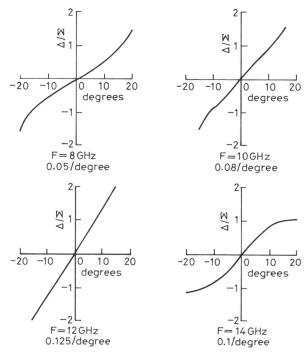

Fig. 7.24 *Measured ratio $\Delta/\sum$ between difference (Δ) and sum ($\sum$) patterns in terms of frequency. Broad-angular-coverage and large bandwidth antenna*

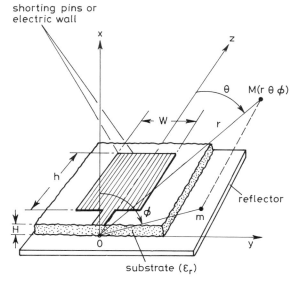

Fig. 7.25 *Half short-circuited microstrip antenna configuration*

Mutual impedance and coupling: Explicit formulas are presented for the mutual impedance and coupling between two parallel or co-linear short-circuited flat dipoles at resonance in air medium in terms of three dimensions with respect to wavelength [31]. Experimental and theoretical results are in good agreement and show that these small microstrip antennas are particularly well uncoupled, and therefore suitable for incorporation in a phased array with a steering beam inside a large angular sector. Thus, for the following values (Fig. 7.25):

$$W/\lambda_0 = 0.35 \qquad h/\lambda_0 = 0.185 \qquad H/\lambda_0 = 0.055 \qquad \varepsilon_r = 1$$

the coupling factor in the E- and H-planes is lower than $-20\,\text{dB}$ between short-circuited flat dipoles separated by a distance $0.5\,\lambda_0$.

Radiated fields and polarisation: The far field radiated by this source has been calculated [32]. With the same notation as in Fig. 7.25 we obtained:

The Normalised electric field in the E-plane ($\phi = 0$)

$$(E_\theta)_N = \frac{(\varepsilon_r - 1)\cos\theta}{\varepsilon_r - \cos^2\theta}\left[\cos\theta + j\sqrt{\varepsilon_e}\,\exp(j\frac{\pi}{2\sqrt{\varepsilon_e}}\cos\theta)\right]$$

$$+ \frac{\sqrt{\varepsilon_e}\sin^2\theta}{\varepsilon_e - \cos^2\theta}\left[\sqrt{\varepsilon_e} + j\cos\theta\exp(j\frac{\pi}{2\sqrt{\varepsilon_e}}\cos\theta)\right]$$

$$- j\sqrt{\varepsilon_e}\,\cos\theta\exp(j\frac{\pi}{2\sqrt{\varepsilon_e}}\cos\theta) \qquad (7.15)$$

The normalised electric field in the H-plane ($\theta = \pi/2$)

$$(E_{\frac{\pi}{2}})_N = \cos\phi\,\frac{\sin(k_0\frac{W}{2}\sin\phi)}{k_0\frac{W}{2}\sin\phi} \qquad (7.16)$$

7.2.2 Array designs

These short-circuited patches acting at a quarter-wave resonance (section 7.2.1) have been used in several arrays.

7.2.2.1 Phased array with steering beam: Effect of mutual coupling: This section concerns a Ku-band phased array acting in a large angular [33] at around 15 GHz. The array has 64 short-circuited patches in air medium, located on a 8 × 8 square lattice. The element spacing is 10 mm ($0.5\,\lambda_0$), such that the coupling coefficient in the E- and H-planes is lower than $-20\,\text{dB}$. The measured results on each radiating element at around 15 GHz are as follows:

Bandwidth for a VSWR ⩽ 2:17%
Pattern in E-plane: ±1 dB in an angular sector of 160°
Beamwidth in H-plane at 3 dB:80°

Feedthrough is used to excite each radiating element from the output of the associated phase shifter by means of a short coaxial line. This 3 bit digital phase shifter was recently described [34]. A total of only eight PIN diodes is required

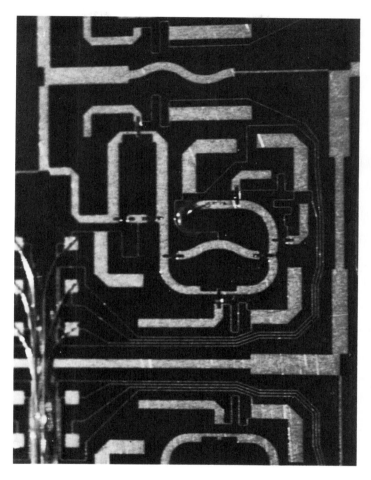

Fig. 7.26 *3 bit digital phase shifter with eight PIN diodes in Ku band*

for the whole phase shifter (Fig. 7.26) instead of ten for a conventional type. It is arranged in such a way that no series capacitance is necessary to separate the different bias [35]. So only three wires are used to apply the DC bias to the eight PIN diodes. For each of the eight states, the phase-shifter insertion losses is equal to 3 ± 1 dB in a bandwidth of 15% at about 15 GHz. The 64 phase shifters are photo-etched on four fused-quartz substrate plates each of 5.08×5.08 cm (Fig. 7.27). So for every deflected beam we can form sum and difference patterns. The feeding structure, which is composed of splitters,

380 Wideband flat dipole and short-circuit microstrip

branch lines, corporate feeds and DC bias, is also photoetched according to a microstrip technique. Four plugs are connected by pliant and thin conductors to the etched wires for DC bias feeding. The square feeding RF network is divided into four parts. The adjacents parts are deduced from each other by rotation symmetry (RS) so that the four outputs 1, 2, 3, 4 are located on the four array sides (Figs. 7.29 and 7.27). This rotation symmetry is only related to the four parts of the square feed RF network, and does not concern the short-

Fig. 7.27 *Phased array with steering beam in Ku band: 64 3 bit digital phase shifters and feeding arrangements*

circuited radiating dipoles which are linearly polarised (Fig. 7.28). The parasitic signals at the input of the four quadrants, which are due to the radiating-dipole aerial alternately coupling in the *E* or *H* planes, are out of phase. This concept

Wideband flat dipole and short-circuit microstrip 381

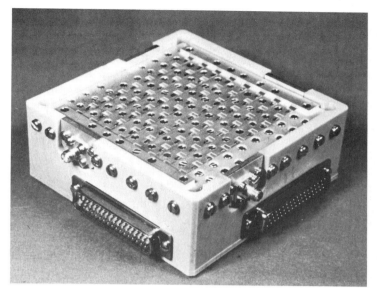

Fig. 7.28 *Phased array with steering beam in Ku band: 64 radiating sources*

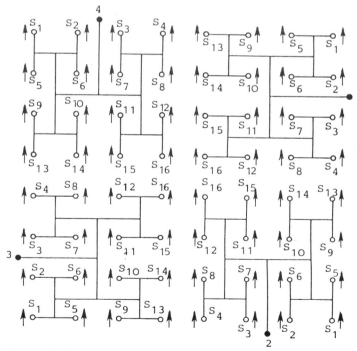

Fig. 7.29 *Square feeding network (From Reference 36)*
 S: output towards source ↑:source electrical moment

[36] has advantages as compared with an axial-symmetry network (AS). For instance, in Fig. 7.30 we show the theoretical sum patterns in the H-plane deflected by 45° and related to (RS) and (AS). Furthermore, because of the compensation of aerial mutual coupling, the risk of scan blindness is reduced when the (RS) concept is used. Fig. 7.31 shows measured sum gains in the E, H or diagonal planes and Fig. 7.32 shows the ratio Δ/Σ between the difference and sum patterns for different deflection angles α. The measured efficiency $\eta\%$ is

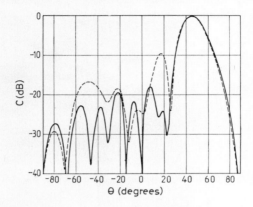

Fig. 7.30 *Deflected sum patterns in H-plane at f = 15 GHz (From reference 36)*
——— (RS)
- - - - (AS)

deduced from the following expressions:

$$\eta\% = 100/10^{0.1\,(D_M - G_M)}$$

$$D_M = 10\log_{10}\left[\left(\frac{4\pi A}{\lambda_0^2}\right)\cos\alpha\right]$$

G_M is the measured maximum isotropic linear gain (dB) and A is the antenna geometrical area.

In conclusion, major advances have been made in lowering side-lobe level [36], in the reduction of phase-shifter complexity [35] as well as in the manufacturing procedures developed for building the array.

7.2.2.2 Omnidirectional radiation array for a mobile telecommunication station: This section discusses a very flat antenna with circular or elliptical polarised directional radiation over a large conical angular sector, and linear omnidirectional radiation in the reflector plane [37, 38]. It can be advantageously used for communication between mobile stations. The antenna comprises four short-circuited half dipoles (section 7.2.1) acting at quarter-wavelength resonance and fed in phase quadrature (Fig. 7.33 *a* and *b*). The ends of the four short-circuited half dipoles (1, 2, 3, 4) are screwed (7) on the edges of a cavity

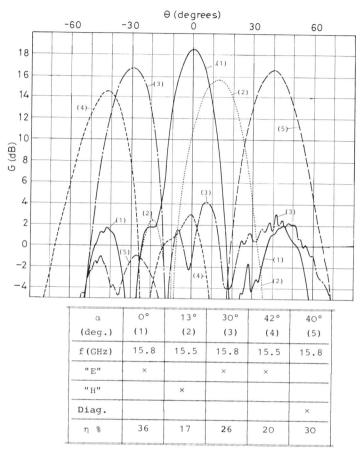

α (deg.)	0° (1)	13° (2)	30° (3)	42° (4)	40° (5)
f (GHz)	15.8	15.5	15.8	15.5	15.8
"E"	×		×	×	
"H"		×			
Diag.					×
η %	36	17	26	20	30

Fig. 7.31 *Measured sum gains G(dB) for different deflection angles α (degrees)*

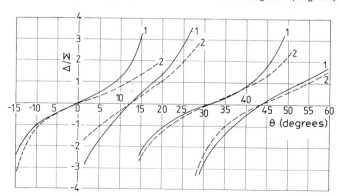

Fig. 7.32 *measured ratio $\Delta/\sum$ between difference (Δ) and sum ($\sum$) patterns for different deflection angles α*
1: E-plane
2: H-plane

384 Wideband flat dipole and short-circuit microstrip

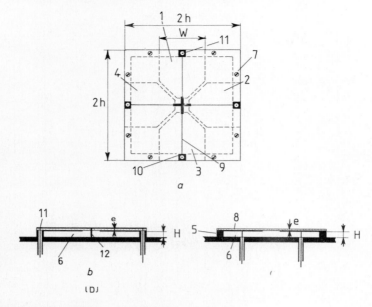

Fig. 7.33 Omnidirectional radiation array configuration

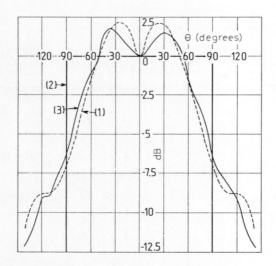

Fig. 7.34 Linear directivity in E-plane (f = 1·2 GHz)
　　　　(1): GTD
　　　　(2): GO
　　　　(3): Measurements (mm) $2h = 115/W = 50/H = 7·5/e = 1·6$

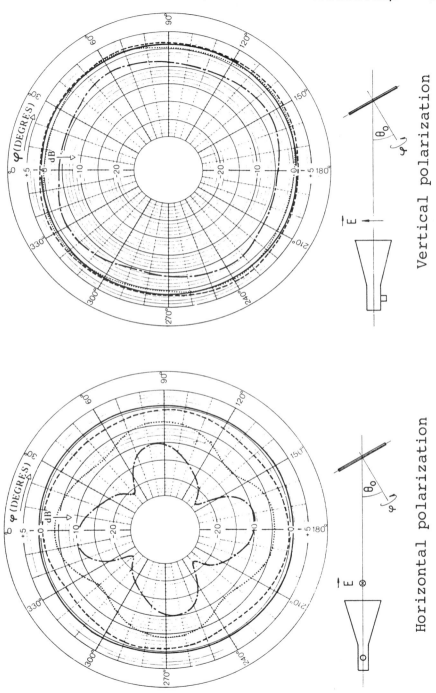

Fig. 7.35 *Linear isotropic gain measured for different site angles (θ_o)*

in air medium (6). A printed circuit (8) of thickness e supports, on one face, the four half dipoles and, on the other face, the feed microstrip lines (10), with quarter-wave transformers (9) which are connected to coaxial lines (11). The four quarter-wave-transformer ends are joined and short-circuited to the bottom of the cavity by 12. Another feed concept is possible (Fig. 7.33 c) by which the feed-point location is chosen to match the radiation resistance of the coaxial line.

With the following dimensions (in mm): $2h = 115$, $W = 50$, $H = 7.5$, $e = 1.6$, $2d = 300$ ($2d$ is the dimension of the square reflector); the bandwidth is equal to 4·35% for mismatch losses lower than 0·5 dB, while the theoretical bandwidth given by eqn. 7.13, is equal to 4·3% for a VSWR $\leqslant 2$ and $\varepsilon_r = 1$. The mean frequency is equal to 1·2 GHz.

Fig., 7.34 shows the linear directivity in the E-plane. The E-plane is the symmetrical plane related to the two opposite short-circuited half dipoles (Fig. 7.33 b or c) which are linearly polarised. Curve (1) gives the directivity when GTD on the reflector edges is taken into account. Fig. 7.35 shows the linear isotropic gain measured for different site angles θ_0 with horizontally or vertically polarised waves. We can see that the antenna is circularly polarised along its axis ($\theta_0 = 0°$), which is perpendicular to its plane, and quasi-linearly polarised in all directions ($\theta_0 = 90°$) of the reflector plane.

Another configuration is shown in Fig. 7.36. The antenna, which is vertically polarised with omnidirectional radiation, is composed of four short-circuited half-dipoles acting at quarter-wavelength resonance in air medium and fed in phase [39]. To match the radiation resistance of each half-dipole, a correct feed-point location along the symmetrical axis is found [40] and soldered to the central conductor of a coaxial line. The four coaxial lines are fed in parallel by means of a standard divider. A model was calculated and tested with the following parameters (in mm): $\varepsilon_r = 1$, $l = 40.5$, $W = 52$, $H = 2.6$, $h = 34.5$, $2d = 400$. The measured resonance frequency was found to be 1·90 GHz while the theoretical one (f_r given by eqn. 7.17) is equal to 1·93 GHz.

$$f_r^{-1} = 4\mu_0^{1/2} \varepsilon_0^{1/2} [h + H + 0.72 H(W/H + 0.26)(W/H + 0.81)^{-1}]$$

(7.17)

The measured bandwidth for a VSWR of less than 2 is equal to 4·7% while the theoretical bandwidth, given by eqn. 7.13, is 4·62%. Fig. 7.37 shows the measured E-plane, co-polar and cross-polar gain patterns, and the theoretical one when diffraction corrections, obtained with GTD applied to the square reflector edges, are taken into account. Maximum isotropic linear directivity of 6·46 dB occurs for $\theta = 37°$, while the measured maximum gain of the antenna with its divider is equal to 5·8 dB. For a constant angle θ around the antenna, the measured radiated far field is practically constant, with variations of ± 1 dB when $0 \leqslant \theta \leqslant 90°$. The bandwidth can be improved when the height H is increased, and the maximum site directivity angle may be altered by when adjusted the l parameter. This model is particularly thin since H/λ_0 is equal to 0·017.

7.2.2.3 Log-periodic array: This antennas is a microstrip travelling-wave antenna of large bandwidth [19, 41, 42, 43]. The log-periodic array is composed of flat short-circuited half dipoles acting at quarter-wave resonance and fed in series by means of a high-characteristic-resistance microstrip line, R_a (Fig. 7.38).

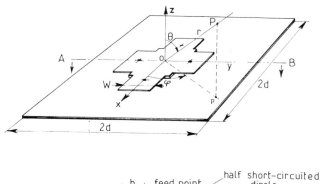

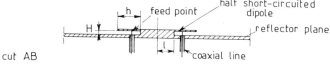

Fig. 7.36 *Vertically polarised antenna with omnidirectional radiation*

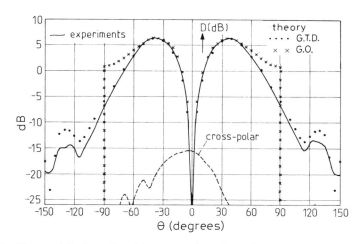

Fig. 7.37 *Measured E-plane directivity pattern (copolar and cross-polar)*
$\phi = \pi/2$, Θ: variable
$\varepsilon_r = 1$, $l = 40.5$ mm, $W = 52$ mm, $H = 2.6$ mm/$h = 34.5$ mm, $2d = 400$ mm

Each half dipole has a large *E*-plane pattern and weak coupling, especially in the parallel position (*H*-plane), and so the log-periodic array can be used for a progressive wave, when a squinted beam in the *H*-plane and large pattern

388 *Wideband flat dipole and short-circuit microstrip*

coverage in the E-plane are suitable. Fig. 7.39 shows the geometrical parameters W_n, H_n, and h_n for the half-dipole (D_n) of the nth order, acting at a quarter-wave resonance in air medium with $h_n + H_n \simeq (4f_n\sqrt{\mu_0\varepsilon_0})^{-1}$, where f_n is the resonance frequency. The various radiating parts are located on one metallic face

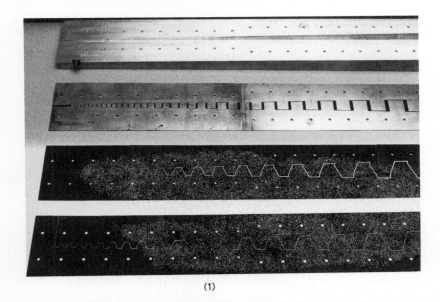

(1)

(2)

Fig. 7.38 *Log-periodic array of short-circuited patches*
    (1): Printed circuits and complete array
    (2): Cross-section

Wideband flat dipole and short-circuit microstrip

of a printed circuit (I). They are fed in series across a gap (G_n) by means of a stripline (M) etched on a second printed-circuit sheet (J). The classical log-periodic parameters are:

$$\sigma \simeq 2\frac{d_n}{\lambda_n} \text{ and } \tau = \frac{H_{n-1}}{H_n} = \frac{l_{n-1}}{l_n} = \frac{d_n-1}{d_n} < 1$$

l_{n+1} being the distance along the stripline between the gaps (G_n) and (G_{n+1}) of (D_n) and (D_{n+1}) separated by a distance d_{n+1}. To limit the variation of the VSWR in the frequency band, it is necessary to increase the expansion parameter τ. The distance l_n must be lower than $\lambda_g/2$ to compensate for the various reflections appearing at each gap. The following parameters have been adopted: $\tau = 0.95$, $\sigma = 0.4$, $d_n/\lambda_n = 0.2$, $H_n/\lambda_n = 0.1$, $l_n/\lambda_n = 0.32$, $W_n/\lambda_n = 0.166$, $R_a = 180\,\Omega$. With these parameters, a 50° 3 dB beamwidth and 45° deflected beam are calculated. With 50 radiating sources the antenna dimensions are $1.1 \times 0.1 \times 0.07$ m, which corresponds to a theoretical frequency band of 0.9 – 6 GHz bounded by a VSWR ≤ 2 and an efficiency of 95%.

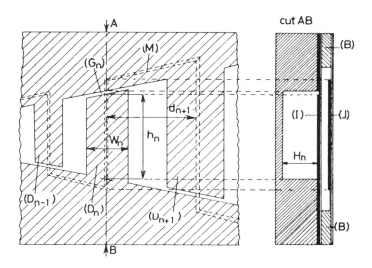

Fig. 7.39 *Geometrical parameters of the log-periodic structure*

The antenna performs well between 0.75 and 4.5 GHz. Fig. 7.40 shows the measured *H*-plane radiation patterns at three frequencies. The average values are 42° for the deflection angle and 50° for the 3 dB beamwidth over all the experimental bandwidth.

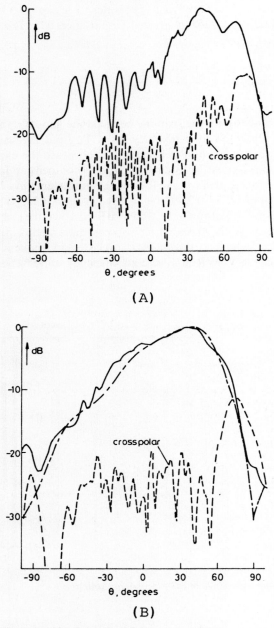

Fig. 7.40 *A Measured H-plane directivity patterns at 4.5 GHz*
B Measured H-plane directivity patterns
– · – Copolar at 0·8 GHz
——— Copolar at 2·9 GHz
– – – Cross-polar at 2·9 GHz

7.3 References

1. DUBOST, G., and ZISLER, S.: 'Antennes à large bande. Théories et applications' (Masson, Paris, 1976)
2. DUBOST, G., and VINATIER, C.: 'Doublet replié symetrique en plaques fonctionnant à très hautes fréquences et à large bande'. Brevet Européen 0044779 B1, 13 Nov. 1985; USA, 284 702, 20 July 1981; Japan, 114 526, 23 July 1981
3. DUBOST, G.: 'Flat radiating dipoles and applications to arrays' (John Wiley, 1981)
4. DUBOST, G. BEAUQUET, G., and VINATIER, C.: 'Theoretical radiation admittance of a large bandwidth flat symmetrical folded dipole', *Electron. Lett*, 1984, **20**, pp. 252–253
5. DUBOST, G., and RABBAA, A.: 'Analysis of a slot microstrip antenna', IEEE Trans., 1986, **AP-34**, pp. 155–163
6. DUBOST, G., and VINATIER, C.: 'Réseau de doublets repliés symétriques en plaques, à large bande autour de 12 GHz', *L'Onde Electrique*, 1981, **61**, pp. 34–41
7. DUBOST, G., and GUEHO, S.: 'Impédance mutuelle et couplage entre deux doublets repliés plans parallèles en fonction de leur écartement', *CR Acad. Sci. Paris*, 1985, **301**, sér. II, pp. 79–82
8. DUBOST, G.: 'Mutual coupling between flat folded dipoles in terms of frequency'. Int. Symposium Antennas and EM Theory, Beijing, China, Aug. 1985, pp. 706–711
9. DUBOST, G., and FRIN, R.: 'Antenne plaque à double polarisation croisée'. (Brevet 36 05 990, 23 April 1986
10. DUBOST, G.: 'Large bandwidth dual polarized multilayer microstrip antenna'. AP-S Intern. Symp., Philadelphia, USA, June 1986, pp. 455–458
11. PALANISAMY, V., and GARG, R.: 'Analysis of circularly polarized square ring and crossed-strip microstrip antennas', *IEEE Trans.*, 1986, **AP-34**, pp. 1340–1345
12. DUBOST, G.: 'Broadband circularly polarized flat antenna'. Int. Symp. Ant. & Prop., Japan, 1978, pp. 89–92
13. CHIBA T., *et al.*: 'Suppression of higher modes and cross polarized component for microstrip antennas'. IEEE AP-S, 1982, p. 285
14. HUANG, J.: 'CP microstrip array with wide axial ratio bandwidth and single feed LP elements'. IEEE AP-S, 1985, pp. 705–708
15. DUBOST, G., FRIN, R.: 'Antenne à double polarisation associée à des directeurs'. Brevet Europe, USA and Japan, April 1987
16. DUBOST, G., and RABBAA, A.: 'Synthèse de l'antenne plaque à double fente', *L'Onde Electrique*, 1987, **67**, pp. 72–79
17. DUBOST, G., and RABBAA, A.: 'Substrate influence on flat folded dipole bandwidth', *Electron. Lett.*, 1985, **21**, pp. 426–427
18. JAMES, J. R., HALL, P. S., and WOOD, C.: 'Microstrip Antenna' (Peter Peregrinus 1981) IEE Electromaghehe Waves Series 12.
19. DUBOST, G.: 'Comparison between flat radiating source bandwidth'. IEE Coloquim on Antenna bandwidth extension techniques, London, 28 Oct. 1985.
20. DUBOST, G., and VINATIER, C.: 'Large bandwidth and high grain array of flat folded dipoles acting at 12 GHz' 3rd Int. Conf. on Ant. and Propag. ICAP, Norwich, April 1983
21. DUBOST, G.: 'Improvements in printed-circuit radiating sources and arrays'. IEE Colloquim on receiving antennas for satellite broadcasting', London, April 1984
22. DUBOST, G., BEGUIN, D., CHAPUIS, E., AURIOL, A.: 'Réseau plat à grand gain et à faibles lobes secondaires à 5 GHz'. Conférence Internationale sur le Radar, Paris, May 1984
23. MARCHAND, M.: 'Antenne plane à réseau en bande X', *Rev. Tech. Thomson–CSF (France)*, 1985, **17**, pp. 83–109
24. DUBOST, G., POTIER, P.: 'Réseau plat à commutation électronique de faisceaux dans la bande des 12 GHz', *L'Onde electrique*, 1985, **65**, pp. 56–61
25. DUBOST, G., and NICOLAS M.: 'A braod angular coverage and large bandwidth antenna'. 17th European Microwaves Conf., Rome, Sept. 1987

26 DUBOST G., NICOLAS, M., VALLEE, P.: 'Antenne à pas de balayage réduit dans un large secteur angulaire'. Brevet 85 07 348, 15 May 1985
27 DUBOST, G.: 'Antenne à symétrie de révoloution associant une lentille diélectrique à une source plaque'. IEEE Symposium AP-S, Vancouver, Canada, June 1985, pp. 587–590
28 DUBOST, G.: 'Flat linear radiating array applied on a cylindrical lens'. Melecon'85 Mediterranean Electrotechnical Conference, Madrid, Oct. 1985, pp. 215–218
29 DUBOST, G.: 'Short-or open-circuited dipole parallel to perfect reflector plane and embedded in substrate and acting at resonance', *Electron. Lett.*, 1981, **17**, pp. 914–916
30 DUBOST, G.: 'Transmission-line model analysis of a lossy rectangular microstrip patch' *Electron. Lett.* 1982, **18**, pp. 282–282
31 DUBOST, G., and RABBAA, A.: 'Mutual impedance between two short-circuited flat resonant dipoles', IEEE Trans., 1981, **AP-29**, pp. 668–671
32 DUBOST, G.: 'Far field radiated by short-circuited microstrip antenna acting at a quarter wavelength resonance' *Electron. Lett. 1983*, **19**, pp. 737–739
33 DUBOST, G., GUEHO, S., and BEGUIN, D.: 'Ku band phased array in a large angular sector', 5th Int. Conf. on Antennas and Propag., ICAP 87, 1987, University of York
34 DUBOST, G., and GUEHO, S.: 'A 3 bits digital phase shifter in Ku band for microstrip phased array'. 8th Coloquim on Microwave communication, Aug. 1986, Budapest, Hungary
35 DUBOST, G., GUEHO, S., BEGUIN, D.: 'Déphaseur élémentaire en ligne microruban et déphaseur à commande numérique en faisant application', Brevet 86 11 923, 21 Aug. 1986
36 DUBOST, G., GUEHO, S., and BEGUIN, D.: 'Antenne réseau carré, monopulse à balayage électronique'. Brevet 87 20774, 13 Jun. 1987
37 DUBOST, G., ALEXIS, R.: 'Antenne formée à partir d'une cavité résonnante comportant une face rayonnate'. Brevet 84 08.392, 19 May 1984
38 DUBOST, G.: 'Antennes plaques pour les télécommunications entre stations mobiles' *L'Onde Electrique,* 1985), **65**, pp. 41–49
39 DUBOST, G.: 'Vertically polarized flat antenna with omnidirectional radiation". Int. Symposium on Antennas and Propagation, ISAP, Aug. 1985, Kyoto, Japan, pp. 109–112
40 DUBOST, G.: 'Influence of feed-point location on radiation resistance of a short-circuited flat dipole', *Electron. Lett.*, 1984, **20**, pp. 980–981
41 DUBOST, G., BIZOUARD, A.: 'Antenne périodique plane'. Brevet 83.19.924 13 Dec. 1983
42 DUBOST, G., GUEHO, S., and BIZOUARD, A.: 'Log-periodic short-circuited dipole array with a squinted beam', *Electron. Lett.*, **20**, pp. 411–413
43 DUBOST, G., and GUEHO, S.: 'Theory of a large bandwidth microstrip plane array with a deflected beam'. IEEE Int. Symposium on Ant. Prop., June 1984, Boston, Mass., USA

Chapter 8

Numerical analysis of microstrip patch antennas

J.R. Mosig, R.C. Hall and F.E. Gardiol

8.1 Introduction

8.1.1 General description

Microstrip patch antennas are thin and lightweight radiating elements, formed by a substrate, including one or several dielectric layers, backed by a metallic sheet (the ground plane).

Thin metallic patches (the radiating elements) are located on the air–substrate interface and, possibly, between the dielectric layers. Microstrip antennas are manufactured by the photolithographic process developed for printed circuits. Their low profile, low weight and mechanical ruggedness make them an ideal choice for aerospace applications. They can be mass-produced, and could thus provide inexpensive receiver antennas for direct reception of microwave signals from satellites (television, mobile communications). Finally, they are ideally suited to be combined in large arrays, the individual patches sharing the same substrate. Thus directive antennas can be obtained in spite of the inherent low directivity of a single patch.

The remarkable practical advantages offered by microstrip antennas are offset, to some extent, by their inhomogeneous nature, and a rigorous analysis was long considered to be a hopeless task.

An accurate model should take into account the three inhomogeneities of a microstrip structure (Fig. 8.1):

(*a*) The presence of at least two dielectrics (often air and substrate)
(*b*) The boundary conditions on the interfaces between different layers are inhomogeneous since thin metallic plates making up the radiating elements and feeding the structure can partially fill the interfaces
(*c*) Any microstrip structure is finite in dimensions; i.e. its ground plane and its dielectric substrate are bounded in the transverse directions. The edges may, however, be located at a very large distance, in which case this third inhomogeneity may be neglected (the structure is then assumed, mathematically, to extend to infinity).

Models used to study microstrip patch antennas range from very simplified ones, such as the transmission-line model, through cavity models, planar circuit analysis, segmentation techniques, and up to quite sophisticated approaches based on an integral-equation formulation. In the framework of the integral-equation model, many different approaches exist, depending on the use of spectral or space quantities and on the geometries to be included.

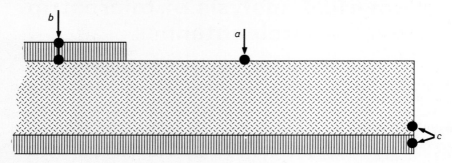

Fig. 8.1 *The three inhomogeneities of a microstrip structure.*
 a Dielectric media of unequal permittivity,
 b Infinitely thin conductors introducing surface current between the dielectric media
 c Finite transverse dimensions of substrate and ground plane

Detailed surveys of these models are available (e.g. Reference 1), and several of them are treated elsewhere in this book.

8.1.2 The integral equation model
The purpose of this Chapter is to provide a rigorous treatment of microstrip antennas, free from over-simplifying assumptions. Among its principal features, the proposed model is able to handle patches of arbitrary shapes where no educated guess of the surface-current distribution is possible. Also there is no limitation in frequency and substrate thickness. The model automatically takes into account mutual coupling between elements and can predict the performance of a patch embedded in an array environment. Surface waves are included as well as dielectric and ohmic losses. Thus the model allows accurate prediction of quasistatic behaviour, dominant and higher modes of resonance, and input impedances, coupling coefficients, radiation patterns, gain and efficiency at any frequency.

The model relies upon the identification of a microstrip antenna as a particular case of a stratified medium. The pioneer work on electromagnetic-wave propagation in stratified media must be ascribed to A. Sommerfeld, who investigated the radio-wave propagation above a lossy ground as early as 1909.

The microstrip antenna is modelled by an integral equation where the main unknown is the electric surface current density on the patches. The Green's functions forming the kernel of this equation include the effects of the layers, and are obtained in the form of inverse Hankel transforms by the systematic use of stratified media theory. The first Sections of this Chapter are devoted to the construction of the integral equation and of the pertinent Green's functions. Considerable attention is paid to the development of efficient numerical techniques for evaluating the Green's functions.

The integral equation is directly formulated in the space domain using a vector and a scalar potential. The resulting mixed potential integral equation (MPIE), similar to that obtained for wire antennas, is better suited for numerical analysis than the customary electric-field integral formulation.

The MPIE is solved by a method of moments. In the general case, rooftop subsectional-basis functions are used. For particular geometries, it can be more efficient to use entire domain-basis functions corresponding to the eigenvalues of the equivalent cavity. Finally, standard circuit analysis is used to deal with multiport antennas, loaded antennas and arrays.

The final Sections of the Chapter present numerical results for the input impedance, coupling coefficients and radiation patterns of several microstrip antennas and arrays of practical interest.

8.2 Model based on the electric surface current

8.2.1 Geometry of the model and boundary conditions
In order to present the theory in the clearest possible manner, the electric surface-current model will be developed for the simple microstrip structure of Fig. 8.2 with a single dielectric layer and a metallic patch. The generalisation to multilayer antennas and to multiple patches (arrays) will be considered later.

The substrate is assumed to extend to infinity in the transverse directions and is made of a nonmagnetic, isotropic, homogeneous material which can be lossy. The ground plane also has infinite dimensions, and the upper conductor (the metallic patch) has zero thickness. Both the ground plane and patch are allowed to have ohmic losses.

The direction perpendicular to the ground plane is selected as the z-axis (Fig. 8.2). The patch extends over part of the $z = 0$ plane, denoted by the surface S_0. The remainder of the $z = 0$ plane, i.e. the surface separating the two dielectric media, is denoted S and called hereinafter the interface.

Indexes 1 and 2 are associated, respectively, with the infinite dielectric extending above the substrate, usually the air, and with the substrate itself. Thus, we have

$$\varepsilon_1 = \varepsilon_0, \quad z > 0 \tag{8.1}$$

$$\varepsilon_2 = \varepsilon_0 \varepsilon_r = \varepsilon_0 \varepsilon_r'(1 - j\tan\delta) \quad -h < z < 0$$

where h is the substrate thickness and

$$\mu_1 = \mu_2 = \mu_0 \tag{8.2}$$

everywhere.

The excitation is provided by a time-harmonic electromagnetic field. Complex phasor notation is used throughout this Chapter. Any complex scalar quantity C represents an instantaneous quantity $C(t)$ given by:

$$C(t) = \sqrt{2}\,\text{Re}[C\exp(j\omega t)] \tag{8.3}$$

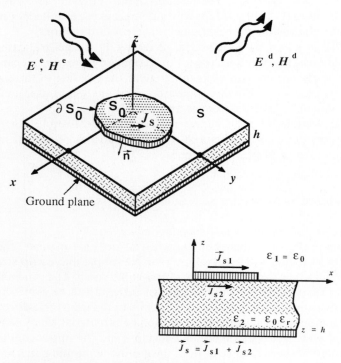

Fig. 8.2 *General view of a microstrip antenna and vertical cut in the $y = 0$ plane*
Superscripts e and d refer to the incident fields (excitation) and to the scattered (diffracted) fields. For a infinitely thin patch the currents J_{s1} and J_{s2} can no longer be distinguished and the meaningful quantity is the total current $J_s = J_{s1} + J_{s2}$ flowing on the patch.

The excitation fields are denoted by E^e and H^e. They can be the fields of a plane wave coming from infinity (receiving antenna) or the local fields created by a finite source located within the microstrip structure (transmitting antenna).

In either case, the excitation fields induce surface currents on the upper side of the ground plane and on both sides of the metallic patch. However, since the patch is assumed to have zero thickness, the model cannot differentiate between the currents flowing on its upper and lower side. Indeed, the patch is modelled as a sheet of current J_s whose value at any point is the algebraic sum of the upper and lower surface currents J_{s1} and J_{s2} existing at $z = 0^+$ and $z = 0^-$ (Fig. 8.2).

The induced currents in turn create diffracted or scattered electromagnetic fields. These fields, denoted E^d, H^d, add to the excitation fields to yield the total fields E, H existing in the entire space.

On the air–dielectric interface (the plane $z = 0$, excluding the surface of the patch S_0) the boundary conditions are:

$$e_z \times (E_1 - E_2) = 0 \tag{8.4}$$

$$e_z \times (H_1 - H_2) = 0 \tag{8.5}$$

On the metallic surfaces the boundary conditions will be inhomogeneous owing to the presence of the currents. Assuming that the patch and the ground plane are perfect conductors (this restriction will be removed later) we have on the upper side of the patch $S_0(z = 0^+)$:

$$e_z \times E_1 = 0; \quad e_z \times H_1 = J_{s1} \tag{8.6}$$

and, similarly on the lower side $z = 0^-$:

$$e_z \times E_2 = 0; \quad e_z \times H_2 = -J_{s2} \tag{8.7}$$

Combining the above pairs of equations, we obtain

$$e_z \times (E_1 - E_2) = 0 \tag{8.8}$$

$$e_z \times (H_1 - H_2) = J_{s1} + J_{s2} = J_s \tag{8.9}$$

which apply to the fields on both sides of the patch. Eqns. 8.4–8.9 are written in terms of the total fields. However, since the excitation fields are assumed to be continuous, the boundary conditions (eqns. 8.4–8.5) and (eqns. 8.8–8.9) also hold for the diffracted fields. Hence the diffracted tangential electric field is continuous across the patch, while the jump in the diffracted tangential magnetic field equals the total surface current on the patch.

Finally, the boundary conditions on the ground plane $z = -h$ are

$$e_z \times E_2 = 0 \tag{8.10}$$

$$e_z \times H_2 = J_s \tag{8.11}$$

It must be pointed out that the microstrip antenna problem can be completely solved without actually knowing the exact distribution of surface current on the ground plane. In fact, image theory can be used to remove the ground plane and the boundary conditions (eqns. 8.10–8.11) will then be included automatically in the Green's functions.

8.2.2 Potentials for the diffracted fields

Since no volume sources are considered in this model, the diffracted fields satisfy the homogeneous Maxwell's curl equations:

$$\nabla \times E^d = -j\omega\mu_0 H^d \tag{8.12}$$

$$\nabla \times H^d = j\omega\varepsilon E^d \tag{8.13}$$

The resolution of antenna problems can in many cases be simplified by introducing the scalar and the vector potentials for the diffracted fields:

$$H^d = (1/\mu_0)\nabla \times A \tag{8.14}$$

$$E^d = -j\omega A - \nabla V \tag{8.15}$$

subject to Lorentz's gauge:

$$\nabla \cdot A + j\omega\mu_0\varepsilon V = 0 \tag{8.16}$$

Introducing the above expressions into Maxwell's equations 8.12 and 8.13 and combining, we obtain two homogeneous Helmholtz's equations for the potentials:

$$(\nabla^2 + k_i^2)A = 0 \tag{8.17}$$

$$(\nabla^2 + k_i^2)V = 0 \tag{8.18}$$

where $k_i = \omega(\mu_0\varepsilon_i)^{1/2}$ is the wavenumber in medium i.

It must be recalled here that the choice of the couple A, V is not unique [2]. Indeed, any vector $A^* = A + \operatorname{grad} \Psi$ can be used as a new vector potential provided that the scalar potential is replaced by $V^* = V - j\omega\Psi$. Moreover, if Ψ is a solution of the homogeneous Helmholtz's equation, the new potential will still satisfy Lorentz's gauge (eqn. 8.16). We shall discuss later some convenient choices for the potentials.

8.2.3 Green's functions

Let us consider an arbitrarily oriented Hertz dipole of moment $I\,dl$ located at the point r' (Fig. 8.3). In general, the vector potential at the point r due to this dipole is given by the linear relationship

$$A(r) = \bar{\bar{G}}_A(r|r') \cdot I\,dl \tag{8.19}$$

where $\bar{\bar{G}}_A$ is a three-dimensional dyadic Green's function. The physical meaning of $\bar{\bar{G}}_A$ is evident: the scalar component G_A^{st} gives the s-component of the vector potential existing at the point r created by a t-directed Hertz dipole located at the point r'.

If the source and the observer are surrounded by an infinite homogeneous medium, the dyadic $\bar{\bar{G}}_A$ is diagonal and can be expressed as the product of a scalar Green's function G_A times the unit dyadic $\bar{\bar{U}}$. In this case, the vector potential is always colinear with the source dipole.

For a microstrip antenna it is possible to use a scalar free-space Green's function of the vector potential. However, if we do, we then need to use fictive electric and magnetic surface currents on the air-dielectric interface in order to satisfy the boundary conditions. These currents are also unknowns in the integral-equation formulation of the problem and add to the complexity of the numerical solution. The preferred solution is to include in the Green's functions effects of the dielectric substrate and of the ground plane. Therefore, the Green's functions must satisfy the boundary conditions (eqns. 8.4–8.5) and (eqns. 8.10–8.11). Moreover, the boundary condition for the magnetic field (eqn. 8.9) is automatically built into the formulation of the Green's functions. These functions are dyadics, and unfortunately do not have a closed analytical expression. However, once they are formulated and numerically evaluated, the only unknown which remains is the true electric surface-current distribution on the conducting patch.

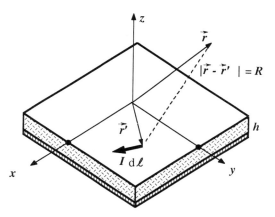

Fig. 8.3 *Horizontal electric dipole (HED) on a microstrip substrate*
The fields and potentials of such an elementary source give the Green's functions associated with a microstrip antenna.

Keeping in mind the linearity of Maxwell's equations, the vector potential of a given current distribution can be written as a superposition integral involving the corresponding dyadic Green's function

$$A(r) = \int_{S_0} \bar{\bar{G}}_A(r|r') \cdot J_s(r') \, dS' \tag{8.20}$$

The scalar potential V is obtained by introducing the above expression in the Lorentz gauge with the result:

$$j\omega\mu\varepsilon \, V(r) = -\int_{S_0} \nabla \cdot \bar{\bar{G}}_A(r|r') \cdot J_s(r') \, dS' \tag{8.21}$$

The Green's function G_V associated with the scalar potential must be carefully defined. In fact, the uniqueness of G_V is guaranteed only if the divergence of $\bm{G}_A$ is an irrotational vector. Thus we can write [3]

$$\nabla \cdot \bar{\bar{G}}_A(r|r') = \mu\varepsilon \nabla G_V(r|r') = -\mu\varepsilon \nabla' G_V(r|r') \tag{8.22}$$

where ∇' acts on the primed co-ordinates. Expression 8.21 is now easily transformed into

$$V(\mathbf{r}) = -\int_{S_0} G_V(\mathbf{r}|\mathbf{r}') \frac{\nabla' \cdot \mathbf{J}_s(\mathbf{r}')}{j\omega} dS' + \int_{\partial S_0} G_V(\mathbf{r}|\mathbf{r}') \mathbf{J}_s(\mathbf{r}') \cdot \mathbf{n}\, dS' \quad (8.23)$$

where ∂S_0 is the perimeter of the patch and $\mathbf{n}$ is the outwards-pointing normal unit vector (Fig. 8.2). The edge condition guarantees that the normal component of the surface current vanishes on the perimeter of the patch. Hence, the line integral in eqn. 8.23 can be eliminated.

We can now introduce the associated surface-charge density q_s via the continuity equation:

$$\nabla \cdot \mathbf{J}_s + j\omega q_s = 0 \quad (8.24)$$

Finally, we can express the scalar potential as:

$$V(\mathbf{r}) = \int_{S_0} G_V(\mathbf{r}|\mathbf{r}') q_s(\mathbf{r}')\, dS \quad (8.25)$$

It is worth mentioning that the edge condition can be applied even if S_0 is a portion of a larger patch. Such a situation may arise when solving the problem with a method of moments using subsectional-basis functions. In this case the line integral in eqn. 8.23 can still be eliminated, but since $\mathbf{J}_s$ does not necessarily vanish on the boundary of S_0, the continuity equation must be interpreted according to the theory of distributions, and delta functions corresponding to line charges in the boundary of S_0, will appear in the expression for q_s.

The Green's function G_V can be viewed as the scalar potential created by a point charge, even if isolated time-varying point charges do not exist in the real world. Thus, owing to the lack of a sound physical interpretation, it is better to consider G_V only as a useful mathematical device. Only when the frequency goes to zero, does this function become the familiar electrostatic potential of a point charge.

8.2.4 Mixed potential integral equation (MPIE)

The diffracted fields derived from the potentials of eqns. 8.20 and 8.25 satisfy Maxwell's equations and the boundary conditions of the problem (eqns. 8.4–8.5) and (eqns. 8.8–8.9). The last step is now to relate these fields to the excitation fields via conditions (eqns. 8.6–8.7). If the total tangential electric field is forced to vanish on the patch surface, we get the standard electric field integral equation. This equation can be slightly modified to account for the ohmic losses on the patch. The total tangential electric field is now proportional to the total surface current, and we can write

$$\mathbf{e}_z \times (\mathbf{E}^e + \mathbf{E}^d) = \mathbf{e}_z \times Z_s \mathbf{J}_s \quad (8.26a)$$

or, introducing the potentials

$$\mathbf{e}_z \times (j\omega \mathbf{A} + \nabla V + Z_s \mathbf{J}_s) = \mathbf{e}_z \times \mathbf{E}^e \quad (8.26b)$$

where the proportionality constant Z_s is a surface impedance (measured in ohms) which accounts for the finite conductivity of the patch. An accurate value for Z_s can only be obtained by measurement since Z_s must include technological data such as the thickness and roughness of the metallic patch. However, in most cases the patch is thick compared with the skin depth δ, and the classical expression

$$Z_s = (1 + j)/\sigma^*\delta \qquad (8.27)$$

represents a good approximation. In the above expression σ^* is an effective conductivity that includes roughness effects and can be several times lower than the values of conductivity found in standard tables.

Finally, introducing the integral form of the potentials (eqns. 8.20–8.21) in eqn. 8.26 we get the final expression for the mixed potential integral equation (MPIE):

$$e_z \times (j\omega \int_{S_0} \bar{\bar{G}}_A \cdot J_s dS' + \nabla \int_{S_0} G_V q_s dS' + Z_s J_s) = e_z \times E^e \qquad (8.28)$$

The validity of this equation depends on the possibility of defining the Green's function G_V by eqn. 8.22.

Eqn. 8.28 is a Fredholm integral equation of the second kind. However, the term $Z_s J_s$ is usually small and the iterative techniques commonly used for Fredholm integral equations of the second kind that arise, for example, when using the magnetic field integral equation [4], do not apply here.

The unknowns in the integral equation 8.28 are the surface current J_s and the surface charge q_s. However, they are not independent, and are related through the continuity equation.

8.2.5 Sketch of the proposed technique

The successive steps in solving the microstrip antenna problem are now clear. We start with the theoretical determination of the required Green's functions G_A and G_V. In general, the Green's functions are available as definite integrals over semi-infinite intervals and they must be numerically evaluated for distances ranging from zero to the maximum linear dimension of the patch. The construction of accurate numerical integration algorithms to evaluate the Green's functions is a crucial step of the overall problem.

Once the Green's functions are computed, the unknown surface current is expanded over a set of basis functions and the integral equation is tested against a set of test functions using the so-called method of moments (MoM). Here, the correct choice of these sets of functions is essential for the quality of the final results. In this way, the integral equation is discretised and transformed into a set of linear equations. The complex eigenvalues of the matrix equation provide the unperturbed resonant frequencies of the patch and its unloaded quality factor.

The next step is the construction of the excitation fields. These fields depend strongly on the physical nature of the excitation. In many cases (coaxial pin, microstrip line) the computation of the excitation fields calls for the same or related Green's functions that were calculated for the MoM matrix. Testing the excitation fields yields the independent vector of the matrix equation.

The solution of the matrix equation gives a numerical estimation of the unknown surface current. Computing the voltages at the excitation points allows the determination of the impedance and scattering matrices of the antenna. Standard circuit analysis may be used to account for any load or for multiple excitations. Resonant frequencies and loaded quality factors are easily derived from the input impedance.

Table 8.1 *Essential steps (−) and main results (∗) of the proposed numerical technique*

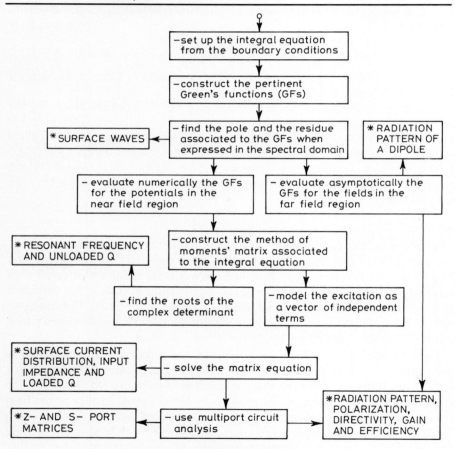

To calculate the radiation pattern we need to go back to the Green's functions and obtain their asymptotic forms in the far field. This can be done analytically and the radiation pattern of the antenna, including polarisation and phase information, is obtained by using a superposition integral over the patch.

Finally, integration of the far fields over the upper half-space will give the directivity of the antenna and an estimation of its efficiency and gain.
These steps are summarised in Table 8.1.

8.3 Horizontal electric dipole (HED) in microstrip

The construction of the Green's functions requires the determination of the fields created by a horizontal electric dipole (Hertzian dipole) located on the air–dielectric interface of a microstrip structure (Fig. 8.3).

The first investigations of a HED embedded within a stratified medium are due to Sommerfeld, who published in 1909 the exact solution for a dipole over an imperfect ground. The Hankel integral transforms which appear within such a problem are often known as Sommerfeld integrals.

The vertical dipole over a conducting plane covered with a dielectric layer was studied by Lo and Brick in two articles which appeared almost simultaneously [5, 6]. The problem is quite similar to the one arising in microstrip antennas, except for the dipole orientation. However, even though the first microstrip lines were introduced around the time these articles were published, no connection between the two fields was made – microstrip antennas were to be developed some 20 years later. This may explain the different approaches to the two problems, in particular the absence of a detailed study of the near field, which is essential when the source and the observer are both located on the surface of the substrate.

The general theory of dipoles – either electric or magnetic, horizontal or vertical, located within an arbitrary stratified medium – was developed later, mainly by Brekhovskikh [7], Wait [8], Felsen and Marcuvitz [9], and Kong [10]. However, as was done in previous publications, the emphasis was put on the study of radiated fields, for which approximate asymptotic analysis is sufficient.

For the accurate study of microstrip radiation, however, precise knowledge of the surface currents on the patch, and hence of the near fields on the dielectric interface, are required. For this reason, the fields created by a HED located on the air – dielectric interface will be determined.

8.3.1 The vector potential
Let us consider a HED of moment Idx equal to unity placed at the point $\mathbf{r}' = 0$ as indicated in Fig. 8.4.

In order to ease the mathematical development required for solving the Helmholtz equation 8.17 satisfied by the vector potential, we will replace the space variables x, y by their spectral counterparts k_x, k_y according to the double Fourier transform:

$$\tilde{f}(k_x, k_y) = \mathscr{F}[f(x,y)] = \frac{1}{2\pi} \iint_{-\infty}^{\infty} f(x,y) \exp(-jk_x x - jk_y y)\, dx\, dy$$

(8.29)

$$f(x, y) = \mathcal{F}^{-1}[\tilde{f}(k_x, k_y)]$$

$$= \frac{1}{2\pi} \iint_{-\infty}^{\infty} \tilde{f}(k_x, k_y) \exp(jk_x x + jk_y y) dk_x dk_y \qquad (8.30)$$

In the spectral domain the homogeneous Helmholtz equation becomes:

$$\left(\frac{d^2}{dz^2} + u_i^2\right) \tilde{A} = 0$$

where

$$u_i^2 = k_x^2 + k_y^2 - k_i^2 = k_\varrho^2 - k_i^2 \qquad (8.31)$$

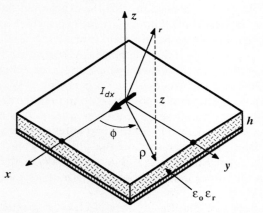

Fig. 8.4 *Co-ordinate system for the study of an x-directed horizontal electric dipole (HED) on a microstrip substrate*

The air–dielectric interface is at $z = 0$ and the ground plane is at $z = -h$. The continuity equation implies that there are two point charges of value $\pm I\, dx/j\omega$ at both extremities of the dipole

According to eqn. 8.1 we shall adopt from now on the shorthand notation $u_1^2 = u_0^2 = k_\varrho^2 - k_0^2$ and $u_2^2 = u^2 = k_\varrho^2 - \varepsilon_r k_0^2$.

The general solution of eqn. 8.31 for a cartesian component of A is ($s = x, y, z$):

$$\tilde{A}_s = a \exp(-u_i z) + b \exp(+u_i z) \qquad (8.32)$$

where the unknown factors a, b may be functions of the spectral variables.

The fields are obtained by using eqns. 8.14–8.15 and the Lorentz gauge, which also holds in the spectral domain. Since no external excitation fields are considered here, the fields of eqns. 8.14–8.15 are total fields. These fields must satisfy the boundary conditions eqns. 8.4–8.5 on the interface and eqn. 8.10 on the ground plane. In particular, the HED is equivalent to a surface current density in the plane $z = 0$ given by:

$$J_s = e_x \delta(x)\delta(y) \Rightarrow \tilde{J}_s = e_z(1/2\pi) \qquad (8.33)$$

If the dipole is embedded in an infinite homogeneous medium of permittivity ε_0, the vector potential is parallel to the dipole and exhibits a spherical symmetry:

$$A = e_x \frac{\mu}{4\pi} \frac{\exp(-jk_0 r)}{r} \Rightarrow \tilde{A} = e_x \frac{\mu}{4\pi} \exp(-u_0|z|)/u_0 \qquad (8.34)$$

On the other hand, it is well known [11] that two cartesian components of A are needed to satisfy the boundary conditions in an inhomogeneous structure such as a microstrip antenna. Here, we shall adopt Sommerfeld's choice and postulate an additional vertical component for the vector potential. Hence $A = e_x A_x + e_z A_z$.

Choosing now for A_x and A_z general expressions of the form of eqn. 8.32, we obtain after satisfying the boundary conditions the expressions:

$$\tilde{A}_x = \frac{\mu_0}{2\pi} \begin{Bmatrix} \exp(-u_0 z)/D_{TE} \\ \sinh u(z+h)/(D_{TE} \sinh uh) \end{Bmatrix} \qquad (8.35)$$

$$\tilde{A}_z = \frac{\mu_0}{2\pi} (\varepsilon_r - 1) jk_x \begin{Bmatrix} \exp(-u_0 z)/(D_{TE} D_{TM}) \\ \dfrac{\cosh u(z+h)}{(D_{TE} D_{TM} \cosh uh)} \end{Bmatrix} \qquad (8.36)$$

where $D_{TE} = u_0 + u \coth uh$, $D_{TM} = \varepsilon_r u_0 + u \tanh uh$, and the upper and lower expression inside the symbol { } correspond, respectively, to the upper semi-infinite medium ($z > 0$) and to the substrate ($-h < z < 0$).

It can be easily shown that if $\varepsilon_r = 1$ and $h \to \infty$, the vertical component A_z vanishes and A_x becomes the free-space vector potential given by eqn. 8.34.

8.3.2 Scalar potential and the fields
The continuity equation applied to an electric dipole implies the existence of two point charges $q = \pm I/j\omega$ at both ends of the dipole.

Since the product Idx has been assumed to be equal to unity, the moment of this pair of charges is $qdx = 1/j\omega$. The scalar potential associated with the dipole is given directly by the Lorentz gauge. Introducing eqns. 8.35 and 8.36 in eqn. 8.16 we get:

$$\tilde{V} = -\frac{jk_x}{2\pi j\omega\varepsilon_0} \begin{Bmatrix} N\exp(-u_0 z)/(D_{TE} D_{TM}) \\ N\sinh u(z+h)/(D_{TE} D_{TM} \sinh uh) \end{Bmatrix} \qquad (8.37)$$

with $N = u_0 + u \tanh uh$.

In the space domain, the scalar potential V of an electrostatic dipole of moment $1/j\omega$ is related to the scalar potential V_q of a single unit point charge by the well known expression:

$$V = -\frac{1}{j\omega} \frac{\partial V_q}{\partial x} \qquad (8.38)$$

or, in the spectral domain by

$$\tilde{V} = -\frac{jk_x}{j\omega}\tilde{V}_q \tag{8.39}$$

Comparing eqn. 8.39 with eqn. 8.37, we can deduce by analogy that the potential of a unit point charge on the air–dielectric interface of a microstrip antenna is given by

$$\tilde{V}_q = \frac{1}{2\pi\varepsilon_0}\begin{Bmatrix} N\exp(-u_0 z)/(D_{TE}D_{TM}) \\ \dfrac{N\sinh u(z+h)}{D_{TE}D_{TM}\sinh uh} \end{Bmatrix} \tag{8.40}$$

Now, the construction of the fields is straightforward, for we have in the spectral domain:

$$\begin{aligned}
\tilde{E}_x &= -j\omega\tilde{A}_x - jk_x\tilde{V} & \mu\tilde{H}_x &= jk_y\tilde{A}_z \\
\tilde{E}_y &= -jk_y\tilde{V} & \mu\tilde{H}_y &= (\partial\tilde{A}_x/\partial z) - jk_x\tilde{A}_z \\
\tilde{E}_z &= j\omega\tilde{A}_z - (\partial\tilde{V}/\partial z) & \mu\tilde{H}_z &= -jk_y\tilde{A}_x
\end{aligned} \tag{8.41}$$

It can be easily demonstrated that the vertical component E_z shows the expected jump discontinuity when crossing the air–dielectric interface.

8.3.3 Surface waves and the spectral plane k_ϱ
It can be shown [12] that the equations $D_{TE} = 0$, $D_{TM} = 0$ are the characteristic equations for the TE and TM surface-wave modes propagating on a dielectric-coated conducting plane. Hence the zeros of D_{TE} and D_{TM} give the phase constant of the surface waves existing on a microstrip structure.

The terms D_{TE} and D_{TM} depend on k_x and k_y only through the radial spectral variable $k_\varrho^2 = k_x^2 + k_y^2$. For functions exhibiting such a kind of dependence, the inverse Fourier transform 8.30 can be written as

$$\mathscr{F}^{-1}[\tilde{f}(k_\varrho)] = \int_0^\infty J_0(k_\varrho \varrho)k_\varrho \tilde{f}(k_\varrho)\,dk_\varrho \tag{8.42a}$$

and

$$\mathscr{F}^{-1}[jk_x\tilde{f}(k_\varrho)] = -\cos\varphi \int_0^\infty J_1(k_\varrho \varrho)k_\varrho^2 f(k_\varrho)\,dk_\varrho \tag{8.42b}$$

where ϱ, ϕ are polar co-ordinates and J_0 is the Bessel function of zeroth order and first kind. The replacement of the double Fourier transform by Hankel transforms is of great utility when performing the numerical evaluation of the fields and potentials. For instance, if the observation point is also on the interface ($z = 0$) we have:

$$A_x = \frac{\mu_0}{2\pi}\int_0^\infty J_0(k_\varrho\varrho)\frac{k_\varrho}{D_{TE}}\,dk_\varrho \tag{8.43a}$$

$$A_z = -\frac{\mu_0}{2\pi}\cos\varphi\,(\varepsilon_r - 1)\int_0^\infty J_1(k_\varrho\varrho)\frac{k_\varrho^2}{D_{TE}D_{TM}}\,dk_\varrho \tag{8.43b}$$

$$V_q = \frac{1}{2\pi\varepsilon_0} \int_0^\infty J_0(k_\varrho \varrho) \frac{k_\varrho}{D_{TE}} dk_\varrho \tag{8.44}$$

Hence, the surface waves appear as poles of the integrands in the complex plane $k_\varrho = \lambda + jv$. It can be shown [13] that D_{TE} has no zeros if $k_0 h(\varepsilon_r' - 1)^{1/2} < \pi/2$ and D_{TM} has only one corresponding to the dominant zero-cutoff TM surface wave. This condition is equivalent to the restriction:

$$f[\text{GHz}] \leqslant 75/\{h[\text{mm}]\sqrt{(\varepsilon_r' - 1)}\} \tag{8.45}$$

For the sake of simplicity we shall assume from now on that this inequality holds. Higher frequencies would add new poles, but the analysis made for the single-pole case remains qualitatively valid.

For substrates with moderate losses the pole $\lambda_r + jv_p$ lies slightly below the real axis ($v_p < 0$) and its real part is bounded by $1 < \lambda_p/k_0 < \varepsilon_r^{1/2}$ (Fig. 8.5).

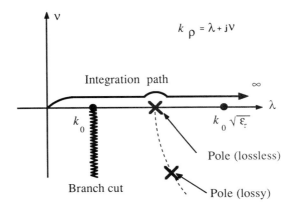

Fig. 8.5 *Topology of the complex plane k_ϱ with the original integration path from zero to infinity*
The Figure also shows the geometrical locus of the pole as a function of dielectric losses.

More precisely, a good approximation for electrically thin substrates [14] is

$$\frac{\lambda_p}{k_0} = 1 + (k_0 h)^2 \frac{(\varepsilon_r' - 1)^2}{2\varepsilon_r'^2} \tag{8.46}$$

$$v_p = -(\varepsilon_r' - 1)\tan\delta \left(\frac{k_0 h}{\varepsilon_r'}\right)^2$$

The integration path of the Sommerfeld integral eqn. 8.43–8.44 is, in general, the real positive axis λ. But, if we consider the theoretical case of a lossless substrate, then $v_p = 0$ and the pole is on the real axis. Since, by continuity, the integration path must remain above the pole, the integral from zero to infinity in the lossless case is interpreted as:

$$\int_0^\infty = \fint_0^\infty - j\pi R \qquad (8.47)$$

where the symbol $\fint$ stands for Cauchy's principal value and R is the residue of the integral at the pole $k_\varrho = \lambda_p$.

Finally, it should be mentioned that the function $u_0^2 = k_\varrho^2 - k_0^2$ introduces a branch point at $k_\varrho = k_0$ (Fig. 8.5). However, the integral remains bounded here and no deformation of the integration path is needed.

8.3.4 Far-field approximations

Far-field approximations are essential for the evaluation of the radiation pattern of a microstrip antenna. They can be obtained by using standard asymptotic techniques, such as the steepest-descent method [9]. We shall outline briefly the application of these techniques to Sommerfeld integrals. We begin by replacing the Bessel functions in these integrals by Hankel functions. This is done recalling the identity

$$\int_0^\infty J_0(k_\varrho\varrho) k_\varrho \tilde{f}(k_\varrho) dk_\varrho = \frac{1}{2} \int_{-\infty}^\infty H_0^{(2)}(k_\varrho\varrho) k_\varrho \tilde{f}(k_\varrho) dk_\varrho \qquad (8.48)$$

which holds if $\tilde{f}$ is an even function of k_ϱ [11]. In this way, the integration path C in the complex plane k_ϱ closes at infinity (Fig. 8.6) while the topology of the plane remains unchanged except for an additional branch point introduced by the Hankel function (Fig. 8.6). For the sake of clarity, the pole $k_\varrho = \lambda_p$ is located on the real axis (lossless substrate, see Fig. 8.5).

Let us consider now a typical integral in the air:

$$I = \int_C H_n^{(2)}(k_\varrho\varrho) f(k_\varrho) \exp(-u_0 z) dk_\varrho \qquad (8.49)$$

where C is the path shown in Fig. 8.6.

To obtain an asymptotic expansion for I, the plane k_ϱ is transformed into a new complex plane w defined by the relation

$$k_\varrho = k_0 \sin w \qquad (8.50)$$

The transformed path C^* in the plane w is shown in Fig. 8.7. The pole is now located at $w_p = \pi/2 + j\operatorname{arcosh}(\lambda_p/k_0)$ and the branch cuts associated with the points $k_\varrho = \pm k_0$ disappear owing to the new choice of variables. Introducing spherical co-ordinates r, θ ($z = r\cos\theta$, $\varrho = r\sin\theta$) the argument of the Hankel function becomes

$$k_\varrho \varrho = k_0 r \sin\theta \sin w \qquad (8.51)$$

Since we want an expression useful in the far field ($k_0 r \gg 1$), we can replace the Hankel function by its first-order asymptotic expression. The integration path can be deformed far from the origin $w = 0$ in order that $\sin w$ never vanishes.

On the other hand, it can be shown [9] that this replacement will give correct results even for the broadside direction $\theta = 0°$.

The final expression of the integral in the w-plane is

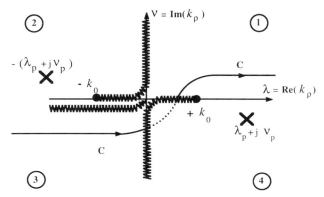

Fig. 8.6 *The integration path C closing at infinity in the k_ϱ plane and the new branch cut introduced by the Hankel function*
A portion (dotted line) of the path C is in the lower Riemann sheet

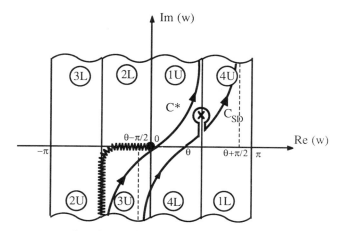

Fig. 8.7 *The new complex plane w ($k_\varrho = k_0 \sin w$) showing the transformed path C^*, the steepest descent path C_{SD} and the new location of the pole*
Also the correspondence with the four quadrants of the plane k_ϱ is given, either in the upper (U) or in the lower (L) Riemann sheet.

$$I = \int_{C^*} F(w) \exp[\Omega q(w)] \, dw \qquad (8.52)$$

$$F(w) = j^n \left(\frac{2j}{\pi \Omega \sin \theta \sin w} \right)^{1/2} k_0 \cos w f(k_0 \sin w)$$

with $\Omega = k_0 r$ and $q(w) = -j \cos(w - \theta)$.

The saddle point w_s of function q is given by $w_s = \theta$. The steepest-descent path C_{SD} through w_s is defined by $\text{Im}(q) = -1$. Its particularities are easily derived (Fig. 8.7):

(i) C_{SD} is at a 45° angle with the real axis.
(ii) C_{SD} intersects the line $\text{Re}(w) = \pi/2$ at the level $\text{Im}(w) = \cosh^{-1}(1/\sin\theta)$.
(iii) As a result, the pole w_r is located between C_{SD} and the positive real axis only when $\theta > \theta_p = \sin^{-1}(k_0/\lambda_p)$.
(iv) C_{SD} possesses two vertical asymptotes at $\text{Re}(w) = \theta \pm \pi/2$.

Contributions from the integrals joining the two paths C^* and C_{SD} at infinity can be eliminated since the term $\exp(\Omega q)$ vanishes in these regions. The path C^* can then be deformed into the path C_{SD}, provided the contribution of the pole is added for angles greater than the critical angle θ_p.

This is written symbolically as

$$\int_C = \int_{C_{SD}} + U(\theta - \theta_p) \int_{C_p} \tag{8.53}$$

where U is the Heaviside unit step function and C_p is a patch surrounding the pole w_p (Fig. 8.7).

The first-order approximation for the integral along C_{SD} can now be obtained by standard techniques [9]. The integral around the pole is evaluated using the residue theorem and is expressed in the original k_ϱ plane.

Finally, the asymptotic expression of the integral I (eqn. 8.49), valid in the far field region, is

$$I \approx 2j^{n+1} \cotan\theta f(k_0 \sin\theta) \frac{\exp(-jk_0 r)}{r}[1 + 0(r^{-1})] - U(\theta - \theta_p)$$

$$\times 2\pi j R H_n^{(2)}(\lambda_p \varrho) \exp[-z\sqrt{\lambda_p^2 - k_0^2}][1 + 0(r^{-1})] \tag{8.54}$$

where the residue R is given by

$$R = \lim_{k_\varrho \to \lambda_p}(k_\varrho - \lambda_p) f(k_\varrho) \tag{8.55}$$

and the Landau notation $0(r^n)$ indicates the behavior at infinity of the first term neglected in the asymptotic expansion.

It must be noted here that the asymptotic approximation eqn. 8.54 is only valid if the pole λ_p is located far enough from the saddle point $\lambda = k_0 \sin\theta$, i.e. $(k_0 \sin\theta - \lambda_p)r \gg 1$. Otherwise, the contributions of the saddle point and of the pole cannot be separated and a modified saddle-point method must be used [9].

From the asymptotic point of view, the fields and potentials are the sum of two terms. The first term, due to the saddle point, represents a spatial wave with a complex factor depending on angle θ and corresponds to the geometrical optical fields. The second term, due to the pole at λ_p, represents a cylindrical wave decreasing exponentially away from the substrate that corresponds to the surface wave.

The surface wave is only relevant for angles $\theta > \theta_p$, i.e. near the interface. However, its field dependence on $\varrho^{-1/2}$ can make it the dominant term of eqn. 8.54 when fields on the substrate surface at $z = 0$ are evaluated.

Radiated electric field: An immediate application of the asymptotic relationship eqn. 8.54 is the computation of the radiated field. It is assumed here that $\theta < \theta_p$, so that the surface wave can be neglected. In a real situation, a substrate always has finite dimensions and the surface wave can be observed directly only close to the substrate. For angles near $\theta = \pi/2$ (grazing angles) but far from the substrate, diffraction effects of the surface wave on the edges become significant.

The radiation field is obtained by transforming, for $z > 0$, the rectangular components of $\boldsymbol{E}$ (eqn. 8.41) into spherical components and then applying eqn. 8.54 to the resulting integrals. The final expressions are

$$E_\theta = -j(Z_0/\lambda_0)\cos\phi g_\theta(\theta)\exp(-jk_0 r)/r$$
$$E_\phi = j(Z_0/\lambda_0)\sin\phi g_\phi(\theta)\exp(-jk_0 r)/r \qquad (8.56)$$
$$E_r \sim 0$$

where Z_0 is the free-space impedance, λ_0 the free-space wavelength, not to be confused with the spectral variable $\lambda = \mathrm{Re}[k_\varrho]$, and

$$g_\theta(\theta) = T\cos\theta/(T - j\varepsilon_r\cos\theta\cot an\, k_0 hT)$$
$$g_\phi(\theta) = \cos\theta/(\cos\theta - jT\cot an\, k_0 hT)$$
$$T = (\varepsilon_r - \sin^2\theta)^{1/2}$$

These asymptotic expansions have also been derived, with a different approach, by several other authors [15, 16]. The result for E_r shows that this component decreases faster with distance than $1/r$, and is thus not a radiated component.

Figs. 8.8 and 8.9 give the polar radiation patterns, respectively, in the E-plane and H-plane. In each Figure, four substrate thickness $k_0 h = 0.05, 0.1, 0.2$ and 0.5 have been considered, and, for each thicknesses, curves corresponding to four dielectric constants $\varepsilon_r = 1, 2, 5$ and 10 have been plotted.

The presence of a dielectric layer increases, in general, the half-power beamwidth in the E-plane, especially for thin substrates. On the other hand, the H-plane pattern is almost independent of the substrate parameters, except for very thick substrates just on the threshold for the generation of a second surface wave.

Potentials at the interface: The Green's functions appearing in the integral equation 8.28 can be obtained from the potentials A_x and V_q. Solving the integral equation requires the knowledge of these potentials only at the interface. If we apply eqns. 8.54 to eqns. 8.43 and 8.44, transformed according to eqn. 8.48, we obtain with $z = 0$ and $\theta = \pi/2$:

$$\frac{4\pi}{\mu} A_x \sim 0(\varrho^{-2}) \qquad (8.57)$$

$$4\pi\varepsilon_0 V_q \sim -2\pi j R H_0^{(2)}(\lambda_p \varrho) \tag{8.58}$$

where the residue R is given by eqn. 8.55 with $f = k_\varrho N/D_{TE}D_{TM}$. It is here apparent that the ϱ^{-1} contribution of the saddle point vanishes in both expressions. If higher-order terms in ϱ^{-2} are desired for A_x, which has no surface-wave component, the whole C_{SD} integration path in the complex plane k_ϱ must be considered.

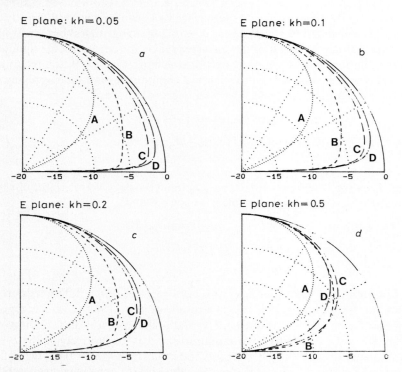

Fig. 8.8 *Polar plot of the E-plane radiation pattern (in dB) of a HED on a microstrip substrate for four normalised substrate thicknesses*
(a) $k_0 h = 0.05$ (b) $k_0 h = 0.10$
(c) $k_0 h = 0.20$ (d) $k_0 h = 0.50$
A: $\varepsilon_r = 1$ B: $\varepsilon_r = 2$ C: $\varepsilon_r = 5$ D: $\varepsilon_r = 10$

It can be shown that the asymptotic behaviour of the integral is mainly determined by the discontinuity of the derivative of the integral at the point $k_\varrho = k_0$. The dominant term in the asymptotic expansion is

$$\frac{4\pi}{\mu_0} A_x \approx \left(\frac{\tan \Delta}{\Delta}\right) \frac{2j}{k_0} \frac{\exp(-jk_0\varrho)}{\varrho^2} \tag{8.59}$$

with the parameter Δ defined as $\Delta = k_0 h(\varepsilon_r - 1)^{1/2}$.

It is possible to replace eqn. 8.58 by a uniform asymptotic development valid in all situations, but the calculations are rather complex, requiring error functions of complex arguments [9], and they will not be carried out here.

Asymptotic expressions 8.58 and 8.59 will be used in the following development to check the results obtained using numerical integration of the potentials.

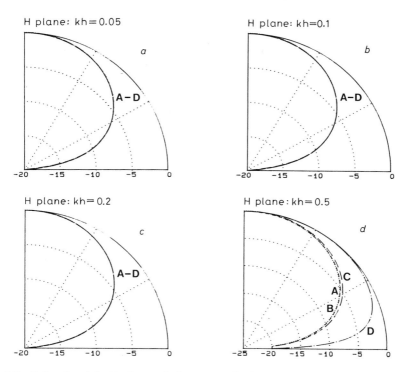

Fig. 8.9 *Polar plot of the H-plane radiation pattern (in dB) of a HED on a microstrip substrate for four normalised substrate thicknesses*
(a) $k_0 h = 0.05$ (b) $k_0 h = 0.10$
(c) $k_0 h = 0.20$ (d) $k_0 h = 0.50$
A: $\varepsilon_r = 1$ B: $\varepsilon_r = 2$ C: $\varepsilon_r = 5$ D: $\varepsilon_r = 10$

8.3.5 Radiation resistance and antenna efficiency

If the cylindrical components of the fields are expressed in terms of the potentials by transforming eqns. 8.41, it can be shown that the term D_{TM} appears only in the denominator of components E_ϱ, E_z and H_ϕ. Therefore, these are the components that include a surface-wave term, and thus the Poynting vector S associated with the surface wave has a radial and a vertical component, respectively, $S_\varrho = -E_z H_\phi$ and $S_z = E_\varrho H_\phi$. Consideration of the general expression 8.54 shows that the vertical component decreases exponentially with z, and consequently does not contribute to the radiated power. On the other hand, the radial component can be written according to expression 8.54 as

$$S_\varrho = (Z_0/2\pi k_0)(\varepsilon_r - 1)\cos^2\phi R^2 (Idl)^2 \lambda_p^4 F(z)/\varrho \qquad (8.60)$$

with

$$F(z) = \begin{cases} \exp(-u_0 z)/D_{TE} \\ \cosh u(z+h)/D_{TE}\cosh uh \end{cases}$$

where the asymptotic expansion of the Hankel function has been used, and R is the residue of $1/D_{TM}$ at the pole.

It can be shown that the integral of this component over a cylindrical surface of radius ϱ extending from $z = -h$ to infinity has a non-zero value independent of ϱ. Hence, there is a net amount of power carried away by the surface wave. The surface integral must, in general, be numerically evaluated, essentially because there is no analytical formula for computing the pole λ_p. However, for thin substrates we can use the approximation given in eqn. 8.46 and estimate the residue appearing in eqn. 8.60 as:

$$R \approx \frac{\varepsilon_r - 1}{\varepsilon_r^2} k_0 h, \qquad (8.61)$$

Then we define a radiation resistance R_{surf} associated with the surface wave as

$$R_{surf} \approx 8\pi^5 Z_0 \left(\frac{dl}{\lambda_0}\right)^2 \left(\frac{\varepsilon_r - 1}{\varepsilon_r}\frac{h}{\lambda_0}\right)^3 \qquad (8.62)$$

In a similar fashion, we can introduce the radiation resistance R_{sp} associated with the space wave. Starting with the asymptotic expansions 8.56 for the fields and using the fact that in the far-field zone we have $Z_0 \mathbf{H} = \mathbf{e}_r \times \mathbf{E}$ we can demonstrate that the Poynting vector is radial, its modulus being given by

$$S_r = Z_0 I^2 \left(\frac{dl}{\lambda_0}\right)^2 \frac{1}{r^2} (|g_\theta|^2 \cos^2\varphi + |g_\varphi|^2 \sin^2\varphi) \qquad (8.63)$$

Integrating this expression over the upper ($z > 0$) half-sphere of radius r and equating the resulting power to $I^2 R_{sp}$ we obtain the expression of the radiation resistance of the space wave:

$$R_{sp} = \pi Z_0 \left(\frac{dl}{\lambda_0}\right)^2 \int_0^{\pi/2} (|g_\theta|^2 + |g_\varphi|^2) \sin\theta \, d\theta \qquad (8.64)$$

As before, this surface integral cannot be analytically evaluated except for the case $\varepsilon_r = 1$ and $h = \infty$, where we recover the classical result for the radiation resistance of a Hertzian dipole radiating into free space.

However, for thin substrates we can again estimate the surface integral 8.64 by using the approximations:

$$g_\phi = \frac{jk_0 h(\varepsilon_r - \sin^2\theta)}{\varepsilon_r}; \quad g_\varphi = jk_0 h\cos\theta, \qquad (8.65)$$

and we get

$$R_{sp} = 4\pi^3 Z_0 \left(\frac{dl}{\lambda_0}\right)^2 \left(\frac{h}{\lambda_0}\right)^2 \left(\frac{4}{3} - \frac{4}{3\varepsilon_r} + \frac{8}{15\varepsilon_r^2}\right) \quad (8.66)$$

Here again, the above formula becomes for $\varepsilon_r = 1$ the well known radiation resistance of a horizontal dipole above a ground plane [40]. It is worthwhile to compute numerically the integral 8.64 and plot the values of the radiation resistance R_{sp}, normalised to $\pi Z_0 (dl/\lambda_0)^2$, as a function of the parameter $\Delta = k_0 h \sqrt{\varepsilon_r - 1}$, which is proportional to the substrate thickness and to the frequency. This has been done and the results are given in Fig. 8.10 for three values of the dielectric constant. Initially, the radiation resistance increases with the square of the thickness, as predicted by eqn. 8.66. But as the normalised thickness increases, the values of the resistance oscillate and show a discontinuous derivative at the points where Δ is an odd multiple of $\pi/2$. These values correspond to the generation of higher-order surface waves [17].

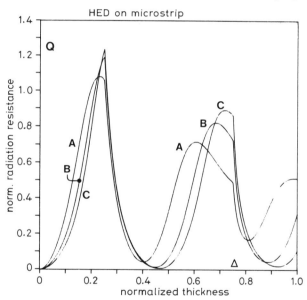

Fig. 8.10 *Normalised radiation resistance $Q = R_{sp}/[\pi Z_0 (dl/\lambda_0)^2]$ of a HED on a microstrip substrate as a function of the parameter $\Delta = k_0 h \sqrt{(\varepsilon_r - 1)}$. $Z_0 \approx 120\pi$ is the free-space wave impedance.*
A: $\varepsilon_r = 2$
B: $\varepsilon_r = 3$
C: $\varepsilon_r = 4$

We can now evaluate the ratio between the power carried away by the surface wave and the power radiated by the space wave as:

$$q = \frac{\text{Power (surface wave)}}{\text{Power (spatial wave)}} = \pi^2 \frac{(\varepsilon_r - 1)^3 h/\lambda_0}{\frac{2}{3}\varepsilon_r^2(\varepsilon_r - 1) + \frac{4}{15}\varepsilon_r} \quad (8.67)$$

which is proportional to the normalised thickness and shows a rather complicated dependence on the permittivity of the substrate. Finally, the radiation efficiency of a HED on a thin lossless microstrip substrate is given by

$$\eta = 1/(1 + q) \tag{8.68}$$

Numerical tests have shown eqn. 8.67 to be accurate for $h \leqslant 0{\cdot}05\,\lambda_0$. Fig. 8.11 gives the theoretical values of the efficiency for several dielectric constants.

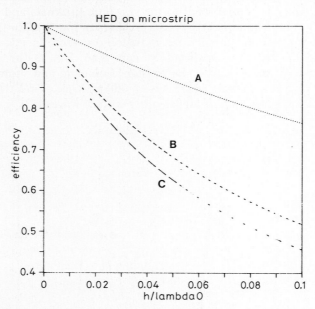

Fig. 8.11 *Radiation efficiency (space-wave radiated-power/total radiated-power) of a HED on a lossless microstrip substrate as a function of the normalised substrate thickness*
A: $\varepsilon_r = 2$
B: $\varepsilon_r = 5$
C: $\varepsilon_r = 10$

Finite size patches: The above considerations refer to an elementary Hertzian dipole, but can be easily extended to finite-size microstrip patches by using superposition. The patch is excited by a unit current, and the input impedance Z_{IN} of the antenna is obtained with techniques to be described in the following Sections. Once the radiated fields are known, a radiation resistance R_{sp} is calculated using the techniques outlined in this Section. The overall antenna efficiency is given by the ratio $R_{sp}/\text{Re}\,(Z_{IN})$. If the antenna has been analysed assuming a lossless substrate and a perfect conductor, the conservation of power implies that $R_{sp} + R_{surf} = \text{Re}\,(Z_{IN})$. This is a useful check on the accuracy of the numerical calculations. Otherwise $\text{Re}\,(Z_{IN})$ is greater than $R_{sp} + R_{surf}$ and the difference gives the power dissipated in the antenna in the form of ohmic and dielectric losses.

The gain of the patch can now be defined in a customary way, as the product of the efficiency times the directivity.

8.4 Numerical techniques for Sommerfeld Integrals

When a microstrip antenna is analysed by an integral-equation technique, it is necessary to evaluate the interaction between points separated by distances ranging from zero to several wavelengths. For most of these distances the accuracy of near- and far-field approximations is not sufficient and the potentials must be numerically evaluated. For a single-layer microstrip antenna the source and the observation point are both on the interface. Hence $z = 0$ in the integral expressions for the fields and the potentials, and the exponential function which ensures convergence of the integrands disappears. This is the most difficult case numerically, and we will concentrate on it in this Section.

Even though many deformations of the original path C of Fig. 8.6 have been tried [14, 18], we feel that the integration along C (the real positive axis λ of the complex plane k_ϱ) provides the most efficient algorithm for evaluating the Sommerfeld integrals appearing in microstrip problems.

8.4.1 Numerical integration on the real axis
The functions to be integrated oscillate on the real axis due to the Bessel functions. The square root $u_0 = (\lambda^2 - k_0^2)^{1/2}$ introduces a discontinuity of the derivative at $\lambda = k_0$ that corresponds to a branch point in the complex plane. If the integrand contains D_{TM} in the denominator there is a pole just below the real axis (or on the axis itself for a lossless substrate) that produces very strong variations of the integrands. Finally, many of the oscillating integrands have an envelope which converges very slowly (A_x, V_q) or even diverges at infinity (A_z, V). When the envelope diverges, the integral diverges in the Riemann sense since the area under the curve representing the integrand fails to converge to a finite value as the upper bound goes to infinity. However, the integral can be interpreted in the Abel sense as

$$\int_0^\infty F(\lambda)d\lambda = \lim_{z \to 0} \int_0^\infty F(\lambda)\exp(-u_0 z)\,d\lambda \qquad (8.69)$$

and the exponential guarantees convergence. This means, physically, that the potentials at the interface can be considered as the limiting case of the potentials in the air when the height of the observation point above the interface goes to zero.

All these facts are clearly depicted in Fig. 8.12, which shows the integrand of the scalar potential V_q for $\varepsilon_{r1} = 2.55$, $\tan \delta = 0$, $k_0 h = 0.3\,\pi$, and $k_0 \varrho = 3$. In this Figure, the substrate has been chosen quite thick for the sake of pictorial clarity. Electrically thinner, more common substrates will exhibit a pole very close to the branch point.

The integration interval is decomposed into three subintervals $[0, k_0]$, $[k_0, k_0\sqrt{\varepsilon_r}]$ and $[k_0\sqrt{\varepsilon_r}, \infty]$. In the region $[0, k_0]$ the infinite derivative in k_0 is eliminated with a change of variables $\lambda = k_0 \cos t$. The resulting smoother function is integrated numerically. In the interval $[k_0, k_0\sqrt{\varepsilon_r}]$, the singularity is first extracted if the integrand has D_{TM} in the denominator. By writing the function under the integral sign in the form $F(\lambda) = J_n(\lambda \varrho) f(\lambda)$, we have

$$F(\lambda) = [J_n(\lambda \varrho) f(\lambda) - F_{sing}(\lambda)] + F_{sing}(\lambda) \tag{8.70}$$

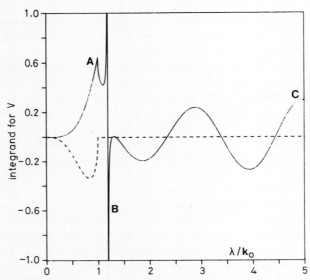

Fig. 8.12 *Normalised values of the integrand associated with the scalar potential V of a HED on microstrip*
$\varepsilon_r = 2\cdot 55 \quad k_0 h = 0\cdot 3\pi \quad k_0 \varrho = 3$
A: Discontinuities in the derivative
B: Sharp peaks due to the pole
C: Oscillatory and divergent behaviour at infinity
——— Real part   - - - - Imaginary part

where

$$F_{sing}(\lambda) = \frac{R}{\lambda - (\lambda_p + jv_p)}$$

Here $\lambda_p + jv_p$ is the complex pole ($v_p \leq 0$) and R the residue of F at the pole. The function F_{sing} is integrated analytically as

$$\begin{aligned}I_{sing} = \int_{k_0}^{k_0\sqrt{\varepsilon_r}} F_{sing}\, d\lambda &= \frac{R}{2} \ln \frac{(k_0\sqrt{\varepsilon_r} - \lambda_p)^2 + v_p^2}{(\lambda_p - k_0)^2 + v_p^2} \\ &+ jR \arctan \frac{k_0\sqrt{\varepsilon_r} - \lambda_p}{v_p} \\ &+ jR \arctan \frac{\lambda_p - k_0}{v_p}\end{aligned} \tag{8.71}$$

It is worth mentioning that, in the lossless case ($v_p = 0$), the above integral becomes

$$I_{sing} = R \ln \frac{k_0\sqrt{\varepsilon_r} - \lambda_p}{\lambda_p - k_0} - j\pi R = R \fint_{k_0}^{k_0\sqrt{\varepsilon_r}} \frac{d\lambda}{\lambda - \lambda_p} - j\pi R \qquad (8.72)$$

and therefore the principal-value formulation (eqn. 8.47) of the lossless case is included as a limiting case in this numerical technique.

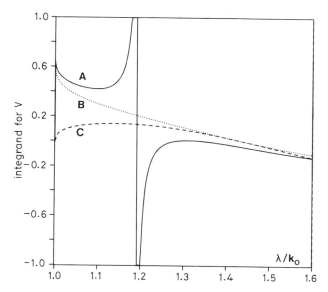

Fig. 8.13 *Real part of the integrand of Fig. 8.12 for the lossy case in the interval $[k_0, k_0\sqrt{(\varepsilon_r)}]$*
A: Before the extraction of the singularity
B: After the extraction of the singularity
C: After the change of variables: $\lambda = k_0 \cosh t$

Fig. 8.13 depicts the real part of the original function $F(\lambda)$ (solid line, A) and the difference $F(\lambda) - F_{sing}(\lambda)$ (dotted line, B) after the singularity has been extracted. There is still an infinite derivative in the curve B at $\lambda = k_0$; however, with a change of variables $\lambda = k_0 \cosh(t)$ one finally obtains a very smooth integrand (the dashed line C in Fig. 8.13), which is integrated by a Gaussian quadrature. The same procedure is applied to the imaginary part of $F(\lambda)$ to eliminate in a similar way the sharp peak and the infinite derivative.

Finally, in the region $[k_0\sqrt{\varepsilon_r}, \infty]$ we first extract the static term defined by $F(\lambda, k_0 = 0)$. Fig. 8.14 depicts the integrand $F(\lambda, k_0)$ (curve A) and the difference $F(\lambda, k_0) - F(\lambda, 0)$ (curve B). It can be shown that the static term has the form

$$F(\lambda, k_0 = 0) = C J_n(\lambda \varrho) \lambda^m \qquad (8.73)$$

and hence it can be integrated analytically.

The remaining part is a slowly convergent oscillating function over a semi-infinite interval that is integrated with specially tailored techniques.

8.4.2 Integrating oscillating functions over unbounded intervals

Sommerfeld integrals, as given by eqn. 8.42, can be grouped in a more general class of integrals defined by:

$$I(\varrho) = \int_a^\infty g(\lambda\varrho)f(\lambda)\,d\lambda \qquad (8.74)$$

where

(a) $g(\lambda\varrho)$ is a complex function whose real and imaginary parts oscillate with a strictly periodic behaviour (sin, cos), or behave asymptotically as the product of a periodic function and a monotonic function. A typical example of this class of functions, which will be termed from now on as quasi-periodic, are the Bessel functions of the first kind.

(b) $f(\lambda)$ is a smooth, non-oscillating function which behaves asymptotically as $\lambda^\alpha \exp(-\lambda z)$. Therefore, for points on the interface ($z = 0$), the function $f(\lambda)$ decreases very slowly or even diverges at infinity if $\alpha > 0$. Here, we shall discuss the most interesting case, $z = 0$, which is also the most difficult from a numerical point of view. For the sake of simplicity, $f(\lambda)$ is assumed to be real. Complex functions can be handled by working alternately with their real and imaginary parts.

(c) The lower integration bound a has been chosen conveniently to ensure that the interval $[a, \infty)$ is far enough from any possible singularities of $f(\lambda)$. For instance in our problem, we shall take $a = \sqrt{\varepsilon_r}$.

It is worth mentioning that the general expression 8.74 includes many integral transforms such as Fourier and Hankel transforms. Hence, the following techniques can be applied to many other problems in numerical analysis.

The classical problem involving Sommerfeld integrals is the problem of radio-wave propagation above a lossy ground, where the comprehensive monograph of Lytle and Lager is the classical reference [19]. These authors have found an iterative Romberg integration satisfactory, since here the integrand displays an exponential convergence and its poles have been removed from the real axis. In microstrip problems, Romberg integration has also been used, but its effectiveness decreases considerably in the absence of a well-behaved integrand.

In recent years, there has been a considerable amount of work published on the numerical evaluation of Fourier transforms, which are included in eqn. 8.74 as a particular case. The involved techniques can be classified in three groups.

(a) The decomposition $[a, \infty] = [a, A] + [A, \infty]$. Here, Filon's algorithm is applied to the finite interval $[a, A]$, while an asymptotic expression of the integrand is used to estimate the integral's value over $[A, \infty]$ [20]. The most serious drawbacks of this approach are the choice of A and the analytical work required – two features which are difficult to incorporate in an automatic computation routine.

(b) Another approach applies if $g(\lambda\varrho)$ is a strictly periodic function. The following decomposition is then used:

$$\int_a^\infty g(\lambda\varrho)f(\lambda)\,d\lambda = \int_a^{a+P} g(\lambda\varrho) \sum_{n=0}^{\infty} f(\lambda + nP)\,d\lambda \tag{8.75}$$

where P is the period of the function g. The infinite sum under the integration sign can be evaluated using standard techniques such as Euler's transformation. Recently, a more involved technique using theoretical Fourier-transform concepts has been described in connection with a problem on quantum-mechanics impact cross-sections [21]. These methods work very well for large values of ϱ and an exponentially decreasing integrand. Unfortunately, their extension to quasi-periodic diverging integrands seems problematic.

(c) The third group of techniques, introduced by Hurwitz and Zweifel [22], are defined by the decomposition

$$\int_a^\infty g(\lambda\varrho)f(\lambda)\,d\lambda = \sum_{n=0}^{\infty} \int_{a+nP/2}^{a+(n+1)P/2} g(\lambda\varrho)f(\lambda)\,d\lambda \tag{8.76}$$

The integration over each half-cycle is performed prior to the series' summation. As before, an accelerating device, such as the nonlinear transformations of Shanks [23] and Sidi [24] can be used to sum the infinite series.

We feel that the decomposition 8.76 is particularly well suited for the Sommerfeld integrals encountered in microstrip problems and we have used it extensively throughout this work. However, instead of the sophisticated nonlinear techniques mentioned above, we have devised a new simple technique based on the concept of a weighted average between the half-cycle integrals [14]. This accelerating device has proved to be faster and more accurate for these kind of integrals.

8.5 Construction of the Green's functions

Once the potentials of a HED are known, the potentials created by an arbitrary surface current J_s existing on the plane $z = 0$ can be determined by using the superposition integrals 8.20 and 8.25. The Green's functions arising in these expressions are closely related to the potentials of a HED and can be easily obtained if the symmetry properties of microstrip structures are taken into account.

We shall restrict ourselves here to the case where the source and observer are both on the air–dielectric interface ($z = z' = 0$). Then, according to the translational invariance of the microstrip structure in any plane perpendicular to the z-axis (eqn. 8.43), we have

$$\begin{vmatrix} G_A^{xx}(r|r') = \dfrac{\mu_0}{2\pi} \int_0^\infty J_0(\lambda R) \dfrac{\lambda}{D_{TE}} d\lambda \\ G_A^{yx}(r|r') = 0 \\ G_A^{zx}(r|r') = -\dfrac{\mu_0}{2\pi}(\varepsilon_r - 1)\cos\alpha \int_0^\infty J_1(\lambda R) \dfrac{\lambda^2}{D_{TE}D_{TM}} d\lambda \end{vmatrix} \quad (8.77)$$

where the angle α is given by $\alpha = \tan^{-1}\{(y - y')/(x - x')\}$. In short, we can say that the components G_A^{xx} and G_A^{zx} are given by the Sommerfeld integrals 8.43 with the polar co-ordinates ϱ, φ replaced by $R = |r-r'|$ and α.

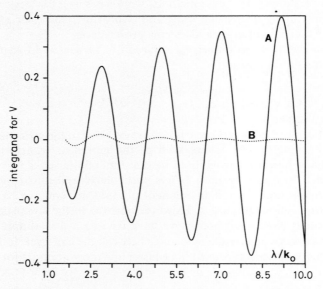

Fig. 8.14 *Real part of the integrand of Fig. 8.12 for the lossy case in the interval $[k_0\sqrt{(\varepsilon_r)}, \infty]$*
A: Before the extraction of the static term
B: After the extraction of the static term

In a similar way, since a microstrip substrate shows a symmetry of revolution around the z-axis, we can write

$$\begin{vmatrix} G_Z^{yy}(r|r') = G_A^{xx}(r|r') \\ G_A^{xy}(r|r') = 0 \\ G_A^{zy}(r|r') = -\dfrac{\mu_0}{2\pi}(\varepsilon_r - 1)\sin\alpha \int_0^\infty J_1(\lambda R) \dfrac{\lambda^2}{D_{TE}D_{TM}} d\lambda \end{vmatrix} \quad (8.78)$$

To evaluate G_A^{xz}, G_A^{yz}, G_A^{zz} we would need the potentials of a vertical electric dipole. However, since only horizontal surface currents are considered we do not need to compute these expressions.

It is now a matter of straightforward algebra to show that the transverse divergence of the dyadic

$$\bar{\bar{G}}_A = \sum_{i=x,y,z} \sum_{j=x,y,z} e_i G_A^{ij} e_j \tag{8.79}$$

is given by

$$\nabla_t \cdot \bar{\bar{G}}_A = e_x \left(\frac{\partial G_A^{xx}}{\partial x} + \frac{\partial G_A^{zx}}{\partial z} \right) + e_y \left(\frac{\partial G_A^{yy}}{\partial y} + \frac{\partial G_A^{zy}}{\partial z} \right) \tag{8.80}$$

and, correspondingly, is derived from a scalar potential according to eqn. 8.22. Therefore, provided that no vertical currents are considered, it is possible to define a Green's function associated with the scalar potential as

$$G_V(r|r') = \text{eqn. 8.44 with } \varrho \text{ replaced by } |r - r'| = R. \tag{8.81}$$

This concludes the derivation of the Green's functions needed to solve the integral equation 8.28. A method-of-moments solution, described briefly in the next section, is used to numerically solve the equation and calculate the microstrip antenna parameters of interest.

8.6 Method of moments

The integral equation 8.28 will be numerically tackled with a method of moments (MoM). This technique [25] transforms the integral equation into a matrix algebraic equation which can be easily solved on a computer. The method of moments is among the most widespread numerical techniques in electromagnetics and a detailed account of the underlying principles will not be given here. For the problem of the microstrip antenna two particular versions of the MoM deserve attention: the subsectional-basis functions approach and Galerkin's method with entire domain-basis functions.

8.6.1 Rooftop (subsectional)-basis functions

If no a priori assumptions about the shape of the patches are made, a successful technique must decompose the patch into small elementary cells and define simple approximations for the surface current on each cell. The most commonly used shapes for the elementary cells are the triangle [26] and the rectangle. Even though the triangular shape is more flexible, rectangular cells involve simpler calculations and suffice for many microstrip antenna problems. Concerning the basis functions to be defined on these rectangular cells, a comparison of available possibilities [13] led to the selection of rooftop functions for the surface current J_s, that have been successfully used in similar problems [27]. To implement these functions, the patch's boundary is replaced by a Manhattan-type

424 Numerical analysis of microstrip patch antennas

polygonal line (Fig. 8.15). As most commonly used antennas exhibit this kind of geometry anyway, this requirement is easily satisfied.

The patch's surface is then divided into rectangular cells, called charge cells, which, for the sake of clarity, will be assumed to be of identical size. This is not an essential requirement for the theory of the algorithm, but the use of different cell sizes considerably increases the computation time.

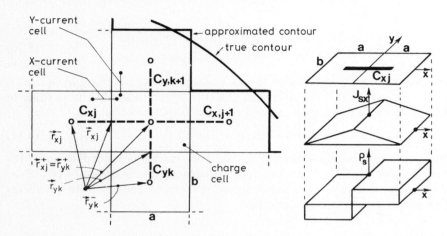

Fig. 8.15 *Decomposition of the upper conductor in elementary cells showing the discretisation of the current and the test segments. After [38].*

Two adjacent charge cells, sharing a common border perpendicular to the x-direction (y-direction) will form an x-directed (y-directed) current cell (Fig. 8.16). An automatic overlapping of current cells is obtained in this manner so that a particular charge cell may belong to up to four different current cells. The number of charge cells is thus related to the number of current cells, though the relationship is not a simple one, since it depends on the shape of the patch. However, for rectangular patches with $m \times n$ charge cells, the number of x-directed current cells is $M = (m - 1)n$, and there are $N = m(n - 1)$ y-directed current cells.

Every current cell supports one rooftop basis function and there is one associated test segment joining the centres of the two charge cells making up the current cell. The centre of the segment C_{xj} associated with the jth x-directed current will be denoted by the vector r_{xj}, and its ends by r_{xj}^- and r_{xj}^+ (Fig. 8.16). These three vectors are related through

$$r_{xj}^{\pm} = r_{xj} \pm e_x(a/2) \qquad j = 1, 2, \ldots M \qquad (8.82)$$

A similar relationship is written for y-directed segments C_{yj} ($j = 1, 2, \ldots N$).

Basis functions: The Cartesian components of the surface current are expanded over a set of basis functions T_x, T_y:

$$J_{sx} = \frac{1}{b} \sum_{j=1}^{M} I_{xj} T_x(\mathbf{r} - \mathbf{r}_{xj})$$
$$J_{sy} = \frac{1}{a} \sum_{j=1}^{N} I_{yj} T_y(\mathbf{r} - \mathbf{r}_{yj})$$
(8.83)

where the basis functions are rooftop-type functions shown in Fig. 8.16:

$$T_x(\mathbf{r}) = \begin{cases} 1 - |x|/a & |x| < a, |y| < b/2 \\ 0 & \text{elsewhere} \end{cases}$$
(8.84)

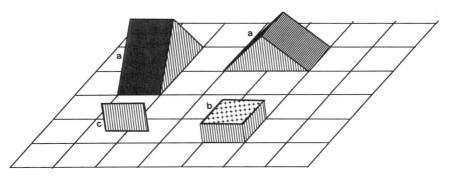

Fig. 8.16 *Subsectional basis functions defined over pairs of adjacent cells (a), associated constant charge distribution on each cell (b), and razor testing functions (c)*

A similar expression is obtained for T_y by interchanging $a \leftrightarrow b$ and $x \leftrightarrow y$ in the above equation.

The introduction of factors $1/a$ and $1/b$ in eqn. 8.83 yields unknown coefficients I_{xj} and I_{yj} having the dimensions of a current. Moreover, every coefficient gives the total current flowing across the common boundary of two charge cells.

The associated surface charge density is obtained from eqn. 8.83 by using the continuity equation, yielding

$$q_s = \frac{1}{j\omega ab} \left\{ \sum_{j=1}^{M} I_{xj} [\Pi(\mathbf{r} - \mathbf{r}_{xj}^+) - \Pi(\mathbf{r} - \mathbf{r}_{xj}^-)] \right.$$
$$\left. + \sum_{j=1}^{N} I_{yj} [\Pi(\mathbf{r} - \mathbf{r}_{yj}^+) - \Pi(\mathbf{r} - \mathbf{r}_{yj}^-)] \right\}$$
(8.85)

where $\Pi(\mathbf{r})$ is a two-dimensional unit pulse function defined over a rectangle of dimensions $a \times b$, centered at $\mathbf{r} = 0$.

The charge density within every elementary cell remains constant, justifying the name charge cell. For the charge cell of Fig. 8.15, with four test segments ending at its centre, the surface charge density is simply given by

$$q_s = \frac{1}{j\omega ab} [I_{x,j+1} - I_{x,j} + I_{y,k+1} - I_{y,k}]$$
(8.86)

The charge density is discontinuous on the borders between charge cells. However, the scalar potential remains bounded, while the electric field becomes singular, since q_s does not satisfy a Hölder condition [13]. This means that the test functions must be selected carefully, avoiding the locations where the electric field is singular.

Discrete Green's functions: The notation and the computational task can be simplified by introducing 'discrete Green's functions', that have as a source a complete basis function, rather than the traditional elementary point source.

The vector potential Γ_A is created by a rooftop distribution of surface current, whereas Γ_V is the scalar potential resulting from a rectangular distribution of unit surface charge. In practice it is convenient to deal with dimensionless quantities and in a normalised space where physical lengths are replaced by electrical lengths. The following dimensionless expressions are therefore introduced that define the discrete Green's functions:

$$\Gamma_A^{xx}(r|r_{xj}) = \int_{S_{xj}} \frac{1}{\mu_0 k_0} G_A^{xx}(r|r') T_x(r' - r_{xj})(k_0^2 dS') \tag{8.87}$$

$$\Gamma_V(r|r_{0j}) = \int_{S_{0j}} \frac{\varepsilon_0}{k_0} G_V(r|r') \Pi(r' - r_{0j})(k_0^2 dS') \tag{8.88}$$

A similar expression may be written for Γ_A^{yy}. In these formulas $r_{xj}(r_{0j})$ denotes the centre of a test segment and $S_{xj}(S_{0j})$ the surface of a current (charge) cell.

The discrete Green's functions exhibit the same properties of translational invariance and symmetry as the conventional Green's functions. In general, the surface integrals in eqns. 8.87 and 8.88 must be evaluated numerically. When the observation point r belongs to the source cell, some difficulties arise in the integration process. It is then recommended that the singular part of the Green's function which corresponds to the dominant term of their static value be extracted, i.e. $G = G_s + (G - G_s)$ where the static value G_s is given by:

$$G_A^{xx}(r|r') = \frac{\mu_0}{4\pi|r - r'|} \tag{8.89}$$

for the vector potential and

$$G_V(r|r') = \frac{1}{4\pi\varepsilon_0|r - r'|} \tag{8.90}$$

for the scalar potential.

The singular part G_s can be analytically integrated over the cell's surface. For instance, the singular part of eqn. 8.88 is

$$2\pi(\varepsilon_r + 1)\Gamma_V(0|0) \simeq 2k_0 a \ln \tan\left(\frac{\alpha}{2} + \frac{\pi}{4}\right) - 2k_0 b \ln \tan(\alpha/2) \tag{8.91}$$

with $\tan \alpha = b/a$.

When the observer is located many cells away from the sources, the sources can be concentrated at the centre of the cell. The following approximations may then be used:

$$\Gamma_A^{xx}(r|r_{xi}) \simeq \frac{1}{\mu_0 k_0} G_A^{xx}(r|r_{xi})(k_0 a)(k_0 b)$$

$$\Gamma_V(r|r_{0i}) \simeq \frac{\varepsilon_0}{k_0} G_V(r|r_{0i})(k_0 a)(k_0 b)$$

(8.92)

Test functions: The last step of the solution with the moment method is the selection of a suitable test function. Previous work [13] has shown that the best choice, compatible with the basis functions selected, is the use of unidimensional rectangular pulses. The use of these test functions, also called razor test functions (see Fig. 8.16), is equivalent to integrating the boundary condition (eqn. 8.26) along the segments linking the centres of adjacent cells (test segments), and therefore the testing procedure yields equations of the type:

$$j\omega \int_{C_{xi}} A_x dx + V(r_{xi}^+) - V(r_{xi}^-)$$

$$+ Z_s \int_{C_{xi}} J_{sx} dx = \int_{C_{xi}} E_x^{(e)} dx = V_{xi}^{(e)}$$

(8.93)

where C_{xi} is the x-directed test segment extending from r_{xi}^- to r_{xi}^+ and $V_{xi}^{(e)}$ is the excitation (impressed) voltage along the segment. A similar relationship is obtained for y-directed test segments. It is worth mentioning that this choice eliminates the need for computing field values near the edges where field singularities can adversely affect the performance of the moment method.

Eqn. 8.93 is well suited for numerical treatment since all derivatives have been removed. The integration of J_{sx} can be done easily using the expansion given by eqn. 8.83 with the result

$$\int_{C_{xi}} J_{sx} dx = \frac{a}{4b}[2I_{xi} + I_{xi+1} + I_{xi-1}] \simeq \frac{a}{b} I_{xi}$$

(8.94)

The last approximation is valid for a reasonably smooth current distribution.

The matrix equation: Introducing the expansions 8.83 and 8.85 in eqn. 8.93 and using the discrete Green's functions defined above, the following matrix equation is obtained:

$$\begin{pmatrix} C^{xx} & | & C^{xy} \\ \hline C^{yx} & | & C^{yy} \end{pmatrix} \begin{pmatrix} I_x \\ I_y \end{pmatrix} = \frac{1}{jZ_0} \begin{pmatrix} V_x^{(e)} \\ V_y^{(e)} \end{pmatrix}$$

(8.95)

The elements in the submatrices are given by

$$C_{ij}^{xx} = \frac{1}{k_0 a k_0 b}[-\Gamma_V(r_{xi}^+|r_{xj}^-) - \Gamma_V(r_{xi}^-|r_{xj}^+)$$
$$+ \Gamma_V(r_{xi}^+|r_{xj}^+) + \Gamma_V(r_{xi}^-|r_{xj}^-)]$$
$$- \frac{1}{k_0 b} \int_{C_{xi}} \Gamma_A^{xx}(r|r_{xj}) k_0 dx + j\frac{Z_s}{Z_0}\frac{a}{b}\delta_{ij}$$
$$i = 1\ldots M, j = 1\ldots M \qquad (8.96)$$

$$C_{ij}^{xy} = \frac{1}{k_0 a k_0 b}[-\Gamma_V(r_{xi}^+|r_{yj}^-) - \Gamma_V(r_{xi}^-|r_{yj}^+)$$
$$+ \Gamma_V(r_{xi}^+|r_{yj}^+) + \Gamma_V(r_{xi}^-|r_{yj}^-)]$$
$$i = 1\ldots M, j = 1\ldots N \qquad (8.97)$$

where δ_{ij} is the Kronecker delta. The expression for C_{ij}^{yy} is obtained by interchanging the couples (x, y), (a, b) and (M, N) within eqn. 8.96. Finally, it is easily shown that $C_{ij}^{yx} = C_{ji}^{xy}$.

For distances $|r_{xi} - r_{xj}|$ much greater than the dimensions of a cell, the integrals in eqn. (8.96) can be replaced by

$$\int_{C_{xi}} \Gamma_A^{xx}(r|r_{xj}) k_0 dx \simeq k_0 a \Gamma_A^{xx}(r_{xi}|r_{xj}) \qquad (8.98)$$

In principle, this approximation is not valid for short distances between cells. For these situations, however, the contribution of the vector potential is overshadowed by that of the scalar potential, so that the approximation of eqn. 8.98 still suffices. As a matter of fact, eqn. 8.98 may be used everywhere but on the diagonal terms. This approximation has been confirmed by extensive numerical tests.

A last point worth mentioning concerns the number of discrete Green's functions that must be calculated. For a rectangular patch with $m \times n$ charge cells, the number of matrix elements is $(M + N)^2$, with $M = (m - 1)n$ and $N = (n - 1)m$. When all the cells have identical sizes, only $m \times n$ values of Γ_V, M values of Γ_A^{xx} and N values of Γ_A^{yy} are needed in order to completely fill the matrix. This is the great advantage of using cells of equal size, and it is generally more convenient to use a larger number of identical cells rather than fewer cells of different sizes.

Interpolation among Green's functions: The evaluation of the matrix in eqn. 8.95 requires a large amount of computation. For a rectangular patch divided into 10×10 cells, the order of the matrix is 180; hence the number of elements in the matrix is $180^2 = 32400$. Even when a simple 4×4 Gaussian quadrature is used to evaluate the discrete Green's functions, the number of Sommerfeld integrals that need to be evaluated would exceed half a million.

Numerical analysis of microstrip patch antennas

Fortunately, for a given structure these integrals depend only upon the distance from source to observer. It is thus possible to tabulate the integrals for a small number of distances, and then to interpolate between the tabulated values. The distances to be considered range from zero to the maximum linear dimension of the patch. Several interpolation schemes have been tried [13]. The best solution was obtained by separating the Green's functions according to eqn. 8.89, and then using a simple parabolic Lagrange interpolation for the regular part.

For a square patch with 10 × 10 cells, at frequencies for which the patch's length is less than a free-space wavelength, the error obtained when interpolating from 25 tabulated values is hardly noticeable: less than 0·5%, even though the computation time was reduced by a factor of 100!

8.6.2 Entire domain basis functions

If the microstrip patch has a simple regular shape (circular or rectangular) we can consider the equivalent electromagnetic cavity obtained when the patch is enclosed by a lateral magnetic wall. If the eigen modes have a simple analytical expression it is reasonable to use them as a set of entire domain basis functions. For thin substrates, the surface-current distribution on a microstrip patch at resonance follows closely the behaviour of the corresponding eigenmode except for a slight disturbance due to the antenna's excitation. Therefore, meaningful results can frequently be obtained by using a very small number of entire domain basis functions. This fact enables the analysis of microstrip arrays having hundreds of elements. The size of the linear system to be solved will be equal to two or three times the number of patches.

It is clear that if only one basis function is allowed per patch, we cannot use a poor testing procedure such as point matching and match the boundary condition only at the centre of the patch. The best alternative is provided by Galerkin's method where the test functions are identical to the basis functions and the inner product includes a surface integration over the patches [25].

To be clear, let us consider a single rectangular patch of dimensions L, W. The eigen modes of the corresponding cavity are [28]

$$f_j = e_x \sin \frac{m\pi x}{L} \cos \frac{n\pi y}{W} + e_y \cos \frac{m\pi x}{L} \sin \frac{n\pi y}{W} \qquad (8.99)$$

where the co-ordinate origin is in the lower left corner of the patch. The vectors f_j correspond to the modes TM_{mn0} of the equivalent cavity. The choice of the modes TM_{mn0} in the expansion of J_s is rather arbitrary and depends on the problem considered. For instance, they can be ordered according to their resonant frequencies, or we can consider only the TM_{m00} subset if variations along the co-ordinate y can be neglected. In any case, the relation between the integers m, n and the ordinal number j in eqn. 8.99 should be clearly defined.

We assume now for the unknown surface current density the following expansion:

$$J_s = \sum_j \alpha_j f_j \tag{8.100}$$

and consequently

$$q_s = \sum_j \alpha_j \left(\frac{-\nabla \cdot f_j}{j\omega} \right) = \sum_j \alpha_j h_j \tag{8.101}$$

Notice that here the unknown coefficients α_i are interpreted as being amplitudes of the surface current distribution [A/m] while, when using subsectional basis, the unknowns were the total currents [A] flowing across contiguous cells.

If we test now the boundary condition 8.26 in the Galerkin sense, we get the set of equations

$$j\omega \int_{S_0} f_i \cdot A \, ds + \int_{S_0} f_i \cdot \nabla V \, ds + Z_s \int_{S_0} f_i \cdot J_s \, ds = \int_{S_0} f_i \cdot E^{(e)} \, ds \tag{8.102}$$

The integral involving the scalar potential can be rewritten as

$$\int_{S_0} f_i \cdot \nabla V \, ds = \frac{1}{j\omega} \int_{S_0} h_i V \, ds \tag{8.103}$$

where the identity $\nabla \cdot (f_i V) = f_i \cdot \nabla V + V \nabla \cdot f_i$ has been used.

Finally, introducing expansions 8.100 and 8.101 into eqn. 8.102 we get the linear system

$$\sum_j c_{ij} \alpha_j = b_i \quad \text{or} \quad C\alpha = b \tag{8.104}$$

where

$$c_{ij} = j\omega \int_{S_0} f_i \cdot \int_{S_0} \bar{\bar{G}}_A \cdot f_j \, dsds' + \frac{1}{j\omega} \int_{S_0} h_i \int_{S_0} G_v h_j \, dsds'$$

$$+ Z_s \int_{S_0} f_i \cdot f_j \, ds \tag{8.105}$$

and

$$b_i = \int_{S_0} f_i \cdot E^{(e)} \, ds \tag{8.106}$$

The last term in the expression of c_{ij} vanishes if $i \neq j$ due to the orthogonality of the eigenmodes f_i.

It can be seen that, essentially, each matrix element requires the computation of two fourfold integrals. Fortunately, two of the four integrations can be performed analytically, by introducing the new variables $u = x - x'$, $u' = x + x'$, $v = y - y'$, $v' = y + y'$. As in the subsectional basis case, the elements c_{ii} will include a singularity when $r = r'$ in the Green's functions. Again, a decomposition of the type of eqn. 8.89 or a change to polar co-ordinates may be used to eliminate the singularity.

Numerical analysis of microstrip patch antennas

The above techniques can easily be generalised to the analysis of an array of patches. In this case, the domains of the ith and jth modes do not necessarily coincide. If the distance between the centres of the patches is greater than the largest linear dimension of a single patch, we can use the approximation

$$\int_{S_0} g_i \int_{S_j} G(r|r') g_j \, ds \, ds' \simeq G(r_i|r_j) \int_{S_i} g_i \, ds \int_{S_j} g_i \, ds' \qquad (8.107)$$

where r_i, r_j denote the centres of the patches.

More accurate approximations can be obtained by expanding the Green's functions in a Taylor series. These mathematical tools and powerful numerical integration routines enable the total computation time to be kept within reasonable limits.

8.7 Excitation and loading

In practice, microstrip antennas are excited by many different techniques. A good survey of these has been recently given by Pozar [29]. This Section will present a brief description of those most commonly found, keeping in mind that the method of moments will be used to analyse the antenna.

8.7.1 Several microstrip-antenna excitations

A very common technique used for feeding a microstrip antenna consists of a microstrip line directly connected to the patch. A thorough treatment of this excitation requires the analysis of the incident and reflected quasi-TEM current waves existing on the line, which is assumed to extend from the patch to infinity. A more realistic model assumes that the microstrip line has a finite length and introduces a mathematical excitation (for instance, a filament of vertical current or a series voltage generator) at the end of the line. The vertical filament of current is a good model for the coaxial excitation and on a physical basis is preferred to the series voltage generator. The microstrip line is then cut at a point where uniform line conditions can be assumed. In this way the discontinuity created by the line-patch junction, and possibly discontinuities of the line itself, are included in the analysis. If a vertical filament of unit current is used as excitation, the input impedance is simply the voltage at the insertion point. When using other mathematical excitations, the section of line included in the analysis must be long enough to support a standing-wave pattern that can be used to estimate the input impedance of the patch.

Finally, it is worth mentioning that replacing the microstrip line by a coaxial probe at the edge of the patch is a first-order approximation that gives surprisingly good results in many practical cases.

Microstrip line under the patch: A more sophisticated excitation, designed to minimise the spurious radiation coming from the feed line, uses a microstrip line under the patch to couple energy to it electromagnetically [29, 30]. The presence of two dielectric layers adds an additional degree of freedom to the design. The comments made above for the line directly connected to the patch also apply here. The microstrip line can be cut far from the patch and terminated in a coaxial probe or excited by some other mathematical device.

Slot in the ground plane: A slot in the ground plane can be used to couple energy to the patch from, presumably, a triplate transmission line [31]. The mathematical treatment replaces the slot by an equivalent distribution of surface magnetic current. The excitation field $\boldsymbol{E}^{(e)}$ results from this current. In the numerical procedure, this modifies only the independent terms b_i (eqn. 8.106).

8.7.2 Coaxial excitation and input impedance

A coaxial line attached to the bottom of the patch (Fig. 8.17) is also a practical way of feeding microstrip patches. From a theoretical point of view, the coaxial excitation is of great interest because simple yet accurate mathematical models are available. While the models presented here are constant current or voltage sources and not constant power sources, that are actually used in practice, the calculated values of the input impedance agrees closely with measured results.

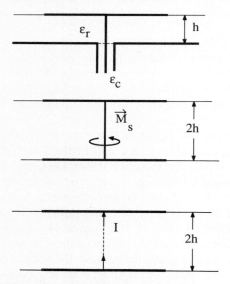

Fig. 8.17 *Microstrip patch antenna excited by a coaxial probe*
The most rigorous model of this configuration considers the probe as belonging to the patch and the whole structure excited by a frill of magnetic current. Simpler models reduce the coaxial line to a filamentary current entering the patch.

The most accurate treatment of a coaxial probe [32, 33] assumes that the portion of the inner coaxial conductor embedded in the substrate belongs to the patch. The whole structure is then excited by a frill of magnetic current existing between the inner and outer conductors of the coaxial line on the ground plane (Fig. 8.17). For thin substrates, however, replacing the coaxial probe by a vertical filament of current gives results accurate enough for engineering purposes.

If the Galerkin technique is used, we can model the probe as a filament of zero diameter ending in a point charge at the junction with the patch. The calculation of the excitation fields would normally require a knowledge of the fields created by vertical electric dipoles embedded in the substrate; however, the reciprocity theorem allows the evaluation of the terms b_i (eqn. 8.106) using only formulas related to horizontal electric dipoles. Thus, we have

$$b_i = \int_{S_0} \boldsymbol{f}_i \cdot \boldsymbol{E}^e \, ds = \int_{V_e} \boldsymbol{J}^e \cdot \boldsymbol{E}_i \, dv \qquad (8.108)$$

where $\boldsymbol{E}_i$ is the field created by the surface current density $\boldsymbol{J}_s = \boldsymbol{f}_i$ of eqn. 8.99 and $\boldsymbol{E}^e$ is the field produced by the excitation current density $\boldsymbol{J}^e$ [A/m^2] entering the feed point (x_e, y_e). The total excitation current is normalised to 1A, i.e.

$$\boldsymbol{J}^e = \boldsymbol{e}_z \delta(x - x_e)\delta(y - y_e) \qquad -h < z < 0 \qquad (8.109)$$

and its domain V_e reduces to a segment of length h in the case of zero-diameter filament. Consequently:

$$b_i = \int_{-h}^{0} \boldsymbol{E}_i \cdot \boldsymbol{e}_z \, dz \qquad (8.110)$$

Finally, in terms of the Green's functions we have

$$b_i = -\int_{S_0} G_v(\boldsymbol{r}_e|\boldsymbol{r}')h_i \, ds' - j\omega \int_{-h}^{0} \boldsymbol{e}_z \cdot \int_{S_0} \bar{\bar{\boldsymbol{G}}}_A(\boldsymbol{r}_e|\boldsymbol{r}') \cdot \boldsymbol{f}_i \, ds' \qquad (8.111)$$

an expression which can be cast in an easier form as

$$b_i = -\int_{S_0} G_v^*(\boldsymbol{r}_e|\boldsymbol{r}')h_i \, ds' \qquad (8.112)$$

with the modified scalar Green's function being given by

$$G_v^* = u_0 \tanh uh/u \, D_{TM}.$$

Now, once the vector of excitation terms $\boldsymbol{b} = (b_i)$ is known, the amplitudes of the basis functions are obtained by solving the linear system of equations (eqn. 8.104). The input impedance is finally given by

$$Z_{IN} = -\sum_i \alpha_i \int_{-h}^{0} \boldsymbol{E}_i(\boldsymbol{r}_e) \cdot \boldsymbol{e}_z \, dz = -\boldsymbol{b} \cdot \boldsymbol{C}^{-1} \cdot \boldsymbol{b} \qquad (8.113)$$

When the subsectional basis functions are used together with razor test functions, the coaxial probe must be modelled more carefully in order to avoid mathematical singularities in the excitation terms. The total current entering the probe must be spread over a region of the patch surrounding the insertion point. A simplified attachment mode in which the current is spread over a cell with a linear dependence has been developed and successfully tested [38]. In the framework of this model, the expression 8.108 for the excitation terms is still valid but the excitation E^e is created by the currents belonging to the probe and to the attachment mode. Also, the expression 8.113 for the input impedance can still be used, but the effect of the attachment mode on itself must be added in order to obtain accurate predictions for the reactance values. Finally, owing to the discretisation inherent in this approach, the excitation point (x_e, y_e) can only be located at the centre of a charge cell.

The reactance of the probe: Expression 8.113 gives the impedance at the patch level, i.e. $z = 0$. To obtain the impedance at the ground plane level, $z = -h$, we need to add in series the self-impedance of the coaxial probe. Assuming now that the inner conductor of radius r_c carries a current I evenly distributed on its surface, we have

$$J_s = e_z \frac{I}{2\pi r_c} \tag{8.114}$$

For thin substrates, we can approximate the magnetic potential due to these currents by

$$A_z = \frac{\mu_0}{4\pi} \frac{1 - jk_0|\mathbf{r} - \mathbf{r}'|}{|\mathbf{r} - \mathbf{r}'|} \tag{8.115}$$

and the self impedance is given by

$$Z_{coax} = 60 \left[(k_0 h)^2 \sqrt{\varepsilon_r} + jk_0 h \arcsin \frac{2h}{r_c} + \frac{r_c - \sqrt{r_c^2 + 4h^2}}{2h} \right] \text{ohms} \tag{8.116}$$

This term is mainly inductive. Thus, finally, a better estimate of the input impedance of a coaxial-fed antenna is

$$Z_{in} = \text{eqn. 8.113} + Z_{coax} \tag{8.117}$$

8.7.3 Multiport analysis

In many practical situations the microstrip antenna is excited simultaneously at M points, for instance, in the case of a microstrip array. In this case the antenna can be considered as an M-port device and standard circuit theory may be applied to completely characterize the antenna.

The first step is to solve the linear system $C\alpha = b$, obtained by application of the method of moments, M times for M different excitation vectors b_j. These vectors correspond to a physical situation in which a unit current is entering the jth port while the remaining M-1 ports are open-circuited. After solving the matrix equation, we get the vector $\alpha_j = C^{-1} b_j$ containing the amplitudes of the N basis functions. Then, by computing the voltage at each port we get the quantities $Z_{ij}, i = 1, 2 \ldots M$, which is the jth column of the impedance matrix of the M-port. Therefore, the complete determination of the impedance matrix requires the solution of M linear systems, but, fortunately, the matrix C of the system remains unchanged when exciting each port.

The elements Z_{jj} have been termed 'input impedances' previously (eqns. 8.113 and 8.117); however, it must be pointed out that these are input impedances corresponding to a single-port excitation since the remaining M-1 ports are open-circuited. These impedances may be quite different from the true input impedances for which an expression will be given now.

Let us consider that each port is connected to a voltage generator U_i with an internal impedance Z_{Li} (Fig. 8.18). This arrangement includes the case of a passive load Z_{Li} by allowing $U_i = 0$.

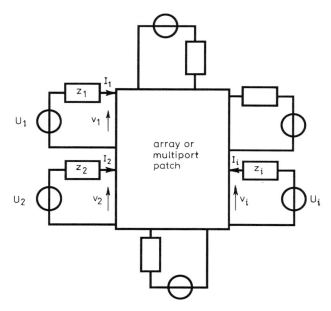

Fig. 8.18 *Equivalent circuit of a microstrip antenna array considered as a multiport device*
The voltage generators are replaced by short circuits at the ports terminated with a passive load

We define a vector U with elements U_i and a diagonal load matrix Z_L with elements Z_{Li}. The equations relating currents I_i and voltages V_i at each port (Fig. 8.18) are

$$U = Z_L I + V = (Z_L + Z)I \qquad (8.118)$$

where Z is the matrix of impedances previously calculated. The vector of port currents is then given by

$$I = (Z_L + Z)^{-1} U \qquad (8.119)$$

and the vector of port voltages is $V = Z(Z_L + Z)^{-1} U$. The new vector

$$\alpha^* = \sum_{i=1}^{N} I_i \alpha_i \tag{8.120}$$

contains the amplitudes of the basis functions for the real working conditions of the antenna, i.e., with all the loads and excitations simultaneously present. This is the vector to be used in the computations of the radiation pattern and when studying the surface current distribution. Finally, the input impedance at each port is

$$Z_{IN}|_{port\,i} = \frac{V_i}{I_i} = \frac{Z(Z_L + Z)^{-1} U]_i}{(Z_L + Z)^{-1} U]_i} \tag{8.121}$$

It is clear that, for a single-port antenna, this input impedance equals the parameter Z_{11} which is directly given by eqns. 8.113 or 8.117.

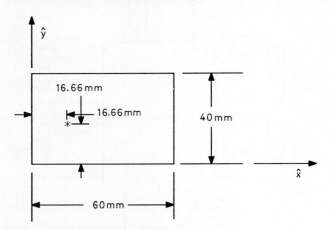

Fig. 8.19 Geometry of a single rectangular patch
$\varepsilon_r = 2\cdot55$ $\tan \delta = 0\cdot002$
$h = 1\cdot28$ mm $Z_0 = 0\cdot9e{-}7$

8.8 Single rectangular patch antenna

In previous Sections the mathematical theory and numerical procedures have been developed for the analysis of general microstrip structures. This Section will concentrate on a single, rectangular, coaxial-fed patch to illustrate how the theory is applied, present computed results for a simple common structure and answer some of the questions that remain. Remaining questions include the convergence of the method-of-moments procedure and the advantages and disadvantages of the various choices of basis functions presented in Section 8.6. Results will be presented for a single patch as shown in Fig. 8.19 showing the input-impedance loci on the Smith chart, the surface current distribution at several resonances and the far-field radiation pattern.

8.8.1 Entire-domain versus subdomain basis functions
Section 8.6 presented several choices of basis functions that could be used in a MoM procedure. Among the choices were wide-triangle or rooftop subdomain basis functions used with razor testing and entire-domain cosine basis functions used in a Galerkin procedure, i.e. tested with cosine testing functions. Computer programs have been written using these two choices of basis and testing functions, and a comparison of the results obtained will be presented in this Section.

The calculation of the far-field radiation pattern will be considered first. If subdomain basis functions are used the patch is reduced to an array of horizontal electric dipoles (HED). In this sense, each rooftop basis function is equivalent to a HED whose moment is given by the product of the total current flowing across the common border of two cells times the distance between the centres of the cells.

Now, the far field for a horizontal dipole, which can be thought of as the element pattern in this procedure, is multiplied by the array factor resulting from the segmentation of the patch to give the total far field pattern. Mathematically, the far field is thus given by

$$E_\alpha = G_E^{\alpha x}(r|0) \sum_{i=1}^{M} aI_{xi} \exp(jk_0 e_r \cdot \varrho_i')$$
$$+ G_E^{\alpha y}(r|0) \sum_{j=1}^{N} bI_{yj} \exp(jk_0 e_r \cdot \varrho_j') \qquad (8.122)$$

where $\alpha = \theta, \varphi$, I_{xi} and I_{yi} are the MoM current coefficients and G_E represents the far fields due to a HED. The pattern can then be integrated to calculate the directivity, gain and efficiency.

When properly chosen entire domain basis functions are used, the far field for each basis function can be calculated analytically, and the total far field is a simple sum of the fields generated by each basis function. For other choices of basis functions costly numerical integration techniques may be required. Fig. 8.20 shows the far-field pattern for a single rectangular patch. The subdomain rooftop and entire-domain cosine basis functions yield identical co-polar patterns at resonance while only the rooftop basis functions yield an estimate of the cross-polar pattern. The cross-polar pattern is due mostly to currents on the patch in phase with the excitation spreading out from the coaxial probe, and to the currents in the probe. In general, the cross-polarised pattern is very difficult to calculate accurately and is sensitive in practice to the size of the ground plane; an item not included in the numerical model. The model can be used, however, to study the effect of the placement of the coaxial probe on the cross-polar pattern. Figs. 8.20a and 8.20b show the far-field pattern of a single patch when excited by a coaxial probe located at different points.

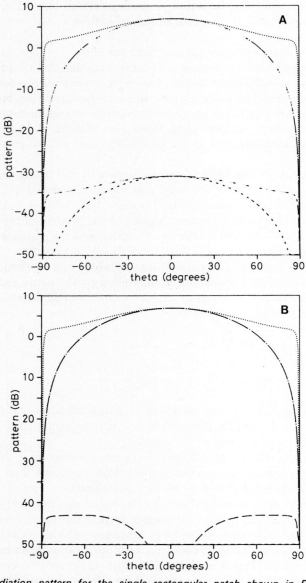

Fig. 8.20 *Radiation pattern for the single rectangular patch shown in Fig. 8.19. Frequency = 1·565 GHz.*
A Coaxial feed at x = 16·66 mm, y = 16·66 mm
B Coaxial feed at x = 16·66 mm, y = 20 mm
····· E-plane co-polar
—·—·— H-plane co-polar
– – – – E-plane cross-polar
– – – H-plane cross-polar

Numerical analysis of microstrip patch antennas 439

The entire domain functions could, in theory, model accurately the patch at frequencies away from resonance and its cross-polar behaviour, however, in practice, this is costly. Near resonance only one or two entire domain functions are needed to model surface current, and are therefore very well suited to standard geometries near resonance. The cost to include additional basis functions needed away from resonance or to calculate the cross-polar pattern is, however, relatively high compared to the subdomain basis functions. Thus it may be more efficient to use subdomain basis functions when studying a single patch or small array away from resonance or when cross-polar pattern is needed.

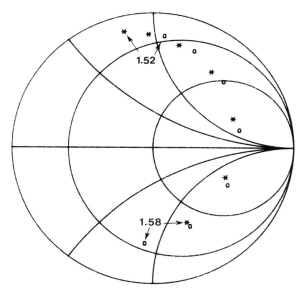

Fig. 8.21 *Input impedance for the single rectangular patch shown in Fig. 8.19.*
Frequency range: 1·52–1·58 GHz.
Frequency increases clockwise with a 0·01 GHz step
* Rooftop basis functions
O Entire-domain cosine basis functions

The input impedance for a single patch is given on the Smith chart of Fig. 8.21. The two choices of basis functions yield approximately the same results with a slight shift in frequency. The resonant frequencies obtained differ by 0·77%, which is often much less than what arises due to uncertainties in the manufacturing process and material parameters. Fig. 8.22 shows the variation of the real part of the input impedance as a function of position for three choices of width/aspect ratios. In each case, the coaxial line was centered in the non-resonant direction on the patch and then moved inward from the edge of the patch to the centre. Note that the results were calculated at the resonant frequency for each patch.

8.8.2 Convergence using subsectional basis functions

The question of convergence must always be dealt with when using a moment method. If a numerical result has not converged there is virtually no hope of its being correct. The factors affecting convergence include: the choice of basis and testing functions, frequency, antenna shape, the dielectric used and even the numerical precision of the computer. Since this list includes nearly all the parameters of the antenna and affects nearly all of the decisions made during the development of the computer program to some degree, it is difficult to give rules that guarantee that a particular result has converged. However, several rules of thumb are applicable and can be used as a base when studying convergence. This Section will briefly demonstrate how the MoM solution using subsectional basis functions converges.

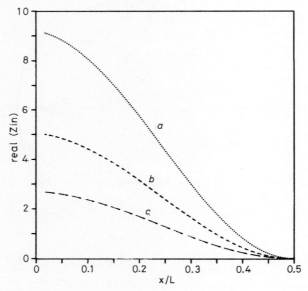

Fig. 8.22 *Real part of the input impedance as a function of the position on the patch. The coaxial is centered along the non-resonant direction and moved from the edge of the patch ($x/L = 0$) to the centre of the patch ($x/L = 0.5$).*
For case (a) the antenna parameters are given in Fig. 8.19 where $L = 60$ mm and $W = 40$ mm. For cases (b) and (c) the width W is varied.
(a) ····· $L/W = 1.5$ $f_{resonance} = 1.555$ GHz
(b) ---- $L/W = 1.0$ $f_{resonance} = 1.543$ GHz
(c) --- $L/W = 0.667$ $f_{resonance} = 1.535$ GHz

The general rule of thumb given in the published literature is that, when using subsectional basis functions, of the order of 10 basis functions per wavelength are needed to obtain good results. This rule also holds for microstrip antennas operating near the first few resonances when calculating the input impedance. It is interesting to note, though, that the number of basis functions in the direction transverse to the resonance direction, i.e. parallel to the H plane at the first resonant frequency, may be reduced without a significant penalty, resulting in large savings in computation time. Fig. 8.23 shows how the input impedance of a rectangular patch converges when the number of basis functions in the H-plane direction is varied at frequencies near the first resonance. Note the resonant frequency changes by only 0·3% when using three basis functions in the transverse direction as opposed to using 7, however, the input impedance changes by approximately 20%. Thus, a rough study of the antenna's resonant frequency and radiation pattern can be performed quickly at low cost. The final analysis can then be performed using additional basis functions.

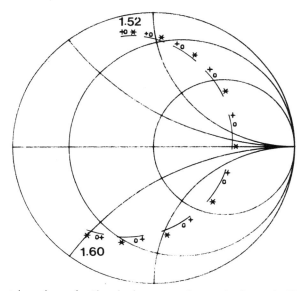

Fig. 8.23 *Input impedance for the single rectangular patch shown in Fig. 8.19 with the coaxial centered vertically (y = 20 mm) and at x = 16·66 mm versus the number of basis functions in the H-plane direction, i.e. along the y-axis*
Frequency range is from 1·52 GHz increasing clockwise to 1·60 GHz with a step of 0·01 GHz
+: 9 by 7 cells
0: 9 by 5 cells
*: 9 by 3 cells

8.8.3 Surface currents

The subsectional basis functions can be used to model virtually any current distribution owing to their flexibility. As an example, a rectangular patch was analysed and measured at the first four resonances, TM_{100}, TM_{010}, TM_{110} and TM_{200}. The patch dimensions are 60 × 40 mm and the dielectric is a standard low-frequency printed-circuit substrate with high losses in the microwave range ($\varepsilon_r = 4·34$ and $\tan \delta \approx 0·02$). The excitation point has been selected at

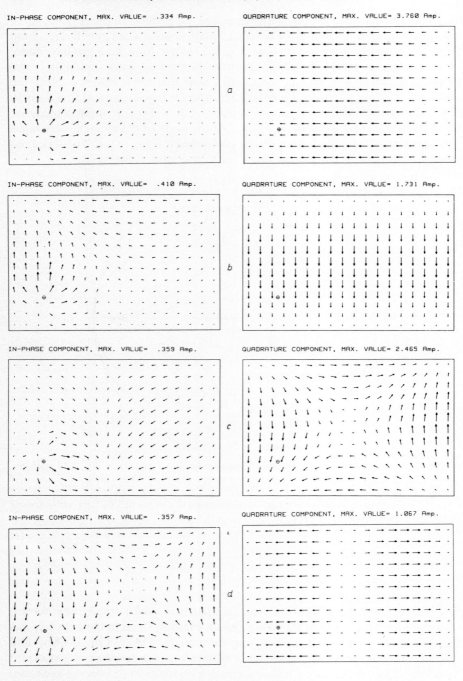

Fig. 8.24 *In-phase and quadrature-phase current distributions for a rectangular patch (60 × 40 mm, h = 0·8 mm, ε_r = 4·34, tan δ = 0·02). After [38].*
  (a) TM_{100} mode; frequency = 1·206 GHz
  (b) TM_{010} mode; frequency = 1·783 GHz
  (c) TM_{110} mode; frequency = 2·177 GHz
  (d) TM_{200} mode; frequency = 2·405 GHz

$x = y = 10$ mm (taking a corner as origin of co-ordinates). This location provides reasonable impedance levels for the first four resonances. The theoretical predictions for the resonant frequencies were, respectively, 1·206, 1·783, 2·177 and 2·405 GHz. The accuracy of these results, when compared with measurements, was always better than 1%.

Fig. 8.24 shows the current distribution calculated using (9 × 6) subsectional basis functions. For each resonance, two plots give the real and the imaginary part of the currents, respectively, in phase and out of phase (in quadrature) with the coaxial excitation. The numerical values given in each Figure correspond to the peak value of the current (longest arrow). The imaginary parts of the current are dominant at resonance and their peak value can be many times greater than the unit current used as excitation. The distribution of the imaginary part is independent of the coaxial position and follows closely the shape of the eigen modes of the equivalent cavity. On the other hand, there is a real part whose peak value is always less than unity and whose distribution is mainly controlled by the position of the excitation.

These results justify the use of entire-domain basis functions as a first approximation of the true current distribution. However, an entire-domain basis cannot account for the part of the current depending directly on the excitation (the real part). Therefore, the input impedance values will not be very accurate for weak resonances, and predictions for cross-polar radiation will fail for the patch. However if the impedance level is high, a single entire-domain basis function will yield engineering accuracy at the first few resonances where the antenna is most likely to be operated.

Fig. 8.25 gives the measured and computed values of the input impedance for the two first resonances. Good agreement is observed in each case.

8.9 Microstrip arrays

Microstrip patch antennas are inherently low-gain antennas and this fact partially offsets the advantages they offer in terms of cost and ease of fabrication. However, microstrip patch antennas are particularly well suited from a technological point of view to be grouped in large arrays to obtain high gain. Typically, the feed lines are integrated on the same substrate. It may also be possible to integrate active devices for phase shifters and pre-amplifiers on the same substrate, or on a lower layer supporting the active elements that has a different permittivity better suited to the active elements. An integrated array constructed in this manner is compact, light weight and reliable. This Section will be concerned not with the feed network or the active elements but with the actual array of microstrip patch antennas and how the numerical techniques presented in the previous Sections can be applied to the study of arrays.

Antennas at the heart of an array interact among one another, often in a manner that degrades the overall array performance. Each active antenna induces currents in its neighbours affecting the element's radiation pattern and input impedance. Thus, elements in an array environment have to be studied in the actual array environment to properly account for mutual coupling [34, 35]. However, when the coupling between array elements is less than 20 or 30 dB, it may be possible to neglect mutual coupling and still obtain acceptable results.

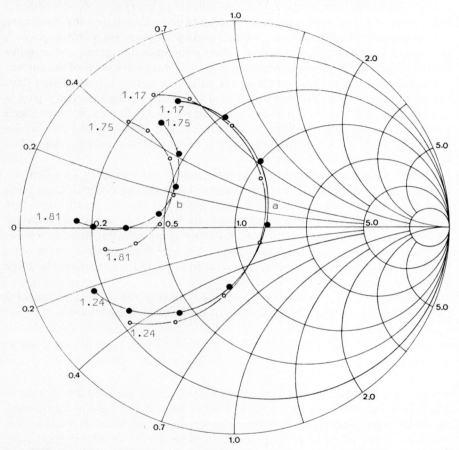

Fig. 8.25 *Input impedance near the two first resonances of the patch of Fig. 8.24. After [38].*
○ Theory (9 × 6 cells)
● Measurements.

This section will present results for several small arrays and show how the element factor and input impedance are affected when an isolated element is incorporated in an array.

8.9.1 Array modelling
To accurately model an array the model should not assume any particular

current distribution on the elements. The method of moments is well suited for the task because any current distribution can be calculated to the desired precision and coupling coefficients are easily calculated. However, practical limitations exist when applying the MoM to anything but small arrays. The use of subsectional basis functions, while very flexible in modelling arbitrary geometries and current distributions, comes with a high price in the number of unknowns. To accurately model an element of the array approximately 50 unknowns are needed; thus for a linear array 10 elements long, 500 unknowns are needed. It can be seen that the capacity of even the largest currently available computer is quickly surpassed.

There are, however, several techniques that can be used to study larger arrays without simply using a larger computer or more computer time. These techniques include:

(a) Entire domain basis functions
(b) Infinite array techniques

Additionally, specialised numerical techniques may be applied with success in certain cases. The method of conjugate gradients has been proposed as a method that would allow the solution of larger systems of linear equations [36, 37]. However, normally the MoM matrix is not sparse and the slow convergence of iterative routines applied to fully populated linear systems precludes their use.

Of the two techniques discussed above only the use of entire-domain basis functions will be discussed since infinite array techniques are included as a full Chapter of this handbook. Using entire-domain basis function only one or two basis functions are typically needed at resonance per element; so arrays having up to several hundred elements can be studied easily with today's computers. However, the study of circularly polarised elements or the cross-polarised fields requires the use of additional higher-order basis functions. This considerably increases the computation time and reduces the size of largest array that can be studied.

8.9.2 Mutual coupling

A 2 × 2 element array was build on a lossy, inexpensive substrate and the four elements of the scattering matrix were measured and compared with the results obtained using the numerical procedures presented previously in this Chapter. The scattering matrix is defined by

$$S = (Z - I)(Z + I)^{-1} \qquad (8.123)$$

where Z is the impedance matrix with elements calculated using eqn. 8.117 and I is the unit dyadic.

The array consisted of four identical patches, each 60 mm along the E-plane and 40 mm along H-plane. The substrate thickness was 0·8 mm and the dielectric constant was 4·34 with a loss tangent of 0·02. The elements were coaxially fed with the coaxial line centered along the H-plane and located 10 mm from the

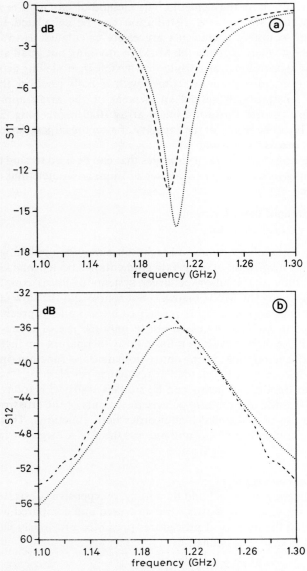

Fig. 8.26 *Scattering parameters for a 2 × 2 microstrip array: measured versus theory*
Patch size: 60 mm by 40 mm, h = 0·8 mm
ε_r = 4·34   tan δ = 0·02
E-plane separation 20 mm between patch edges
H-plane separation 16 mm between patch edges
(a) S_{11} (b) S_{12}
(c) S_{13} (d) S_{14}
———— Measured   ····· Theory

Numerical analysis of microstrip patch antennas

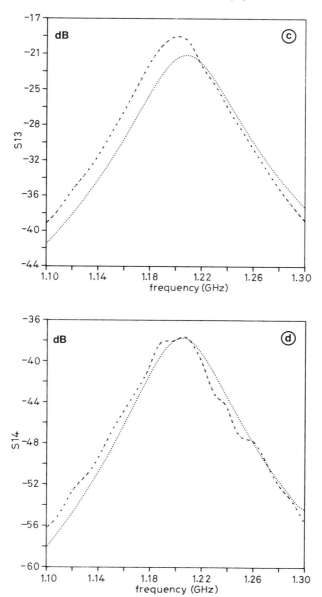

edge. The separation between the patches was 20 mm along the E-plane and 16 mm along the H-plane, corresponding, respectively, to 0·08 and 0·064 free-space wavelengths at the isolated patch's resonance of 1·2 GHz. The results are shown in Fig. 8.26. The two algorithms, which use different basis functions, yield nearly identical results across the frequency band. As can be seen, the agreement between the measured and calculated results is good.

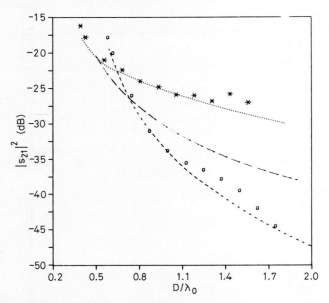

Fig. 8.27 *Measured (Jedlicka, Poe and Carver [35]) and calculated mutual coupling between two coaxial-fed microstrip antennas versus the separation between the patch centres D measured in free-space wavelengths*
W = 105·7 mm   L = 65·5 mm   h = 1·6 mm
ε_r = 2·53 f_r = 1·414 GHz
   * Measured E-plane (From Reference 35)
   O Measured H-plane (From Reference 35)
····· Calculated E-plane
---- Calculated H-plane
--- Calculated 45° plane

Fig. 8.27 gives the E- and H-plane coupling results as a function of the distance between two patches measured by Jedlicka, Poe and Carver [35], and compares these results with those calculated using the theory presented above. Entire-domain basis functions were used. In addition, the calculated coupling results between two antennas located along a diagonal are also presented. It is readily apparent that the H-plane coupling is stronger for small separations.

This is the kind of coupling found in standard microstrip coupled line filters, and it is mainly due to quasi-static terms and to the space wave. However, the

Numerical analysis of microstrip patch antennas

H-plane coupling decreases faster than the E-plane coupling, which is sustained by the surface wave and becomes dominant for greater separations.

It must be finally pointed out that the values of the coupling (s_{21} parameter) are dependent on the input impedance (z_{11} parameter) and are usually given for a couple of patches matched to 50 Ω. The small differences with measurements in Fig. 8.27 can be due to a slight mismatch of the patches.

8.9.3 Linear array of four patches

The last example in this Chapter will illustrate, from a theoretical point of view, the relevance of mutual coupling in the computation of input impedances and mutual coupling. The selected configuration is shown in Fig. 8.28 and consists of four rectangular identical patches fed by coaxial probes and working in the lowest-order mode. The substrate parameters are $h = 0.787$ mm, and $\varepsilon_r = 2.23$. The patches are coupled by their non-resonant sides (H-plane coupling) and they are excited uniformly. On the other hand, their spacing is non-uniform in order to obtain a lower first-sidelobe level [39].

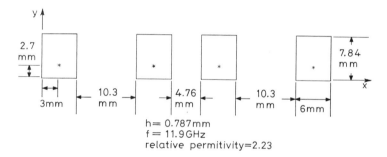

Fig. 8.28 *Linear array with four identical patches and nonuniform spacing*
The substrate has a permittivity of $\varepsilon_r = 2.23$ and a thickness of $h = 0.787$ mm. The nominal resonant frequency is $f_0 = 11.9$ GHz.

Fig. 8.29 shows the normalised real and imaginary parts of the input impedance presented by an inner patch. For each of the quantities, two curves are given, corresponding to theoretical calculations without mutual coupling (the patch is considered as an isolated element) and with mutual coupling (the patch is embedded in an array environment with the three other patches terminated by 50 Ω loads). It can be seen that, for this array, mutual coupling raises the maximum resistance from 130 Ω to 154 Ω. This significant change (18%) shows clearly that mutual coupling should be taken into account for properly matching a microstrip array.

The influence of mutual coupling on radiation patterns is even stronger. Fig. 8.30 shows the H-plane radiation patterns of an isolated patch and of an inner patch in an embedded configuration. The asymmetries in the geometrical environment of an inner patch (coupled to two patches on one side but only to one

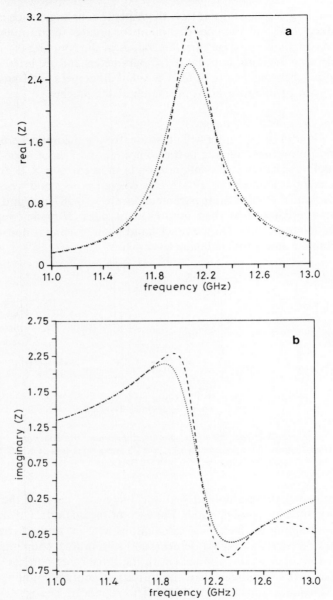

Fig. 8.29 *Calculated isolated-element and embedded-inner-element input impedance versus frequency for the array of Fig. 8.28. Values are normalised to 50Ω*
 a Real part
 b Imaginary part
 ····· Isolated element
 ––– Embedded element

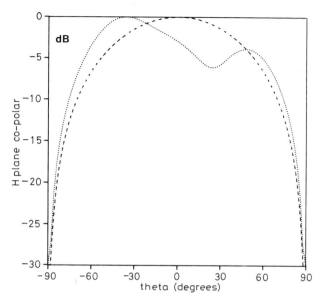

Fig. 8.30 Calculated isolated-element and embedded-inner-element far-field radiation patterns for the array of Fig. 8.28
In the embedded case all the other elements are loaded with 50 Ω
――― Isolated element
· · · · · Embedded element

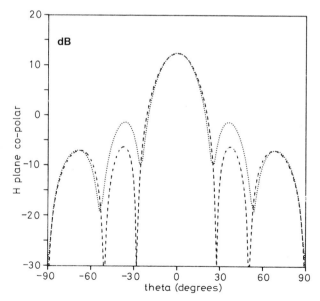

Fig. 8.31 Calculated H-plane far-field pattern of the array shown in Fig. 8.28 with and without mutual coupling
· · · · · With mutual coupling (MoM result)
First sidelobe = −13·8 dB
――― Without mutual coupling
First sidelobe = −18·6 dB

on the other side) are clearly reflected in the asymmetrical pattern for the embedded situation.

Finally, Fig. 8.31 gives the overall radiation pattern of the array in the H-plane. A first theoretical prediction neglects mutual coupling and computes the array pattern as the product of the isolated element pattern times the array factor. The first-order sidelobe is then at -18.6 dB, well below the value of -13.2 dB which can be obtained with four equally spaced, uniformly fed elements [40]. However, if the integral-equation model presented in this Chapter is used, mutual coupling is automatically taken into account. Theoretical predictions show then how mutual coupling deteriorates the radiation pattern, raising the first-sidelobe level to only -13.8 dB.

8.10 Acknowledgments

The authors would like to thank the members of the Laboratoire d'Electromagnétisme et d'Acoustique (LEMA) for their aid and support. Specifically, Anja Skrivervik, Lionel Barlatey and Bertrand Roudot performed many of the calculations and measurements presented in this Chapter. Also special thanks are due to Mrs. Mary Hall for typing the manuscript.

8.11 References

1. BAHL, I. J., and BHARTIA, P.: 'Microstrip antennas' (Artech House, 1980).
2. STRATTON, J. A.: 'Electromagnetic theory' (McGraw-Hill, NY, 1941).
3. MICHALSKI, K. A.: 'On the scalar potential of a point charge associated with a time-harmonic dipole in a layered medium', *IEEE Trans.*, 1988, **AP-36**.
4. POGGIO, A. J., and MILLER, E. K.: 'Integral equation solutions of three-dimensional scattering problems', *In* MITTRA, R. (Ed.), 'Computer Techniques for Electromagnetics' (Pergamon Press, 1973).
5. LO, Y. T.: 'Electromagnetic field of a dipole source above a grounded dielectric slab', *J. Appl. Phys.*, 1954, **25**, p. 733.
6. BRICK, D. B.: 'The radiation of a Hertzian dipole over a coated conductor', *Proc. IEE*, 1955, **102C**, pp. 103–121.
7. BREKHOVSKIKH, L. M.: 'Waves in layered media' (Academic Press, NY, 1960).
8. WAIT, J. R.: 'Electromagnetic waves in stratified media', (Pergamon Press, 1962).
9. FELSEN, L. B., and MARCUVITZ, N.: 'Radiation scattering of waves', (Prentice Hall, New Jersey, 1973).
10. KONG, J. A.: 'Theory of electromagnetic waves', (Wiley, NY, 1975).
11. SOMMERFELD, A.: 'Partial differential equations in physics' (Academic Press, NY 1949).
12. HARRINGTON, R. F.: 'Time harmonic electromagnetic fields' (McGraw-Hill, NY, 1961).
13. MOSIG, J. R., and GARDIOL, F. E.: 'A dynamical radiation model for microstrip structures', *in* HAWKES, P. (Ed.) 'Advances in electronics and electron physics', (Academic Press, NY, 1982) pp. 139–237.
14. MOSIG, J. R., and GARDIOL, F. E.: 'Analytic and numerical techniques in the Green's function treatment of microstrip antennas and scatterers', *IEE Proc.*, 1983, **130H**, pp. 175–182.
15. SHASTRY, S. V. K.; Ph.D. Dissertation, Indian Institute of Science, Bangalore, India, 1979.

16 UZUNOGLU, N. K., ALEXOPOULOS, N. G., and FIKIORIS, J. G.: 'Radiation properties of microstrip dipoles', *IEEE Trans.*, 1979, **AP-27**, pp. 853–858.
17 MOSIG, J. R., and GARDIOL, F. E.: 'Dielectric losses, ohmic losses and surface wave effects in microstrip antennas.' Int. URSI Symposium, Santiago de Compostela, Spain, 1983, pp. 425–428.
18 MICHALSKI, K. A.: 'On the efficient evaluation of integrals arising in the Sommerfeld halfspace problem', *IEE Proc.*, 1985, **132H**, pp. 312–318.
19 LYTLE, R. J., and LAGER, D. L.: 'Numerical evaluation of Sommerfeld integrals'. Report UCRL-52423, Lawrence Livermore Lab., Univ. of California, 1974.
20 PANTIS, G.: 'The evaluation of integrals with oscillatory integrands', *J. Comp. Phys.*, 1975, **17**, pp. 229–233.
21 BORIS, J. P., and ORAN, E. S.: 'Evaluation of oscillatory integrals', *J. Comp. Phys.*, 1975, **17**, pp. 425–433.
22 HURWITZ, H., and ZWEIFEL, P. F.: 'Numerical quadrature of Fourier transform integrals', *Math. Tables Aids Comput.*, 1956, **10**, pp. 140–149.
23 ALAYLIOGLU, A., EVANS, G., and HYSLOP, J.: 'The evaluation of oscillatory integrals with infinite limits,' *J. Comp. Physics*, 1973, **13**, pp. 433–438.
24 SIDI, A.: 'The numerical evaluation of very oscillatory infinite integrals by extrapolation', *Math. Comp.*, 1982, **38**, pp. 517–529.
25 HARRINGTON, R. F.: 'Field computation by moment methods' (MacMillan, NY, 1968).
26 RAO, S. M., WILTON, D. R., and GLISSON, A. W.: 'Electromagnetic scattering by surfaces of arbitrary shape', *IEEE Trans.*, 1982, **AP-30**, pp. 409–418.
27 GLISSON, A. W., and WILTON, D. R.: 'Simple and efficient numerical methods for problems of electromagnetic radiation and scattering from surfaces', *IEEE Trans.*, 1981, **AP-29**, pp. 593–603.
28 COLLIN, R. E.: 'Foundations for microwave engineering', (McGraw-Hill, NY, 1966).
29 POZAR, D. M.: 'New architectures for millimeter wave phased array antennas', Journées Internationales de Nice sur les Antennas (JINA), Nice, 1986, pp. 168–179.
30 KATEHI, P. B., ALEXOPOULOS, N. G.: 'On the modelling of electromagnetic coupled microstrip antennas – The printed strip dipole,' IEEE Trans. Antennas Propagat., **AP-32**, pp. 1179–1186, 1984.
31 SULLIVAN, P. L. and SCHAUBERT D. H., 'Analysis of an aperture coupled microstrip antenna,' IEEE Trans. Antennas Propagat., **AP-34**, pp. 977–984, 1986.
32 POPOVIC, B. D., DRAGOVIC, M. B., and DJORDJEVIC, A. R.: 'Analysis and synthesis of wire antennas', (Wiley, NY, 1982).
33 HALL, R. C., MOSIG, J. R., and GARDIOL, F. E.: 'Analysis of microstrip antenna arrays with thick substrates', 17th European Microwave Conf., Rome, Italy, 1987.
34 ALEXOPOULOS, N. G., and RANA, I. E.: 'Mutual impedance computation between printed dipoles', *IEEE Trans.*, 1981 **AP-29**, pp. 106–111.
35 JEDLICKA, R. P., POE, M. T., and CARVER, K. R.: 'Measured mutual coupling between microstrip antennas', *IEEE Trans.*, 1981, **AP-29**, pp. 147–149.
36 JACOBS, D. A. H.: 'A generalization of the conjugate-gradient method to solve complex systems', *IMAJ of Numerical Analysis*, 1986, **6**, pp. 447–452.
37 PETERSON, A. F., and MITTRA, R.: 'Method of conjugate gradients for the numerical solution of large-body electromagnetic scattering problems', *J. Opt. Soc. Am. A*, 1985, **2**, pp. 971–977.
38 MOSIG, J. R., and GARDIOL, F. E.: 'General integral equation formulation for microstrip antennas and scatterers', *IEE Proc.*, 1985, **132H**, pp. 424–432.
39 GRONAU, G., and WOLFF, I.: 'Spectral domain analysis of microstrip antennas', Proc. of Workshop 'Analytical and numerical techniques for microstrip circuits and antennas', Montreux, Switzerland, March 1988.
40 BALANIS, C. A.: 'Antenna theory: analysis and design', (Harper & Row, NY, 1982).

Chapter 9
Multiport network approach for modelling and analysis of microstrip patch antennas and arrays

K. C. Gupta

9.1 Introduction

The multiport network approach for microstrip patch antennas is based on the use of segmentation and desegmentation methods for analysis of planar structures. Segmentation and desegmentation techniques were developed originally for analysis of two-dimensional (planar) circuit components [1–7]. Since microstrip patch antennas on thin substrates can be treated as two-dimensional planar components, segmentation and desegmentation techniques have been employed for the analysis of microstrip antennas also [8, 9]. This approach has been used successfully for the analysis and design of several types of microstrip patch antennas [10–17] and arrays [18], and promises to be an appropriate methodology for computer-aided design [19, 20] of microstrip patch antennas and arrays in hybrid as well as in monolithic configurations.

This Chapter describes the multiport network approach, segmentation-desegmentation techniques, and their applications to design of microstrip antennas. Relevant aspects of various models for microstrip antennas are presented in Section 9.2. The multiport network model [19], which is an extension of the well known cavity model [21, 22] for microstrip patches, is discussed. Evaluation of multiport impedance matrices from the Green's functions for various types of segments is described in Section 9.3. Modelling of external fields (including fringing, radiation and surface wave) by edge-admittance networks is detailed in Section 9.4.

Segmentation and desegmentation methods for analysis of planar electromagnetic structures (and for multiport networks) are discussed in Section 9.5. Various examples of microstrip antenna configurations, which have been analysed and designed using multiport network approach, are reviewed in Section 9.6. These include various types of single-feed circularly-polarised microstrip patch configurations, such as a diagonally-fed nearly square patch, a square patch with truncated corners, a square patch with a diagonal slot, a pentagonal-shaped patch, a square ring patch, and a cross-shaped patch. The second group of antennas analysed by the multiport network approach consists of broadband multi-resonator microstrip antennas. Three-resonator and five-resonator configurations (coupled to a central patch either by a capacitive-gap coupling or by short sections of microstrip lines) are included in this group. The third category of microstrip antenna configurations, discussed in Section 9.6.3, consists of two-port rectangular, two-port circular patches, and series-fed arrays making use of these two-port patches.

Discussion related to the development of CAD procedures for microstrip patches and arrays is contained in Section 9.7. It is pointed out that multiport network modelling and segmentation/desegmentation methods of analysis are ideally suitable for implementation of CAD procedures for microstrip antennas.

9.2 Models for microstrip antennas

The transmission-line model [23, 24] and the cavity model [21, 22] are the two most widely used network models for analysis of microstrip antennas. We will discuss these models briefly before introducing the multiport network model suitable for implementing segmentation/desegmentation methods.

9.2.1 Transmission-line model

In this model, a rectangular microstrip antenna patch is viewed as a resonant section of a microstrip transmission line. A detailed description of the transmission-line model is given in Chapter 10. The basic concept is shown in Fig. 9.1 which illustrates the transmission-line models for (*a*) and unloaded rectangular patch; (*b*) a rectangular patch with a feed line along the radiating edge; and (*c*) a rectangular patch with a feed line along the non-radiating edge. Z_{0p} is the characteristic impedance of a microstrip line of width W_p, and ε_{rep} is the corresponding effective dielectric constant. B_e and G_e are capacitive and conductive components of the edge admittance Y_e. The susceptance B_e accounts for the fringing field associated with the radiating edge of the width W_p, and G_e is the conductance contributed by the radiation field associated with each edge. Power carried away by the surface wave(s) excited along the slab may also be represented by a lumped loss and added to G_e. In Fig. 9.1*b* and *c*, Z_{0f} and ε_{ref} are the characteristic impedance and the effective dielectric constant for the feeding microstrip line of width W_f. In both of these cases, the parasitic reactances associated with the junction between the line and the patch have not been taken into account.

Transmission-line models may also be developed for two-port rectangular microstrip patches [14]. These configurations are used in the design of series-fed linear (or planar) arrays [25, 18]. Models for two types of two-port rectangular microstrip patches are shown in Fig. 9.2. Fig. 9.2*a* illustrates the equivalent transmission-line network when the two ports are located along the radiating edges, and Fig. 9.2*b* shows the transmission-line model [26] when the two ports are along the non-radiating edges. It has been shown [14, 26] that, when the two ports are located along the non-radiating edges, transmission from port 1 to port 2 can be controlled by suitable choices of distances x_1 and x_2. Again, the two models shown in Figs. 9.2a and b do not incorporate the parasitic reactances associated with the feed-line-patch junctions.

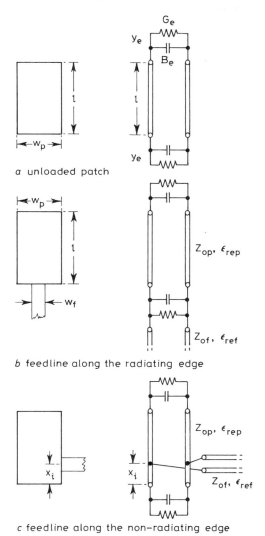

Fig. 9.1 *Transmission-line models for three rectangular microstrip patch configurations*

There are several limitations inherent to the concept of the transmission-line model for microstrip antennas. The basic assumptions include: (i) fields are uniform along the width W_p of the patch; and (ii) there are no currents transverse to the length *l* of the patch. Detailed analysis of rectangular patches has shown [27] that, even at a frequency close to the resonance, field distribution along the radiating edge is not always uniform. Also, the transverse currents are caused by the feeding mechanism and are invariably present. Moreover, the circularly polarised rectangular microstrip antennas (whose operation depends upon the excitation of two orthogonal modes) cannot be represented by the transmission-line model discussed above. Clearly, a more accurate method for modelling of microstrip antennas is needed.

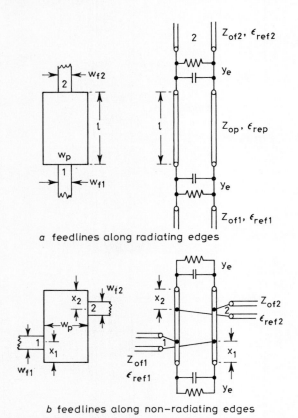

a feedlines along radiating edges

b feedlines along non-radiating edges

Fig. 9.2 *Transmission-line modes for two-port rectangular microstrip patch antennas*

9.2.2 Cavity model

A planar two-dimensional cavity model for microstrip patch antennas [21, 22] offers considerable improvement over the one-dimensional transmission-line model discussed in the previous Section. In this method of modelling, the microstrip patch is considered as a two-dimensional resonator surrounded by a

Table 9.1 Variation of modal fields (ψ_{mn}) and resonant wave numbers (K_{mn}) for various patch geometries analysed by the cavity method (Reproduced from Reference 21 © IEEE, 1979)

Rectangle

$$\psi_{mn} = \cos\frac{m\pi}{a}x \cos\frac{n\pi}{b}y$$

$$k_{mn} = \sqrt{\left(\frac{m\pi}{a}\right)^2 + \left(\frac{n\pi}{b}\right)^2}$$

Circular ring

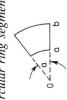

$$\psi_{mn} = [N'_n(k_{mn}a)J_n(k_{mn}\varrho) - J'_n(k_{mn}a)N_n(k_{mn}\varrho)]e^{jn\phi}$$

$$\frac{J'_n(k_{mn}a)}{N'_n(k_{mn}a)} = \frac{J'_n(k_{mn}b)}{N'_n(k_{mn}b)}$$

Circle (disc)

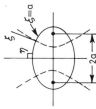

$$\psi_{mn} = J_n(k_{mn}\varrho)e^{jn\phi}$$
$$J'_n(k_{mn}a) = 0$$

Circular ring segment

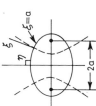

$$\psi_{mn} = [N'_\nu(k_{m\nu}a)J_\nu(k_{m\nu}\varrho) - J'_\nu(k_{m\nu}a)N_\nu(k_{m\nu}\varrho)]\cos\nu\phi$$

$$\nu = n\pi/\alpha$$
$$\frac{J'_\nu(k_{m\nu}a)}{N'_\nu(k_{m\nu}a)} = \frac{J'_\nu(k_{m\nu}b)}{N'_\nu(k_{m\nu}b)}$$

Circular segment

$$\psi_{m\nu} = J_\nu(k_{m\nu}\varrho)\cos\nu\phi$$
$$\nu = n\pi/\alpha, \quad J'_\nu(k_{m\nu}a) = 0$$

Ellipse

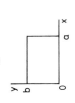

Even-modes
$\psi_{mn} = \text{Re}_m(\xi, \chi e_n) Se_m(\eta, \chi e_n)$
$\text{Re}_m(a, \chi e_n) = 0, \quad \chi e_n = kq$
major axis $= 2q\cosh a$
minor axis $= 2q\sinh a$
Odd-modes:
Replacing e by o in the above

Disk with slot

$$\psi_{mn} = J_{n/2}(k_{mn}\varrho)\cos n\phi/2$$
$$J'_{n/2}(k_{mn}a) = 0$$
$$\alpha \approx 2\pi, \quad \nu = n/2$$

Right isosceles

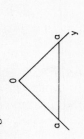

(a) $\psi_m = \cos\dfrac{m\pi}{a}x - \cos\dfrac{m\pi}{a}y$

$k_m = \sqrt{2}\,\dfrac{m\pi}{a}$

(b) $\psi_m = \cos\dfrac{m\pi}{a}x \cos\dfrac{m\pi}{a}y$

$k_m = \sqrt{2}\,\dfrac{m\pi}{a}$

Equilateral triangle

$$\psi_{mn} = \cos\dfrac{2\pi l}{3b}\left(\dfrac{u}{2}+b\right)\cos\dfrac{\pi(m+n)(v-w)}{9b} + \cos\dfrac{2\pi m}{3b}\left(\dfrac{u}{2}+b\right)\cos\dfrac{\pi(n-l)(v-w)}{9b} + \cos\dfrac{2\pi n}{3b}\left(\dfrac{u}{2}+b\right)\cos\dfrac{\pi(l-m)(v-w)}{9b}$$

$l = -(m+n),\ u = \dfrac{\sqrt{3}}{2}x + \dfrac{1}{2}y,$

$v - w = -\dfrac{\sqrt{3}}{2}x + \dfrac{3}{2}y,$

$b = a/2\sqrt{3},$

$k_{mn}^2 = \left(\dfrac{4\pi}{3a}\right)^2 (m^2 + n^2 + mn)$

perfect magnetic wall around the periphery. The fields underneath the patch are expanded in terms of the resonant modes of the two-dimensional resonator. This approach is applicable to a variety of patch geometries. These geometries, the corresponding modal variations denoted by ψ_{mn} and the resonant wave numbers k_{mn} are shown in Table 9.1 (from Reference 21). E and H fields are related to ψ_{mn} by

$$E_{mn} = \psi_{mn}\hat{a} \tag{9.1}$$

$$H_{mn} = \hat{z} \times \nabla_t \psi_{mn} / j\omega\mu \tag{9.2}$$

where $\hat{z}$ is a unit vector normal to the plane of the patch. Resonant wave numbers k_{mn} are solutions of

$$(\nabla_t^2 + k_{mn}^2)\psi_{mn} = 0 \tag{9.3}$$

with

$$\frac{\partial \psi_{mn}}{\partial p} = 0 \tag{9.4}$$

on the magnetic wall (periphery of the patch). ∇_t is the transverse part of the del operator and p is perpendicular to the magnetic wall.

The fringing fields at the edges are accounted for by extending the patch boundary outwards and considering the effective dimensions to be somewhat larger than the physical dimensions of the patch. The radiation is accounted for by considering the effective loss tangent of the dielectric to be larger than the actual value. If the radiated power is estimated to be P_r, the effective loss tangent δ_e may be written as

$$\delta_e = \frac{P_r + P_d}{P_d} \delta_d \tag{9.5}$$

where P_d is the power dissipated in the dielectric substrate and δ_d is the loss tangent for the dielectric medium. The effective loss tangent given by eqn. 9.5 can be modified further to incorporate the conductor loss. The modified loss tangent δ_e is given by

$$\delta_e = \frac{P_r + P_d + P_c}{P_d} \delta_d \tag{9.6}$$

The input impedance of the antenna is calculated by finding the power dissipated in the patch for a unit voltage at the feed port, and is given by

$$Z_{in} = |V|^2 / [P + 2j\omega(W_E - W_M)] \tag{9.7}$$

where $P = P_d + P_c + P_{sw} + P_r$. W_E is the time-averaged electric stored energy, and W_M is the time-averaged magnetic energy. The voltage V equals $E_z d$ averaged over the feed-strip width (d is the substrate thickness). The far-zone field, and radiated power are computed by replacing the equivalent magnetic-current ribbon on the patch's perimeter by a magnetic line current of magnitude Kd on the ground plane (xy plane). The magnetic current source is given by

$$K(x, y) = \hat{n} \times \hat{z}E(x, y) \tag{9.8}$$

where $\hat{n}$ is a unit vector normal to the patch's perimeter and $\hat{z}E(x, y)$ is the component of the electric field perpendicular to the ground plane.

A cavity model for microstrip patch antennas may also be formulated by considering a planar two-dimensional resonator with an impedance boundary wall all around the edges of the patch. A direct form of network analogue (DFNA) method for the analysis of such a cavity model has been discussed in Reference 28.

9.2.3 Multiport network model

The multiport network model of microstrip patch antennas [19, 29] may be considered as an extension of the cavity model discussed above. Electromagnetic fields underneath the patch and outside the patch are modelled separately. The patch itself is analysed as a two-dimensional planar network [1], with a multiple number of ports located all around the edges as shown in Fig. 9.3. Each port represents a small section (of length W_i) of the edge of the patch. W_i is chosen so small that the fields over this length may be assumed to be uniform. Typically, for a rectangular patch, the number of ports along each radiating edge is taken to be 4, and along each non-radiating edge the number is taken to be eight.

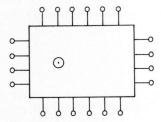

Fig. 9.3 *Multiport representation of a rectangular patch*

Thus, a 24 × 24 matrix is typically adequate for the characterisation of the interior fields of a rectangular patch.

For patches of regular shapes (rectangles, circles, rings, sectors of circles and of rings, and three types of triangles), this multiport planar network model can be analysed by using two-dimensional impedance Green's functions available for these shapes [1]. A multiport Z-matrix characterisation representing the fields underneath the patch can be derived from the Green's function as:

$$Z_{ij} = \frac{1}{W_i W_j} \int_{W_i} \int_{W_j} G(x_i, y_i | x_j, y_j) ds_i ds_j \tag{9.9}$$

where $x_{i,j}$, $y_{i,j}$ denote the locations of the two ports of widths W_i and W_j, respectively. Green's function G is usually a doubly infinite summation with terms corresponding to various modes of the planar resonator (rectangular or circular or triangular) with magnetic walls.

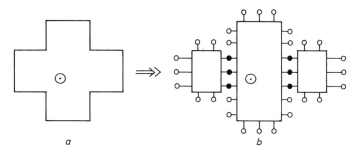

Fig. 9.4 *(a) A cross-shaped microstrip patch. (b) Multiport-network model of the cross-shaped microstrip patch*

For patches of composite shapes (such as a cross shape shown in Fig. 9.4a), a multiport network model can be written by treating the composite shape as a combination of the elementary shapes for which Green's functions are available. The cross shape of Fig. 9.4a can be considered as a combination of three rectangular segments as shown in Fig. 9.4b. Segmentation and desegmentation methods [1] are used for finding the Z-matrix of a composite shape from those of the elementary segments. If the patch or one of the segments of a composite patch is of an irregular shape for which Green's function is not available, a technique called the contour integral method [1] can be used to evaluate the Z-matrix.

In the multiport-network modelling of radiating microstrip patches, the fields outside the patch (namely, the fringing fields at the edges, the surface-wave fields and the radiation field) are incorporated by adding equivalent edge admittance networks (EAN) connected to the various edges of the patch. This representation is shown in Fig. 9.5 for the case of the rectangular patch shown in Fig. 9.3. EANs are multiport networks consisting of parallel combinations of the capacitances C (representing the energy stored in the fringing field) and the conductances G (representing the power carried away by radiation and surface waves) as shown in Fig. 9.6a. Each capacitance–conductance pair is connected to a port of the planar equivalent circuit of the patch. EANs at the non-radiating edges may be simplified to consist of capacitances only, as shown in Fig. 9.6b. Values of capacitance and conductance in the edge-admittance networks may be obtained from the various analyses reported in the literature [30–32].

The flexibility of the multiport network model leads to several advantages when compared with the conventional cavity model discussed in Section 9.2.2. For example, the parasitic reactances at the junction between the feed line and the patch can be incorporated in the multiport network model by considering a small section of the feed line as an equivalent planar circuit connected to the patch at a finite number of (typically five) ports. Solution of this network problem (depicted in Fig. 9.7) is equivalent to the expansion of the fields (in the feed line as well as in the patch) in series of eigen functions and matching the fields at the interface.

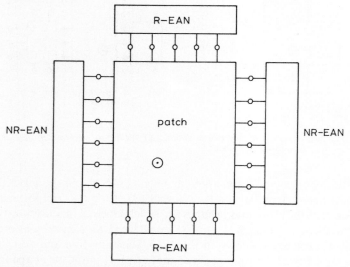

Fig. 9.5 *Edge-admittance networks (EANs) connected to the multiport representation of a rectangular patch*

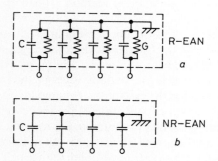

Fig. 9.6 *Edge admittance networks for (a) radiating edges; (b) non-radiating edges*

Also, the multiport network model discussed above can be extended to incorporate the effect of mutual coupling between the two radiating edges [33] by inserting a mutual coupling network (MCN) as shown in Fig. 9.8. The edge-admittance terms associated with various ports at the edges constitute the diagonal terms of the admittance matrix for MCN. The non-diagonal terms of this matrix are obtained from the 'reaction' between the equivalent magnetic current sources at the two corresponding sections of the edges. Similar MCNs may also be included between the non-radiating edges, between a radiating edge and a non-radiating edge, or between two edges of different patches in an array.

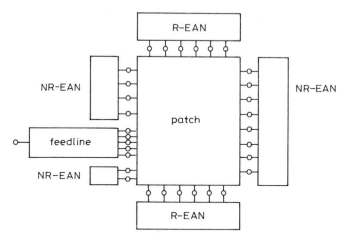

Fig. 9.7 *Incorporation of feed-junction reactance in multiport-network model of a rectangular patch*

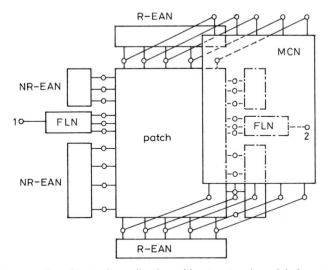

Fig. 9.8 *Incorporation of mutual coupling in multiport-network model of a rectangular patch*

It may be noted that, in the multiport network model, the characterisation of fields underneath the patch is conceptually similar to that used in the conventional cavity model [21, 22]. In both of these models, the fields under the patch

466 Multiport network approach for modelling

are considered two-dimensional with no variations of fields perpendicular to the substrate. For this reason, the limits of the applicability of the technique in terms of substrate permittivity and thickness are similar to that for the cavity model. Also, both of these methods will not be accurate when applied to narrow-width microstrip dipoles rather than to the wide microstrip patches discussed in this Chapter.

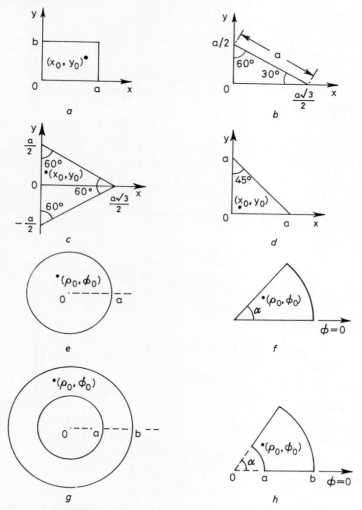

Fig. 9.9 *Various geometries of the planar segments for which Green's functions are available*

Green's functions used for evaluating impedance-matrix characterisation for patches of various shapes are discussed in Section 9.3. Derivation of the Z-matrix is also included therein.

9.3 Z-matrix characterisation of planar segments

9.3.1 Green's functions

For practical microstrip antennas, the thickness of the substrate is much smaller than the wavelength. Therefore fields underneath the patch do not vary in the z-direction (perpendicular to the substrate). Electric field has a z-component only. Since $\partial E_z/\partial z = 0$, we may define a voltage $V(x, y)$ given by

$$V(x, y) = -E_z(x, y)d \quad (9.10)$$

where d is the substrate thickness. When a magnetic-wall boundary condition is assumed at the edges of the patch, $V(x, y)$ satisfies the boundary condition given by eqn. 9.4, i.e.

$$\frac{\partial V(x, y)}{\partial p} = 0 \quad (9.11)$$

If we consider a z-directed electric current source $J_x(x_0, y_0)$ located at (x_0, y_0), the voltage $V(x, y)$ is related to the source current through a two-dimensional impedance Green's function $G(x, y|x_0, y_0)$ defined by

$$V(x, y) = \int_D \int G(x, y|x_0, y_0) J_z(x_0, y_0) \, dx_0 dy_0 \quad (9.12)$$

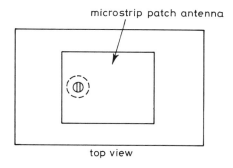

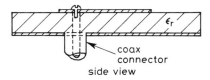

Fig. 9.10 *Microstrip patch antenna with a probe feed perpendicular to the substrate*

where the source current J_z is distributed over a region D in x, y plane. These Green's functions are known [1] for several regular shapes shown in Fig. 9.9. Expressions for these Green's functions are listed in Appendix 9.8.

When a microstrip antenna is excited by a probe feed perpendicular to the substrate, as shown in Fig. 9.10, the current density may be related to the axial current through the z-directed probe. For patches excited by a microstrip line feed, the current J_{in} flowing into the patch can be expressed as an equivalent z-directed electric current sheet J_z as follows. At the magnetic wall surrounding the patch (as shown in Fig. 9.11),

$$J_z = \hat{p} \times H_t \qquad (9.13)$$

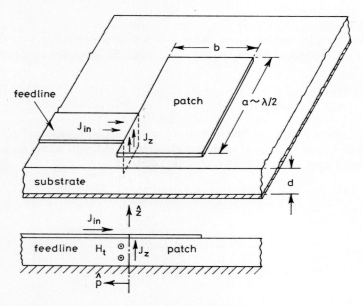

Fig. 9.11 *Equivalence between the port current and the z-directed fictitious current density at the junction between a microstrip-line feed and a patch*

and for the planar waveguide model of the microstrip line feeding the patch

$$J_{in} = \hat{z} \times H_t \qquad (9.14)$$

Thus $|J_z| = |J_{in}|$. If the effective width of the microstrip line is W_j (for the jth port), the input current at the port j may be written as

$$I_j = W_j |J_z| \qquad (9.15)$$

9.3.2 Evaluation of Z-matrix from Green's functions

Green's functions discussed above may be used to find the Z-matrix characterisation of various planar segments of the shapes shown in Fig. 9.9 with respect to specified locations of external ports. These external ports may be either of the probe-feed type (Fig. 9.10) or the microstrip-feed type (Fig. 9.11) or a combination of these. Evaluation of the Z-matrix is based on relation 9.9

and involves integration of $G(x, y|x_0, y_0)$ over the extent of the two-ports corresponding to the specific element of the Z-matrix. For a microstrip line feed with an effective width W_i, the integral is carried out over the width W_i. For a probe-feed type of external port, the integration is carried out over a circular path corresponding to the cylindrical surface of the probe. Alternatively, the circular probe may be replaced by an equivalent strip and the integration carried over this equivalent width.

Z-matrix for rectangular segments: Green's functions for various geometries, discussed in Appendix 9.8, appear as double-infinite summations. In numerical computations of the Z-matrix elements, order of integration and summations could be interchanged. For rectangular segments, the integrals involved may be carried out analytically. When sides of the rectangle are oriented along x- and y-axes and for the two ports (say, port p and port q), the impedance-matrix element Z_{pq} may be written in the following form [37]:

$$Z_{pq} = \frac{j\omega\mu d}{ab} \sum_{m=0}^{\infty} \sum_{n=0}^{\infty} \sigma_m \sigma_n \phi_{mn}(x_p, y_p)$$
$$\cdot \phi_{mn}(x_q, y_q)/(k_x^2 + k_y^2 - k^2) \qquad (9.16)$$

where, for ports oriented along the y-direction,

$$\phi_{mn}(x, y) = \cos(k_x x)\cos(k_y y)\operatorname{sinc}\left(\frac{k_y w}{2}\right) \qquad (9.17)$$

and for ports oriented along the x-direction

$$\phi_{mn}(x, y) = \cos(k_x x)\cos(k_y y)\sin\left(\frac{k_x w}{2}\right) \qquad (9.18)$$

The function $\operatorname{sinc}(z)$ is defined as $\sin(z)/z$, and

$$k_x = \frac{m\pi}{a}, \quad k_y = \frac{n\pi}{b}$$

$$\sigma_m = \begin{cases} 1, & m = 0 \\ 2, & m \neq 0 \end{cases}$$

$$k^2 = \omega^2 \mu \varepsilon_0 \varepsilon_r (1 - j\delta)$$

$$\delta = \text{loss tangent of the dielectric}$$

The length of rectangle is a, its width is b, and height of the substrate is d. The points (x_p, y_p) and (x_q, y_q) denote the locations of the p and q ports, respectively.

It has been shown [37] that the doubly infinite series in (eqn. 9.16), along with eqns. 9.17 and 9.18, can be reduced to a singly infinite series by summing the inner sum. The choice of summation over n or m depends on the relative locations of the ports p and q, and also on the aspect ratio of the rectangular segment. We consider two different cases

Case 1: When both the ports (p and q) are oriented along the same direction (x or y). We may write Z_{pq} as

$$Z_{pq} = -CF \sum_{l=0}^{L} \sigma_l \cos(k_u u_p) \cos(k_u u_q)$$
$$\cdot \cos(\gamma_l z_>) \cos(\gamma_l z_<)$$
$$\cdot \frac{\operatorname{sinc}\left(\frac{k_u w_p}{2}\right) \operatorname{sinc}\left(\frac{k_u w_q}{2}\right)}{\gamma_l \sin(\gamma_l F)} - jCF$$
$$\cdot \sum_{l=L+1}^{\infty} \cos(k_u u_q) \cos(k_u u_p) \operatorname{sinc}\left(\frac{k_u w_p}{2}\right)$$
$$\cdot \operatorname{sinc}\left(\frac{k_u w_q}{2}\right) \frac{\exp(-j\gamma_l (v_> - v_<))}{\gamma_l} \qquad (9.19)$$

where

$$(v_>, v_<) = \begin{cases} (y_>, y_<) & l = m \\ (x_>, x_<) & l = n \end{cases}$$

and

$$C = j\omega\mu d/(ab)$$

When the two ports are oriented in the y-direction we choose $l = n$, and when they are along x-direction l is put equal to m. Also,

$$F = \begin{cases} b, & l = m \\ a, & l = n \end{cases}$$

$$(u_p, u_q) = \begin{cases} (x_p, x_q), & l = m \\ (y_p, y_q), & l = n \end{cases}$$

$$\gamma_l = \pm \sqrt{k^2 - k_u^2}$$

$$k_u = \begin{cases} \dfrac{m\pi}{a}, & l = m \\ \dfrac{n\pi}{b}, & l = n \end{cases}$$

and

$$(z_>, z_<) = \begin{cases} (y_> - b, y_<), & l = m \\ (x_> - a, x_<), & l = n \end{cases}$$

The sign of γ_l is chosen so that Im (γ_l) is negative. w_p and w_q are widths of ports p and q, respectively. Also, we use

$$y_> = \max(y_p, y_q) \qquad y_< = \min(y_p, y_q)$$

and a similar notation for $x_>$ and $x_<$ when $l = n$. The choice of the integer L in eqn. 9.19 becomes a trade-off between fast computation and accuracy. A compromise is to select L so that $(\gamma_l F)$ is less than or equal to 100.

Case 2: When the two ports (p and q) are oriented in different directions (x and y), various elements of the Z-matrix may be

$$Z_{pq} = -CF\frac{1}{\eta}\sum_{l=0}^{L}\sigma_l \cos(k_u u_p)\cos(k_u u_q)\cos(\gamma_l z_<)$$

$$\cdot \cos(\gamma_l z_>) \frac{\text{sinc}\left(k_u \frac{w_i}{2}\right)\text{sinc}\left(\frac{\gamma_l w_j}{2}\right)}{\gamma_l \sin(\gamma_l F)}$$

$$- CF\frac{1}{\eta}\sum_{l=L+1}^{\infty}\cos(k_u u_p)\cos(k_u u_q)$$

$$\cdot \text{sinc}\left(\frac{k_u w_i}{2}\right)\frac{\exp\left(-j\gamma_l\left(v_> - v_< - \frac{w_j}{2}\right)\right)}{\gamma_l^2 w_j} \qquad (9.20)$$

Choice of l is made by noting that, for convergence of the last summation in the above equation, we need

$$(v_> - v_< - w_j/2) > 0 \qquad (9.21)$$

We choose the index of the inner summation so that this condition is satisfied. This condition may be written more explicitly as

$$l = m, \quad \text{if}\{\max(y_p, y_q) - \min(y_p, y_q) - w_j/2\} > 0 \qquad (9.22)$$

and

$$l = n, \quad \text{if}\{\max(x_p, x_q) - \min(x_p, x_q) - w_j/2\} > 0 \qquad (9.23)$$

When both of these conditions are satisfied, any choice of l will ensure convergence.

If $l = n$, w_i corresponds to the port oriented along y-direction and w_j corresponds to the port along the x-direction. On the other hand if $l = m$, w_i is for the port along the x-direction and w_j for the port along the y-direction.

Z-matrix for circular segments: For circular-shaped patches, the impedance Green's function is given by eqns. (A9.10) and (A9.11) in Appendix 9.8. When ports are located along the circumference of the circle (as shown in Fig. 9.12), various elements of the Z-matrix [38] may be written as follows. For any port i, the Z-matrix element Z_{ii} may be written as

$$Z_{ii} = \frac{2j\omega\mu da^2}{\pi W_i^2} \sum_{n=0}^{\infty} \sum_{m=1}^{\infty} \frac{\sigma_n\{1 - \cos[2n\sin^{-1}(W_i/2a)]\}}{n^2 \left(a^2 - \dfrac{n^2}{k_{nm}^2}\right)(k_{nm}^2 - k^2)} \quad (9.24)$$

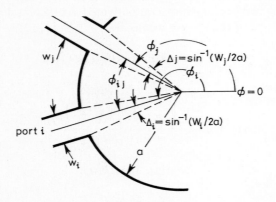

Fig. 9.12 *Various parameters for ports located at the circumference of a circular segment*

Off-diagonal terms of the impedance matrix are found to be

$$Z_{ij} = \frac{2j\omega\mu da^2}{\pi W_i W_j} \sum_{n=0}^{\infty} \sum_{m=1}^{\infty} \frac{\sigma_n}{n^2 \left(a^2 - \dfrac{n^2}{k_{nm}^2}\right)(k_{nm}^2 - k^2)}$$

$$\cdot \{\cos[n(\Delta_i - \Delta_j)] - \cos[n(\Delta_i + \Delta_j)]\} \cos(n\phi_{ij}) \quad (9.25)$$

where

$$\Delta_{i,j} = \sin^{-1}(W_{i,j}/2a)$$

Similar expressions for computation of the Z-matrices for planar segments of other geometries shown in Fig. 9.9 have not been reported so far.

9.3.3 Z-matrices for segments of arbitrary shape

When we come across segments of arbitrary shapes, for which Green's functions do not exist, the impedance matrix can be found by a method known as contour integral method [39]. The contour integral method is based on the Green's theorem in cylindrical co-ordinates. The RF voltage at any point $M(s)$ inside the periphery of an arbitrarily shaped planar segment shown in Fig. 9.13a is given by

$$4jv(s) = \oint_C \left\{ H_0^{(2)}(kr) \frac{\partial v(s_0)}{\partial n} - v(s_0) \frac{\partial H_0^{(2)}(kr)}{\partial n} \right\} ds_0 \quad (9.26)$$

where $H_0^{(2)}$ is the zeroth-order Hankel function of the second kind and r is the straight-line distance between the point $M(s)$ and the source point on the periphery (given by $L(s_0)$). The integral on the right-hand side of eqn. 9.26 is carried out over the entire periphery. The RF voltage at any point *just* inside the periphery can be derived from the above relationship. We obtain

$$2jv(s) = \oint_C \{k\cos\theta H_1^{(2)}(kr)v(s_0) + j\omega\mu d J_n(s_0)H_0^{(2)}(kr)\}\,ds_0 \qquad (9.27)$$

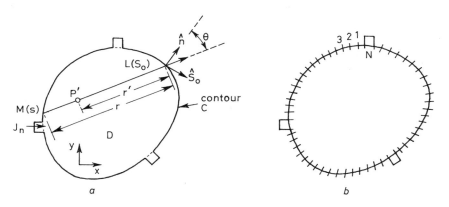

Fig. 9.13 *(a) Configuration of a planar segment for analysis by contour integral method; (b) Division of the periphery in N sections for the analysis*

where $H_1^{(2)}$ is the first-order Hankel function of the second kind, and J_n denotes line current density flowing into the segment at s_0. The variables s and s_0 denote distances along the contour C and r is the distance between the two points M and L (specified by s and s_0) as shown in Fig. 9.13a. The angle θ is the angle made by the straight line joining points M and L with the normal to the periphery at L. Line current density J_n, flowing into the segment at a coupling port, is given by

$$J_n = \frac{1}{j\omega\mu d}\frac{\partial v}{\partial n} \qquad (9.28)$$

For the numerical calculation of the impedance matrix, we divide the periphery into N sections having arbitrary widths $W_1, W_2 \ldots W_N$ as shown in Fig. 9.13b. The periphery is divided in such a manner that each coupling port contains an integral number of such sections. For greater accuracy, wider coupling ports may be divided into a multiple number of sections. We set N sampling points, one at the centre of each section, and assume that each section is a straight edge. It is further assumed that the widths of the sections are so small that magnetic and electric fields can be considered constant over each section. Under the assumptions outlined above, the line integral in eqn. 9.27 can be replaced by summation over the N sections. The resulting expression is given by

$$2jv_l = \sum_{m=1}^{N} \{kv_m G_{lm} + j\omega\mu d\, i_m F_{lm}\} \tag{9.29}$$

where v_l is the voltage over the lth section and $i_m(=J_m W_m)$ is the total current flowing into the mth section. The matrix elements G_{lm} and F_{lm} are given as

$$G_{lm} = \begin{cases} \int\!\int_{W_m} \cos\theta H_1^{(2)}(kr)\,ds_0, & \text{if } l \neq m \\ 0, & \text{otherwise} \end{cases} \tag{9.30}$$

and

$$F_{lm} = \begin{cases} \dfrac{1}{W_m}\int_{W_m} H_0^{(2)}(kr)\,ds_0, & \text{if } l \neq m \\ 1 - \dfrac{2j}{\pi}\left(\ln\dfrac{kW_l}{4} - 1 + \gamma\right), & \text{otherwise} \end{cases} \tag{9.31}$$

In eqn. 9.31, $\gamma(=0.5772\ldots)$ denotes the Euler's constant. In the above discussion we assume that the current can be fed into the planar circuit from all the N sections and i_m denotes the current fed from the mth section. This yields the impedance matrix for the N-port circuit. This matrix can be used to obtain the impedance matrix for any specified number and location of ports on the planar circuit being analysed. Eqn. 9.29 is written for each section l on the periphery of the planar circuit. All these equations combined together in matrix form become

$$Av = Bi \tag{9.32}$$

where v and i are the voltage and the current vectors at each section. A and B denote N by N matrices, determined by the shape of the circuit. The elements of these matrices, obtained from eqn. 9.29, are

$$a_{lm} = -kG_{lm} \quad \text{for} \quad l \neq m \tag{9.33}$$

$$a_{ll} = 2j$$

and

$$b_{lm} = j\omega\mu d\, F_{lm} \tag{9.34}$$

From eqn. 9.32, the impedance matrix for the N sections, considered as ports, is obtained as

$$Z_N = A^{-1}B \tag{9.35}$$

In practice, the coupling ports are connected to only a few of the N sections. Rows and columns corresponding to the sections which are open-circuited can be deleted from Z_N. If each coupling port covers only one section, the matrix thus obtained (after deleting rows and columns corresponding to the open sections from Z_N) is the required impedance matrix. If some coupling ports extend to more than one section, the sections in these coupling ports are like sub-ports and the procedure detailed in Reference 1 can be used to obtain the overall admittance matrix at the coupling ports (and hence impedance matrix, if desired).

9.4 Edge-admittance and mutual-coupling networks

As discussed in Section 9.2.3, the multiport network modelling approach assumes that the fields outside the patch may be represented by an equivalent network model. The concept of edge admittance associated with the radiating edges has been widely used in conjunction with the transmission-line model [23, 24, 26]. The same concept can be extended to the multiport network modelling approach. For the multiport network model, the two-terminal edge admittance (used in the transmission-line model) is replaced by the multiport network shown in Fig. 9.6a or b. If an edge of the patch is divided into n sections, the edge-admittance network (EAN) is an n-port network with the common terminal being the ground return.

In this Section, we discuss various methods for evaluating parameters of edge-admittance networks. Also, the network modelling approach can be extended to incorporate the effect of mutual coupling between the radiating edges of a single patch, as well as among the edges of adjacent patches in an array environment.

9.4.1 Edge-admittance networks

Edge admittance associated with a radiating microstrip patch consist of two components: (i) a susceptance representing the energy stored in the fringing field associated with the edge; and (ii) a conductance G representing the power transmitted to the radiation field as well as the power carried away by the surface waves excited along the dielectric substrate. For incorporating edge-admittance networks in the multiport network model for microstrip patch antennas, a capacitance–conductance pair is connected to each of the ports of the equivalent planar multiport representation of the patch. The conductance G consists of two parts: a radiation conductance G_r and a surface-wave conductance G_s. Radiation conductance associated with an edge of a microstrip patch is defined as an ohmic conductance (distributed or lumped), which, when connected to the edge (continuously or at discrete ports), will dissipate a power equal to that radiated by the edge (or by an equivalent magnetic current source) for the same voltage distribution. If the edge has width W and the power radiated for a uniform voltage distribution is P_{rad}, the radiation conductance per unit length of the edge is given by $2P_{rad}/W$, where P_{rad} is calculated for a unit voltage at the edge. When the voltage distribution along an edge is given by $f(s)$, the radiation conductance per unit length is obtained as:

$$\frac{G_r}{W} = \frac{2P_{rad}}{\int_0^W f^2(s)\,ds} \qquad (9.36a)$$

where s denotes the distance along the edge of the patch. If we select n uniformly spaced ports (each representing a section of length W/n along the edge, the conductance connected to each of the ports is taken as G_r/n. This concept of edge conductance can be implemented when $f(s)$ is known *a priori*. In most of the cases, microstrip antennas are operated near resonance of the patch and $f(s)$ is known at least approximately. For more accurate results, iterative computations may be needed. Starting from an approximate $f(s)$, an analysis based on the MNM model is carried out to evaluate the voltages at the n ports on the edge. This computed voltage distribution is used as modified $f(s)$ and the computation of G_r is repeated.

The surface-wave conductance G_s is defined in a similar manner. We write:

$$\frac{G_s}{W} = \frac{2P_{sur}}{\int_0^W f^2(s)\,ds} \qquad (9.36b)$$

where P_{sur} is the power coupled to the surface waves (on the substrate) excited by the voltage distribution at the edges. For thin substrates without any cover layer, G_s is much smaller than G_r and may be neglected.

The edge susceptance accounts for the reactive energy associated with the fringing field at the edge. As in the case of the edge conductance, the edge susceptance is also distributed uniformly over the n ports.

Values of the capacitances and the conductances in the edge-admittance networks for edges with uniform voltage distribution, may be obtained from the various analyses reported in the literature. Available results are discussed in this Subsection.

Edge conductance: Various available formulations for the radiation conductance associated with the radiating edge of a rectangular microstrip patch are summarised in this Section.

The most widely used formula has been obtained by considering radiating edges as slots radiating in half free space. As given by James *et al.* [40], this formula may be expressed as:

$$G_R = \begin{cases} \dfrac{b_e^2}{90\,\lambda_0^2} & b_e \leqslant 0\cdot 35\,\lambda_0 & (9.37a) \\[1em] \dfrac{b_e}{120\,\lambda_0} - \dfrac{1}{60\,\pi^2} & 0\cdot 35\,\lambda_0 \leqslant b_e \leqslant 2\,\lambda_0 & (9.37b) \\[1em] \dfrac{b_e}{120\,\lambda_0} & 2\,\lambda_0 < b_e & (9.37c) \end{cases}$$

where λ_0 is the free-space wavelength and b_e is the effective width of the rectangular resonator of width b. The effective width is defined as the width of an equivalent parallel-plate waveguide corresponding to a microstrip line of width b. G_R is the total radiation conductance (in mhos) associated with the radiating edge.

A more precise formula based also on the slot-radiator model of the edge is given by Van de Capelle et al. [31] and may be expressed as

$$G_R = \frac{1}{\pi\eta_0}\left\{\left[k_0b_e\,\text{si}(k_0b_e) + \cos(k_0b_e) + \frac{\sin(k_0b_e)}{k_0b_e} - 2\right]\right.$$
$$\left.\times\left(1 - \frac{(k_0d)^2}{24}\right) + \frac{(k_0d)^2}{12}\cdot\left[\frac{1}{3} + \frac{\cos(k_0b_e)}{(k_0b_e)^2} - \frac{\sin(k_0b_e)}{(k_0b_e)^3}\right]\right\}$$

(9.38)

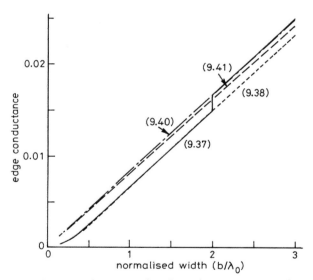

Fig. 9.14 *Edge conductance of a rectangular microstrip antenna versus the normalised width (From Reference 27)*

where η_0 = free-space wave impedance = 120π ohms, $k_0 = 2\pi/\lambda_0$ and d is the height of the substrate. The function $\text{si}(x)$ is defined as

$$\text{si}(x) = \int_0^x \frac{\sin(u)}{u}\,du \qquad (9.39)$$

Relation 9.38 is based on the assumption that the radiation from a rectangular antenna (Fig. 9.11) can be modelled by two rectangular slots with dimensions b_e and d in an infinite ground plane. Results based on these formulas are plotted in Fig. 9.14.

Two other formulas, useful for design and based on Wiener–Hopf characterisation of an infinitely wide microstrip patch edge, are given by Kuester et al. [32] and Gogoi et al. [30].

The formula of Kuester et al. (32) for electrically thin substrates ($k_0 d \ll 1$) may be written as

$$G_R = \text{Re}\left(\frac{1 - e^{j\chi(0)}}{1 + e^{j\chi(0)}}\right) \tag{9.40}$$

where

$$\chi(0) = \frac{2k_0 d}{\pi\sqrt{\varepsilon_r}}\left\{\left[\ln(jk_0 d) + \gamma - 1\right] + \varepsilon_r\left[2Q_0(-\delta_\varepsilon) - \ln(2\pi)\right]\right\}$$

$$\delta_\varepsilon = \frac{\varepsilon_r - 1}{\varepsilon_r + 1}$$

$$Q_0(z) = \left(\frac{z}{1-z}\right)^2\left[\ln 2 - Q_2(z)\right]$$

$$Q_2(z) = \sum_{m=1}^{\infty} z^m \ln\left(\frac{(m+1)^2}{m(m+2)}\right), \qquad |z| < 1$$

and

$$\gamma = 0.57721 \text{ (Eurler's constant)}$$

The formula given by Gogoi et al. [30] is as:

$$G_R = b_e \frac{7.75 + 2.2k_0 d + 4.8(k_0 d)^2}{1000\lambda_0}\left\{1 + \frac{(\varepsilon_r - 2.45)(k_0 d)^3}{1.3}\right\} \tag{9.41}$$

Accuracy of eqn. 9.41 is 1·1% for $0.05 \leqslant k_0 d < 0.6$ and $2.45 < \varepsilon_r < 2.65$. These two formulas are accurate for wide patches (large value of b in Fig. 9.11). Results based on these formulas are also plotted in Fig. 9.14.

Fig. 9.14 shows a comparison of the formulas 9.37–9.41 for the following set of data: frequency $f = 7.5\,\text{GHz}$; dielectric constant $\varepsilon_r = 2.48$; and thickness of substrate, $d = 1/32\,\text{in}$. It is inferred from Fig. 9.14 that formulas 9.37 and 9.38 yield close results as expected. The difference between formulas 9.37 and 9.41 increases with decreasing value of the width. This is because of the fact that formulas 9.40 and 9.41 are valid only for wide patches. Formulas 9.40 and 9.41 are both based on the Wiener–Hopf formulation and therefore yield identical results for all values of the width. From this limited discussion, it seems reasonable to use formula 9.37 owing to its simplicity.

Although the power coupled to surface waves is very small compared to the radiated power, the conductance corresponding to the surface waves should be added to G_R. This conductance may be expressed as [30]

$$G_S = k_0 d[20.493 + 65.167 k_0 d + 104.333 (k_0 d)^2] 10^{-4} \cdot b_e$$
$$[1 + 3.5(\varepsilon_r - 2.45)(k_0 d)^3]/\lambda_0 \,\text{mhos/m} \tag{9.42}$$

Accuracy of formula 9.42 is 2·6% for $0·2 < k_0 d < 0·6$ and $2·45 < \varepsilon_r < 2·65$.

More rigorous formulations for edge conductance of open-ended microstrips have been reported by James and Henderson [41], and more recently by Katehi and Alexopoulos [42] and by Jackson and Pozar [43]. Analysis techniques and numerical results for a limited set of parameters have been given in References 41–43. James and Henderson [41] have pointed out that their rigorous results agree with eqn. 9.37a and a previous estimation by Lewin [44] for $d/\lambda_0 < 0·09$ when $\varepsilon_r = 2·32$, and for $d/\lambda_0 < 0·03$ when $\varepsilon_r = 10$. Results based on full-wave analysis using the method of moments [43] agree with results in Reference 41 for substrate thicknesses up to $0·1 \lambda_0$.

Edge susceptance: As in the case of edge conductance, several different results are available for edge susceptance also. One of the formulas for edge susceptance B is based on the parallel-plate waveguide model of a microstrip and is given by Reference 45 as:

$$B = \pi f \left\{ \frac{\sqrt{\varepsilon_{re}(a, d, t, \varepsilon_r)}}{c Z_0(a, d, t, \varepsilon_r)} - \frac{\varepsilon_0 \varepsilon_r a}{h} \right\} b_e \quad (9.43)$$

where Z_0 and ε_{re} are the characteristic impedance and effective dielectric constant of a microstrip line of width a. Expressions for Z_0 and ε_{re} are well known [1]. c is the velocity of waves in free space ($= 3 \times 10^8$ m/s). Another formula for B, which is based on open-end capacitance of a microstrip line [46] is given by

$$B = 0·01668 \frac{\Delta a}{d} \frac{b_e}{\lambda_0} \varepsilon_{re} \quad (9.44)$$

where

$$\frac{\Delta a}{d} = 0·412 \frac{(\varepsilon_{re} + 0·3)(w/d + 0·264)}{(\varepsilon_{re} - 0·258)(w/d + 0·8)} \quad (9.45)$$

Other formulas, which are based on the Wiener–Hopf formulation and which can be used for wide patches, are given in References 30 and 32. From Reference 32,

$$B = \text{Im} \left\{ \frac{1 - \exp(j\chi(0))}{1 + \exp(j\chi(0))} \right\} \quad (9.46)$$

where $\chi(0)$ is as given earlier for eqn. 9.40. According to Reference 30,

$$B = 0·01668 \frac{\Delta a}{d} \frac{b_e}{\lambda_0} \varepsilon_{re} \quad (9.47)$$

where

$$\frac{\Delta a}{d} = \frac{0·95}{1 + 0·85 k_0 d} - \frac{0·075(\varepsilon_r - 2·45)}{1 + 10 k_0 d} \quad (9.48)$$

The accuracy of expression 9.48 is 2% for $0.1 < k_0 d < 0.6$ and $2.45 < \varepsilon_r < 2.65$. Formulas 9.43–9.48 are compared in Fig. 9.15. Expressions 9.43, 9.44 and 9.47 give close results for all practical values of the resonator width ($0.25\lambda_0 < w < 0.6\lambda_0$). Eqn. 9.44 predicts an end-susceptance value of one-half of that computed using other formulas. For all practical values of interest, formula 9.43 may be used, since there is no restriction on the width of the patch for this formula. When the width is large, both expressions 9.46 and 9.47 can be used.

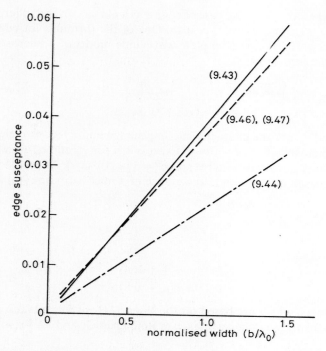

Fig. 9.15 *Edge susceptance of a rectangular microstrip antenna versus normalised width (From Reference 27)*

More rigorous characterisation of microstrip open-edge susceptance has been reported in [41, 43, 47, 48]. Results of the analysis in References 47 and 48 are available [49] as a closed form expression obtained by curve fitting of numerical data. Normalised outward extension of the radiating edge ($\Delta a/d$ in eqn. 9.44) is given by

$$\Delta a/d = (\xi_1 \xi_3 \xi_5 / \xi_4)$$

with

$$\xi_1 = 0.434907 \frac{\varepsilon_{re}^{0.81} + 0.26 (w/d)^{0.8544} + 0.236}{\varepsilon_{re}^{0.81} - 0.189 (w/d)^{0.8544} + 0.87}$$

$$\xi_2 = 1 + \frac{(w/d)^{0.371}}{2.358\varepsilon_r + 1}$$

$$\xi_3 = 1 + \frac{0.5274 \arctan[0.084(w/d)^{1.9413/\xi_2}]}{\varepsilon_{re}^{0.9236}}$$

$$\xi_4 = 1 + 0.0377 \arctan[0.067(w/d)^{1.456}]$$
$$\times \{6 - 5\exp(0.036(1 - \varepsilon_r))\}$$

$$\xi_5 = 1 - 0.218 \exp(-7.5w/d) \tag{9.49}$$

Expressions in eqns. 9.49 make use of the effective dielectric constant formula of Reference 50, i.e.

$$\varepsilon_{re}(u, \varepsilon_r) = \frac{\varepsilon_r + 1}{2} + \frac{\varepsilon_r - 1}{2}\left(1 + \frac{10}{u}\right)^{-a(u)b(\varepsilon_r)}$$

$$a(u) = 1 + \frac{1}{49}\ln\frac{u^4 + (u/52)^2}{u^4 + 0.432} + \frac{1}{18.7}\ln\left[1 + \left(\frac{u}{18.1}\right)^3\right]$$

$$b(\varepsilon_r) = 0.564 \left(\frac{\varepsilon_r - 0.9}{\varepsilon_r + 3}\right)^{0.053} \tag{9.50}$$

where $u = w/d$.

Accuracy of results given by eqn. 9.49 is claimed to be better than 2·5% for the range of normalised widths $0.01 \leqslant w/d \leqslant 100$ and $\varepsilon_r \leqslant 50$.

Non-radiating edges: For rectangular patches radiating linearly polarised electromagnetic waves, radiating and non-radiating edges can be distinguished clearly. As shown in Fig. 9.6*b*, the non-radiating edges can be modelled by EAN consisting of capacitances only. An equivalent approach is to extend the width of the patch by moving the non-radiating edges outwards so that the edge capacitance is accounted for by the increased capacitance of the wider patch. The concept of radiating and non-radiating edges has been studied [59] by studying the total and partial reflections from the end of a parallel-plate waveguide with an extended dielectric slab. It has been pointed out [60] that the radiating or non-radiating nature of the edge depends on the angle at which the wave underneath the patch is incident on the edge. For grazing incidence there is no radiation, whereas for normal incidence the edge radiates. In between there is a critical angle where the transition from non-radiation to radiation takes place.

It has been recognised that the so-called non-radiating edges of a rectangular patch (operating at the resonance of 1,0 mode) do contribute to a cross-polarised radiation field. Compared to a single radiating edge, cross-polarised radiation from a single non-radiating edge is typically 10–15 dB lower. When we combine the radiated fields (in the broadside direction) from the two non-radiating edges, the total field is much smaller. This is caused by the fact that equivalent magnetic currents corresponding to the fringing fields at the two non-radiating edges are in the opposite direction and tend to cancel each other. However, the non-radiating edge-admittance network (NR-EAN), shown in Fig. 9.6b can be used only in cases where approximate results (obtained by ignoring the cross-polarised radiation from NR edges) are considered satisfactory. In more general cases, especially when the antenna operation is not at 1, 0 mode resonance (as for the circular polarised radiator discussed in Section 9.6.1), the EANs at all edges are similar to that shown in Fig. 9.6a.

In spite of the numerous results for edge admittance that have been reviewed in this Section, the lack of an accurate characterisation for the edge admittance is one of the major shortcomings in the design information required for precise design of microstrip patches and arrays.

9.4.2 Mutual-coupling network

As shown in Fig. 9.8, the mutual interaction between the fringing fields associated with any two edges can be expressed in terms of a network model. The basis idea of using an admittance element to represent mutual coupling between two radiating edges was initially presented in Reference 51 in conjunction with the transmission-line model of microstrip antennas. The reader is referred to Chapter 10 for details of this concept. The multiport mutual-coupling network shown in Fig. 9.8 is an extension of this concept suitable for incorporating in the multiport network model of microstrip patch antennas discussed in this Chapter.

Evaluation of mutual-coupling networks: For computation of external mutual coupling between various edges of a microstrip patch antenna on a thin substrate, the field at the edge may be modelled by equivalent line sources of magnetic current placed directly on the ground plane at the location of the edges. This is illustrated in Fig. 9.16. Magnetic currents M are given by

$$M = -\hat{n} \times Ed \tag{9.51}$$

where d is the height of the substrate. Product $(-Ed)$ is the voltage $V(x, y)$ defined in eqn. 9.10. $\hat{n}$ denotes a unit vector normal to the ground plane as shown. For the E-field in the direction indicated in this Figure, both magnetic current sources M are in y-direction, i.e. directed out of the plane of the paper.

The coupling between two magnetic current line sources is evaluated by dividing each of the line sources in small sections, each of length dl. The magnetic field produced by each of these sections (on the line source 1) at the locations of the various sections on the line source 2 can be written by using fields of a magnetic current dipole in free space [52]. The configuration and

Multiport network approach for modelling 483

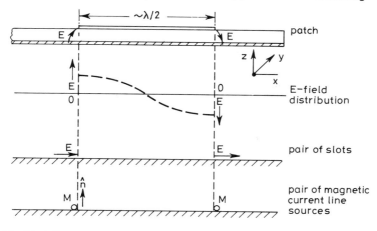

Fig. 9.16 *Modelling of the fringing field at the patch edges in terms of magnetic-current line sources*

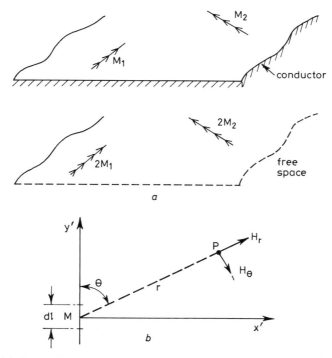

Fig. 9.17 *(a) Two arbitrarily spaced magnetic current elements; and (b) the co-ordinate system used for computation of fields*

co-ordination system is shown in Fig. 9.17. We have

$$H_\theta = j\frac{k_0 M dl \sin\theta}{4\pi\eta_0 r}\left\{1 + \frac{1}{jk_0 r} - \frac{1}{(k_0 r)^2}\right\}e^{-jk_0 r} \quad (9.52)$$

$$H_r = \frac{M dl \cos\theta}{2\pi\eta_0 r^2}\left\{1 + \frac{1}{jk_0 r}\right\}e^{-jk_0 r} \quad (9.53)$$

Here k_0 is the free-space wave number and r is the distance between the point P and the magnetic current element Mdl. When the two edges (say i and j) are oriented arbitrarily, as shown in Fig. 9.18a, the magnetic field H at (x_j, y_j) produced by a source $dl_i M$ at (x_i, y_i) may be written as

$$H = \hat{x}H_x + \hat{y}H_y \quad (9.54)$$

with

$$H_y = H_{y'}\cos\theta_i - H_{x'}\sin\theta_i \quad (9.55)$$

$$H_x = H_{y'}\sin\theta_i + H_{x'}\cos\theta_i \quad (9.56)$$

where

$$H_{x'} = H_\theta \cos\theta + H_r \sin\theta \quad (9.57)$$

$$H_{y'} = -H_\theta \sin\theta + H_r \cos\theta \quad (9.58)$$

Co-ordinate systems (x, y) and (x', y') are illustrated in Fig. 9.18b. Mutual admittance between sections j and i may be written in terms of the electric current density J_j induced in the upper surface of the edge segment j. We have

$$J_j = \hat{n} \times H = (-H_x \cos\delta_j - H_y \sin\delta_j)\hat{j} \quad (9.59)$$

The current density induced on the surface of the edge section j underneath the patch is $-J_j$. The mutual admittance between sections i and j is given by the negative of the current flow into section j (underneath the patch) divided by the voltage at section i, i.e.

$$Y_{ji} = -(-J_j \, dl_j/M_i) \quad (9.60)$$

The second minus sign in eqn. 9.60 accounts for the fact that the direction of current for defining the admittance matrix of a multiport network is directed into the network as shown in Fig. 9.19. The two edges shown in Fig. 9.18a may be the edges of the same radiating patch or those of the different patches in an array environment. When coupling between two adjacent patches in an array environment is being computed, several individual edges of the two patches contribute to the mutual coupling network (MCN). An MCN configuration taking the four radiating edges into account is shown in Fig. 9.20. Here, the MCN is connected to four ports along each radiating edge. In practice, the number of sections considered on each edge for mutual-coupling calculations is usually larger (typically 12). However, while using MCN for antenna analysis (by segmentation), a small number of ports along radiating edges (typically 4) is sufficient. Thus the original mutual-admittance matrix (48 × 48 for 12 ports

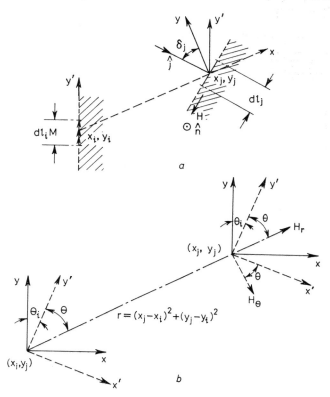

Fig. 9.18 *(a) Configuration showing sections i and j of two patch edges; and (b) two different co-ordinate systems used for mutual coupling calculations*

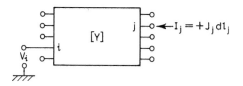

Fig. 9.19 *Representation of the mutual coupling by an admittance matrix*

along each edge) is reduced to a smaller size (16 × 16 as shown) by paralleling the ports in subgroups of three each. Contributions of non-radiating edges can also be incorporated in MCN. This is not shown in the Figure. Detailed computations [53] point out that the mutual-coupling contribution by non-radiating edges is usually much smaller and may be ignored as a first-order approximation.

Mutual coupling computations based on the above formulation have been verified [53] by comparison with the available experimental results [54]. Some of

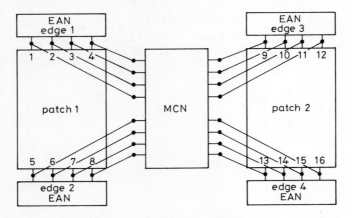

Fig. 9.20 *A mutual coupling network (MCN) representing the coupling between two adjacent patches in an array*

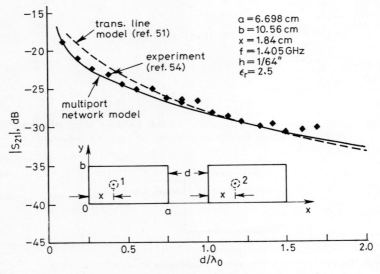

Fig. 9.21 *Comparison of theoretical and experimental results for E-plane mutual coupling between two rectangular microstrip patches (Reproduced from Reference 53)*

these results are shown in Fig. 9.21 for E-plane coupling and in Fig. 9.22 for H-plane coupling between two probe-fed rectangular patches. A very good agreement between the computation (solid line) and experimental results (diamond points) is seen. Also shown in these Figures are results of Van Lil *et al.* [51], based in transmission-line theory.

It may be noted that the preceding method of evaluating mutual coupling (based on the equivalent magnetic current model shown in Fig. 9.16) is valid only for electrically thin substrates where the effect of surface waves along the substrate is negligible. This modelling approach has been recently extended to microstrip patches on thin substrates, but covered by a relatively thicker dielectric cover layer [61, 62]. The equivalent magnetic current model used in this case

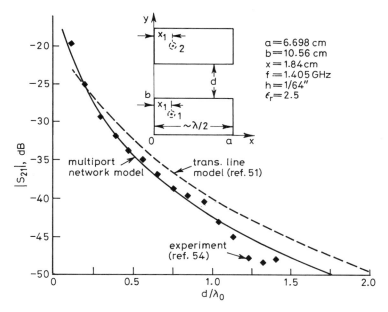

Fig. 9.22 *Comparison of theoretical and experimental results for H-plane mutual coupling between two rectangular microstrip patches (Reproduced from Reference 53)*

is shown in Fig. 9.23. The basic approach is similar to that for the case without a cover layer discussed earlier. Eqns. 9.52 and 9.53 are replaced by H_θ and H_r in the presence of the cover layer. These field components are now dominated by the effect of surface waves in the thicker cover layer.

When the substrate thickness is increased, equivalent magnetic current models for Figs. 9.16 and 9.23 become more and more inaccurate. Conceptually, multiport-network modelling of mutual coupling between two patches is still possible if more rigorous analytical/numerical techniques [63, 64] could be extended to arrive at a network representation of mutual coupling.

488 Multiport network approach for modelling

9.5 Analysis of multiport-network model

The most outstanding advantage of the multiport-network model is the fact that various analysis and optimisation techniques available for multiport networks can now be used for analysis and optimisation of microstrip antenna elements and arrays. Most widely used techniques for planar networks are segmentation [1–3, 6–8] and desegmentation [4, 5, 8] methods. These two network-analysis techniques are reviewed in this Section. Examples of various microstrip antenna configurations where these techniques have been used are reviewed in Section 9.6.

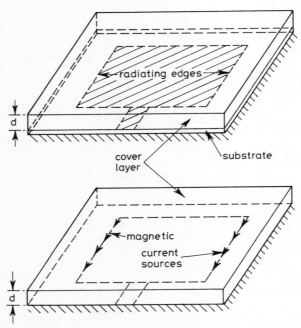

Fig. 9.23 *Magnetic-current source model for computation of mutual coupling in case of microstrip patches on a thin substrate but covered with a thick dielectric layer*

9.5.1 Segmentation method

The name 'segmentation' had been given to this network-analysis method when it was used for planar (two-dimensional) microwave circuits by Okoshi and his colleagues [2, 6, 35, 39]. The basic idea is to divide a single large planar circuit into simpler 'segments' which have regular shapes and can therefore be characterised relatively easily. An example of such a segmentation is shown in Fig. 9.24 where a ring-shaped geometry is broken down into four rectangular segments for which Green's functions are available. For the multiport-network model of a two-port microstrip antenna shown in Fig. 9.8, each of the components (shown by rectangular boxes in the figure) can be considered as a 'segment' for application of the segmentation method. Essentially, the segmentation method gives us overall characterisation or performance of the multiport network, when the characterisation of each of the segments is known. Originally the segmentation method was formulated [2] in terms of S-matrices of individual segments; however, it was found subsequently [3] that a Z-matrix formulation is more efficient for microwave planar circuits (also for microstrip antennas). In this Section, we will describe the procedure based on Z-matrices.

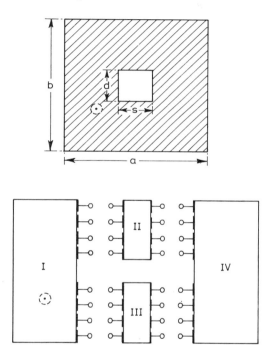

Fig. 9.24 *Segmentation of a ring-shaped structure into four rectangular segments*

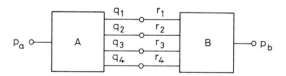

Fig. 9.25 *Two connected multiport networks A and B*

For illustrating the procedure, we consider a multiport network consisting of only two segments A and B, as shown in Fig. 9.25. Various ports of these two segments are numbered as shown. The external (unconnected) ports of segment A are called p_a-ports (which may be more than one). Similarly, the external unconnected ports of segment B are called p_b ports. Connected ports of the segment A are named q-ports and the connected ports of the segment B are designated as r-ports. q- and r-ports are numbered such that q_1 is connected to r_1, q_2 to r_2, and so on. As a result of these interconnections, we can write:

$$V_q = V_r \quad \text{and} \quad i_q = -i_r \tag{9.61}$$

Z-matrices of segments A and B may be written as

$$Z_A = \begin{bmatrix} Z_{p_a} & Z_{p_a q} \\ Z_{q p_a} & Z_{qq} \end{bmatrix}, \quad Z_B = \begin{bmatrix} Z_{p_b} & Z_{p_b r} \\ Z_{r p_b} & Z_{rr} \end{bmatrix} \tag{9.62}$$

where Z_{p_a}, $Z_{p_a q}$, $Z_{q p_a}$, z_{qq}, Z_{p_b}, $Z_{p_b r}$, $Z_{r p_b}$, Z_{rr} are sub-matrices of appropriate dimensions. As we are dealing with reciprocal components

$$Z_{p_a q} = (Z_{q p_a})^t, \quad \text{and} \quad Z_{p_b r} = (Z_{r p_b})^t \tag{9.63}$$

Z-matrices of the segments A and B can be written together as

$$\begin{bmatrix} V_p \\ V_q \\ V_r \end{bmatrix} = \begin{bmatrix} Z_{pp} & Z_{pq} & Z_{pr} \\ Z_{qp} & Z_{qq} & 0 \\ Z_{rp} & 0 & Z_{rr} \end{bmatrix} \begin{bmatrix} i_p \\ i_q \\ i_r \end{bmatrix} \tag{9.64}$$

where

$$V_p = \begin{bmatrix} V_{p_a} \\ V_{p_b} \end{bmatrix}, \quad i_p = \begin{bmatrix} i_{p_a} \\ i_{p_b} \end{bmatrix}$$

and

$$Z_p = \begin{bmatrix} Z_{p_a} & 0 \\ 0 & Z_{p_b} \end{bmatrix}, \quad Z_{pq} = \begin{bmatrix} Z_{p_a q} \\ 0 \end{bmatrix},$$

$$Z_{pr} = \begin{bmatrix} 0 \\ Z_{p_b r} \end{bmatrix},$$

and

$$Z_{qp} = (Z_{pq})^t, \quad Z_{rp} = (Z_{pr})^t$$

Superscript t indicates the transpose of a matrix, 0 denotes a null matrix of appropriate dimensions.

It may be noted that interconnection conditions 9.61 have not been used for writing eqn. 9.64 which represents a rearrangement of individual matrices Z_A and Z_B given in eqn. 9.62. Relations 9.61 can now be substituted in the eqns. 9.64 to eliminate V_p, V_q, i_q and i_r. The resulting expression may be written as $V_p = [Z_{AB}] i_p$, where

$$[Z_{AB}] = \begin{bmatrix} Z_{p_a} & 0 \\ 0 & Z_{p_b} \end{bmatrix} + \begin{bmatrix} Z_{p_a q} \\ -Z_{p_b r} \end{bmatrix} [Z_{qq} + Z_{rr}]^{-1} [-Z_{q p_a}, Z_{r p_b}] \tag{9.65}$$

Multiport network approach for modelling   491

It may be noted that the size of Z_{AB} is $(p_a + p_b) \times (p_a + p_b)$. The second term on the right-hand side is a product of three matrices of the sizes: $(p_a + p_b) \times q$, $q \times q$, and $q \times (p_a + p_b)$, respectively. From the computational point of view, the most time-consuming step is the evaluation of the inverse of a matrix of size $(q \times q)$, where q is the number of interconnected ports.

In order to illustrate the above procedure for combining Z-matrices of two segments together, let us consider an example of two lumped resistive networks connected together as shown in Fig. 9.26a. Z-matrices of the individual components A and B may be written as

$$Z_A = \begin{bmatrix} Z_{11} & | & Z_{13} & Z_{14} \\ \hline Z_{31} & | & Z_{33} & Z_{34} \\ Z_{41} & | & Z_{43} & Z_{44} \end{bmatrix} = \begin{bmatrix} 4 & | & 3 & 3 \\ \hline 3 & | & 7 & 3 \\ 3 & | & 3 & 5 \end{bmatrix} = \begin{bmatrix} Z_{p_a} & | & Z_{p_a q} \\ \hline Z_{q p_a} & | & Z_{qq} \end{bmatrix} \quad (9.66)$$

and

$$Z_B = \begin{bmatrix} Z_{22} & | & Z_{25} & Z_{26} \\ \hline Z_{25} & | & Z_{55} & Z_{56} \\ Z_{62} & | & Z_{65} & Z_{66} \end{bmatrix} = \begin{bmatrix} 5 & | & 3 & 3 \\ \hline 3 & | & 5 & 3 \\ 3 & | & 3 & 7 \end{bmatrix} = \begin{bmatrix} Z_{p_b} & | & Z_{p_b r} \\ \hline Z_{r p_b} & | & Z_{rr} \end{bmatrix} \quad (9.67)$$

In terms of the notations of eqn. 9.64, we have

$$Z_{pp} = \begin{bmatrix} Z_{11} & 0 \\ 0 & Z_{22} \end{bmatrix} = \begin{bmatrix} 4 & 0 \\ 0 & 5 \end{bmatrix}$$

$$Z_{pq} = \begin{bmatrix} Z_{p_a} \\ 0 \end{bmatrix} = \begin{bmatrix} Z_{13} & Z_{14} \\ 0 & 0 \end{bmatrix} = \begin{bmatrix} 3 & 3 \\ 0 & 0 \end{bmatrix}$$

$$Z_{pr} = \begin{bmatrix} 0 \\ Z_{p_b r} \end{bmatrix} = \begin{bmatrix} 0 & 0 \\ Z_{25} & Z_{26} \end{bmatrix} = \begin{bmatrix} 0 & 0 \\ 3 & 3 \end{bmatrix}$$

$$Z_{qq} = \begin{bmatrix} Z_{33} & Z_{34} \\ Z_{43} & Z_{44} \end{bmatrix} = \begin{bmatrix} 7 & 3 \\ 3 & 5 \end{bmatrix}$$

$$Z_{rr} = \begin{bmatrix} Z_{55} & Z_{56} \\ Z_{65} & Z_{66} \end{bmatrix} = \begin{bmatrix} 5 & 3 \\ 3 & 7 \end{bmatrix}$$

$$Z_{qp} = [Z_{q p_a}, 0] = \begin{bmatrix} Z_{31} & 0 \\ Z_{41} & 0 \end{bmatrix} = \begin{bmatrix} 3 & 0 \\ 3 & 0 \end{bmatrix}$$

$$Z_{rp} = [0, Z_{r p_b}] = \begin{bmatrix} 0 & Z_{52} \\ 0 & Z_{62} \end{bmatrix} = \begin{bmatrix} 0 & 3 \\ 0 & 3 \end{bmatrix}$$

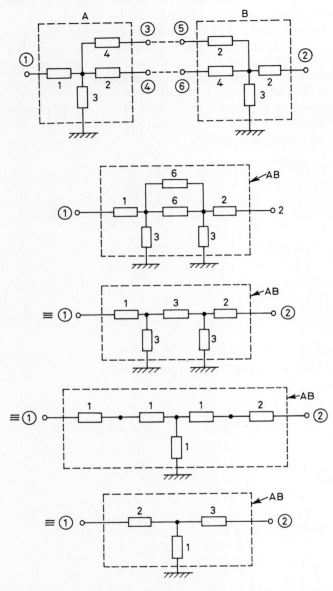

Fig. 9.26 (a) Two lumped networks considered for illustrating the segmentation procedure; (b) Circuit simplification for writing Z-matrix of the combination of the networks A and B

Multiport network approach for modelling   493

Substituting all these sub-matrices in eqn. 9.65 we get

$$[Z_{AB}] = \begin{bmatrix} 4, & 0 \\ 0, & 5 \end{bmatrix} + \begin{bmatrix} 3, & 3 \\ -3, & -3 \end{bmatrix} \begin{bmatrix} 7+5, & 3+3 \\ 3+3, & 5+7 \end{bmatrix}^{-1} \begin{bmatrix} -3, & 3 \\ -3, & 3 \end{bmatrix}$$

which may be evaluated as

$$[Z_{AB}] = \begin{bmatrix} 4, & 0 \\ 0, & 5 \end{bmatrix} + \begin{bmatrix} 3, & 3 \\ -3, & -3 \end{bmatrix} \frac{1}{18} \begin{bmatrix} 2, & -1 \\ -1, & 2 \end{bmatrix} \begin{bmatrix} -3, & 3 \\ -3, & 3 \end{bmatrix}$$

$$= \begin{bmatrix} 4, & 0 \\ 0, & 5 \end{bmatrix} + \begin{bmatrix} -1, & +1 \\ -1, & +1 \end{bmatrix} = \begin{bmatrix} 3 & 1 \\ 1 & 4 \end{bmatrix} \qquad (9.68)$$

The resultant matrix Z_{AB} in eqn. 9.68 may be verified by rewriting the circuit shown in Fig. 9.26a as the one shown in Fig. 9.26b.

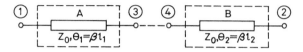

Fig. 9.27 Segmentation as applied to two sections of a transmission line

Let us consider another example. Two transmission-line sections of electrical lengths θ_1 and θ_2 are connected in cascade as shown in Fig. 9.27. Z-matrices of individual sections A and B are given by

$$Z_A = (-jZ_0/\sin\theta_1) \begin{bmatrix} \cos\theta_1 & 1 \\ 1 & \cos\theta_1 \end{bmatrix} \qquad (9.69)$$

and

$$Z_B = (-jZ_0/\sin\theta_2) \begin{bmatrix} \cos\theta_2 & 1 \\ 1 & \cos\theta_2 \end{bmatrix} \qquad (9.70)$$

In terms of notations of eqn. 9.65, we have

$$Z_{pp} = \begin{bmatrix} Z_{p_a} & 0 \\ 0 & Z_{p_b} \end{bmatrix} = \begin{bmatrix} Z_{11} & 0 \\ 0 & Z_{22} \end{bmatrix} = \begin{bmatrix} z_1\cos\theta_1 & 0 \\ 0 & z_2\cos\theta_2 \end{bmatrix}$$

$$Z_{pq} = \begin{bmatrix} Z_{p_a q} \\ 0 \end{bmatrix} = z_1 \begin{bmatrix} 1 \\ 0 \end{bmatrix}, \qquad z_1 = -jZ_0/\sin\theta_1$$

$$Z_{pr} = \begin{bmatrix} 0 \\ Z_{p_b r} \end{bmatrix} = z_2 \begin{bmatrix} 0 \\ 1 \end{bmatrix}, \qquad z_2 = -jZ_0/\sin\theta_2$$

$$Z_{qq} = Z_{33} = z_1 \cos \theta_1$$

$$Z_{rr} = Z_{44} = z_2 \cos \theta_2$$

$$Z_{qp} = [Z_{qp_a}, 0] = [z_1, 0]$$

$$Z_{rp} = [0, Z_{rp_b}] = [0, z_2]$$

Substituting all these submatrices in eqn. 9.65 we get

$$Z_{AB} = \begin{bmatrix} z_1 \cos \theta_1 & 0 \\ 0 & z_2 \cos \theta_2 \end{bmatrix} + \begin{bmatrix} z_1 \\ -z_2 \end{bmatrix} [z_1 \cos \theta_1 + z_2 \cos \theta_2]^{-1} [-z_1, z_2]$$

$$= \begin{bmatrix} z_1 \cos \theta_1 & 0 \\ 0 & z_2 \cos \theta_2 \end{bmatrix} + \frac{1}{(z_1 \cos \theta_1 + z_2 \cos \theta_2)} \begin{bmatrix} z_1 \\ -z_2 \end{bmatrix} [-z_1, z_2]$$

$$= \begin{bmatrix} z_1 \cos \theta_1 & 0 \\ 0 & z_2 \cos \theta_2 \end{bmatrix} + \frac{1}{(z_1 \cos \theta_1 + z_2 \cos \theta_2)} \begin{bmatrix} -z_1^2 & z_1 z_2 \\ z_1 z_2 & -z_2^2 \end{bmatrix}$$

(9.71a)

Substituting for z_1 and z_2 and using trigonometric formulas for $\sin(\theta_1 + \theta_2)$ and $\cos(\theta_1 + \theta_2)$, eqn. 9.71a may be expressed

$$Z_{AB} = -jZ_0/\sin(\theta_1 + \theta_2) \begin{bmatrix} \cos(\theta_1 + \theta_2) & 1 \\ 1 & \cos(\theta_1 + \theta_2) \end{bmatrix} \quad (9.71b)$$

which is a Z-matrix for a uniform transmission line of length $(\theta_1 + \theta_2)$ and illustrates the validity of eqn. 9.65.

When the segmentation method is applied to the multiport network model of microstrip antennas (such as the one shown in Fig. 9.8), we are interested in Z-matrix with respect to external ports (1 and 2 in Fig. 9.8) and also in the voltages at the ports connecting R–EAN to the patch. This voltage distribution at the radiating edges is expressed in terms of equivalent line source of magnetic current. The radiation field (and associated characteristics like beamwidth, SLL etc.) are obtained from the magnetic current distribution by using far-field ($1/r$ variation) term of eqn. 9.52 and integrating over the various radiating edges.

Referring to Fig. 9.25, voltages at the connected ports (q-ports) may be obtained by (Reference 1, p. 357)

$$V_q = \{Z_{qp} + Z_{qq}(Z_{qq} + Z_{rr})^{-1}(Z_{rp} - Z_{qp})\} i_p \quad (9.72)$$

where i_p is the current vector specifying the input current(s) at the external port(s) of the antenna.

9.5.2 Desegmentation method

There are several configurations of planar components which cannot be analysed by the segmentation method discussed above. For example, the configuration shown in Fig. 9.28 cannot be partitioned into regular segments for which Green's functions are known. In cases like this, an alternative method called desegmentation [4] is useful. The concept of desegmentation can be explained by considering the example of a rectangular patch with a circular hole. Referring to Fig. 9.28, we note that, if a circular disc called segment β (Fig. 9.28c) is added to the configuration of Fig. 9.28a, the resulting configuration γ

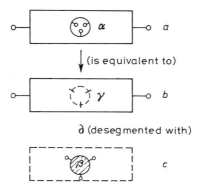

Fig. 9.28 *Concept of desegmentation*

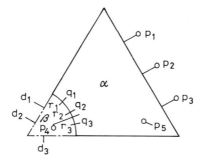

Fig. 9.29 *Port nomenclature used in desegmentation procedure*

is a rectangular segment shown in Fig. 9.28b. Green's functions are known for both the circular (Fig. 9.28c) and rectangular (Fig. 9.28b) shapes, and therefore Z-matrices for characterising both of these components may be derived. The desegmentation method allows us to derive the Z-matrix of the configuration α shown in Fig. 9.28a when the Z-matrices of the rectangular segment in (b) and the circular segment in (c) are known. For deriving a relationship among the Z-matrices of three shapes we consider a generalised configuration (shown in Fig. 9.29). Here, region β is not included in the α segment. Ports p_1, p_2 etc. are external ports of α. Characterisation of α is required with respect to these ports. In general p-ports may also be located on the part of the periphery of α where the segment β is connected. An example of this is port p_4 shown in Fig. 9.29. The

496 *Multiport network approach for modelling*

Z-matrices of β and γ segments are known, and may be written as

$$Z_\beta = \begin{bmatrix} Z_{rr} & Z_{rd} \\ Z_{dr} & Z_{dd\beta} \end{bmatrix} \quad \text{and} \quad Z_\gamma = \begin{bmatrix} Z_{pp\gamma} & Z_{pd} \\ Z_{dp} & Z_{dd\gamma} \end{bmatrix} \qquad (9.73)$$

As in the case of segmentation, ports q (of α) and ports r (of β) are numbered such that q_1 is connected to r_1, q_2 to r_2, etc. Ports d are unconnected (external) port of the segment β. Evaluation of Z_α is simplified when the number of d-ports is made equal to the number of q (or r) ports. The number of q (or r) ports depends upon the nature of field variation along α–β interface and, as in the case of segmentation, is decided by iterative computations. On the other hand, the number of d-ports is arbitrary and can be always made equal to that of q- (or r-) ports after that number has been finalised. Under these conditions, the impedance matrix for the α-segment can be expressed (Reference 5) in terms of the Z-matrices of β- and γ-segments as

$$Z_\alpha = Z_{pp\gamma} - Z_{pd}\{Z_{dd\gamma} - Z_{dd\beta}\}^{-1} Z_{dp} \qquad (9.74)$$

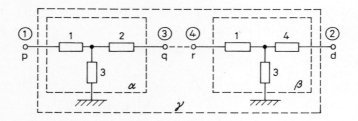

Fig. 9.30 *A resistive network considered for illustrating the desegmentation procedure*

It may be noted that the size of Z_α is $(p \times p)$ since all the specified ports of α segment have been numbered as p-ports. Let us consider an example for illustrating the implementation of eqn. 9.74. Consider the resistive network shown in Fig. 9.30. Let us say that sub-network α is the one whose Z-matrix is to be determined, the Z-matrix of the β segment is known, and the combination of these two is the γ-segment whose Z-matrix is also known. We have

$$Z_\beta = \begin{bmatrix} 4 & 3 \\ 3 & 7 \end{bmatrix} = \begin{bmatrix} Z_{rr} & Z_{dr} \\ Z_{dr} & Z_{dd\beta} \end{bmatrix} \qquad (9.75)$$

If the characterisation of α is to be found with respect to port 1 only, network γ can be considered as a 2-port network with one p-port and one d-port. The Z-matrix of γ is obtained by re-writing the network as in Fig. 9.31. We have

$$Z_\gamma = \begin{bmatrix} 3 & 1 \\ 1 & 6 \end{bmatrix} = \begin{bmatrix} Z_{pp\gamma} & Z_{pd} \\ Z_{dp} & Z_{dd\gamma} \end{bmatrix} \qquad (9.76)$$

Using eqn. 9.74 we get Z_α as

$$Z_\alpha = Z_{pp\gamma} - (Z_{pd}\{Z_{dd\gamma} - Z_{dd\beta}\}^{-1} Z_{dp})$$
$$= 3 - 1\{6 - 7\}^{-1} 1 = 4 \qquad (9.77)$$

which may be verified to the correct value. If the characterisation of the α-network is needed with respect to ports 1 and 3, it becomes necessary to include port 3 in the γ-network also. For this purpose the γ-network is re-written as in Fig. 9.32. Now Z_γ becomes

$$Z_\gamma = \begin{bmatrix} 3 & 4/3 & \vdots & 5 \\ 4/3 & 20/9 & \vdots & 5/3 \\ \hdashline 1 & 5/3 & \vdots & 6 \end{bmatrix} = \begin{bmatrix} Z_{pp\gamma} & \vdots & Z_{pd} \\ \hdashline Z_{dp} & \vdots & Z_{dd\gamma} \end{bmatrix} \qquad (9.78)$$

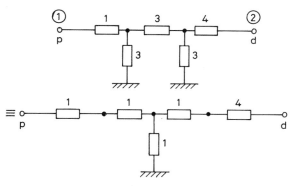

Fig. 9.31 *Reconfiguration of the network in Fig. 9.30 for evaluation of the matrix Z_γ used in desegmentation procedure*

Now Z_α is obtained by using eqn. 9.74 as follows:

$$Z_\alpha = Z_{pp\gamma} - Z_{pd}\{Z_{dd\gamma} - Z_{dd\beta}\}^{-1} Z_{dp}$$
$$= \begin{bmatrix} 3 & 4/3 \\ 4/3 & 20/9 \end{bmatrix} - \begin{bmatrix} 1 \\ 5/3 \end{bmatrix} \{6 - 7\}^{-1} \{1 \; 5/3\}$$
$$= \begin{bmatrix} 3 & 4/3 \\ 4/3 & 20/9 \end{bmatrix} + \begin{bmatrix} 1 & 5/3 \\ 5/3 & 25/9 \end{bmatrix} = \begin{bmatrix} 4 & 3 \\ 3 & 5 \end{bmatrix} \qquad (9.79)$$

which again may be verified to be the correct result.

Let us also consider another example of two transmission-line sections of electrical lengths βl_1 and βl_2 connected in cascade as shown in Fig. 9.27. Say we want to find Z_A when Z_B and Z_{AB} are known. When we need characterisation of A only with respect to port 1, the segment γ is a two-port network with Z_γ

as

$$Z_\gamma = \{-jZ_0/\sin(\theta_1 + \theta_2)\} \left[\begin{array}{c|c} \cos(\theta_1 + \theta_2) & 1 \\ \hline 1 & \cos(\theta_1 + \theta_2) \end{array}\right]$$

$$= \left[\begin{array}{c|c} Z_{pp\gamma} & Z_{pd} \\ \hline Z_{dp} & Z_{dd\gamma} \end{array}\right] \qquad (9.80)$$

and

$$Z_\beta = \left[\begin{array}{c|c} Z_{rr} & Z_{rd} \\ \hline Z_{dr} & Z_{dd\beta} \end{array}\right] = \left[\begin{array}{cc} \cos\theta_2 & 1 \\ 1 & \cos\theta_2 \end{array}\right] \{-jZ_0/\sin\theta_2\} \qquad (9.81)$$

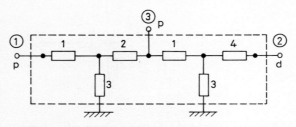

Fig. 9.32 Modification of the γ-network for two-port characterisation of the α-segment in desegmentation procedure

The Z-matrix (1 port) of segment A may now be written using eqn. 9.74 as

$$\begin{aligned}
Z_\alpha &= Z_{pp\gamma} - Z_{pd}\{Z_{dd\gamma} - Z_{dd\beta}\}^{-1} Z_{dp} \\
&= jZ_0 \cot(\theta_1 + \theta_2) - \left(\frac{-jZ_0}{\sin(\theta_1 + \theta_2)}\right)\{-jZ_0 \cot(\theta_1 + \theta_2) \\
&\quad + jZ_0 \cot\theta_2\}^{-1} \left(\frac{-jZ_0}{\sin(\theta_1 + \theta_2)}\right) \\
&= -jZ_0 \cot(\theta_1 + \theta_2) - \left(\frac{(-jZ_0)^2}{\sin^2(\theta_1 + \theta_2)}\right) \\
&\quad \times \left(\frac{1}{-jZ_0[\cot(\theta_1 + \theta_2) - \cot\theta_2]}\right) = -jZ_0 \\
&\quad \times \left\{\cot(\theta_1 + \theta_2) - \frac{1}{\sin^2(\theta_1 + \theta_2)[\cot(\theta_1 + \theta_2) - \cot\theta_2]}\right\}
\end{aligned}$$

(9.82)

This can be simplified to

$$Z_\alpha = -jZ_0 \cot \theta_1 \qquad (9.83)$$

These two examples illustrate the applications of the desegmentation eqn. 9.74.

It may be noted that, for implementing the desegmentation method, d-ports of the β-segment need not be located on the periphery of the β-segment. In fact, in the case of the rectangular patch with a circular hole (Fig. 9.28a), no region is available on the periphery of the circular β-segment for locating d-ports. As shown in Fig. 9.33, d-ports may be located inside the circular region. In this case three d-ports d_1, d_2, d_3 are shown located inside the β-segment. Since the Green's functions are valid for any point on the segment (or on the periphery), the desegmentation procedure remains unchanged.

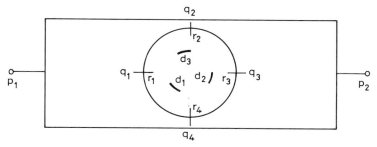

Fig. 9.33 *A configuration where the desegmentation procedure requires location of d-ports inside the β-segment*

9.6 Examples of microstrip antenna structures analysed by multiport-network approach

The multiport-network modelling and analysis approach discussed above has been used for the analysis, design and optimisation of a variety of microstrip radiators [8, 10–18]. We will discuss these applications in three groups: (i) circularly polarised microstrip patches; (ii) broad-band multiresonator microstrip antennas; and (iii) multiport microstrip antennas and series-fed arrays.

9.6.1 Circularly polarised microstrip patches
Single-feed circularly polarised microstrip patches analysed by multiport-network approach include: diagonal-fed nearly square patch [10], truncated-corners square patch [10], square patch with a diagonal slot [10], a pentagonal-shaped patch [8], square ring patch [17], and cross-shaped patch [17]. The desegmentation method has been used for analysing a truncated-corners square patch and a square patch with a diagonal slot. Implementation of the desegmentation procedure in these two cases is illustrated in Fig. 9.34. For a truncated-corners patch, the α-segment (c.f. Fig. 9.29) is shown in Fig. 9.34a. For

500 Multiport network approach for modelling

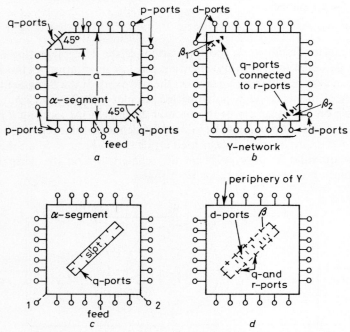

Fig. 9.34 *(a) and (b) Desegmentation method applied to a corners-truncated antenna; (c) and (d) desegmentation method applied to a square antenna with a diagonal slot*

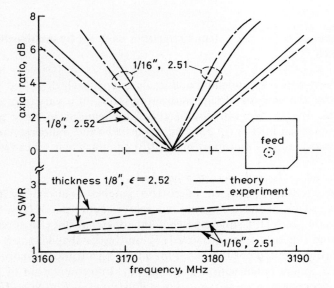

Fig. 9.35 *Theoretical and experimental results for axial ratio and input VSWR for a truncated-corner square antenna (Reproduced from Reference 10 © IEEE 1983)*

implementation of the desegmentation procedure in this case, we use two triangular-shaped β-segments β_1 and β_2. Addition of β_1 and β_2 to α-segments results in a perfect square-shaped γ-segment shown in Fig. 9.34b. Configuration of the α-segment for a square with a diagonal slot is shown in Fig. 9.34c. In this case the β-segment is a rectangle of the size of the slot, and d-ports are located inside the rectangle as shown in Fig. 9.34d (this configuration may be compared with Fig. 9.33). Detailed results for these two configurations are given in

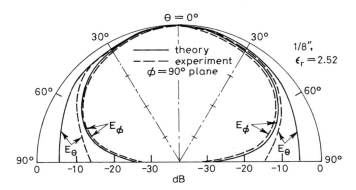

Fig. 9.36 *Radiation pattern for circularly polarized truncated-corners square antenna (Reproduced from Reference 10 © IEEE 1983)*
Thickness: 1/8 in; ε_r = 2·52; frequency 3·176 GHz

Reference 10. Results based on multiport-network analysis have been compared [10] with experimental results. The type of agreement observed for the case of a truncated-corner square antenna is illustrated in Figs. 9.35 and 9.36. Theoretical and experimental values of VSWRs and axial ratios are plotted in Fig. 9.35, whereas radiation patterns are compared in Fig. 9.36. A more quantitative comparison of the theoretical and experimental performance is contained in Table 9.2 (from Reference 10). Reasonable agreement between the theoretical and experimental results verifies the validity of the multiport-network approach. Corresponding results for a square antenna with a diagonal slot are depicted in Figs. 9.37 and 9.38 and in Table 9.3 (again from Reference 10). The multiport-network approach is seen to perform equally well in this case also.

Square-ring patch and cross-shaped patches have been analysed [17] by the segmentation method. Segmentation of a square ring in four rectangular segments is shown in Fig. 9.24. Segmentation of a cross-shaped patch in three rectangular segments is illustrated in Fig. 9.39. Detailed results for these two configurations are given in Reference 17. Comparison for five different shapes (square ring, crossed strip, almost square patch, corner-chopped square patch, and square with a diagonal slot) indicates that the maximum axial-ratio bandwidth (about 5·2% for f = 3·0 GHz, ε_r = 2·5, and h = 0·159 cm) is obtained by using a square-ring configuration.

Table 9.2 Performance of corners-chopped square patch antennas (Reproduced from Reference 10 © IEEE 1983)

I	Parameters	Antenna I		Antenna II	
	1 Thickness, ε_r	1/8", 2·52		1/16", 2·51	
	2 Dimensions $a \times a\,\text{cm}^2$ } (see Fig. 9.34)	2·73 × 2·73		2·86 × 2·86	
	3 Truncation b/a	0·04578		0·0573	

II	Performance	Theoretical	Experimental	Theoretical	Experimental
	1 Centre frequency f_c (GHz)	3·1758	3·1750	3·1756	3·1753
	2 Resonant frequencies of orthogonal modes (GHz)	3·1340	3·1325	3·1370	3·1343
		3·2155	3·2125	3·2340	3·2298
	3 Axial ratio at centre frequency (dB)	0·02	0·0	0·12	0·15
	4 Bandwidth (MHz) for axial ratio <6dB	26·4 (0·831%)	29·4 (0·925%)	14·0 (0·44%)	14·4 (0·4535%)
	5 Input VSWR at centre frequency	2·26	2·26	1·6	1·8
	6 Beamwidth for 3 dB difference between $|E_\theta|$ and $|E_\phi|$	129°	152°	129°	138°

Table 9.3 *Performance of square-patch antenna with a diagonal slot (Reproduced from Reference 10 © IEEE 1983)*

	Theoretical	Experimental
1 Centre frequency f_c (GHz)	3·130	3·130
2 Resonance frequency of orthogonal modes (GHz)	3·063	3·060
	3·212	3·210
3 Axial ratio at f_c	0·198	0·2
4 Bandwidth for axial ratio less than 6 dB	35·5 MHz	38·0 MHz
	(1·134%)	(1·214%)
5 Input VSWR at chosen feed location	2·9	2·9
6 Beamwidth for 3 dB difference between $|E_\theta|$ and $|E_\phi|$	116°	124°

Substrate thickness = 1/8 in, ε_r = 2·52. Dimensions of square patch = 2·602 cm × 2·602 cm
Dimensions of slot = 2·89 cm × 0·47 cm.

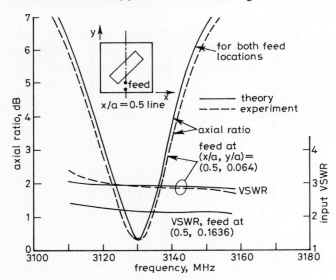

Fig. 9.37 *Theoretical and experimental results for a square antenna with a diagonal slot (Reproduced from Reference 10 © IEEE 1983)*
Thickness = 1/8 in; ε_r = 2·52; frequency = 3·130 GHz (Reproduced from Reference 10 © IEEE 1983)

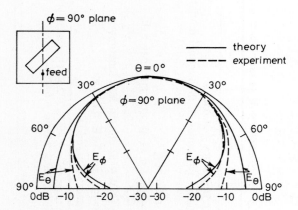

Fig. 9.38 *Radiation pattern for a circularly polarised square antenna with a diagonal slot*
Thickness = 1/8 in; ε_r = 2·52; frequency = 3·130 GHz (Reproduced from Reference 10 © IEEE 1983)

Another interesting antenna configuration analysed by the multiport-network approach is a pentagonal-shaped patch originally proposed in References 55 and 28, and analysed in Reference 8. The antenna configuration is shown in Fig. 9.40a and two different methods of analysis are depicted in Fig. 9.40b and c. Desegmentation with two triangular segments β_1 and β_2 yields a 90°-60°-30°

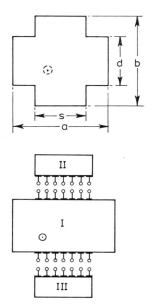

Fig. 9.39 *Segmentation of a cross-shaped patch into three rectangular segments*

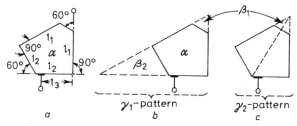

Fig. 9.40 *(a) A pentagonal-shaped microstrip patch*
(b) Desegmentation procedure applied to the pentagonal shape
(c) Combination of segmentation and desegmentation procedure for the pentagonal patch

triangular segment for which Green's function is available. The second approach illustrated in Fig. 9.40c employs desegmentation with one segment β_1 to yield a kite-shaped geometry. The kite shape is then segmented into two identical 90°-60°-30° triangles as shown. Both of these approaches yield identical results.

506 Multiport network approach for modelling

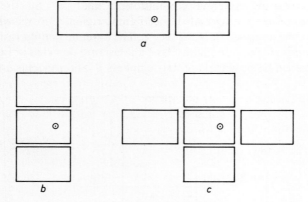

Fig. 9.41 *(a) Configuration of a radiating-edge gap-coupled microstrip antenna (REGCOMA)*
(b) Configuration of a non-radiating-edge gap-coupled microstrip patch antenna (NEGCOMA)
(c) Four-edges gap-coupled microstrip patch antenna (FEGCOMA)

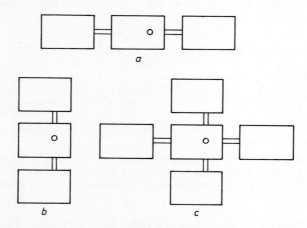

Fig. 9.42 *(a) Configuration of a radiating-edge directly coupled microstrip patch antenna (REDCOMA)*
(b) Configuraton of a non-radiating-edge directly coupled microstrip patch antenna (NEDCOMA)
(c) Configuration of a four-edge directly coupled microstrip antenna (FEDCOMA)

9.6.2 Broadband multiresonator microstrip antennas

Another group of microstrip antenna configurations which have been analysed using the multiport-network approach are broadband microstrip antennas using coupled resonators [11-13]. All these configurations use multiple resonators with slightly different resonant frequencies. These configurations are shown in Figs. 9.41 and 9.42. Different resonators in any of these configurations are coupled to each other, and only one (usually the central one) is connected to the feedline. Two different coupling mechanisms have been used. The three configurations shown in Fig. 9.41 use capacitive coupling across the gaps between the closely spaced edges, whereas the three configurations of Fig. 9.42 employ short sections of microstrip lines for providing the necessary coupling.

Analysis procedure for a gap-coupled multiresonator antenna configuration is illustrated in Fig. 9.43. Coupling gaps are modelled by the multiport lumped *RC* network shown in Fig. 9.43*b*. Values of C_1, C_2 and C_g are obtained from coupled microstrip transmission-line analysis. *G* represents the radiation conductance and is obtained by treating the gap fields as a line source of equivalent magnetic current. Since the feed point is located along the centre line XX (Fig. 9.43*a*), symmetry of the configuration may be used to simplify the computations and only one-half of the antenna configuration, shown in Fig. 9.43*c*, need be analysed. The multiport-network model is shown in Fig. 9.43*d*. RCs represent edge-admittance networks and GCs are two networks modelling the coupling gaps. Mutual coupling networks are not shown in this Figure because the effect of mutual coupling was not incorporated in the results presented in Reference 11. Component REs are planar network models for the three patches. Ports 1, 2, and 3 on the central patch are the three locations investigated for locating the probe feed. Experiments were performed to verify the theoretical results obtained by using the multiport-network approach, and a comparison is shown in Fig. 9.44. In this case, a 0·159 cm-thick copper-clad substrate ($\varepsilon_r = 2\cdot55$) was used. The experimental bandwidth of the antenna is 225 MHz (6·9% at centre frequency $f_0 = 3\cdot27\,\text{GHz}$), which is slightly more than the theoretical value (207 MHz), possibly because of the dielectric, conductor and surface-wave losses ignored in the computations of the results reported in Reference 11. For comparison, the corresponding bandwidth of a single patch is 65 MHz. Thus the microstrip antenna configuration shown in Fig. 9.44 yields a bandwidth nearly 3·5 times that of a single patch.

A multiport-network for a directly coupled three-resonator antenna configuration is shown in Fig. 9.45. In this case also, one can make use of geometrical symmetry, and only one-half of the antenna configuration (with a magnetic wall placed along the plane XX) needs to be analysed. The multiport-network model is drawn in 9.45*c*. Interconnecting microstrip line sections are also modelled by two planar rectangular segments RE_L. We have nine edge-admittance networks denoted by RCs. The segmentation formula 9.65 is used for finding the input impedance at the feed port and eqn 9·72 for evaluating the voltage distribution at the edges of the radiating patches.

Fairly wide impedance bandwidth for microstrip antennas can be achieved by using the multiple resonator configurations shown in Figs. 9.41 and 9.42. Typical values for the six configurations fabricated on substrates with $\varepsilon_r = 2\cdot55$

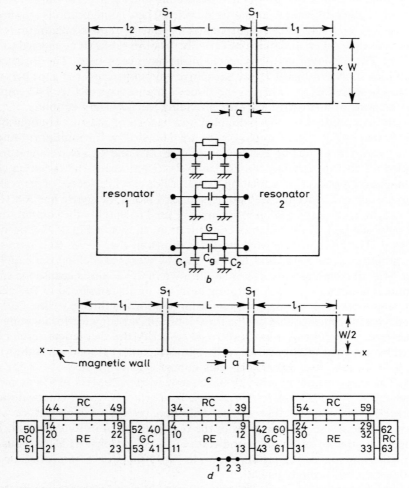

Fig. 9.43 (a) Three-patch gap-coupled antenna configuration
(b) Multiport-network modelling of the gap between two patches
(c) Half-section of the antenna configuration with a magnetic wall along the plane of symmetry XX
(d) Multiport network model of the half-section of the antenna

are summarised in Table 9.4. Various acronyms (REGCOMA etc.) are defined in Figs. 9.41 and 9.42. The factor M gives the bandwidth BW as a multiple of the corresponding value for a single rectangular patch antenna; f_0 is the centre frequency and d is the thickness of the substrate.

The concept of the non-radiating-edge gap-coupled microstrip antenna is ideally suited for overcoming the narrow bandwidth limitation of microstrip antenna. An extension of this concept called the multiple-coupled microstrip

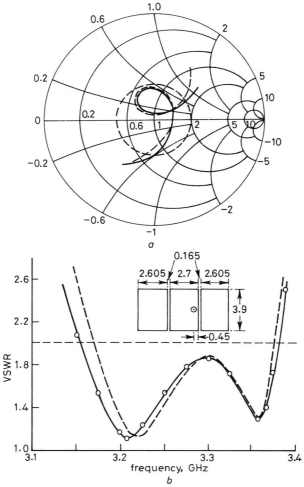

Fig. 9.44 *Theoretical (- - - -) and experimental (———) results for a gap-coupled triple-resonator antenna (Reproduced from Reference 11 © IEEE 1984)*
(a) Impedance locus on Smith Chart
(b) VSWR variations.

antenna (configuration shown in Fig. 9.46) is being currently developed as a compact broadband microstrip patch [57].

9.6.3 Multiport microstrip patches and series-fed arrays
Series-fed linear arrays of microstrip patches employ two-port radiators as basic

510 Multiport network approach for modelling

Table 9.4 *Typical impedance bandwidth values for microstrip antennas using multiple coupled resonators (Based on Reference 56)*

Configuration	d(cm)	f(GHz)	BW(MHz)	BW(%)	M
REGCOMA	0·159	3·29	331	10·0	5·3
NEGCOMA	0·318	3·11	480	15·4	4·0
FEGCOMA	0·318	3·16	815	25·8	6·7
REDCOMA	0·318	3·20	548	17·1	5·0
NEDCOMA	0·318	3·31	605	18·3	5·5
FEDCOMA	0·318	3·38	810	24·0	7·36

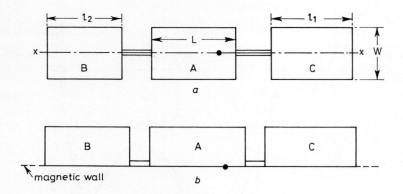

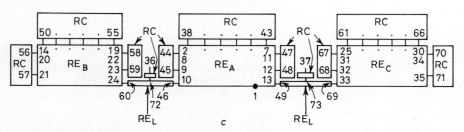

Fig. 9.45 *(a) Three-patch antenna configuration with direct (microstrip-line) coupling between the patches*
(b) Half-section of the antenna configuration in (a) with a magnetic wall along the plane of symmetry XX
(c) Multiport-network model of the antenna half-section shown in (b)

building blocks. For this application, both two-port rectangular patches [14, 29] as well as two-port circular patches [15] have been analysed by using the multiport-network modelling approach.

Two-port rectangular patch: The multiport-network model of a rectangular patch with two microstrip-line ports along the non-radiating edges is shown in Fig. 9.8. Segments labelled FLN (feed line network) are rectangular planar segments representing small sections (typically $\lambda/8$ long) of microstrip lines connected to the two ports. Widths of FLNs are equal to the effective widths of

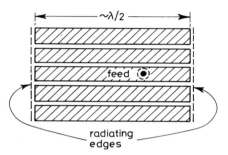

Fig. 9.46 *Configuration of a broadband coupled microstrip line radiating patch*

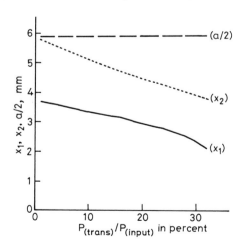

Fig. 9.47 *Variations of the power transmitted to the port 2 (as a percentage of the input power) with the changes in the locations of the two ports (Reproduced from Reference 27)*

the two lines, respectively. Multiple interconnections between FLNs and the patch ensure that the parasitic reactances associated with the feed-line-patch junctions are taken into account. Radiating edge admittance networks (R-EAN)

512 Multiport network approach for modelling

and non-radiating edge-admittance networks (NR-EAN) are obtained by modelling fringing fields at the edges as discussed in Section 9.4.1. The mutual coupling network (MCN) represents the external interaction between two radiating edges as discussed in Section 9.4.2.

As mentioned earlier, for two-port patches with ports along the non-radiating edges, transmission from port 1 to port 2 can be controlled by suitable location of the ports (distance x_1 and x_2 in Fig. 9.2b). An example of this feature is presented in Fig. 9.47. This Figure shows the variation of the power transmitted to port 2 with the relative locations of external ports. Values of x_1 are chosen to ensure match at the input port ($S_{11} \simeq 0$). When the port locations are altered, the associated change in the junction reactances causes the patch resonance frequency to shift slightly. The corresponding change in the resonant dimension a is also plotted in this Figure. The results shown are for a substrate with $\varepsilon_r = 2.48, d = 1/32$ in, $\tan\delta = 0.002$ and for a resonant frequency of 7.5 GHz. A comparison of theoretical and experimental results for S_{21} of a two-port patch is shown in Fig. 9.48. Design parameters of the two-port patch are also listed in this Figure. Apart from the magnitude of S_{21} which determines the amplitude

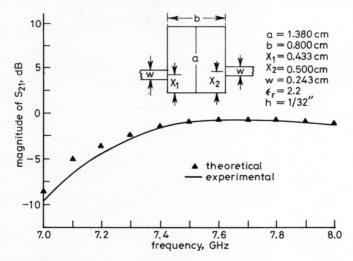

Fig. 9.48 *Comparison of the theoretical and experimental results for transmission coefficient of a two-port rectangular patch (Reproduced from Reference 29)*

variation along the linear series-fed array, another parameter of interest in the two-port patches is the phase angle of the transmission coefficient from port 1 to port 2. Theoretical computations of the phase angle of S_{21} (based on the multiport-network approach) have been compared with experimental values obtained from measurements using an automatic network analyser. Fig. 9.49 shows these results. The excellent agreement obtained demonstrates the usefulness of multiport-network model for computations of the phase angles also.

Thus we conclude that the multiport-network model and the analysis approach discussed here are well suited for S-parameter characterisation of the radiating patches.

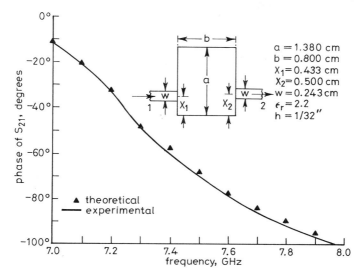

Fig. 9.49 *Comparison of the theoretical and experimental values of the transmisson phase angle for a two-port rectangular patch (Reproduced from Reference 29)*

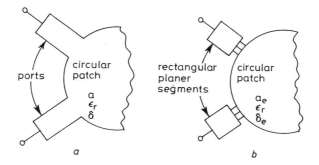

Fig. 9.50 *Analysis of a circular two-port microstrip patch antenna by combining multiport network-modelling approach with the cavity method*

Two-port circular patches: Circular microstrip patches with two ports located along the circumference have been analysed [15] by using a cavity model for the circular patch and a multiport modelling approach for the input/output microstrip feed-line junction. This approach is illustrated in Fig. 9.50. The physical radius a of the disc and its loss tangent δ are replaced by effective values a_e and δ_e. The effective radius a takes into account the fringing capacitance around the circumference [45]. The effective loss tangent δ_e includes 'loss' due to the power radiated from the patch. Relation 9.5 or 9.6 may be used for this purpose. Power radiated P_r, power dissipated in the dielectric P_d and the power lost because of finite conductor conductivity P_c may be evaluated as illustrated in Reference 46, pp. 92–94.

Approximate results [15], using the dominant mode only and ignoring the feed-junction reactances, point out that, for a match at the input port ($S_{11} = 0$), the impedance Z_0 of the feed line at the input port is related to the Z_{11} element of the Z-matrix by

$$Z_0 = Z_{11} \sin(\phi_{12}) \tag{9.84}$$

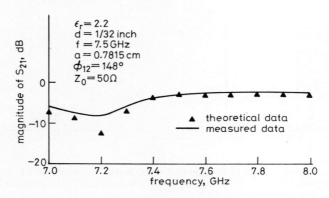

Fig. 9.51 *Comparison of the theoretical and experimental results for transmission coefficient of a two-port circular patch (Reproduced from Reference 58)*

where ϕ_{12} is the angular separation between the two ports and

$$Z_{11} = 1\cdot 674 \left(\frac{d}{\lambda_0}\right) \eta_0/\delta_e \tag{9.85}$$

where d is the substrate thickness. The corresponding transmission coefficient S_{21} is given by

$$S_{21} = \cos\phi_{12}/(1 + \sin\phi_{12}) \tag{9.86}$$

Eqn. 9.86 suggests that, for the dominant mode, the transmission coefficient S_{21} varies from 1 to 0 as the angular separation ϕ_{12} is changed from 0 to 90°. Also we note that, for high values of S_{21} close to unity, the input-port characteristic impedance becomes very small and makes the design impracticable.

Results based on the above method have been verified experimentally and a sample comaprison of theoretical and measured S_{21} values is presented in Fig. 9.51. Also, a comparison of the transmission phase in this case is shown in Fig. 9.52. Again, a reasonably good agreement is obtained. It may be recalled that, for this example of a two-port circular patch, the method of analysis followed is not exactly the multiport-network approach discussed in Section 9.2.3 but is a hybrid combination of the conventional cavity method (Section 9.2.2) and the multiport-analysis technique. The multiport-network approach itself can, in principle, be applied to circular patches also, but no such efforts have been reported to date.

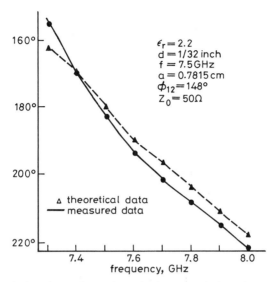

Fig. 9.52 *Theoretical and experimental results for transmission phase angle of a two-port circular patch (Reproduced from Reference 58)*

Series-fed microstrip arrays: The two-port rectangular or circular patches discussed above may be cascaded together to form a series-fed linear array as shown schematically in Fig. 9.53a and b. The multiport-network modelling approach has been used [18] for design and sensitivity analysis of series-fed arrays. Each unit cell of the array is characterised in terms of a 2-port Z-matrix. A two-port representation of a typical unit cell of a series-fed array is shown in Fig. 9.54. This is a simplification of the multiport-network representation shown in Fig. 9.8. Edge-admittance networks associated with non-radiating edges are not shown. Multiple interconnections between various components are shown symbolically by a single multiconnection symbol defined in the inset of the Figure. Ports 1 and 2 are two external ports of the unit cell. $e1$ and $e2$ are the two radiating edges. Typically, there are four interconnections (at each of these edges) among the R-EAN, the patch, and the mutual coupling network (MCN). Analysis based on the segmentation procedure allows us to determine voltages (or currents) at the input/output ports of each of the cells of the array. This information is used in conjunction with another set of impedance matrices (called Z_v matrices) for each cell. The Z_v matrices (obtained from eqn. 9.72) relate the voltages at various ports on radiating edges $e1$ and $e2$ to the input currents at the input/output ports 1 and 2 of each cell. Thus we have a multiport-network representation of the series-fed array. This representation may be extended to include MCNs (mutual coupling networks) representing the coupling between adjacent cells. In this case, the two-port representation of each cell

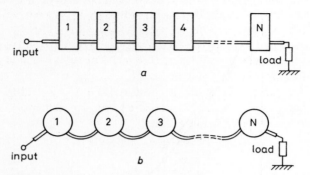

Fig. 9.53 (a) Series-fed linear array of rectangular patches
(b) Series-fed linear array of circular patches

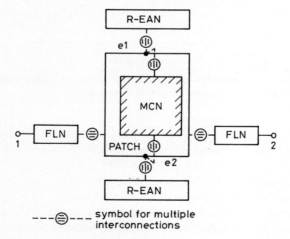

Fig. 9.54 Multiport network representation of a typical unit cell of a series-fed linear array

needs to be extended to multiple n-port representation, the additional $(n - 2)$ ports being the fictitious ports to account for the mutual coupling. A schematic network representation of such a unit cell of a series-fed array is shown in Fig. 9.55. The network MCN2 denotes a multiport mutual coupling network representing the coupling between the two adjacent patches (i.e. between the patch shown and the next patch on the right-hand side). In the model shown, the coupling to the third patch (on either side) has been ignored. The n-port network shown here represents a unit cell of the series-fed array when a non-negligible mutual coupling is present between the adjacent cells. Such a network representation can be used to evaluate the effect of mutual coupling on the array performance and to iteratively modify the array geometry to compensate (as far as possible) for the undesirable effects of the mutual coupling.

For analysing multiport-network models of large arrays, a circuit analysis technique called the sub-network growth method [1] is very convenient. In this method, only two adjacent components are combined together (at any stage in the iterative loop) to form a larger sub-network. Consequently, the size of the matrices to be processed is restricted to the number of ports in the two components and does not increase proportionately to the size of the array.

At present, research efforts are in progress for using multiport-network modelling techniques for the automated design of one-dimensional and two-dimensional arrays of microstrip patches.

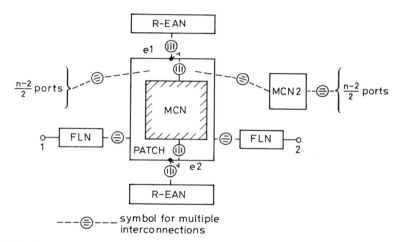

Fig. 9.55 *Incorporation of mutual coupling between adjacent patches in the network representation of a unit cell shown in Fig. 9.54*

9.7 CAD of microstrip patch antennas and arrays

Increasing interest in the use of microstrip antenna technology in phased-array systems, and the potential of fabricating millimeter-wave arrays monolithically on GaAs wafers, have made it necessary to develop CAD techniques for the accurate design of microstrip patches and arrays. The multiport-network approach presented in this Chapter is well suited for CAD (i.e. for modelling, analysis and optimisation) of microstrip patch antennas and arrays [20].

Basic aspects of CAD methodology are well known [1] and are common to the computer-aided-design process in various other disciplines. A generic flow chart for CAD is shown in Fig. 9.56. Starting with a given set of specifications, synthesis methods and available designs (pre-stored in computer) help us to arrive at the initial design. A model of this initial design is analysed by a computer-aided-analysis (simulation and performance evaluation) package.

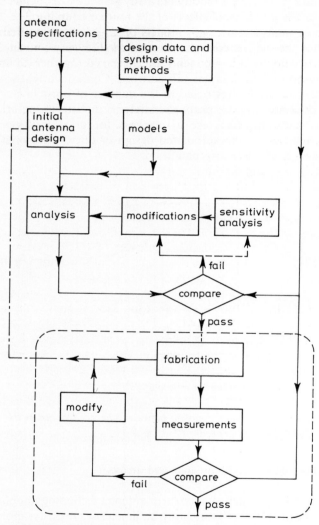

Fig. 9.56 *Typical flow chart for computer-aided design methodology for microstrip antennas*

The performance characteristics obtained are compared with the given specifications. When the specifications are not met, the designable parameters of the antenna configuration are modified and the analysis is repeated. These analysis, modification and comparison steps constitute a single iteration in the optimisation loop. Several optimisation strategies include sensitivity analysis of the design for calculating the changes in the designable parameters. Iterations in the optimisation loop are carried out until the specifications are met or the optimum performance of the design (within the given constraints) is achieved. The antenna design so obtained is now fabricated and experimental measurements are carried out. As indicated in the lower part of Fig. 9.56 (portion inside the dashed rectangle), some modifications may still be necessary if the modelling and/or analysis has not been sufficiently accurate. The modifications, hopefully, should be small and the aim of CAD is to minimise these experimental iterations as far as is practicable.

The three main aspects of the computer-aided-design process are modelling, analysis and optimisation. Modelling and analysis approaches suitable for CAD of microstrip patches and arrays have been discussed in this Chapter. The most attractive features of the multiple-network approach is the extension of the network-analysis methods to microstrip patches and antennas. Also, the techniques for sensitivity analysis and optimisation, which have been developed extensively for multiport networks [1], can now be extended to network models of microstrip patches and arrays. Use of gradient optimisation techniques involves calculation of gradients of antenna performance with respect to various designable parameters. The adjoint network method of sensitivity analysis has been used extensively for calculating gradients for circuit optimisation. This method can now be used for sensitivity analysis of microstrip patch antenna configurations also. Since the network models involved are passive and reciprocal, the adjoint network is identical to the original network model itself, and thus a single network analysis is sufficient to yield the sensitivity information also. Among other things, this would yield the sensitivity of the voltage distribution along the radiating edges with respect to the various designable parameters of the antenna configuration. Sensitivity of the radiation characteristics with respect to the antenna parameters can be calculated therefrom.

It is expected that the techniques reported in this Chapter will lead to implementation of accurate CAD procedures for microstrip patches and arrays.

9.8 Appendix: Green's functions for various planar configurations

In this Section Green's functions for some planar shapes, shown in Fig. 9.9, are given. In the expressions that follow, σ_i is given by

$$\sigma_i \stackrel{\Delta}{=} \begin{cases} 1, & \text{if } i = 0 \\ 2, & \text{otherwise} \end{cases} \tag{A9.1}$$

(a) *A rectangle:* The Green's function for the rectangle shown in Fig. 9.9a is given as [2]

$$G(x, y | x_0, y_0) = \frac{j\omega\mu d}{ab} \sum_{n=0}^{\infty} \sum_{m=0}^{\infty} \frac{\sigma_m \sigma_n \cos(k_x x_0) \cos(k_y y_0) \cos(k_x x) \cos(k_y y)}{k_x^2 + k_y^2 - k^2} \tag{A9.2}$$

where

$$k_x = \frac{m\pi}{a} \quad \text{and} \quad k_y = \frac{n\pi}{b}$$

(b) A 30°–60° right-angled triangle: The Green's functions for the triangle shown in Fig. 9.9b is given as [34]:

$$G(x, y|x_0, y_0) = 8j\omega\mu d \sum_{m=-\infty}^{\infty} \sum_{n=-\infty}^{\infty}$$

$$\frac{T_1(x_0, y_0)T_1(x, y)}{16\sqrt{3}\pi^2(m^2 + mn + n^2) - 9\sqrt{3}a^2k^2} \tag{A9.3}$$

where

$$T_1(x, y) = (-1)^l \cos\left(\frac{2\pi lx}{\sqrt{3}a}\right) \cos\left[\frac{2\pi(m-n)y}{3a}\right]$$

$$+ (-1)^m \cos\left(\frac{2\pi mx}{\sqrt{3}a}\right) \cos\left[\frac{2\pi(n-l)y}{3a}\right]$$

$$+ (-1)^n \cos\left(\frac{2\pi nx}{\sqrt{3}a}\right) \cos\left[\frac{2\pi(l-m)y}{3a}\right] \tag{A9.4}$$

with the condition that

$$l = -(m+n) \tag{A9.5}$$

(c) An equilateral triangle: The Green's function for the equilateral triangle shown in Fig. 9.9c is given as [34]

$$G(x, y|x_0, y_0) = 4j\omega\mu d \sum_{m=-\infty}^{\infty} \sum_{n=-\infty}^{\infty}$$

$$\frac{T_1(x_0, y_0)T_1(x, y) + T_2(x_0, y_0)T_2(x, y)}{16\sqrt{3}\pi^2(m^2 + mn + n^2) - 9\sqrt{3}a^2k^2} \tag{A9.6}$$

where $T_1(x, y)$ is given by eqn. A9.4 and

$$T_2(x, y) = (-1)^l \cos\left(\frac{2\pi lx}{\sqrt{3}a}\right) \sin\left[\frac{2\pi(m-n)y}{3a}\right]$$

$$+ (-1)^m \cos\left(\frac{2\pi mx}{\sqrt{3}a}\right) \sin\left[\frac{2\pi(n-l)y}{3a}\right]$$

$$+ (-1)^n \cos\left(\frac{2\pi nx}{\sqrt{3}a}\right) \sin\left[\frac{2\pi(l-m)y}{3a}\right] \tag{A9.6}$$

As for $T_1(x, y)$, the integer l in $T_2(x, y)$ is given by eqn. A9.5.

(d) A right-angled isosceles triangle: The Green's function for the right-angled isosceles triangle shown in Fig. 9.9d is given by

$$G(x, y | x_0, y_0) = \frac{j\omega\mu d}{2} \sum_{m=0}^{\infty} \sum_{n=0}^{\infty} \frac{\sigma_m \sigma_n T(x_0, y_0) T(x, y)}{(m^2 + n^2)\pi^2 - a^2 k^2} \quad (A9.8)$$

where

$$T(x, y) = \cos \frac{m\pi x}{a} \cos \frac{n\pi y}{a} + (-1)^{m+n} \cos \frac{n\pi x}{a} \cos \frac{m\pi y}{a} \quad (A9.9)$$

(e) A circle: The Green's function for the circle shown in Fig. 9.9e is given by [35]

$$G(\varrho, \phi | \varrho_0, \phi_0) = \frac{d}{j\omega\varepsilon\pi a^2} + j\omega\mu d \sum_{n=0}^{\infty} \sum_{m=1}^{\infty}$$

$$\times \frac{\sigma_n J_n(k_{mn}\varrho_0) J_n(k_{mn}\varrho) \cos[n(\phi - \phi_0)]}{\pi(a^2 - n^2/k_{mn}^2)(k_{mn}^2 - k^2) J_n^2(k_{mn}a)} \quad (A9.10)$$

where $J_n(\cdot)$ represents Bessel's function of the nth order, and k_{mn} satisfies

$$\frac{\partial}{\partial p} J_n(k_{mn}\varrho)|_{\varrho = a} = 0 \quad (A9.11)$$

The subscripts m in k_{mn} denotes the mth root of eqn. A9.11. For the zeroth-order Bessel's function, the first root of eqn. A9.11 is taken to be the non-zero root.

(f) A circular sector: The Green's function for circular sectors are available only when the sector angle α is a sub-multiple of π. For the circular sector shown in Fig. 9.9f for which $\alpha = \pi/l$, the Green's function is given as [36]

$$G(\varrho, \phi | \varrho_0, \phi_0) = \frac{2ld}{j\omega\varepsilon\pi a^2} + 2jl\omega\mu d \sum_{n=0}^{\infty} \sum_{m=1}^{\infty}$$

$$\times \frac{\sigma_n J_{n_i}(k_{mn_i}\varrho_0) J_{n_i}(k_{mn_i}\varrho) \cos(n_i\phi_0) \cos(n_i\phi)}{\pi[a^2 - n_i^2/k_{mn_i}^2](k_{mn_i}^2 - k^2) J_{n_i}^2(k_{mn_i}a)} \quad (A9.12)$$

where $n_i = nl$, and k_{mn_i} are given by

$$\frac{\partial}{\partial p} J_{n_i}(k_{mn_i}\varrho)|_{\varrho = a} = 0 \quad (A9.13)$$

(g) An annual ring: The Green's function for the annular ring shown in Fig. 9.9g is given as [36]

$$G(\varrho, \phi | \varrho_0, \phi_0) = \frac{d}{j\omega\varepsilon\pi(b^2 - a^2)} + j\omega\mu d \sum_{n=0}^{\infty} \sum_{m=1}^{\infty}$$

$$\frac{\sigma_n F_{mn}(\varrho_0) F_{mn}(\varrho) \cos[n(\phi - \phi_0)]}{\pi[(b^2 - n^2/k_{mn}^2) F_{mn}^2(b) - (a^2 - n^2/k_{mn}^2) F_{mn}^2(a)](k_{mn}^2 - k^2)} \quad (A9.14)$$

where

$$F_{mn}(\varrho) = N'_n(k_{mn}a)J_n(k_{mn}\varrho) - J'_n(k_{mn}a)N_n(k_{mn}\varrho) \qquad (A9.15)$$

and k_{mn} are solutions of

$$\frac{J'_n(k_{mn}a)}{N'_n(k_{mn}a)} = \frac{J'_n(k_{mn}b)}{N'_n(k_{mn})b} \qquad (A9.16)$$

In the above relations $N_n(\cdot)$ denotes Neumann's function of order n and $J'_n(\cdot)$ and $N'_n(\cdot)$ denote first derivatives with respect to the arguments.

(*h*) *An annular sector:* As in the case of circular sectors, the Green's function for annular sectors are available only if the sector angle α is a submultiple of π. For the annular sector shown in Fig. 9.9*h* for which $\alpha = \pi/l$, the Green's function is given as [36]

$$G(\varrho, \phi | \varrho_0, \phi_0) = \frac{2ld}{i\omega\varepsilon\pi(b^2 - a^2)} + 2jl\omega\mu d \sum_{n=0}^{\infty} \sum_{m=1}^{\infty}$$

$$\times \frac{\sigma_n F_{mn_i}(\varrho) F_{mn_i}(\varrho_0) \cos(n_i\phi_0) \cos(n_i\phi)}{\pi[b^2 - n_i^2/k_{mn_i}^2)F_{mn_i}^2(b) - (a^2 - n_i^2/k_{mn_i}^2)F_{mn_i}^2(a)](k_{mn_i}^2 - k^2)}$$

$$(A9.17)$$

where $n_i = nl$, and $F_{mn_i}(\cdot)$ is defined in eqn. A9.15. The values of k_{mn_i} are obtained from eqn. A9.16.

9.9 Acknowledgments

Most of the material discussed in this Chapter is based on the theses and publications of several of my students and colleagues: Rakesh Chadha, P. C. Sharma, Girish Kumar, Yinggang Tu and Abdelaziz Benalla. Their contributions are gratefully acknowledged.

9.10 References

1. GUPTA, K. C., *et al.*: 'Computer-aided design of microwave circuits', (Artech House, USA, 1981) chaps 8 and 11
2. OKOSHI, T., and TAKEUCHI, T.: 'Analysis of planar circuits by segmentation method', *Electron. Commun. Japan*, 1975, **58-B**, pp. 71–79
3. CHADHA, R., and GUPTA, K. C.: 'Segmentation method using impedance-matrices for analysis of planar microwave circuits', *IEEE Trans.*, 1981, **MTT-29**, pp. 71–74
4. SHARMA, P. C., and GUPTA, K. C.: 'Desegmentation method for analysis of two-dimensional microwave circuits', *IEEE Trans.*, 1981, **MTT-29**, pp. 1094–1098
5. SHARMA, P. C., and GUPTA, K. C.: 'An alternative procedure for implementing desegmentation method', *IEEE Trans.*, 1984, **MTT-32**, pp. 1–4

6 OKOSHI, T.: 'Planar circuits for microwaves and lightwaves' (Springer-Verlag, 1985) chap. 5
7 GUPTA, K. C., et al.: 'Two-dimensional analysis for stripline/microstrip circuits'. 1981 IEEE MTT-S International Microwave Symp. Digest, pp. 504–506
8 GUPTA, K. C., and SHARMA, P. C.: 'Segmentation and desegmentation techniques for analysis of two-dimensional microstrip antennas'. 1981 IEEE AP-S International Antennas and Propagation Symp. Digest, pp. 19–22
9 GUPTA, K. C.: 'Two-dimensional analysis of microstrip circuits and antennas', *J. Inst. Electron. Telecommon. Engrs. (India)*, 1982, **28**, pp. 346–364
10 SHARMA, P. C., and GUPTA, K. C.: 'Analysis and optimised design of single feed circularly polarised microstrip antennas', *IEEE Trans.*, 1983, **AP-31**, pp. 949–955
11 KUMAR, G., and GUPTA, K. C.: 'Broadband microstrip antennas using additional resonators gap-coupled to radiating edges', *IEEE Trans.*, 1984, **AP-32**, pp. 1375–1379
12 KUMAR, G., and GUPTA, K. C.: 'Non-radiating edges and four-edges gap-coupled multiple resonator, broadband microstrip antennas', *IEEE Trans.*, 1985, **AP-33**, pp. 173–178
13 KUMAR, G., and GUPTA, K.C.: 'Directly coupled multiple resonator wideband microstrip antennas', *IEEE Trans.*, 1985, **AP-33**, pp. 588–593
14 GUPTA, K. C.: 'Two-port transmission characteristics of rectangular microstrip patch radiators'. 1985 IEEE AP-S International Antennas Propagat. Symp. Digest, pp. 71–74
15 GUPTA, K. C., and BENALLA, A.: 'Two-port transmission characteristics of circular microstrip patch antennas'. 1986 IEEE AP-S International Symp. Antennas Propagat. Digest, pp. 821–824
16 PALANISAMY, V., and GARG, R.: 'Analysis of arbitrary shaped microstrip patch antennas using segmentation technique and cavity model', *IEEE Trans.*, 1986, **AP-34**, pp. 1208–1213
17 PALANISAMY, V., and GARG, R.: 'Analysis of circularly polarised square ring and crossed-strip microstrip antennas', *IEEE Trans.*, 1986, **AP-34**, pp. 1340–1346
18 BENALLA, A., and GUPTA, K. C.: 'A method for sensitivity analysis of series-fed arrays of rectangular microstrip patches'. National Radio Science Meeting (URSI), Boulder (CO), USA, Jan. 1987, Digest, p. 65
19 GUPTA, K. C.: 'Multiport-network modelling approach for computer-aided design of microstrip patches and arrays'. 1987 IEEE AP-S International Symp. Antennas Propagat., Blacksburg (VA), USA, June 1987
20 GUPTA, K. C., and BENALLA, A.: 'Computer-aided design of microstrip patches and arrays'. Int. Microwave Symp./Brazil, July 1987, Symp. Proc. Vol. 1, pp. 591–596
21 LO, Y. T., et al.: 'Theory and experiment on microstrip antennas', *IEEE Trans.*, 1979, **AP-27**, pp. 137–145
22 RICHARDS, W. F., et al.: 'An improved theory for microstrip antennas and applications', *IEEE Trans.*, 1981, **AP-29**, pp. 38–46
23 MUNSON, R. E.: 'Conformal microstrip antennas and microstrip phased arrays', *IEEE Trans.*, 1974, **AP-22**, pp. 74–78
24 DERNERYD, A. G.: 'Microstrip array antenna'. Proc. 6th European Microwave Conf., 1976, pp. 339–343
25 CAMPI, M.: 'Design of microstrip linear array antennas by computer'. Proc. Antenna Applications Symp., Robert Alerton Park, Univ. of Illinois, Urbana, USA, Sept. 1981
26 BENALLA, A., and GUPTA, K. C.: 'Transmission line model for 2-port rectangular microstrip patches with ports at the non-radiating edges', *Electron. Lett.*, 1987, **23**, pp. 882–884
27 BENALLA, A., and GUPTA, K. C.: 'Two-dimensional analysis of one-port and two-port microstrip antennas'. Electromagnetics Laboratory, Scientific Rept. 85, Univ. of Colorado, May 1986, p. 48
28 COFFEY, E. L., and LEHMAN, T. H.: 'A new analysis technique for calculating the self and mutual impedance of microstrip antennas'. Proc. Workshop Printed Circuit Antenna Technology, New Mexico State Univ., 1979, pp. 31.1–31.21
29 BENALLA, A., and GUPTA, K. C.: 'Multiport-network model and transmission characteristics of two-port rectangular microstrip antennas', IEEE Trans., Oct 1988, AP-36, pp. 1337–42

30 GOGOI, A., and GUPTA, K. C.: 'Wiener–Hopf computation of edge admittances for microstrip patch radiators', *AEU*, 1982, **36**, pp. 247–251
31 VAN DE CAPELLE, A., et al.: 'A simple accurate formula for the radiation conductance of a rectangular microstrip antenna'. 1981 IEEE AP-S International Symp. Antennas Propagat., Digest, pp. 23–26
32 KUESTER, E. F., et al.: 'The thin-substrate approximation for reflection from the end of a slab-loaded parallel plate waveguide with application to microstrip patch antenna', *IEEE Trans.*, 1982, **AP-30**, pp. 910–917
33 GUPTA, K. C., and BENALLA, A.: 'Effect of mutual coupling on the input impedance and the resonant frequency of a rectangular microstrip patch antenna'. National Radio Science Meeting (URSI), Boulder, Jan. 1986, Digest, p. 226
34 CHADHA, R., and GUPTA, K. C.: 'Green's functions for triangular segments in planar microwave circuits', *IEEE Trans.*, 1980, **MTT-28**, pp. 1139–1143
35 OKOSHI, T., et al.: 'Planar 3-dB hybrid circuits', *Electron. Commun. Japan*, 1975, **58-B**, pp. 80–90
36 CHADHA, R., and GUPTA, K. C.: 'Green's functions for circular sectors, annular rings and annular sectors in planar microwave circuits', *IEEE Trans.*, 1981, **MTT-29**, pp. 68–71
37 BENALLA, A., and GUPTA, K. C.: 'Faster computation of Z-matrices for rectangular segments in planar microstrip circuits', *IEEE Trans.*, 1986, **MTT-34**, pp. 733–736
38 GUPTA, K. C., and ABOUZAHRA, M. D.: 'Analysis and design of four-port and five-port microstrip disc circuits', *IEEE Trans.*, 1985, **MTT-33**, pp. 1422–1428
39 OKOSHI, T., and MIYOSHI, T.: 'The planar circuit – An approach to microwave integrated circuitry', *IEEE Trans.*, 1972, **MTT-20**, pp. 245–252
40 JAMES, J. R., et al.: 'Microstrip antenna theory and design', (Peter Peregrinus, 1981), p. 23
41 JAMES, J. R., and HENDERSON, A.: 'High-frequency behaviour of microstrip open-end terminations', *IEE Microwaves, Optics & Acoustics*, 1979, **3**, pp. 205–218
42 KATEHI, P. B., and ALEXOPOULOS, N. G.: 'Frequency-dependent characteristics of microstrip discontinuities in millimeter-wave integrated circuits', *IEEE Trans.*, 1985, **MTT-33**, pp. 1029–1035
43 JACKSON, R. W., and POZAR, D. M.: 'Full-wave analysis of microstrip open-end and gap discontinuities', *IEEE Trans.*, 1985, **MTT-33**, pp. 1036–1042
44 LEWIN, L.: 'Radiation from discontinuities in stripline', *Proc. IEE*, 1960, **107C**, pp. 163–170
45 WOLFF, I., and KNOPPIK, N.: 'Rectangular and circular microstrip disk capacitors and resonators', *IEEE Trans.*, 1974, **MTT-22**, pp. 857–864
46 BAHL, I. J., and BHARTIA, P.: 'Microstrip antennas' (Artech House, 1980) chap. 2
47 JANSEN, R. H., and KOSTER, N. H. L.: 'Accurate results on the end effect of single and coupled microstrip lines for use in microwave circuit design', *AEU*, 1980, **34**, pp. 453–459
48 JANSEN, R. H.: 'Hybrid mode analysis of the end effects of planar microwave and millimeter-wave transmission lines', *Proc. IEE*, 1981, **129**, pp. 77–86
49 KIRSCHNING, M., et al.: 'Accurate model for open end effect of microstrip lines', *Electron. Lett.*, 1981, **17**, pp. 123–125
50 HAMMERSTAD, E., and JENSEN, O.: 'Accurate models for microstrip computer-aided design', 1980 IEEE MTT-S Int. Microwave Symp. Digest, Washington, 1980, pp. 407–409
51 VAN LIL, E. H., and VAN DE CAPELLE, A. R.: 'Transmission line model for mutual coupling between microstrip antennas', *IEEE Trans.*, 1984, **AP-32**, pp. 816–821
52 BALANIS, C. A.: 'Antenna theory analysis and design', (Harper and Row, 1982), p. 169
53 BENALLA, A., and GUPTA, K. C.: 'Multiport network approach for modelling mutual coupling effects in microstrip patch antennas and arrays', IEEE Trans., Feb 1989, **AP-37**, pp. 148–52
54 JEDLICKA, R. P., et al.: 'Measured mutual coupling between microstrip antennas', *IEEE Trans.*, 1981, **AP-29**, pp. 147–149
55 WIENCHEL, H. D.: 'A cylindrical array of circularly polarised microstrip antennas'. IEEE-APS Int. Symp. Antennas Propagation Digest, 1975, pp. 177–180

56 KUMAR, G., and GUPTA, K. C.: 'Broadband microstrip antennas using coupled resonators'. 1983 IEEE AP-S Int. Antennas Propagat. Symp. Digest, pp. 67–70
57 GUPTA, K. C., and BANDHAUER, B.: 'Coupled line model for multiresonator wide band microstrip antennas'. National Radio Science Meeting (URSI), Boulder, Jan. 1988
58 BENALLA, A.: Unpublished experimental results, 1986
59 CHANG, D. C., and KUESTER, E. F.: 'Total and partial reflection from the end of a parallel-plate waveguide with an extended dielectric slab', *Radio Sci.*, 1981, **16**, pp. 1–13
60 CHANG, D. C.: 'Analytical theory of an unloaded microstrip patch', *IEEE Trans.*, 1981, **AP-29**, pp. 54–62
61 TU, Y.: 'Edge admittance and mutual coupling in rectangular microstrip patch antennas with a dielectric cover layer', Ph.D. Thesis, Univ. of Colorado, 1987, pp. 62–84
62 TU, Y., GUPTA, K. C., and CHANG, D. C.: 'Mutual coupling computations for rectangular microstrip patch antennas with a dielectric cover layer', URSI National Radio Science Meeting, Boulder, Jan. 1987, Digest p. 66
63 POZAR, D. M.: 'Input impedance and mutual coupling of rectangular microstrip antennas', *IEEE Trans.*, 1982, **AP-30**, pp. 1191–1196
64 JACKSON, D. R., *et al.*: 'An exact mutual coupling theory for microstrip patches', 1987 IEEE AP-S Int. Symp. Antennas Propag. Digest, Vol. 2, 1987, pp. 790–793

Chapter 10
Transmission-line model for rectangular microstrip antennas

A. Van de Capelle

List of sumbols

L = length of patch
W = width of patch
t = thickness of patch or co-planar strip conductor
σ_p = conductivity of patch or co-planar strip conductor
Δ_p = RMS surface error of patch or co-planar strip conductor
L_s = length of substrate
W_s = width of substrate
h = thickness of substrate
ε_r = relative permittivity of substrate
δ_s = loss tangent of substrate
σ_g = conductivity of ground plane
Δ_g = RMS surface error of ground plane
t_g = thickness of ground plane
W_m = width of strip conductor of a microstrip line
L_m = length of microstrip line
$Y_{c,m}$ = characteristic admittance of microstrip line
γ_m = propagation constant of microstrip line
Y_c = characteristic admittance of transmission line representing a rectangular microstrip antenna
γ_p = propagation constant of the above
Y_s = self-admittance representing the open-end terminations of a microstrip antenna
G_s = real part of Y_s, self-conductance
B_s = imaginary part of Y_s, self-susceptance
Δ_l = extra length of microstrip line by open-end effect
y_s = admittance per unit length of a slot with infinite length
g_s = real part of y_s
b_s = imaginary part of y_s
E_a, H_a = electric, magnetic field in equivalent slot apertures

V_i = excitation voltage of a slot i
$\mathbf{i}_x, \mathbf{i}_y, \mathbf{i}_z$ = unit vectors of x, y, z co-ordinates
S = width of equivalent slots
$\mathcal{E}_a, \mathcal{H}_a$ = Fourier transform of $\mathbf{E}_a, \mathbf{H}_a$
k_x, k_y, k_z = components of $\mathbf{k}$
$\mathbf{k}$ = propagation vector
p = real part of the complex radiated power per unit length
q = imaginary part of the complex radiated power per unit length
η = wave impedance in half-space above antenna
k = propagation constant in half-space above antenna
μ = permeability of half-space above antenna
ε = permittivity of half-space above antenna
y_m = mutual admittance per unit length
g_m = real part of y_m
b_m = imaginary part of y_m
J = Bessel function of the first kind
Y = Bessel function of the second kind
s = normalised slot width
C_e = Euler's constant
ε_{eff} = effective relative permittivity
W_{eff} = effective width
α = attenuation constant
β = phase constant
η_0 = wave impedance of free space
k_0 = propagation constant in free space
λ_0 = free-space wavelength
W_e = length of equivalent slot
w = normalised slot length
L_e = centre distance between equivalent slots
F_g = auxiliary coupling function for the mutual conductance
E_p = relative error on the radiation conductance G_r
F_b = auxiliary coupling function for the mutual susceptance
Y_m = mutual admittance
G_m = real part of Y_m
B_m = imaginary part of Y_m
K_g = correction function for the mutual conductance
K_b = correction function for the mutual susceptance
D = directivity in broadside direction
G = gain in broadside direction
q = antenna efficiency
Q = antenna quality factor
BW = impedance bandwidth

10.1 Introduction

Microstrip antennas have a physical structure derived from microstrip transmission lines. Therefore a transmission-line model is the first and most obvious choice for the analysis and the design of microstrip antennas. However, the transmission-line model is often regarded as a simplified and somewhat dated theory. This is true for the original, simple transmission-line model; but the accuracy of the improved transmission-line model is comparable to that of other more complicated methods. Even mutual coupling between rectangular microstrip antennas can be calculated in a fairly accurate and very efficient way with the transmission-line approach.

The practical design of a microstrip antenna or a microstrip array, including matching and feeding networks, has to be done by means of a CAD software package. Existing programs represent the network components by equivalent transmission lines. If the antenna elements are modelled by the same transmission-line approach, the incorporation in the available CAD software is straightforward.

The concept of the transmission-line model can be applied to any microstrip antenna configuration for which separation of variables is possible. In this Chapter we will devote our attention entirely to rectangular (and square) microstrip antennas.

The transmission-line model does not include surface waves. Therefore, the application is limited to antenna configurations where the thickness and the substrate permittivity are sufficiently small to avoid considerable excitation of those surface waves. But, in practice, this is not a severe limitation. However, research is going on to also include surface waves in the transmission-line model.

10.2 Simple transmission-line model

10.2.1 Description of the transmission-line model

The transmission-line model will be discussed for rectangular (and square) microstrip antennas. The antenna consists of a conducting patch, a dielectric substrate and a conducting ground plane. The antenna is fed by a microstrip line (as shown in Fig. 10.1) or by a coaxial probe (Fig. 10.2).

The patch is characterised by the resonant length L (resonant for the fundamental mode), the width W, the thickness t, the conductivity σ_p and the RMS surface error Δ_p.

In the analysis, the dielectric substrate is supposed to have infinite dimensions in the plane of the patch. In practice, it has a length L_s, a width W_s and a thickness h. Electrically, it is characterised by a relative permittivity ε_r and a loss tangent δ_s. It is supposed that the substrate consists of one homogeneous layer. A multilayer substrate can be replaced by an equivalent homogeneous layer with equivalent relative permittivity and loss tangent.

The conducting ground plane has the same dimensions as the substrate: L_s, W_s in practice, infinite of extent for the analysis. It is further characterised by a conductivity σ_g, a RMS surface error Δ_g and a thickness t_g.

In the case of an antenna fed by a co-planar microstrip line, the strip conductor has a width W_m and a length L_m. The other parameters of the microstrip line $(t, h, t_g, \sigma_p, \sigma_g, \Delta_p, \Delta_g, \varepsilon_r, \delta_s)$ are the same as for the antenna. The cross-sectional geometry of the microstrip line is characterised by the aspect ratio W_m/h. Likewise, the microstrip antenna can be considered as a microstrip line with a very large aspect ratio W/h.

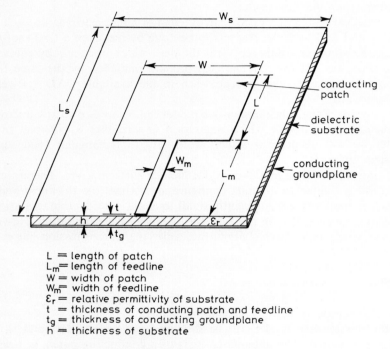

L = length of patch
L_m = length of feedline
W = width of patch
W_m = width of feedline
ε_r = relative permittivity of substrate
t = thickness of conducting patch and feedline
t_g = thickness of conducting groundplane
h = thickness of substrate

Fig. 10.1 *Rectangular microstrip antenna fed by a microstrip line*

In the case of a microstrip antenna fed by a microstrip line (Fig. 10.1), the introduction of the transmission-line model is straightforward (Fig. 10.3):

● The microstrip feed line is represented by a transmission line with a characteristic admittance $Y_{c,m}$ (mainly determined by the aspect ratio W_m/h and the relative permittivity ε_r), a propagation constant γ_m and a physical length L_m.

● The rectangular microstrip antenna is represented by a transmission line with a characteristic admittance Y_c (mainly determined by the aspect ratio W/h and the relative permittivity ε_r), a propagation constant γ_p and a physical length L.

Transmission-line model for rectangular microstrip antennas 531

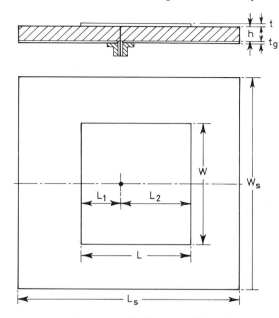

Fig. 10.2 *Rectangular microstrip antenna fed by a coaxial probe*

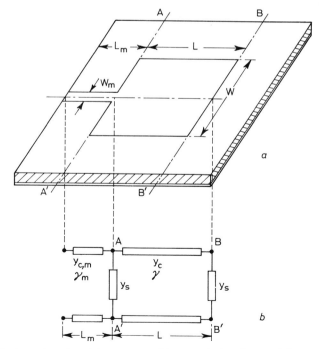

Fig. 10.3 *Rectangular microstrip antenna with transmission-line model*

532 *Transmission-line model for rectangular microstrip antennas*

● At the cross-sections AA′ and BB′, in Fig. 10.3a, the microstrip line with aspect ratio W/h has an open-ended termination, which can be represented by a parallel admittance ($Y_s = G_s + jB_s$).

In the case of excitation with a coaxial probe, the equivalent transmission-line model has to be modified as shown in Fig. 10.4.

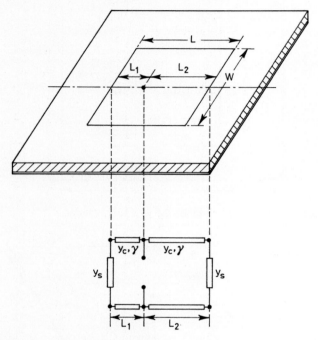

Fig. 10.4 *Rectangular microstrip antenna with coaxial feed and equivalent transmission-line model*

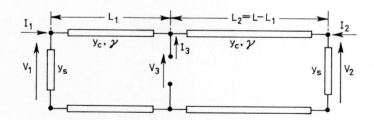

Fig. 10.5 *General three-port equivalent transmission-line model*

A general transmission-line model, which can be applied in both cases (microstrip line feed or coaxial feed), consists of a three-port circuit (Fig. 10.5). The general three-port model has to be completed with a piece of transmission line at port 1 and port 3 has to be left open (Fig. 10.3) in the case of a microstrip line feed. In the case of a coaxial-feed probe the model has to be completed with an inductance at port 3.

The main step in the modelling of a microstrip antenna by a transmission-line equivalent, is the representation of the open-ended terminations by a parallel admittance Y_s. An open-ended microstrip line does not perform as a perfect open circuit (see Fig. 10.6):

● The field lines do not stop abruptly at the end of the strip conductor: there is a stray field extending beyond the end of the strip; this can be interpreted as an electrical lengthening Δl of the line, which implies an amount of stored energy; on the other hand, the stray field is also source of power radiated in the space above the antenna and launched as surface waves along the substrate;
● The real part G_s of the parallel admittance Y_s represents the radiation effect (and surface waves), and the imaginary part B_s models the stored energy in the extra line length.

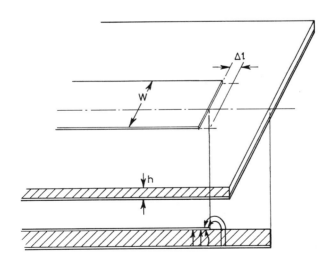

Fig. 10.6 *Open-ended microstrip line with aspect ratio W/h*

10.2.2 Expressions for G_s and B_s

The accuracy of the transmission-line model depends strongly on the choice of expressions for G_s and B_s. In the original transmission-line model, proposed by Munson [1], a simple but very approximate expression for Y_s has been proposed:

$$Y_s = W y_s \tag{10.1}$$

with y_s = admittance per unit length of a uniformly excited slot with infinite length and width h, in an infinite, perfectly conducting plane.

This expression is not accurate enough, but it is important because of the concept behind it:

- The radiation of a rectangular microstrip antenna can be explained as originating from the tangential electric field in the plane of the patch.
- In the fundamental mode, only the contribution from the two open ends is important.
- The source of radiation can be limited to two narrow zones along the two open ends of the patch.
- The field in these two narrow zones can be thought of as the field of two rectangular slots in an infinite, perfectly conducting plane.
- For the fundamental mode of the microstrip antenna the tangential field in these two slots can be considered to be uniformly distributed.
- A slot with a uniform excitation field can be considered as a cut from an infinitely long, uniformly excited slot.

The idea of representing the microstrip antenna by equivalent slots in an infinite, perfectly conducting plane is very powerful. The inaccuracy of eqn. 10.1 is mainly due to the last simplification, where the edge effects of finite-length slots are neglected.

We now have two concepts available to explain the radiation of a microstrip antenna:

- the open end concept
- the equivalent slot concept.

These two concepts can be used to derive expressions for the parameters in the transmission-line model. In the following Sections the equivalent, slot concept will be applied where possible, and the open-end concept where necessary.

In this Section we want to derive suitable expressions for the parameters of the simple transmission-line model of Fig. 10.5. We proceed with eqn. 10.1 and derive expressions for the real part g_s and the imaginary part b_s of y_s.

A configuration of two equivalent slots, as shown in Fig. 10.7, is considered. The slots are pieces of length W, taken from infinitely long, uniformly excited slots. The tangential electric field in the slot apertures can be written as:

$$E_a = \begin{cases} \dfrac{V_1}{S} i_y & \text{for} \quad \dfrac{L_e - S}{2} \leqslant y \leqslant \dfrac{L_e + S}{2} \\ \dfrac{V_2}{S} i_y & \text{for} \quad \dfrac{L_e - S}{2} \leqslant -y \leqslant \dfrac{L_e + S}{2} \\ 0 & \text{elsewhere} \end{cases} \quad (10.2)$$

where i_x, i_y, i_z = unit vectors of the x, y, z co-ordinates; V_1, V_2 = excitation voltage of slot 1 and slot 2, respectively; $S = h$ = width of the equivalent slots; $L_e = L + h$ = centre distance between the equivalent slots.

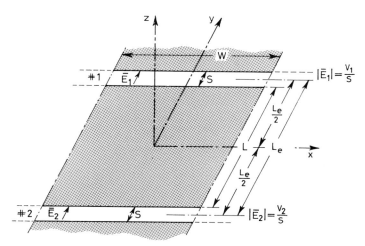

Fig. 10.7 *Two-slot model*

The spatial Fourier transform, with respect to y, of this aperture is given by

$$\mathscr{E}_a = \int_{-\infty}^{+\infty} E_a e^{jk_y y} \, dy \tag{10.3}$$

where k_x, k_y, k_z = components of the propagation vector $\mathbf{k}$; $|\mathbf{k}| = k = \omega(\mu\varepsilon)^{1/2}$ = propagation constant.
As the field of eqn. 10.2 has only a y-component, the Fourier transform is

$$\mathscr{E}_a = \mathscr{E}_y i_y \tag{10.4}$$

with

$$\mathscr{E}_y = jS \frac{\sin(k_y S/2)}{k_y S/2} (V_1 e^{jk_y L_e/2} + V_2 e^{-jk_y L_e/2}) \tag{10.5}$$

The complex radiated power ($p + jq$) per unit slot length, in terms of the spatial Fourier transform, is given by [Reference 2 pp. 61–68]:

$$p = \frac{k}{4\pi\eta} \int_{-k}^{+k} |\mathscr{E}_y|^2 \frac{dk_y}{\sqrt{k^2 - k_y^2}} \tag{10.6}$$

$$q = \frac{-k}{4\pi\eta} \left(\int_{-\infty}^{-k} + \int_{k}^{+\infty} |\mathscr{E}_y|^2 \frac{dk_y}{\sqrt{k_y^2 - k^2}} \right) \tag{10.7}$$

where $\eta = (\mu/\varepsilon)^{1/2}$ = wave impedance.

From a network point of view, the two-slot configuration of Fig. 10.7 can be considered as a symmetrical two-port with a self-admittance per unit length $y_s (= g_s + jb_s)$ and a mutual self-admittance per unit length $y_m (= g_m + jb_m)$. Expressed in these quantities, the complex radiated power per unit length is given by

$$p + jq = \frac{1}{2} y_s^* (|V_1|^2 + |V_2|^2) + y_m^* \operatorname{Re}(V_1 V_2^*) \tag{10.8}$$

Taking $V_2 = 0$, g_s and b_s follow from eqns. 10.5–10.8:

$$g_s = \frac{k}{\pi \eta} \int_0^k \frac{\sin^2(k_y S/2)}{(k_y S/2)^2} \frac{dk_y}{\sqrt{k^2 - k_y^2}} \tag{10.9}$$

$$b_s = \frac{k}{\pi \eta} \int_k^\infty \frac{\sin^2(k_y S/2)}{(k_y S/2)^2} \frac{dk_y}{\sqrt{k_y^2 - k^2}} \tag{10.10}$$

The single integrals in eqns. 10.9 and 10.10 can be written (see Reference 3 appendix) as double integrals of a Bessel function of the first kind J, and the second kind Y, respectively:

$$g_s = \frac{k}{\eta s^2} J_0^{ii}(s) \tag{10.11}$$

$$b_s = -\frac{k}{\eta s^2} Y_0^{ii}(s) \tag{10.12}$$

where $s = kS$ is the normalised slot width

$$J_0^{ii}(s) = \int_0^s \int_0^u J_0(v) \, dv \, du \tag{10.13}$$

$$Y_0^{ii}(s) = \int_0^s \int_0^u Y_0(v) \, dv \, du \tag{10.14}$$

By twice integrating the series expansion of J_0 and Y_0 [4], the following series are obtained

$$J_0^{ii}(s) = \frac{s^2}{2} \left(1 - \frac{s^2}{24} + \frac{s^4}{960} \cdots \right) \tag{10.15}$$

$$Y_0^{ii}(s) = \frac{s^2}{\pi} \left\{ \left(X - \frac{3}{2} \right) - \frac{s^2}{24} \left(X - \frac{19}{12} \right) \right.$$
$$\left. + \frac{s^4}{960} \left(X - \frac{28}{15} \right) \cdots \right\} \tag{10.16}$$

where $X = \ln(s/2) + C_e$; $C_e =$ Euler's constant $= 0.577216$.

From eqns. 10.9–10.17 the following expressions for g_s and b_s can be derived

$$g_s \approx \frac{k}{2\eta}\left(1 - \frac{s^2}{24}\right) \quad (10.18)$$

$$b_s \approx -\frac{k}{\pi\eta}\left\{\left(\ln\frac{s}{2} + C_e - \frac{3}{2}\right)\left(1 - \frac{s^2}{24}\right) + \frac{s^2}{288}\right\} \quad (10.19)$$

where the terms in s^4, s^6 etc. have been neglected. The maximum truncation error of eqns. 10.18 and 10.19 is not larger than 0·1% for $s \leqslant 1$. Expressions 10.18 and 10.19, combined with expression 10.1, completely determine the parallel admittance Y_s.

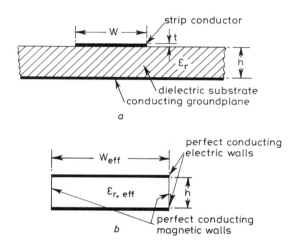

Fig. 10.8 *Planar-waveguide model for microstripline*
 a Cross-section of microstripline with aspect ratio W/h
 b Cross-section of planar-waveguide model

10.2.3 Expressions for the line parameters
To derive expressions for the characteristic admittance Y_c and the propagation constant $\gamma_p(=\alpha_p + j\beta_p)$ of the equivalent transmission line representing the antenna, and for the characteristic admittance Y_{cm} and the propagation constant γ_m of the microstrip feed line, the planar-waveguide model is used, see [5].

A microstrip line with aspect ratio W/h (cross-section see Fig. 10.8a) and with a dielectric substrate of relative permittivity ε_r is modelled by a planar waveguide (cross-section see Fig. 10.8b). The top and bottom walls of this planar waveguide are electrically conducting, while the side walls are perfect magnetic conductors. The guide is of height h equal to the microstrip substrate height, but it has an effective width W_{eff}, larger than the physical width W of the strip. The guide is filled with a dielectric that has an effective relative permittivity $\varepsilon_{\mathit{eff}}$.

The characteristic impedance Z_c and the phase constant β of the fundamental mode propagating in the microstrip line are given in terms of the planar-waveguide parameters:

$$Z_C = \frac{\eta_0}{\sqrt{\varepsilon_{eff}}} \frac{h}{W_{eff}} \tag{10.20}$$

$$\beta = k_0 \sqrt{\varepsilon_{eff}} \tag{10.21}$$

where $\eta_0 = \sqrt{\mu_0/\varepsilon_0}$ = wave impedance in free space; $k_0 = \omega\sqrt{\mu_0\varepsilon_0}$ = propagation constant in free space.

In the quasi-static approximation, Reference 6 gives convenient expressions for

$$W_{eff}(0) = 2\pi h / \ln\{hF/W' + \sqrt{1 + (2h/W')^2}\} \tag{10.22}$$

where

$$F = 6 + (2\pi - 6)\exp\left\{-\frac{4\pi^2}{3}\left(\frac{h}{W'}\right)^{3/4}\right\} \tag{10.23}$$

$$W' = W + \frac{t}{\pi}\{1 + \ln(4/\sqrt{(t/h)^2 + (1/\pi)^2/(W/t + 1\cdot 1)^2}\} \tag{10.24}$$

and for

$$\varepsilon_{eff}(0) = \frac{1}{2}\{\varepsilon_r + 1 + (\varepsilon_r - 1)G\} \tag{10.25}$$

where

$$G = (1 + 10h/W)^{-AB} - \frac{\ln 4}{\pi}\frac{t}{\sqrt{Wh}} \tag{10.26}$$

$$A = 1 + \frac{1}{49}\ln\left\{\frac{(W/h)^4 + W^2/(52h)^2}{(W/h)^4 + 0\cdot 432}\right\} + \frac{1}{18\cdot 7}\ln\left\{1 + \left(\frac{W}{18\cdot 1h}\right)^3\right\} \tag{10.27}$$

$$B = 0\cdot 564 \exp\left(-\frac{0\cdot 2}{\varepsilon_r + 0\cdot 3}\right) \tag{10.28}$$

The attenuation α and the frequency dependence of Z_c and β are neglected in the simple transmission-line model. Indeed, as will be explained in the improved transmission-line model, the simple model has important shortcomings, so that it makes no sense to take into account these second-order effects.

10.3 Improved transmission-line model

10.3.1 Description of the improved transmission-line model
The simple transmission line model has important shortcomings:

(i) The expressions for Y_s are inaccurate for the usual patch widths (i.e. for $W \leqslant \lambda_0$; λ_0 = free-space wavelength);

(ii) The mutual coupling between the two equivalent slots is neglected;
(iii) The radiation from the side walls is not taken into account.

Derneryd [7, 8] has partly eliminated the first two shortcomings:

(i) To determine $G_s = \text{Re}(Y_s)$, he considers the two main slots with an identical excitation and a negligible width. He finds an integral expression for G_s, for which an approximate analytical solution has been derived by Lier [9]. Derneryd's model corrects the first two shortcomings of [Reference 1] for the real part of Y_s, but it still neglects the influence of the side walls on G_s.
(ii) To determine the susceptance $B_s = \text{Im}(Y_s)$, Derneryd makes this parameter equal to the open-end self-susceptance of the microstrip line formed by the patch. This corrects the first shortcoming of [Reference 1] for the imaginary part of Y_s.

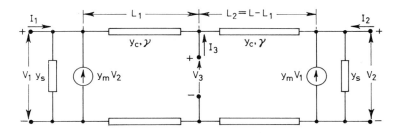

Fig. 10.9 *Improved transmission-line model represented as a three-port*

An improved transmission-line model, proposed by Pues and Van de Capelle [10], will be discussed here. This model corrects the three shortcomings of [Reference 1] for the real as well as for the imaginary part, and has a broad range of validity. The circuit representation of the present model is shown in Fig. 10.9. In this network Y_s is the self admittance of the open-end terminations of the patch, and Y_m is their mutual (radiation) admittance. The mutual coupling is formally taken into account by voltage-dependent current sources.

The admittance matrix of this three-port model is given by

$$[Y] = \begin{bmatrix} Y_s + Y_c \coth(\gamma_p L_1) & -Y_m & -Y_c \operatorname{csch}(\gamma_p L_1) \\ -Y_m & Y_s + Y_c \coth(\gamma_p L_2) & -Y_c \operatorname{csch}(\gamma_p L_2) \\ -Y_c \operatorname{csch}(\gamma_p L_1) & -Y_c \operatorname{csch}(\gamma_p L_2) & Y_c \coth(\gamma_p L_1) + Y_c \coth(\gamma_p L_2) \end{bmatrix}$$

(10.29)

where $\coth z$ and $\operatorname{csch} z$ are the complex hyperbolic cotangent and cosecant functions of argument z, respectively. The copper and dielectric losses of the antenna are taken into account by the attenuation constant α_p, the real part of the complex propagation constant γ_p.

If there is only one feed point, an input admittance can be defined. Assuming $I_1 = I_2 = 0$, it follows from eqn. 10.29 that

$$Y_{in} = 2Y \left[\frac{Y_c^2 + Y_s^2 - Y_m^2 + 2Y_s Y_c \coth(\gamma_p L) - 2Y_m Y_c \operatorname{csch}(\gamma_p L)}{Y_c^2 + Y_s^2 - Y_m^2)\coth(\gamma_p L) + (Y_c^2 - Y_s^2 + Y_m^2)\cosh(2\gamma_p \Delta)\operatorname{csch}(\gamma_p L) + 2Y_s Y_c} \right]$$

(10.30)

where

$$\Delta = \left| \frac{L}{2} - L_1 \right| = \left| \frac{L}{2} - L_2 \right| \tag{10.31}$$

L_1 and L_2 are defined in Fig. 10.9.

In the case of a microstrip-line-fed antenna, this corresponds to $I_2 = I_3 = 0$. It follows from eqns. 10.29 and 10.30 that

$$Y_{in} = \frac{Y_c^2 + Y_s^2 - Y_m^2 + 2Y_s Y_c \coth(\gamma_p L) - 2Y_m Y_c \operatorname{csch}(\gamma_p L)}{Y_s + Y_c \coth(\gamma_p L)}$$

(10.32)

To model the parasitic effects of the feed line on the antenna behaviour, the self-admittance of the open-end termination facing the feed line is reduced by a factor

$$r = 1 - \frac{W_m}{W_{eff}} \tag{10.33}$$

where W_m = width of feed line; W_{eff} = effective width of patch.

This reduction takes into account the partial covering of the open-end termination by the feed line. The reduction of the self-admittance at terminal 1 can be considered as an addition of a parallel admittance

$$Y_F = (r - 1) Y_s \tag{10.34}$$

The antenna input admittance is given by

$$\begin{aligned} Y'_{in} &= Y_{in} + Y_F \\ &= rY_s + \frac{Y_c^2 - Y_m^2 + Y_s Y_c \coth(\gamma_p L) - 2Y_m Y_c \operatorname{csch}(\gamma_p L)}{Y_s + Y_c \coth(\gamma_p L)} \end{aligned}$$

(10.35)

The accuracy of the improved transmission-line model depends strongly on the accuracy of the expressions for the model parameters. Therefore much effort has

Transmission-line model for rectangular microstrip antennas 541

been spent in comparing available formulas and deriving new ones where needed. It was a primary goal to combine accuracy with numerical efficiency. We have tried to obtain analytical expressions for all the model parameters.

The imaginary part of Y_s, the self-susceptance B_s, is determined by means of the open-end-effect concept. Indeed, in the equivalent-slot concept, the self-susceptance depends strongly on the aperture field and there is no information available on an appropriate choice of this field. The real part of Y_s, the self-conductance G_s, is modelled as the radiation conductance of an equivalent slot. The mutual admittance Y_m is also determined from the equivalent-slot concept. For the line parameters, the attenuation and the frequency dependence of Y_c and γ are included.

10.3.2 Expression for the self-susceptance B_s

For the self-susceptance B_s, the correct transmission-line formula is used:

$$B_s = Y_c \tan(\beta \Delta l) \tag{10.36}$$

where Y_c, β, Δl are, respectively, the characteristic admittance, the phase constant and the open-end extension of a microstrip line with aspect ratio W/h, as formed by the patch.

The most appropriate expression for Δl is given in Reference 11:

$$\Delta l = h\xi_1 \xi_3 \xi_5 / \xi_4 \tag{10.37}$$

where

$$\xi_1 = 0.434907 \frac{\varepsilon_{\it eff}^{0.81} + 0.26}{\varepsilon_{\it eff}^{0.81} - 0.189} \frac{(W/h)^{0.8544} + 0.236}{(W/h)^{0.8544} + 0.87} \tag{10.38}$$

$$\xi_2 = 1 + \frac{(W/h)^{0.371}}{2.358 \varepsilon_r + 1} \tag{10.39}$$

$$\xi_3 = 1 + \frac{0.5274 \arctan\{0.084(W/h)^{1.9413/\xi_2}\}}{\varepsilon_{\it eff}^{0.9236}} \tag{10.40}$$

$$\xi_4 = 1 + 0.0377 \arctan\{0.067(W/h)^{1.456}\}$$
$$\{6 - 5\exp[0.036(1 - \varepsilon_r)]\} \tag{10.41}$$

$$\xi_5 = 1 - 0.218 \exp(-7.5W/h) \tag{10.42}$$

Expressions for Y_s, β and $\varepsilon_{\it eff}$ are given in Section 10.3.6.

10.3.3 Expression for the self-conductance G_s

For the real part of Y_s, the self-conductance G_s, the equivalent-slot concept explained in Section 10.2.2 is applied. The model is similar to that of Derneryd [7], except for the dimensions of the equivalent slot. The open-end terminations of the patch are replaced by uniformly TE-excited narrow rectangular slots of length $W_e = W_{\it eff}$ (instead of W in Reference 7) and width Δl (instead of h in Reference 7).

Transmission-line model for rectangular microstrip antennas

To calculate the self-conductance G_s, one such equivalent slot is considered, as shown in Fig. 10.10. The electric field in the slot aperture is assumed to be uniform:

$$E_a = \frac{V_s}{S} i_y \quad \text{for} \quad |y| \leq \frac{S}{2}; \; |x| \leq \frac{W_e}{2} \tag{10.43}$$

where V_s = excitation voltage of the equivalent slot; $S = \Delta l$ = width of the equivalent slot.

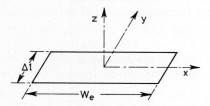

Fig. 10.10 *Equivalent slot radiator in an infinite, perfectly conducting plane*

The spatial Fourier transform of the aperture field is defined as

$$\mathscr{E}_a = \int_{-\infty}^{+\infty} \int_{-\infty}^{+\infty} E_a e^{jk_x x} e^{jk_y y} \, dx \, dy \tag{10.44}$$

The aperture field has only a y-component, so that the Fourier transform

$$\mathscr{E}_a = \mathscr{E}_y i_y \tag{10.45}$$

where

$$\mathscr{E}_y = V_s W_e \frac{\sin(k_x W_e/2)}{(k_x W_e/2)} \frac{\sin(k_y S/2)}{(k_y S/2)} \tag{10.46}$$

The complex power radiated by this slot may be found by integrating the complex Poynting vector over the aperture surface A:

$$P + jQ = \frac{1}{2} \iint_A E_a \times H_a^* \cdot i_z \, dx \, dy \tag{10.47}$$

with H_a the magnetic field in the slot aperture A.
Expressed in terms of the Fourier-transformed aperture field

$$P + jQ = \frac{1}{8\pi^2 \eta k} \int_{-\infty}^{+\infty} \int_{-\infty}^{+\infty} \{(k_z^2)|\mathscr{E}_y|^2 + |k_y \mathscr{E}_y|^2\} \frac{dk_x dk_y}{k_z^*} \tag{10.48}$$

where

$$\left. \begin{array}{l} k_z = +(k^2 - k_x^2 - k_y^2)^{1/2} \quad \text{for} \quad k^2 \geq k_x^2 + k_y^2 \\ = -j(k_x^2 + k_y^2 - k^2)^{1/2} \quad \text{for} \quad k^2 < k_x^2 + k_y^2 \end{array} \right\} \tag{10.49}$$

Transmission-line model for rectangular microstrip antennas 543

This complex power can also be written in terms of network parameters:

$$P + jQ = \frac{1}{2} Y_s^* |V_s|^2 = \frac{1}{2}(G_s - jB_s)|V_s|^2 \qquad (10.50)$$

Equating expressions 10.48 and 10.50, expressions for G_s and B_s follow

$$G_s = \frac{1}{\pi^2 \eta k |V_s|^2} \int_0^k \left\{ \int_0^{\sqrt{k^2-k_x^2}} (k^2 - k_x^2)|\mathscr{E}_y|^2 \frac{dk_y}{\sqrt{k^2 - k_x^2 - k_y^2}} \right\} dk_x \qquad (10.51)$$

$$B_s = \frac{1}{\pi^2 \eta k |V_s|^2} \int_0^k \left\{ \int_{\sqrt{k^2-k_x^2}}^{\infty} (k^2 - k_x^2)|\mathscr{E}_y|^2 \frac{dk_y}{\sqrt{k_x^2 + k_y^2 - k^2}} \right\} dk_x$$

$$+ \frac{1}{\pi^2 \eta k |V_s|^2} \int_k^{\infty} \left\{ \int_0^{\infty} (k^2 - k_x^2)|\mathscr{E}_y|^2 \frac{dk_y}{\sqrt{k_x^2 + k_y^2 - k^2}} \right\} dk_x \qquad (10.52)$$

Using expression 10.46 for $\mathscr{E}_y$, we obtain for G_s:

$$G_s = \frac{4}{\pi^2 \eta k} \int_0^k \frac{k^2 - k_x^2}{k_x^2} \sin^2(k_x W_e/2)$$

$$\times \left\{ \int_0^{\sqrt{k^2-k_x^2}} \frac{\sin^2(k_y S/2)}{(k_y S/2)^2} \frac{dk_y}{\sqrt{k^2 - k_x^2 - k_y^2}} \right\} dk_x \qquad (10.53)$$

The inner integral can be written as a double integral of the Bessel function of the first kind and order zero [3]. Expansion of the Bessel function in a Maclaurin series and double integration term by term give:

$$\int_0^{\sqrt{k^2-k_x^2}} \frac{\sin^2(k_y S/2)}{(k_y S/2)^2} \frac{dk_y}{\sqrt{k^2 - k_x^2 - k_y^2}} = \frac{\pi}{2} \left\{ 1 - \frac{1}{24} (\sqrt{k^2 - k_x^2} S)^2 + \frac{1}{960} (\sqrt{k^2 - k_x^2} S)^4 \ldots \right\} \qquad (10.54)$$

The first two terms of this series expression are used in eqn. 10.53 to obtain finally

$$G_s \approx \frac{1}{\pi \eta} \left\{ [w \, \text{Si}(w) + \frac{\sin w}{w} + \cos w - 2] \right.$$

$$\left. \times \left(1 - \frac{s^2}{24}\right) + \frac{s^2}{12}\left(\frac{1}{3} + \frac{\cos w}{w^2} - \frac{\sin w}{w^3}\right) \right\} \qquad (10.55)$$

where $w = kW_e$ = normalised slot length; $s = kS$ = normalised slot width;

$$\text{Si}(x) = \int_0^x \frac{\sin u}{u} du$$

544 *Transmission-line model for rectangular microstrip antennas*

As explained before, expression 10.52 is not used to calculate B_s, as it is impossible to define a suitable aperture field. One has to fall back on the open-end-effect concept (see Section 10.3.2).

10.3.4 Expression for the mutual conductance G_m

The expression for the mutual conductance G_m of finite-length slots will be derived from the mutual conductance between infinite slots. Therefore, an auxiliary coupling function is defined

$$F_g = g_m/g_s \tag{10.56}$$

where g_s and g_m are the per-unit-length self-conductance and mutual conductance, respectively, of two infinite-length TE-excited slots in a perfectly conducting infinite ground plane, as shown in Fig. 10.7. The aperture field was given in eqn. 10.2. An analytical expression has been derived for g_s in Section 10.2.2:

$$g_s \approx \frac{k}{2\eta}\left(1 - \frac{s^2}{24}\right) \tag{10.18}$$

To obtain an expression for g_m, the complex radiated power per unit slot length $(p + jq)$ has to be expressed in terms of the Fourier-transformed aperture field (eqns. 10.6 and 10.7), and in terms of the slot voltages V_1 and V_2 (eqn. 10.8). Setting $V_1 = V_2$ and equating the real part p of the radiated power, the following expression for g_m is found:

$$g_m = \frac{k}{\pi\eta}\int_0^k \frac{\sin^2(k_y S/2)}{(k_y S/2)^2}\cos(k_y L_e)\frac{dk_y}{\sqrt{k^2 - k_y^2}} \tag{10.57}$$

where $L_e = L + \Delta l =$ centre distance between the two slots; $S = \Delta l =$ width of the equivalent slots.

Notice that

$$\sin^2\left(k_y \frac{S}{2}\right)\cos(k_y L_e) = \frac{1}{2}\sin^2\left(k_y \frac{L_e + \Delta l}{2}\right)$$

$$+ \frac{1}{2}\sin^2\left(k_y \frac{L_e - \Delta l}{2}\right) - \sin^2\left(k_y \frac{L_e}{2}\right) \tag{10.58}$$

Similarly to the derivation of eqn. 10.11 from eqn. 10.9, we obtain the following expression for eqn. 10.57 using eqn. 10.58:

$$g_m = \frac{k}{2\eta s^2}\{J_0^{ii}(l + s) + J_0^{ii}(l - s) - 2J_0^{ii}(l)\} \tag{10.59}$$

with $J_0^{ii}(s)$ as defined in eqn. 10.13 and $l = kL_e$. Expanding $J_0^{ii}(l \pm s)$ in a Taylor series around l leads to

$$g_m = \frac{k_0}{\eta s^2}\sum_{n=1}^{\infty}\frac{s^{2n}}{(2n)!}J_0^{ii(2n)}(l) \tag{10.60}$$

Transmission-line model for rectangular microstrip antennas 545

where the superscript $(2n)$ denotes the $2n$th derivative. Truncation of these series, maintaining the first two terms, gives

$$g_m \simeq \frac{k}{\eta} \left\{ \frac{1}{2!} J_0(l) + \frac{s^2}{4!} J_0^{(2)}(l) \right\} \tag{10.61}$$

Using the identity

$$J_0^{(2)}(l) = \frac{1}{2} \{J_2(l) - J_0(l)\} \tag{10.62}$$

we finally obtain for g_m:

$$g_m \simeq \frac{k}{2\eta} \left\{ \left(1 - \frac{s^2}{24}\right) J_0(l) + \frac{s^2}{24} J_2(l) \right\} \tag{10.63}$$

The maximum truncation error of eqn. 10.63 is about 0·1% of g_s for $s \leqslant 1$.

Using eqn. 10.18 found for g_s and eqn. 10.63 for g_m the auxiliary coupling function $F_g = g_m/g_s$ can be expressed as

$$F_g \simeq J_0(l) + \frac{s^2}{24 - s^2} J_2(l) \tag{10.64}$$

The auxiliary coupling function F_g has been introduced to calculate a first approximation of the mutual conductance G_m of the finite-length slots by putting

$$G_m = G_s F_g \tag{10.65}$$

with G_s the self-conductance as given by eqn. 10.55.

The results of this approximation are compared to the following reference: the radiation conductance of the four-slot equivalent system shown in Fig. 10.11. This four-slot system consists of two main slots and two side slots. The main slots have a length $W_e = W_{\mathit{eff}}$ and a width Δl, as used before, and have a centre distance $L_e = L + \Delta l$. The side slots have a length L_e, a width Δl and a centre distance W_e. The tangential electric field in the aperture plane $z = 0$ is:

$$E_a = \begin{cases} \dfrac{V_s}{\Delta l} i_y & \text{for} \quad |x| \leqslant \dfrac{W_e}{2}; \dfrac{L_e - \Delta l}{2} \leqslant |y| \leqslant \dfrac{L_e + \Delta l}{2} \\[2mm] \dfrac{V_s}{\Delta l} \sin\left(\dfrac{\pi y}{L_e}\right) i_x & \text{for} \quad |y| \leqslant \dfrac{L_e}{2}; \dfrac{W_e - \Delta l}{2} \leqslant x \leqslant \dfrac{W_e + \Delta l}{2} \\[2mm] \dfrac{-V_s}{\Delta l} \sin\left(\dfrac{\pi y}{L_e}\right) i_x & \text{for} \quad |y| \leqslant \dfrac{L_e}{2}; \dfrac{W_e - \Delta l}{2} \leqslant -x \leqslant \dfrac{W_e + \Delta l}{2} \\[2mm] 0 & \text{elsewhere} \end{cases}$$

$$\tag{10.66}$$

This aperture field is an acceptable approximation of the true tangential electric field in the plane $z = 0$ of the microstrip antenna excited in the fundamental mode. As shown in Reference 12, it allows an accurate computation of the far field and the radiation conductance. The computation is straightforward using

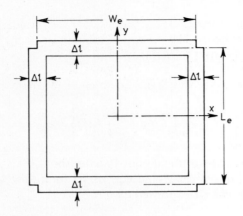

Fig. 10.11 *Four-slot radiation model*

the plane-wave spectral method. The resulting integral expression for the radiation conductance is too complicated for analytical integration, but it can be evaluated numerically without any difficulty. We call this numerically evaluated quantity, the reference conductance G_r^{ref} and we will use it to verify the accuracy of the radiation conductance G_r^{mod} as predicted by the transmission-line model (Fig. 10.9). The conductance G_r^{mod} is given by

$$G_r^{mod} = 2(G_s + G_m) \tag{10.67}$$

where G_s is calculated from eqn. 10.55 and G_m is expressed to a first approximation by eqn. 10.65. Poor correspondence between G_r^{mod} and G_r^{ref} is expected; therefore we add a correction function K_g to compensate for eventual influence of the side slots and of the finite length of the main slots. Instead of eqn. 10.65, we use:

$$G_m = G_s F_g K_g \tag{10.68}$$

The correction function K_g has to be determined by comparison of G_r^{mod} and G_r^{ref}:

$$G_r^{mod} = 2(G_s + G_m) = 2G_s(1 + F_g K_g) \tag{10.69}$$

To obtain a good correspondance between G_r^{mod} and G_r^{ref}, the correction function K_g has to be a good approximation to the numerical reference quantity

$$K_g^{ref} = \left(\frac{G_r^{ref}}{2G_s} - 1\right) \bigg/ F_g \tag{10.70}$$

Extensive numerical investigation of this quantity for a large number of parameter values in the ranges $w \geqslant 0{\cdot}1$, $l \leqslant 3{\cdot}2$ and $s \leqslant 1$ shows the surprising result

$$K_g^{ref} \approx 1 \qquad (10.71)$$

Therefore, the simple expression $G_m = G_s F_g$ can be considered as valid within the given parameter ranges. The validity of this expression for both small ($\geqslant 0{\cdot}1$) and large values of w can be understood as some kind of compensation: the influence of the side slots (which is not taken into account in the calculation of G_s and G_m) and the influence of the finite length of the main slots (which is neglected in the calculation of the factor F_g in G_m) appear to cancel each other out almost perfectly. To illustrate this effect, Table 10.1 lists the quantities G_r^1 ($= 2G_s$, i.e. 2 × self-conductance of a main slot), G_r^2 ($=$ radiation conductance of the two-slot system consisting of the two main slots, see Fig. 10.12), G_r^4 ($= G_r^{ref}$, i.e. the radiation conductance of the four-slot system consisting of the two main slots and the two side slots, see Fig. 10.11) and G_r^{mod} (i.e. the radiation conductance found with the transmission-line model of Fig. 10.9 and calculated from expression 10.69), as a function of w for $l = 2$ and $s = 0$. The influence of the side slots can be deduced from a comparison between G_r^2 and G_r^4; the influence of the finite length of the main slots can be seen from a comparison between G_r^{mod} and G_r^2. In Reference 9 it is argued that the influence of the side slots on the radiation conductance can be neglected, but according to Reference 10 a distinctly better correspondence with experiment is obtained if the side slots are taken into account as described above.

Table 10.1 *Radiation conductance for l = 2 and s = 0*

w	G_r^1 (mS)	G_r^2 (mS)	G_r^4 (mS)	G_r^{mod} (mS)
1	0·55	0·75	0·69	0·68
2	2·11	2·84	2·63	2·58
3	4·40	5·86	5·48	5·38
4	7·80	9·33	8·79	8·67
5	9·87	12·84	12·23	12·08
6	12·61	16·19	15·58	15·43
7	15·26	19·39	18·82	18·68
8	17·86	22·54	22·00	21·86

The accuracy of expression 10.65 can also be shown in a systematic way by computing the relative error:

$$E_p = \frac{G_r^{ref} - G_r^{mod}}{G_r^{ref}} \qquad (10.72)$$

in the above mentioned range of parameters. It appears that E_p is:

- Always positive
- A decreasing function of w (if l and s are constant)
- An increasing function of l (if w and s are constant)
- A decreasing function of s (if w and s are constant).

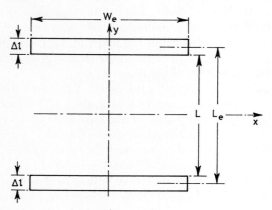

Fig. 10.12 *Two-slot model*

The maximum error for a fixed value of l occurs for $w \blacktriangleright 0$ and $s \blacktriangleright 0$. This case is shown in Fig. 10.13. If the antenna is working in its fundamental mode, the normalised distance l between the main slots is about a half wavelength along the microstrip line formed by the patch:

$$l \approx \frac{\pi}{\sqrt{\varepsilon_{\text{eff}}}} \approx \frac{\pi}{\sqrt{\varepsilon_r}} \tag{10.73}$$

For substrates with $\varepsilon_r > 2$, the error E_p is less than 2·5%.

10.3.5 Expression for the mutual susceptance B_m

The derivation of an expression for B_m is similar to the procedure followed for G_m. First an auxiliary coupling function for the susceptances is defined:

$$F_b = \frac{b_m}{b_s} \tag{10.74}$$

where b_s and b_m are the per-unit-length self-susceptance and the mutual susceptance, respectively, of two infinite-length TE-excited slots in an infinite, perfectly conducting plane, as shown in Fig. 10.7. The electric field in the aperture plane was given in eqn. 10.2. The complex radiated power per unit slot length has been expressed in terms of the Fourier-transformed aperture field (eqns. 10.6 and 10.7), and in terms of the slot voltages V_1 and V_2 (eqn. 10.8). Setting $V_1 = V_2$ and

Transmission-line model for rectangular microstrip antennas

equating the imaginary part q of the radiated power, the following expression for b_m is found

$$b_m = \frac{k}{\pi\eta} \int_k^\infty \frac{\sin^2(k_y S/2)}{(k_y S/2)^2} \cos(k_y L_e) \frac{dk_y}{\sqrt{k_y^2 - k^2}} \tag{10.75}$$

where $S = \Delta l$ = slot width; $L_e = L + \Delta l$ = centre distance between the two slots. Similarly, as for g_m, this integral can be written as

$$b_m = -\frac{k}{2\eta s^2} \{Y_0^{ii}(l+s) + Y_0^{ii}(l-s) - 2Y_0^{ii}(l)\} \tag{10.76}$$

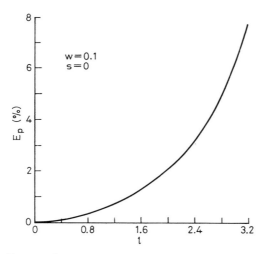

Fig. 10.13 *Error E_p versus l*

with $Y_0^{ii}(s)$ as defined in eqn. 10.14. Expanding $Y_0^{ii}(l \pm s)$ in a Taylor series around l leads to

$$b_m = -\frac{k}{\eta s^2} \sum_{n=1}^\infty \frac{s^{2n}}{(2n)!} Y_0^{ii(2n)}(l) \tag{10.77}$$

where the superscript $2n$ denotes the $2n$th derivative. Truncation of these series, maintaining the first two terms, gives

$$b_m \approx -\frac{k}{\eta} \left\{ \frac{1}{2!} Y_0(l) + \frac{s^2}{4!} Y_0^{(2)}(l) \right\} \tag{10.78}$$

Using the identity

$$Y_0^{(2)}(l) = \frac{1}{2} \{Y_2(l) - Y_0(l)\} \tag{10.79}$$

we finally obtain for b_m

$$b_m \simeq -\frac{k}{2\eta}\left\{\left(1 - \frac{s^2}{24}\right) Y_0(l) + \frac{s^2}{24} Y_2(l)\right\} \tag{10.80}$$

The maximum truncation error of eqn. 10.80 is about 0·1% of b_s if $s \leqslant 1$.

In Section 10.2.2 a closed-form expression for b_s was derived:

$$b_s \simeq -[\frac{k}{\pi\eta}\left\{\left(\ln\frac{s}{2} + C_e - \frac{3}{2}\right)\left(1 - \frac{s^2}{24}\right) + \frac{s^2}{288}\right\} \tag{10.19}$$

Using eqns. 10.80 and 10.19 in eqn. 10.74 enables to write the auxiliary coupling function F_b as

$$F_b = \frac{b_m}{b_s} \simeq \frac{\pi}{2} \frac{Y_0(l) + \frac{s^2}{24 - s^2} Y_2(l)}{\ln\left(\frac{s}{2}\right) + C_e - \frac{3}{2} + \frac{s^2/12}{24 - s^2}} \tag{10.81}$$

We consider the product $(B_s F_b)$ as a first approximation of the mutual susceptance B_m and introduce a correction function K_b so that:

$$B_m = B_s F_b K_b \tag{10.82}$$

One could expect to apply the same method to determine K_b as is used for K_g; i.e. numerical evaluation of the aperture susceptance of the four-slot system of Fig. 10.11 and to consider this quantity as a reference for deriving a suitable expression for K_b. But following problems occur:

(a) It is not clear how a reference susceptance evaluated for a four-slot system has to be related to the quantity K_b of the transmission-line model. We cannot assume conservation of reactive power passing through two different reference planes: the aperture plane $z = 0$ in the equivalent slot model and the input port plane (e.g. port 1 or 2) in the transmission-line model.

(b) The susceptance of a radiating aperture is much more sensitive to the precise form of the aperture-field distribution than is the conductance. Consequently the approximate field distribution of eqn. 10.66 is not appropriate for computing a reference susceptance. Besides, this field distribution does not even meet the required edge behaviour.

(c) It is much more difficult to compute the susceptance of a radiating aperture than its conductance. Using the plane-wave spectral method, the conductance is given by a surface integral over a finite part of the wave-number plane, whereas the susceptance requires a surface integration over an infinite domain.

Because of problem (a) we are unable to use a four-slot system to determine the correction function K_b, but must use the two-slot system of Fig. 10.12 consisting of the main slots. Hence we neglect the influence of the side slots on the susceptance.

Because of problem (*b*), an aperture field that meets the edge conditions [3] is required: the tangential components of the electric field perpendicular to the edge must decrease with distance *d* from the edge, and the components parallel to the edge must decrease with d^2 from the edge. There is no further information available to select the appropriate field distribution. Different distributions give different values for the aperture susceptance. But a detailed numerical evaluation shows that the dependence of the aperture susceptance on the aperture distribution comes from the dependence of the self-susceptance, not from the mutual susceptance. For the configuration of Fig. 10.12, it can be concluded that the uniform aperture-field distribution predicts the mutual susceptance within 1% of the values obtained using appropriate tapered distributions. Hence our reference correction function K_b^{ref} is defined as

$$K_b^{ref} = B_m^{ref}/(b_m W_e) \tag{10.83}$$

where B_m^{ref} is the mutual susceptance of the two-slot system of Fig. 10.12 with a uniform aperture field, and $b_m W_e$ is the mutual susceptance of an equivalent system of length W_e taken from two infinite-length slots having the same width Δl and the same centre distance L_e. Numerical investigation shows that K_b^{ref} is nearly independent of *s* or *l*. It can be concluded that an expression having only the variable *w* can represent the correction function K_b:

$$K_b = 1 - \exp(-0.21w) \tag{10.84}$$

This equation has the correct asymptotic behaviour for $w \blacktriangleright \infty$, as the influence of the finite length of the main slots disappears and K_b has to approach unity.

10.3.6 Expressions for the line parameters

In Section 10.2.3 we discussed the planar waveguide model and the corresponding expressions for the line parameters. Eqns. 10.20–10.28 describe the characteristics of a microstrip line in a quasi-static approach. The frequency dependence of the model parameters can be taken into account through frequency dependence of ε_{eff} and W_{eff}.

A convenient expression for ε_{eff} can be found in References 13 and 14:

$$\varepsilon_{eff}(f) = \varepsilon_r - \frac{\varepsilon_r - \varepsilon_{eff}(0)}{1 + P} \tag{10.85}$$

where $\varepsilon_{eff}(0)$ is given in eqns. 10.25–10.28 and

$$P = P_1 P_2 \{(0.1844 + P_3 P_4) f_n\}^{1.5763} \tag{10.86}$$

$$P_1 = 0.27488 + \{0.6315 + 0.525/(1 + 0.0157 f_n)^{20}\} u$$
$$\quad - 0.065683 \exp(-8.7513 u) \tag{10.87}$$

$$P_2 = 0.33622\{1 - \exp(-0.03442 \varepsilon_r)\} \tag{10.88}$$

$$P_3 = 0.0363 \exp(-4.6 u)\{1 - \exp[-(f_n/38.7)^{4.97}]\} \tag{10.89}$$

$$P_4 = 1 + 2\cdot751\{1 - \exp[-(\varepsilon_r/15\cdot916)^8]\} \tag{10.90}$$

$$f_n = fh[\text{in GHz mm}] = 47\cdot713\,kh \tag{10.91}$$

$$u = \{W + (W' - W)/\varepsilon_r\}/h \tag{10.92}$$

For $W_{\text{eff}}(f)$ an expression has been proposed in Reference 15:

$$W_{\text{eff}}(f) = W + \frac{W_{\text{eff}}(0) - W}{1 + K(f/f_{c1})^2} \tag{10.93}$$

with $W_{\text{eff}}(0)$ as given in eqns. 10.22–10.24; according to Reference 15, K can be equal to 1 and

$$f_{c1} \simeq \frac{c_0}{2\sqrt{\varepsilon_{\text{eff}}(0)}\,W_{\text{eff}}(0)} \tag{10.94}$$

where c_0 is the free space velocity of light.

However, it has been shown by Pues and Van de Capelle [16] that a better asymptotic behaviour for $\varepsilon_r \blacktriangleright 1$ is obtained if

$$K = \frac{\varepsilon_{\text{eff}}(f) - 1}{\varepsilon_{\text{eff}}(f)} \tag{10.95}$$

and a better accuracy, particularly for high frequencies, if

$$f_{c1} = \frac{c_0}{2\sqrt{\varepsilon_{\text{eff}}(f)}\,W_{\text{eff}}(f)} \tag{10.96}$$

Substitution of eqns. 10.95 and 10.96 in eqn. 10.93 gives a cubic equation from which $W_{\text{eff}}(f)$ has to be solved. There is one real solution:

$$W_{\text{eff}}(f) = \frac{W}{3} + (R_w + P_w)^{1/3} - (R_w - P_w)^{1/3} \tag{10.97}$$

where

$$P_w = \left(\frac{W}{3}\right)^3 + \frac{S_w}{2}\left[W_{\text{eff}}(0) - \frac{W}{3}\right] \tag{10.98}$$

$$Q_w = \frac{S_w}{3} - \left(\frac{W}{3}\right)^2 \tag{10.99}$$

$$R_w = (P_w^2 + Q_w^3)^{1/2} \tag{10.100}$$

$$S_w = \frac{c_0^2}{4f^2[\varepsilon_{\text{eff}}(f) - 1]} \tag{10.101}$$

The attenuation constant α can be divided into dielectric losses in the substrate (α_d), conducting losses in the strip conductor (α_{cs}) and in the ground plane (α_{cg}):

$$\alpha = \alpha_d + \alpha_{cs} + \alpha_{cg} \tag{10.102}$$

The dielectric losses are given in Reference 17 as

$$\alpha_d = 0.5\beta \frac{\varepsilon_r}{\varepsilon_{eff}(f)} \frac{\varepsilon_{eff}(f) - 1}{\varepsilon_r - 1} \tan\delta \qquad (10.103)$$

The conducting losses are given in References 17 and 18 as

$$\alpha_{cs} = \alpha_n R_{ss} F_{\Delta s} F_s \qquad (10.104)$$

$$\alpha_{cg} = \alpha_n R_{sg} F_{\Delta g} \qquad (10.105)$$

with

$$R_{ss} = \sqrt{\pi f \mu_0 / \sigma_s} \qquad (10.106)$$

$$R_{sg} = \sqrt{\pi f \mu_0 / \sigma_g} \qquad (10.107)$$

$$\alpha_n = \begin{cases} \dfrac{1}{4\pi h Z_c(0)} \dfrac{32 - (W'/h)^2}{32 + (W'/h)^2}; & W'/h < 1 \quad (10.108) \\[6pt] \dfrac{\sqrt{\varepsilon_{eff}(0)}}{2\eta_0 W_{eff}(0)} \left(W'/h + \dfrac{0.667 W'/h}{W'/h + 1.444} \right); & W'/h \geq 1 \quad (10.109) \end{cases}$$

$$F_{\Delta s} = 1 + \frac{2}{\pi} \arctan\{1.4 (R_{ss} \Delta_s \sigma_s)^2\} \qquad (10.110)$$

$$F_{\Delta g} = 1 + \frac{2}{\pi} \arctan\{1.4 (R_{sg} \Delta_g \sigma_g)^2\} \qquad (10.111)$$

$$F_s = 1 + \frac{2h}{W'} \left(1 + \frac{\partial W'}{\partial t} \right) \qquad (10.112)$$

which gives, with eqn. 10.24 for W',

$$F_s \simeq 1 + \frac{2h}{W'} \left(1 - \frac{1}{\pi} + \frac{W' - W}{t} \right) \qquad (10.113)$$

10.4 Application of the improved transmission-line model

10.4.1 Analysis and design of rectangular microstrip antennas

All parameters of the improved transmission-line model have been given in terms of closed-form expressions. This enables one to program the model very easily for analysis as well as for design purposes.

The input admittance Y_{in} is expressed in eqn. 10.30 and the resonance condition is defined as

$$\text{Im}(Y_{in}) = 0 \qquad (10.114)$$

For the case of a microstrip-line-fed antenna the input admittance is given in eqn. 10.35. The usefulness of the transmission-line model is illustrated by the simple expressions that can be derived for several important antenna characteristics. For the radiation conductance we find

$$G_r = G_s(r + |v|^2) - 2G_m \operatorname{Re}(v) \tag{10.115}$$

where r is defined in eqn. 10.33; v = voltage-excitation ratio of the main slots

$$v = \frac{V_2}{V_1} = \frac{Y_m + Y_c \operatorname{csch}(\gamma_p L)}{Y_s + Y_c \operatorname{coth}(\gamma_p L)} \tag{10.116}$$

The antenna efficiency follows from

$$q = \frac{G_r}{\operatorname{Re}(Y_{in})} \tag{10.117}$$

The directivity in the broadside direction is given by

$$D = \frac{|r - v|^2 w^2}{\pi \eta_0 G_r} \approx \frac{4w^2}{\pi \eta_0 G_r} \tag{10.118}$$

and the antenna gain by

$$G = qD \tag{10.119}$$

The resonant input conductance is defined as

$$G_{res} = \operatorname{Re}(Y_{in})|_{f=f_{res}} \tag{10.120}$$

where f_{res} is the resonant frequency which follows from eqn. 10.114.

The antenna input admittance, given by eqn. 10.35, can be modelled fairly accurately by the resonant input conductance G_{res} connected in parallel with a lossless open-ended half-wavelength transmission line with characteristic admittance Y_s:

$$Y'_{in} \approx G_{res} + jY_c \tan\left(\pi \frac{f}{f_{res}}\right) \tag{10.121}$$

Consequently, the unloaded antenna quality factor is given by

$$Q = \frac{\pi}{2} \frac{Y_c}{G_{res}} \tag{10.122}$$

and the impedance bandwidth by using Reference 19,

$$BW = \frac{100\%}{Q} \frac{S - 1}{S} \tag{10.123}$$

where S is the maximum value of the voltage standing-wave ratio that is allowed on the feed line.

10.4.2 Comparison with other methods

To verify the usefulness of the improved transmission-line model, we compare it with other published results, theoretical as well as experimental. Fig. 10.14 shows the input impedance of a rectangular microstrip antenna excited by a microstrip line. The Figure compares measured results of Lo et al. [20], calculated results published by Deshpande and Bailey [21] and calculated results obtained with the improved transmission-line model. Eqn. 10.35 was evaluated for: $W = 144$ mm, $L = 76$ mm, $W_m = 4.3$ mm, $h = 1.59$ mm, $\varepsilon_r = 2.62$, $\tan \delta = 0.001$, $t = 0.035$ mm, $\sigma_s = \sigma_g = 0.556 \times 10^5$ S/mm, $\Delta_s = \Delta_g = 0.0015$ mm.

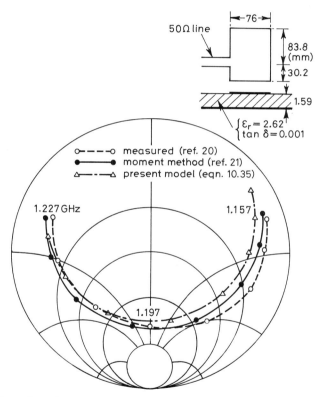

Fig. 10.14 *Input impedance of a rectangular microstrip antenna fed by a microstrip line*

The moment-method results of Reference 21 agree somewhat better with the experimental results [20], than does the transmission-line model. However, detailed comparison with Reference 21 proves that the transmission-line model is more accurate than the calculated results of Lo et al. obtained with a cavity model [20] and of Newman and Tulyathan obtained by a moment method [22]. The discrepancy between the transmission-line model and experiment can be explained by the tolerances on the structural parameters. For example, almost perfect agreement with experimental results was obtained using $\varepsilon_r = 2\cdot 64$ instead of $2\cdot 62$, and if the losses were somewhat less. Observe that copper losses are neglected in the calculations of Deshpande and Bailey [21].

10.4.3 Comparison with experimental results

The improved transmission-line model is used to analyse a square microstrip antenna, shown in Fig. 10.15. The antenna is matched to $50\,\Omega$ by a quarter-wave transformer. The structure has been photo-etched on a RT/Duroid 5880 substrate of $0\cdot 031\,\text{in} = 0\cdot 787\,\text{mm}$ thickness and connected to an OSM–215–3 connector. The measured dimensions of the copper pattern are:

$L = 33\cdot 147\,\text{mm}$, $W = 33\cdot 165\,\text{mm}$, $W_m = 0\cdot 473\,\text{mm}$, $L_m = 18\cdot 713\,\text{mm}$, $W_F = 2\cdot 403\,\text{mm}$, $L_F = 20\,\text{mm}$

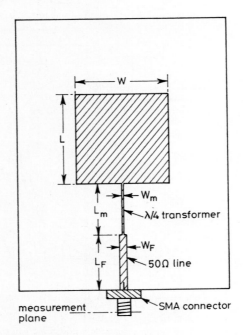

Fig. 10.15 *Square microstrip antenna fed with quarter-wave transformer*

The parameters of the substrate are:

$h = 0\cdot 7874\,\text{mm}$, $\varepsilon_r = 2\cdot 20$, $\tan\delta = 0\cdot 0009$, $t = 0\cdot 018\,\text{mm}$,
$\sigma_g = \sigma_s = 0\cdot 556 \times 10^5\,\text{S/mm}$, $\Delta s = \Delta g = 0\cdot 5 \times 10^{-3}\,\text{mm}$

The antenna has been analysed completely by a cascade of transmission-line models (see Fig. 10.16):

- Model for the coaxial–microstripline transition [23],
- Transmission-line representation of the feed line
- Model for the step discontinuity in the microstrip lines
- Transmission-line representation of the quarter-wave transformer
- Improved transmission-line model for the microstrip antenna

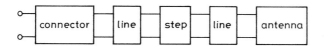

Fig. 10.16 *Schematic representation of complete analysis model for the antenna of Fig. 10.15*

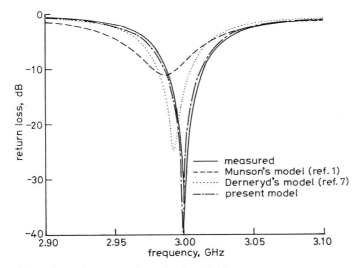

Fig. 10.17 *Return loss of antenna shown in Fig. 10.15*

Fig. 10.17 shows the measured reflection diagram and three calculated curves. To model the antenna element, we have used eqn. 10.35, but with different values of Y_s and Y_m. To simulate Derneryd's model [7] we set $Y_m = 0$; and to simulate Munson's model [1] we set $Y_m = 0$ and $Y_s = W_e\, y_s$. One can clearly observe the effect of neglecting the mutual coupling between the equivalent slots (by comparing Derneryd's model with the improved transmission-line model) and of neglecting the influence of the finite length of the slots (by comparing Munson's model with Derneryd's model).

10.4.4 Design application

The design of microstrip antennas by the improved transmission-line model is

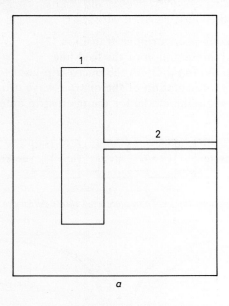

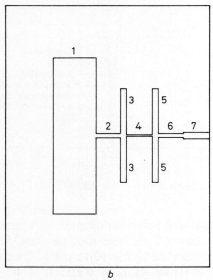

Fig. 10.18 *Rectangular microstrip antenna*
  a Fed by a 50Ω microstrip line
  b Fed by a co-planar impedance-matching network

very powerful if the model is combined with CAD packages for microstriplines or striplines. To illustrate this application we discuss the design of a rectangular microstrip antenna, combined with a broadband impedance-matching network.

The antenna shown in Fig. 10.18a has the following dimensions: $W_1 = 119.83$ mm, $L_1 = 31.68$ mm, $W_2 = 4.87$ mm, $L_2 = 83.32$ mm. It has been etched on a RT/Duroid 5880 substrate of 200 mm × 150 mm × 1·5748 mm. The other parameters (ε_r, tan δ, t, σ_s, σ_g, Δ_s, Δ_g) are the same as in the previous Section. Fig. 10.19 shows the calculated and the measured return loss of this antenna. The best match occurs at 3·025 GHz (-21.5 dB) and the improved transmission-line model predicts best match at 3·040 GHz (-26.96 dB).

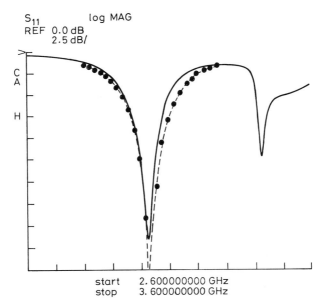

Fig. 10.19 *Return loss of the antenna of Fig. 10.18a*
——— measured
---- calculated

The impedance bandwidth can be increased with a reactive matching network [24]. A broadband-matching design procedure developed by Pues (US Patent 4445122) can be applied. The design has been carried out combining the improved transmission line model for the antenna element with suitable design models for the microstrip-network components. The design result is shown in Fig. 10.18b. The dimensions are as follows:

$$W_1 = 119.83 \text{ mm}, L_1 = 31.68 \text{ mm}, W_2 = 3.14 \text{ mm}, L_2 = 20.38 \text{ mm},$$
$$W_3 = 4.34 \text{ mm}, L_3 = 35.98 \text{ mm}, W_4 = 0.70 \text{ mm}, L_4 = 23.15 \text{ mm},$$

$W_5 = 4.46\,\text{mm}$, $L_5 = 35.95\,\text{mm}$, $W_6 = 2.65\,\text{mm}$, $L_6 = 20.50\,\text{mm}$,
$W_7 = 4.87\,\text{mm}$, $L_7 = 19.30\,\text{mm}$

Both prototypes (Figs. 10.18a and b) have been realised by the same etch process on pieces of substrate cut from the same sheet. The return loss of the impedance-matched antenna is shown in Fig. 10.20. Within the band of operation, the worst match occurs at 3·035 GHz (−8·8 dB). The bandwidth at this reflection level (VSWR = 2·14) has been increased by a factor of 3·2 up to a value of 275 MHz

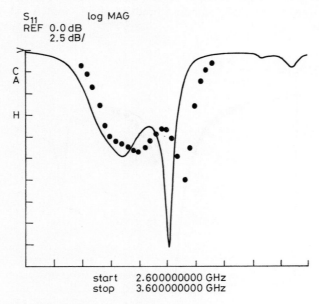

Fig. 10.20 *Return loss of the antenna of Fig. 10.18b*
——— measured
- - - - calculated

or 9·1%, whereas the theoretical maximum bandwidth-enlargement factor for this degree of matching equals 4·0. For further illustration of this application, Fig. 10.21 compares the transmission performance of the two antennas, which is proportional to the realised gain. The impedance-matched antenna is a more efficient radiator, including dissipation losses in the matching network, over the 2·832–2·988 GHz and the 3·055–3·174 GHz band, whereas the antenna without impedance matching is more efficient between both frequency bands. The maximum gain difference equals 0·61 dB and occurs at 3·026 GHz. The co- and cross-polar radiation patterns have been recorded in *E*- and *H*-plane at three different frequencies: 2·9, 3·0 and 3·1 GHz. The results are shown in Fig. 10.22 for the unmatched antenna and in Fig. 10·23 for the impedance-matched antenna. It can be observed that the co-planar matching network does not disturb the radiation characteristics.

This design procedure can also be applied to combine the antenna element with a stripline matching network in a multi-layer structure [25].

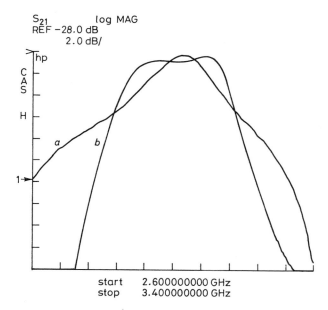

Fig. 10.21 *Comparison of transmission performance of the antennas of Fig. 10.18*
  a Antenna of Fig. 10.18*a*
  b Antenna of Fig. 10.18*b*

10.5 Transmission-line model for mutual coupling

10.5.1 Description of the model

Although more rigorous methods have been developed to calculate the mutual coupling between microstrip antennas, the transmission-line model provides a numerically efficient alternative. The mutual coupling is caused by the simultaneous effect of:

- Interaction through free-space radiation
- Interaction through surface waves.

The influence of surface waves can be neglected if we confine our attention to antennas with substrates of small electrical thickness and low permittivity. In practice, this limitation is not very restrictive.

To develop the transmission-line model for mutual coupling, the following procedure is used [26]:

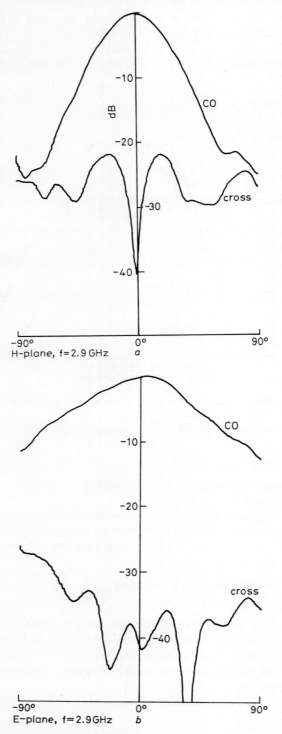

Fig. 10.22 *Radiation patterns of the antenna of Fig. 10.18a*

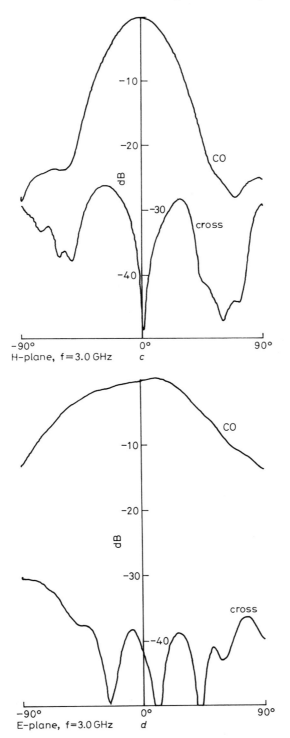

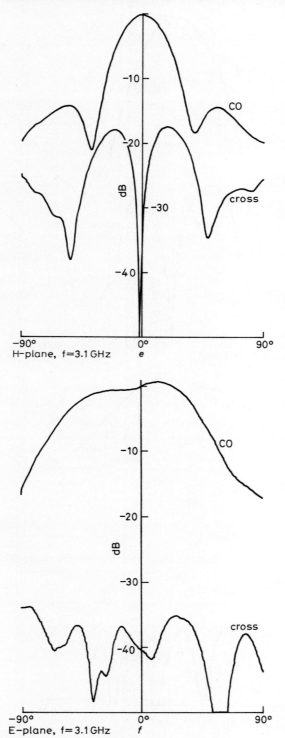

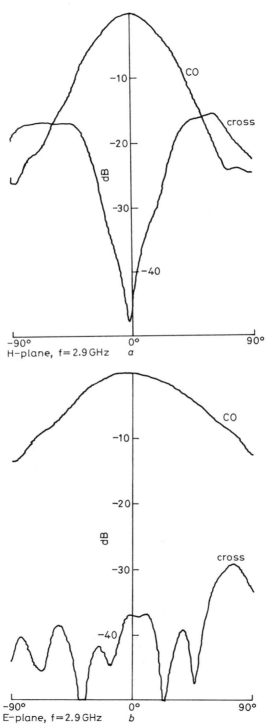

Fig. 10.23 *Radiation patterns of the antenna of Fig. 10.18b*

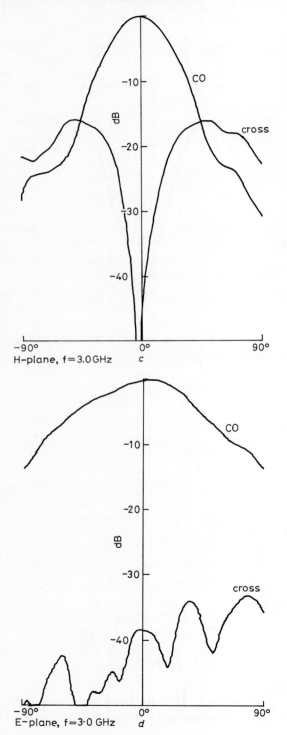

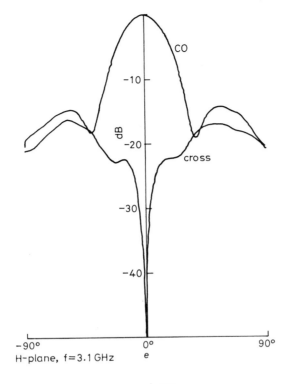
H-plane, f = 3.1 GHz e

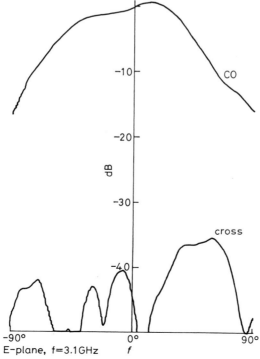
E-plane, f = 3.1 GHz f

568 Transmission-line model for rectangular microstrip antennas

- Each microstrip antenna is represented by its improved transmission-line model as described in Section 10.3.
- To model the mutual coupling between different antennas, each antenna is replaced by a two-slot system, consisting of the two main slots (Fig. 10.24b); hence the influence of the side slots on the mutual coupling is neglected.
- The aperture field in the equivalent slots is assumed to be uniform; the slots have length $W_e = W_{eff}$, width $S = \Delta l$ and a centre distance $L_e = L + \Delta l$.
- The transmission line model of each antenna is completed with voltage-dependent current sources representing the mutual coupling between equivalent slots of different antennas (Fig. 10.25).

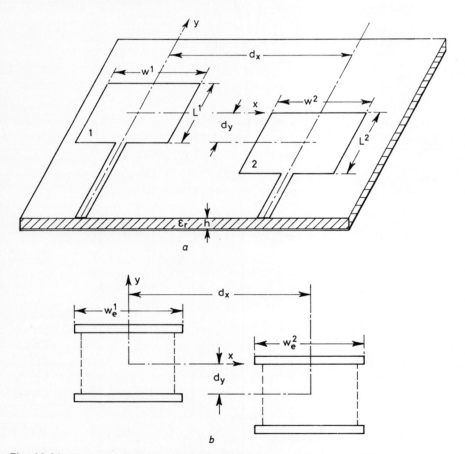

Fig. 10.24 *Two-slot model for the mutual coupling between rectangular microstrip antennas*

Fig. 10.25 shows the complete transmission-line model, including mutual coupling, for the case of two antennas (1 and 2). The self-admittance in the transmission-line model of each individual antenna is denoted by Y_s^1 for antenna 1,

Y_s^2 for antenna 2. The mutual admittance between equivalent slots within one antenna is given by Y_m^1 and Y_m^2, respectively. The mutual admittances between equivalent slots of different antennas are denoted by Y_{me}, Y_{mr}, Y_{mf}, Y_{ms}, respectively.

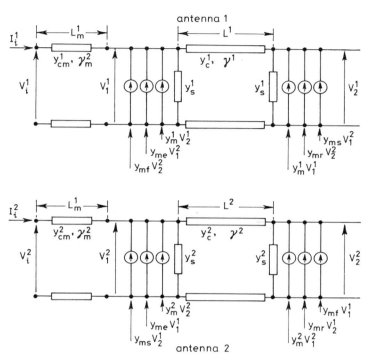

Fig. 10.25 *Transmission-line model for mutual coupling between rectangular microstrip antennas*

Consider the simplest case of feeding directly at the edge of each microstrip antenna (i.e. $L_m^1 = L_m^2 = 0$ in Fig. 10.25), and assume port 1 of each antenna is the respective input port. The input admittances Y^{11} of antenna 1, Y^{22} of antenna 2 and the mutual admittances $Y^{12} = Y^{21}$ between antennas 1 and 2 are obtained through elimination of the voltages (V_2^1 and V_2^2) at ports 2 of both antennas:

$$Y^{11} = \frac{(Y_{st}^1)^2 Y_{st}^2 - Y_{st}^1(Y_{mr})^2 + 2Y_{mt}^1 Y_{mr} Y_{mf} - (Y_{mt}^1)^2 Y_{st}^2 - (Y_{mf})^2 Y_{st}^1}{N}$$

(10.124)

$$Y^{22} = \frac{(Y_{st}^2)^2 Y_{st}^1 - Y_{st}^2(Y_{mr})^2 + 2Y_{mt}^2 Y_{mr} Y_{ms} - (Y_{mt}^2)^2 Y_{st}^1 - (Y_{ms})^2 Y_{st}^2}{N}$$

(10.125)

$$Y^{12} = Y^{21} = \frac{Y_{me} Y_{st}^1 Y_{st}^2 - Y_{me}(Y_{mr})^2 + Y_{mt}^1 Y_{mt}^2 Y_{mr}}{N}$$
$$- \frac{Y_{mt}^1 Y_{st}^2 Y_{ms} - Y_{mf} Y_{mt}^2 Y_{st}^1 + Y_{mf} Y_{ms} Y_{mr}}{N} \quad (10.126)$$

with

$$N = Y_{st}^1 Y_{st}^2 - (Y_{mr})^2 \quad (10.127)$$
$$Y_{st}^1 = Y_s^1 + Y_c^1 \coth(\gamma_p^1 l^1) \quad (10.128)$$
$$Y_{mt}^1 = Y_m^1 - Y_c^1 \operatorname{csch}(\gamma_p^1 l^1) \quad (10.129)$$
$$Y_{st}^2 = Y_s^2 + Y_c^2 \coth(\gamma_p^2 l^2) \quad (10.130)$$
$$Y_{mt}^2 = Y_m^2 - Y_c^2 \operatorname{csch}(\gamma_p^2 l^2) \quad (10.131)$$

In these expressions a superscript (e.g. i in Y^i) denotes the number of the antenna ($i = 1$ or 2); an exponent is denoted as $(Y^i)^2$, which means Y^i to the power 2.

10.5.2 Calculation of the model parameters

The transmission-line model for mutual coupling (Fig. 10.25) contains:

- Parameters depending on only one antenna
- Parameters expressing the mutual admittance between equivalent slots of different antennas

For the first kind of parameters, the expressions derived in Section 10.3 are valid:

Self conductance G_s^i = expression 10.55
Self-conductance B_s^i = expression 10.36
Mutual conductance G_m^i = expression 10.65
Mutual susceptance B_m^i = expression 10.82
Line parameters, see Section 10.2.3 and 10.3.6

For the second kind of parameters, the mutual admittance between equivalent slots of different microstrip antennas has to be calculated. The geometrical configuration of two arbitrarily chosen slots i and j is shown in Fig. 10.26. The slots have length W_e^i, W_e^j respectively; width Δl^i and Δl^j, respectively; centre distance Δx^{ij} in the x-direction, and centre distance Δy^{ij} in the y-direction.

We start from the expression for the complex radiated power:

$$P + jQ = \frac{1}{2} \iint_A \mathbf{E} \times \mathbf{H}^* \cdot \mathbf{i}_z \, dx \, dy \quad (10.132)$$

where A is the aperture surface of the slots i and j, and $\mathbf{E}$ and $\mathbf{H}$ are the electric and magnetic fields, respectively, in the aperture plane.

The complex radiated power can be written in terms of the impressed voltages

V_i and V_j impressed on the equivalent slots i and j, respectively, and in terms of the self and mutual admittances:

$$P + jQ = \frac{1}{2}(V_i I_i^* + V_j I_j^*) = \frac{1}{2} Y_{si}^* V_i V_i^* + \frac{1}{2} Y_{sj}^* V_j V_j^*$$
$$+ \frac{1}{2} Y_{ij}^* V_i V_j^* + \frac{1}{2} Y_{ji}^* V_j V_i^* \qquad (10.133)$$

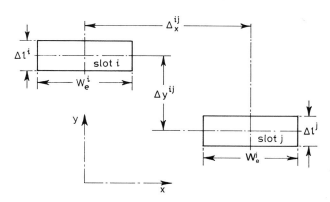

Fig. 10.26 *Geometrical configuration of two arbitrarily chosen slots i and j*

where Y_{si}, Y_{sj} are the self-admittances of slots i and j, respectively. Y_{ij}, Y_{ji} is the mutual admittance between slot i and slot j. Owing to the reciprocity theorem the mutual admittances are equal:

$$Y_{ij} = Y_{ji} \qquad (10.134)$$

Eqn. 10.133 can be developed in terms of the fields E, H or in terms of the Fourier-transformed fields $\mathscr{E}$, $\mathscr{H}$. The last one is suitable for deriving analytical series expressions (as used in Section 10.2.2, 10.3.3, 10.3.4 and 10.3.5). The integral expression in terms of the fields E, H is more suitable for direct numerical evaluation. Both methods will be considered.

To develop eqn. 10.132 in terms of the E, H fields, the aperture fields have to be expressed in their impressed and induced (by mutual coupling) field components:

$$E_i = E_{ai} + E_{ij} \qquad (10.135)$$
$$H_i = H_{ai} + H_{ij} \qquad (10.136)$$
$$E_j = E_{aj} + E_{ji} \qquad (10.137)$$
$$H_j = H_{aj} + H_{ji} \qquad (10.138)$$

where E_i, H_i, E_j, H_j = total tangential fields, slots i, j; E_{ai}, H_{ai}, E_{aj}, H_{aj} = impressed tangential fields, slots i, j; E_{ij}, H_{ij}, E_{ji}, H_{ji} = induced tangential fields, slots i, j. Eqn. 10.133 is written in terms of impressed voltages; consequently eqn. 10.132 has to be developed in terms of impressed electric fields in order to enable identification of terms. Consequently:

$$E_{ij} = 0 \tag{10.139}$$
$$E_{ji} = 0 \tag{10.140}$$

and

$$\begin{aligned} P + jQ &= \frac{1}{2} \iint_{A_i} \boldsymbol{E}_{ai} \times (\boldsymbol{H}_{ai}^* + \boldsymbol{H}_{ij}^*) \cdot \boldsymbol{i}_z \, dx \, dy \\ &+ \frac{1}{2} \iint_{A_j} \boldsymbol{E}_{aj} \times (\boldsymbol{H}_{aj}^* + \boldsymbol{H}_{ji}^*) \cdot \boldsymbol{i}_z \, dx \, dy \\ &= \frac{1}{2} \iint_{A_i} \{ (\boldsymbol{i}_z \times \boldsymbol{E}_{ai}) \cdot \boldsymbol{H}_{ai}^*) + (\boldsymbol{i}_z \times \boldsymbol{E}_{ai}) \cdot \boldsymbol{H}_{ij}^* \} \, dx \, dy \\ &+ \frac{1}{2} \iint_{A_j} \{ (\boldsymbol{i}_z \times \boldsymbol{E}_{aj}) \cdot \boldsymbol{H}_{aj}^* + (\boldsymbol{i}_z \times \boldsymbol{E}_{aj}) \cdot \boldsymbol{H}_{ji}^* \} \, dx \, dy \end{aligned} \tag{10.141}$$

Equating eqns. 10.133 and 10.141, and identification of terms, gives:

$$Y_{si}^* V_i V_i^* = \iint_{A_i} (\boldsymbol{i}_z \times \boldsymbol{E}_{ai}) \cdot \boldsymbol{H}_{ai}^* \, dx \, dy \tag{10.142}$$

$$Y_{sj}^* V_j V_j^* = \iint_{A_j} (\boldsymbol{i}_z \times \boldsymbol{E}_{aj}) \cdot \boldsymbol{H}_{aj}^* \, dx \, dy \tag{10.143}$$

$$Y_{ij}^* V_i V_j^* = \iint_{A_i} (\boldsymbol{i}_z \times \boldsymbol{E}_{ai}) \cdot \boldsymbol{H}_{ij}^* \, dx \, dy \tag{10.144}$$

$$Y_{ji}^* V_j V_i^* = \iint_{A_j} (\boldsymbol{i}_z \times \boldsymbol{E}_{aj}) \cdot \boldsymbol{H}_{ji}^* \, dx \, dy \tag{10.145}$$

Eqns. 10.142 and 10.143 determine the self-admittances Y_{si} and Y_{sj}. The mutual admittance $Y_{ij} = Y_{ji}$ can be calculated from eqn. 10.144 or 10.145:

$$Y_{ij}^* = Y_{ji}^* = \frac{1}{V_i V_j^*} \iint_{A_i} (\boldsymbol{i}_z \times \boldsymbol{E}_{ai}) \cdot \boldsymbol{H}_{ij}^* \, dx \, dy \tag{10.146}$$

$$= \frac{1}{V_j V_i^*} \iint_{A_j} (\boldsymbol{i}_z \times \boldsymbol{E}_{aj}) \cdot \boldsymbol{H}_{ji}^* \, dx \, dy \tag{10.147}$$

We proceed with eqn. 10.147. The vector product

$$\boldsymbol{i}_z \times \boldsymbol{E}_{aj} = -\boldsymbol{K}_{mj} \tag{10.148}$$

where $\boldsymbol{K}_{mj}$ is an equivalent magnetic surface current.

The induced magnetic field $\boldsymbol{H}_{ji}$ in slot j is caused by the impressed field $\boldsymbol{E}_{ai}$ in slot i. This induced field can be expressed as

$$\boldsymbol{H}_{ji} = \iint_{A_i} \boldsymbol{h}_{ji} \cdot \boldsymbol{K}_{mi} \, dx \, dy \tag{10.149}$$

where h_{ji} is a dyadic Green's function giving the magnetic field at a position r_j on slot j caused by a unit magnetic current source located at position r_i on slot i; and K_{mi} is the equivalent magnetic surface current replacing the aperture field E_{ai} by a current source in free space:

$$K_{mi} = -2i_z \times E_{ai} \quad (10.150)$$

The impressed aperture fields E_{ai} and E_{aj} occurring in the two-slot system of Fig. 10.24b have only a -y-component. Consequently, K_{mi} and K_{mj} have only an x-component:

$$K_{mi} = -2i_z \times i_y E_{ai} = 2E_{ai} i_x \quad (10.151)$$

$$K_{mj} = -i_z \times i_y E_{aj} = E_{aj} i_x \quad (10.152)$$

Applying eqns. 10.148–10.152 in eqn. 10.147 enables to write

$$Y^*_{ij} = \frac{-2}{V_j V^*_i} \iint_{A_j} E_{aj} \cdot i_x \left(\iint_{A_i} i_x \cdot h_{ji} \cdot i_x E_{ai} \, dx \, dy \right) dx \, dy \quad (10.153)$$

We denote the xx-component of the dyadic Green's function as h^{xx}_{ji}. This depends only on the distance $|r_j - r_i|$ between observation point and source point. The impressed field distributions were assumed to be uniform in the equivalent two-slot system of Fig. 10.26; consequently:

$$Y^*_{ij} = \frac{-2E_{ai}E_{aj}}{V_j V^*_i} \iint_{A_j} \left(\iint_{A_i} h^{xx}_{ji}(r_j - r_i) \, dx \, dy \right) dx \, dy \quad (10.154)$$

After substitution:

$$u = r_j - r_i \quad (10.155)$$

$$v = r_j + r_i \quad (10.156)$$

the integration with respect to v can be performed analytically and the four-dimensional integral can be reduced to a two-dimensional one, which improves the numerical efficiency and makes this expression very suited for direct numerical evaluation.

In order to avoid numerical evaluation of integral expressions, and to further improve the efficiency, analytical series expressions have been derived [27]. These expressions are obtained starting from the expressions for the mutual admittance in terms of the Fourier-transformed aperture fields.

10.5.3 Comparison with other methods

In order to check the validity of the transmission-line model for mutual coupling, comparison with published theoretical and experimental results will be discussed.

First we compare with the experimental results published by Jedlicka and Carver [28]. It deals with the mutual coupling between two identical microstrip antennas in the E-plane (Fig. 10.27) and in the H-plane (Fig. 10.28). The coupling is measured as a function of the distance between the patch edges, which is normalised with respect to the free-space wavelength λ_0. Neither the dimensions or the location of the feed probe, nor the permittivity of the substrate are mentioned in Reference 28. As the microstrip antenna is matched at 1·405 GHz and the permittivity was estimated in Reference 29 to be 2·50, we obtain a probe diameter of 0·3 mm and a distance between the feed point and the edge of the patch equal to 20·0 mm. To account for the probe inductance, Harrington's formula (Reference 30 pp. 378), has been used. The calculated results obtained with the transmission-line model are shown by the solid line in Figs. 10.27 and 10.28. In the *E*-plane, the correspondence between the transmission-line model and the experimental results is quite good; the largest difference occurs at a small distance of $0·2\,\lambda_0$. In the *H*-plane, the correspondence is less good, as can be expected from neglecting the side slots.

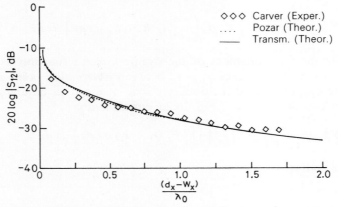

Fig. 10.27 *Mutual coupling between two rectangular microstrip antennas in the **E**-plane*
$W_x = 66$ mm $W_y = 105.6$ mm $f = 1.405$ GHz
$h = 1.5785$ mm $\varepsilon_r = 2.50$ $W_{x1} = 20.0$ mm

In order to compare the transmission-line model with more complicated models, Figs. 10.27 and 10.28 also show by dotted line the results published by Pozar [29]. It is a moment-method solution, which uses the grounded-dielectric-slab Green's function to account for the presence of the substrate and surface waves. This method gives, as expected, excellent agreement with experiment, even though the calculations were made with at most three expansion functions.

Another theoretical method, due to Penard [31], is based on a cavity model where the mutual coupling is considered as the coupling between two current loops around the patch surface. This method takes into account all the equivalent slots around the patch, but neglects the width of the slots, the surface waves, the variation of the ideal field distribution in the slots versus frequency and the higher-order cavity modes. Figs. 10.29 and 10.30 compare Penard's results with the transmission-line model for a severe test case, i.e. mutual distances smaller than $0·5\,\lambda_0$. Because of the presence of the side slots in Penard's model, it sometimes gives better agreement with experiment than the transmission-line model, except for very small distances ($<0.1\lambda_0$), where Penard's model does not give satisfactory results owing to neglecting the slot width.

In Fig. 10.31 the transmission-line model is compared with theoretical and experimental results published by Malkomes [32]. In this case the mutual

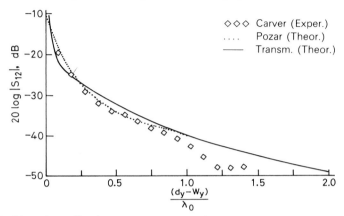

Fig. 10.28 *Mutual coupling between two rectangular microstrip antennas in the **H**-plane*
$W_x = 66$ mm $W_y = 105.6$ mm $f = 1.405$ GHz
$h = 1.5785$ mm $\varepsilon_r = 2.50$ $W_{x1} = 20.0$ mm

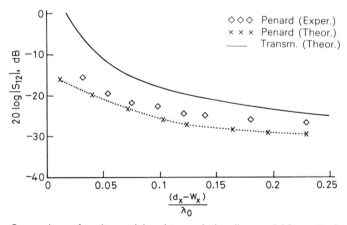

Fig. 10.29 *Comparison of cavity model and transmission-line model for mutual coupling in the **E**-plane*
$W_x = 40$ mm $W_y = 60$ mm $f = 1.548$ GHz
$h = 1.52$ mm $\varepsilon_r = 2.55$ $W_{x1} = 12.5$ mm

coupling is given as a function of the centre distance between the microstrip antennas. The antennas are fed by microstrip lines and the measurements are done at a frequency of 4·77 GHz. Malkomes' theoretical results have been obtained by taking into account all the cavity modes and the four equivalent slots around the patch. Fig. 10.31 shows that, for distances ranging from 0·95 to 4·45 λ_0, the transmission-line model offers the same accuracy as the more complicated and time-consuming method of Malkomes.

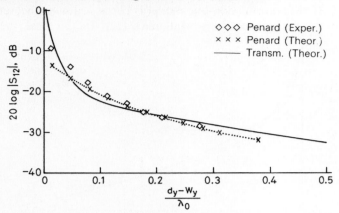

Fig. 10.30 *Comparison of cavity model and transmission-line model for mutual coupling in the **H**-plane*
W_x = 40 mm W_y = 60 mm f = 2.250 GHz
h = 1.52 mm ε_r = 2.55 W_{x1} = 0.0 mm

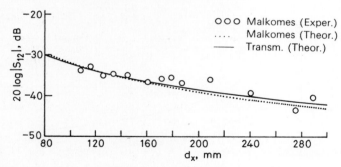

Fig. 10.31 *Comparison of transmission-line model with results of Malkomes (Reference 32) for mutual coupling in the **E**-plane*
W_x = 20 mm W_y = 29 mm f = 4.77 GHz
h = 0.5 mm ε_r = 2.32 W_{x1} = 0.0 mm

A final comparison with results provided by Daniel [33] is given in Fig. 10.32. It shows the mutual coupling between two rectangular microstrip antennas in an eight-element array, designed to work near 21 GHz. Matching of each element was obtained by a proper choice of the feed location. The feed is a coaxial feed-through connection with a diameter of 0·3 mm for the centre conductor. In Fig. 10.32 the experimental results are compared with those obtained with Penard's model and with the transmission-line model. In this case, the agreement of the transmission-line model is even better than for the cavity model. To illustrate the influence of the feed inductance on the mutual coupling, the transmission-line-model results are also given for a centre-conductor diameter of 0·1 mm.

These comparisons show that the improved transmission-line model calculates the mutual coupling in a fairly accurate and very efficient way. It offers a powerful tool for the practical analysis and design of microstrip antennas and arrays, as it can easily be combined with existing microwave CAD packages.

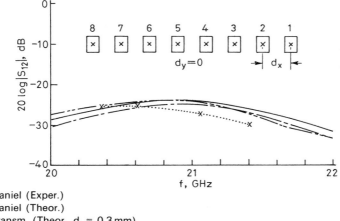

— — — Daniel (Exper.)
× × × Daniel (Theor.)
———— Transm. (Theor., $d_f = 0.3$ mm)
– – – Transm. (Theor., $d_f = 0.1$ mm)

Fig. 10.32 *Mutual coupling between elements in an eight-element linear microstrip array*
$W_x = 4.5$ mm $W_y = 5.96$ mm $d_x = 10$ mm
$h = 0.38$ mm $\varepsilon_r = 2.17$ $W_{x1} = 1.06$ mm

10.6 Acknowledgments

The author acknowledges the National Fund for Scientific Research of Belgium for the support of his research on microstrip antennas. He also expresses his gratitude to Dr. H. Pues, Dr. B. Nauwelaers and Dr. E. Van Lil, who have largely contributed to the development of the material presented in this Chapter.

10.7 References

1. MUNSON, R.: 'Conformal microstrip antennas and microstrip phased arrays' *IEEE Trans.*, 1974, **AP-22**, pp. 74–78
2. COLLIN, R., and ZUCKER, F. Eds.: 'Antenna theory', (McGraw-Hill, NY, 1969)
3. RHODES, D.: 'Synthesis of planar antenna sources', (Oxford University Press, 1974)
4. ABRAMOWITZ, M., and STEGUN, I.: 'Handbook of mathematical functions', (Dover, NY)
5. KOMPA, G., and MEHRAN, R.: 'Planar waveguide model for calculating microstrip components', *Electron. Lett.*, 1975, **11**, pp. 459–460
6. HAMMERSTAD, E., MELHUS, I., JENSEN, O., and BEKKADAL, F.: 'Simulation of microwave components'. ELAB Report. STF44 F80127, Trondheim, Norway, Dec. 1980
7. DERNERYD, A.: 'Linearly polarised microstrip antennas', *IEEE Trans.*, 1976, **AP-24**, pp. 846–851
8. DERNERYD, A.: 'A theoretical investigation of the rectangular microstrip antenna element', *IEEE Trans.*, 1978, **AP-26**, pp. 532–535
9. LIER, E.: 'Improved formulas for input impedance of coax-fed microstrip patch antennas', *IEE Proc H.*, 1982m **129**, pp. 161–164

10 PUES, H., and VAN DE CAPELLE, A.: 'Accurate transmission-line model for the rectangular microstrip antenna', *IEE Proc.*, 1984, **131H**, pp. 334–340
11 KIRSCHNING, M., JANSEN, R., and KOSTER, N.: 'Accurate model for open end effect of microstrip lines', *Electron. Lett.*, 1981, **17**, pp. 123–125
12 HAMMER, P., VAN BOUCHAUTE, D., VERSCHRAEVEN, D., and VAN DE CAPELLE, A.: 'A model for calculating the radiation field of microstrip antennas', *IEEE Trans.*, 1979, **AP-27**, pp. 267–270
13 KIRSCHNING, M., and JANSEN, R.: 'Accurate model for effective dielectric constant of microstrip with validity up to millimeter-wave frequencies', *Electron. Lett.*, 1982, **18**, pp. 272–273
14 WHEELER, H.: 'Transmission-line properties of a strip on a dielectric sheet on a plane', *IEEE Trans.*, 1977, **MTT-25**, pp. 631–647
15 OWENS, R.: 'Predicted frequency dependence of microstrip characteristic impedance using the planar-waveguide model', *Electron. Lett.* 1976, **12**, pp. 269–270
16 PUES, H., and VAN DE CAPELLE, A.: 'Approximate formulas for frequency dependence of microstrip parameters', *Electron. Lett.*, 1980, **16**, pp. 870–872
17 HAMMERSTADT, E., and BEKKADAL, F.: 'Microstrip handbook', ELAB Report STF44 A74169, Trondheim, Norway, Feb. 1975
18 BAHL, I., and GUPTA, K.: 'Average power-handling capability of microstrip lines', *IEE J. Microwaves, Optics & Antennas*, 1979, **3**, pp. 1–4
19 VANDESANDE, J., PUES, H., and VAN DE CAPELLE, A.: 'Calculation of the bandwidth of microstrip resonator antennas'. Proc. 9th European Microwave Conf., Brighton, Sept. 1979, pp. 116–119
20 LO, Y., SOLOMON D., and RICHARDS, W.: 'Theory and experiment on microstrip antennas', *IEEE Trans.*, 1979, **AP-27**, pp. 137–145
21 DESHPANDE, M., and BAILEY, M.: 'Input impedance of microstrip antennas', *IEEE Trans.*, 1982, **AP-30**, pp. 645–650
22 NEWMAN, E., and TULYATHAN, P.: 'Analysis of microstrip antennas using moment methods', *IEEE Trans.*, 1981, **AP-29**, pp. 47–53
23 PUES, H., and VAN DE CAPELLE, A.: 'Computer-aided experimental characterisation of microstrip-to-coaxial transitions'. Proc. 14th European Microw. Conf., Liège, Sept. 1984, pp. 137–141
24 PUES, H., VANDENSANDE, J., and VAN DE CAPELLE, A.: 'Broadband microstrip resonator antennas'. Int. IEEE/AP-S Ant. & Prop. Symp. Digest, Washington, May 1978, pp. 268–271
25 PUES, H., VAN LIEBERGEN, H., THISSEN, L., NAUWELAERS, B., and VAN DE CAPELLE, A.: 'Broadband multi-layer microstrip antenna'. Proc. MIOP '87 Conf., Wiesbaden, May 1987
26 VAN LIL, E., and VAN DE CAPELLE, A.: 'Transmission line model for mutual coupling between microstrip antennas', *IEEE Trans.*, 1984, **AP-32**, pp. 816–821
27 NAUWELAERS, B., VAN DE CAPELLE, A.: 'Formulas for the calculation of mutual coupling between rectangular microstrip antennas'. Proc. Int. Conf. Ant. & Prop., April 1985, Coventry, pp. 99–102
28 JEDLICKA, R., and CARVER, K.: 'Mutual coupling between microstrip antennas'. Proc. Workshop on Printed Circuit Antenna Technology, Las Cruces, pp. 4-1/4-19, Oct. 1979
29 POZAR, D.: 'Input impedance and mutual coupling of rectangular microstrip antennas', *IEEE Trans.*, 1982, **AP-30**, pp. 1191–1196
30 HARRINGTON, R.: 'Time-harmonic electromagnetic fields', (McGraw-Hill, NY, 1961)
31 PENARD, E., and DANIEL, J.: 'Mutual coupling between microstrip antennas', *Electron. Lett.*, 1982, **18**, pp. 605–607
32 MALKOMES, M.: 'Mutual coupling between microstrip patch antennas', *Electron. Lett.*, 1982, **18**, pp. 520–522
33 DANIEL, J., VAN DE CAPELLE, A., and FORREST, J.: 'Microstrip patch arrays for satellite communications'. ESA/COST 204 Phased Array Antenna Workshop, ESTEC, Noordwijk, June 1983, pp. 9–14

Chapter 11
Design and technology of low-cost printed antennas

J.P. Daniel, E. Penard and C. Terret

11.1 Introduction

In the last decade printed-array antennas have received increasing attention for applications in various communication and navigation systems. Microstrip patches can be very efficient candidates for inexpensive antennas when narrow bandwidth (typically less than 5%) and medium gain are required (15–25 dB). However, divergence in substrate parameters and manufacturing tolerances means that a wider frequency bandwidth and a better control of radiation characteristics are necessary in the mass production of printed antennas.

Thus simple but accurate investigations of radiating elements are necessary to obtain the design requirements. Analysis of normal-shaped patches and slots can be developed using both known models (transmission-line or cavity models) or more elaborate theory (spectral-domain-approach).

On the other hand, the design of planar arrays requires a thorough knowledge of typical properties of printed linear sub-arrays such as directivity versus spacing, mutual-coupling effects, losses etc; for this purpose, simple formulas and analyses have been developed. Two-dimensional arrays, with non-identical sub-arrays, such as cross-fed structures, are well suited for low-cost antennas. Design equations and curves are included.

Microstrip patches exhibit different E-plane and H-plane radiation patterns; analytical synthesis methods (Fourier, Chebӯshev etc.) are not always suitable for a small number of sources or when the pattern is specified by a given outline. Two new numerical synthesis methods, taking into account the directivity pattern of sources with equal or unequal spacings, are proposed.

Numerical programs can be implemented on conventional personal computers.

Cost reduction will necessitate common microstrip laminates or new polymer substrates exhibiting good mechanical and electrical properties (typically a low dielectric constant of about 2–2·5 and losses of $\tan \delta \approx 10^{-3}$. A new low-cost polypropylene whose fabrication process is quite simple has been developed at

CNET Lannion, France. This substrate can be made as a multi-layer structure or with thick metal backing.

11.2 Analysis of simple patches and slots

The microstrip antenna designer needs a method of analysis (not too time consuming) to calculate, as nearly as possible, the parameters of interest: resonant frequency, Q-factor, input impedance, pattern etc. Moreover, he should be able to evaluate the surface-wave effects and to take into account the superstrate applied to the antenna as a protective layer.

11.2.1 Rectangular and circular patches

Two simple techniques can be used for the analysis of microstrip antennas: transmission-line model [1, 2, 3] and cavity model. More sophisticated analysis will be presented for multi-layered structures.

11.2.1.1 Cavity-model analysis: Cavity-model analysis [4] allows modal field description and gives good results depending on the effective parameters obtained from previous microstrip line formulas, as long as the substrate thickness is thin compared with the wavelength. This method has often been used with magnetic walls to calculate the electrical properties of open microstrip antennas (OMA) with simple shapes [5, 6] (cf. Chapter 3). Its application to the rectangular hybrid microstrip antenna (HMA) [7] permits a comprehensive comparison between OMA (Fig. 11.1a) and HMA (Fig. 11.1b) antennas [8]. It can also be easily applied to multi-port microstrip antennas [9, 10].

In the cavity method, according to the Lo and Richards model, the total interior field with a unit z-directed excitation current at (x_0, x_0) is given by

$$E_z(x, y) = j\omega\mu_0 \sum_m \sum_n \frac{\psi_{mn}(x, y)\psi_{mn}(x_0, y_0)}{k^2 - k_{mn}^2}$$

$$\times J_0\left(\frac{m\pi d}{2a}\right) \frac{\varepsilon_{0m}\varepsilon_{0p}}{ab} \qquad (11.1)$$

with

$$k^2 = k_0^2 \varepsilon_r (1 - j\delta_{eff})$$

$$J_0(x) = \frac{\sin(x)}{x}$$

$$\varepsilon_{0p} = \begin{cases} = 1 & p = 0 \\ = 2 & p \neq 0 \end{cases}$$

k_{mn} is the wave resonant number. The eigen functions ψ_{mn} are solutions of the Helmholtz equation with different boundary conditions:

- OMA with four magnetic walls (as in Fig. 11.1a)
- HMA with three magnetic walls and one electric wall (as in Fig. 11.1b)

The treatment considers perfect magnetic or electric walls and groups all antenna losses together in an effective dielectric loss tangent δ_{eff} determined by an iterative process (Fig. 11.2). In this process it is necessary to calculate the electric and magnetic energies W_e and W_h stored in the cavity.

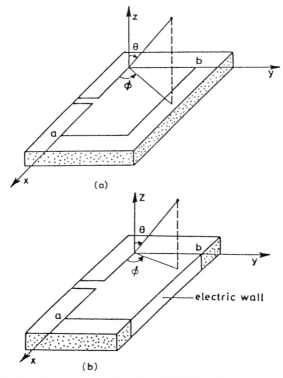

Fig. 11.1 *(a) OMA with four magnetic walls. (b) HMA with three magnetic walls and one electric wall*

These quantities are, of course, evaluated by integrating the electric E and magnetic H fields over the cavity volume V. In fact, the exact expression of W_e and W_h can be obtained in a very simple manner without performing any integration at all, as explained in the following [11]. The total power P is injected into the cavity as

$$P = VI_z^* = V = -tE_z(x_0, y_0) \tag{11.2}$$

where V is the voltage at the source point. On the other hand, the same power can be expressed as a function of W_e and W_h:

$$P = 2\omega\delta_{eff}W_e + j2\omega(W_h - W_e) \tag{11.3}$$

Which leads to the following relationships:

$$W_e = -\frac{t}{2\omega\delta_{eff}}\operatorname{Re}[E_z(x_0, y_0)] \tag{11.4}$$

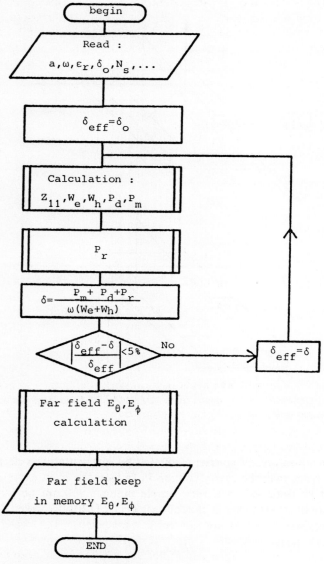

Fig. 11.2 *Flow chart of the cavity method*

$$W_h = W_e - \frac{t}{2\omega} \text{Im}\,[E_z(x_0, y_0)] \tag{11.5}$$

Where Re and Im stand for real and imaginary parts, respectively.

Note that W_h is determined without using the H field. Eqns. 11.4–11.5 are obtained independently of the mathematical method employed to derive the expressions for the E_z field or of the geometry of the cavity section, i.e. the patch shape. Thus eqns. 11.4 and 11.5 are general formulas valid for planar microstrip antennas of arbitrary shape. However, the cavity thickness t must be quite small compared to λ_e in order to ensure the validity of expression 11.2.

The analysis yields input impedance, resonant frequency and Q-factor; the radiation pattern is found from the knowledge of the interior field. Two methods are available:

(a) Electric-current method
The radiated field is found from the electric currents J_s flowing on the patch and on the electric walls. If there is a dielectric, volume polarisation currents J_p must be added to the cavity region (Fig. 11.3) [12, 13, 14, 15].

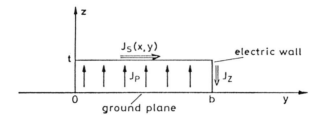

Fig. 11.3 *Electric currents and polarisation currents in an HMA*

(b) Aperture-radiation method
One application of the equivalence theorem [16] is to choose a perfect closed electric surface S with null field inside and equivalent magnetic sources outside: $M_s = E \times n$; n being the normal outward from S. S defines the limit of an homogeneous exterior medium and must be carefully selected.

As the rectangular OMA structure has been extensively described, only the HMA is considered here; e.g., the choice of S for HMA is indicated in Fig. 11.4.

The surface S_1 around the cavity (Fig. 11.4a) is not permitted, except in the case of a vacuum medium ($\varepsilon_r = 1$). It seems more convenient to choose S_2 (Fig. 11.4b) lying on the $z = t$ plane, because it is always available whatever the medium, i.e. $\varepsilon_r = 1$ or $\varepsilon_r \neq 1$.

In practice, electric walls are not perfect conductors, but the electric-field distribution in the vicinity of the electric wall (Fig. 11.4c) suggests that antenna radiation occurs mainly from the magnetic walls. This assumption is well confirmed by experiment. It may be assumed that HMA radiation is entirely due to the magnetic walls, i.e. three-slots array. Experimental and theoretical results for an HMA are given in Fig. 11.5.

The impedance loci in the case of the dominant mode (0, 1) are given in Fig. 11.5a. All the modes have been taken into account in the computation. The antenna was edge-fed by a coaxial-line probe at a position $x_0 = 3 \cdot 5$ cm, $y_0 = 0$. Two methods were used for the radiated field (equivalence theorem and electric current plus polarisation current) and both agree well with experimental data. The experimental patterns in planes $\phi = 0°$ and $\phi = 90°$ are plotted in Figs. 11.5b and 11.5c.

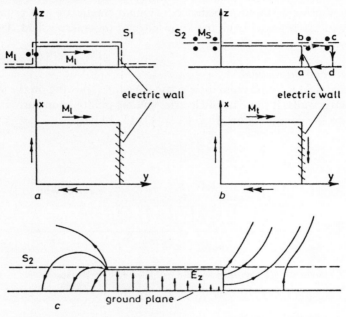

Fig. 11.4 *Application of the equivalence theorem to HMA*
a First choice of surface S_1
b Second choice of surface S_2
c Local geometry of the electric field of an HMA near the short circuit

The uniform aperture lying along the Ox-axis gives an omnidirectional pattern in the plane $\phi = 90°$ (Fig. 11.5c), the cross-polarisation level being very low (< 30 dB). The discrepancy between the patterns computed from the electric currents alone (broken line) and from the aperture (large broken line) appears clearly in this Figure; it confirms the importance of the dielectric slab. Polarisation currents must be included in the far-field calculation. They are z-directed and therefore contribute only to the E_θ components. Therefore, the electric-current approach gives a useful physical insight into the radiation properties, particularly regarding the major influence of the dielectric constant on the E_θ component.

Design and technology of low-cost printed antennas 585

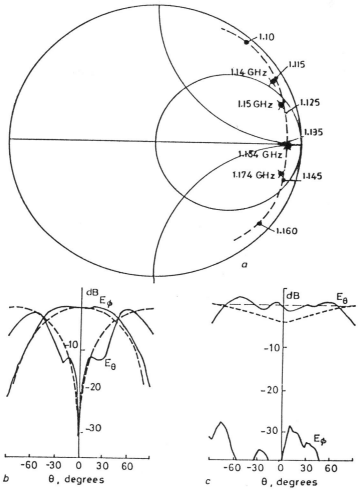

Fig. 11.5 *Experimental and theoretical results in the case of an HMA*
$a = 6 \cdot 00$ cm, $b = 4 \cdot 00$ cm, $\varepsilon_r = 2 \cdot 56$
$t = 0 \cdot 146$ cm, $\tan \delta = 0 \cdot 001$, mode(0, 1)
(a) Impedance loci
  Feed point $x_0 = 3 \cdot 5$ cm, $y_0 = 0$
  ●—● experiment
  ○   magnetic currents
  X   electric currents and polarisation currents
(b) Plane $\phi = 0°$
  ——— experiment
  ---- theory
(c) Plane $\phi = 90°$
  ——— experiment
  ---- theoretical pattern: electric currents only
  ——— theoretical pattern: magnetic-currents method

In the plane $\phi = 0°$, the radiated field of the two slots lying along the y-axis leads to a high level of cross-polarisation radiation with a null in the broadside direction (the magnetic currents are each 180° out of phase). The diffraction [17] at the edge of the ground plane appears particularly in the oscillations of the principal component E_θ in the planes $\phi = 0°$ (notably for $\theta = 30°$ and $\theta = 90°$). Nevertheless, the theoretical results are again well confirmed by experiment.

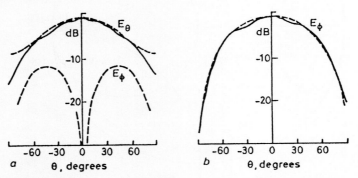

Fig. 11.6 *Experimental and theoretical patterns in for the HMA with one electrical wall*
$a = 6.00\,\text{cm}$, $b = 4.00\,\text{cm}$, $\varepsilon_r = 2.56$, $t = 0.148\,\text{cm}$, mode(1.1)
——— experiment
- - - - theory
(a) Plane $\phi = 0°$
(b) Plane $\phi = 90°$

The high-level cross-polarisation can be reduced when the exciting mode is (1, 1). The radiation pattern in the $\phi = 0°$ plane (Fig. 11.6a) is therefore due to an array of two slots directed along the y-axis and has a cosine field distribution. The cross-polarisation due to the x-slot is smaller than for mode (0, 1) because of the phase inversion along the x-apertures, and is cancelled in the plane $\phi = 90°$ (Fig. 11.6b, theoretical results only). The important reduction of beamwidth is worthy of note here.

Feed position: As in the OMA case, it is possible to match the HMA by properly selecting the feed position along the y-axis.

When considering the dominant mode alone and its resonant frequency (given by $k^2 = k^2 mn$), the input impedance $Z(= R_i + jX_i)$, which becomes real at resonance, can be simply expressed as a function of the feed position x_0, y_0. For the HMA with one electrical wall

$$R_i(x_0, y_0) = R_i(0, 0)\cos^2\left(\frac{mx_0}{a}\right)\cos^2\left(\frac{ny_0}{2b}\right) \qquad (11.6)$$

where $R_i(0, 0)$ is the input impedance for $x_0 = 0$, $y_0 = 0$ (corner feed).

For example, in the case of the one-electric-wall antenna and the mode (1, 1),

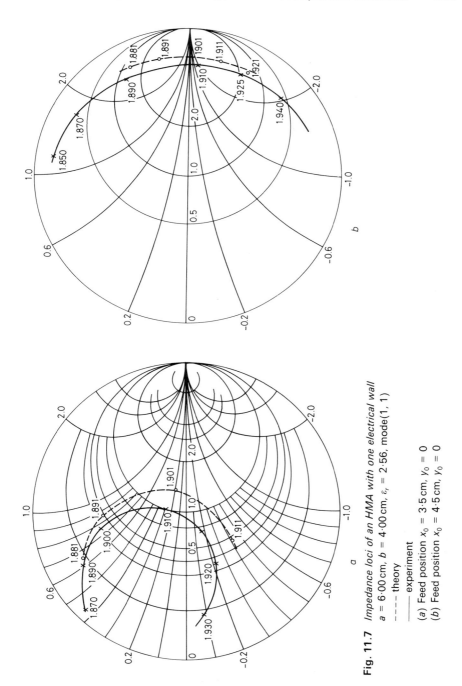

Fig. 11.7 *Impedance loci of an HMA with one electrical wall*
$a = 6.00$ cm, $b = 4.00$ cm, $\varepsilon_r = 2.56$, mode(1, 1)
---- theory
—— experiment
(a) Feed position $x_0 = 3.5$ cm, $y_0 = 0$
(b) Feed position $x_0 = 4.5$ cm, $y_0 = 0$

the impedance loci are given in Figs. 11.7a and 11.7b for two feed positions $x_0 = 3.5$ cm, $y_0 = 0$ and $x_0 = 4.5$ cm, $y_0 = 0$. It can be seen that the impedance decreases as the feed moves along the x-axis toward the centre of the edge. The discrepancy between measurement and the theoretical results is due first to the experimental error in the feed position and secondly to the feed model, which does not take into account the disturbance of the internal field distribution near the probe.

(c) Comparison between HMA and OMA characteristics

The different parameters, Q-factor, input impedance and radiation patterns of the OMA and the HMA, can be compared for the same electrical dimensions. Both act in the dominant mode (0, 1) and at the same resonant frequency. The wave number k_0 in free space is given by

$$k_0 \sqrt{\varepsilon_r} = k_{01} = \frac{\pi}{b'} = \frac{\pi}{2b}$$

Table 11.1 *Comparison of theoretical impedance and Q-factor*

	HMA		OMA	
	Experiment	Theory	Experiment	Theory
R, Ω	355·8	467·4	388	406
Q-factor	34·16	41	61	69·5
f_{r0}, MHz	1740·00	1763·00	1753	1773·7

Table 11.2 *Comparison of cavity method with Wood's method [7]*

	OMA		HMA	
	Wood	Cavity method	Wood	Cavity method
f_{r0}, MHz	1275	1257·2	1275·0	1278
Δf, MHz	26	22·4	22·5	23

with b' the OMA dimension along the y-axis and b the HMA dimension along the y-axis; therefore $b' = \lambda_\varepsilon/2$ and $b = \lambda_\varepsilon/4$ (λ_ε is the wavelength in the dielectric).

Vacuum medium, $\varepsilon_r = 1$: The theoretical impedance and Q-factor are compared in Table 11.1 with experimental data for both antennas. In this case the Q-factor of the HMA is smaller but the impedances have approximately the same value. The slight discrepancy between experiment and theory is due to the failure of the magnetic-wall model when $\varepsilon_r = 1$.

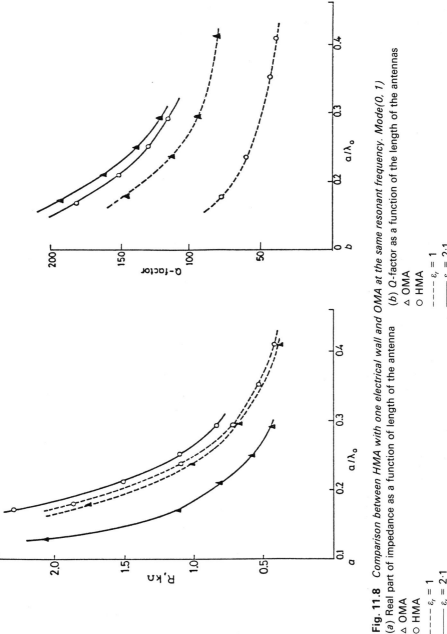

Fig. 11.8 *Comparison between HMA with one electrical wall and OMA at the same resonant frequency. Mode(0, 1)*
(a) Real part of impedance as a function of length of the antenna
(b) Q-factor as a function of the length of the antennas

△ OMA
○ HMA

---- $\varepsilon_r = 1$
——— $\varepsilon_r = 2\cdot 1$

Dielectric medium case, $\varepsilon_r = 2\cdot 5$: Theoretical results for the bandwidth and resonant frequency are compared with Wood's measurements [7] in Table 11.2. There is an important difference compared with the previous case: here the bandwidths of the two kinds of antenna are very similar.

(d) Theoretical results for different antenna lengths a/λ_0

Q-factor, input impedance: In Fig. 11.8a the variation of the input resistance at resonance for the HMA and OMA is given as a function of the parameter a/λ_0 for $\varepsilon_r = 1$ and $\varepsilon_r = 2\cdot 1$. The Q-factor variation is given in Fig. 11.8b; for all the different cases considered it can be concluded that, for $\varepsilon_r = 1$ the Q-factor of the HMA is lower than that of the OMA (the bandwidth is therefore larger). However, the impedances of OMA and HMA are very similar.

For $\varepsilon_r = 2\cdot 1$ the HMA and OMA have approximately the same Q-factor (and therefore the same bandwidth), but the input impedance of the HMA is twice that of the OMA. These important differences between the vacuum medium and the dielectric medium can be explained by the contribution of the polarisation currents. Their effect should be more important for the HMA than for the OMA owing to the cosine distribution of the internal field, which has no change of phase along the y-axis (Fig. 11.4c). For a full comparison of the two antennas, the cross-polarisation level must also be considered.

Cross-polarisation level as a function of a/λ_0: Theoretical results are given in Fig. 11.9 in the plane $\phi = 0°$ for different values of θ ($\theta = 20°, 60°$) and for $\varepsilon_r = 2\cdot 1$. The cross-polarisation C_p is defined here by

$$C_p = 20 \log_{10} \left| \frac{E_\theta}{E_\phi} \right| \tag{11.7}$$

It may be seen that, for the HMA, the cross-polarisation remains very high, and therefore this antenna is not suitable for use in an electronically scanned array (except when very small scanning angles are used).

In conclusion, the HMA is shorter ($\lambda_\varepsilon/4$ long) than the OMA ($\lambda_\varepsilon/2$ long), each having the same resonant frequency for the mode (0, 1). A closer comparison of these two kinds of antenna shows that the dielectric plays a fundamental role in the far-field pattern, the input impedance and the bandwidth.

When $\varepsilon_r = 1$, the impedance of the HMA and OMA have the same values, whereas the HMA bandwidth is twice as broad. On the other hand, with the usual dielectric ($\varepsilon_r \approx 2$), the bandwidths have the same value, but the impedance of the HMA is twice that of the OMA.

11.2.1.2 Spectral-domain approach (SDA): As the frequency of operation is increased, more complex analyses are necessary to rigorously account for the effect of the dielectric substrate, which may couple surface waves. Recently, quite efficient approaches based on integral formulations and numerical resolu-

Design and technology of low-cost printed antennas 591

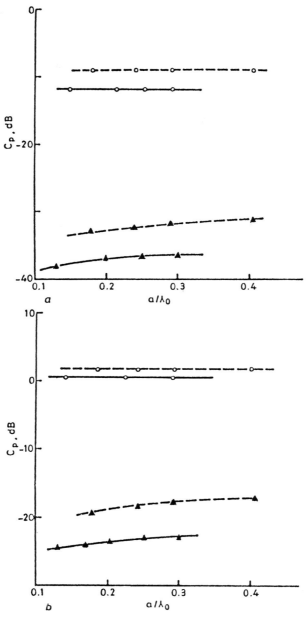

Fig. 11.9 Comparison of the cross-polarisation level of the OMA and HMA for different values of λ_0/a. Results are given for the plane $\phi = 0°$
△ OMA
○ HMA
---- $\varepsilon_r = 1$
—— $\varepsilon_r = 2\cdot1$
(a) $\theta = 20°$
(b) $\theta = 60°$

tions were proposed [18–23]. These methods use the exact Green's function for the grounded dielectric slab, and hence the results depend on the numerical techniques used to calculate these functions accurately.

(a) Resonant frequency and quality factor
Among these approaches, the SDA [20] is particularly suitable for determining the resonant frequency and Q-factor of patch antennas embedded in dielectric substrates. The structure and the co-ordinate system employed are shown in Fig. 11.10. In SDA, the Fourier transform of the dielectric field $E_x(\alpha, \beta)$, $E_y(\alpha, \beta)$ at $z = d_1$ in region 2 is related to the current distributions $J_x(\alpha, \beta)$, $J_y(\alpha, \beta)$ on the

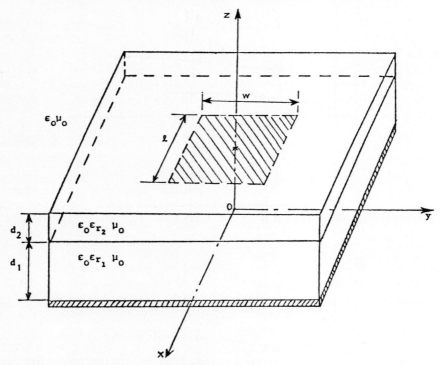

Fig. 11.10 *Patch antenna with dielectric layer*
 d_2 = protective dielectric-layer thickness
 d_1 = dielectric-substrate thickness
 (W, l) = width and length of the patch

conducting patch by

$$\begin{bmatrix} E_x \\ E_y \end{bmatrix} = \frac{j}{\omega\varepsilon_0 \, \text{dte dtm}} \bar{\bar{M}}(\gamma_i, d_i) \begin{bmatrix} J_x \\ J_y \end{bmatrix} \quad (11.8)$$

where γ_i is the propagation constant in the z-direction in the ith region and

$\varepsilon_i = \varepsilon_0 \varepsilon_{ri}$ ($i = 0, 1, 2$) is the complex permittivity
$$\gamma_i^2 = \alpha^2 + \beta^2 - k_i^2; \quad k_i = \varepsilon_{ri} k_0^2 = \omega^2 \mu_0 \varepsilon_0 \varepsilon_{ri}$$
($a \exp j\omega t$ time variation is assumed)

$$\begin{cases} \text{dte} = \varepsilon_{r2}\left[\gamma_2 \text{ch}\gamma_2 d_2 \left[\gamma_0 + \frac{\gamma_1}{\text{th}\gamma_1 d_1}\right] + \gamma_2 \text{sh}\gamma_2 d_2 \left(\gamma_2 + \frac{\gamma_0 \gamma_1}{\gamma_2 \text{th}\gamma_1 d_1}\right)\right] \\ \text{dtm} = \gamma_2 \text{sh}\gamma_2 d_2 \left[\gamma_2 \varepsilon_{r1} + \frac{\varepsilon_{r2}^2 \gamma_0}{\gamma_2} \gamma_1 \text{th}\gamma_1 d_1\right] + \varepsilon_{r2}(\gamma_2 \text{ch}\gamma_2 d_2)(\gamma_0 \varepsilon_{r1} + \gamma_1 + \gamma_1 \text{th}\gamma_1 d_1) \end{cases}$$

(11.9)

The zeros of dte and dtm define the surface-wave poles in the composite layer, the dominant mode of which, TM_0, is always above cut off regardless of slab thicknesses.

The matrix M can easily be obtained in terms of an equivalent transmission-line circuit as presented in Fig. 11.11, by generalising the method of Reference 20. Using the moments method and a modal representation of J, a matrix equation is derived:

$$\sum_{m=1}^{M} K_{pm}^{xx} C_m + \sum_{n=1}^{N} K_{pn}^{xy} d_n = 0 \quad p = 1, 2, \ldots M$$

$$\sum_{m=1}^{M} K_{qm}^{yx} C_m + \sum_{n=1}^{N} K_{qn}^{yy} d_n = 0 \quad q = 1, 2, \ldots N$$

(11.10)

where the elements of matrix K are given by double integrals on α and β. With the transformation $\alpha = k_\varrho \sin\theta$, $\beta = k_\varrho \cos\theta$, we have, for example

$$K_{pn}^{xy}(\omega) = 4 \int_0^{\pi/2} \int_0^{+\infty} J_{xp}(k\varrho, \theta) M_{xy}(k\varrho, \theta, \omega) J_{yn}(k\varrho, \theta) k_\varrho dk_\varrho d\theta \quad (11.11)$$

The condition $\det[K] = 0$ is only satisfied by a complex frequency $f = f' + jf''$ that leads to the resonant frequency f' and the quality factor $Q = f'/2f''$ of this radiating open resonator. In this free regime, k_i is complex with $\text{Im}[k_i] > 0$, and a proper Riemann surface must be defined for evaluation of $\gamma_i = (k_\varrho^2 - k_i^2)^{1/2}$. The sign of γ_1, γ_2 does not affect the value of the integrals, as the terms involving the radicals are even functions of γ_1, γ_2. For this reason, only the branch-cut contribution by the radical γ_0 has to be considered, the branch cuts being shown in Fig. 11.12. It can be shown that this condition involves an integration path in the k_ϱ complex plane and not only on a real axis as in Reference 20. Furthermore, it is worthy of note that the surface-wave poles lie within the range $k_0 \to \max[k_1, k_2]$.

A possible position of an integration path in the k_ϱ-plane is shown in Fig. 11.12, with $\Delta \gg \text{Im}[k_0]$, taking into account the presence of the poles.

In Table 11.3 and Fig. 11.13, the present method of analysis is compared with available theoretical and experimental results on resonant frequency and quality factor.

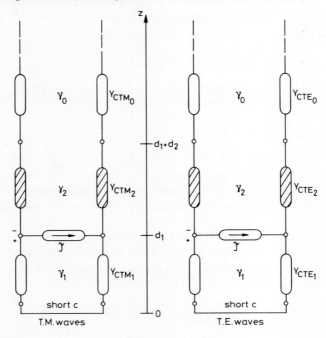

Fig. 11.11 *Transmission-line model in the spectral domain*

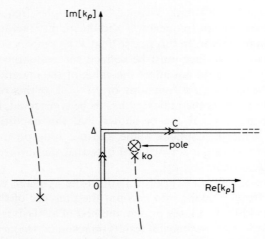

Fig. 11.12 *Branch-cuts position and path integration in the k_ϱ plane*

Design and technology of low-cost printed antennas   595

Generally, the resonant frequency f_0 of a microstrip antenna is defined as the frequency giving the input reactance X equal to zero. For electrically thin substrates, f_0 is also very close to the frequency f_r where the input resistance R

Table 11.3 *Measured and predicted f_r and Q of a rectangular patch antenna: TM_{01} and TM_{10} modes*

Method	f_r	Q	f_r	Q
	TM_{10}		TM_{01}	
SDA [24]	1·5458	102·7	2·2616	44·68
Cavity [25]	1·548	112	2·272	47
Dubost [3]	1·52	105	2·3	36·2
Exp [25]	1·53	97 ± 10	2·237	44 ± 6

$W = 60$ mm; $l = 40$ mm; $d_1 = 1\cdot46$ mm; $\varepsilon_r = 2\cdot56$; $\mathrm{tg}\,\Delta_1 = 10^{-3}$ $(d_2 = 0)$

Table 11.4 *Comparison between the present method and results predicted by Bahl [26]*

d_2 (mm)		f_r GHz	Q	$\frac{\Delta f_r}{f_r}$ (%)
0	SDA	4·092	33·68	0
	Experience	4·104		0
0·8	SDA	3·986	34·52	2·62
	Experience	4·008		2·34
1·59	SDA	3·9336	33·49	3·9
	Experience	3·934		4·14
3·18	SDA	3·8767	31·82	5·09
	Experience	3·895		5·09

$W = 19$ mm; $l = 22\cdot9$ mm; $d_1 = 1\cdot59$ mm; $\varepsilon_{r1} = 2\cdot32$ $\mathrm{tg}\,\Delta_1 = 10^{-3}$; $\varepsilon_{r2} = 2\cdot32$; $\mathrm{tg}\,\Delta_2 = 10^{-3}$

$\frac{\Delta f_r}{f_r} = \frac{f_{r0} - f_r}{f_{r0}}$ f_{r0} = resonant frequency without protection

reaches a maximum, R_{max}. In coaxial-feed excitation, as the substrate thickness increases, f_0 lies farther from f_r owing to an inductive shift of the reactance curve. Furthermore, the bandwidth, defined in terms of impedance, is affected by this inductive shift. In order to remove the disturbance of the feed, and to compare measured and predicted results, the resonant frequency and Q-factor were taken

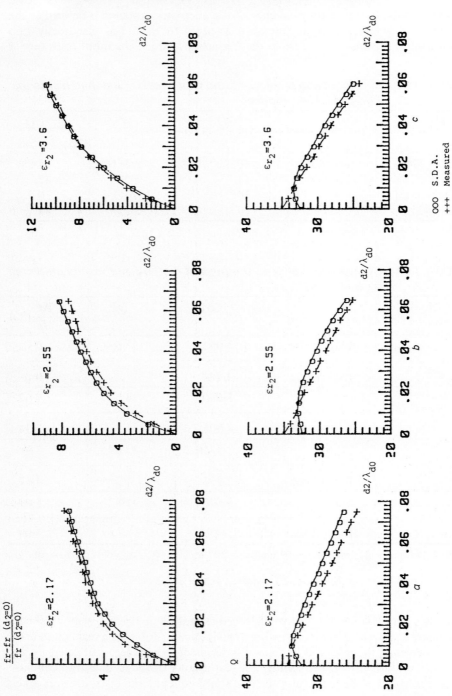

as

$$f_r = f(R = R_{max})$$

and

$$Q = \frac{f_r}{2R_{max}} \left[\frac{\partial X(f)}{\partial f} \right]_{f=f_r} \quad (11.12)$$

It should be noted that the SDA method gives

$$f = f' + jf''$$

with

$$f' \cong f_r = f_0 \sqrt{1 - \left(\frac{1}{2Q}\right)^2} \quad \text{and} \quad f'' = \frac{f_0}{2Q}$$

The effect of a protective layer on the resonant frequency and Q-factor of a rectangular microstrip antenna is shown in Table 11.4. In this case the Q-factor varies slightly, while the resonant frequency clearly decreases.

Additional experiments were carried out on a rectangular patch antenna for several different thicknesses of superstrate and three different values of the superstrate dielectric constant: $\varepsilon_{r2} = 2\cdot17, 2\cdot55$ and $3\cdot6$. The results are presented in Fig. 11.13. It should be noted that, for $d_2/\lambda_{d0} \leqslant 0\cdot015$, the theoretical Q-factor increases slightly. This corresponds to a better matching of the antenna to free space and a maximum efficiency of radiation. As d_2 increases, so does the coupling of energy with the superstrate by the surface waves, and radiation efficiency decreases while the bandwidth increases. It is worthy of note that, for a given thickness, $\Delta f_r/f_r$ and bandwidth increase with increasing ε_{r2}.

(b) Input impedance

The input impedance can be determined using Richmond's reaction equation in spectral domain. If the real axis is used as the integration path, the residue contribution of the surface-wave poles has to be taken into account [21]. In our approach this is avoided by using the previous integration path in the complex plane (Fig. 11.12). Fig. 11.14 shows our theoretical results compared with those of Bailey *et al.* for an antenna without (with) dielectric cover.

Fig. 11.13 *(a) Effect of dielectric loading on resonant frequency and Q-factor of a rectangular microstrip antenna*
$l = 22\cdot9$ mm, $W = 19$ mm, $\varepsilon_{r1} = 2\cdot17$, $d_1 = 1\cdot58$ mm
$\tan \Delta_1 = 6\cdot0 \times 10^{-4}$, $\varepsilon_{r2} = 2\cdot17$, $\tan \Delta_2 = 6\cdot0 \times 10^{-4}$
(b) Effect of dielectric loading on f_r and Q-factor of a rectangular patch antenna
$l = 22\cdot9$ mm, $W = 19$ mm, $\varepsilon_{r1} = 2\cdot17$, $d_1 = 1\cdot58$ mm
$\tan \Delta_1 = 6\cdot0 \times 10^{-4}$, $\varepsilon_{r2} = 2\cdot55$, $\tan \Delta_2 = 10^{-3}$
(c) Effect of dielectric loading on f_r and Q-factor of a rectangular patch antenna
$l = 22\cdot9$ mm, $W = 19$ mm, $\varepsilon_{r1} = 2\cdot17$, $d_1 = 1\cdot58$ mm
$\tan \Delta_1 = 6\cdot0 \times 10^{-4}$, $\varepsilon_{r2} = 3\cdot6$, $\tan \Delta_2 = 2 \times 10^{-3}$

(c) Radiation efficiency [27, 28]

The SDA can also be used to calculate the influence of surface waves on the radiation efficiency of a rectangular patch antenna. Once the problem is solved for the resonant frequency f_r, far-field radiation may be obtained by using the inverse Fourier transform. Hence the electric-field components in each point of

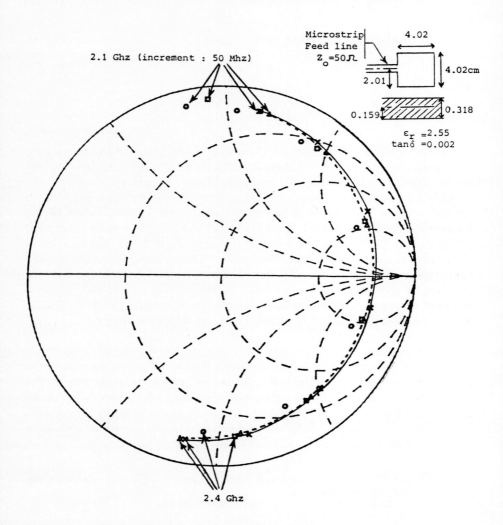

Fig. 11.14 *Input impedance of microstrip antenna without (with) dielectric cover compared with Bailey's results (Reference 21)*

●——● Bailey *et al.*
X   X Our results
△——△ Bailey *et al.*
O   O Our results

region 0 are given by

$$E_q(x, y, z) = \frac{1}{2\pi} \int_{-\infty}^{+\infty} \int_{-\infty}^{\infty} E_q(\alpha, \beta) \exp(-\gamma_0 \times (z-d)) \exp(-j(\alpha x + \beta y)) \, d\alpha \, d\beta$$

$$q \equiv (x, y, z) \qquad (d_2 = 0, d_1 = d)$$

with

$$\gamma_0 = \sqrt{\alpha^2 + \beta^2 - k_0^2} \quad \text{Re}[\gamma_0] > 0, \ \text{Im}[\gamma_0] > 0, \ k_0 = \omega\sqrt{\mu_0 \varepsilon_0} \quad (11.13)$$

and ω real.

To compute the far field, eqn. 11.13 is first transformed into spherical coordinates and secondly transformed to specify the (α, β) plane in terms of a spherical-polar angle (u, v). Branch points of γ_0 have thus disappeared, and the resulting integral can be evaluated by the saddle-point method, provided that the contribution of the integrand poles is correctly taken into account.

The integral along the steepest-descent path corresponds to the space-wave component, while the summation of the residues of these poles is related to the TM- and TE-mode surface waves. Only the dominant lowest-order TM mode which may be excited for the usual values of d/l_0 and ε_r has been considered here. After integration, the electric far field may be expressed as

$$E_\theta(r, \theta, \phi) = E_{\theta 1} + E_{\theta 2}$$
$$E_\phi(r, \theta, f) = E_{\phi 1} \qquad (11.14)$$

where $E_{\theta 1}$, $E_{\phi 1}$ and $E_{\theta 2}$ are expressions for the space and surface waves, respectively. Note that the space-wave components are the Fourier-transformed interface $(z = d)$ fields, and may be expressed as

$$E_{\theta 1} \propto \sin\phi E_{y0}(\alpha, \beta) + \cos\phi E_{x0}(\alpha, \beta)$$
$$E_{\phi 1} \propto \cos\theta[\cos\phi E_{y0}(\alpha, \beta) - \sin\phi E_{x0}(\alpha, \beta)] \qquad (11.15)$$

with

$$\alpha = k_0 \sin\theta \cos\phi$$
$$\beta = k_0 \sin\theta \sin\phi$$

The efficiency of space-wave launching can be calculated as the ratio of radiated power P_{sp} to total (radiated plus surface-wave) power:

$$\eta\% = 100 \times \frac{P_{sp}}{P_{sp} + P_{su}} \qquad (11.16)$$

with

$$P_{sp} = \frac{1}{2z_0} \int_0^{2\pi} d\phi \int_0^{\pi/2} [(E_{\theta 1})^2 + (E_{\phi 1})^2] r^2 \sin\theta \, d\theta$$

$$P_{su} = \frac{\omega \varepsilon_0}{\xi_p} \int_0^{2\pi} d\phi \int_0^\infty \varrho |E_z|^2 \, dz$$

$$z_0 = \sqrt{(\mu_0/\varepsilon_0)} \quad \text{and} \quad \xi_p = \sqrt{\alpha_p^2 + \beta_p^2} = k_0 \sin up$$

$$\varrho = r \sin \theta \qquad \qquad \text{pole in the } \xi \text{ plane}$$

E_z having been deduced from $E_{\theta 2}$. Furthermore, the gain may be defined by

$$G = 4\pi r^2 \frac{[|E_{\theta 1}|^2 + |E_{\phi 1}|^2]_{\theta=0}}{P_{sp} + P_{su}} \tag{11.17}$$

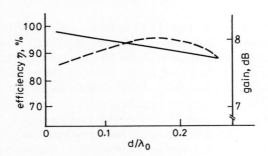

Fig. 11.15 *Radiation efficiency and gain against d/λ_0 for rectangular patch antenna*
$W = 15\,mm;\ l = 10\,mm,\ d = 1\cdot 58\,mm,\ \varepsilon_r = 2\cdot 35,\ f_r = 8\cdot 58\,GHz$

Fig. 11.15 shows numerical results for η and G against dielectric thickness d normalised to λ_0 for a rectangular patch antenna fed along the x-direction. An optimum value of gain for $d \approx 0\cdot 15\lambda_0$ with $\eta \approx 95\%$ is observed. For $d > 0\cdot 15\lambda_0$ the increase of P_{su} causes a slight reduction of the gain, but the first TE-mode surface waves excited for $d \approx 0\cdot 216\lambda_0$ are not considered here. Identical results have been obtained for microstrip disc antennas [27] for large values of dielectric constant.

(d) Surface-wave effects on radiation patterns [24, 28, 29]

Radiation into the horizon: The radiating E_θ^{sp} field for printed antennas with an infinite dielectric layer tends to zero along the horizon ($\theta \to \pi/2$). However, in many cases the dielectric layer should be considered truncated after a certain distance a (Fig. 11.16), and the surface waves radiate some of their energy when they reach this discontinuity. If the radius a is sufficiently large, the far-field E_z of the surface wave may be used to calculate the corresponding far-field E_θ^{su}.

Fig. 11.17 shows the radiation pattern in the E-plane ($\phi = 0°$) for the total ($E_\theta^{sp} + E_\theta^{su}$) electric field of the rectangular patch antenna with a truncated dielectric radius $a \approx 3\cdot 5\lambda_0$. Note that, for $\theta = 90°$, the far field is due only to the components of the surface wave. In the H-plane ($\phi = 90°$), $E_\phi^{su} = 0$, and the total field varies as $\cos \theta$.

11.2.2 Conical antennas

In some cases, the narrow bandwidth of microstrip patch antennas continues to be the main constraint. The widest bandwidths are likely to be achieved with thicker substrates [30], but higher modes and surface waves are limiting factors. Replacing a conventional circular disc by a solid conical patch of the same radius, a new type of 'microstrip antenna' [31] is obtained. This structure

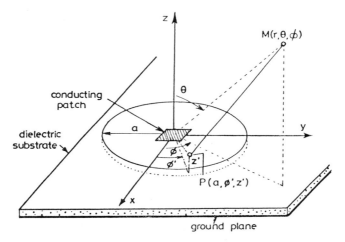

Fig. 11.16 *Microstrip patch antenna with a truncated dielectric layer of radius a*

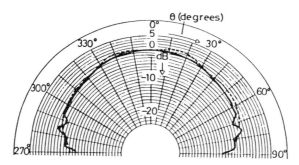

Fig. 11.17 *Radiation pattern in the E-plane for the total electric field $E_\theta = E_\theta^{sp} + E_\theta^{su}$ of a rectangular patch antenna with a truncated dielectric layer*
$W = 15$ mm, $l = 10$ mm, $d = 1.58$ mm, $\varepsilon_r = 2.35$, $f_r = 8.58$ GHz, $a = 3.5\lambda_0$
---- Theory
——— Experiment

facilitates radiation owing to better matching of the internal field of the microstrip cavity to free space, and it provides a broader bandwidth [32].

11.2.2.1 Resonant frequency: The cavity model is applied to the volume bounded by the ground plane, the cone surface and the spherical magnetic wall

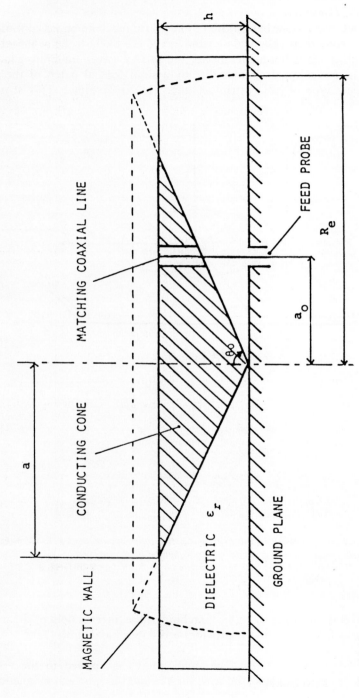

Fig. 11.18 *Geometry of conical patch antenna*

(Fig. 11.18). By analogy with the circular disc, we can define an effective spherical radius Re given by

$$\text{Re} = \frac{a}{\sin\theta_0}\left[1 + \frac{2}{\pi\varepsilon_r}\cot\theta_0 \ln\left(\frac{\pi}{2}\tan\theta_0\right) + 1\cdot7726\right]^{1/2} \quad (11.18)$$

where θ_0 and a are, respectively, the angle and the base radius of the conducting cone, and ε_r is the relative dielectric constant. The fields within the cavity corresponding to radial TE_{mp}-modes may be derived from a scalar electric potential $\psi(r, \theta, \phi)$ which must satisfy the wave equation

$$(\Delta + k^2)\frac{\psi}{r} = 0 \quad (11.19)$$

where $k = \omega\sqrt{\mu\varepsilon}$.

Table 11.5 *Theoretical and experimental resonance frequencies of the TE_{11}-mode for several values of θ_0 and ε_r*

ε_r	1	1	2·1	2·1
θ_0	74°	85°	72°	75°
f_r (GHz) (theoretical)	4·79	5·48	3·50	3·61
f_{rexp}	4·86	5·46	3·56	3·56
$\frac{f_{th} - f_{exp}}{f_{exp}}$	1·5%	1·5%	−1·7%	−1·4%

For θ_0 near $\pi/2$, applying boundary conditions on the conducting walls, ψ can be written as

$$\psi(r, \theta, \phi) \cong \frac{\sqrt{k\pi r}}{2} J_{\alpha+1/2}(kr)\cos m\phi$$

where $J_{\alpha+1/2}$ is the Bessel function of the first kind and

$$\alpha = \alpha_m \cong \frac{\sqrt{4m^2+1}-1}{2} \quad (11.20)$$

Then the resonance frequency can be obtained from the boundary condition on the magnetic wall:

$$\left\{\frac{\partial}{\partial r}\left[\frac{\sqrt{k\pi r}}{2} J_{\alpha_m+1/2}(kr)\right]\right\}_{r=\text{Re}} = 0 \quad (11.21)$$

The radiating mode of interest is the TE_{11}-mode corresponding to $\alpha_1 \approx 0\cdot618$. Solving eqn. 11.21 numerically, we find

$$f_r = \frac{2\cdot303c}{2\pi\text{Re}\sqrt{(\varepsilon_r)}}$$

where c is the velocity of light in free space. In Table 11.5 the theoretical and experimental resonance frequencies of the TE_{11}-mode are presented for several values of θ_0 and ε_r; the radius of the cone base is fixed at 18·5 mm, while the substrate thickness is $h = 6$ mm.

11.2.2.2 Quality factor: The resonant frequencies, Q-factor and corresponding bandwidth (BW) of conical-patch and circular-disc antennas with the same parameters a, and ε_r are given in Table 11.6. It should be noted that in an air-dielectric ($\varepsilon_r = 1$), the same conical patch leads to: $f_r = 4\cdot 86$ GHz, $Q = 5$ and a $BW \approx 20\%$.

Table 11.6 *Resonant frequencies and bandwidths of conical-patch and circular-disc antennas*

Antenna	f_r (GHz)	Q	BW
Disc	2·8	12	8%
Conical structure	3·56	6	16%

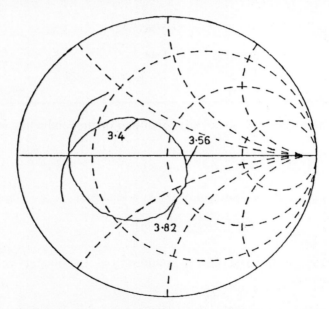

Fig. 11.19 *Input-impedance locus of matched conical patch antenna*

11.2.2.3 Impedance matching: As the substrate is relatively thick ($h \approx 0\cdot 1\lambda_\varepsilon$), the effect of the feed probe introduces an effective reactance and causes a clockwise rotation on the Smith chart of the input impedance seen at the feed port. An open coaxial transmission line situated inside the conducting cone

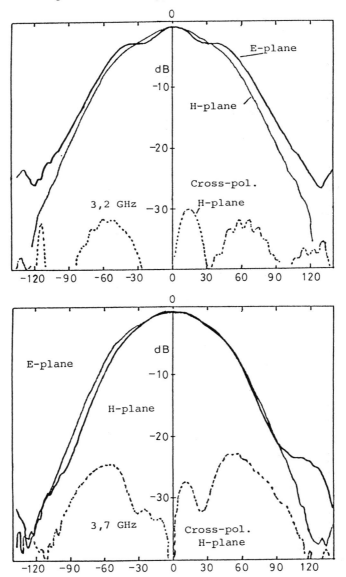

Fig. 11.20 *Radiation pattern of a conical patch with two excitations*

provides series-reactance compensation of the feed probe. Furthermore, by shifting the probe away to $a_0 = 10$ mm, we obtain the curve of Fig. 11.19 and a bandwidth (VSWR $\leqslant$ 2) of 12%.

11.2.2.4 Radiation patterns:
They are found to be relatively stable for linear polarisation over a wide-frequency range. Dissymetries in the *E*-plane and cross-polarised components in the *H*-plane, due to the generation of higher modes, can be suppressed as usual by using two radially opposite feed probes (Fig. 11.20). The necessary 0–180° phase difference can be obtained with a rat-race (3 dB coupler).

11.2.2.5 Circularly polarised conical-patch antennas:
It can be observed in Fig. 11.20, contrary to the case of the classical disc antenna, that the patterns are identical in both *E*- and *H*-planes. Thus, by feeding the conical patch as shown in Fig. 11.21, a good broadband, circularly polarised antenna can be realised. The axial ratio versus elevation angle θ is presented.

It can be concluded that flat conical-patch antennas are more advantageous in some respects than traditional microstrip antennas. Infact, they present a broader bandwidth, and their structure lends itself to a simple way of impedance matching. Moreover, they are well adapted to circular polarisation.

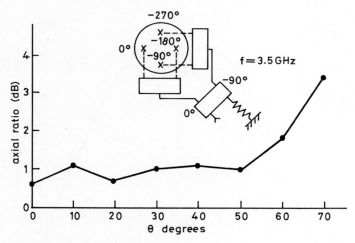

Fig. 11.21 *Axial ratio of a microstrip conical patch measured at 3·5 GHz versus elevation angle θ*

11.2.3 Linear and annular slots
Printed slots have received somewhat less attention than patch antennas in array design, since care must be taken to suppress undesired modes such as the parallel-plate mode excited between the ground planes in a typical stripline-fed slot [33–35]. However, the feeding and matching techniques remain very simple.

11.2.3.1 Linear slot [36, 37]:
A microstrip line offers the possibility of

Design and technology of low-cost printed antennas

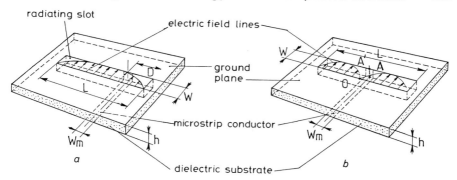

Fig. 11.22 *(a) and (b): Printed slot fed by a microstrip line at first resonance and second resonance*

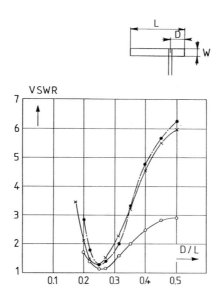

Fig. 11.23 *VSWR versus D/L of MSA at first resonance*
──── $L = 13$ cm $\quad F = 1{\cdot}15$ GHz
XXXX $L = 15{\cdot}5$ cm $\quad F = 0{\cdot}94$ GHz
OOO $L = 9$ cm $\quad F = 1{\cdot}53$ GHz
$\varepsilon_r = 4{\cdot}4 \quad w = 5{\cdot}0$ mm $\quad h = 11{\cdot}6$ mm

608 *Design and technology of low-cost printed antennas*

feeding the slot antenna at the first or second resonance without excessively disturbing the field distribution in the slot (Fig. 11.22). The strip conductor is connected through the dielectric substrate to the edge of the slot.

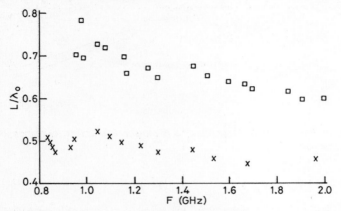

Fig. 11.24 *Radiating slot length normalised to λ_0 versus frequency measured at the first and second resonance*
$W = 5$ mm, $\varepsilon_r = 4\cdot 4$, $h = 1\cdot 6$ mm
XXX   first resonance
OOO   second resonance

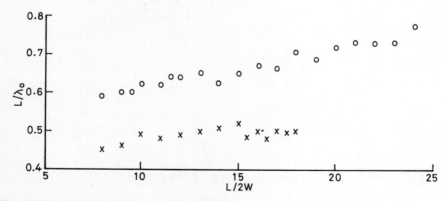

Fig. 11.25 *Influence of W on the first and second resonant length of a MSA*
$W = 5$ mm, $\varepsilon_r = 4\cdot 4$, $h = 1\cdot 6$ mm
XXX   first resonance
OOO   second resonance

A good match to the 50 Ω microstrip line can be obtained by choosing a centre-fed slot or an offset-fed slot for second-resonance or first-resonance operations, respectively. Fig. 11.23 gives the VSWR of the microstrip slot antenna (MSA) at first-resonance operation as a function of D/L, where L is the

Table 11.7 Bandwidth at resonance

1st resonance	$BW = 14\%$ for $L/2W \cong 16$
2nd resonance	$BW = 20\%$ for $10 < L/2W < 19$

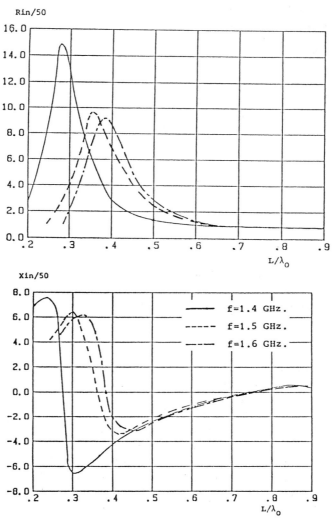

Fig. 11.26 *Input impedance of MSA fed by a microstrip line versus L/λ_0 at different frequencies [40]*
$W = 0.6$ mm, $\varepsilon_r = 2.17$, $h = 0.8$ mm

length normalised to λ_0 of the slot of width $W \ll \lambda_0$. D is the distance between the centre of the strip conductor of width W_m and the small side of the slot, and λ_0 is the free-space wavelength. From the Figure it can be seen that the slot is matched to the feed line for $D/L \approx 0.25$.

Fig. 11.24 shows the length normalised to λ_0 of a radiating slot of 0·5 cm width, measured at the first and second resonance, versus frequency. The influence of the width W on the first- and second-resonant length is shown in Fig. 11.25. At the first resonance $L = 0.49\lambda_0$, where λ_0 is the free-space wavelength, while at the second resonance L increases with $L/2W$. These variations are identical to those obtained for a cylindrical dipole [38, 39]. For the previous slot antennas, an optimum value of bandwidth (VSWR < 2) has been obtained (Table 11.7].

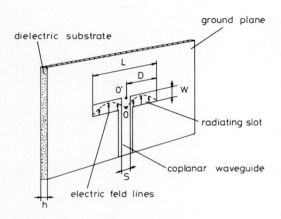

Fig. 11.27 *Printed slot fed by co-planar waveguide*

The large bandwidth obtained at the second resonance is due to the weak variation of the input impedance of the MSA, as shown in Fig. 11.26 [40]. Despite various attempts, an exact theoretical study of the input impedance of the MSA with limited ground plane does not really exist [34, 41].

An expression for the complex admittance at the first resonance of a radiating slot in the ground plane of a microstripline was found by Das [42] from the complex radiated power and discontinuity in the modal voltage. The concept of a complementary dipole [43] in an uniform medium of effective permittivity $\varepsilon_{re} = 2\varepsilon_r/(1 + \varepsilon_r)$ has been used to calculate the radiated power.

Experimental studies have been carried out over the frequency range 0·8–2 GHz using a 50 Ω coplanar waveguide [37] (Fig. 11.27); a bandwidth (VSWR < 2) of about 30% has been obtained. Nesic [44, 45] used coplanar-fed slots in primary radiators in a phase-scanned antenna.

11.2.3.2 Annular slots (Fig. 11.28)

(a) Model of an annular slot (without reflector plane) [46]: Analysis of radiating slots has been developed using a lossy transmission-line model (Fig. 11.29); it requires the computation of a propagation constant $\alpha + j\beta$ and a characteristic impedance Z_c. β and Z_c are obtained using Cohn's method [47], and α is the solution of the numerical equation $P_l(\alpha) = P_r(\alpha)$ where $P_l(\alpha)$ is the power delivered to the lossy loop and $P_r(\alpha)$ is power radiated from the annular slot.

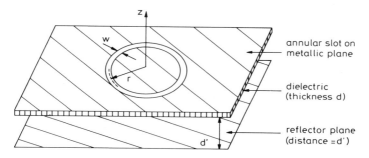

Fig. 11.28 *Printed annular slot above a reflector plane*

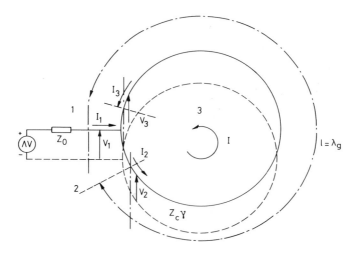

Fig. 11.29 *Transmission-line model of the annular slot*

(ai) Guided wavelength and impedance [48]

Theory: To date, no result of Cohn's method for calculating the slot parameters on a low-permittivity substrate has appeared in the literature. The use of Cohn's formula, for magnetic walls parallel to the slot and a relative permittivity of 2·17, gives the ratio of guided wavelength, λ_g, to wavelength in free space, λ, and the characteristic impedance defined in Reference 47. The results of computation are plotted in Fig. 11.30 in the manner they were presented earlier. A simple formula for λ_g/λ has been found like that in Reference 51:

$$\frac{\lambda_g}{\lambda} = 0.856 + 0.042 \frac{W}{d} - \log_{10}\left[100 \frac{d}{\lambda}\right]\left[0.035 + 0.017 \frac{W}{d}\right]$$

(11.22)

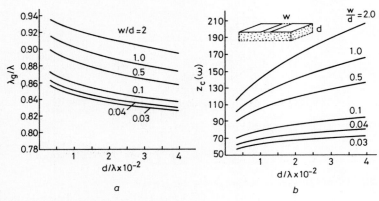

Fig. 11.30 (a) λ_g/λ versus d/λ and w/d
(b) Characteristic impedance versus d/λ and w/λ
$\varepsilon_r = 2.17$

Table 11.8 Some theoretical results of λ_g/λ (from Reference 49) and Cohn's method

Frequency (GHz)	Cohn's method	Spectral-domain technique [4]
2	0.8890	0.8885
2.5	0.8840	0.8834
3.0	0.8796	0.879
3.5	0.8756	0.875
4.0	0.8718	0.871

$\varepsilon_r = 2.55$; $d = 1.57$ mm; $w/d = 1.335$

The following set of parameters is considered:

$\varepsilon_r = 2.17$

$0.01 \leq w/d \leq 2.0$

$0.01 \leq d/\lambda < 0.1$

The previous formula fits the theoretical results with better than 2% accuracy. The upper limit of d/λ corresponds to appreciable excitation of surface waves

[52]. Comparison of that method with theoretical results obtained by the spectral-domain technique in a recent paper [49] gives good agreement as shown in Table 11.8.

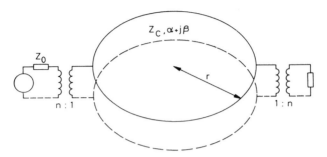

Fig. 11.31 *Resonant-ring transmission-line model*

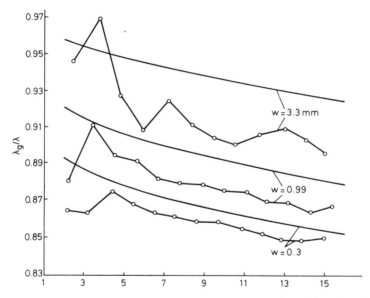

Fig. 11.32 *Theoretical and experimental results for guided wavelength for slots of different width w*
—o—experiment (ε_r = 2·17, substrate thickness = 0·78 mm)
- - - - theory

Experiment: Experiments were carried out to verify Cohn's method. The method of the resonant slot ring was chosen, because it has given very good results on substrates of higher permittivity (ε_r = 9·6) [50]. Three rings were etched. The outer radius of the rings was equal to 38 mm. The measured widths were 0·3, 0·99 and 3·3 mm. The thickness of the substrate was 0·78 mm and the relative dielectric constant was 2·17. Measurements were performed on an automatic network analyser HP 8510. The output and input lines were cut until the signal level weakened at about $-40\,\text{dB}$ below the reference set to $10\,\text{dBm}$ (Fig. 11.31).

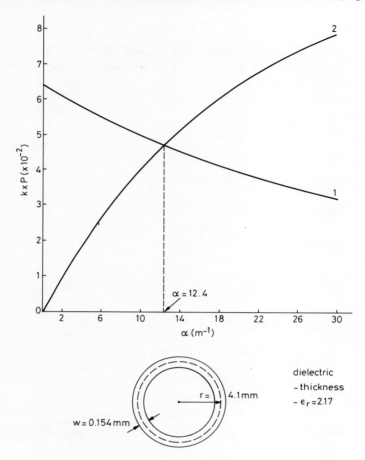

Fig. 11.33 *α solution of $P_r(\alpha) = P_1(\alpha)$ for X-band annular slot*

The difference between the theoretical and experimental curves of Fig. 11.32 is due to ohmic, dielectric and radiation losses. In the case of a slot on a dielectric substrate of low permittivity, the radiation loss is important: it increases when the radius of curvature decreases and remains significant for a straight slot. It can be noted that, for quasi-straight slots, the experimental ratio λ_g/λ always appears lower than the theoretical one, as in Reference 49.

(aii) Attenuation coefficient and antenna design
The method which is used to find α requires numerical computation.

- P_1 is the power delivered to the lossy loop (analytically known)
- P_r is the radiated power.

then α is the solution to the equation

$$P_r(\alpha) = P_1(\alpha)$$

It will be noticed that P_1 increases with α while P_r decreases; thus the solution is unique. An example is given in Fig. 11.33; the loop was designed at X-band from the dimensions given in Fig. 11.33. The variation of α near the resonant frequency is plotted in Fig. 11.34, and it appears that α has a linear variation.

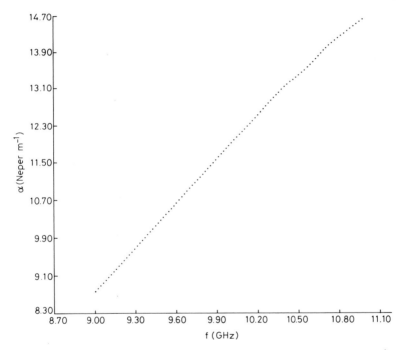

Fig. 11.34 α *variation versus frequency (near the resonant frequency)*
$r = 4\cdot14$ mm, $W = 154\,\mu$m

The design of antennas matched to $50\,\Omega$ has been considered using the following model shown in Fig. 11.35; a quarter-wavelength line (length l_1) is used to feed the loop and a second transformer (length l_2) is used as a matching section. The equivalent transformer from the microstrip feed line and slot are taken into account. Impedance curves are drawn in Fig. 11.36 for frequencies varying from 9 to 11 GHz. The theoretical results agree well with the experimental values.

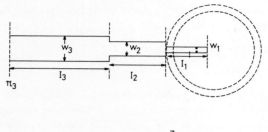
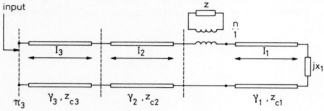

Fig. 11.35 *Annular slot and its equivalent circuit*

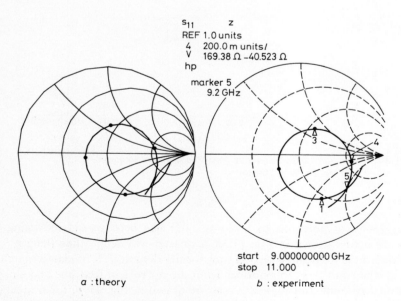

a : theory   b : experiment

Fig. 11.36 *Impedance of the printed annular slot*
r = 4·1 mm, *W* = 140 μm, W_1 = 0·76 mm

Design and technology of low-cost printed antennas 617

To facilitate the design of antennas, analytical formulas for α have been derived in the vicinity of the resonant frequency:

$$\alpha = \frac{A}{\lambda} + \frac{B}{d} \tag{11.23}$$

where

α = attenuation, N/m
λ = free-space wavelength
d = thickness of dielectric

$$A \times 1000 = 25{\cdot}81 \left(\log_{10} \left(100 \frac{w}{d} \right) \right)^2 \left(\frac{d}{\lambda_{res}} \right)^{2/3}$$
$$+ 79{\cdot}15 \frac{w}{d} \left(\frac{d}{\lambda_{res}} \right)^{2/3} - 1300{\cdot}31 \left(\frac{w}{d} \right)^{1/2} \frac{d}{\lambda_{res}}$$
$$- 251{\cdot}44 \frac{d}{\lambda_{res}} + 0{\cdot}82 + 9{\cdot}36 \left(\frac{w}{d} \right)^{1/2} \log_{10} \left(120 \frac{d}{\lambda_{res}} \right)$$
$$- 94{\cdot}83 \frac{w}{\lambda_{res}}$$

$$B \times 1000 = 183{\cdot}8 \left(\log_{10} \left(100 \frac{w}{d} \right) \right)^{5/4} + 459{\cdot}8 + 3076{\cdot}5 \frac{d}{\lambda_{res}}$$
$$- 18005{\cdot}3 \left(\frac{d}{\lambda_{res}} \right)^2 + 570{\cdot}8 \left(\log_{10} 200 \frac{d}{\lambda_{res}} \right)^{4/3} \left(\frac{w}{d} \right)^{3/4}$$

The accuracy of the above formula is better than 2%; it can be used for the following parameters:

$$\varepsilon_r = 2{\cdot}17;\ 0{\cdot}01 \leq \frac{w}{d} \leq 2;\ 0{\cdot}01 \leq \frac{d}{\lambda_{res}} \leq 0{\cdot}1$$

(b) Annular slots with reflector planes (Fig. 11.28)

The analysis is very similar to the previous one; a modification of Cohn's method makes it possible to obtain the two main parameters, guided wavelength and characteristic impedance of the slot line in front of the reflector plane Figs. 11.37a and b show the variation of λ_g and Z_c for the two distances d' ($d'/d = 3$ and $d'/d = 11$) as a function of d/λ. The dielectric constant is 2·17 and the slot is very thin ($W/d = 0{\cdot}01$). It appears that λ_g and Z_c do not differ from the values for slot lines without a reflector when it is sufficiently displaced. Detailed analysis of the mathematical expressions [46] shows that it is possible to find the limit distance d'_{lim}, which leads to identical values of λ_g and Z_c with or without

618   Design and technology of low-cost printed antennas

a reflector plane:

$$\frac{d'_{lim}}{\lambda} > \frac{0.3}{\pi\left[\left(\frac{\lambda}{\lambda_g}\right)^2 - 1\right]^{1/2}} \tag{11.24}$$

The following assumptions are then necessary to obtain the value of α.

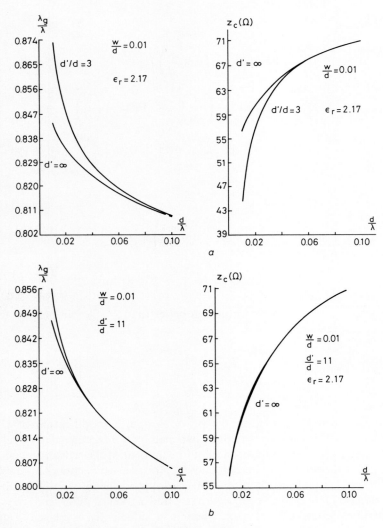

Fig. 11.37 (a) Variation of λ_g/λ and Z_c for slot above reflector plane ($d'/d = 3$)
(b) Variation of λ_g/λ and Z_c for slot above reflector plane ($d'/d = 11$)

(a) Radiation occurs only in the upper half space.
(b) No energy couples to the space between the two metallic planes (slot plane and ground plane).

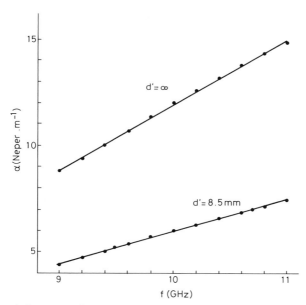

Fig. 11.38 α variation versus frequency for slot above reflector plane
$r = 4\cdot 41$ mm, $w = 154\,\mu$m

Table 11.9 Dimensions of the two annular slots of Figs. 11.40a and b

Antenna	r (mm)	W (μm)	l_1 (mm)	l_2 (mm)	l_3 (mm)	W_1 (μm)	W_2 (μm)	W_3 (μm)
1	4·065	140	5·45	5·85	22·6	760	760	2435
2	4·090	154	6·28	5	22	373	373	2310

Then, for a EMF given generator, the total radiated power is half the value obtained without a reflector plane; α is also reduced by a factor of 2, and the impedance is twice that without a reflector plane. Fig. 11.38 shows the frequency variation of α with or without the reflector plane.

To avoid guided waves between the two metallic planes, four cylindrical metallic posts are positioned as shown in Fig. 11.39a; the post spacing equals one half-wavelength, and their height is 8·5 mm. In Fig. 11.39b impedance curves have been plotted either for theoretical results (assuming radiation in the upper half-space only) or for experimental results with and without posts.

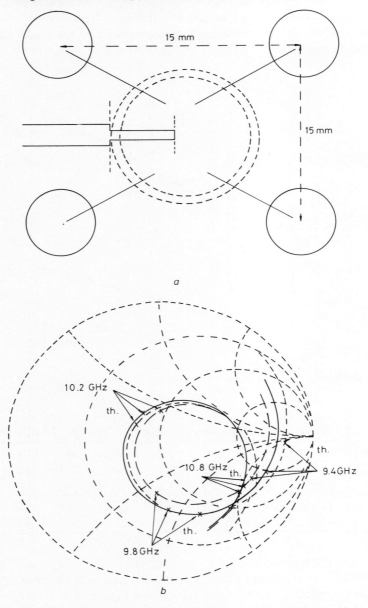

Fig. 11.39 (a) Location of the metallic posts near the annular slot
(b) Input impedance of the annular slot (reference plane π_3)
---- theory
—— experience with posts
– – – experience without posts

Design and technology of low-cost printed antennas

Theory and experiment agrees well when the posts are taken into account.

Input impedances have been computed and measured for two different slots and feed lines with the same ground-plane distance $d' = 8.5$ mm (Fig. 11.40); the reference plane is π_3 (Fig. 11.35) and the dimensions are given in Table 11.9.

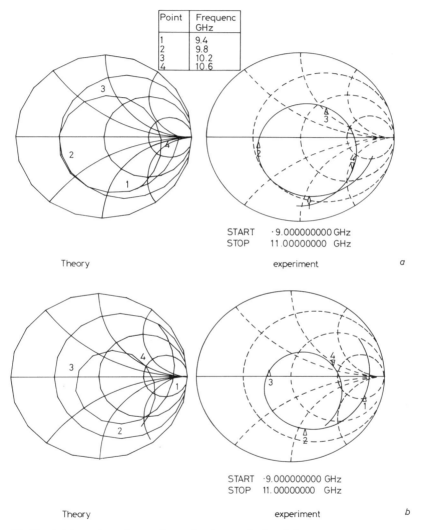

Fig. 11.40 (a) Input impedance of annular slot no. 1: $d' = 8.5$ mm
(b) Input impedance of annular slot no. 2: $d' = 8.5$ mm

Conclusion: It has been proved that the transmission-line model of annular slots yields good results in terms of impedance variation versus frequency. However, some problems occur when a ground plane is added to produce a directional antenna; further work is necessary to understand the parasitic effects of guided waves between the ground plane and the metallic plane of the antenna.

11.3 Design of planar printed arrays

Photo-etching techniques offer great flexibility in designing one- or two-dimensional arrays of microstrip antennas. Many parameters could be considered; e.g. position of elements, spacings, amplitude and phase distribution, feeding lines (on the same side or behind), connection with active devices etc. In fact, each structure is developed in response to a given class of problem. Only the usual design parameters and the realisation of pencil beam antennas are developed here; shaped-beam pattern design will be considered, together with the synthesis methods, in Section 11.4.

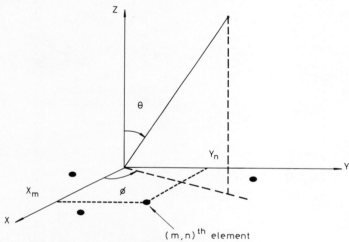

Fig. 11.41 Co-ordinate system and location of the (m, n)th element

11.3.1 Design parameters

With the assumption of identical elements, the array gain is decomposed into the product of array factor, isolated element gain pattern, and a factor displaying the impedance effects of mutual coupling. When mutual coupling is negligible, the overall radiation pattern becomes [53]

$$F(\theta, \phi) = f(\theta, \phi) \, g(\theta, \phi) \tag{11.25}$$

where

$g(\theta, \phi)$ = radiation pattern of one element

$f(\theta, \phi)$ = array factor

$$= \sum_m \sum_n a_{m,n} \, e^{jk_0(x_n \sin\theta\cos\phi \, + \, Y_n \sin\theta\sin\phi)} \tag{11.26}$$

Design and technology of low-cost printed antennas 623

and $a_{m,n}$ is the excitation coefficient (relative to a reference element) of the (m, n) source located at (x_m, y_n) (Fig. 11.41). When the radiating elements are regularly spaced along two orthogonal directions (Fig. 11.42), simple ex-

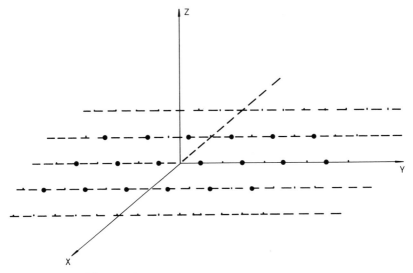

Fig. 11.42 *Array of linear sub-arrays*

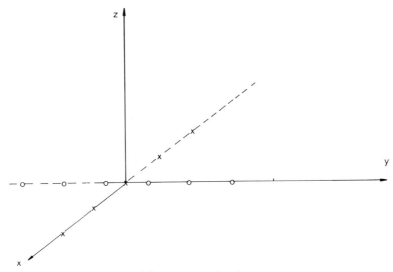

Fig. 11.43 *Equivalent orthogonal linear arrays of a planar array*

pressions can be derived for the two main planes, $\phi = 0°$ and $\phi = \pi/2$. For any ϕ-direction, the planar array can be analysed using the pattern-multiplication method if the various sub-arrays are identical (same number of elements and same distribution); then the planar-array factor remains the product of two linear-array factors (Fig. 11.43).

As planar arrays and linear arrays have much in common, some basic results for linear series microstrip antenna arrays are given first: (i) directivity estimation taking into account the dielectric substrate, (ii) line losses depending on the element impedance value, and (iii) beam width and limitation on gain.

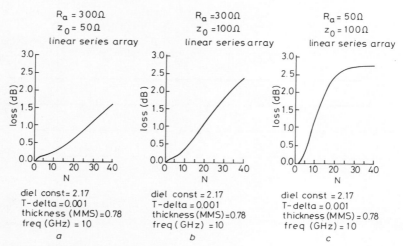

Fig. 11.44 Line losses of linear series array
(a) $R_a = 300\,\Omega$, $Z_0 = 50\,\Omega$
(b) $R_a = 300\,\Omega$, $Z_0 = 100\,\Omega$
(c) $R_a = 50\,\Omega$, $Z_0 = 100\,\Omega$

11.3.1.1 Line losses, efficiency of linear series array: Typical linear resonant arrays of microstrip antennas can be considered as a transmission line loaded periodically by shunt resistances whose values depend on the elements themselves (rectangular, circular, triangular with proper feeding points), or on the equivalent sub-array attached to the main line. For the sake of simplicity, let us consider a transmission line with a characteristic impedance Z_0 and a complex propagation constant $\gamma = \alpha + j\beta$. The periodic loads (radiating elements) are located at each guided wavelength in order to get a uniformly excited array. The resonant structure is then represented by a cascade of lossy transmission-line sections. The radiated power P_r and input power P_{in} are readily computed, and efficiency can be deduced:

$$\eta = \frac{P_r}{P_{in}} = \frac{R_{in}}{R_a} \sum_{j=1}^{N} |v_j|^2 \qquad (11.27)$$

where R_{in} = input resistance; R_a = resistance of each radiating element and V_j = potential at node j.

Losses have been plotted (Fig. 11.44a, b, c) on a dB scale (loss (dB) = P_{indB} − P_{rdB}) for a number N varying from 2 to 40. Three main parameters have been considered: Z_0, the characteristic impedance with typical values of 50 Ω and 100 Ω; dielectric loss tangent, typically 10^{-3}; and R_a equal to 300 or 50 Ω. In each case the usual dielectric-constant value for printed circuits of Teflon fibre substrate is equal to 2·2, and the different curves have been obtained for a frequency of 10 GHz. When N increases, the efficiency is reduced, as might be expected; the highest value of 3 dB is obtained only for a large number of elements. However, the slope of the curves depends on the antenna resistance and characteristic impedance; it will be noticed that small R_a (typically 50 Ω) and large Z_0 (here 100 Ω) lead to high losses, even with a moderate number of elements.

In order to reduce line losses, the designer should choose a high input impedance R_a and a low characteristic impedance Z_0. However, the last constraint often conflicts with the necessarily small width of the input feed line compared with the width of the antenna itself.

11.3.1.2 Directivity of uniform linear arrays: The simplest type of linear array is the uniform one. Analytical expressions for directivity have been proposed for typical elements such as ideal isotropic sources, co-linear short dipoles and parallel short dipoles [53–56]. Rectangular printed antennas radiate from fringing fields around the edges; the fields along the two radiating edges are approximately uniform. Thus each antenna can be considered as an array of two uniformly excited identical slots; its directivity is about 5–8 dB depending on the dielectric substrate. Two linear-array structures are to be considered, namely the H-plane and E-plane (Fig. 11.45); the slots of each patch are defined by their length W and their spacing d_0; d equals the distance between each element. The array radiates in the half-space defined by $x > 0$. Elements are located along Oz to get simple expressions for slot patterns and simple integrals of directivity.

(a) Directivity expression
The element pattern of each patch $g(\theta, \phi)$ can be easily expressed in H- and E-plane configuration as:

H-plane: $g(\theta, \phi) = \sin\theta \cos\left(k \dfrac{d_0}{2} \sin\theta \sin\phi\right)$

E-plane $g(\theta, \phi) = (1 - \sin^2\theta \sin^2\phi)^{1/2} \cos\left(k \dfrac{d_0}{2} \cos\theta\right)$

and the directivity expression of an N-element array in the broadside direction may be written

$$D = \dfrac{2}{\dfrac{1}{4\pi} \int_{4\pi} f^2(\theta) g^2(\theta, \phi)\, d\Omega} = \dfrac{2}{\dfrac{\Omega_A}{4\pi}} \qquad (11.28)$$

where

$$f(\theta) = \frac{1}{N}\left[\frac{\sin(N\psi/2)}{\sin(\psi/2)}\right]$$

$$\psi = kd\cos\theta$$

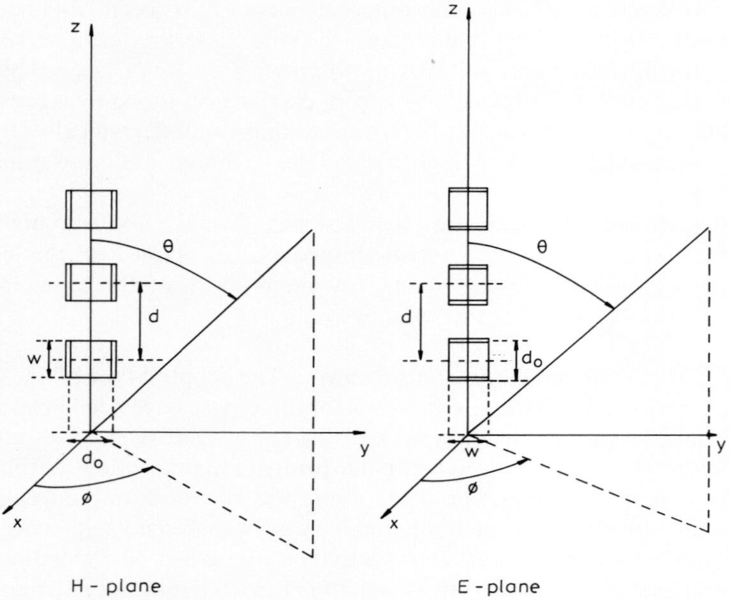

Fig. 11.45 *E-plane and H-plane linear arrays of equivalent slots of microstrip antenna*

The coefficient 2 arises because the antenna radiates in the half-space, $X > 0$. With $d\Omega = \sin\theta\, d\theta\, d\phi$ and using the finite-series expression [55, 59] of $f^2(\theta)$:

$$f^2(\theta) = \frac{1}{N} + \frac{2}{N^2}\sum_{m=1}^{N-1}(N-m)\cos m\psi$$

the results are

H-plane configuration:

$$D = \frac{2}{\frac{2}{3}\frac{1}{N} + \frac{2}{N^2}\sum_{m=1}^{N-1}2(N-m)SC - aI_1 + bI_2} \qquad (11.29)$$

E-plane configuration:

$$D = \frac{2}{\frac{1}{2}\left[\frac{1}{N}\left(\frac{2}{3} + S_0 - SC_0\right) + \frac{1}{N^2}\sum_{m=1}^{N-1}(N-m)[2S + S_1 + S_2 - (2SC + SC_1 + SC_2)]\right]}$$

(11.30)

where

$$S_i = \frac{\sin \alpha_i}{\alpha_i} \quad SC_i = \frac{1}{\alpha_i}\left(\frac{\sin \alpha_i}{\alpha_i^2} - \frac{\cos \alpha_i}{\alpha_i}\right) \quad SC = \frac{1}{\alpha}\left(\frac{\sin \alpha}{\alpha^2} - \frac{\cos \alpha}{\alpha}\right)$$

$$I_1 = \alpha_0^2\left[\frac{4}{15}\frac{1}{N} + \frac{8}{N^2}\sum_{m=1}^{N-1}\left(\frac{N-m}{\alpha^2}\right)\left(S\left(\frac{3}{\alpha^2}-1\right) - \frac{3\cos\alpha}{\alpha^2}\right)\right]$$

$$I_2 = \alpha_0^4\left[\frac{8}{35}\frac{1}{N} + \frac{24}{N^2}\sum_{m=1}^{N-1}\left(\frac{N-m}{\alpha^3}\right)\left[S\left(\frac{15}{\alpha^3}-\frac{6}{\alpha}\right) + \frac{\cos\alpha}{\alpha}\left(1-\frac{15}{\alpha}\right)\right]\right]$$

and

$$\alpha_0 = kd_0 \qquad a = 0\cdot24774379$$
$$\alpha = mkd \qquad b = 0\cdot01294148$$
$$\alpha_1 = mkd + kd_0$$
$$\alpha_2 = mkd - kd_0$$

The a and b constants which occur in the H-plane-configuration expression appear in the θ integration of Ω_A

$$\Omega_A = \int_0^{2\pi}\int_0^{\pi} \cos^2\left(\frac{kd_0}{2}\sin\theta\sin\phi\right) g^2(\theta) f^2(\theta) \sin\theta \, d\theta \, d\phi$$

$$\Omega_A = \int_0^{\pi} \pi(1 + J_0(kd_0 \sin\theta)) g^2(\theta) f^2(\theta) \sin\theta \, d\theta$$

The kernel of the Bessel function J_0 remains between 0 and 2·5 for most printed antennas ($\varepsilon_r \geq 2$); then $J_0(x)$ can be approximated by a simple polynomial expression:

$$J_0(x) \approx 1 - ax^2 + bx^4$$

The use of two points $x_1 = 1$ and $x_2 = 2\cdot402$ leads to the previous values a and b. $J_0(x)$, and the approximate polynomial are plotted in Fig. 11.46. There is good agreement within $(x_1 x_2)$.

(b) Results

Figs. 11.47 and 11.48 show the variations of D for linear arrays of printed antennas in H- and E-plane configurations for two kinds of dielectric: PTFE ($\varepsilon_r = 2\cdot2$ and thickness $= 1\cdot6$ mm) and alumina ($\varepsilon_r = 9\cdot8$ and thickness $= 0\cdot625$ mm). The dimensions of the antennas were computed using the following expressions:

$$d_0 = \frac{\lambda_0}{2\sqrt{\varepsilon_{\text{eff}}}} - 2\Delta d_0 \quad \text{and} \quad \Delta d_0 = \text{length extension [6]}$$

$$W = \frac{\lambda_0}{2}\left(\frac{\varepsilon_r + 1}{2}\right)^{-1/2}$$

The curves show that D is an increasing function of d/λ when d/λ is less than 1 and N is large. The optimum distance d for large directivity and no grating lobes lies between $0.7\lambda_0$ and $0.9\lambda_0$. For usual substrates ($\varepsilon_r \approx 2$ to 2.5) and with

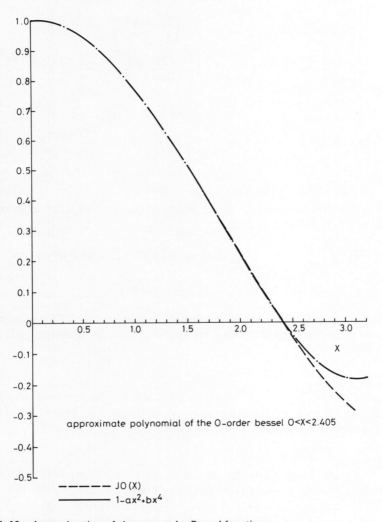

Fig. 11.46 *Approximation of the zero-order Bessel function*

one-λ_g-long straight feeding lines between neighbouring elements, the distance d is approximately $0.75\lambda_0$. For small values of N, the E-plane arrays exhibit a larger directivity than H-plane ones. On the other hand, the H-plane arrays present similar directivity variations for $\varepsilon_r = 2.2$ and 9.8.

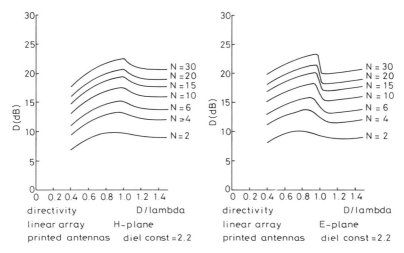

Fig. 11.47 *Directivity of an N-element linear array of printed antennas as a function of uniform element spacing*
Dielectric constant = 2·2.

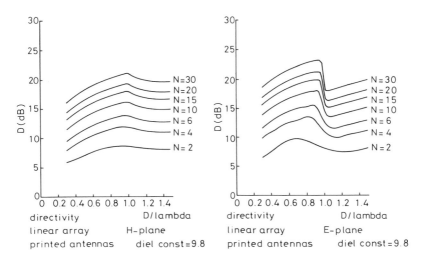

Fig. 11.48 *Directivity of an N-element linear array of printed antennas as a function of uniform element spacing*
Dielectric constant = 9·8.

11.3.1.3 Non-uniform linear arrays:

The reduction of the sidelobe level requires tapering of the amplitude distribution; some care must be taken in the element spacing; the optimum distance $d \approx 0.75\lambda_0$ for printed antennas on a PTFE substrate does not always fulfil the spacing requirement of the Chebyshev design for small number N and low sidelobes. Fig. 11.49 gives $(d/\lambda_0)_{max}$ versus sidelobe level for various N; and example of an array factor with $N = 6$ and

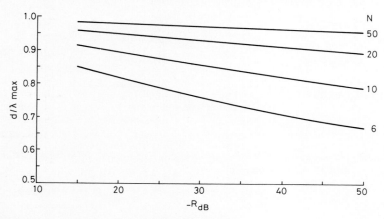

Fig. 11.49 $(d/\lambda)_{max}$ versus sidelobe level for Chebýchev taper

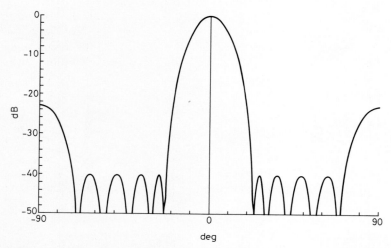

Fig. 11.50 Chebýshev pattern of a six-isotropic-source linear array (Spacing = $0.75\lambda_0$)

$(d/\lambda_0) = 0.75$, designed for $-40\,dB$ sidelobe level, is shown in Fig. 11.50. A strong effect appears on the sidelobe level, which reaches $-23\,dB$ at $\theta = \pm 90°$ instead of $-40\,dB$. On the other hand, directivity D decreases and the half-power beamwidth (HPBW) increases. Directivity can be computed for isotropic elements using the following formula [54]:

$$D = \frac{\left(\sum_{n=1}^{N} A_n\right)^2}{\sum_{n=1}^{N}\sum_{m=1}^{M} A_n A_m \frac{\sin(n-m)kd}{(n-m)kd}} \quad (11.31)$$

where A_n is the normalised amplitude of the nth element. However, the excitation efficiency $\eta_e = D/D_0$ (where D_0 is the directivity of uniform array of isotropic elements) is available either for Chebȳshev, Taylor or Villeneuve distribution [53, 54, 57, 58, 60]; assuming that η_e of printed antennas remains very similar to η_e of isotropic sources, a quick estimate of D can be obtained from D_0 given previously.

11.3.2 Cavity-model analysis of mutual coupling

Since the experimental work of Jedlicka and Carver [61, 62], many theoretical efforts have been made to calculate the mutual interaction between microstrip antennas [25, 63–68]. Mutual-coupling effects are caused by radiation through free space and by surface waves which propagate along a dielectric substrate. The theoretical method proposed by Penard and Daniel [65, 25] is restricted to the first effect; this is acceptable as long as the dielectric constant is low and the thickness is small compared to the wavelength [67]. The method uses the cavity model of the antenna, and it is assumed that mutual coupling does not disturb the field distribution in the cavity. In the cavity method, by using the equivalence principle, one can relate the internal field in the cavity to the magnetic-current loop radiating in the upper half-space. Then the mutual-coupling coefficients can be derived from the interaction between magnetic current loops.

The mutual impedance Z_{ij} between two elements is deduced from the reaction theorem

$$Z_{ij} = \oint_{c_j} \frac{H_i \cdot M_j}{I_i I_j} dl \quad (11.32)$$

where C_j = contour of antenna j; M_j = magnetic line source; H_i = magnetic field set up by antenna i on antenna j; I_i and I_j = current feeds of the two patches.

Only the fundamental mode is usually considered in evaluating integral 11.32 in order to reduce the computation time. The coupling coefficient S_{ij} can easily be calculated by applying the impedance-matrix equations of network theory. Provided the internal electric field is known, printed antennas of any shape can be considered.

11.3.2.1 Rectangular patch antennas: In the H-plane configuration (Fig. 11.51), the mutual impedance between identical OMA (open microstrip antennas) is given by [65]

$$z_{12} = \frac{jt^2 A_0^2}{240\pi^2 k_0} (R_1 + R_2 + R_3) \tag{11.33}$$

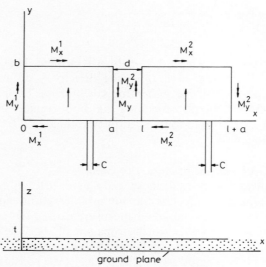

Fig. 11.51 *Geometry of two rectangular microstrip antennas in H-plane configuration*

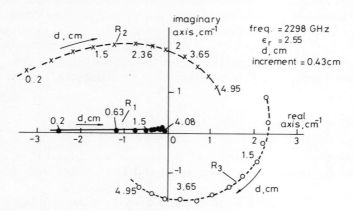

Fig. 11.52 *Contribution of OX edges and OY edges in mutual-impedance calculation in H-plane configuration; mode(0, 1)*

where t is the substrate thickness with

$$A_0 = k\omega\mu \frac{C}{a} \frac{1}{\beta_0 \tan(\beta_0 b)}$$

Design and technology of low-cost printed antennas

where C is the width of the excitation and

$$\beta_0^2 = k^2(1 - jtg\Delta)$$
$$k = k_0 \sqrt{\varepsilon_r}$$

The expression takes all the equivalent slots around the patch into account. The contribution of the two $0X$-edge radiating slots appears in the integral R_3.

$$R_3 = -2k^2 \int_0^a \int_l^{1+a} \frac{e^{-jk_0|x-x'|}}{|x-x'|} + \frac{e^{-jk_0\sqrt{(x-x')^2+b^2}}}{\sqrt{[(x-x')^2+b^2]}} \, dx \, dx'$$

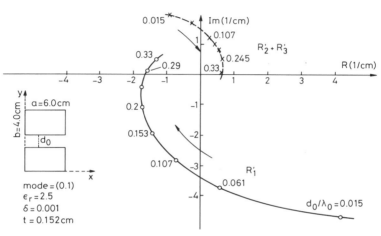

Fig. 11.53 Contribution of $0X$ edges and $0Y$ edges in mutual-impedance calculation in E-plane configuration; mode(0, 1)

The effect of the $0Y$ slots appears in R_1 and R_2 and cannot be ignored, as shown in Fig. 11.52.

$$R_1 = \left(\left(\frac{\pi}{b}\right)^2 - k^2\right) \int_0^b \int_0^b \cos\frac{\pi}{b}y' \cos\frac{\pi}{b}y \left[2\frac{e^{-jk_0\sqrt{l^2+(y-y')^2}}}{\sqrt{l^2+(y-y')^2}}\right.$$
$$\left. - \frac{e^{-jk_0\sqrt{d^2+(y-y')^2}}}{\sqrt{d^2+(y-y')^2}} - \frac{e^{-jk_0\sqrt{(l+a)^2+(y-y')^2}}}{\sqrt{(l+a)^2+(y-y')^2}}\right] dy \, dy'$$

$$R_2 = 2\left(\frac{\pi}{b}\right) \int_0^b \sin\frac{\pi}{b}y \left[-\frac{e^{-jk_0\sqrt{d^2+y^2}}}{\sqrt{(d^2+y^2)}} + 2\frac{e^{-jk_0\sqrt{l^2+y^2}}}{\sqrt{(l^2+y^2)}}\right.$$
$$\left. - \frac{e^{-jk_0\sqrt{(l+a)^2+y^2}}}{\sqrt{(l+a)^2+y^2}}\right] dy$$

It should be noted that R_1 is cancelled out when the dielectric slab is replaced by a vacuum medium. Similar relations can be found for two patches in the

E-plane configurations [25]

$$Z_{12} = \frac{-jA_0^2 t^2}{240\pi^2 k_0} (R'_1 + R'_2 + R'_3) \tag{11.34}$$

In this case, the effect of the two 0X-directed slots (R'_1) is stronger, but the contribution of the longitudinal slots (R'_2 and R'_3) is not negligible (Fig. 11.53). Fig. 11.54 shows calculated values and experimental data of the coupling between two identical patches for various spacing in *E*-plane and *H*-plane. When the distance *d* becomes smaller than $0.08\lambda_0$, the discrepancy between theoretical and experimental results in the *H*-plane increases. In this case, the coupling disturbs the internal field distribution, which clearly appears on the measured value of the input impedance Z_{11} (Fig. 11.55).

Penard [25] has studied the mutual coupling between the hybrid-microstrip antennas (HMA). In this element, the contribution of the apertures situated near the electric walls is very small and can generally be ignored. In Fig. 11.56, the coupling coefficient of HMA and OMA are shown for comparison. In the *H*-plane, S_{12} is smaller for HMA than for OMA, except when the dielectric is replaced by a vacuum medium (Fig. 11.57).

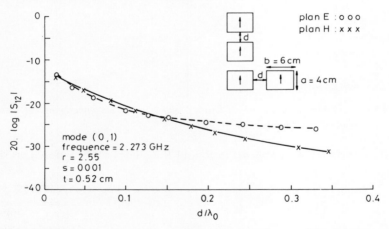

Fig. 11.54 *Mutual-coupling coefficient as a function of the distance $d/\lambda 0$ of the edges in the E-plane and H-plane; mode(0, 1)*

Another interesting result, in view of array design, is that the mutual coupling coefficient slowly increases with the substrate thickness (Fig. 11.58). The effect of the dielectric-constant variations on mutual coupling of short-circuited and open microstrip antennas is presented in Fig. 11.59 and 11.60, respectively.

11.3.2.2 Circular patch antennas: In this case, the circular geometry of the elements allows the separation of the co-ordinate variables, disc centre spacing R_{ij}, disc angular orientation ϕ_{ij} and angular feed positions θ_i, θ_j (Fig. 11.61).

Design and technology of low-cost printed antennas 635

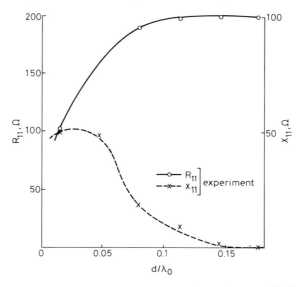

Fig. 11.55 *Experimental results of Z_{11} as a function of the distance d/λ_0 between the two patches; H-plane; mode(0, 1)*

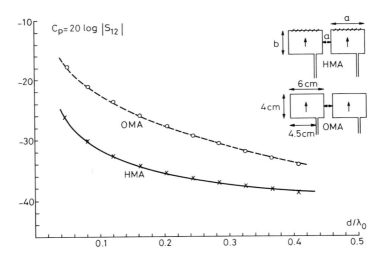

Fig. 11.56 *Comparison of coupling coefficient in H-plane, between HMA (a = 6 cm, b = 2 cm) and OMA (a = 6 cm, b = 4 cm) structures*
$\varepsilon_r = 2\cdot17$, $f_{r0} = 2\cdot45$ GHz, $t = 0\cdot157$ cm, $\tan \Delta = 10^{-4}$

636 Design and technology of low-cost printed antennas

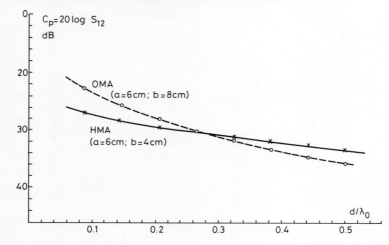

Fig. 11.57 *Comparison of coupling coefficient in H-plane, between HMA and OMA with air–dielectric*
$f_{r0} = 1 \cdot 773 \, \text{GHz}, \, t = 0 \cdot 150 \, \text{cm}$

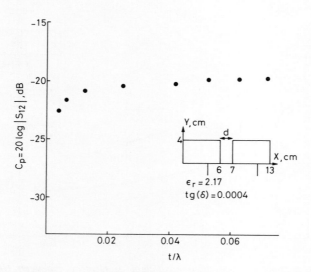

Fig. 11.58 *Theoretical coupling coefficient as a function of the dielectric thickness t/λ; mode(0, 1)*

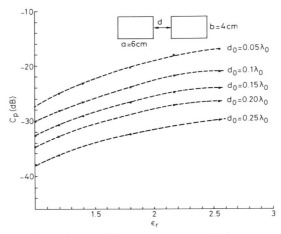

Fig. 11.59 *Theoretical coupling coefficient between two OMA versus ε_r for different values of d/λ_0*

$\varepsilon_r = 1$ $t/\lambda_0 = 0.017$

$\varepsilon_r = 1.4$ $t/\lambda_0 = 0.0152$

$\varepsilon_r = 2.17$ $t/\lambda_0 = 0.013$

$\varepsilon_r = 2.55$ $t/\lambda_0 = 0.0115$

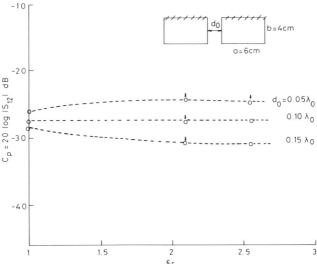

Fig. 11.60 *Theoretical coupling coefficient between two HMA versus ε for different values of d/λ_0*

$\varepsilon_r = 1$ $t/\lambda_0 = 0.0093$ △ experiment

$\varepsilon_r = 2.1$ $t/\lambda_0 = 0.0064$ ○ theory

$\varepsilon_r = 2.55$ $t/\lambda_0 = 0.006$

638 Design and technology of low-cost printed antennas

Mahdjoubi [66] gives a semi-analytical formula for Z_{ij} which permits a great reduction in computation charges:

$$Z_{ij} = \sum_{m=0}^{\infty} Z_i^m \sum_{n=0}^{\infty} Z_j^n [Y_e^{mn}(R_{ij}) \cos m(\phi_{ij} - \theta_i) \cos n(\phi_{ij} - \theta_j)$$
$$+ Y_h^{mn}(R_{ij}) \sin m(\phi_{ij} - \theta_i) \sin n(\phi_{ij} - \theta_j)] \quad (11.35)$$

m and n are the angular mode indices for antennas i and j, respectively.

$$Z_q^p = \frac{j\omega\mu_0 t \sin p\theta_{wq}}{2(1 + \delta_{po}) p\theta_{wq}} \left[Y_p(ka_q) - \frac{Y_p'(ka_q)}{J_p'(ka_q)} J_p(ka_q) \right] J_p(k\varrho_q)$$

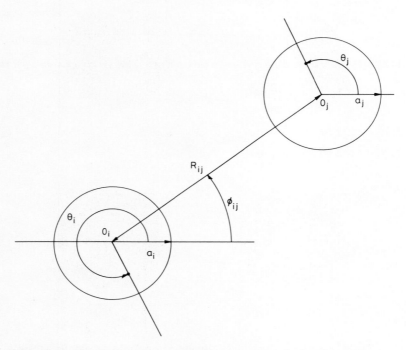

Fig. 11.61 *Geometry of electromagnetically coupled microstrip disc antennas*

= angular width of the feeding probe; ϱ = radial feed position; a = effective radius of the disc antenna; t = substrate thickness; k = propagation constant in the dielectric, J_p and Y_p = Bessel functions of first and second kind of order p.

$Y_e^{mn}(R_{ij})$ and $Y_h^{mn}(R_{ij})$ are the space-wave mutual admittances between mode m of annular slot i and mode n of annular slot j for the E- and H-plane coupling configurations, respectively. For two identical circular discs with only the fundamental modes excited in both antennas ($m = n = 1$), the mutual impedance Z_{12} can be written

$$Z_{12} = Z'_1 Z'_2 Y_{12}(R, \phi, \theta_1, \theta_2) \quad (11.36)$$

where

$$\begin{aligned} Y_{12}(R, \phi, \theta_1, \theta_2) &= Y_e(R)\cos(\phi - \theta_1)\cos(\phi - \theta_2) \\ &+ Y_h(R)\sin(\phi - \theta_1)\sin(\phi - \theta_2) \end{aligned}$$

The Y_e factor corresponds to the E-plane ($\phi = 0, \theta_1 = \theta_2 = 0$) while the Y_h factor corresponds to the H-plane ($\phi = 0, \theta_1 = \theta_2 = 90°$) (see Fig. 11.62). The

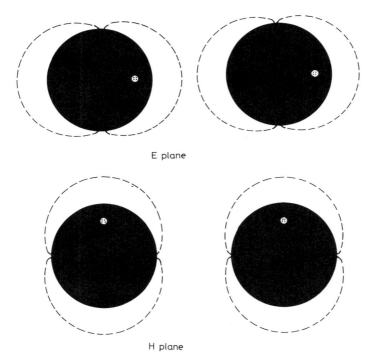

Fig. 11.62 *Magnetic-current distribution along the contour of circular patch for two principal coupling planes*

mutual coupling coefficient S_{12} calculated at $f = 1.4127\,\text{GHz}$, for comparison with experimental data ($f = 1.44\,\text{GHz}$) from Jedlicka *et al.*, are given in Fig. 11.63. Agreement is quite good so long as the frequency shift ($\approx 2\%$) due to the cavity model is taken into account. Whatever the mode, the exact dependence of the coupling phenomena on ϕ is a simple trigonometrical function. Eqn. 11.36 and Fig. 11.64 describe the typical form of the coupling coefficient versus ϕ, normalised to E-plane coupling. The results are compared with those of Bailey and Parks (Reference 5, p. 157).

The curves of coupling coefficient against the angular feed position $\theta_2 (\theta_1 = 0)$ are given in Fig. 11.65 for two edge spacings $d = 0.1\lambda_0$ and $d = \lambda_0$. The E-plane and H-plane of antenna 1 correspond, respectively, to $\phi = 0°$ and $\phi = 90°$. It should be noted that the coupling coefficient varies rapidly near $\theta_2 = 90°$ showing how critical is the precision of the feed angle at this location.

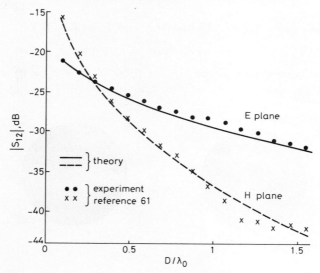

Fig. 11.63 *Coupling coefficient versus distance d between edges*
$\varepsilon_r = 2\cdot5$, $t = 1\cdot575$ mm, patch radius $= 3\cdot85$ cm, probe radial distance $= 1\cdot1$ cm
Frequency:
  Measurement $= 1\cdot44$ GHz
  Theory $= 1\cdot4127$

(a) Input impedance Z_{in}
The input impedance Z_{in} of the elements can be readily calculated by applying the impedance matrix equations of network theory. Fig. 11.66 illustrates Z_{in} variations of the previous two-element array against the edge distance d. It can be observed that, for $d > 0\cdot3\lambda_0$ ($R > 0\cdot68\lambda_0$), the coupling effect on Z_{in} is negligible. In studying Z_{in} as a function of ϕ and frequency, we have chosen a very short distance ($d = 0\cdot1\lambda_0$) in order to observe the coupling influence. Fig. 11.67 shows that in the E-plane ($\phi = 0°$), $Z_{in} \approx Z_{11} = 61\cdot7\angle9°$ ohms, where Z_{11} is the proper impedance of the disc antenna. However, in the H-plane ($\phi = 90°$) where the coupling is stronger, Z_{in} is very different.

(b) Circular sub-array of three identical elements (Fig. 11.68) [68]
In Fig. 11.69 one observes that the coupling effect on the input impedances Z_{in1}, Z_{in2} and Z_{in3} becomes negligible for $d > 0\cdot3\lambda_0$. The dependence of the input impedance on feed angular position θ is presented in Fig. 11.70. Although two of the three impedances can be equal for certain values of θ, nowhere they are identical simultaneously.

Design and technology of low-cost printed antennas   641

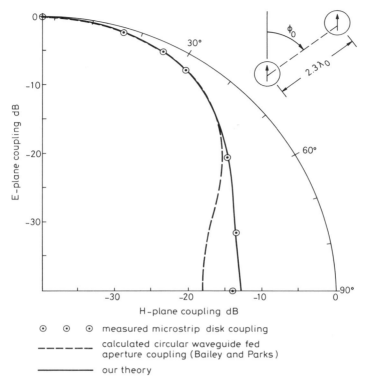

Fig. 11.64 *Coupling coefficient versus orientation ϕ, normalised to the E-plane coupling ($f = 5 \cdot 5\,GHz$)*

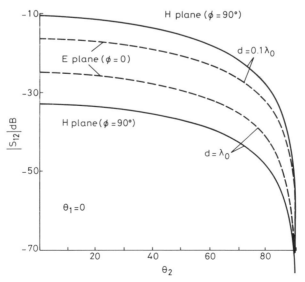

Fig. 11.65 *E- and H-plane coupling coefficients for $m = 1$, $n = 1$, versus angle $\theta_2 (\theta_1 = 0)$*

(c) Conclusion

The cavity method and reaction theorem represent very suitable ways to calculate and predict free-space mutual coupling between array elements of simple shape. According to these results, it seems that the H-plane coupling is stronger for the disc than for the patch (see Figs. 11.54 and 11.63) and identical in the E-plane ($d < 0.3\lambda_0$). Mutual coupling effects become important when the edge-spacing distance d is lower than $0.3\lambda_0$ for an array of few elements.

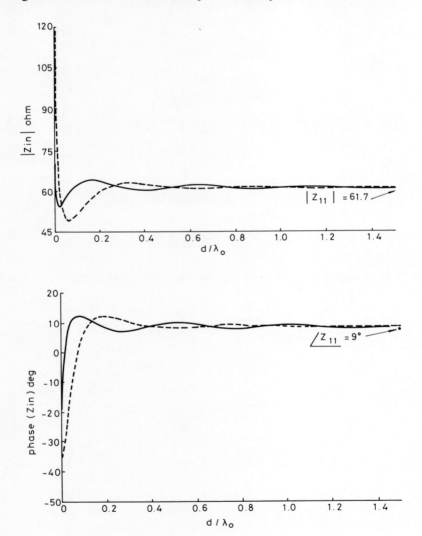

Fig. 11.66 *Input impedance Z_{in} of each array element as a function of d*
  $f = 1.3918\,\text{GHz}$ —— E Plane ---- H Plane

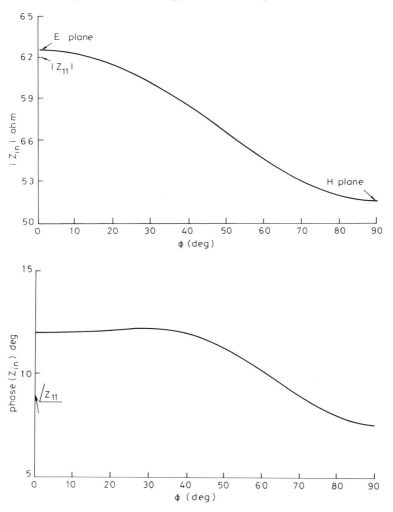

Fig. 11.67 Z_{in} of a two-identical-element array against ϕ for $d = 0\cdot 1\lambda_0$ and $f = 1\cdot 3918\,GHz$

11.3.3 Linear series array of corner-fed square patches [69]

The feed network of microstrip antenna arrays exhibits loses which lead to a limit on the expected gain, and consequently a limited number of elements. The Schelkunoff unit circle, Dolph–Chebÿshev and Villeneuve methods are well suited for the design of fairly small arrays; Taylor's method, sampling line source and Fourier's series expansion are better for large arrays. However, the previous synthesis methods consider only the array factor, while the overall diagram obviously needs the element patterns to be taken into account; for e.g. the $\cos\theta$ variation in the H-plane of most printed antennas has a strong effect for arrays with few elements; the $-40\,dB$ Chebÿshev design example previously mentioned shows a high sidelobe level ($-23\,\text{dB}$) in Fig. 11.50. The sidelobe disappears completely in the H-plane element pattern (Fig. 11.71), and the initial specification is obtained. Two synthesis methods, taking the elementary radiation pattern into account, are developed in Section 11.4.

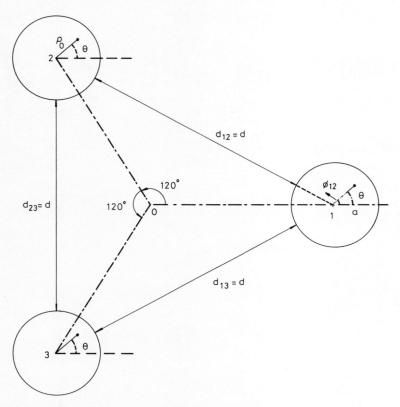

Fig. 11.68 *Geometry of a planar circular array of three microstrip disc antennas*

Only one simple structure using square-shaped microstrip antennas is considered here. The corner-fed square patches are easily excited with a single microstrip line (Fig. 11.72); a tapered distribution is readily obtained using quarter-wavelength transformers along the line. In order to get a broadside pattern, one wavelength spacing is necessary; half-wavelength spacing is also possible, with alternate elements to keep the equi-phase condition.

The corner-fed square patches have been chosen because they provide a high input impedance well suited for series array. It is also very easy to feed each element on the corner. Two new aspects are considered here: reduction of cross-polarisation using altenate location of the elements on the feeding line, and reduction of sidelobe level using simple tapering for a linear series array.

Design and technology of low-cost printed antennas

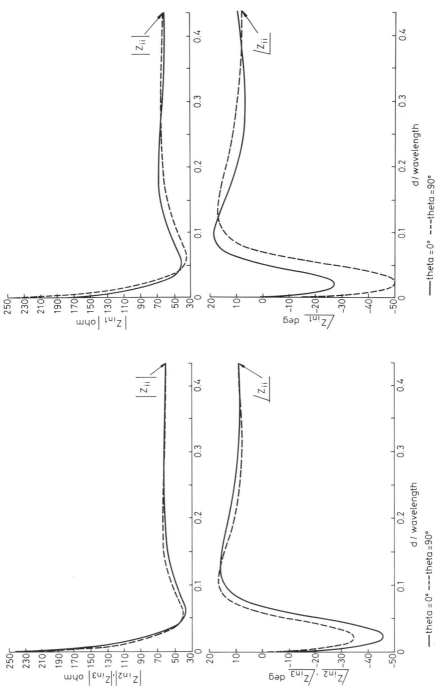

Fig. 11.69 Input impedances Z_{in1}, Z_{in2} and Z_{in3} as a function of edge spacing: $f = f_0$

646 Design and technology of low-cost printed antennas

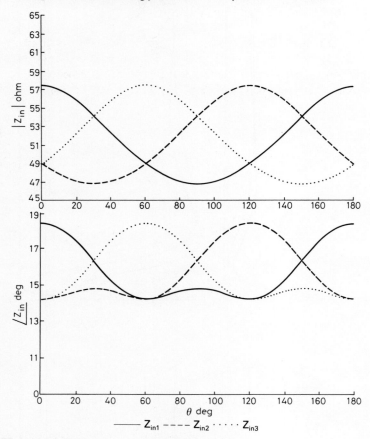

Fig. 11.70 *Input impedance versus frequency: $d = 0.1\lambda_0$*

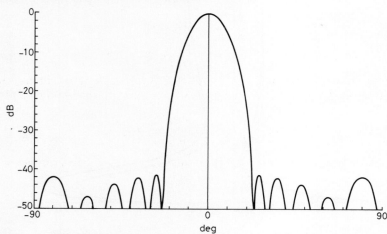

Fig. 11.71 *Chebȳshev pattern of a six-microstrip antenna linear array (H-plane, spacing = $0.75\lambda_0$)*

Design and technology of low-cost printed antennas 647

11.3.3.1 Radiation of corner-fed square patches

(a) Theory: When the patch is excited at one corner (Fig. 11.73a), the cavity model [4, 25] shows that the main part of the internal field is the sum of two degenerate modes with equal amplitudes, i.e. modes (0, 1) and (1, 0). If the higher modes are ignored, the E_x and E_y fields along the edges exhibit the

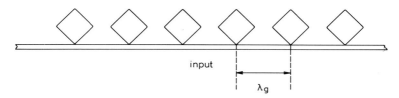

Fig. 11.72 *Corner-fed square-patches array*

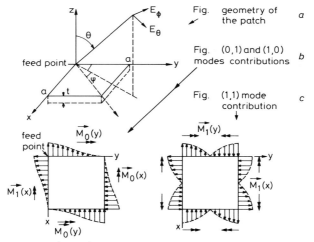

Fig. 11.73 *Geometry and magnetic currents of the corner-fed patch*

variations shown in Fig. 11.73b. The far field is linearly polarised either in the E-plane ($\psi = 0°$) or in the H-plane ($\psi = 45°$). For instance in the H-plane,

$$E_{\phi 0} = -jM_0 \frac{\sin 2C}{2C} \left[1 - \frac{4C^2}{(2C)^2 - \pi^2} \right] \cos\theta \qquad (11.37)$$

where M_0 is proportional to the amplitude of the magnetic current ($x = 0$, $y = 0$)

$$2C = k_0 a \frac{\sqrt{2}}{2} \sin\theta$$

a = side length

As long as the cross-polarised field is needed, higher modes must be considered. The next mode (1, 1) adds a contribution with a magnetic line distribution as shown in Fig. 11.73c. The $E_{\theta 1}$ far-field component is given by

$$E_{\theta 1} = -2M_1 \cos^2 C \frac{2C}{(2C)^2 - \pi^2} \tag{11.38}$$

where M_1 is proportional to the amplitude of the magnetic current ($x = 0$, $y = 0$). It will be noticed that M_1 is much smaller than M_0. The previous formula shows that the cross-polarised component is null for $\theta = 0$ and increases with θ.

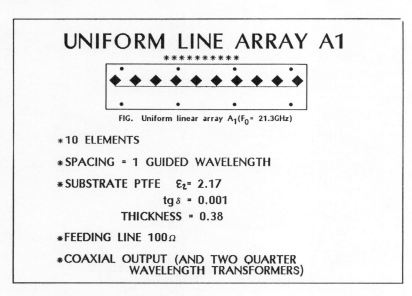

Fig. 11.74 Uniform line array A_1 ($F_0 = 21.3\,GHz$)

(b) Examples of linear arrays

When the spacing along the feeding line equals one guided wavelength, the different elements are uniformly excited (Fig. 11.74). Here the ten-element array is printed on the usual PTFE substrate (dielectric constant 2·17 and thickness 0·38 mm). The quarter-wavelength transformer enables good matching to the 50 Ω coaxial output. As expected, the *H*-plane diagram exhibits the well-known −13 dB first sidelobe; the cross-polarisation component is very low at $\theta = 0$ and then quickly increases with θ to −17 dB (Fig. 11.75). This cross-polarisation component can be reduced in a simple and efficient way. Let us consider each patch located alternatively on each side of the feeding line (Fig. 11.76). Considering half-wavelength spacing, the different co-polar fields ($E_{\theta 0}$ for *H*-plane) add in phase, while the different cross-polar components ($E_{\theta 1}$) maintain

Design and technology of low-cost printed antennas 649

an anti-phase relationship which cancels the different contributions. The diagram is shown in Fig. 11.77. It shows a large reduction of the cross-polar level, which is close to $-28\,\text{dB}$ instead of $-17\,\text{dB}$ as previously. Obviously, as the length of this array is half that of the first one, the 3 dB beamwidth is larger. To attain the same directivity, 20 elements spaced $\lambda/2$ would be necessary.

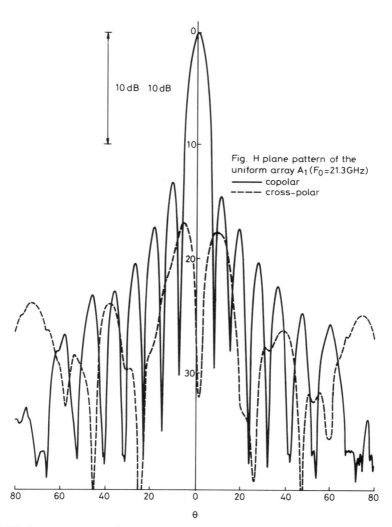

Fig. 11.75 *H-plane pattern of the uniform array A_1 ($F_0 = 21\cdot3\,GHz$)*
——— co-polar
- - - - cross-polar

11.3.3.2 Tapered linear series array

(a) Theory: The previous arrays were uniformly excited; high sidelobes are the consequences of this illumination. The idea was to produce a non-uniform amplitude distribution while keeping the simplicity of the previous series feeding. Let us consider a linear array with wavelength spacing (guided wavelength);

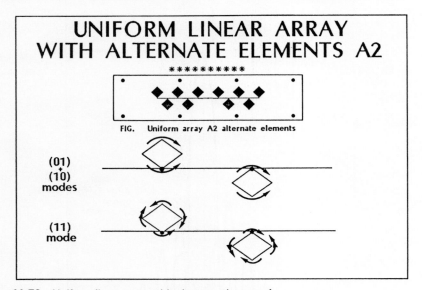

Fig. 11.76 Uniform linear array with alternate elements A_2

as the radiating elements are identical, impedance transformers are necessary to obtain the given amplitude current. To do this, a two-step quarter-wave transformer can be used in each cell (Fig. 11.78). The transformed admittance Y_i in the π'_i plane is given by

$$Y_i = \left(\frac{Y_{L_2}}{Y_{L_1}}\right)^2 Y_{i+1} = n_i^2 Y_{i+1} \tag{11.39}$$

where Y_{i+1} is the admittance of node $(i + 1)$, and Y_{L2} and Y_{L1} are the characteristic admittances of each quarter-wavelength transformer. If necessary, four quarter-wave transformers can be inserted when the spacing equals one wavelength.

When the input voltage leads to a unit current in the first element, the current distribution is readily obtained with the following relations:

$$\begin{aligned} I_0 &= Y_A V = 1 \\ I_1 &= n_1 Y_A V = n_1 \\ I_i &= n_i \cdot n_{i-1} n_1 \end{aligned} \tag{11.40}$$

Design and technology of low-cost printed antennas 651

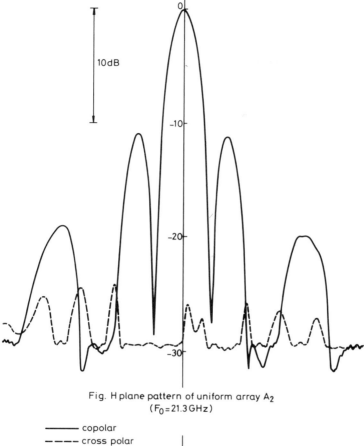

Fig. 11.77 *H-plane pattern of the uniform array A_2 ($F_0 = 21.3\,GHz$)*
——— co-polar
---- cross-polar

However, the various ratio n_i can be deduced step by step from the known values of I_0, I_1, I_i, etc. The input impedance at element 1 is

$$Z_{in} = \frac{Z_A}{1 + n_1^2 + n_1^2 n_2^2 + n_1^2 n_2^2 n_3^2 + \ldots} \qquad (11.41)$$

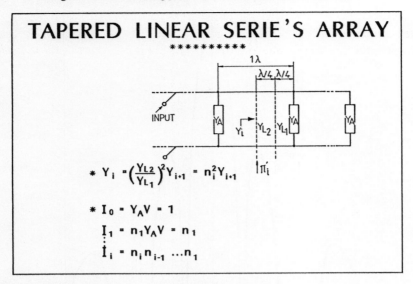

Fig. 11.78 *Tapered linear series array: current distribution*

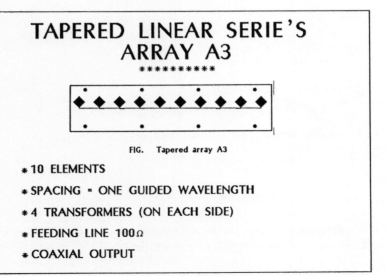

Fig. 11.79 *Tapered linear series array A_3*

(b) Results

A ten-element array has been constructed (Fig. 11.79). The requirement was to get a sidelobe level lower than -20 dB. Only eight transformers were used (four

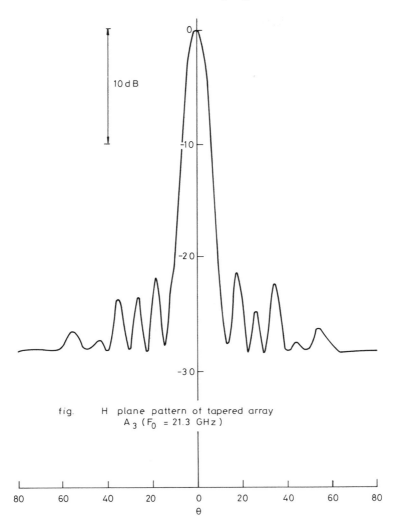

Fig. 11.80 *H-plane pattern of tapered array A_3 ($F_0 = 21.3$ GHz)*

on each side), because Y_{L2} was chosen equal to the characteristic admittance of the half-wavelength following line. Taking the characteristic impedance of the main line as about $100\,\Omega$, the various transformers exhibit impedances between 75 and 95 Ω, which are easily realised with microstrip lines. The experimental H-plane diagram is plotted in Fig. 11.80. It shows that the sidelobe level is lower

654 Design and technology of low-cost printed antennas

than -20 dB while the gain and input-impedance matching remain very similar to those of the uniform array.

A combination of alternate elements and tapered distribution yields a pattern with low sidelobe level and low cross-polarisation over the whole space. Fig. 11.81 shows a ten-element array with quarter-wave transformers. The diagrams

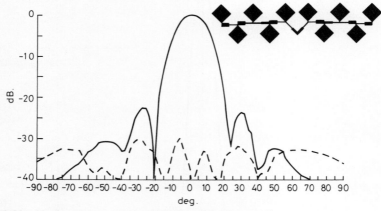

Fig. 11.81 *H-plane pattern of tapered array with alternate elements*

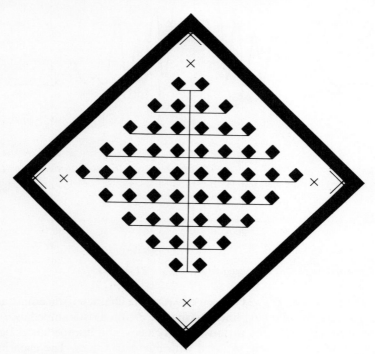

Fig. 11.82 *Typical cross-fed array of square patches (without matching network)*

Design and technology of low-cost printed antennas 655

are plotted in Fig. 11.81, which clearly shows that no degradation occurs to the cross-polarisation level within the frequency band.

11.3.4 Two-dimensional cross-fed arrays [71, 72]

The combination of identical linear sub-arrays leads to a planar array. The feeding network of such structures is developed in Section 11.4. Another simple two-dimensional array named the cross-fed structure can be considered as a combination of non-identical linear sub-arrays.

Cross-fed printed aerials have already been described by Williams [70]. The basic radiating elements were 45° dipoles inclined along the feeding lines. No analysis was proposed; however, the structure appears very attractive owing to its simple feed geometry which avoids having any transformer. Corner-fed patches were chosen because they are easily fed along a straight microstrip line. Moreover, the discontinuities introduced near the corner of each element are symmetrical and identical for all of them. Thus co-polar and cross-polar components remain symmetrical around the broadside direction. Figure 11.82 shows a typical cross-fed array. Matching networks can be added for a coaxial feed. Inter-element spacing equals one guided wavelength. Design equations for a uniform array are given below.

11.3.4.1 Uniform illumination and impedance matching: The overall array is constructed from parallel sub-arrays. The number of elements is reduced from N_i to $N_i - 2$, considering ith and $(i + 1)$th sub-array, respectively. When N is the number along the diagonal, we find:

$N_s = (N - 2)N/4$ elements for the upper or lower group of subarrays

$N_t = N^2/2$ elements for the whole array

The input impedances are different for the half main-line section (R_1) and the upper or lower group of sub-arrays (R_2) (Fig. 11.83). The impedance matching needs one or two quarter-wave-section transformers to get suitable characteristic impedances (Z'_2, Z_2, Z'_1, Z_1). The uniform illumination condition yields a second relation (same voltage V for all the elements). Then the transformer impedance ratios are equal on each side as follows:

$$r = \frac{Z'_2}{Z_2} = \frac{Z'_1}{Z_1} = \frac{1}{N}\left(\frac{2R_A}{R_0}\right)^{1/2} \quad (11.42)$$

where R_A = resistance of a corner-fed antenna; R_0 = desired input impedance; N = number of elements on the main line.

11.3.4.2 Radiation patterns: The total array factor results from the combination of sub-array factors. The expression is

$$S_T = \frac{1}{S_{max}}\left[\frac{\sin N\varphi y/2}{\sin \varphi y/2} + 2\sum_{i=1}^{P} \cos i\varphi_x \frac{\sin(N - 2i)\varphi y/2}{\sin \varphi y/2}\right] \quad (11.43)$$

where $\phi_x = kd_x \cos\phi \sin\theta$; $\varphi_y = kd_y \sin\phi \sin\theta$; d_x and d_y are the inter-element spacing along $0x$ and $0y$; P = number of sub-arrays (above or below the central line); N = number of elements along the central line and $S_{max} = N^2/2$.

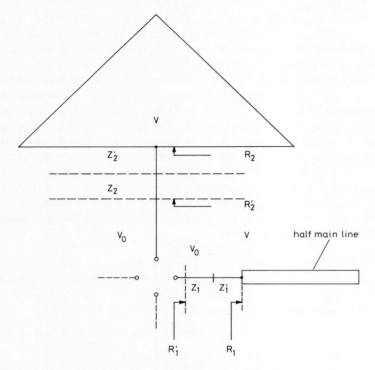

Fig. 11.83 *Input matching transformers of the cross-fed array*

It is interesting to consider this expression for $d_x = d_y = d$ in the two main planes. In the H-plane, S_T is the sum of the usual uniform array factors $S_i = \sin(M_i\varphi_y/2)/\sin(\varphi_y/2)$. However, the nulls of each S_i function have different locations because the sub-arrays have a different number of elements. Then all S_i components are added in phase in the broadside direction while a compensation occurs from the oscillating functions outside $\theta = 0$. The following results are given for usual substrates ($\varepsilon_r \approx 2\cdot2$), and the spacing equals one guided wavelength ($\lambda_g \approx 0\cdot75\lambda_0$).

Fig. 11.84 shows the three components S_0, S_1, S_2 for the case $N = 6$ and $P = 2$ (six diagonal elements and four sub-arrays). Summation then leads to a diagram with a large reduction in the previous oscillations of each of the subarrays. In the E-plane, $\varphi_y = 0$ and S_T is the array factor of an equivalent array exhibiting a linear tapered excitation.

Fig. 11.85a, b and c show the computed patterns in the three main planes,

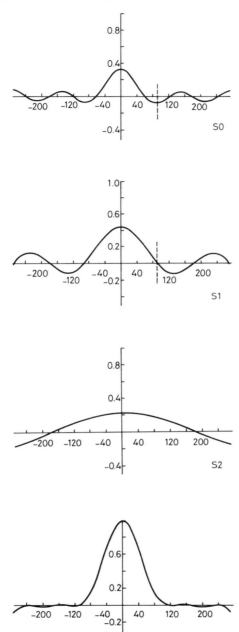

Fig. 11.84 *Sub-array contributions S_i of the global array factor S_T (six diagonal elements and four subarrays)*

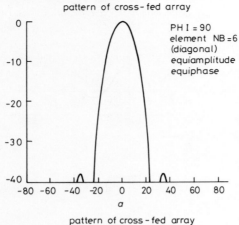
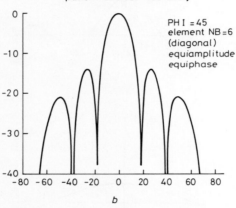
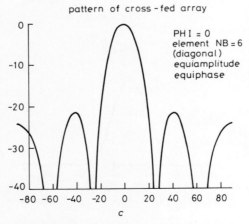

Fig. 11.85 *Computed patterns of the cross-fed array (six diagonal elements, uniform distribution)*
(a) *H*-plane $\phi = 90°$
(b) $45°$-plane $\phi = 45°$
(c) *E*-plane $\phi = 0°$

which clearly show that the sidelobe level (*SLL*) is greatly reduced. However, the $\phi = 45°$ plane yields the -13 dB sidelobe of the uniformly illuminated square structure. On the other hand, the half-power beamwidth θ_3 dB remains very similar in the three planes ($\phi = 0°$, $45°$, $90°$). θ_3 dB and SLL_{max} versus the number N of elements on the diagonal feed line is given in Table 11.10.

Table 11.10 *Half-power beamwidth and sidelobe level of cross-fed array*

	θ_{3dB}			SLL_{max}		
N	$\phi = 0$	$\phi = 45°$	$\phi = 90°$	0	45	90
4	26·4	23	22	−11·5	−15·3	−23·4
6	16·55	15·36	15·35	−21·3	−14·0	−38·0
8	12·14	11·57	11·72	−23·86	−13·7	−31·2
10	9·72	9·23	9·44	−24·88	−13·6	−29·01
12	7·95	7·76	7·79	−25·5	−13·89	−28·16

N = number of elements along the diagonal feed line

It will be noticed that only using the impedance matching condition does not provide a uniform illumination; for instance, if different quarter-wave transformers are used on each arm of the previous structure ($N = 6$ elements on the diagonal arm), the voltages on each element of the upper and lower sub-arrays differ from the voltages of the central line. The computed patterns are plotted in Fig. 11.86 when the voltage equals 1 on the main line and 0·8 on all the other elements. It appears that the *E*-plane is quite transformed, sidelobes reaching -16 dB and θ_3 dB $= 17·8°$.

11.3.4.3 Results: Various arrays have been built either with coaxial or waveguide output. Each of them was printed on a PTFE substrate ($\varepsilon_r = 2·17$, tg$\delta = 10^{-3}$, thickness $= 0·38$ mm) or on polypropylene, as described in Chapter 5. An 18-element cross-fed array designed for 23·5 GHz is considered first. The same quarter-wave transformer was used on each arm to obtain a SWR better than 1·5 at the coaxial output (Fig. 11.87). The measured gain equals 20 dB, while a uniform aperture of the same area yields a 21 dB gain. *E*- and *H*-plane diagrams are given in Fig. 11.88; the sidelobes reach -18 dB in the *E*-plane and are lower than -20 dB in the *H*-plane, and beam widths are 15° (*H*-plane) and 18° (*E*-plane), as expected.

Another 50-element array, printed on polypropylene of the same thickness, has been realised for a frequency near 20 GHz. The number of elements in the upper and lower groups of sub arrays is 40, and the impedance to be matched equals 8 Ω. The transformer then needs three quarter-wave sections as shown in Fig. 11.89 to get equal voltage on each patch and good impedance matching. The radiation patterns are given in Fig. 11.90. No sidelobes larger than -25 dB appear in the *H*-plane while the *E*-plane exhibits -18 dB sidelobes. In both cases the cross-polarisation level remains acceptable. The measured gain is

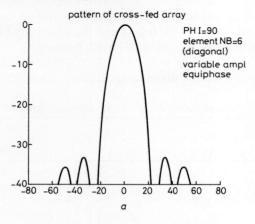

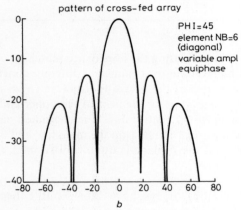

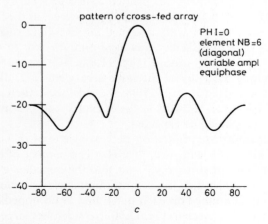

Fig. 11.86 *Computed patterns of the cross-fed array (six-diagonal elements; non-uniform distribution)*

Design and technology of low-cost printed antennas 661

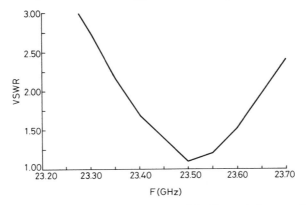

Fig. 11.87 *VSWR of a 18-element cross-fed array (six diagonal elements, $F_0 = 23 \cdot 5\,GHz$)*

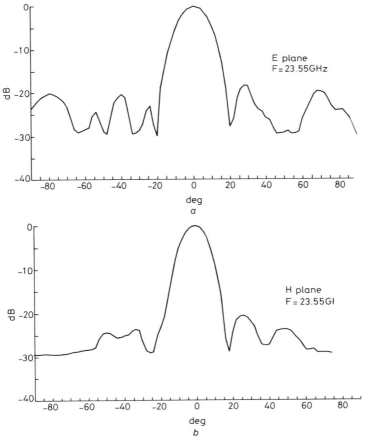

Fig. 11.88 *E-plane and H-plane measured pattern of the 18-element cross-fed array*

23·2 dB; the uniform aperture of the same area would yield 25·48 dB. Losses reach approximately 2·3 dB; the VSWR is lower than 1·8 between 19·5 and 20·4 GHz.

11.4 Synthesis methods for linear arrays [73, 74, 75]

The usual analytical synthesis methods (Fourier, Chebȳshev, Woodward–Lawson) are not always suitable when the directivity pattern is specified by a given outline. Thus, the mean-squared error criterion is a global criterion that does not permit the separation of the main-beam and the sidelobes contribution. Prescribing equi-level sidelobes does not always fit the Chebȳshev requirement.

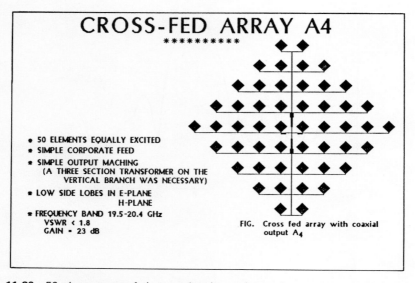

Fig. 11.89 *50-element cross-fed array printed on polypropylene*

Lastly, the Woodward–Lawson sampling method needs a great number of sources, and the choice of sample is sometimes critical.

Numerical methods can take into account the envelope specification, the directivity pattern of the source and the inter-element spacing. Two numerical methods have been studied:

- The relaxation method which enables real excitation coefficients
- The simplex method which uses the Dantzig algorithm and yields symmetrical or non-symmetrical pattern synthesis (real or complex coefficient).

Let $F(\theta)$ be the directivity pattern of a linear array; then

$$F(\theta) = f(\theta)g(\theta)$$

where $f(\theta)$ = array factor; $g(\theta)$ = directivity pattern of the source.
As usual printed antennas have different diagrams in the two main planes (E-plane and H-plane), so it is better to synthesize $F(\theta)$.

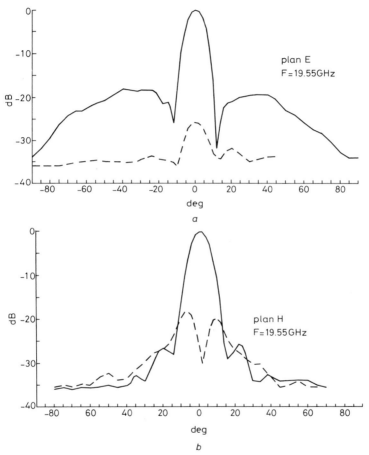

Fig. 11.90 *E-plane and H-plane patterns of a 50-element cross-fed array printed on polypropylene*

11.4.1 Relaxation methods

11.4.1.1 Method: Let us consider a symmetrical linear array of $2N$ elements; $F_d(\theta)$ is the desired directivity pattern and $a = (a_j)_{j=1}^{N}$ is the unknown excitation vector:

$$F_d(\theta) = \sum_{j=1}^{N} a_j \cos\left(2\pi \frac{d_j}{\lambda_0} \sin\theta\right) \times g(\theta) \tag{11.44}$$

d_j = symmetrical-axis relative position of source $\pm j$; λ_0 = free-space wavelength.

The following linear system is obtained using discrete values of θ:

$$C_{(M,N)} a_{(N,1)} = F_{(M,1)}$$

with

$$C = [c_{ij}]_{M,N},\; c_{ij} = \cos\left(2\pi \frac{d_j}{\lambda_0} \sin\theta_i\right),\; a^T = (a_1, a_2 \ldots a_n);$$

$$F^T = (F_d(\theta_1), F_d(\theta_2) \ldots F_d(\theta_M))$$

The functionals to optimise are those like

$$J(a): R^N \to R^M$$

$$a \to J(a) = Ca - F$$

We denote a' and $a \in R^N$ such as $J(a') = \text{Min } J(a)$ in the sense of the chosen criterion. To realise this, a series of vectors a^k is built, such as $J(a^k + 1) < J(a^k)$. The search directions are the co-ordinate axes, each of them being taken periodically. For each component a_j, we realise $\xi = a_j^{k+1}$ such as at the $(k + 1)$th iteration:

$$J\begin{pmatrix} a_1^{k+1} \\ a_j^{k+1} \\ a_j^k \\ a_N^k \end{pmatrix} = \min J \begin{pmatrix} a_1^{k+1} \\ \xi \\ a_j^k \\ a_N^k \end{pmatrix} \text{ in the sense of the choosen criterion}$$

For a given quality criterion and choice of convenient functionals to optimise, the relaxation method provides fast convergence. Some portions of the outline pattern can eventually be preferred. Two examples illustrate the method: sector pattern and directive broadside pattern.

11.4.1.2 Sector-pattern case: Let us define the desired directivity pattern from an outline symmetrical on θ (Fig. 11.91). R_{lim} is the maximum ripple value for $\theta \in [0, \theta_0 - \alpha]$ and $R = 1 - D(\theta)_{min}$. SLL_{lim} is the maximum sidelobe level for $\theta_0 + \alpha < \theta < \pi/2$ and $SLL = D(\theta)_{max}$. The pattern in the transition region is related to α values. An initial value can be chosen, such as $|a_i^\circ| = 1 - \beta(i - 1)/(N-1)\; 0 \leq \beta \leq 1$, or it can be equal to the excitation obtained from classical methods. The functional $J(a)$ is replaced by two functionals $R(a)$ and $SLL(a)$, and min $J(a)$ can be expressed by the following improvement criteria

Design and technology of low-cost printed antennas 665

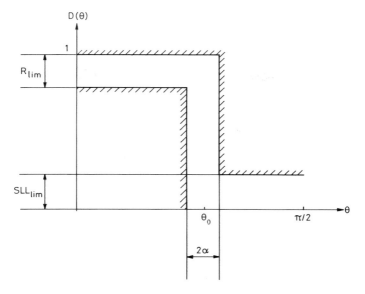

Fig. 11.91 *Outline of desired directivity pattern (example of sector of pattern)*

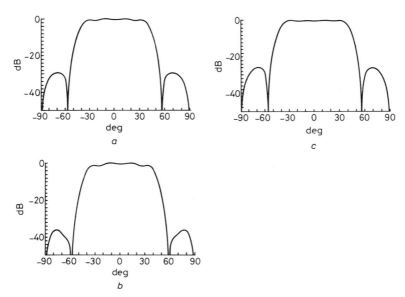

Fig. 11.92 *(a) Sector pattern of a 10-element array (H-plane). Relaxation synthesis:*
 a C_1 criterion
 b C_2 criterion
 c C_3 criterion

c_1, c_2, c_3: let a^+ be the actual vector of the iteration,

c_1: $[R(a^+) < R(a)$ and $SLL(a^+) \leq SLL(a)]$ or $[R(a^+) \leq R(a)$ and $SLL(a^+) < SLL(a)]$

c_2: $[R(a^+) < R(a)$ and $SLL(a^+) \leq SLL_{lim}]$

c_3: $[SLL(a^+) < SLL(a)$ and $R(a^+) \leq R_{lim}]$

The c_1 criterion application improves the ripple and the sidelobe level. c_2 gives the best ripple for a given sidelobe level SLL_{lim}, and c_3 the best sidelobe level for a given ripple R_{lim}. The c_1, c_2, c_3 criteria can be applied successively, depending on the ripple and the sidelobe level requirements.

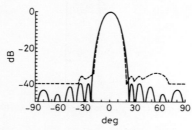

Fig. 11.93 *Directional broadside pattern of a six-element array (H-plane)*
—— Relaxation synthesis
- - - - Experimental results

11.4.1.3 Directive broadside pattern case: For a broadside array of N equispaced elements with equal excitations, the beamwidth between first nulls is equal to $2\lambda_0/Nd$ radians. Applying c_3 with $R_{lim} = 1$ and $\theta_0 > \lambda_0/Nd$ provides a main beam with the lowest sidelobe level, taking d/λ_0 and $g(\theta)$ into account. If $d \leq 0.5\lambda_0$ and $g(\theta) =$ constant, we obtain the Chebyshev excitation. When $d > 0.5\lambda_0$, the Chebyshev method does not always maintain equal sidelobe levels. The relaxation method avoids such limitations, and printed antenna arrays with guided-wavelength spaced sources can be considered.

11.4.1.4 Results: Let us consider a linear printed antenna array on a substrate, having a relative dielectric constant $\varepsilon_r = 2 \cdot 17$ (PTFE). For microstrip transmission lines $\lambda_g = \lambda_0/\sqrt{\varepsilon_{eff}} \approx 0.75\lambda_0$.

(a) Sector pattern
We want to design a 10-source array with a 90° sector pattern and a transition width, $2\alpha = 20°$. If the sources are equally spaced, the array amplitude factor will be zero with any excitation vector for $\sin\theta = \lambda_0/2d$. To avoid a null in the sector region, the first distance to the symmetry axis can be taken as $0.25\lambda_0$.
With $a^{oT} = (1; -0.75; 0.5; -0.25; 0)$, $g(\theta) = \cos(\theta)$ and application of the

c_1 criterion, we obtain $R = -0.8$ dB and $SLL = -29$ dB for a solution vector S (Fig. 11.92a).

With $a^\circ = S$, $SLL_{lim} = -25$ dB and application of c_2, we obtain $R = -0.3$ dB and $SLL = -26$ dB (Fig. 11.92b).

With $a^\circ = S$, $R_{lim} = -1.5$ dB and application of c_3, we obtain $R = -1.5$ dB and $SLL = -36$ dB (Fig. 11.92c).

(b) Directive broadside pattern
On the same substrate, let us design a linear array of six sources with interelement spacings λ_g. For equal excitations, the beamwidth between the first nulls is 25·5° and the first sidelobe is at -13 dB. For $2\theta_0 \approx 35°$, $2\alpha = 10°$ (outline corresponding to a -40 dB sidelobe level, Chebȳshev excitation), $R_{lim} = 1$, $a_1^\circ = 1$, and $a_j^\circ = 0$ $\forall j \neq 1$, $g(\theta) = \cos(\theta)$, the c_3 application provides a -40 dB sidelobe level (Fig. 11.93). The experimental pattern obtained with a linear array of square patches is plotted in Fig. 11.93. The theoretical and experimental outlines agree well if the low-level measurement difficulties are taken into account.

11.4.2 Simplex method
Dantzig's algorithm [76] has been developed for linear programming with a real variable. Under symmetrical amplitude and antisymmetrical phase conditions, it is possible to compute real or complex excitations. The desired diagram is also defined with an envelope specification.

11.4.2.1 Symmetrical pattern: If the pattern is considered at M angular values (without correlation with N), the following inequalities are obtained:

$$F(\theta_2) \leq F_1$$
$$F(\theta_2) \geq F_2 \qquad (11.45)$$
$$F(\theta_M) \leq F_M$$

The problem is to find the ensemble V:

$$V = a \in R^N \left| \begin{cases} c_{11}a_1 + \ldots\ldots\ldots c_{1M}a_N \leq F_1 \\ c_{12}a_1 + \ldots\ldots\ldots c_{2N}a_N \geq F_2 \\ c_{M1}a_1 + \ldots\ldots\ldots c_{MN}a_M \leq F_M \end{cases} \right.$$

c_{ij} has been previously defined, and a functional $J(a) = \sum_{j=1}^{N} r_j a_j$ has to be optimised (minimum or maximum) in order to promote some part of the diagram for instance.

11.4.2.2 Asymmetrical pattern: Choosing $a_j = a_{-j}$ and $\phi_j = -\phi_{-j}$, the array factor can be written in the following form:

$$f(\theta) = \sum_{j=1}^{N} a_j \cos(kd_j \sin\theta + \phi_j)$$

668   Design and technology of low-cost printed antennas

The expansion of the cosine term leads to:

$$F(\theta) = f(\theta)g(\theta) = \sum_{j=1}^{N} (A_j \cos(k_0 d_j \sin \theta) - B_j \sin(k_0 d_j \sin \theta)) g(\theta)$$

(11.46)

A_j and B_j are the real unknowns to be determined from the new $2N$-dimensions linear problem. The algorithm can be used again, and at the end:

$$a_j = (A_j^2 + B_j^2)^{1/2} \qquad \phi_j = \text{tg}^{-1}\left(\frac{B_j}{A_j}\right)$$

The introduction of M difference variables leads to a new linear system of M equations (instead of inequalities) with $N + M$ unknowns associated with the functional $J(a)$ (or $2N + M$ unknowns for an asymmetrical patterns). The Dantzig algorithm shows that only one solution exists (if there is a solution), which is found in a finite number of steps [73, 76].

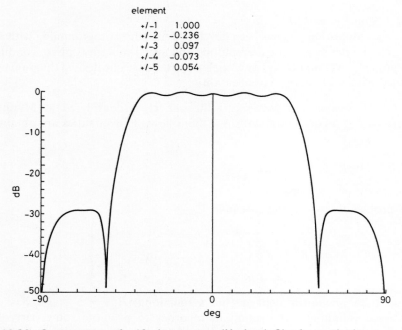

Fig. 11.94 Sector pattern of a 10-element array (H-plane). Simplex synthesis

11.4.2.3 Examples: The second pattern previously mentioned has been computed using the simplex method (Fig. 11.94). It appears that the amplitudes, sidelobe level and ripple are very similar to the relaxation solution. Directional

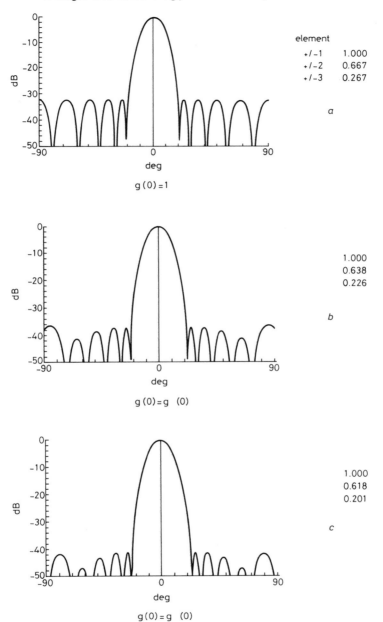

Fig. 11.95 *Directional patterns of a six-element array for various source patterns. Simplex synthesis*
  a Isotropic source
  b E-plane pattern
  c H-plane pattern

patterns are plotted in Fig. 11.95 for an array of six elements spaced $0.75\lambda_0$; three-element patterns have been considered (isotropic, *E*-plane and *H*-plane).

The directional patterns of a linear 32-equi-spaced-element array (spacing = $0.89\lambda_0$) in the *H*-plane are plotted in Fig. 11.96. The sidelobe levels were

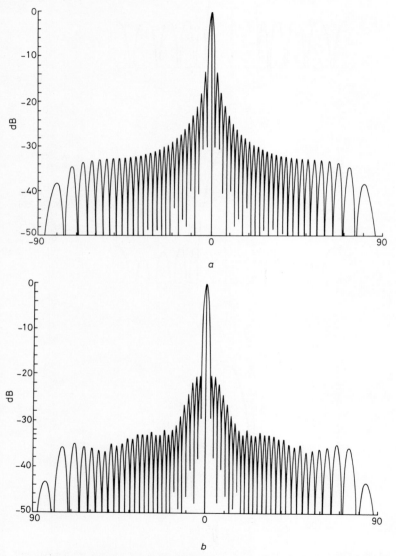

Fig. 11.96 *Computed directional broadside pattern of a 32-linear-element array (H-plane and d/λ = 0·87). Simplex synthesis*
 a Equi-amplitude
 b CCIR–TVRO constraints and −20 dB first sidelobe

Design and technology of low-cost printed antennas 671

constrained by the CCIR-TVRO conditions with an extra limitation of $-20\,\text{dB}$ for the highest one. A cosecant-squared pattern (with a 30° window) was achieved for a 30-element array with half-wavelength spacing. The E-plane-

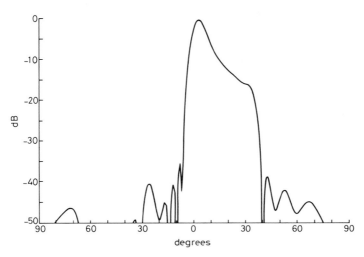

Fig. 11.97 Computed cosecant squared pattern of 30-element linear array (E-plane and $d/\lambda = 0.5$). Simplex synthesis

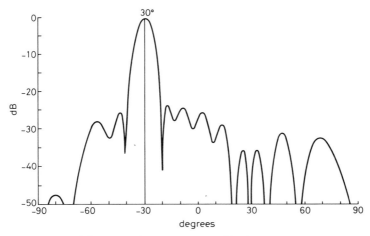

Fig. 11.98 Computed 30° steered-beam pattern of 20-element linear array (H-plane and $d/\lambda = 0.5$). Simplex synthesis

element pattern is considered in Fig. 11.97. A 30° steered beam of a 20-element array (with half-wavelength spacing) in the H-plane, with a sidelobe level lower than $-25\,\text{dB}$, is shown in Fig. 11.98.

672 Design and technology of low-cost printed antennas

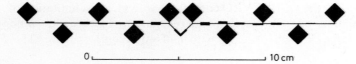

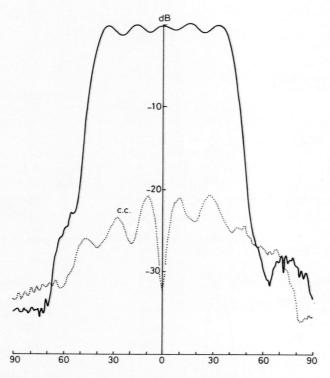

Fig. 11.99 *Measured sector pattern of a 10-element linear array (H-plane and $F_0 = 10.8\,GHz$)*

Table 11.11 *Amplitude and phase distribution along sector pattern array*

Element	Amplitude	Phase
+/−1	1	0
2	0·218	180
3	0·117	0
4	0·082	180
5	0·075	0

11.4.3 Experimental results [73]
Both methods have been used to design sector and directional patterns in the X-bands and K-bands, using corner-fed square patches.

11.4.3.1 Sector pattern: The specification is as follows:

$\theta_0 = 45°$

$2\alpha = 10°$

ripple $= \pm 1\,\text{dB}$

$SLL \leqslant -25\,\text{dB}$

To keep the feed line as simple as possible, the linear series array was chosen. The constraint of straight lines between each element leads to a one-guided-wavelength spacing (or a 0·75 free-space wavelength). However, a sector pattern of 90° beamwidth cannot be obtained with this spacing, because a null occurs in the θ direction (where $\theta = \sin^{-1} \lambda_0/(2\lambda g)$) whatever the amplitudes. To avoid this effect, one solution is to change the spacing. Fig. 11.99 shows the structure which has been used. The two half arrays have been located closer together, then the synthesis methods can perform the amplitude excitations, taking into account the non-identical spacing and the *H*-plane pattern of each patch. Theoretical results are presented in Table 11.11.

table 11.12 *Amplitude distribution along near direction array (in E- and H-plane)*

Element *nb*	*H*-plane (6 elements)	Element *nb*	*E*-plane (8 elements)
+/−1	1	+/−1	1
+/−2	0·627	+/−2	0·735
+/−3	0·212	+/−3	0·377
		+/−4	0·110

The 180° phase shift between two neighbouring elements is easily obtained using alternate positions along the feeding line. Quarter-wave transformers (one or two sections) are used to obtain the amplitude taper. The measured patterns (co- and cross-polarised components) are plotted in Fig. 11.99.

11.4.3.2 Low side-lobe directive array (Fig. 11.100): Two steps were necessary: just to synthesise the *H*-plane pattern (six elements) and secondly to synthesise the *E*-plane pattern (eight elements). The amplitude taper was realised in the usual way, using quarter-wave transformers. Table 11.12 presents the taper values.

Figs. 11.101a and b show the measured patterns in the *E*- and *H*-planes. The

674 Design and technology of low-cost printed antennas

sidelobe levels are higher than the expected − 10 dB. However, mutual coupling has not been taken into account, and the reflectivity of the anechoid chamber reaches − 40 dB.

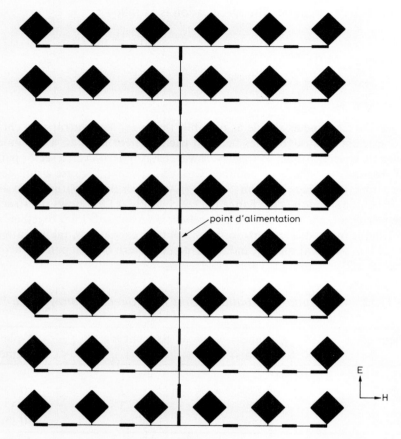

Fig. 11.100 *48-element planar directional array (6 × 8) F_0 = 10·8 GHz*

11.5 New low-cost low-loss substrates [77, 79]

11.5.1 Substrate choice

A large market demand for low-cost printed antennas has emerged from the development of new types of civil communication such as direct-broadcasting satellite reception, data transmission, communications between satellites and mobiles (trucks, ships, etc.) and intruder detectors.

The price of a mass-produced printed antenna is directly related to substrate and connector costs. The choice of the appropriate substrate depends on its

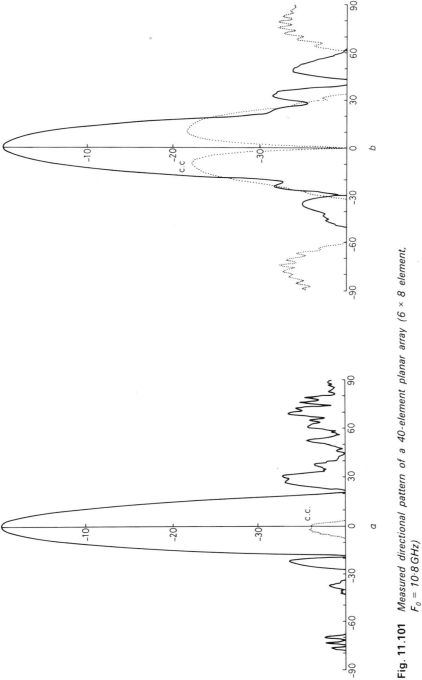

Fig. 11.101 Measured directional pattern of a 40-element planar array (6 × 8 element, $F_0 = 10.8\,GHz$)
(a) E-plane
b H-plane

676   Design and technology of low-cost printed antennas

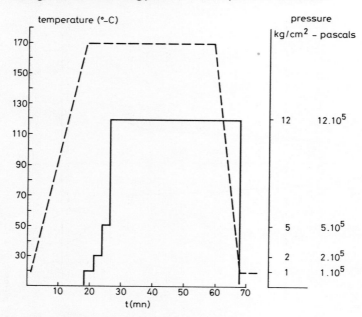

Fig. 11.102 *Pressure and temperature cycles of polypropylene*

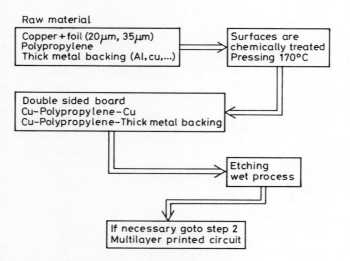

Fig. 11.103 *Steps in manufacture of double-sided printed-circuit board (step 1)*

Table 11.13 Substrate characteristics at X-band

	Polytetrafluoroethylene (unsupported + glass fibre)		Polyetherymide (unsupported + glass fibre)		Cross-linked polystyrene (unsupported + glass fibre)		Polysulfone (unsupported)	Polyphenylene oxide	Modified polyethylene	Polyproylene
Dielectric constant	2·10	2·17	2·9	3·2	2·5	2·6	3	2·6	2·3	2·18
Loss tangent	0·0004	0·002 0·0019	0·0018	0·0022	0·001	0·002	0·0058	0·002	0·0002	0·0003
Maximum temperature °C	260	260	180	180	110	110	150	190	100	150
Cost	high	medium to high	medium	medium	medium	medium	medium	medium	low	low

application, nevertheless, many substrate properties may be involved: dielectric constant, loss, and their variations with temperature and frequency ranges, and mechanical and climatic stresses.

Conventionally, printed antennas require the use of a low-loss low-dielectric-constant substrate. Unfortunately most printed-circuit boards currently used are quite expensive; some are listed in Table 11.13. CNET (Centre National d'Etudes des Telecommunications France) has developed a polypropylene substrate whose characteristics are very similar to commercial substrates, while remaining inexpensive.

Fabrication procedures and evaluation of this new substrate will be discussed in the next Section.

11.5.2 Fabrication procedures
The fabrication of polypropylene printed-circuit boards is a very simple procedure. Two types of board are considered:

● Double-sided printed circuit: Cu–polypropylene–Cu
● Thick metal-backing substrate: thick metal base (Al or Cu) polypropylene–Cu

Polypropylene of different thicknesses (ranging from 0·25 mm to 1·6 mm or more) is manufactured by heating polypropylene granules to the melting point (170°C) and pressing them (Fig. 11.103). Pressure and temperature cycles are

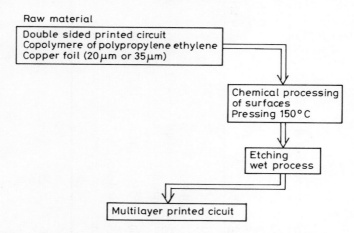

Fig. 11.104 *Steps in manufacture of multi-layer printed-circuit board (step 2)*

detailed in Fig. 11.102. Copper-foil or thick-metal backing is first chemically processed and then laminated to the polypropylene-based dielectric by the same procedure.

Laminated copper is usually selected in preference to electrical-grade copper, which is more lossy. Thicknesses are commonly 20 μm or 35 μm.

Thick-backing construction offers significant advantages over conventional designs [78]:

- It provides high reliability for connector mounting
- For high-power applications heat generated by devices can be dissipated.

Because the substrate is not loaded with glass fibre, it tends to warp when the internal stress between copper and polypropylene is too high; a thick metal cladding can ensure flatness.

The same process can be extended to multilayer structures (Fig. 11.104), but in order not to damage the first layer, a copolymer or polypropylene–ethylene is used.

The electrical characteristics are almost similar, but the melting point is 20° lower (150°C instead of 170°C); applications to triplate feeding network and stacked patches are straightforward as there is no need for bonding the film between different layers. Examples of practical applications are discussed below.

11.5.3 Electrical characteristics

Dispersion measurements were carried out by the ring-resonator technique in triplate technology (in order to avoid unwanted radiation).

Overall losses are plotted against frequency in Fig. 11.105 for different types of printed circuits and are compared with a PTFE substrate (RT Duroid 5880):

A: Cu (20 μm)–polypropylene–Cu (20 μm)
B: Aluminium (Al) (4 mm)–polypropylene–Cu (20 μm)
C: RT Duroid 5880

The dielectric constant remains constant with frequency. Fig. 11.106 shows that polypropylene and the copolymer of polypropylene–ethylene have the same electrical performance.

11.5.4 Environmental tests

11.5.4.1 Damp heat: 95% relative humidity at 40° and for 22 days: Again, overall losses are plotted as a function of frequency in Fig. 11.107 before and after tests, and are compared with the RT Duroid 5880 substrate. The polypropylene does not seem to be very affected by the test conditions, while the PTFE substrate losses are slightly increased. This is probably due to the fact that this substrate is loaded with glass fibre.

11.5.4.2 Thermal shocks: Tests have been performed on a 0·8 mm printed circuit with 4 mm aluminium (Al) backing. The board survived the following tests:

$-30°C \quad +50°C$

680   Design and technology of low-cost printed antennas

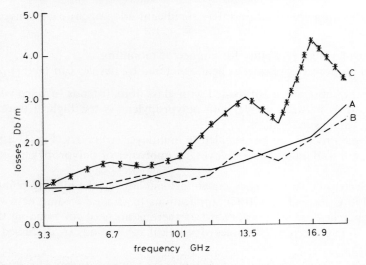

Fig. 11.105 *Losses (dB/m) versus frequency for various printed circuits*
(A) Cu 20 μm–polypropylene–Cu 20 μm
(B) Al (4 mm)-polypropylene–Cu 20 μm
(C) RT-Duroid 58-80 (copper on each side)

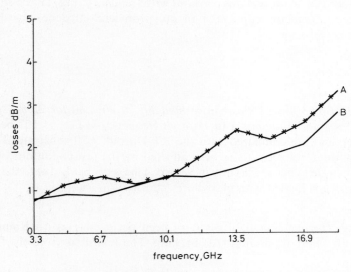

Fig. 11.106 *Comparison of polypropylene (A) and copolymer of polypropylene ethylene (B)*

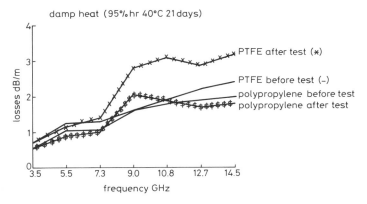

Fig. 11.107 *Losses before and after test of damp heat*

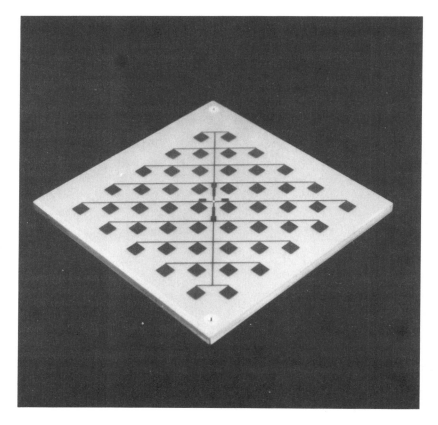

Fig. 11.108 *50-element cross-fed array printed on a 0·4 mm polypropylene substrate. Frequency = 20 GHz*

−40°C   +85°C
−55°C   +100°C

but was destroyed at −65°C +125°C.

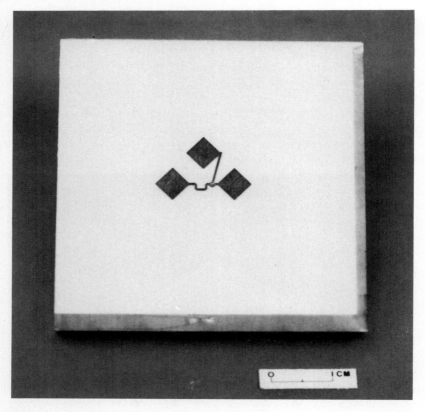

Fig. 11.109 *Three-element array printed on a 0·4 mm polypropylene substrate. Frequency = 23·5 GHz*

11.5.5 Examples of printed antennas on polypropylene substrates
The two examples which are described here give a good insight into the possibilities of this new technology.

11.5.5.1 Single-sided printed antennas: A few types of microstrip antenna arrays have been designed for the 20 GHz band with gain from 11 dB to 23 dB [77]. These antennas are printed on a 0·4 mm polypropylene substrate with 4 mm aluminium backing, which operates as the ground plane of the patches, ensures the flatness of the antenna and eases the connector mounting. A photograph of a 50-element array is shown in Fig. 11.108. This array operates at 19·6 GHz, and its measured gain is 23 dB. The antenna has already been discussed in a previous Section.

The three-element array in Fig. 11.109 was designed for data transmission between mobiles at 23·5 GHz. The required performance was:

E-plane: 3 dB beamwidth = 40°

H-plane: 3 dB beamwidth = 60°

In order to meet the 60° requirement in the *H*-plane, the distance between patches has to be very small, which is the reason why this three-element geometry was chosen. The measured gain is 11 dB because the patches are very close. It is not possible to match the antenna with 0.25λ microstrip line transformer. The problem has been solved by using a coaxial-line transformer machined in the 4 mm aluminium ground plane which provides a VSWR of 1·5. Radiation patterns are given in Fig. 11.110.

11.5.5.2 Multi-layer printed antennas: A printed-slot array was designed in order to test the possibilities of using polypropylene technology for large dimensions. This array was also designed with the aim of replacing a Yagi antenna, whose sidelobes level are very high, by a flat antenna which should have better characteristics and be less expensive. The specifications were:

- 1·38–1·525 GHz
- Gain: 16 dB to 17 dB
- 3 dB beamwidth = 20° to 25°
- Linear polarisation
- Sidelobe level < −20 dB

The gain specification implies dimensions of 60 × 60 cm. The printed-slot element was selected because it can provide wider bandwidth than microstrip patches, and also because it is very easy to feed. The fabrication of the antenna requires few steps. The two copper layers are separated by a 0·8 mm polypropylene layer. The slots are printed on the upper face of the substrate while the feed network is printed on the rear side. The printed-slot plane is covered by a second 0·8 mm polypropylene layer which ensures flatness of the antenna. Photographs of the antenna are given in Fig. 11.111.

The printed slots are electromagnetically coupled to the microstrip lines and the array is suspended over a reflector plane in order to get unidirectional radiation patterns. Since the dielectric thickness is very small compared with the wavelength, the dimensions of the slot are such that $2\pi r = \lambda$. Input impedance is measured in the plane of the outer edge of the slot and then matched to 50 Ω by a matching network. The slot width is $W/\lambda = 0.017$. The VSWR of a single element is plotted against frequency in Figs. 11.112*a* and *b* for two spacings $h/\lambda = 0.12$ and $h/\lambda = 0.25$ of the reflector plane. In the first case, the matching is provided by a quarter-wavelength transformer, while in the second case a

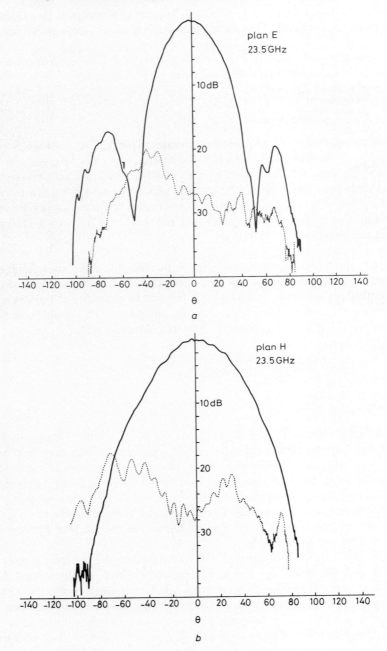

Fig. 11.110 *Radiation patterns of the three-element array (Fig. 11.107)*
  (a) *E*-plane
  (b) *H*-plane

quarter-wavelength transformer and two stubs are required to fit the bandwidth requirement. It appears that the bandwidth decreases as the spacing increases. The 4 × 4-slots array is designed for a 25 dB Chebyshev taper. The results are

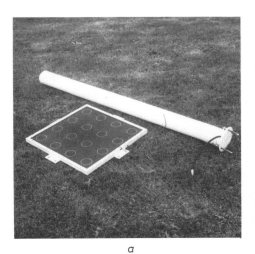
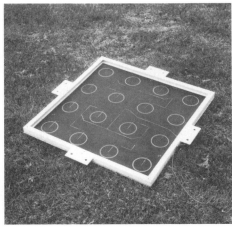

a　　　　　　　　　　　　　　　　　　b

Fig. 11.111　*Printed slot array. Frequency = 1·5 GHz*
　　　　　　a Top view of the printed-slot array beside a Yagi antenna under a radome
　　　　　　b Rear view of the printed-slot array and its feed network

as follows:

$$\text{Slot width} = 0.017\lambda$$
$$\text{Reflector height} = 0.25\lambda$$
$$\text{Gain: } 17.2\,\text{dB}$$
$$\text{Efficiency: } 67\%$$

Radiation patterns are given in Fig. 11.113. Although the radiation patterns and gain are good, a good match over the whole frequency band could not be obtained. The VSWR remains below 2·5 over the whole-frequency band. This is due to the very high coupling between slots, which is increased by the proximity of the reflector plane.

11.6 Concluding remarks

The design of low-cost printed antenna arrays needs accurate analysis of parameters such as resonant frequency, input impedance, mutual coupling and

686   Design and technology of low-cost printed antennas

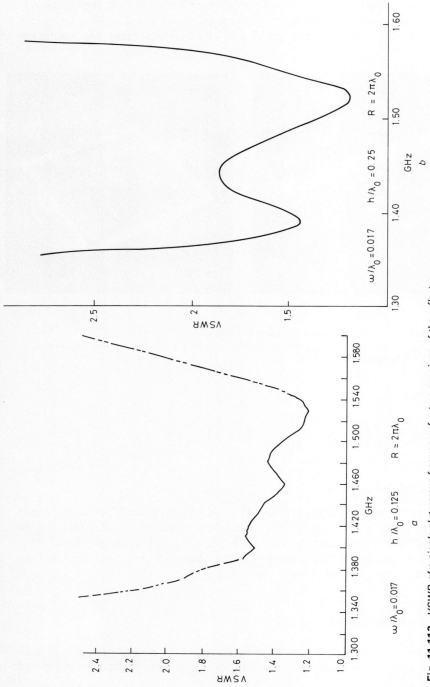

Fig. 11.112 *VSWR of a single slot versus frequency for two spacings of the reflector*
(a) $0.12\lambda_0$
(b) $0.25\lambda_0$

Design and technology of low-cost printed antennas 687

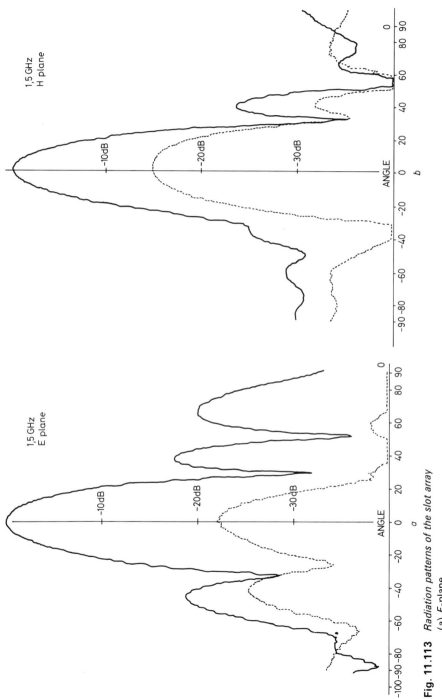

Fig. 11.113 *Radiation patterns of the slot array*
(a) *E*-plane
(b) *H*-plane

sidelobe level with respect to the dimensions, dielectric constant, thickness of protective layer and inter-element spacing. The transmission-line model, the cavity model and the spectral-domain approach are complementary tools which can be used together. In addition, specific synthesis methods have been developed to maintain the flexibility offered by printed structures; namely non-identical element spacing and non-identical E- and H-plane element patterns.

Finally, the polypropylene substrate, whose characteristics are very close to commercial substrates, is very cost competitive in most applications. On the other hand, thick-backing construction using either metal or metallised dielectric, and multi-layer structures without bonding film have been experimented with in order to reduce overall cost.

11.7 References

1 DERNERYD, A. G., and LIND, A. G.: 'Extended analysis of rectangular microstrip resonator antennas', *IEEE Trans.*, 1979, **AP-27**, pp. 846–849
2 VANDESANDE, J., PUES, H., and VAN DE CAPELLE, A.: 'Calculation of the bandwidth of microstrip resonator antennas'. Proc. 9th European Microwave Conference, Brighton, 1979, pp. 116–119
3 DUBOST, G.: 'Linear transmission-line model analysis of arbitrary shape patch antennas', *Electron. Lett.*, 1986, **22**, pp. 798–799
4 LO, Y. T., SOLOMON, D., and RICHARDS, W. F.: 'Theory and experiment on microstrip antennas', *IEEE Trans.*, 1979, **AP-27**, pp. 137–145
5 JAMES, J. C., HALL, P. S., and WOOD, C.: 'Microstrip antenna theory and design' (Peter Peregrinus, 1981)
6 BAHL, I. J., and BHARTHIA, P.: 'Microstrip antennas (Artech House)', USA, 1980)
7 WOOD, C.: 'Improved bandwidth of microstrip antennas using parasite elements', *IEE Proc.*, 1980, **127**, pp. 231–234
8 PENARD, E., and DANIEL, J. P.: 'Open and hybrid microstrip antennas', *IEE Proc.*, 1984, **131H**, pp. 38–44
9 MAHDJOUBI, K., DANIEL, J. P., and TERRET, C.: 'Etude d'antennes imprimées à accès multiples', *Ann. de Telecom*, 1985, **40**, pp. 190–203
10 MAHDJOUBI, K., DANIEL, J. P., and TERRET, C.: 'Dual frequency disc antenna studied by cavity method', *Electron. Lett.*, 1986, **22**, pp. 379–381
11 MAHDJOUBI, K., and TERRET, C.: 'Exact expression for stored energies in the cavity volume of microstrip antennas', *Electron. Lett.*, 1985, **21**, pp. 1221–1222
12 PENARD, E., and DANIEL, J. P.: 'Electric and magnetic currents in microstrip antenna theory'. Int. IEEE/URSI, Albulquerque, New Mexico, 1982
13 PENARD, E., and DANIEL, J. P.: 'Calcul du rayonnement d'antennes microstrip: deux exemples. Journées Nationales Micro-ondes de Toulouse, 1981
14 NEWMAN, E. D., and TULYATHAN, P.: 'Analysis of microstrip antennas using moment method', *IEEE Trans.*, 1981, **AP-29**, pp. 47–53
15 LEWIN, L.: 'Radiation from discontinuities in strip-lines', IEE Monograph 358, 1960
16 HARRINGTON, R. F.: 'Time-harmonic electromagnetic fields' (McGraw-Hill Book, NY, 1961)
17 AAS, J. A., and JAKOBSEN, K.: 'Radiation patterns of rectangular microstrip antennas on finite ground plane'. 12th European Microwave Conference, Helsinki, 1982
18 POZAR, D. M.: 'Finite phase arrays of rectangular microstrip patches', *IEEE Trans.*, 1986, **AP-34**, pp. 658–665

Design and technology of low-cost printed antennas 689

19 POZAR, D. M.: 'Input impedance and mutual coupling of rectangular microstrip antennas', *IEEE Trans.*, 1982, **AP-30**, pp. 1191–1196
20 ITOH, T., and MENZEL, N.: 'A full wave analysis method for open microstrip structures', *IEEE Trans.*, 1981, **AP-29**, pp. 63–68
21 DESHPANDE, M. D., and BAILEY, M. C.: 'Input impedance of microstrip antennas', *IEEE Trans.*, 1982, **AP-30**, pp. 645–650
22 MOSIG, J.: 'Les structures microrubans, analyse au moyen des équations intégrales', D.Sc Thesis, EPFUL, Lausanne, Switzerland, 1984
23 KATEHI, P. B., and ALEXOPOULOS, N. G.: 'On the modelling of electromagnetically coupled microstrip antennas. The printed strip dipoles', *IEEE Trans.*, 1984, **AP-32**, pp. 1179–1186
24 ROUDOT, B.: 'Analyse d'antennes imprimées par une approache dans le domaine spectral', D.Sc Thesis, University of Rennes, France, 1985
25 PENARD, E.: 'Etude d'antennes imprimées par la méthode de la cavité applications au couplage'. D.Sc Thesis, University of Rennes, France, 1982
26 BAHL, I. J., STUCHLY, S. S., and BHARTIA, P.: 'Design of microstrip antennas covered with a dielectric layer', *IEEE Trans.*, 1982, **AP-30**, pp. 314–318
27 DE ASSIS FONSECA, S. B., and GIAROLA, A. J.: 'Microstrip disk antennas. pt. 2: The problem of surface wave radiated by dielectric truncation', *IEEE Trans.*, 1984, **AP-32**, pp. 561–573
28 ROUDOT, B., TERRET, C., DANIEL, J. P., PRIBETICH, P., and KENNIS, P.: 'Fundamental surface-wave effects on microstrip antenna radiation', *Electron. Lett.*, 1985, **21**, pp. 1112–1114
29 DE ASSIS FONSECA, S. B., and GIAROLA, A. J.: 'Influence of surface-wave excitation efficiency of space-wave launching in microstrip disc antennas', *Electron. Lett.*, 1982, **18**, pp. 406–407
30 GRIFFIN, J. M., and FORREST, J. R.: 'Broadband circular disc microstrip antennas', *Electron. Lett.*, 1982, **18**, p. 266–269
31 DAS, N., and CHATTERJEE, J. S.: 'Conically depressed microstrip patch antenna', *IEE Proc.*, 1983, **130H**, pp. 193–196
32 JEDDARI, L., MAHDJOUBI, K., TERRET, C., and DANIEL, J. P., 'Broadband conical microstrip antenna', *Electron. Lett.*, 1985, **21**, pp. 896–898
33 MAILLOUX, R. J.: 'Printed slot arrays with dielectric substrates'. IEEE Symposium on Antenna and Propagation, June 1985
34 OLINER, A. A.: 'The radiation conductance of a series slot in strip transmission line'. IRE Conv. Rec. No. 2, 1954, Pt. 8, pp. 89–90
35 BREITHAUPT, R. W.: 'Conductance data for offset series slot in stripline', *IEEE Trans.*, 1968, **MMT-16**, pp. 969–970
36 TOBARIAS, J.: 'Fente rayonnante bidirectionnelle à la résonance et à l'antirésonance alimentée par une ligne microruban'. 4ème Journées Nationales Micro-ondes, Lannion, June 1984, pp. 236–237
37 TOBARIAS, J., and TERRET, C.: 'Fente rayonnante à la resonance et à l'antirésonance alimentée par une ligne coplanaire'. 4ème Journées Nationales Micro-ondes, Lannion, June 1984, pp. 242–243
38 DUBOST, G., and ZISLER, S.: 'Antennes à large bande', (Masson Editors, France, 1976)
39 WEEKS, W. L.: 'Antennas engineering' (McGraw-Hill, 1968)
40 MARCANO, D., SAILLARD, J., TERRET, C., and DANIEL, J. P.: 'Reseau de fentes à balayage electronique'. 5ème Journées Nationales Micro-ondes, Nice, pp. 265–267
41 YOSHIMURA, Y.: 'A microstrip line slot antenna', *IEEE Trans.*, 1972, **MTT-20**, pp. 760–762
42 DAS, B. N., and JOSHI, K. K.: 'Impedance of a radiating slot in the ground plane of a mirostripline', *IEEE Trans.*, 1982, **AP-30**, pp. 922–926
43 BOOKER, G. G.: 'Slot aerials and their relation to complementary wire aerials', *J. IEE*, 1946, **93**, Pt. IIIA, pp. 620–626

44 NESIC, A.: 'Slotted antenna array excited by a coplanar waveguide', *Electron. Lett.*, 1982, **13**, pp. 404–406
45 NESIC, A.: 'A printed antenna array with slots as primary radiators for phase scanned antenna'. JINA, Nice, Nov. 1986, pp. 281–283
46 DUSSEUX, T.: 'Etude d'antennes fentes annulaires imprimées applications antennas melangeuses, réseaux'. D.Sc Ing. Thesis, University of Rennes, May 1987
47 COHN, S. B.: 'Slot-line on a dielectric substrate, *IEEE Trans.*, 1969, **MTT-17**, pp. 768–778
48 DUSSEUX, T., DANIEL, J. P., and TERRET, C.: 'Theoretical and experimental results of guided wavelength of a slot on a low permittivity substrate', *Electron. Lett.*, 1986, **22**, pp. 589–590
49 JANASWANY, R., and SCHAUBERT, D. H.: 'Dispersion characteristics for wide slot lines on low permittivity substrates', *IEEE Trans.*, 1985, **MTT-33**, pp. 723–726
50 KAWANO, K., and TONIMORO, H.: 'Slot ring resonator and dispersion measurement on slot lines', *Electron. Lett.*, 1981, **17**, pp. 916–917
51 GARG, R., and GUPTA, K. C.: 'Expressions for wavelength and impedance of a slot line', *IEEE Trans.*, 1976, **MTT-24**, p. 532
52 JAMES, J. R., and HENDERSON, A.: 'High-frequency behaviour of microstrip open-circuit terminations', *IEE J. Microwaves, Optics and Acoustics*, 1979, **3**, pp. 205–218
53 ELLIOTT, R. S.: 'The theory of antenna arrays', in HANSEN, R. C. (Ed.): Microwave scanning antennas: Vol. II, (Academic Press, 1986), chap. 1
54 HANSEN, R. C.: 'Linear arrays' and 'Planar arrays' in RUDGE, A. W. *et al.* (Eds): 'The handbook of antenna design: Vol II' (Peter Peregrinus, 1983). chaps. 9 and 10
55 BACH, H.: 'Directivity of basic linear arrays', *IEEE Trans.*, 1970, **AP-18**, pp. 107–110
56 BACH, H., and HANSEN, J. E.: 'Uniformly spaced arrays' in COLLIN, R. E., and ZUCKER, F. J. (Eds): 'Antenna theory. Pt. 1' (McGraw-Hill, NY, 1969)
57 HANSEN, R. C.: 'Aperture efficiency of Villeneuve *n* arrays', *IEEE Trans.*, 1985, **AP-33**, pp. 666–669
58 VILLENEUVE, A. T.: 'Taylor patterns for discrete arrays', *IEEE Trans.*, 1984, **AP-32**, pp. 1089–1093
59 HANSEN, R. C.: 'Comparison of square array directivity formulas', *IEEE Trans.*, 1972, **AP-20**, pp. 100–102
60 WHITTAKER E. T., and WATSON, G. N.: 'A course of modern analysis', Cambridge, London, 1962, p. 170
61 JEDLICKA, R. P., and CARVER, K. R.: 'Mutual coupling between microstrip patch antennas'. Proc. Workshop on printed circuit antenna technology, Oct. 1979
62 JEDLICKA, R. P., POE, M. T., and CARVER, K. R.: 'Measured mutual coupling between microstrip antennas', *IEEE Trans.*, 1981, **AP-29**, (1)
63 MALKOMES, M.: 'Mutual coupling between microstrip patch antennas', *Electron. Lett.*, 1982, **18**, pp. 520–522
64 VAN LIL, E. H., and VAN DE CAPELLE, A. R.: 'Transmission line model for mutual coupling between microstrip antennas', *IEEE Trans.*, 1984, **AP-32**, pp. 816–821
65 PENARD, E., and DANIEL, J. P.: 'Mutual coupling between microstrip antennas', *Electron. Lett.*, 1982, **18**, pp. 605–607
66 MAHDJOUBI, K., PENARD, E., DANIEL, J. P., and TERRET, C.: 'Mutual coupling between circular disc microstrip antennas', *Electron. Lett.*, 1987, **23**, pp. 27–28
67 BHATTACHARYYA, A. K., and SHAFAI, L.: 'Surface wave coupling between circular patch antennas', *Electron. Lett.*, 1986, **22**, pp. 1198–1200
68 MAHDJOUBI, K., PENARD, E., TERRET, C., and DANIEL, J. P.: 'Mutual coupling between microstrip disk antennas'. ICAP '87, late papers, University of York, 1987
69 DANIEL, J. P., PENARD, E., NEDELEC, M., and MUTZIG, J. P.: 'Design of low cost printed antenna arrays'. Proceedings of ISAP'85, Kyoto, Aug. 1985, pp. 121–124
70 WILLIAMS, J. C.: 'Cross fed printed aerial'. Proc. 7th European Microwave Conf. Copenhagen, 1977, p. 292

71 DANIEL, J. P., MUTZIG, J. P., NEDELEC, M., and PENARD, E.: 'Reseaux d'antennes imprimées dans la bande 20/30 GHz'. 4ème Journeés Nationales Microondes, Lannion, France, June 1984, pp. 246–247
72 DANIEL, J. P., MUTZIG, J. P., NEDELEC, M., and PENARD, E.: 'Reseaux d'antennes imprimées dans la bande 20/30 GHz', *L'onde Électrique*, 1985, **65**, pp. 35–41
73 BOGUAIS, M.: 'Contribution à la synthèse de réseaux d'antennes, réalisation en technologie imprimée. D.Sc Thesis, University of Rennes, France
74 BOGUAIS, M., DANIEL, J. P., and TERRET, C.: 'Antenna pattern synthesis using a relaxation method: application to printed antennas', *Electron. Lett.*, 1986, **22**, (7)
75 BOGUAIS, M., DANIEL, J. P., and TERRET, C.: 'Deux methodes de synthèse de réseaux d'antennes, application aux antennes imprimées', JINA, Nice, 1986, pp. 310–311
76 DANTZIG, G. B., and ORCHARD-HAYS, W.: 'The product form for the inverse in the simplex method', *Math. Comp.*, 1959
77 DEMEURE, L.: 'New low cost and low loss substrate: Application to printed antenna', JINA, Nice, France, Nov. 1986
78 BONFIELD, R.: 'Thick metal backing adds value to substrate', *Microwaves and RF*, Feb. 1987
79 Patent 84 402 7078: 'Support métallisé à base de polypropylène et procédé de fabrication de ce support'

Chapter 12
Analysis and design considerations for printed phased-array antennas

D. M. Pozar

12.1 Introduction

Until the last decade or so, phased-array technology generally employed dipole or waveguide radiating elements, with waveguide or coaxial lines for feed networks [1–4]. In more recent years, however, printed or microstrip arrays and feedlines have become quite popular [5–7] owing to features including light weight, conformability, ease of manufacture and, probably most important, potentially low cost. Economics is generally the most critical factor affecting the deployment of phased arrays into more systems, as a variety of applications would benefit from the advantages of a phased-array antenna, which include rapid and selective beam steering, adaptive nulling, and other controlled array-illumination functions. The printed phased array, with its fabrication simplified through the use of photolithographic techniques, offers the promise of lower-cost electronically scanned arrays. This is in spite of some inherent disadvantages of printed antennas, such as low bandwidth and power capacity.

Printed arrays can take many different forms. Radiating elements may be printed dipoles, printed (microstrip) patches or slot elements. Feed circuitry may be in microstripline, or in stripline form. Several combinations then exist for the interconnection of feed lines to radiating elements. One approach is to etch the radiating elements and feed lines in microstrip form on the same substrate, while other approaches use two or more layers to separate the radiating elements from feed circuitry. Phase shifting and other active circuitry functions can be incorporated in hybrid form. At millimetre-wave frequencies the physical size of the array may be small enough so that circuit integration can be carried one step further, resulting in the 'monolithic phased array'. This concept, discussed in more detail in Section 12.3, involves the integration of all active circuitry required for a sub-array module of a millimetre-wave phased array.

This Chapter first considers the rigorous analysis of several canonical printed-array geometries (Section 12.2), and then discusses some design considerations for printed arrays (Section 12.3). During the 1960s, a large analysis effort was carried out for waveguide and dipole phased arrays [1–4]; a corresponding effort for printed phased arrays is still needed, but the present Chapter consolidates some of the solutions which have been completed to date. Section 12.2.2 treats infinite planar arrays of various printed elements. While most of these cases are idealised in some way, they represent a starting point for the analysis and/or design of more practical arrays. In addition, as discussed in more detail in Section 12.2.2.5, the scan performance of phased arrays is often more dependent on substrate parameters and element spacing than on the particular details of the feeding method. In Section 12.3 dealing with design considerations, the monolithic phased-array concept is emphasised, but much of this material is relevant to non-monolithic printed arrays as well.

12.2 Analysis of some canonical printed phased-array geometries

In this Section we present analyses for several types of printed phased-array geometries. These problems are canonical in that the geometries are idealised in some sense, usually in terms of simplifying assumptions about the feed. The solutions here all have a high degree of commonality, being based on the work of the author and his colleagues at the University of Massachusetts. The relevant work of other researchers, however, is noted and discussed in relation to the analyses presented here.

The basic procedure for the analysis of each of the printed-array geometries in this Section is as follows. First, the Green's function for the relevant dielectric-slab geometry is derived in spectral (transform)-domain form, for a single infinitesimal source (electric or magnetic dipole). This result is then extended to an infinite periodic planar array of such sources, with a progressive phase shift for scanning at the desired angle. A moment-method solution is formulated for the unknown current distribution on the antenna element, and an appropriate set of expansion weighting functions is chosen. An impedance matrix results, which can then be used to determine the unknown coefficients of the expansion modes. Because of the periodic nature of the array, the current distributions on all of the elements are the same, except for the imposed progressive phase shift. Thus, formulating the moment-method solution for one 'unit cell' is equivalent to imposing the solution across the entire array. Mutual coupling is implicitly included in the solution. This method has variously been referred to as a 'full-wave solution', or the 'Galerkin method in the spectral domain', and has been applied to a variety of antenna and microwave circuit problems, in both single-element and array form.

After the currents have been determined, other quantities of interest can easily be found. The variation of input impedance with scan angle can be calculated; this result is quite important for matching the array over the desired scan range. A related quantity is the active-element pattern, which also gives information about the scan performance of the array. Other quantities of interest include the

Analysis and design considerations for phased-array antennas 695

cross-polarisation level, and possibly the efficiency of the array. Pattern quantities such as directivity and sidelobe level depend on the size of the array, and so are not very meaningful for infinite arrays (since an infinite array radiates a plane wave, its directivity is infinite, while its sidelobe level is zero).

In the following Subsections, we first present some material that is common to most of the solutions which follow, including a brief derivation of a typical Green's function and a discussion of the scan-blindness effect. We then treat several canonical infinite planar printed arrays, and present results for the scan performance of such arrays. Next, solutions are described for finite arrays of dipoles and rectangular microstrip patches. The finite-array problem is considerably more difficult than the corresponding infinite-array problem, but may be of more practical utility since it includes edge effects.

12.2.1 Some preliminaries

12.2.1.1 Derivation of the Green's function of a grounded dielectric slab:

Central to the solutions that follow is the exact Green's function of the dielectric-slab geometry in spectral, or transform, domain form. Such Green's functions have appeared in a number of recent papers on printed-antenna analysis [8–12], but generally without derivation. Thus it may be useful to present a short derivation of a Green's function, in case the reader is not familiar with the basic procedure. We will derive the Green's function for a grounded dielectric slab, with an infinitesimal electric-current source on its surface. This is one of the most useful Green's function results, being applicable to all of the array geometries below (some additional Green's functions are needed for the slot elements of Section 12.2.2.4). The same procedure, however, can be used to obtain the Green's functions for a number of more general cases, including the following:

- Two (or more) dielectric-layer geometry
- Dielectric layer with a lossy (surface-impedance) ground plane
- Substrate with magnetic properties
- Anisotropic substrates

Fig. 12.1 shows the geometry of a grounded isotropic dielectric slab of thickness d and relative permittivity ε_r. The source is an infinitesimal $\hat{x}$-directed electric dipole, of unit strength and located on the surface of the dielectric slab at (x_0, y_0, d). We desire to find the E_x, E_y and E_z fields generated by this source.

While it is possible, and quite common, to introduce vector potentials, it is actually simpler to work directly with wave equations for E_z and H_z, and find the transverse fields from these field components. Thus, Maxwell's equations,

$$\nabla \times \boldsymbol{E} = -j\omega\mu_0 \boldsymbol{H} \tag{12.1a}$$

$$\nabla \times \boldsymbol{H} = j\omega\varepsilon \boldsymbol{E} \tag{12.1b}$$

can be solved simultaneously for the usual Helmholtz wave equations in a source-free region:

Analysis and design considerations for phased-array antennas

$$\nabla^2 E + k^2 E = 0 \qquad (12.2a)$$

$$\nabla^2 H + k^2 H = 0 \qquad (12.2b)$$

where $k^2 = \omega^2 \mu_0 \varepsilon$, and $\varepsilon = \varepsilon_0 \varepsilon_r$ for $0 < z < d$, and $\varepsilon = \varepsilon_0$ for $z > d$. Anticipating a plane-wave form of solution, with a propagation factor $e^{\pm j k_x x} e^{\pm j k_y y} e^{\pm j k_z z}$, and substituting this form into eqn. 12.2a,b gives the propagation constants in the z-direction as

$$k_1^2 = k_z^2 = \varepsilon_r k_0^2 - \beta^2 \quad \text{for} \quad 0 > z > d \qquad (12.3a)$$

$$k_2^2 = k_z^2 = k_0^2 - \beta^2 \quad \text{for} \quad z > d \qquad (12.3b)$$

with $\beta^2 = k_x^2 + k_y^2$. In the above, the branch of the square-root function should be chosen so that $\text{Im}(k_1) < 0$ and $\text{Im}(k_2) < 0$.

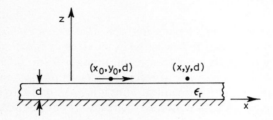

Fig. 12.1 *Geometry of an infinitesimal $\hat{x}$-directed current element on a grounded dielectric slab*

We now define a Fourier-transform pair as

$$E(x, y, z) = \frac{1}{4\pi^2} \int_{-\infty}^{\infty} \int_{-\infty}^{\infty} \tilde{E}(k_x, k_y, z) e^{jk_x x} e^{jk_y y} \, dk_x \, dk_y \qquad (12.4a)$$

$$\tilde{E}(k_x, k_y, z) = \int_{-\infty}^{\infty} \int_{-\infty}^{\infty} E(x, y, z) e^{-jk_x x} e^{-jk_y y} \, dx \, dy \qquad (12.4b)$$

Then, in the transform domain, the transverse fields can be written in terms of $\tilde{E}_z$ and $\tilde{H}_z$ as,

$$\left(\varepsilon_r k_0^2 + \frac{\partial^2}{\partial z^2}\right) \tilde{E}_x = jk_x \frac{\partial}{\partial z} \tilde{E}_z + \omega \mu_0 k_y \tilde{H}_z \qquad (12.5a)$$

$$\left(\varepsilon_r k_0^2 + \frac{\partial^2}{\partial z^2}\right) \tilde{E}_y = jk_y \frac{\partial}{\partial z} \tilde{H}_z - \omega \mu_0 k_x \tilde{H}_z \qquad (12.5b)$$

$$\left(\varepsilon_r k_0^2 + \frac{\partial^2}{\partial z^2}\right) \tilde{H}_x = jk_x \frac{\partial}{\partial z} \tilde{H}_z + \omega \varepsilon_0 \varepsilon_r k_y \tilde{E}_z \qquad (12.5c)$$

$$\left(\varepsilon_r k_0^2 + \frac{\partial^2}{\partial z^2}\right) \tilde{H}_y = jk_y \frac{\partial}{\partial z} \tilde{H}_z + \omega \varepsilon_0 \varepsilon_r k_x \tilde{E}_z \qquad (12.5d)$$

Analysis and design considerations for phased-array antennas 697

These results are for the dielectric region $0 < z < d$, but can be used for the air region $z < d$ by setting $\varepsilon_r = 1$. Similarly, $\partial^2/\partial z^2 = -k_1^2$ or $-k_2^2$, depending on whether $0 < z < d$, or $z > d$, respectively.

From the wave equations of 12.2a and 12.2b, the general solutions for $\tilde{E}_z$ and $\tilde{H}_z$ are:

$$\hat{E}_z = A e^{-jk_2 z} \quad \text{for} \quad z > d \tag{12.6a}$$

$$\hat{H}_z = B e^{-jk_2 z} \quad \text{for} \quad z > d \tag{12.6b}$$

$$\tilde{E}_z = C \cos k_1 z + D \sin k_1 z \quad \text{for} \quad 0 < z < d \tag{12.6c}$$

$$\tilde{H}_z = E \sin k_1 z + F \cos k_1 z \quad \text{for} \quad 0 < z < d \tag{12.6d}$$

where outgoing waves have been assumed in the region $z > d$. With these forms the transverse field components of eqns. 12.5a–d can be rewritten as

$$\tilde{E}_x = \frac{jk_x}{\beta^2} \frac{\partial}{\partial z} \tilde{E}_z + \frac{\omega \mu_0 k_y}{\beta^2} \tilde{H}_z \tag{12.7a}$$

$$\tilde{E}_y = \frac{jk_y}{\beta^2} \frac{\partial}{\partial z} \tilde{E}_z - \frac{\omega \mu_0 k_x}{\beta^2} \tilde{H}_z \tag{12.7b}$$

$$\tilde{H}_x = \frac{jk_x}{\beta^2} \frac{\partial}{\partial z} \tilde{H}_z - \frac{\omega \varepsilon_0 \varepsilon_r k_y}{\beta^2} \tilde{E}_z \tag{12.7c}$$

$$\tilde{H}_y = \frac{jk_y}{\beta^2} \frac{\partial}{\partial z} \tilde{H}_z + \frac{\omega \varepsilon_0 \varepsilon_r k_x}{\beta^2} \tilde{E}_z \tag{12.7d}$$

which also apply to the region $z > d$ after ε_r is set to unity. Applying eqns. 12.7a and b to eqns. 12.6c and d to enforce the boundary condition that $E_x = E_y = 0$ at $z = 0$ yields $D = F = 0$. There then remains four constants (A, B, C, E) to be evaluated by the continuity of E_x, E_y and H_x at $z = d$ (the dielectric–air interface), and a jump condition in H_y at $z = d$ (due to the current source). After some straightforward algebraic manipulation, the results for $\tilde{E}_z$ and $\tilde{H}_z$ for $0 < z < d$ are

$$\tilde{E}_z = \frac{Z_0 k_2 k_x}{k_0 T_m} \cos k_1 z \, e^{-jk_x x_0} e^{-jk_y y_0} \tag{12.8a}$$

$$\tilde{H}_z = \frac{-jk_y}{T_e} \sin k_1 z \, e^{-jk_x x_0} e^{-jk_y y_0} \tag{12.8b}$$

where

$$T_m = \varepsilon_r k_2 \cos k_1 d + jk_1 \sin k_1 d \tag{12.9a}$$

$$T_e = k_1 \cos k_1 d + jk_2 \sin k_1 d \tag{12.9b}$$

and $Z_0 = \sqrt{\mu_0/\varepsilon_0}$. The zeros of the T_m and T_e functions correspond to the TM and TE surface-wave poles of the grounded dielectric slab.

Using eqns. 12.7a–d and eqns. 12.8a and b and taking the inverse transform in eqn. 12.4a allows the transverse electric fields at $z = d$ to be evaluated as

$$E_x^0(x, y, d) = \frac{1}{4\pi^2} \int_{-\infty}^{\infty} \int_{-\infty}^{\infty} G_{xx}^{EJ}(k_x, k_y) e^{jk_x(x-x_0)} e^{jk_y(y-y_0)} dk_x dk_y$$

(12.10a)

$$E_y^0(x, y, d) = \frac{1}{4\pi^2} \int_{-\infty}^{\infty} \int_{-\infty}^{\infty} G_{yx}^{EJ}(k_x, k_y) e^{jk_x(x-x_0)} e^{jk_y(y-y_0)} dk_x dk_y$$

(12.10b)

where the following quantities have been defined:

$$G_{xx}^{EJ} = \frac{-jZ_0}{k_0} \frac{(\varepsilon_r k_0^2 - k_x^2) k_2 \cos k_1 d + jk_1(k_0^2 - k_x^2) \sin k_1 d}{T_e T_m} \sin k_1 d$$

(12.11a)

$$G_{yx}^{EJ} = \frac{jZ_0}{k_0} \frac{k_x k_y \sin k_1 d [k_2 \cos k_1 d + jk_1 \sin k_1 d]}{T_e T_m}$$

(12.11b)

In eqns. 12.10a and b the notation E^0 has been used to denote the field due to a single source. The above field expressions are directly applicable to the analysis of isolated antennas printed on the surface of a dielectric slab. Note that the results for E_x and E_y of eqns. 12.10a and b satisfy reciprocity, as an interchange of x, x_0 and y, y_0 does not change the result.

12.2.1.2 Extension to an infinite array: We now show how the Green's function of the previous Section for a single infinitesimal electric dipole can be generalised to an infinite phased array of such sources. Fig. 12.2 shows the geometry of an infinite periodic array of infinitesimal sources, with spacing a in the E-plane (x) direction, and spacing b in the H-plane (y) direction. The m, n th source is thus located at

$$x_m = x_0 + ma \qquad (12.12a)$$

$$y_n = y_0 + nb \qquad (12.12b)$$

where m, n are integer indices with $-\infty < m, n < \infty$. Now for scanning at the angle θ, ϕ the currents on the m, n th source must be phased as

$$e^{-jk_0(mau + nbv)} \qquad (12.13)$$

where

$$u = \sin\theta \cos\phi \qquad (12.14a)$$

$$v = \sin\theta \sin\phi \qquad (12.14b)$$

Analysis and design considerations for phased-array antennas

are the direction cosines. Then by superposition, eqns. 12.10a and b can be used to find the total field from this infinite array (after replacing m, n, k_x and k_y with $-m$, $-n$, $-k_x$ and $-k_y$, respectively):

$$E_x(x, y, d) = \frac{1}{4\pi^2} \sum_{m=-\infty}^{\infty} \sum_{n=-\infty}^{\infty} e^{jk_0(mau+nbv)} \int_{-\infty}^{\infty} \int_{-\infty}^{\infty} G_{xx}^{EJ}(k_x, k_y)$$

$$e^{jk_x(x_0 - x - ma)} e^{jk_y(y_0 - y - nb)} dk_x dk_y \quad (12.15a)$$

$$E_y(x, y, d) = \frac{1}{4\pi^2} \sum_{m=-\infty}^{\infty} \sum_{n=-\infty}^{\infty} e^{jk_0(mau+nbv)} \int_{-\infty}^{\infty} \int_{-\infty}^{\infty} G_{yx}^{EJ}(k_x, k_y)$$

$$e^{jk_x(x_0 - x - ma)} e^{jk_y(y_0 - y - nb)} dk_x dk_y \quad (12.15b)$$

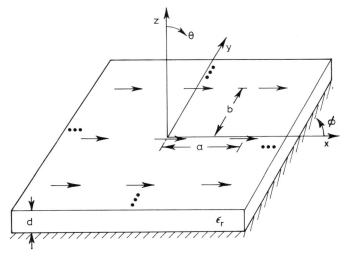

Fig. 12.2 *Geometry of an infinite periodic array of $\hat{x}$-directed infinitesimal current elements on a grounded dielectric slab*

Eqns. 12.15a and b give the transverse fields at (x, y, d) due to the infinite array of dipoles and may be thought of as the Green's function of the infinite array. Observe, however, that E_x and E_y of eqns. 12.15a and b do not satisfy reciprocity upon interchange of x, x_0 and y, y_0. This is due to the asymmetry introduced by the phasing of eqn. 12.13.

Eqns. 12.15a and b are rigorous expressions, but clearly not in a very usable form from a computational viewpoint. The Poisson-sum formula can be applied to greatly simplify the result. Consider an expression of the form

$$g(x) = \frac{1}{2\pi} \sum_{m=-\infty}^{\infty} \int_{-\infty}^{\infty} e^{jk_0 mau} q(k_x) e^{jk_x(x-ma)} dk_x \quad (12.16)$$

The Poisson sum formula can be written as

$$\sum_{m} e^{jm\omega_0 t} F(m\omega_0) = T \sum_{m} f(t + mT) \quad (12.17)$$

where $T = 2\pi/\omega_0$, and $f(t)$ and $F(\omega)$ form a Fourier transform pair:

$$F(\omega) = \int_{-\infty}^{\infty} f(t) e^{-j\omega t} dt \qquad (12.18a)$$

$$f(t) = \frac{1}{2\pi} \int_{-\infty}^{\infty} F(\omega) e^{j\omega t} dt \qquad (12.18b)$$

Now let $f(t) = h(t) e^{j\omega_1 t}$. Then $F(\omega) = H(\omega - \omega_1)$, and eqn. 12.17 becomes

$$\sum_m e^{jm\omega_0 t} H(m\omega_0 - \omega_1) = T \sum_m h(t + mT) e^{j\omega_1 (t + mT)}$$

or

$$\sum_m e^{jm\omega_0 t} \int_{-\infty}^{\infty} h(t') e^{-j(m\omega_0 - \omega_1)t'} dt' = T \sum_m h(t + mT) e^{j\omega_1 (t + mT)} \quad (12.19)$$

Now compare eqns. 12.19 and 12.16, and let $t = k_0 u$, $\omega_0 = a$, $t' = k_x$, $\omega_1 = x$, $T = 2\pi/a$ and $h = q$, to get

$$g(x) = \frac{1}{a} \sum_m q(k_x) e^{jk_x x} \qquad (12.20)$$

with

$$k_x = \left(\frac{2\pi m}{a} + k_0 u \right) \qquad (12.21)$$

So we see that the Poisson-sum formula can be used to eliminate the infinite integration of eqn. 12.16. This result can be applied twice (for k_x and k_y) to eqns. 12.15a and b), to give the following results:

$$E_x(x, y, d) = \frac{1}{ab} \sum_{m=-\infty}^{\infty} \sum_{n=-\infty}^{\infty} G_{xx}^{EJ}(k_x, k_y) e^{-jk_x(x-x_0)} e^{-jk_y(y-y_0)} \quad (12.22a)$$

$$E_y(x, y, d) = \frac{1}{ab} \sum_{m=-\infty}^{\infty} \sum_{n=-\infty}^{\infty} G_{yx}^{EJ}(k_x, k_y) e^{-jk_x(x-x_0)} e^{-jk_y(y-y_0)} \quad (12.22b)$$

where the variables k_x and k_y take on the discrete values

$$k_x = \frac{2\pi m}{a} + k_0 u \qquad (12.23a)$$

$$k_y = \frac{2\pi n}{b} + k_0 v \qquad (12.23b)$$

Note the similarity between eqns. 12.22 and 12.10; eqn. 12.22 can be considered as a discretised version of the continuous integration in eqn. 12.10.

12.2.1.3 The scan-blindness effect: Scan blindness refers to a condition where, for certain scan angles, no real power can be transmitted (or received) by a phased array. This effect has been experimentally and theoretically observed in a number of different types of arrays [2, 4, 8, 13], and is generally related to the resonance of some type of trapped or guided mode of the array structure. For example, waveguide arrays with dielectric plugs or dielectric cover layers have exhibited the scan-blindness phenomenon [4]. Printed phased arrays, because of the presence of a dielectric slab, also show scan-blindness effects. Scan blindness is total only in infinite arrays (or waveguide simulators), but in large arrays the effect can be severe enough to seriously degrade performance. Thus it is important to both understand the scan-blindness effect and to be able to predict the occurrence of this effect in printed phased arrays. Unless otherwise stated, the discussion below refers to an infinite array.

One way of observing scan blindness is to look at the reflection-coefficient magnitude at one element of an infinite phased array. Assuming a reasonable impedance match at broadside scan ($\theta = 0°$) the reflection-coefficient magnitude will be small there. As the scan angle increases towards endfire ($\theta = 90°$), the reflection-coefficient magnitude must increase to unity, since an infinite phased array does not transmit any real power away from the face of the array at endfire scan. A scan blindness, however, will show its presence by unity (or near unity) reflection-coefficient magnitude at some scan angle before endfire. This means that each element of the transmitting array is reflecting all the power incident on it, and so the array is 'blind' at this scan angle.

The active-element pattern of the array provides another way of looking at the scan-blindness effect. The active-element pattern is defined as the radiation pattern of an array obtained when one element is driven and all other elements are terminated in matched loads [3]. If no grating lobes are present, it can be shown [3] that the active-element (power) pattern $F(\theta, \phi)$ of an array is related to the active reflection coefficient $R(\theta, \phi)$ by

$$F(\theta, \phi) = (1 - |R(\theta, \phi)|^2)\cos\theta \qquad (12.24)$$

The active-element pattern is significant because it is relatively easy to measure (no power-divider/phase-shifter network is required), and it provides information about the scan performance of the array. A scan-blindness condition will show up as a null in the active-element pattern, owing to the unity reflection coefficient at the blind spot. Thus, some workers refer to scan blindnesses as nulls in the active-element pattern.

One can discuss the scan-blindness effect from several viewpoints. If a specific array geometry is being considered, a rigorous analysis of the scan performance of the array will allow the prediction of the blindness effect. This will be done in the following Sections for several printed arrays of practical interest. In addition, certain canonical problems can be posed which are complete enough to yield data on the scan performance of printed arrays, including blindness angles, in a more general sense. This has been done with the current sheet model discussed in Section 12.2.2.5, and with the infinite array of infinitesimal dipoles discussed below.

If no details of the array geometry are available, or if one wishes to look at the blindness effect from a different point of view, an analysis based on mutual coupling (measured or calculated) can be used. Thus, consider an infinite planar array, with the elements indexed as in eqn. 12.12. Let the reference element be the $m = n = 0$ element, and assume each element is fed with a unit amplitude voltage source having a phase given by eqn. 12.13. If the scattering matrix coefficient between the m, n th element and the 0, 0 reference element is S_{mn}, then the reflection coefficient is

$$R = \sum_{m=-\infty}^{\infty} \sum_{n=-\infty}^{\infty} S_{mn} e^{-jk_0(mau+nbv)}$$

$$= S_{00} + \sum_{\substack{m=-\infty \\ (m \neq 0)}}^{\infty} \sum_{\substack{n=-\infty \\ (n \neq 0)}}^{\infty} S_{mn} e^{-jk_0(mau+nbv)} \qquad (12.25)$$

This result shows that the reflected wave at any given element is due to the mismatch of the isolated element (S_{00}), plus contributions from all the neighbouring elements. The effect of the coupling from the neighbouring elements depends on the strength of the coupling, and on the scan angle. For coupling coefficients of a certain magnitude and phase, it is possible for an in-phase accumulation of coupled power to lead to total reflection at certain scan angles. This discussion shows how scan blindness can occur from a mutual-coupling point of view, but drawing more specific conclusions is difficult unless data on the coupling coefficients is available, or can be assumed [14].

We now look at a specific printed-array geometry – an infinite planar array of infinitesimal dipoles – to obtain more information about the scan-blindness effect. This is an idealised case, of course, but is complete enough to show some of the essential blindness mechanisms that occur in this array and in other infinite printed arrays. The preliminary derivations of the Green's function and the extension to an infinite array of infinitesimal $\hat{x}$-directed currents of Sections 12.2.1.1 and 12.2.1.2 can be applied directly.

We assume that the $m = n = 0$ element of the array is located at $x = y = 0$, so that $x_0 = y_0 = 0$ in eqn. 12.22a. We now compute the complex power leaving the unit cell centered around the 0, 0 element from

$$P = -\int_{x=-\frac{a}{2}}^{a/2} \int_{y=-\frac{b}{2}}^{b/2} E_x J_{sx} \, dx \, dy \qquad (12.26)$$

where E_x is given by eqn. 12.22a, and J_{sx} is the electric-surface-current density of the sources at $z = d$:

$$J_{sx} = \sum_{m=-\infty}^{\infty} \sum_{n=-\infty}^{\infty} e^{-jk_0(mau+nbv)} \delta(x - ma) \delta(y - nb) \qquad (12.27)$$

In eqn. 12.26, the power is evaluated at $z = d^+$, and is the same for each cell in the array. Substituting eqns. 12.27 and 12.22a into eqn. 12.26 and performing the integration gives

$$P = \frac{-1}{ab} \sum_{m=-\infty}^{\infty} \sum_{n=-\infty}^{\infty} G_{xx}^{EJ}(k_x, k_y) \qquad (12.28)$$

where $G_{xx}^{EJ}(k_x, k_y)$ is given by eqn. 12.11a. Recall that k_x and k_y are functions of m, n and the scan angle θ, ϕ, as given by eqn. 12.23.

The result in eqn. 12.28 shows that the complex power leaving the face of the array is a superposition of the powers contained in each of the Floquet modes forming the field solution. Now if we assume that the array spacing is such that $a \leqslant \lambda/2$ and $b \leqslant \lambda/2$, so that no grating lobes are present for any scan angle, then from eqn. 12.23 it is easy to see that

$$|k_x| < k_0 \quad \text{only for} \quad m = n = 0$$

$$|k_y| < k_0 \quad \text{only for} \quad m = n = 0$$

From eqn. 12.3b, these conditions imply that k_2 is purely imaginary except for the $m = n = 0$ Floquet mode, and a study of the G_{xx}^{EJ} function then leads to the conclusion that only the $m = n = 0$ term of eqn. 12.28 contributes a real part to P; all the other terms are purely imaginary. Thus the $m = n = 0$ Floquet mode is the only mode carrying power away from the array face, and all the other Floquet modes are evanescent, storing energy near the array surface (or carrying power across the surface of the array).

Fig. 12.3 shows the complex power P_{mn} for three Floquet modes ($m = -1$, $n = 0$; $m = n = 0$; $m = 1$, $n = 0$) versus E-plane scan angle for an infinite array of infinitesimal dipoles, with $a = b = \lambda_0/2$ and a substrate with $\varepsilon_r = 12\cdot 8$ and $d = 0\cdot 06\lambda_0$. The $m = n = 0$ mode is the only Floquet mode with a non-zero real part as discussed above. All the modes have imaginary contributions, which are generally well behaved with scan angle. The exception is the $m = -1$, $n = 0$ Floquet mode, which can be seen from Fig. 12.3 to have a singularity in its imaginary part near a scan angle of about 45°. This results in a scan blindness at this angle.

The singularity in Im(P_{-10}) can be traced to a zero of the T_m function in the denominator of G_{xx}^{EJ}. The zeros of this function correspond to TM surface waves of the unloaded dielectric slab. As the scan angle approaches the blindness angle, the propagation constant of the $m = -1$, $n = 0$ Floquet mode approaches that of the TM surface-wave mode of the dielectric slab, and resonates this mode. The surface wave propagates along the surface of the array, and so does not carry power away from the array. In this situation k_1 is real, meaning that waves are propagating (up and down) inside the dielectric slab, while k_2 is imaginary, meaning that the field above the surface of the array is evanescent. Thus, such a wave is sometimes called a 'trapped mode'.

This is not unlike the case of total reflection of a plane wave, in a region having a low dielectric constant, incident on a region having a higher dielectric constant. For incidence angles greater than the critical angle, all power is reflected but a surface wave field is excited in the higher dielectric-constant region. This surface wave propagates along the interface, and exponentially decays away from the interface. Since this surface wave field cannot exist in the absence of the incident plane wave, some authors refer to it as a forced surface wave, while other authors feel that it is not a 'true' surface wave field at all, but is only 'surface-wave-like' [13].

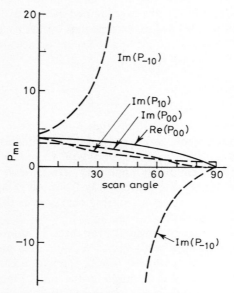

Fig. 12.3 *Complex powers radiated by the $m = -1, n = 0$; $m = n = 0$; and the $m = 1, n = 0$ Floquet modes for an infinite array of infinitesimal dipoles, versus E-plane scan angle*
$\varepsilon_r = 12.8$, $d = 0.06\lambda_0$, $a = b = \lambda_0/2$

Another point to be noted from Fig. 12.3 is that the real part of the power (in the $m = n = 0$ mode) is well-behaved with scan angle, and is finite and non-zero at the blindness angle. Thus blindness occurs because the imaginary part of the input impedance becomes extremely large, leading to severe impedance mismatch, even though the real part may be non-zero. Some authors [15] have been led to erroneous conclusions in this regard.

Scan blindnesses of this type can be predicted by comparing the propagation constants of the surface wave of the dielectric slab and the various Floquet modes. Let β_{sw} be the propagation constant of the first (TM) surface-wave mode of the unloaded dielectric slab where $k_0 < \beta_{sw} < \sqrt{\varepsilon_r}k_0$. (In practice, substrates are usually thin enough so that only the lowest-order TM mode propagates.) Then a surface wave resonance will occur when β_{sw} matches a particular Floquet-mode propagation constant. Mathematically, this condition can be ex-

Analysis and design considerations for phased-array antennas

pressed as:

$$(\beta_{sw}/k_0)^2 = (k_x/k_0)^2 + (k_y/k_0)^2 = \left(\frac{m}{a/\lambda_0} + u\right)^2 + \left(\frac{n}{b/\lambda_0} + v\right)^2 \quad (12.29)$$

where λ_0 is the free-space wavelength.

As an example, for the $\varepsilon_r = 12\cdot 8$, $d = 0\cdot 06\lambda_0$ dielectric slab of Fig. 12.3, $\beta_{sw}/k_0 = 1\cdot 28582$. In the E-plane, $u = \sin\theta$ and $v = 0$, so for half-wave spacing eqn. 12.29 reduces to

$$(1\cdot 28582)^2 = (2m + \sin\theta_{sw})^2 + (2n)^2$$

Clearly the only solution occurs for $m = -1$ and $n = 0$; so

$$\theta_{sw} = \sin^{-1}|1\cdot 28582 - 2| = 45\cdot 6°$$

A useful graphical technique, referred to as a surface-wave circle diagram [8, 16], can be arrived at by noting that eqn. 12.29 describes a set of circles in the u–v plane. Fig. 12.4 shows such a diagram. The solid-line circles represent the usual grating-lobe circles, with centres at $u = -m\lambda_0/a$, $v = -n\lambda_0/b$, and unit

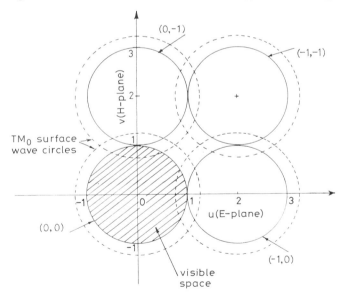

Fig. 12.4 *Surface-wave circle diagram for an infinite phased array with $a = b = 0\cdot 5\lambda_0$, $\varepsilon_r = 12\cdot 8$, $d = 0\cdot 06\lambda_0$*

radius. For half-wave spacing, the edges of these circles just touch, indicating that no grating lobes are visible. The grating-lobe circle centred at the origin represents visible space, since, within this region, u and v are such that θ and ϕ

correspond to real scan angles. The dotted circles in Fig. 12.4 are solutions to eqn. 12.29, and so are called surface-wave circles. When a surface-wave circle intersects visible space, a scan blindness can occur at those scan angles, unless the scan angle is such that a polarisation mismatch occurs between the fields of the relevant Floquet mode and the surface wave of the slab. This occurs, for example, in the H-plane scan of the infinitesimal dipole array because the polarisation of the x-directed dipoles cannot couple to a surface wave propagating along the y-axis. Mathematically, an inspection of G_{xx}^{EJ} in eqn. 12.11a shows that the T_m function in the denominator is cancelled by an identical term in the numerator when $k_x = 0$ (H-plane scan). Thus, there is no blind spot at 45·6° in the H-plane.

The surface-wave circle diagram is a very convenient way to study the effect of grid spacing and substrate parameters (ε_r and d) on the potential blindness angle. For example, the diagram shows that it is possible to completely eliminate a scan blindness by decreasing the element spacing, since this has the effect of moving the grating-lobe and surface-wave circles further apart. The diagram also shows that, for half-wave spacings or greater, there will always be a scan blindness, and that it will occur closest to broadside for E-plane scanning. In practice, however, for electrically thin substrates β_{sw} will be close to k_0, so the scan blindness will occur close to endfire. Decreasing the substrate dielectric constant will also move the blindness angle towards endfire; as $\varepsilon_r \to 1$, the blindness angle approaches 90°.

The surface-wave-type blindnesses discussed above seem to occur in any type of printed array. There are other types of resonances that are also possible, however, depending on the type of array element being used. Patch elements, for example, may in certain circumstances load the dielectric slab enough to support 'leaky wave' modes [15].

12.2.2 Infinite-planar-array solutions

12.2.2.1 Printed dipoles: In this Section we consider the analysis of an infinite planar array of printed dipoles. The dipoles are assumed to be thin, and to be fed with idealised delta-gap generators at their mid-points. Isolated printed dipoles and mutual coupling between pairs of such dipoles have been studied by several workers [9, 17, 18]. Infinite arrays of printed dipoles have been analysed in References 8, 19 and 20. The solution presented here follows that of Reference 8.

Fig. 12.5 shows the geometry; the dipoles are of length L and width W. A rectangular grid is shown, but a triangular grid can be treated by replacing k_y in the following solution by

$$k_y \leftarrow \left(\frac{2\pi n}{b} - \frac{2\pi m}{a \tan \alpha}\right) + k_0 v \qquad (12.30)$$

where α is the skew angle of the vertical columns measured from the x-axis ($\alpha = 90°$ then reduces to the rectangular grid).

Analysis and design considerations for phased-array antennas 707

Using the results of Section 12.2.1.2, the moment method can be applied to a dipole in a single unit cell; by periodicity, all dipoles in the infinite array and their mutual interactions are then accounted for. The $\hat{x}$-electric-surface-current density on the dipole is expanded in a set of piecewise-sinusoidal (PWS) modes:

$$J_{sx}(x, y_0) = \sum_{i=1}^{N} I_i J_i(x_0, y_0) \qquad (12.31)$$

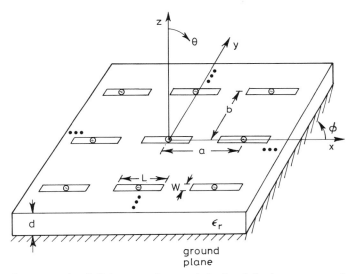

Fig. 12.5 *Geometry of an infinite array of centre-fed printed dipoles on a grounded dielectric substrate*

where

$$J_i(x_0, y_0) = f_p(x_0, x_i) f_u(y_0) \qquad (12.32)$$

is the ith expansion mode, and

$$f_p(x_0, x_i) = \frac{\sin k_e(h - |x_0 - x_i|)}{\sin k_e h} \text{ for } |x_0 - x_i| < h \qquad (12.33)$$

is a piecewise-sinusoidal expansion mode with terminals at x_i, and a half-length of h, and

$$f_u(y_0) = \frac{1}{W} \text{ for } |y_0| < W/2 \qquad (12.34)$$

is a uniform distribution representing the current variation across the width of the dipole. An edge condition could be incorporated here, but past experience has shown that this is generally not worth the trouble. In eqn. 12.32, the wave number k_e can be arbitrarily chosen; here it is set to $k_e = k_0 \sqrt{(\varepsilon_r + 1)/2}$, which has been found to give good results.

708 Analysis and design considerations for phased-array antennas

After the expansion modes have been selected, a Galerkin moment-method procedure can be applied to the electric-field integral equation that enforces $E_{tan} = 0$ over the surface of a dipole in one unit cell. The other boundary conditions at the ground plane and dielectric–air interface are guaranteed to be satisfied through the use of the Green's function of eqn. 12.22a. A general impedance matrix element, representing the coupling of expansion mode j to weight mode i, is defined as

$$Z_{ij} = - \int_{S_i} J_i(x, y) \int_{S_j} E_x(x, y, d) J_j(x_0, y_0) \, dx_0 \, dy_0 \, dx \, dy \qquad (12.35)$$

where $E_x(x, y, d)$ is the field at (x, y, d) due to a periodic array of infinitesimal dipoles, as given by eqn. 12.22a. The space integratons in eqn. 12.35 can be evaluated as the Fourier transforms of the expansion and weighting modes to give the following:

$$Z_{ij} = \frac{-1}{ab} \sum_{m=-\infty}^{\infty} \sum_{n=-\infty}^{\infty} F_{p_i}(k_x) F_u(k_y) G_{xx}^{EJ}(k_x, k_y) F_{p_j}^*(k_x) F_u^*(k_y) \qquad (12.36)$$

where F_{p_i} and F_u are the Fourier transforms of the f_p and f_u functions, as defined by

$$F(k_x) = \int_{-\infty}^{\infty} f(x) e^{-jk_x x} \, dx \qquad (12.37)$$

Evaluating eqn. 12.37 for f_p of eqn. 12.33 and f_u or eqn. 12.34 gives

$$F_{p_i}(k_x) = \frac{2k_e[\cos k_x h - \cos k_e h]}{\sin k_e h (k_e^2 - k_x^2)} e^{-jk_x x_i} \qquad (12.38)$$

$$F_u(k_y) = \frac{\sin k_y W/2}{k_y W/2} \qquad (12.39)$$

If there are N (odd) PWS expansion modes used on each dipole, the voltage-vector elements can be defined as

$$V_i = \begin{cases} 1 & \text{for } i = (N+1)/2 \\ 0 & \text{otherwise} \end{cases} \qquad (12.40)$$

Then in matrix form the expansion coefficients I_i can be found from

$$[Z][I] = [V] \qquad (12.41)$$

The input impedance at any dipole in the array is then

$$Z_{in}(\theta, \phi) = V_k / I_k \qquad (12.42)$$

where $k = (N+1)/2$ is the index of the mode at the dipole terminals. The active reflection coefficient is calculated as

$$R(\theta, \phi) = \frac{Z_{in}(\theta, \phi) - Z_{in}(0, 0)}{Z_{in}(\theta, \phi) + Z_{in}^*(0, 0)} \qquad (12.43)$$

so that the array is conjugate-matched to its broadside scan impedance.

For typical grid spacings a, b of the order of $\lambda/2$, the series in eqn. 12.36 converges for upper limits of about $m, n = \pm 60$. This makes computational efficiency an issue that requires some attention. One thing that helps is to recognise the Toeplitz-like symmetry in the $[Z]$ matrix. Since the Green's function of eqn. 12.22a is not reciprocal in terms of interchanging x_0, y_0 with x, y the $[Z]$ matrix of eqn. 12.36 is not symmetric. But if the N PWS expansion modes are laid out uniformly on the dipole, other symmetries of the form $Z_{ij} = Z_{i-1, j-1}$, for $1 < i \leqslant j$, allow the entire matrix to be filled by computing only the first row and the first column of the matrix ($2N - 1$ elements), rather than all N^2 elements. Additional time savings can be obtained by careful writing of the computer code. In particular, it is relatively easy to write the program so that the G_{xx}^{EJ} function of eqn. 12.36 is computed only once for a new set of k_x and k_y values, and not recalculated for each impedance-matrix element. The Fourier-transform functions in eqn. 12.36 can similarly be handled. Such techniques result in an efficient computer program: a given matrix element takes less than 4 s of CPU time, and the input impedance using three PWS modes can be calculated for one scan angle in less than 30 s. (These CPU times are for a Micro VAX/II computer.) The infinite-array solution is considerably faster than a full-wave solution for a single dipole.

It is interesting to note that the solution for an isolated dipole can be recovered from the infinite-array solution by an integration over all scan angles defined by a unit square on the grating-lobe diagram ($|u| < 1$, $|v| < 1$), for $a = b = \lambda_0/2$.

We now present some results for the scan performance of infinite dipole arrays. Fig. 12.6 shows the magnitude of the reflection coefficient of a dipole array with grid spacings $a = b = 0.5\lambda_0$, on a substrate with $d = 0.19\lambda_0$ and $\varepsilon_r = 2.55$. This substrate is thicker than is usually used in practice, but serves to clearly illustrate some of the important scan effects. The dipole is matched at broadside scan ($\theta = 0°$) with an input impedance of $75 + j0\Omega$, and curves are shown for E-plane scan ($\phi = 0°$), H-plane scan ($\phi = 90°$) and a diagonal (D-plane) scan ($\phi = 45°$). Note that all curves tend to unity as $\theta \to 90°$, and that a scan blindness exists in the E-plane at $\theta = 45.8°$. The substrate supports a surface wave with $\beta_{sw}/k_0 = 1.283$. Two solutions to eqn. 12.29 are possible in the principal planes:

(a) $m = -1, n = 0; u = 0.717, v = 0$

(b) $m = 0, n = -1; u = 0, v = 0.717$

Solution (a) leads to the blindness seen in the E-plane at $\theta = 45.8°$. Solution (b) would lead to a blindspot in the H-plane at $\theta = 45.8°$; but $k_x = 0$, so the TM surface-wave pole is cancelled, as discussed in Section 12.2.1.3. Other solutions to eqn. 12.29, however, are possible off the principal planes. Fig. 12.7 shows a contour plot of the reflection-coefficient magnitude in the u/v scan plane for this same array. Note the two semi-circular loci of the unity reflection-coefficient magnitude, with one starting in the E-plane at $\theta = 45.8°$ and leaving visible space at $\phi = 32.7°$; and the other entering visible space at $\phi = 45.8°$. The

710 Analysis and design considerations for phased-array antennas

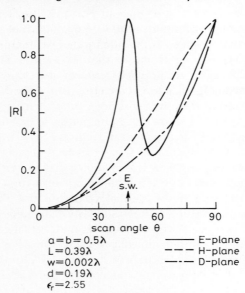

Fig. 12.6 *Reflection-coefficient magnitude versus scan angle for an infinite printed dipole array*
$a = b = 0.5\lambda_0$, $d = 0.19\lambda_0$, $\varepsilon_r = 2.55$, $L = 0.39\lambda_0$, $W = 0.002\lambda_0$

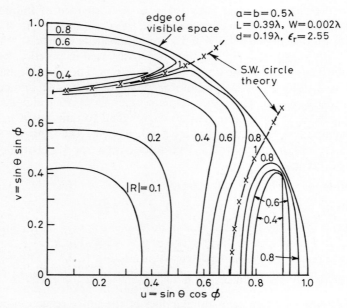

Fig. 12.7 *Contour plot of the reflection-coefficient magnitude in the u/v plane for the array of Fig. 12.6*

region of large reflection gets smaller and smaller as the H-plane is approached, however, until the unity-magnitude region vanishes entirely at the H-plane. Another way of viewing this effect is with the surface-wave circle diagram shown in Fig. 12.8. This simply-obtained diagram predicts quite accurately the scan blindnesses seen in Fig. 12.6 and 12.7. The reader is referred to Reference 8 for further examples, including the effects of grid spacing, substrate thickness and dielectric constant, and a triangular grid.

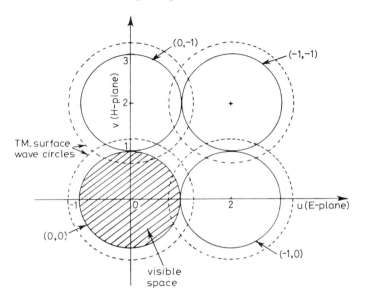

Fig. 12.8 *Surface-wave circle diagram for the array of Fig. 12.6*

The above theory has been experimentally verified by using a waveguide simulator. Waveguide simulators provide a convenient way to test an element in an infinite-array environment, and have been used extensively for waveguide arrays [4, 21]. Here, two printed dipole elements were used in a simulator to test the above theory, and to provide experimental evidence of scan blindness in infinite arrays of printed dipoles.

The simulator was made from a piece of C-band waveguide, with inner dimensions of 2·22 cm × 4·75 cm. Two printed monopoles of length 1·02 cm were laid on a glass-fibre epoxy substrate of thickness $d = 0·95$ cm, with a permittivity $\varepsilon_r = 4·35$. The monopoles were fed with SMA coaxial connectors through the broad wall of the waveguide. The element spacings were $a = 4·43$ cm and $b = 2·38$ cm. By varying the frequency from about 4·0 to 6·3 GHz, the simulator scanned in the H-plane to an angle given by

$$\sin \theta = \lambda/4b$$

Over this frequency range, only the TE_{10}-mode was propagating in the waveguide. A ground plane backed the glass-fibre substrate, and a standard waveguide-matched load was used on the array end. The two printed-antenna elements are fed in phase, with a Wilkinson-type power divider, even though the array is effectively scanning off broadside. This is because the waveguide mode corresponds to two propagating plane waves, one at angle θ and the other at $-\theta$. For proper simulator operation, the array must generate both of these waves which, by superposition, results in in-phase feed voltages.

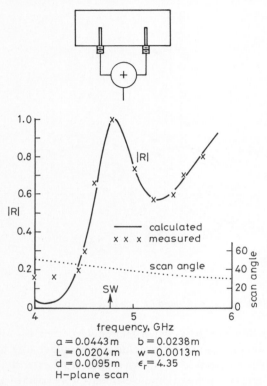

Fig. 12.9 *Measured and calculated active-reflection-coefficient magnitude for a waveguide simulation of an infinite array of printed dipoles*

Fig. 12.9 shows the measured reflection-coefficient magnitudes compared with the calculated values. This reflection coefficient is based on the input impedance of a printed monopole and a 50 Ω system. The agreement is generally quite good; the measured results at 4·0 and 4·2 GHz are somewhat higher than calculated owing to some residual mismatch at the power divider. Since the original measurements were made [8], these results have been improved.

The blind spot is clearly seen at 4·76 GHz, with a measured reflection-coefficient magnitude of 0·96; it is presumed that a non-unity value resulted because of copper and dielectric loss. The blindness in this case is caused by coupling of the TM_0 surface wave to the $m = \pm 1, n = 0$ Floquet mode in the H-plane, as can be seen from the surface-wave circle diagram for the simulator, shown in Fig. 12.10. The diagram shows that, because of the large element spacing in the E-plane and the large diameter of the surface-wave circle, an intersection of the TM_0 surface-wave circle occurs for H-plane scan, and does not result in cancellation because k_x is not zero (since $u = 0$, but $m = \pm 1$).

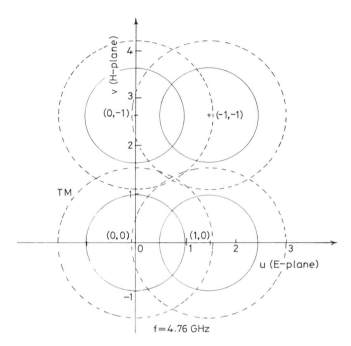

Fig. 12.10 *Surface-wave circle diagram for the dipole array simulator of Fig. 12.9*

A slightly different solution for the infinite array of printed dipoles has been reported in Reference 20. Results from that work, which include several waveguide simulator measurements, have been favourably compared with those from the present solution. The analysis in Reference 20 uses a singularity subtraction technique in the space domain to improve computational efficiency, and may be limited to substrates that are not too thin.

In Reference 19, a solution is described for a planar array of printed dipoles with a superstrate (cover layer). The method of analysis is similar to that presented above. Also treated in Reference 19 is an infinite layer array of dipoles proximity-coupled to microstrip feed lines.

12.2.2.2 Rectangular probe-fed patches: We now consider the analysis of an infinite array of probe-fed rectangular microstrip patches; the geometry is shown in Fig. 12.11. This solution is based on the work reported in Reference

22, and assumes an idealised probe feed model. This feed model uses a constant-current filament to mode the probe, and does not attempt to model the rapid variation of surface current near the probe–patch junction. Such a model has been found to work quite well for single patches on thin substrates [10, 18, 23, 24], since the patch Q in this case is relatively large, so that the resonant mode current dominates the total current. Reference 25 shows a plot of the currents on a probe-fed patch which graphically illustrates this effect.

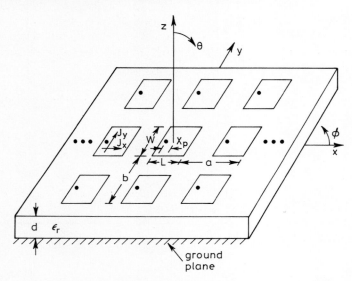

Fig. 12.11 *Geometry of an infinite array of probe-fed rectangular microstrip patches*

The use of this idealised feed model considerably simplifies the analysis, yet provides useful information on the scanning performance of the array and the effects of parameters such as substrate thickness and permittivity, and grid spacings. While the absolute values of active impedance obtained from this solution may only be valid for substrate thicknesses on the order of $0.02\lambda_0$ or less, the active reflection coefficient, being a normalised quantity, has a greater range of validity. In addition, comparisons of patch arrays and printed-dipole arrays with the same substrate parameters and grid spacings show significant similarity in terms of reflection coefficient versus scan angle, suggesting that the element type or feed arrangement is not a dominant factor in the scan performance or active-element patterns. This idea can be pursued further with the current-sheet model discussed in Section 12.2.2.5.

The moment-method theory of the analysis of the probe-fed patch array will be developed below, as an extension of the dipole-array solution of Section 12.2.2.1. Then several calculated results will be presented for the scan performance of infinite patch arrays, followed by some measurements from a waveguide simulator.

Analysis and design considerations for phased-array antennas

For the patch problem it is necessary to use both x and y expansion currents; so in the interest of clarity and conciseness it becomes useful to define dyads representing x and y field components as

$$\overline{\overline{G}}^{EJ} = \hat{x} G_{xx}^{EJ} \hat{x} + \hat{x} G_{xy}^{EJ} \hat{y} + \hat{y} G_{yx}^{EJ} \hat{x} + \hat{y} G_{yy}^{EJ} \hat{y} \tag{12.44}$$

$$\overline{\overline{E}} = \frac{1}{ab} \sum_{m=-\infty}^{\infty} \sum_{n=-\infty}^{\infty} \overline{\overline{G}}^{EJ}(k_x, k_y) e^{-jk_x(x-x_0)} e^{-jk_y(y-y_0)} \tag{12.45}$$

Thus, consistent with the notation of Section 12.2.1.1, G_{pq}^{EJ} represents the $\hat{p}$-component of the electric field due to a $\hat{q}$-directed electric-current source. G_{xx}^{EJ} and G_{yx}^{EJ} are given by eqns. 12.11a and b and G_{yy}^{EJ} and G_{xy}^{EJ} can be found from eqns. 12.11a and b by interchanging k_x and k_y. The results are

$$G_{yy}^{EJ} = \frac{-jZ_0}{k_0} \frac{(\varepsilon_r k_0^2 - k_y^2) k_2 \cos k_1 d + jk_1(k_0^2 - k_y^2) \sin k_1 d}{T_e T_m} \sin k_1 d \tag{12.46a}$$

$$G_{xy}^{EJ} = \frac{jZ_0}{k_0} \frac{k_x k_y \sin k_1 d [k_2 \cos k_1 d + jk_1 \sin k_1 d]}{T_e T_m} \tag{12.46b}$$

Note that $G_{xy}^{EJ} = G_{yx}^{EJ}$. We will also require the E_z field due to $\hat{x}$ and $\hat{y}$ currents; from Section 12.2.1.1 these can be derived as

$$G_{zx}^{EJ} = \frac{Z_0 k_x k_2 \sin k_1 d}{k_0 k_1 T_m} \tag{12.47a}$$

$$G_{zy}^{EJ} = \frac{Z_0 k_y k_2 \sin k_1 d}{k_0 k_1 T_m} \tag{12.47b}$$

By reciprocity, $G_{xz}^{EJ} = -G_{zx}^{EJ}$, and $G_{yz}^{EJ} = -G_{zy}^{EJ}$. The E_z fields represented by eqns. 12.47a and b have been integrated over z for $0 \leq z \leq d$, since this is the form in which they will be used later.

The electric-field integral equation, representing the boundary condition that the total tangential electric field must vanish on the patch conductor, can be written as

$$E_{tan}^{inc} = -E_{tan}^{scat} = -\int_S \overline{\overline{E}} \cdot J_s \, ds \tag{12.48}$$

where E_{tan}^{inc} is the tangential ($\hat{x}, \hat{y}$) component of the incident electric field due to the probe source evaluated at $z = d$, J_s is the total vector electric surface-current density on the patch (the sum of the currents on the top and bottom patch surfaces), S is the patch surface, and E_{tan}^{scat} is the tangential field scattered by the patch. The surface-current density J is now expanded in a set of basis functions as

$$J_s(x_0, y_0) = \sum_j I_j J_j(x_0, y_0) \tag{12.49}$$

where J_j is an expansion mode representing current flow in either the $\hat{x}$ or $\hat{y}$ direction, and I_j is the unknown coefficient. Substitution of eqn. 12.49 into eqn. 12.48, multiplication by a weighting function J_i and integration over s yields, for $i = 1, 2, 3 \ldots N$,

$$\int_s J_i \cdot E_{tan}^{inc} ds = -\sum_j I_j \int_s \int_s J_i \cdot \bar{\bar{E}} \cdot J_j \, ds \, ds \tag{12.50}$$

We can now define an impedance matrix element as

$$Z_{ij} = -\int_s \int_s J_i \cdot \bar{\bar{E}} \cdot J_j \, ds \, ds \tag{12.51}$$

and a voltage-vector element as

$$V_i^t = \int_s J_i \cdot E_{tan}^{inc} \, ds \tag{12.52}$$

where the superscript t indicates that this voltage-vector term is based on a test, or weighting, mode. These matrix elements can be written, using eqn. 12.45, as

$$Z_{ij} = \frac{-1}{ab} \sum_{m=-\infty}^{\infty} \sum_{n=-\infty}^{\infty} F_i(k_x, k_y) \cdot \bar{\bar{G}}^{EJ}(k_x, k_y) \cdot F_j^*(k_x, k_y) \tag{12.53}$$

$$V_i^t = \frac{-1}{ab} \sum_{m=-\infty}^{\infty} \sum_{n=-\infty}^{\infty} F_i(k_x, k_y) \cdot [\hat{x} G_{xz}^{EJ}(k_x, k_y) + \hat{y} G_{yz}^{EJ}(k_x, k_y)]$$

$$\times e^{jk_x x_p} e^{jk_y y_p} \tag{12.54}$$

where x_p, y_p are the co-ordinates of the feed probe, and F_i represents the Fourier transform of the J_i expansion mode, defined as

$$F_i(k_x, k_y) = \int_s \int_s J_i(x_0, y_0) e^{-jk_x x_0} e^{-jk_y y_0} dx_0 \, dy_0 \tag{12.55}$$

(*Note*: Because this definition of the Fourier transform differs from that of Reference 22 in the signs of the exponential terms, some of the above results differ from those of Reference 22 by conjugation.)

The unknown expansion coefficients I_j can then be found as solutions to the following set of linear equations:

$$\sum_j Z_{ij} I_j = V_i^t, \quad \text{for all } i \tag{12.56}$$

Note that $[Z]$ is not a symmetric matrix. The input impedance at the probe can be calculated as

$$Z_{in} = \frac{-1}{I_p^2} \int_0^h \hat{z} \cdot E^{scat} \, dz \tag{12.57}$$

where I_p is the current on the probe which, if $d \ll \lambda_0$, is assumed to be uniform along the probe, and will be chosen as 1 A. Then from eqn. 12.57, 12.49, 12.48 and 12.45 we can write

$$Z_{in} = -\sum_j I_j V_j^e \tag{12.58}$$

Analysis and design considerations for phased-array antennas 717

where a voltage-vector element based on expansion mode j has been defined as

$$V_j^e = \frac{-1}{ab} \sum_{m=-\infty}^{\infty} \sum_{n=-\infty}^{\infty} [G_{zx}^{EJ}(k_x, k_y)\hat{x} + G_{zy}^{EJ}(k_x, k_y)\hat{y}] \cdot \boldsymbol{F}_j^*(k_x, k_y)$$
$$\times e^{-jk_x x_p} e^{-jk_y y_p} \qquad (12.59)$$

In words, V_i^t represents a voltage based on the field from the probe integrated over a surface-test mode, while V_j^e represents a voltage based on the field from a surface-expansion mode integrated over the probe. Thus $V_j^e \neq V_j^t$ in general.

The above solution may be described as a Galerkin solution in the spectral domain, since the matrix elements are expressed in terms of the Fourier transforms of the fields and currents. The probe self-reactance has been ignored here.

The next step is to choose the expansion/weighting functions. Because of their correspondence with the cavity model, entire domain modes of the following form were used:

$$J_i(x, y) = \hat{x} \sin \frac{k\pi}{L} (x + L/2) \cos \frac{l\pi}{W} (y + W/2) \qquad (12.60a)$$

for $\hat{x}$ currents, and

$$J_i(x, y) = \hat{y} \cos \frac{k\pi}{L} (x + L/2) \sin \frac{l\pi}{W} (y + W/2) \qquad (12.60b)$$

for $\hat{y}$ currents, where k and l are integer indices accounting for the number of variations in the x and y directions, respectively. The Fourier transforms of these modes can be easily calculated analytically from eqn. 12.55.

Through numerical convergence checks it was found that, for E-plane scanning, the $(k, l) = (1, 0), (3, 0), (5, 0), (7, 0)$ $\hat{x}$-directed currents and the $(k, l) = (0, 2)$ $\hat{y}$-directed current gave a fairly stable solution, while for off E-plane scan the $(0, 1)$ $\hat{y}$-directed current mode should also be included. The edge condition could be incorporated into the above currents, but it has been found that this is an unnecessary expense in a Galerkin-type solution for patches [10, 23].

Fig. 12.12 shows the reflection-coefficient magnitude versus scan angle for an infinite array of probe-fed microstrip patches on a substrate with $\varepsilon_r = 2.55$ and $d = 0.06\lambda_0$, but the E-plane spacing is $a = 0.51\lambda_0$ which leads to a grating lobe at $\theta = 73°$. The blindspot position in the E-plane then occurs at $68.8°$. Because of the presence of $\hat{y}$-directed currents, the H-plane also shows a blind angle, at $\theta = 76.4°$; this blindness has a much higher Q than the E-plane blindness since these $\hat{y}$-directed currents are highly reactive and radiate little power. In practice, any loss, probe radiation, or random-error effects would probably 'wash-out' this spurious H-plane blindness.

Fig. 12.13 shows the behaviour of an infinite patch array on a thin high-dielectric-constant substrate having $d = 0.02\lambda_0$ and $\varepsilon_r = 12.8$. Because of the thinness of the substrate, the blind spots in the E- and H-planes now occur at

718   Analysis and design considerations for phased-array antennas

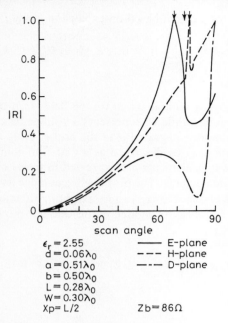

Fig. 12.12 *Reflection-coefficient magnitude versus scan angle for an infinite array of rectangular microstrip patches. $d = 0.06\lambda_0$, $\varepsilon_r = 2.55$*

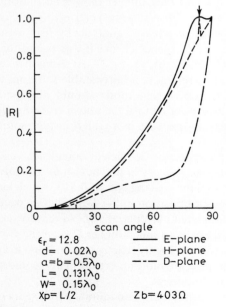

Fig. 12.13 *Reflection-coefficient magnitude versus scan angle for an infinite array of rectangular microstrip patches. $d = 0.02\lambda_0$, $\varepsilon_r = 12.8$*

$\theta = 82 \cdot 9°$. This result shows that, even though a scan blindness may be present in the visible range of an array, it may have little effect on the scan range if it is sufficiently close to endfire. The reader is referred to Reference 22 for further examples.

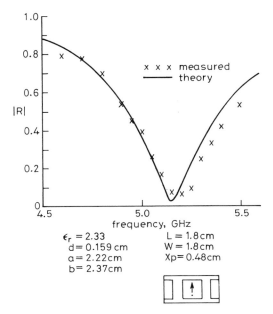

Fig. 12.14 *Measured and calculated reflection-coefficient magnitude of a microstrip array in a waveguide simulator. The centre patch is fed*

The above theory has been verified with several waveguide simulator measurements, and a typical example is shown in Fig. 12.14. Unlike the dipole simulator described in Section 12.2.2.1, this simulator uses only one fed element centered in the guide, with two half-patches at the sides. These half-patches carry zero feed current. Good agreement with theory is obtained, although it should be pointed out that the data of Fig. 12.14 is dominated by the high-Q resonance of the patch, and differs only slightly from the frequency-dependent behaviour of the isolated patch element. It should also be noted that, although the patches image correctly across the waveguide simulator walls, the feed probes do not. This is felt to be a negligible factor for electrically thin substrates.

Another solution for the probe-fed patch array has been reported in Reference 15. This work uses a singularity subtraction technique for the patch current near the feed point, and so overcomes the main drawback of the solution in Reference 22, which is limited to thin substrates because of the idealised feed model. Reference 15 presents some useful design data for patch arrays, and interprets the behaviour of such arrays in terms of surface-wave and leaky-wave effects. It erroneously presumes, however, that the solution of Reference 22 is missing leaky-wave effects, which come about because of the patches loading the dielectric slab. The solution of Reference 22 is full-wave and accounts for the presence of the patches on the dielectric slab through the moment-method procedure, and comparisons with some of the results in Reference 15 show the same effects which are labeled as 'leaky waves' in Reference 15. A recent paper [47] describes experimental confirmation of surface-wave scan blindnesses in large arrays of microstrip patches, in agreement with the theory of Reference 22.

12.2.2.3 Circular probe-fed patches: Like the rectangular patch, the circular microstrip antenna is often used as an array element, and so we describe here the extension of the previous analysis to an infinite array of circular patches. This case clearly illustrates the versatility of the Green's-function/moment method (or 'Galerkin's method in the spectral domain,' as some authors have referred to it) which has been presented in the preceding Sections. As we will see, the only major change needed to treat circular patches is to use the appropriate expansion modes and Fourier transforms of those modes [26].

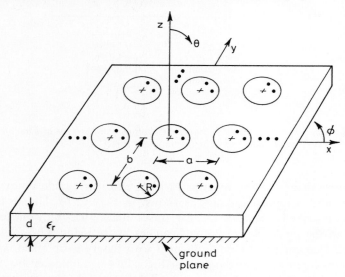

Fig. 12.15 *Geometry of an infinite array of circular probe-fed microstrip patches*

The geometry of the infinite circular patch array is shown in Fig. 12.15. A rectangular grid is assumed, but a triangular grid can be easily treated. As in the rectangular-patch case, the probe-feed model here is also idealised in that it does not attempt to mode the singularity in patch current near the probe, thus limiting the solution to thin substrates. The arguments presented in Section 12.2.2.2, concerning the utility and justification for such an approximation, then apply here, as well.

Analysis and design considerations for phased-array antennas

The analysis for the infinite array of circular patches follows identically the solution for the rectangular-patch case of Section 12.2.2.2, until the expansion modes are chosen. Analogous to the rectangular-patch case, we look to the cavity model and select the TM_{pq} circular waveguide modes as expansion functions. If we assume the feed point of the reference patch lies on the $\phi = 0$ line, the ith expansion mode can be written in cylindrical co-ordinates as

$$J_{si}(\varrho_0, \phi_0) = \hat{\varrho}_0 [\beta_{p_i q_i} J'_{p_i}(\beta_{p_i q_i} \varrho) \cos p_i \phi]$$

$$- \hat{\phi} \left[\frac{p_i}{\varrho} J_{p_i}(\beta_{p_i q_i} \varrho) \sin p_i \phi \right] \quad (12.61)$$

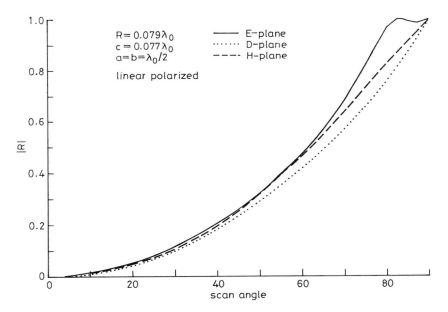

Fig. 12.16 *Reflection-coefficient magnitude versus scan angle for an infinite array of probe-fed circular microstrip patches*
$d = 0{\cdot}02\lambda_0$, $\varepsilon_r = 12{\cdot}8$. Single-probe feed.

where $\beta_{pq} R$ is the qth zero of $J'_p(x)$, where $J_p(x)$ is the Bessel function of order p, and R is the radius of the circular patch elements. In eqn. 12.61, the single index i is used to form a one-dimensional sequence of the TM_{pq} modes. To apply eqns. 12.51–12.59, the Fourier transform of the above expansion modes is needed. These expressions can be derived in closed-form [26], but are too lengthy to list here.

Fig. 12.16 shows a typical result for the scan performance of an infinite circular-patch array. The grid spacing ($a = b = \lambda_0/2$) and substrate parameters ($d = 0{\cdot}02\lambda_0$, $\varepsilon_r = 12{\cdot}8$) are the same as the rectangular-patch case shown in

Fig. 12.13, and it is interesting to observe that the results are practically identical. Only in the diagonal plane is there much difference between the rectangular- and circular-patch results.

Multiple-probe-fed patches are also of practical interest. Fig. 12.17 shows three common feeding circuits for circular patches, and these techniques are also relevant for rectangular (or square, for circular polarisation) microstrip patches. Fig. 12.17a shows the single probe-fed patch which has already been treated. The two-probe case of Fig. 12.17b is fed with a 180° hybrid, which reduces the amount of cross-polarised radiation. In Fig. 12.17c the two feed probes are in orthogonal planes and are fed with a quadrature hybrid to generate circular polarisation. The theory which has been presented above can easily be extended to handle the two-probe feed cases of Figs. 12.17b and c.

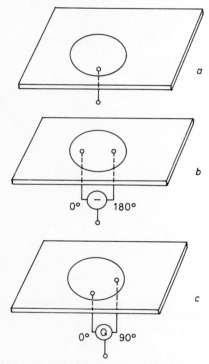

Fig. 12.17 *Three common feed circuits for patch antennas*
    a Single probe feed
    b Balanced 180° hybrid feed for reduced cross-polarisation
    c Quadrature hybrid feed for circular polarisation

A given element in the infinite-array environment can be treated as a two-port network, with an open-circuit 'port' impedance matrix of the form

$$[Z^p] = \begin{bmatrix} Z^p_{11} & Z^p_{12} \\ Z^p_{21} & Z^p_{22} \end{bmatrix} \quad (12.62)$$

Analysis and design considerations for phased-array antennas 723

where the matrix elements can be determined from

$$Z^p_{kl} = -\int_{v_k} \boldsymbol{E}_k \cdot \boldsymbol{J}_l \, dv_k \tag{12.63}$$

which gives the open-circuit voltage induced on the k th feed probe due to the fields excited by a unit current on the l th feed probe. After the matrix of eqn. 12.62 is found (which is a generalisation of the single-port input impedance of eqn. 12.57), an equivalent circuit which models the feed network of either Fig. 12.17b or c can be used to find the reflection coefficients R_1 and R_2, seen looking into the antenna element ports. If the hybrids have isolation, then, in general, some of the reflected power from the antenna element will be dissipated in the hybrid and some will pass back through the hybrid. Thus the reflection coefficient R at the input of the hybrid does not account for this lost (non-radiated) power. A better indication is to plot the active-element gain pattern, including the efficiency of the feed network, defined as

$$G(\theta, \phi) = (1 - |R(\theta, \phi)|^2)\eta(\theta, \phi)\cos\theta \tag{12.64}$$

where η is the feed-network efficiency:

$$\eta(\theta, \phi) = 1 - \frac{|R_1 + R_2 \, e^{-2j\alpha}|^2}{4(1 - |R|^2)} \tag{12.65}$$

and α is the phase angle between the two feed ports (either 90° or 180°). Since R_1 and R_2 in eqn. 12.65 vary with scan angle, η also varies with scan angle.

Fig. 12.18 shows such an active-element gain pattern for a circularly polarised circular patch array. The element and substrate geometry is the same as that of Fig. 12.16. The resulting axial ratio is shown in Fig. 12.19. Observe that, while the single-probe-fed array of Fig. 12.16 shows a scan blindness of about $\theta = 83°$ in the E-plane, the corresponding circularly polarised array with the two feed probes per element of Fig. 12.18 does not show a blindness at this angle. This is because the reflection coefficient at the feed probe which drives E-plane currents (probe at $\phi = 0$) may have a unity reflection-coefficient magnitude at $\theta = 83°$ in the E-plane, but power can still be delivered to the cross-polarised currents fed by the other feed probe (at $\phi = 90°$). This polarisation, being H-plane directed, is decoupled from the E-plane surface wave. The axial ratio, however, becomes infinite at $\theta = 83°$, as shown in Fig. 12.18. Because of symmetry, this argument applies to both E- and H-plane scan for the circularly polarised circular-patch array.

The above results for circular-patch arrays are preliminary; theoretical work on this problem and experimental verifications are continuing.

12.2.2.4 Aperture-coupled patches: The next type of printed array to be considered is one using aperture-coupled rectangular microstrip patches. The aperture-coupled patch element [27] consists of two substrates, with a ground plane in between. As shown in the geometry for a single aperture-coupled patch in Fig. 12.20, a microstrip feed line is printed on the bottom (feed) substrate,

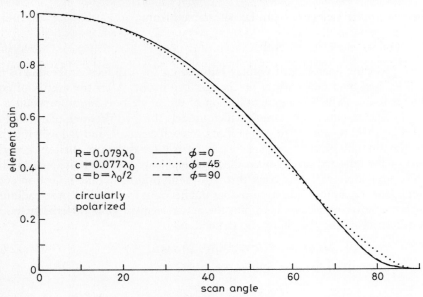

Fig. 12.18 *Active element gain for an infinite array of probe-fed circular patches. $d = 0.02\lambda_0$, $\varepsilon_r = 12.8$*
Two-probe feed with a quadrature hybrid for circular polarisation

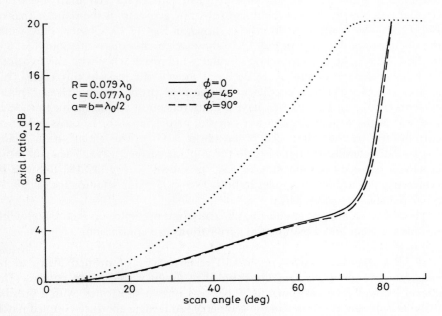

Fig. 12.19 *Axial ratio versus scan angle for the circularly polarised array of Fig. 12.18*

while the patch element is printed on the top (antenna) substrate. Coupling between the feed line and the radiating element is through a small slot in the ground plane below the patch. As will be discussed further in Section 12.3.2, this type of element has a number of attractive features when used in a phased-array configuration.

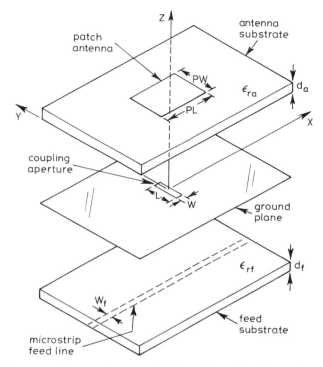

Fig. 12.20 *Geometry of an aperture-coupled microstrip patch antenna element*

The single aperture-coupled patch has been theoretically analysed in References 28 and 29, using Green's-function/moment-method procedures. In Reference 28, the currents on the feed line were expanded in terms of travelling-wave and piecewise sinusoidal expansion modes, following the method of Reference 30. Since this method requires modelling of the feed line over at least several wavelengths, it cannot be applied directly to an infinite phased array without the feed lines running into neighbouring patches or other feed lines. In contrast, the method of Reference 29 first treats the coupling of the aperture fields to the patch, then relates this to an equivalent series impedance seen by the microstrip feed line. This makes the method of Reference 29 a viable approach for treating the infinite array of aperture-coupled patches, since the feed-line interaction is avoided. The solution thus accounts for mutual coupling between the patches, and between the slots, but assumes the feed lines do not couple to each other. This should be a safe assumption for practical microstrip networks.

Aperture coupling is a more complicated way of feeding the patch, as compared to probe feeds or edge feeds (via microstrip line), but it is interesting to note that it is easier to do the analysis of the former case in a more rigorous manner than the latter. This is because the aperture-coupled patch is proximity coupled, without a direct contact, eliminating the difficulty of the patch-current singularity that occurs with probe or edge feeds.

The present analysis follows that of Reference 29, as extended to the infinite array. The $\hat{x}$ and $\hat{y}$ surface currents J_s are expanded in a set of entire domain-basis functions, as in eqns. 12.49 and 12.60a and b. The electric-field integral equation of eqn. 12.48 is then applied to the patch surface, where E^{inc} is the field radiated by the equivalent magnetic current

$$M_{sy}(x_0, y_0) = -e_x^a(x_0, y_0) \tag{12.66}$$

where $e_x^a(x_0, y_0)$ is the unknown aperture electric field. Since the coupling aperture is electrically small, a good approximation is to model the field distribution with a single PWS mode:

$$e_x^a(x_0, y_0) = f_u(x_0)f_p(y_0, 0) \tag{12.67}$$

where f_u denotes a uniform distribution of field across the width (narrow dimension) of the coupling aperture, as defined in eqn. 12.34, and f_p denotes a PWS distribution across the length (long dimension) of the aperture, as defined in eqn. 12.33. If there is some compelling reason to do so, additional PWS modes could be included, as discussed in Reference 29.

The electric-field integral equation 12.48 then can be reduced to matrix form as,

$$[V^M] = [Z][I] \tag{12.68}$$

where $[I]$ is the column vector of unknown patch expansion mode coefficients, $[Z]$ is the impedance matrix of the patch with elements given by eqn. 12.53, and $[V^M]$ is the voltage vector due to an excitation of magnetic current in the aperture. The elements of $[V^M]$ are

$$\begin{aligned} V_i^M &= \int_s \boldsymbol{J}_i \cdot \boldsymbol{E}^{inc}\, ds \\ &= \frac{1}{ab} \sum_{m=-\infty}^{\infty} \sum_{n=-\infty}^{\infty} \boldsymbol{F}_i(k_x, k_y) \cdot \boldsymbol{G}_y^{EJ}(k_x, k_y) F_u^*(k_x) F_p^*(k_y) \end{aligned}$$

$$(12.69)$$

where $\boldsymbol{F}_i$, F_u, and F_p are previously defined Fourier transforms, and $\boldsymbol{G}_y^{EM} = \hat{x} G_{xy}^{EM} + \hat{y} G_{yy}^{EJ}$ is a Green's function representing the $\hat{x}$ and $\hat{y}$ electric field at $z = d$ due to a $\hat{y}$-directed infinitesimal magnetic current element at

$z = 0$. The individual components are [29]:

$$G_{xy}^{EM} = \frac{jk_x^2(\varepsilon_r - 1)\sin k_1 d}{T_e T_m} + \frac{k_1}{T_e} \qquad (12.70a)$$

$$G_{yy}^{EM} = \frac{-jk_x k_y(\varepsilon_r - 1)\sin k_1 d}{T_e T_m} \qquad (12.70b)$$

In eqn. 12.69 s represents the patch surface.

Now an aperture admittance Y^p can be defined as the reaction of the aperture field and the field scattered by the patch:

$$Y^p = \int_{S_a} e_x^a(x, y) H_y(x, y)\, ds_a = [V^J]^t[I] \qquad (12.71)$$

where H_y is the magnetic field at the aperture due to the currents on the patch, S_a represents the aperture surface, and the elements of the voltage vector $[V^J]$ defined as

$$V_i^J = \int_{S_a}\int_S e_x^a(x, y) H_y \cdot J_i\, ds\, ds_a$$

$$= \frac{1}{ab}\sum_{m=-\infty}^{\infty}\sum_{n=-\infty}^{\infty} F_p(k_y) F_u(k_x) G_y^{HJ}(k_x, k_y) \cdot F_i^*(k_x, k_y) \qquad (12.72)$$

where $G_y^{HJ} = G_{yx}^{HJ}\hat{x} + G_{yy}^{HJ}\hat{y}$ is a Green's function representing the H_y field at $z = 0$ due to an $\hat{x}$ or $\hat{y}$-directed infinitesimal electric dipole at $z = d$. The individual components are [29]:

$$G_{yx}^{HJ} = \frac{-jk_x^2(\varepsilon_r - 1)\sin k_1 d}{T_e T_m} + \frac{k_1}{T_e} \qquad (12.73a)$$

$$G_{yy}^{HJ} = \frac{-jk_x k_y(\varepsilon_r - 1)\sin k_1 d}{T_e T_m} \qquad (12.73b)$$

The coupling aperture also has a self-admittance Y^s caused by the direct radiation of the aperture on either side of the ground plane. For an assumed aperture field of the form of eqn. 12.67, we have

$$Y^s = \int_{S_a} e_x^a(x, y) H_y^M\, ds_a$$

$$= \frac{1}{ab}\sum_{m=-\infty}^{\infty}\sum_{n=-\infty}^{\infty} F_u^2(k_x) F_p^2(k_y)[G_{yy}^{HM+}(k_x, k_y) + G_{yy}^{HM-}(k_x, k_y)] \qquad (12.74)$$

where $G_{yy}^{HM\pm}$ refers to the Green's function representing the H_y field radiated on either the top $(+)$ side or the bottom $(-)$ side of the ground plane, due to a $\hat{y}$-directed magnetic current in the aperture. These two terms are similar in form, but may have different values if different substrates are used on the two sides of the ground plane. From Reference 29, the contribution from one side of the ground plane is given as

$$G_{yy}^{HM} = \frac{1}{k_0 Z_0} \left[\frac{(\varepsilon_r k_0^2 - k_y^2) k_1 \cos k_1 d + j k_2 \varepsilon_r \sin k_1 d)}{k_1 T_m} - \frac{k_1 k_y^2 (\varepsilon_r - 1)}{T_e T_m} \right] \quad (12.75)$$

Then, as derived in Reference 29, the slot-coupled patch antenna appears to the microstrip feed line as a series impedance Z, where Z is given by

$$Z = Z_c \frac{\Delta v^2}{Y^p + Y^s} \quad (12.76)$$

where Y^p and Y^s are given by eqns. 12.71 and 12.74, Z_c is the characteristic impedance of the feed line, and Δv is a modal voltage due to the discontinuity of the slot. From Reference 29, Δv is given by

$$\Delta v = \int_{S_a} e_x^a(x, y) h_y(x, y) \, ds_a \quad (12.77)$$

where $h_y(x, y)$ is the normalised magnetic field of the quasi-TEM microstrip-line mode [29]:

$$h_y(x, y) = \frac{1}{2\pi \sqrt{Z_c}} \int_{-\infty}^{\infty} F_u(k_y) G_{yx}^{HJ}(k_x = -\beta_m, k_y) e^{-j\beta_m x} e^{jk_y y} \, dk_y \quad (12.78)$$

where β_m is the propagation constant of the microstrip line. The equivalent circuit is shown in Fig. 12.21. A tuning stub is generally used to terminate the feed line and to adjust the impedance match of the antenna; this is easily treated via the equivalent circuit of Fig. 12.21.

Figs. 12.22 and 12.23 show results for an aperture-coupled array geometry with an antenna substrate having $\varepsilon_r = 2.55$ and $d = 0.02\lambda_0$, and a feed substrate having $\varepsilon_r = 12.8$ and $d = 0.05\lambda_0$. The feed substrate was intentionally made thicker than usual to show surface-wave resonances; a thinner substrate would move the feed-substrate resonance closer to endfire. Fig. 12.22 shows the reflection-coefficient magnitude versus scan angle, where surface wave blind spots are seen to occur at $\theta = 62°$ and at 86° in the E-plane. The former is due to surface-wave excitation on the feed substrate (from the coupling slots), while the latter is due to surface-wave excitation on the antenna substrate (from the slots and the patches). The individual contributions of the patch and slot to this phenomenon are shown in Fig. 12.23, where the real and imaginary components of the patch (Y^p) and slot (Y^s) admittances as seen by the microstrip feed line at the coupling slot are plotted against E-plane scan angle. The patch admittance, $Y^p = G^p + jB^p$, is seen to have a resonance with a near-zero real part at about $\theta = 85°$. The slot looks into both the antenna substrate and the feed substrate, and so the slot admittance Y^s has been separated into an antenna substrate component, $G^{sa} + jB^{sa}$, and a feed substrate component, $G^{sf} + jB^{sf}$.

The former has a resonance with an infinite susceptance at $\theta = 86°$, while the latter has a resonance with an infinite susceptance at about $\theta = 62°$. Thus, surface-wave resonances are possible on both the feed and the antenna substrate, but these blind spots can be moved closer to endfire by making the substrates thinner.

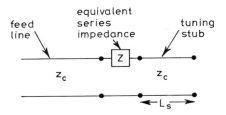

Fig. 12.21 *Equivalent circuit of a stub-tuned aperture-coupled microstrip patch element*

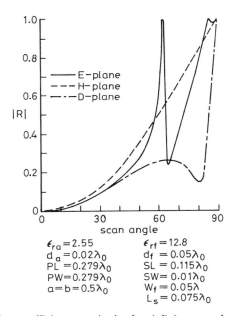

$\epsilon_{ra} = 2.55$ $\epsilon_{rf} = 12.8$
$d_a = 0.02\lambda_0$ $d_f = 0.05\lambda_0$
$PL = 0.279\lambda_0$ $SL = 0.115\lambda_0$
$PW = 0.279\lambda_0$ $SW = 0.01\lambda_0$
$a = b = 0.5\lambda_0$ $W_f = 0.05\lambda$
  $L_s = 0.075\lambda_0$

Fig. 12.22 *Reflection-coefficient magnitude of an infinite array of aperture-coupled patches*

The results from this solution, which are seen to be qualitatively similar to those of the probe-fed patch solutions of Sections 12.2.2.2 and 12.2.2.3, lend credibility to the feed-model approximations used in those solutions.

12.2.2.5 Other geometries: We have analysed several of the most popular types of printed phased arrays in the preceding Sections, but some other results could not be presented here because of space limitations. We will briefly discuss these results, and refer the interested reader to the literature for more details.

After studying a number of different phased-array geometries of dipoles and patches, it becomes evident that many of the dominant characteristics of printed phased arrays are controlled by the element spacing and substrate parameters,

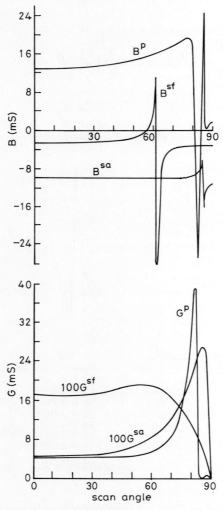

Fig. 12.23 *Active slot admittances of the patch and slot elements for the array of Fig. 12.22*

as opposed to the specific type of radiating element or feeding technique. This observation led to a current-sheet model of a printed phased array [31], based on an extension of some early work by Wheeler [32]. The purpose of this model was not to generate a solution or detailed data that was specific for any printed

Analysis and design considerations for phased-array antennas

phased array, but to develop a simple analysis that could predict the major trends in the scanning performance of a general printed phased array. Such a model can enhance our understanding of the operation of such antennas. A variety of results, given in Reference 31, show how the relatively simple current-sheet model can be used to predict many of the characteristics, such a reflection-coefficient variation with scan angle and scan blindnesses, of various types of phased arrays.

Another type of phased-array geometry is analysed in Reference 33, where elements are printed on a semi-infinite substrate (half-space). Both dipole and slot elements were considered, and the analysis employed a Green's-function/moment-method procedure similar to the above work. The motivation for such a study was the possible integration of array elements and circuitry at the surface of an electrically large dielectric lens. Several problems with this configuration, including poor scan performance, have precluded it from further consideration for general applications.

12.2.3 Finite-array solutions

Of necessity, all practical phased arrays are finite in size, and so it is important to determine the efficacy of the infinite-array assumption. If it is very large, the central elements of a finite array are generally modelled quite well by the infinite-array approximation. The infinite-array solution, however, does not account for edge effects in a finite array, and it is generally not known *a priori* how 'big' a finite array has to be before it can be reasonably modelled as infinite. Thus, the analysis of finite arrays may be of more practical utility than either the analysis of isolated elements (no mutual coupling), or infinite arrays (no edge effects).

Another reason for considering finite arrays of printed antennas concerns the role of surface waves. It has been shown [6, 9, 12, 18] that a single printed antenna element can convert a significant fraction of its input power into surface waves, as opposed to radiated power. On the other hand, surface waves cannot exist on infinite phased arrays except at blindness angles, where all input power is converted to surface-wave power, and no radiation leaves the surface of the array. The question then arises as to the effect of array size on the generation of surface-wave power. Does a finite array of printed dipoles, for example, excite more or less surface-wave power than a single printed dipole on the same substrate? And how does this power vary with array size, and scan angle? The analysis of finite printed arrays provides answers to these questions, and clearly shows the relation of surface-wave excitation to the scan-blindness phenomenon.

Finite arrays, however, are considerably more difficult to analyse than either single elements or infinite arrays. Using the 'element-by-element' approach [3], the mutual coupling between each pair of elements in the array must be calculated, and matrices of order equal to the number of elements in the array (or larger, if there is more than one expansion mode per element) must be inverted.

The size of arrays that can be handled by this method is thus quite limited. It is important to realise that, even though mutual coupling is calculated between pairs of open-circuited elements, the complete solution includes the effect of terminations, and is completely rigorous in the moment-method sense.

In the following Section we present the solution for a finite array of printed dipoles [35]. The key step in this analysis is the efficient and accurate calculation of mutual coupling between pairs of dipoles, which is carried out with a moment-method procedure using the Green's function results of Section 12.2.1.1. Quantities such as the active input impedance, reflection coefficient and element patterns can then be calculated. Section 12.2.3.2 gives a brief discussion of the analysis of finite probe-fed rectangular patch arrays [36], and presents calculated and measured results for mutual coupling between microstrip patches, and some active element patterns.

Besides the element-by-element method used here, a techique called the 'finite periodic structure approach' [34] has recently been developed, and appears capable of treating large arrays. It is based on a modification of the infinite-array solution, and is similar to a technique that has been applied to finite waveguide arrays [4].

12.2.3.1 Printed dipoles: Fig. 12.24 shows the geometry of a finite array of printed dipoles. Each dipole is assumed to have a length L, a width W, and to be uniformly spaced from its neighbours by distances a in the x-direction and b in the y-direction. The solution can treat rectangular arrays of arbitrary size, but in the interest of simplicity only square arrays are considered here. The dipoles are assumed to be thin, so only $\hat{x}$-directed currents are used. The appropriate Green's function is then given by eqn. 12.10a. The current on the dipoles is expanded in a set of piecewise sinusoidal (PWS) modes, as defined in eqns. 12.32 and 12.33. The dipoles are assumed to be centre-fed with idealised delta-gap generators with series impedance Z_T. The equivalent circuit of the fed is shown in Fig. 12.25. Using a Galerkin procedure the electric-field integral equation reduces to

$$\{[Z] + [Z_T]\}[I] = [V] \tag{12.79}$$

where $[Z]$ is the impedance matrix representing the mutual coupling between all the PWS modes on the dipoles, $[Z_T]$ is the generator terminating impedance matrix (a diagonal matrix), $[I]$ is the unknown vector of expansion-mode coefficients, and $[V]$ is the excitation vector of generator voltages. The mn th element of the impedance matrix is given by

$$Z_{mn} = \frac{-1}{4\pi^2} \int_{-\infty}^{\infty} \int_{-\infty}^{\infty} F_{p_m}^*(k_x) F_u^*(k_y) G_{xx}^{EJ}(k_x, k_y) F_{p_n}(k_x) F_u(k_y) \, dk_x \, dk_y \tag{12.80}$$

where F_{p_m} is the Fourier transform of the m th PWS expansion mode given by eqn. 12.38, and F_u is the Fourier transform of the uniform y-variation of current given by eqn. 12.39. Note that reciprocity is satisfied, so that $Z_{mn} = Z_{nm}$. The efficient numerical evaluation of eqn. 12.80 is discussed in Reference 23.

Now consider an $N \times N$ planar array of printed dipoles with M PWS expansion modes on each dipole. Then the order of the linear system of equations in eqn. 12.79 is $N \times N \times M$. Thus, for example, an 11×11 square dipole array with three expansion modes per dipole requires an impedance matrix of size 363×363, and the order increases as the square of N. It is therefore very important to minimize the number of basis functions used. The PWS mode of eqn. 12.32, with a wave number given by $k_e = k_0\sqrt{(\varepsilon_r + 1)/2}$, was found to give quite good results for resonant dipoles, even when only one mode was used [8, 18, 34]. So, for the majority of calculations in Reference 35, and the results presented here, one PWS mode was used per dipole ($M = 1$).

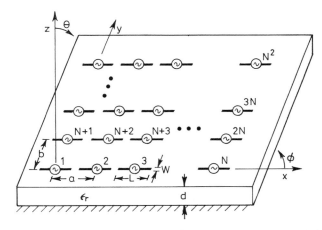

Fig. 12.24 *Geometry of an $N \times N$ planar array of dipoles printed on a grounded dielectric slab*

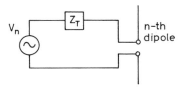

Fig. 12.25 *Equivalent circuit of each dipole in the finite array*

The issue of a complete modal expansion is an important one. Discussions of current expansions for free-space dipole arrays can be found in Reference 3 and 37. In Reference 37 it was found that for arrays of thin dipoles near resonance, not spaced too closely together, the dipole currents were practically identical. For printed dipoles, the situation should then be even better, because the printed-dipole resonance has a much higher Q than the dipole in free space. A judicious choice of a single basis function can thus give a very good approximation to the true current.

The best justification for a single-mode approximation, however, is a comparison with results computed using more than one expansion mode per dipole. Thus, for the example shown in Figs. 12.26–12.28, the input impedance, reflection coefficient and radiation efficiency were computed against scan angle using one and three PWS modes per dipole, for array sizes up to 9 × 9. The input impedance differed by about 10%, and the reflection-coefficient magnitude and radiation efficiency (which are normalised quantities) differed by less than 5%.

It should be emphasised that the solution presented above is capable of handling any number of PWS expansion modes per dipole, and that it is desired to use only one mode per dipole in order to analyse larger arrays. The presence of all dipoles in the array and their mutual coupling is accounted for in the solution. In addition, the solution can handle both the 'forced excitation' ($Z_T = 0$) case, as well as the 'free excitation' ($Z_T \neq 0$) case.

If one expansion mode per dipole is used, the voltage-vector elements can be written as

$$V_m = e^{-jk_0(ux_m + vy_m)} \tag{12.81}$$

where x_m, y_m are the co-ordinates of the centre of the mth dipole and u, v are direction cosines for scanning, as given by eqn. 12.14. Then after the matrix eqn. 12.79 is solved for the currents, the input impedance at the nth dipole can be computed as

$$Z_{in}^n(\theta, \phi) = V_n/I_n \tag{12.82}$$

Note that the input impedance at a dipole of the finite array is dependent on the location of that dipole, as opposed to the infinite-array case where the input impedance would be the same for all dipoles. The active reflection coefficients at the nth dipole can then be calculated according to eqn. 12.43.

A quantity of interest for the finite-array case is the radiation efficiency e based on the power lost to surface waves:

$$e = 1 - \frac{P_{sw}}{P_{in}} \tag{12.83}$$

where P_{in} is the total input power to the array, and P_{sw} is the surface-wave power excited by the array. These quantities can be calculated as [18]

$$P_{in} = \text{Re}\left\{\sum_m \sum_n I_m^* Z_{mn} I_n\right\} \tag{12.84a}$$

$$P_{sw} = \text{Re}\left\{\sum_m \sum_n I_m^* Z_{mn}^{sw} I_n\right\} \tag{12.84b}$$

where Z_{mn} is given by eqn. 12.80, and Z_{mn}^{sw} is the surface-wave contribution (from the residue of the surface-wave pole or poles) to the impedance Z_{mn}. The

Analysis and design considerations for phased-array antennas

active-element pattern can also be calculated, as discussed in Reference 35; this Reference also discusses some useful techniques for improving the computational efficiency of the finite-array solution.

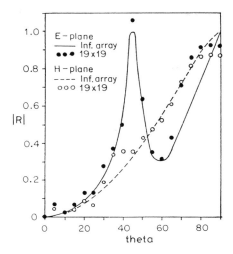

Fig. 12.26 *Reflection-coefficient magnitude versus scan angle (E- and H-plane) for a finite (19 × 19 centre element) printed dipole array and an infinite array*
$\varepsilon_r = 2\cdot 55$, $d = 0\cdot 19$, $a = b = 0\cdot 5\lambda_0$, $L = 0\cdot 39\lambda_0$, $W = 0\cdot 01\lambda_0$

Fig. 12.26 shows the reflection coefficient magnitude of a 19 × 19 printed dipole array on an $\varepsilon_r = 2\cdot 55$ substrate, compared with the result for an infinite array of similar dipoles. The reflection coefficient of the finite array is computed at the centre element of the array, and is matched at broadside scan. Note that the 19 × 19 array is sufficiently large that its reflection-coefficient magnitude versus scan angle follows that of the infinite array relatively closely. This array shows a scan blindness at $\theta = 45\cdot 8°$ in the E-plane. At this scan angle, the reflection-coefficient magnitude of the infinite array is unity, but that of the centre element of the finite array is actually greater than unity. This means that the centre dipole is delivering power back to its generator and load. This power, of course, is being transferred from other ports, and does not violate any conservation laws. The input impedance across the finite array is thus non-uniform. Fig. 12.27 illustrates this variation, showing the reflection-coefficient magnitude as a function of element position across the E-plane (x-direction) of the 19 × 19 array of Fig. 12.26, for various scan angles. The $\theta = 0$ (broadside) case is symmetrical about the centre of the array, and the data shows that the centre element (no. 10) is perfectly matched, but that other elements are slightly mismatched. For $\theta = 30°$ (scanning to the right of the Figure), the mismatch is greater and is asymmetrical. The $\theta = 45°$ data shows that a number of dipole ports on the right-hand side of the array have reflection-coefficient magnitudes greater than unity; it appears from the data that the left-hand elements are absorbing power from the generators and delivering it to the right-hand elements.

Fig. 12.28 shows the efficiency of this array, based on power lost to surface waves as defined in eqn. 12.83, versus E-plane scan angle, for various array sizes. This is a particularly interesting result because it shows the role of surface waves in the transition from a single-element printed antenna, to a finite array, and to an infinite array.

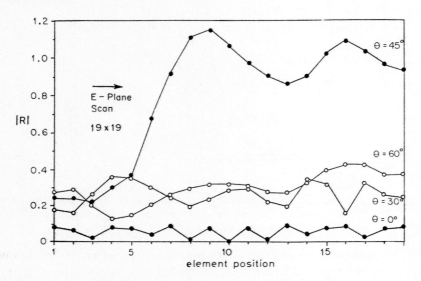

Fig. 12.27 *Reflection-coefficient magnitude versus element position across the E-plane of the 19 × 19 finite array of Fig. 12.26*

For a single dipole (1 × 1), about 22% of the input power is converted to surface-wave power (with the remainder going into space-wave radiation); and this ratio, of course, does not vary with scan angle. For arrays, however, a significant variation of efficiency occurs with scanning. The general trend is that the efficiency improves rapidly for even modest-sized arrays, and increases with array size at all scan angles except those near 45·8°, at which angle the efficiency decreases (more surface-wave power) with increasing array size. This is precisely the angle at which the infinite array has a scan blindness. If the efficiency of the infinite array were plotted in Fig. 12.28, it would be unity at all scan angles except 45·8°, where it would be zero. Since there is no scan blindness in the H-plane of the infinite array, the efficiency of finite arrays for H-plane scan is near unity.

This effect can be explained as the destructive or constructive interference of the surface wave of the unloaded dielectric slab with the radiation of the array. As the array becomes larger, the periodicity and phasing of the array tend to cancel the surface wave at all scan angles except at the scan-blindness angle. At

Analysis and design considerations for phased-array antennas

this angle, the periodicity and phasing of the array are such as to reinforce, or resonate, the surface wave.

As a practical matter, the data of Fig. 12.28 show that the scan-blindness phenomenon can be a problem for even relatively small arrays, and that a prudent array design should probably limit the maximum scan range to about 10° less than the blindness angle.

Reference 35 shows the active-element patterns for the array of Fig. 12.26, as well as examples of other arrays.

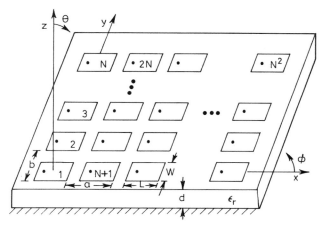

Fig. 12.28 *Radiation efficiency (power loss to surface waves) of the finite dipole array of Fig. 12.26, versus E-plane scan angle for various array sizes*

12.2.3.2 Microstrip patches: The above analysis for finite arrays of printed dipoles can be readily extended to finite arrays of rectangular microstrip patches, as reported in Reference 36. This solution uses the idealised probe-feed model discussed in Section 12.2.2.2 and has been verified by mutual coupling and active-element patterns for patches on thin substrates [23, 24, 36].

The geometry of the finite patch array is shown in Fig. 12.29. As in Section 12.2.2.2, PWS expansion modes are also used here, and impedance matrix elements can be defined as in eqn. 12.80. It must be realised, however, that the mutual impedance defined by eqn. 12.80 are not the same as those seen at the inputs to the probe feeds of the patches. That is, unlike the dipole case, we must make a distinction between the moment-method impedance matrix $[Z]$ and the 'port' impedance matrix $[Z^p]$ defined at the probe terminals. The current flow on the patch is related to the voltage excited at the probe terminals by the modal voltage V_0 [23]:

$$V_0 = \frac{1}{4\pi^2} \int_{-\infty}^{\infty} \int_{-\infty}^{\infty} F_p^*(k_x) F_u^*(k_y) G_{zx}^{EJ}(k_x, k_y) e^{-jk_x x_p} e^{-jk_y y_p} \, dk_x \, dk_y$$

(12.85)

where G_{zx}^{EJ} is given by eqn. 12.47a.

738 Analysis and design considerations for phased-array antennas

The computation of Z_{mn} and V_0 as given in eqns. 12.80 and 12.85 constitutes the bulk of the computational effort for the finite patch-array solution, and so it is important that these terms be evaluated in an efficient manner. References 23 and 36 discuss this issue. In addition, it is possible and desirable to use only one x-directed expansion mode on each patch. This allows a smaller matrix size to be used for a given array, and the arguments presented in Section 12.2.2.1 for the one-mode approximation can be used here as well. Reference 36 shows a result for the reflection-coefficient magnitude versus scan angle for a 7×7 patch array, computed using one and three expansion modes. The results are in good agreement, except for about 10% error in the H-plane scan near endfire. Thus, although the solution can accommodate more than one expansion mode per patch, it appears that in many cases this is not necessary, which then allows the treatment of larger arrays.

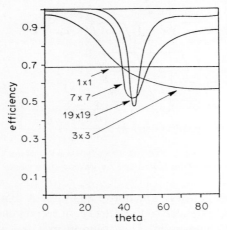

Fig. 12.29 Geometry of a finite array of rectangular microstrip patches

For scanning at the angle θ, ϕ, the probe (port) currents should be driven as

$$I_m^p = e^{-jk_0(ux_m + vy_m)} \tag{12.86}$$

where x_m, y_m are the co-ordinates of the probe feed on the mth patch, and u, v are direction cosines given by eqn. 12.14. The patch-current amplitudes are then given by the column vector [I] and [23]

$$[I] = V_0[Y][I^p] \tag{12.87}$$

where $[Y] = [Z]^{-1}$ is the inverse of the moment-method impedance matrix. The active input impedance at the mth patch is then [23]

$$Z_{in}^m(\theta, \phi) = \frac{-V_0 I_m}{I_m^p} \tag{12.88}$$

Analysis and design considerations for phased-array antennas

The active reflection coefficient can then be calculated from eqn. 12.43, and a radiation efficiency calculated as in the dipole-array case.

To talk about mutual coupling between the probe feed 'ports' of the array, we must define a port impedance matrix $[Z^p]$ as

$$[V^p] = [Z^p][I^p] \tag{12.89}$$

where $[Z^p]$ is found from

$$[Z^p] = [Y^p]^{-1} = -V_0^2[Y] \tag{12.90}$$

Scattering matrix elements can then be calculated directly from $[Z^p]$:

$$[S] = \{[Z^p] - [Z_c]\}\{[Z^p] + [Z_c]\}^{-1} \tag{12.91}$$

where $[Z_c]$ is a diagonal matrix with elements Z_c, the characteristic impedance of the connecting transmission lines.

The active-element pattern can be calculated as follows. From eqns. 12.87, 12.89 and 12.90, the patch currents $[I]$ due to a set of port voltages $[V^p]$ can be calculated as

$$[I] = V_0[Y][Y^p][V^p] = \frac{-1}{V_0}[V^p] \tag{12.92}$$

Now define $[I^s]$ as the driving-current source vector for the active-element pattern of the jth element. Then all elements of $[I^s]$ are zero except for the jth element, which may be set to unity. The port voltages due to $[I^s]$ are found from

$$\{[Y^p] + [Y^T]\}[V^p] = [I^s] \tag{12.93}$$

where $[Y^T]$ is a square diagonal matrix with elements $1/Z_T$, and where Z_T is the termination impedance at each patch port. Then from eqns. 12.92 and 12.93, the patch currents for the active-element pattern are

$$[I] = \frac{-1}{V_0}\{[Y^p] + [Y^T]\}^{-1}[I^s] \tag{12.94}$$

The active-element pattern of the jth element is then computed as

$$E^j(\theta, \phi) = E^0(\theta, \phi) \sum_m I_m e^{-jk_0(ux_m + vy_m)} \tag{12.95}$$

where $E^0(\theta, \phi)$ is the pattern of a single PWS mode [35]. The active-element gain is then

$$G^j(\theta, \phi) = \frac{4\pi|E^j(\theta, \phi)|^2}{Z_0 P_{in}} \tag{12.96}$$

This definition does not include power lost in the terminating impedance of the fed element.

An intermediate result that can easily be obtained from this analysis is the mutual coupling between two microstrip patches. Mutual coupling has been calculated or measured by several authors [18, 23, 24, 38–41], with a wide variety of analytical methods. Fig. 12.30 shows data for the *E*- and *H*-plane mutual coupling between two rectangular patches using the above formulation. Observe that the magnitude of the *H*-plane coupling decays much faster than the *E*-plane coupling. It can be shown that the *H*-plane coupling decays as $1/r^2$, while the

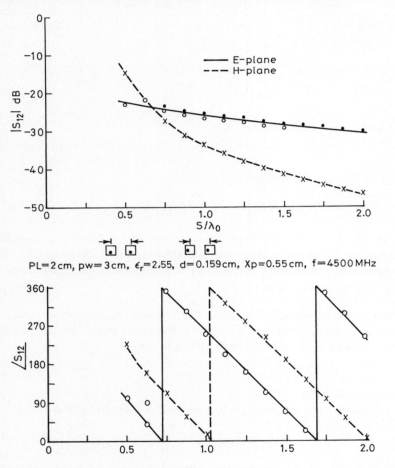

Fig. 12.30 *Calculated mutual coupling (magnitude and phase) between two rectangular microstrip patches*

E-plane coupling decays much slower owing to surface-wave interaction. It is also interesting to note from the phase data of Fig. 12.30 that there is essentially an $e^{-jk_0 r}$ phase dependence with distance for both *E*- and *H*-plane coupling. Even though the *E*-plane coupling is dominated by a surface-wave field, the surface-wave propagation constant of the thin substrate is close to k_0.

As in the case of printed dipoles, the reflection coefficient, efficiency and active-element patterns can be calculated for finite patch arrays. The patch-array. Scan range is also constrained by the scan-blindness effect, which is losely related to element spacing and substrate parameters, as discussed in Section 12.2.1.3.

Table 12.1 *Measured and calculated S-parameters*

i, j	S_{ij} (measured)	S_{ij} (calculated)
25, 24	-12.5 dB $/-147°$	-13.4 dB $/-140°$
25, 26	-12.5 dB $/-145°$	-13.4 dB $/-140°$
25, 23	-21.0 dB $/44°$	-21.5 dB $/59°$
25, 27	-21.0 dB $/49°$	-21.5 dB $/59°$
24, 26	-21.5 dB $/45°$	-21.8 dB $/57°$
25, 18	-24.5 dB $/113°$	-26.0 dB $/122°$
25, 32	-25.0 dB $/112°$	-26.0 dB $/122°$
25, 11	-29.5 dB $/-113°$	-29.8 dB $/-105°$
25, 39	-30.0 dB $/-128°$	-29.8 dB $/-105°$
18, 32	-30.0 dB $/-115°$	-29.8 dB $/-105°$

7×7 element array; $\varepsilon_r = 2.55$; $d = 0.16$ cm; $a = b = 3.45$ cm; $L = 2.0$ cm; $W = 3.0$ cm; $X_p = 0.55$ cm, $Y_p = 0$, $f = 4.35$ GHz

Measurements were made on a 7×7 patch array on a thin substrate to verify the theory. With all ports terminated in $50\,\Omega$, S-parameters were measured and calculated for various element pairs at different locations in the array; typical data are shown in Table 12.1. The elements are numbered across the H-plane, as in Fig. 12.29. Because of the difficulty in obtaining an accurate phase reference, it is estimated that the measured phase data in Table 12.1 may be in error by about $10°$. Also note that the mutual-coupling data in Table 12.1 are between two patch elements in the presence of all the other (terminated) elements, as opposed to the data of Fig. 12.30, which is for two isolated elements.

Element patterns were also measured for the above array, by terminating all but the centre element. Figs. 12.31a and b show the measured and calculated patterns. Agreement is generally within 1 dB or so, although it is clear that the evident in the measured patterns. Asymmetry in the E-plane could possibly be due to feed-probe radiation, which was neglected but could easily be included in the solution.

12.3 Design considerations for printed phased arrays

In this Section we will discuss a variety of considerations for the design and development of printed phased arrays. Much of this material has appeared in the literature [42–44], in relation to the monolithic phased-array concept, but is also relevant for a broader class of printed phased-array antennas. Since printed and integrated phased arrays are still very much in the development stages, we unfortunately cannot be completely thorough in this discussion.

A phased-array antenna offers a number of desirable features to the systems designer, such as rapid beam scanning, pattern control and compatibility with

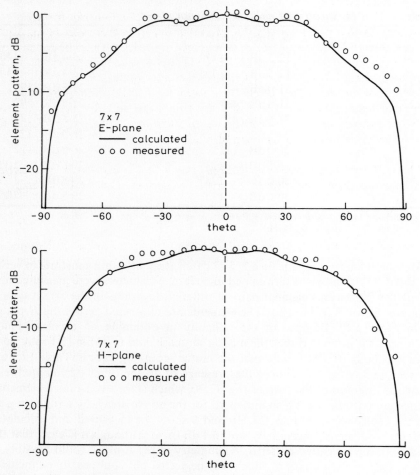

Fig. 12.31 *(a) E-plane, (b) H-plane measured and calculated active element patterns (centre element) of a 7 × 7 rectangular microstrip patch array*
$\varepsilon_r = 2\cdot55$, $d = 0\cdot159$ cm, $a = b = 3\cdot45$ cm, $L = 2\cdot0$ cm, $W = 3\cdot0$ cm, $X_p = 0\cdot55$ cm, $Y_p = 0$, $f = 4\cdot35$ GHz

adaptive and beam-forming systems. The limiting factor in the deployment of phased-array systems, however, is cost, and the cost of such systems seems to be increasing. There exists, then, a strong interest in the integrated phased array, as such a design would use the technology of integration to (hopefully) lower the cost of phased arrays.

The integrated phased array is a general concept that refers to an antenna that takes advantage of photolithographic techniqes and microwave integrated circuitry (MIC) for the radiating elements, feed network and active (phase-shifter/amplifier/switching) circuitry. The logical extension of this concept is the monolithic phased array, where the radiating elements, active circuitry, and feed networks are all integrated on one substrate, (or in sub-array form on one substrate). Such a purely monolithic phased array is far from realisation at the present time and, for reasons discussed below, may not even be desirable from a technical viewpoint. Thus it has become more common to speak of an integrated phased array that is as monolithic as possible.

The following Section will discuss some general factors affecting the design of integrated arrays. Section 12.3.2 will then describe and discuss the relative merits of a variety of array geometries, or architectures, that may be suitable for various levels of phased-array integration.

12.3.1 Design considerations: Design criteria for integrated phased arrays may be categorised according to electrical or mechanical considerations:

(a) Electrical considerations

Type of substrate: To achieve a high level of integration, a semiconductor (high ε_r) substrate (e.g. GaAs) is desirable for active devices and circuitry, but a low-ε_r substrate is preferable for the antenna elements, to enhance bandwidth and scan range.

Maximum scan range: The maximum scan range and the desire to avoid grating lobes controls the element spacing, and hence packing density, of the array. Scan range is also constrained by the scan-blindness effect, which is closely related to element spacing and substrate parameters, as discussed in Section 12.2.1.3.

Bandwidth: The substrate permittivity and thickness, and the element type, all affect the bandwidth of the array. Thick substrates with low permittivity are generally preferred for improved bandwidth.

Type of polarisation: This basically affects the complexity of the array. Linear polarisation is the easiest to obtain; circular polarisation usually requires a quadrature hybrid, and switched polarisation requires a switching network. Dual polarisation is probably the most complicated, as it requires two separate sets of circuitry for each element.

Spurious radiation: Radiation from the feed network and/or active circuitry may degrade the sidelobe level, polarisation or gain of the array.

(b) Mechanical considerations

Number of elements: A typical phased array may require from 10^3 to 10^5 elements. The array architecture must be able to accommodate this number of elements and the requisite feed and control circuitry.

Substrate area: Substrate 'real estate' must exist for radiating elements, feed networks, active circuitry and bias control lines.

Heat transfer: The efficiency of most active devices (particularly FETs) is low. Thus heat removal is often a necessity, especially at millimetre-wave frequencies.

Modularity: To facilitate the reliable fabrication and repair of an integrated phased array, some type of modularity is needed.
A number of the above electrical problems arise from the apparent requirement of using a high-dielectric-constant substrate for both the radiating elements and the active circuitry. For example, microstrip antennas have better bandwidth and less surface-wave excitation for low-dielectric-constant substrates, but the likely semiconductor substrates have a relatively high dielectric constant. In a sense, then, it is a conflicting requirement to have a single substrate for the distinct functions of radiation (loosely bound fields) and circuitry (tightly bound fields). As will be seen in the next Section, a number of new printed-antenna feed methods have been developed to resolve this basic problem by using separate substrates for the radiating elements and the active circuitry.

Substrate space is another prime concern, since a scanning array requires RF power-distribution networks, control and bias circuits, phase-shifter circuits, and possibly amplifier circuits, in addition to radiating elements. The amplifiers may be needed to compensate for increased circuit losses at millimetre-wave frequencies. As will be discussed below, a number of array configurations use more than a single substrate to provide more space, as well as some other advantages. In such cases, a method is needed to couple from one substrate to another. Via holes (plated-through holes) can sometimes be used, but in general it is desirable to avoid such direct connections because of very low yields, and because such connections are usually very inductive at high frequencies. As an alternative, some proximity coupling schemes are discussed below.

A large integrated phased array will probably consist of a number of sub-arrays. Such sub-arrays, for example, might be fabricated on a single 'chip', perhaps with one phase-shifter/amplifier circuit feeding all the antenna elements associated with that chip. All the subarray 'chips' could then be mounted on a 'mother board' to supply RF, bias and control lines. Interconnections here also pose a problem.

Circular or switchable (dual) polarisation is required for a number of applications, and, of course, such requirements complicate the design. Dual polarisation is probably the most difficult case to accommodate, as this essentially requires two separate orthogonally polarised elements, or at least a single element (such as a square microstrip patch) that can be switched between two polarisation states. Circular polarisation is somewhat easier to obtain, by using a circularly polarised element or a polariser to convert linear polarisation to circular.

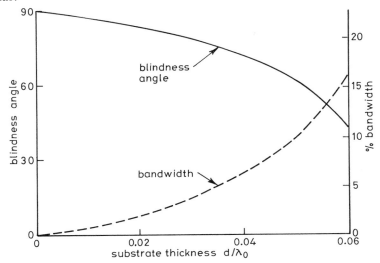

Fig. 12.32 *Scan-blindness angle and bandwidth for a patch array with $\lambda_0/2$ spacing on a GaAs substrate versus substrate thickness*

As discussed in Section 12.2.1.3, the scan-blindness effect can limit the scan range of printed phased arrays. As the substrate is made electrically thicker (as a result of higher frequency, dielectric constant or thickness), the angle at which scan blindness occurs moves closer to broadside. This blindness angle thus effectively limits the scanning range of the array. Fig. 12.32 shows the blindness angle of an infinite microstrip patch array on a GaAs substrate versus substrate thickness. Such a substrate $0.04\lambda_0$ thick, for example, would have a blindness angle of about 60°, which would probably limit the useful scan range of the array to less than 50° owing to the rapid increase of the reflection coefficient near the blindness angle. The data of Fig. 12.32 assumes an element spacing of $\lambda_0/2$ – the blindness angle moves closer to broadside for larger spacings. Also shown in Fig. 12.32 is the approximate bandwidth of the patch element, which shows that a trade-off exists between the bandwidth of the array and its maximum scan range.

12.3.2 Array architectures

In this Section we will discuss several types of printed-phased-array geometries, and their relative merits. Several of these configurations correspond to specific canonical arrays analysed in Section 12.2.

12.3.2.1 Single-layer substrate: The type of geometry that probably first comes to mind when considering an integrated phased array is the single-layer substrate shown in Fig. 12.33, where radiating patches, active circuitry and the necessary feed networks are all contained on the same substrate. A major problem with this approach is that there may not be enough room on the substrate for all of the components. To avoid grating lobes, antenna elements must be spaced no more than about $\lambda_0/2$ apart; so if the phase-shifter circuitry, RF feed network and bias lines can be fitted in at all, the spurious coupling between these components may be severe. Another problem with this geometry is the scan-blindness/bandwidth-trade-off which was discussed above. Scan blindness will always occur at some scan angle for a printed array, but, for thin substrates, the blind angle will be closer to endfire. Fig. 12.13, for example, shows a calculated result for the reflection-coefficient magnitude versus scan angle in the three planes, for an infinite array of microstrip patches on a $0.02\lambda_0$-thick GaAs substrate. The blindness angle is seen to occur at about 82° in the *E*-plane (unity reflection-coefficient magnitude), although the reflection-coefficient magnitude is still about 0·5 at 60° scan in the principal planes. If the substrate thickness is increased, because of higher-frequency operation or a desire for more bandwidth, the blindness angle will move closer to broadside, as indicated in Fig. 12.32.

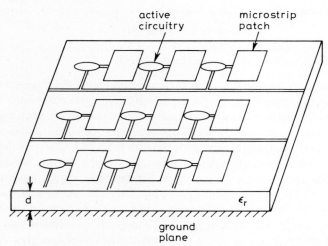

Fig. 12.33 *Geometry of an array of microstrip patches on a single-layer substrate*

This geometry is also susceptible to spurious radiation from the active circuitry and/or the feed network, which can degrade sidelobe levels or polarisation.

12.3.2.2 Two-sided geometry: Fig. 12.34 shows a two-sided substrate design that eliminated many of the problems encountered with the single-layer case by going to the root cause of those problems, and using two separate substrates for the distinct functions of radiation and circuitry. A substrate with a low dielectric constant holds the radiating microstrip patches, while a parallel semiconductor substrate contains active circuitry and feed networks. The two substrates are separated by a ground plane, and apertures in this ground plane are used to couple RF power from the feeds to the radiating elements.

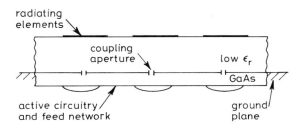

Fig. 12.34 *Cross-sectional view of a two-sided integrated-array geometry with aperture-coupled patch radiators*

This design thus matches the substrate to the electrical function, resulting in improved blindness/bandwidth performance. For example, with an $\varepsilon_r = 2{\cdot}55$ antenna substrate, the thickness would have to be about $0{\cdot}05\lambda_0$ for a blindness at 80°, and the situation would be even better for a lower-dielectric-constant substrate. Since we have two substrates, much more space is available than in the single-layer case. In addition, the ground plane effectively isolates the active circuitry and feed network from the radiating elements to reduce spurious coupling and radiation.

This array configuration is dependent on the aperture-coupled microstrip antenna, which has been described in detail in Reference 27 and theoretically analysed as a single element in Reference 28 and 29, and as an array in Section 12.2.2.4. Fig. 12.20 shows the geometry of a single aperture-coupled patch antenna, fed by a microstripline on the feed substrate. The feed line is usually terminated in an open-circuited stub for tuning. The aperture is smaller than resonant size, so very little radiation occurs in the back region. Models have been successfully fabricated and tested at frequencies from 2 to 20 GHz.

A final feature of the two-sided array, and the array configurations to follow, is the fact that it offers better radiation 'hardening' from lightning or EMP effects compared with the single-layer design, owing to the shielding effect of the ground plane. The coupling of the sensitive active circuitry to the outside world must take place through the microstrip antennas and coupling apertures, which present a two-pole (or more) filter response to signals outside their bandwidth.

12.3.2.3 Perpendicular feed substrates: Another design that uses separate substrates for the radiating elements and active circuitry is shown in Fig. 12.35. In this case, a vertical substrate holding the radiating elements is fed by a number of parallel-feed substrates. Coupling is again through apertures in the ground plane of the antenna substrate. This design also allows the use of a low-dielectric-constant substrate for the radiating elements and a separate semiconductor substrate for the active circuitry, similar to the two-sided geometry, and so has the same advantages in relation to scan-blindness/bandwidth performance and shielding of spurious radiation or coupling. In addition, this architecture has a number of other advantages.

First is the fact that the feed substrate can be of virtually unlimited size, since there is no immediate restriction on the 'depth' dimension away from the vertical-antenna substrate. Waveguide phased arrays usually use this depth dimension to a similar advantage. The geometry also permits a modular construction, where feed modules could conceivably be plugged into receptacles on the antenna substrate.

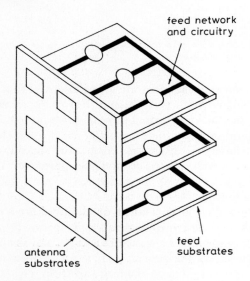

Fig. 12.35 *An integrated phased-array configuration using a feed substrate perpendicular to the radiating-element substrate*

This design also allows efficient heat transfer from the ground plane of the feed substrate. At millimetre-wave frequencies, low device efficiency requires efficient heat transfer from active circuitry. The unobstructed ground plane of the feed substrates allows much heat removal to take place, while the embedded ground plane of the two-sided design makes heat removal more difficult.

Finally, such a geometry would lend itself well to space-fed phased-array lens designs, which may be of interest for some applications. This could be implemented by having antenna substrates at both ends of the feed substrates. It does not appear, however, that this geometry would be useful if dual polarisation were required.

Analysis and design considerations for phased-array antennas 749

The array with perpendicular-feed substrates depends on the feasibility of feeding a single patch through an aperture with a microstrip line on a perpendicularly oriented substrate. Such a geometry is shown in Fig. 12.36, and has been discussed in more detail in Reference 45. This design has been verified experimentally, but no theory has been developed beyond the simple arguments presented in Reference 45.

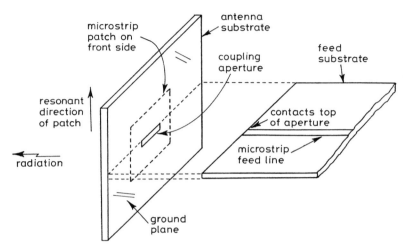

Fig. 12.36 *Geometry of a microstrip antenna fed through an aperture with a microstrip feed line on a perpendicularly oriented feed substrate*

The geometry in Fig. 12.36 shows a direct connection from the feed line to the top of the aperture; the two ground planes are also in electrical contact. Another version of the perpendicularly fed antenna excites the aperture by proximity coupling, eliminating the need for a direct connection of the feed line, as shown in Fig. 12.37. Other variations, including the use of a co-planar waveguide feed, are also possible.

12.3.2.4 Endfire elements: The previously discussed integrated-phased-array designs all used microstrip patches or printed dipoles which radiate normal to the substrate on which they are printed. An alternative to this approach is to use elements which radiate endfire to the substrate, as shown in Fig. 12.38. This example shows the use of tapered-slot antennas, but other elements capable of endfire (to the substrate) radiation, such as dipoles, could be used as well.

This type of geometry then uses a single substrate for both active circuitry and radiating elements, but in a rather different manner from the single-layer design discussed earlier. A lot of substrate space is available for feed networks and circuitry, and the design can readily be used for space-fed lens arrays. The individual substrates can be made in modular form, and heat transfer should not be a problem.

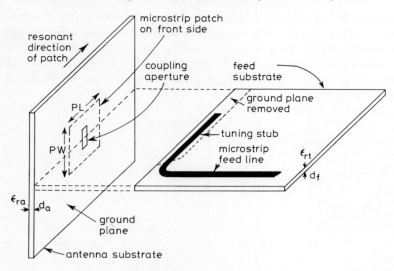

Fig. 12.37 *Geometry of a microstrip antenna fed through an aperture which is proximity coupled to a microstrip feed line on a perpendicularly oriented feed substrate*

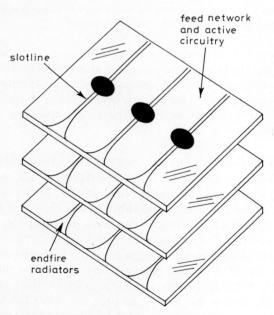

Fig. 12.38 *An integrated phased-array geometry using tapered-slot elements that radiate in the endfire direction*

Although surface waves can still be excited on the substrates, an additional problem is the possibility of scan blindness caused by surface waves on the protruding grid of dielectric slabs; such effects have been observed in similar arrays with protruding dielectrics. In addition, this configuration would probably not be useful if circular polarisation is desired.

The tapered-slot element has been discussed in Reference 7, and may be constructed with either a linear taper or a curved slot. The slot antenna can be proximity fed with a microstrip-line/slot-line transition, in Reference 7, or the slot line could be directly fed from the active circuitry. In this regard, it is interesting to note that slot line has a number of distinct advantages over microstrip in such millimetre-wave integrated-circuit applications [46].

12.4 Conclusion

This Chapter has discussed the analysis and design of printed phased arrays. Analytical techniques were outlined, and applied to several canonical infinite and finite printed arrays. General considerations for the design of integrated arrays were also discussed.

This Chapter has summarised most of the work to date on the anaysis of printed arrays, but there is much yet to be done. Some topics include proximity-coupled elements, the use of wide-angle impedance-matching layers, the development of improved probe-feed models, and the effect of substrate anisotropy.

12.5 Acknowledgment

The author would like to thank his graduate students, James Aberle and Fran Harackiewicz, for reviewing the manuscript and making valuable suggestions regarding the consistency of notation.

12.6 References

1. OLINER, A. A., and KNITTEL, G. H.: 'Phased array antennas'. *in* Proc. Phased Array Antenna Symposium', (Artech House, 1972)
2. STARK, L.: 'Microwave theory of phased array antennas – A review', *Proc. IEEE*, 1974, **62**, pp. 1661–1701
3. HANSEN, R. C. (Ed.): 'Microwave scanning antennas', (Academic Press, NY, 1966)
4. AMITAY, N., GALINDO, V., and WU, C. P.: 'Theory and analysis of phased array antennas', (Wiley Interscience, NY, 1972)
5. MAILLOUX, R. J., McILVENNA, J. G., and KERNWEIS, N. P.: 'Microstrip array technology', *IEEE Trans.*, 1981, **AP-29**, pp. 25–37
6. JAMES, J. R., HALL, P. S., and WOOD, C.: 'Microstrip antenna theory and design', (Peter Peregrinus, 1982)

7   BAHL, I. J., and BHARTIA, P.: 'Microstrip antennas (Artect House, 1980)
8   POZAR, D. M., and SCHAUBERT, D. H.: 'Scan blindness in infinite phased arrays of printed dipoles, *IEEE Trans.*, 1984, **AP-32**, pp. 602–610
9   RANA, I. E., and ALEXOPOULOS, N. G.: 'Current distribution and input impedance of printed dipoles', *IEEE Trans.*, 1981, **AP-29**, pp. 99–105
10 DESHPANDE, M. D., and BAILEY, M. C.: 'Input impedance of microstrip antennas', *IEEE Trans.*, 1983, **AP-31**, pp. 740–747
11 MOSIG, R., and GARDIOL, F. E.: 'A dynamical radiation model for microstrip structures *in* Advances in electronic and electron physics: Vol. 59' (Academic Press, 1982) pp. 139–237
12 PERLMUTTER, P., SHTRIKMAN, S., and TREVES, D.: 'Electric surface current model for the analysis of microstrip antennas with application to rectangular elements', *IEEE Trans.*, 1985, **AP-33**, pp. 301–311
13 KNITTEL, G. H., HESSEL, A., and OLINER, A. A.: 'Element pattern nulls in phased arrays and their relation to guided waves', *Proc. IEEE*, 1968, **56**, pp. 1822–1836
14 LECHTRECK, L. W.: 'Effects of coupling accumulation in antenna arrays', *IEEE Trans.*, 1968, **AP-16**
15 LIU, C. C., HESSEL, A., and SHMOYS, J.: 'Performance of probe-fed microstrip-patch element phased arrays'. Phased Arrays Symposium, Bedford, MA, 1985
16 FRAZITA, R. F.: 'Surface-wave behavior of a phased array analyzed by the grating-lobe series', *IEEE Trans.*, 1967, **AP-15**, pp. 823–824
17 ALEXOPOULOS, N. G., and RANA, I. E.: 'Mutual impedance computation between printed dipoles', *IEEE Trans.*, 1981, **AP-29**, pp. 106–111
18 POZAR, D. M.: 'Considerations for millimeter wave printed antennas', *IEEE Trans.*, 1983, **AP-31**, pp. 740–747
19 CASTANEDA, J., and ALEXOPOULOS, N. G.: 'Infinite arrays of microstrip dipoles with a superstrate (cover) layer'. IEEE AP-S International Symposium Digest, Vancouver, Canada, 1985, pp. 713–717
20 WRIGHT, S. M., and LO, Y. T.: 'Efficient analysis for infinite microstrip dipole arrays', *Electron. Lett.*, 1983, **19**, pp. 1043–1045
21 WHEELER, H. A.: 'A survey of the simulator technique for designing a radiating element', *in* OLINER, A. A., AND KNITTEL, G. H., (Eds.): 'Phased array antennas' (Artech House, 1972)
22 POZAR, D. M., and SCHAUBERT, D. H.: 'Analysis of an infinite array of rectangular microstrip patches with idealized probe feeds', *IEEE Trans.*, 1984, **AP-32**, pp. 1101–1107
23 POZAR, D. M.: 'Input impedance and mutual coupling of rectangular microstrip antennas', *IEEE Trans.*, 1982, **AP-30**, pp. 1191–1196
24 NEWMAN, E. H., RICHMOND, J. H., and KWAN, B. W.: 'Mutual impedance computation between microstrip antennas', *IEEE Trans.*, 1983, **MTT-31**, pp. 941–945
25 MOSIG, J. R., and GARDIOL, F. E.: 'General integral equation formulation for microstrip antennas and scatterers', *Proc. IEE*, 1985, **132H**, pp. 424–432
26 ABERLE, J. T., and POZAR, D. M.: 'Analysis of infinite arrays of one- and two-probe-fed circular patches', *IEEE Trans.*, **AP**. (Accepted for publication)
27 POZAR, D. M.: 'A microstrip antenna aperture coupled to a microstrip line', *Electron. Lett*, 1985, **21**, pp. 49–50
28 SULLIVAN, P. L., and SCHAUBERT, D. H.: 'Analysis of an aperture coupled microstrip antenna', *IEEE Trans.*, 1986, **AP-34**, pp. 977–984
29 POZAR, D. M.: 'A reciprocity method of analysis for printed slot and slot-coupled microstrip antennas', *IEEE Trans.*, 1986, **AP-34**, pp. 1439–1446
30 JACKSON, R. W., and POZAR, D. M.: 'Full-wave analysis of microstrip open-end and gap discontinuities', *IEEE Trans.*, 1985, **MTT-33**, pp. 1036–1042
31 POZAR, D. M.: 'General relations for a phased array of printed antennas derived from infinite current sheets', *IEEE Trans.*, 1985, **AP-33**, pp. 498–504
32 WHEELER, H. A.: 'Simple relations derived from a phased-array antenna made of an infinite current sheet', *IEEE Trans.*, 1965, **AP-13**, pp. 506–514

33 KOMINAMI, M., POZAR, D. M., and SCHAUBERT, D. H.: 'Dipole and slot elements and arrays on semi-infinite substrates', *IEEE Trans.*, 1985, **AP-33**, pp. 600–607
34 ISHIMARU, A., COE, R. J., MILLER, G. E., and GEREN, W. P.: 'Finite periodic structure approach to large scanning array problems', *IEEE Trans.*, 1985, **AP-33**
35 POZAR, D. M.: 'Analysis of finite phased arrays of printed dipoles', *IEEE Trans.*, 1985, **AP-33**, pp. 1045–1053
36 POZAR, D. M.: 'Finite phased arrays of rectangular microstrip antennas', *IEEE Trans.*, 1986, **AP-34**, pp. 658–665
37 KING, R. W. P., MACK, R. B., and SANDLER, S. S.: 'Arrays of cylindrical dipoles' (Cambridge University Press, 1968)
38 JEDLICKA, R. P., POE, M. T., and CARVER, K. R.: 'Measured mutual coupling between microstrip antennas', *IEEE Trans.*, 1981, **AP-29**, pp. 147–149
39 PENARD, E., and DANIEL, J. P.: 'Mutual coupling between microstrip antennas', *Electron. Lett.*, 1982, **18**, pp. 605–607
40 MALKOMES, M.: 'Mutual coupling between microstrip patch antennas', *Electron. Lett.*, 1982, **18**, pp. 520–522
41 VAN LIL, E., and VAN DECAPELLE, A.: 'Comparison of models for calculating mutual coupling in microstrip arrays', IEEE AP-S Symposium Digest, Boston, 1984, pp. 745–748
42 POZAR, D. M., and SCHAUBERT, D. H.: 'Comparison of architectures for monolithic phased array antennas', *Microwave J.*, 1986, **29**, pp. 93–104
43 POZAR, D. M.: 'Phased arrays of printed antennas', ISAP Symposium, Kyoto, Japan, 1985
44 POZAR, D. M.: 'New architectures for millimeter wave phased array antennas'. JINA International Symposium on Antennas, Nice, France, 1986
45 BUCK, A. C., and POZAR, D. M.: 'An aperture coupled microstrip antenna with a perpendicular feed', *Electron. Lett.*, 1986, **22**, pp. 125–126
46 JACKSON, R. W.: 'Coplanar versus microstrip for millimeter wave integrated circuits'. Microwave Theory and Techniques Symposium, Baltimore, 1986
47 SCHUSS, J. J., HANFLING, J. D., and MORROW, R. E.: 'Observation of Scan Blindness Due to Surface Wave Resonance in an Array of Printed Circuit Patch Radiators'. 1987 International IEEE Antennas and Propagation Symposium, Blacksburg, VA, 1987

Chapter 13
Circularly polarised antenna arrays

K. Ito, T. Teshirogi and S. Nishimura

Introduction

This chapter is concerned with various techniques for circularly polarised microstrip arrays. Circular polarisation is effective for many radio systems, such as communications, remote sensing, navigation and radar systems. In particular, at present, mobile-satellite-communication and direct-broadcasting-satellite systems use circular polarisation, because they do not need polarisation tracking.

In these systems, it is desirable for each ground terminal to have a low-profile and lightweight antenna. Also, it is a requirement to achieve a specific gain which cannot be obtained by a single radiating element. In addition to high gain, multiple functions, such as electronic beam scanning, beam shaping and low sidelobe-radiation patterns, are often required. For these reasons, interest in circularly polarised microstrip-array antennas is increasing rapidly.

Circularly polarised microstrip arrays are classified into three major categories. The first group includes arrays which are composed of circularly polarised (sometimes linearly or elliptically polarised) microstrip patches. This type of array is the most common and widely used, and it includes many variations. The second type of array is composed of composite elements, which consist of electric- and magnetic-current-source elements. The third type are travelling-wave arrays which utilise radiation due to suitable discontinuities in travelling-wave transmission lines. These techniques will be discussed in this chapter.

13.1 Various types of circularly polarised arrays

13.1.1 Arrays of patch radiators
The conventional method of obtaining a circularly polarised array is to arrange circularly polarised microstrip patches with appropriate feeding.

(a) Circularly polarised radiating elements: There are various types of circularly polarised patches or resonators, and these are described in Chapter 4. In

756 Circularly polarised antenna arrays

Fig. 13.1 typical circularly polarised patches which can be used as array elements are shown.

The most direct approach for obtaining circular polarisation is to excite a square or circular patch with two orthogonal modes of equal amplitude and a differential phase shift of $\pm 90°$, by using a 90° hybrid as shown in Fig. 13.1a and b. The antenna can be excited from a single feed point by use of a dual-feed device, such as a 90° hybrid or power splitter with the necessary phase shift [1, 2]. For wider bandwidth applications, four-probe feeds with 0°, 90°, 180° and 270° phase differentials are used which can suppress higher-order modes formed in the thick substrate [3], as shown in Fig. 13.1c.

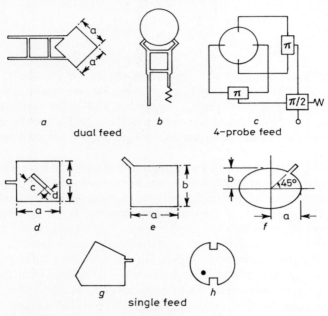

Fig. 13.1 *Circularly polarised microstrip patch antennas*

Several methods have been proposed to provide circular polarisation without the complexities inherent in dual-feed devices. One approach is to attach a single feed point at a location so as to excite two equal-amplitude degenerate orthogonal modes, and then to introduce some asymmetry into the cavity so that the degeneracy of the modes is removed. Examples of this technique are the square microstrip patch with a tilted slot [4], the corner-fed rectangular patch [5, 6], the slightly elliptical patch [7, 8], the pentagon-shaped patch [9], and the circular disc with perturbation element [10], as shown in Fig. 13.1d–h, respectively. However, these perturbation techniques for generating circular polarisation have very limited axial-ratio bandwidth — generally of the order of 1%.

(b) Feed methods and array configuration: There are several feed methods for a linear or planar circularly polarised array, and the detail and the corresponding array configuration will be described in the next Section.

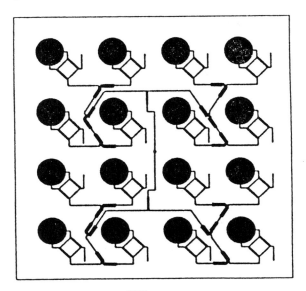

Fig. 13.2 *Example of co-planar arrays [67]*

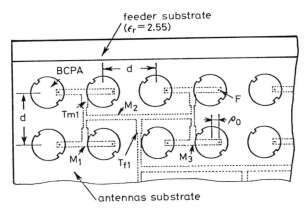

Fig. 13.3 *Rear-feed microstrip array [47]*

The most simple is the corporate (or parallel) feed system which splits the power between n output ports with a prescribed distribution while maintaining equal path length from the input to output ports. The bandwidth of this type of array is essentially wide. In practice, it will primarily be limited by the match of the radiating elements. Figs. 13.2 and 13.3 show examples of corporate feeds.

Fig. 13.2 is a co-planar array in which the array and the corporate feed system are formed on the same plane. Although this array configuration is simpler to manufacture, the radiation from the microstrip feed line deteriorates the overall radiation characteristic of the array.

The rear feed system, in which the feed network is located behind the array, is effective for shielding the spurious radiation from the feed lines and devices. One example, as shown in Fig. 13.3, is a circularly polarised array, composed of circular discs with perturbation notches, each of which is driven from the corporate feed circuit in the rear side through a feed probe.

Another simple form of feed system is a series feed in which the circularly polarised radiating patches are attached periodically to a transmission line. In this configuration, the phase of the radiating elements is determined by their spacing along the transmission line; therefore, as the frequency is altered, a progressive phase shift results down the array, which causes the main beam direction to change and the beam to squint.

Generally, beam squint with frequency is a particular disadvantage of travelling-wave arrays. However, this beam squint can be eliminated by equalising the path lengths between the input and each element. A compact form of such squintless array has been developed by Rodgers [11].

(c) *Circularly polarised array composed of linearly or elliptically polarised elements*: A circularly polarised array can also be realised even by using linearly or elliptically polarised elements. In general, since microstrip antennas, and particularly single-feed-type antennas, have a narrow-ellipticity bandwidth, techniques for obtaining an array which is composed of linearly or elliptically polarised elements, but radiates circular polarisation over a wide frequency band, are useful. Details of these wideband techniques will be described in Section 13.4.

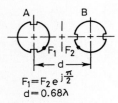

Fig. 13.4 *Arrangement and excitation of a pair of single-feed-type microstrip antenna [12]*
(© 1982 IEEE)

In this Section, these techniques are just introduced as a type included in the category of circularly polarised arrays of patch radiators.

There are three kinds of array. The first is when an array is composed of sets of pairs, in each of which the differential orientation angle and phase shift between two antennas is 90°, as shown in Fig. 13.4 [12]. Although each element has a narrow axial-ratio bandwidth, and radiates a heavily elliptical or almost

Circularly polarised antenna arrays 759

linear polarisation at frequencies off resonance, the cross-polarisations of the paired elements cancel each other out and the array can maintain good polarisation characteristics over a wide bandwidth ($\geqslant 10\%$).

In the second case, circular polarisation is achieved by having a basic 2 × 2 sub-array composed of single-feed linearly polarised elements with unique angular and phase arrangements of the elements, as shown in Fig. 13.5a and b [13]. Both the angular orientation and feed phase of the element are arranged in a 0°, 90°, 0°, 90°, or 0°, 90°, 180°, 270° fashion.

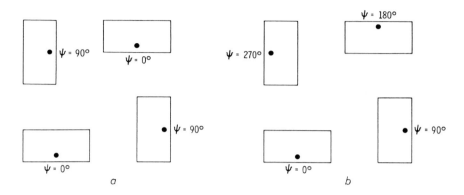

Fig. 13.5 *Circularly polarised 2 × 2 sub-array of linearly polarised elements [13] (© 1986 IEEE)*
  a 0°, 90°, 0°, 90° arrangement
  b 0°, 90°, 180°, 270° arrangement

The third array, which is called a sequential array, is a more generalised configuration [14]. A differential orientation angle and phase shift are provided sequentially for each element of the array. A schematic arrangement of an N-element sequential array is shown in Fig. 13.6. The sequential array provides not only excellent circular polarisation in the boresight, but also low VSWR over the wide frequency band.

13.1.2 Arrays of composite elements

In general, a proper combination of electric and magnetic radiating elements will be able to produce circular polarisation when the two radiation fields from the elements are perpendicular to each other and have 90° phase difference. Such a combination of elements is sometimes referred to as a composite element.

Several kinds of composite elements for circular polarisation have been reported, including a combination of a slot and two parasitic dipoles [15], as shown in Fig. 13.7a, and a combination of a strip and a slot [16], as shown in Fig. 13.7b.

Fig. 13.7a shows the configuration of an array element composed of an

760   *Circularly polarised antenna arrays*

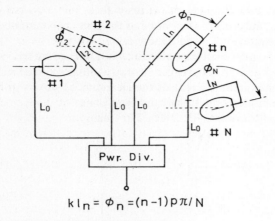

$$kl_n = \phi_n = (n-1)p\pi/N$$

Fig. 13.6   *Configuration of a sequential array [14]*

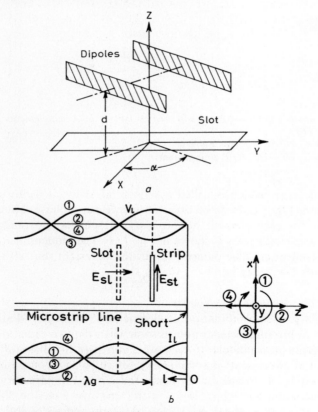

Fig. 13.7   *Two types of composite elements for circular polarisation*
   a Slot and two parasitic dipoles [15]
   b Basic elements and working principle of CP-PASS [16] (© IEE)

Circularly polarised antenna arrays

excited slot and two parasitic flat dipoles [15]. The slot can be excited by a microstrip line. The dipoles are placed above the slot at a distance d and angle α. The two parameters, d and α, are determined so as to produce circular polarisation. The array element was analysed using the technique of reaction matching. A planar array consisting of four identical elements, as shown in Fig. 13.8, was designed, and its various characteristics were simulated taking into account the inter-element coupling effects. The simulation showed that the planar array could produce a gain of more than 13 dB and the 1·5 dB bandwidth of the axial ratio was about 6%. An actual array can be constructed by using a proper substrate and by modifying the design method.

Fig. 13.7b shows the fundamental structure of a circularly polarised printed array composed of strips and slots (CP–PASS). It consists of a strip on a thin substrate, a slot in the ground plane and a microstrip feed line. The strip and the slot — basic radiating elements — are almost half a wavelength long and the spacing between them is a quarter of a guide wavelength λ_g along the line. The strip and the slot are excited by the electric and magnetic fields propagating along the microstrip line, and radiate electric fields E_{st} and E_{sl}, respectively, shown in Fig. 13.7b in the broadside direction.

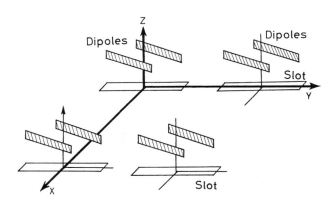

Fig. 13.8 *Configuration of array (four-element sets) [15]*

Fig. 13.7b also shows the voltage and current distributions V_l and I_l along the microstrip line when the line is short-circuited at $l = 0$. If the strip and the slot are located at the maximum points of V_l and I_l, respectively, the elements can produce circular polarisation efficiently in the broadside direction.

Fig. 13.9 shows a typical configuration of a linear-array-type CP–PASS, which consists of three sets (i.e. six pairs) of elements. Each element pair has a strip element (the term 'strip element' means a combination of a strip dipole and a window) and a slot. The window, a kind of wide and long slot, is located in the ground plane in order to effectively increase the gain and bandwidth of the strip dipole [17].

762 Circularly polarised antenna arrays

Tapered window edges are also excited slightly by the magnetic field along the line, and radiate unwanted waves. However, as shown in Fig. 13.9, the strip dipoles are placed at the voltage maxima along the line with a half-wavelength spacing, so that the unwanted radiation from the windows can be effectively suppressed. The details of unwanted radiation will be discussed in Section 13.3.2.

Additionally, it is quite easy to control aperture distribution along the feed line by adjusting the coupling gaps between the elements and the feed line. A design procedure for CP–PASS and a design example are given in Section 13.2.2.

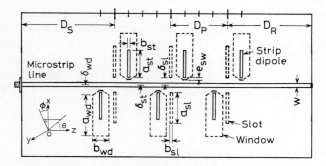

Fig. 13.9 *Configuration of CP–PASS (three-element sets) [38] (© IEE)*

13.1.3 Travelling-wave arrays

It is a well known fact that discontinuities in a microstrip line produce radiation [18]. The microstrip-line antenna utilises this phenomenon: the strip conductor of a microstrip line, which is bent periodically like a meander, forms a circularly polarised travelling-wave array.

Fig. 13.10 shows some circularly polarised microstrip-line antennas. These are:

(a) Rampart-line antenna [19, 20]
(b) Chain antenna [21]
(c) Square-loop-type microstrip-line antenna [22]
(d) Crank-type microstrip-line antenna [23]

These antennas will radiate right-handed circularly polarised waves when the power is fed from the left-hand end, and the right-hand end is terminated in a matched load. On the other hand, when the power comes from the right-hand end and the matched load terminates the left-hand end, each antenna radiates left-handed circularly polarised waves.

The circularly polarised radiation of the four types of antennas (a)–(d) can be explained by the magnetic-current source method [24] for (a), and the line electric-current source method for (b)–(d). The magnetic-current source method assumes that electromagnetic waves radiate in the main from the bent parts of the microstrip line. The line electric-current source method [25] assumes that the source of the radiation is the line electric current (of uniform amplitude) along the central line. These two radiation mechanisms, though based on different principles, give the results that are in good agreement.

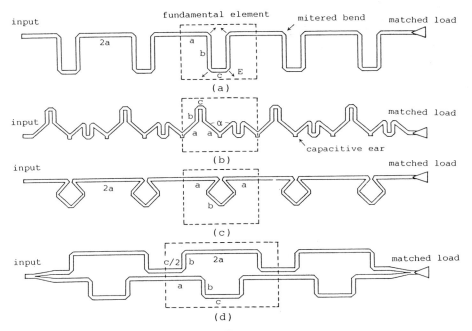

Fig. 13.10 *Circularly polarised microstrip-line antennas*
  a Rampart-line antenna
  b Chain antenna
  c Square-loop-type microstrip-line antenna
  d Crank-type microstrip-line antenna

The rampart-line antenna shown in Fig. 13.10a is built from sets of four right-angled bends. The right-angled bends have to be chamfered in order to reduce the right-angled-discontinuity susceptance [26]. The radiation arises from the bends, and the polarity has the direction shown in Fig. 13.10a. Let us consider the fundamental element surrounded by a dotted line in Fig. 13.10a. When the lengths a, b and c are appropriately chosen, the radiated field from the four bends will produce circular polarisation. For example, when $a = 3\lambda_g/8$, $b = \lambda_g/2$ and $c = \lambda_g/4$ (λg is the guide wavelength of the microstrip line), the antenna will become a circularly polarised travelling-wave-array antenna [20].

The chain antenna and one of its fundamental elements are shown in Fig. 13.10b. Each fundamental element is built from a V-shaped circularly polarised radiating element and a U-shaped phase shifter. A U-shaped phase shifter is inserted between two V-shaped teeth of the zigzag line. Each arm of the V-shaped radiating element has the length a, and the angle between the two arms is α. For practical purposes, the optimum ranges of the parameters are: $a = 0.25\lambda_g - 0.5\lambda_g$, $\alpha = 90°–150°$. To suppress reflection from the angled parts, a capacitive ear is added to each V-shaped radiating element

The square-loop-type microstrip-line antenna, shown in Fig. 13.10c, is formed from a series of fundamental elements. Each fundamental element is made of a square loop, which radiates circular polarisation, and a straight feeder. When the perimeter of the square loop is λ_g, the loop will radiate circular polarisation. In this case, the linear portion will operate as a feeder, and its effective length will be λ_g.

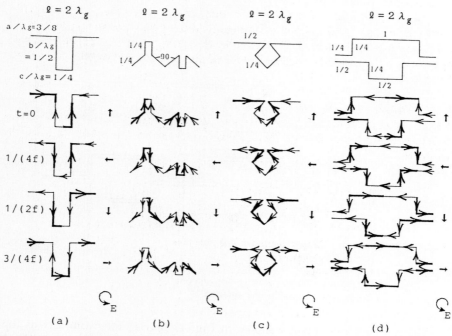

Fig. 13.11 *Instantaneous current distribution and polarity*

Fig. 13.10d shows a crank-type microstrip-line antenna which gives better frequency characteristics for the axial ratio and radiation pattern, compared with a single rampart-line antenna. The antenna is made of two parallel meander (rampart) lines of the same dimensions. One of the meander lines is shifted for one-half of its period. In order to reduce the susceptance, every bend is chamfered, but in special cases the bends need not be chamfered [23]. When the lengths of each segment of the fundamental element are selected properly, the fundamental element will radiate circular polarisation. The method of selection for the lengths is described in detail in Section 13.2.3.

Using the line electric-current source method, we will explain now how these four types of antenna radiate circular polarisation. Suppose that the power is fed from the left-hand end. The instantaneous current distribution is shown by the arrows in Fig. 13.11. The arrows point in the opposite direction every half guide wavelength λ_g. Fig. 13.11a shows the fundamental element of a rampart-line antenna. At time $t = 0$, the total radiation field from each segment consists only of the radiation from the vertical segments, because the radiated fields of the horizontal segments cancel each other out. When $t = 1/(4f)$, where f is the frequency, the total radiation field will be of horizontal left-oriented polarity. Similarly, when $t = 1/(2f)$ or $t = 3/(4f)$, the polarity of the total radiation field will be oriented downwards or to the right, respectively. As shown in Fig. 13.11a–d, the polarity of the total field of the radiated electromagnetic waves (normal to the surface of the paper), rotates counterclockwise, and completes one cycle in time $1/f$. Thus the fundamental element operates as a right-handed circularly polarised antenna. Similarly, the three other types (b), (c) and (d) also operate as right-handed circularly polarised antennas, as can be seen from Figs. 13.11b–d.

13.1.4 Others types of arrays
Several other types of circularly polarised microstrip or printed arrays have been developed and reported. In the following, some typical arrays will be introduced.

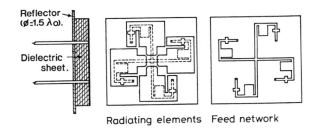

Fig. 13.12 *Broadband flat radiating element [27]*

Dubost [27] has proposed a broadband circularly polarised flat antenna, as shown in Fig. 13.12, which consists of two flat radiating elements that are linearly polarised and placed orthogonally. Each radiating element is a symmetrically fed flat folded dipole, and it is separated from the reflector by a dielectric sheet. Dubost *et al.* [28] have described a cylindrical array composed of four such flat antennas producing circular polarisation and omnidirectional radiation.

Ito *et al.* [29] have proposed a travelling-wave-type circularly polarised array as shown in Fig. 13.13a. Inclined half-wavelength printed dipoles are arranged along both sides of a microstrip feed line terminated in a matched load. The spacing D_S between adjacent dipoles is a quarter of a guide wavelength, and the spacing D_p and the angle α are determined from the desired main-beam direction. The frequency bandwidth was as narrow as those of other simple microstrip antennas.

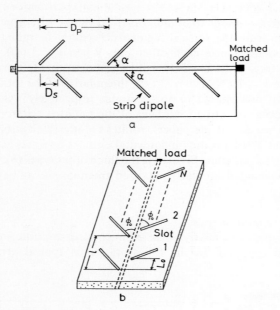

Fig. 13.13 *Travelling-wave printed arrays*
  a Dipole array [29]
  b Slot array [30]

Nakaoka *et al.* [30] have proposed another type of circularly polarised travelling-wave array composed of inclined slots, as shown in Fig. 13.13*b*. The slot arrangement can be determined in a similar way to the printed dipole array in Fig. 13.13*a*. Although the slot array requires a reflector under the substrate in practical use, its frequency bandwidth could be much wider than that of the printed-dipole array.

The crossed printed-dipole array [31], shown in Fig. 13.14*a*, or the crossed slot array [32], shown in Fig. 13.14*b*, both fed from microstrip lines, can be used as circularly polarised printed arrays. Compared with microstrip patch arrays, such arrays have relatively wide frequency bandwidths.

An alternative method of constructing a circularly polarised array is to place a circular polariser over a linearly polarised microstrip array. As a typical example, Henderson and James [33] have proposed a DBS reception array shown in Fig. 13.15. A parallel-plate polariser is overlaid on a comb-line array antenna [34] with a suitable spacer.

13.2 Design of circularly polarised arrays

13.2.1 Arrays of patch radiators

The generalised design method for circularly polarised microstrip patch arrays can be divided into two steps: the design of circularly polarised patch radiators themselves and the design of the appropriate feed network. The former is elaborated in Chapter 4.

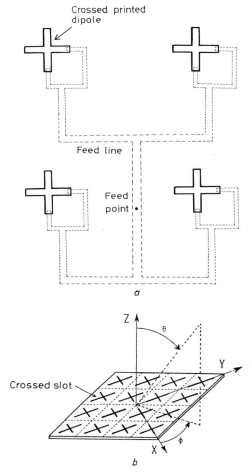

Fig. 13.14 *Crossed printed-element arrays*
 a Dipole array [31]
 b Slot array [32]

As is well known, typical feed systems for microstrip patch arrays are series feeding and parallel or corporate feeding [35]. Such feed systems will be described more fully in Chapter 14.

Comb-line array
E⊥
E∥
45°
Polariser

Fig. 13.15 *Linear polarised array and polariser [33] (© IEE)*

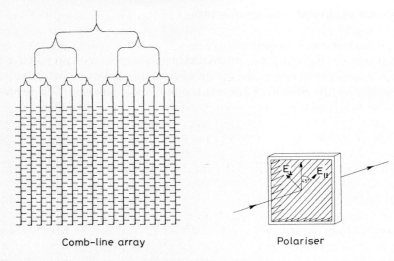

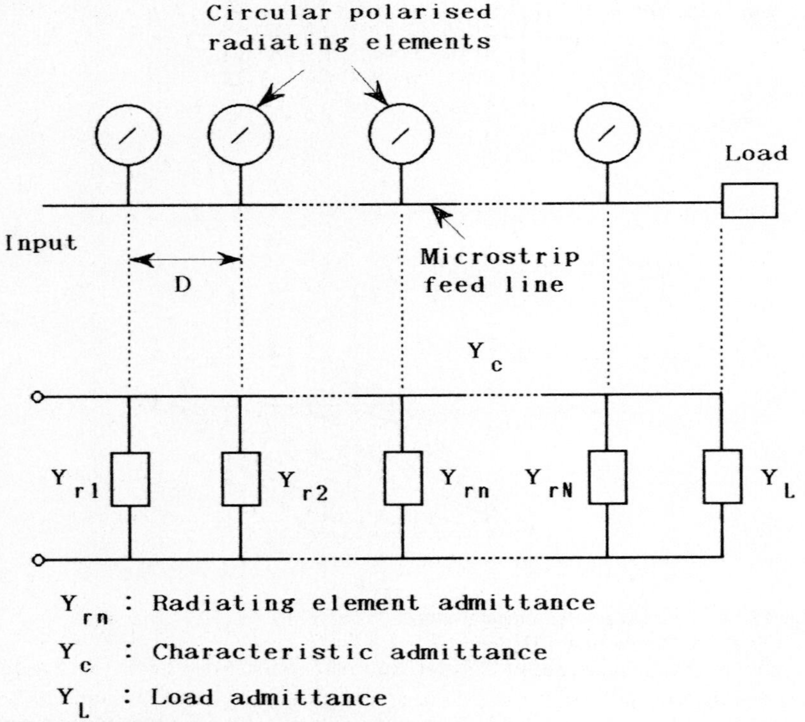

Circular polarised radiating elements
Load
a
Input
D
Microstrip feed line
Yc
b
Yr1
Yr2
Yrn
YrN
YL
Yrn : Radiating element admittance
Yc : Characteristic admittance
YL : Load admittance

Fig. 13.16 *Series-fed patch array and its equivalent circuit*
a Circularly polarised series-fed linear array
b Equivalent circuit

Circularly polarised antenna arrays 769

A typical series-fed linear array, as shown in Fig. 13.16a, consists of identical patch radiators producing circular polarisation, a microstrip feed line and its terminal load. Various shapes of patch radiators, as shown in Fig. 13.1, could be candidates for the array. Fig. 13.16a illustrates circular patch radiators with a single feed point. These radiators should be designed to produce circular polarisation and to meet the required conditions mentioned in Chapter 4.

For the design of such a series-fed linear array, an equivalent circuit model, as shown in Fig. 13.16b, is employed using transmission-line theory [36]. Normally, the patch radiators are attached to the feed line with equal spacing D. If the load admittance Y_L is equal to the line characteristic admittance Y_C, a travelling-wave array is formed [35]. On the other hand, if Y_L is equal to zero or infinity, a resonant array is formed. All the radiating-element admittances Y_{rn} ($1 \leqslant n \leqslant N$) can be determined from the conditions of input-impedance matching and aperture distribution required. A more generalised design method, using such an equivalent circuit, will be elaborated in the next Section.

In the array configuration, the excitation phase of the patch radiators is determined by their element spacing along the feed line. Therefore, the main beam direction will change with frequency variation. A series-fed linear array can minimise the feed-line losses at the sacrifice of frequency bandwidth.

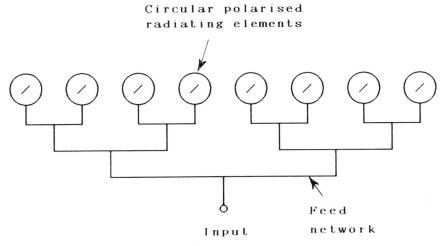

Fig. 13.17 *Example of corporate-fed linear array*

A typical example of the other feed system, parallel or corporate feeding, is shown in Fig. 13.17 feeding a circularly polarised linear array. In general, the patch radiators will be designed independently of the feed system in order to produce circular polarisation and to meet the required conditions mentioned above. The feed system, in this case, splits the input power between the output ports with a prescribed distribution, while maintaining equal electrical path

770 Circularly polarised antenna arrays

lengths from the input to output ports. Therefore, the frequency bandwidth of such an array could be essentially wide. In other words, the frequency bandwidth will be limited by the patch radiators. However, the feed-line losses would be quite large. Detailed design methods for the feed system will be described in Chapter 14.

In a practical patch array, the antenna performance with regard to such matters as radiation pattern and axial ratio could be deteriorated by mutual coupling between the radiating elements or unwanted radiation from, for example, its feed network. These practical design problems will be disussed in Section 13.3.

In practice, the frequency bandwidth of a microstrip patch antenna is relatively narrow. Section 13.4 will describe several design techniques for wideband circularly polarised patch arrays.

13.2.2 Arrays of composite elements

This Section will concentrate on the design of a circularly polarised printed array composed of strip dipoles and slots (CP–PASS) as a typical design method for series-fed circularly polarised arrays.

Fig. 13.9 shows a configuration of CP–PASS consisting of M radiating element sets which are fed in series from a microstrip feed line. For the design of a series-fed linear array, in general, an equivalent-circuit model and transmission-line theory are employed to determine the element spacing and input immittances [36].

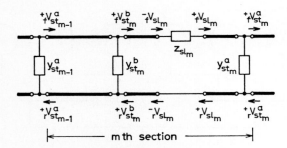

Fig. 13.18 *Equivalent circuit of mth element set of circularly polarised printed array of strips and slots (CP–PASS) [37] (© IEE)*

In an equivalent-circuit model, a strip dipole and a slot can be approximately represented by a shunt and a series element, respectively, to a transmission line [37]. The effect of the windows placed in the ground plane can be neglected at a design frequency of about f_c. In a practical design, inter-element coupling effects should be included in the determiantion of the element input immittances. Therefore, an equivalent circuit of the mth element set at f_c can be expressed as shown in Fig. 13.18, where z_{slm}, y_{stm}^a and y_{stm}^b are the normalised element input immittances and ${}_f^{\pm} V_{stm}^a$ etc. are the travelling-wave voltages (subscripts f and r indicate forward and reflected waves, respectively, and superscripts $+$ and $-$ indicate the load and generator sides of each element, respectively).

For simplicity in the design, we will set

$$y_{stm}^{a} = y_{stm}^{b} = y_{stm} = g_{stm} + jb_{stm} \qquad (13.1)$$

and also

$$z_{slm} = r_{slm} + jx_{slm} \qquad (13.2)$$

The relationship between z_{slm} and y_{stm} may be derived from the design conditions of both circular polarisation and input-impedance matching [37]. As a result, the design conditions for the mth element set will be written as

$$z_{slm} = 2y_{stm}^{*} \quad (\text{* is complex conjugate}) \qquad (13.3a)$$

$$b_{stm} = -g_{stm} \tan(\pi h \sqrt{\varepsilon_r}/\lambda_c) \qquad (13.3b)$$

where h and ε_r are the thickness and the relative dielectric constant of the substrate and λ_c is the free-space wavelength at f_c.

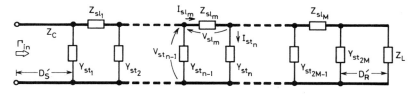

Fig. 13.19 *Equivalent circuit of M-set CP-PASS [38] (© IEE)*

Fig. 13.19 shows an equivalent circuit of an M-set CP–PASS at f_c, where Y_{stn} and Z_{slm} are the nth input admittance of a strip element and the mth input impedance of a pair of slots, respectively, and Z_L is a terminal load (normally it is a short or open circuit). Y_{stn} and Z_{slm} can be determined from the design conditions mentioned above and the aperture distribution required.

A generalised design procedure for linear-array-type CP–PASS is briefly described as follows [38]:

(a) At f_c, evaluate the propagation constant $\gamma_g (= \alpha_g + j\beta_g)$ of the microstrip line on a specific substrate. The substrate used throughout this Subsection is mentioned in Table 13.1.
(b) Calculate all the input immittances of the elements required for a specific CP–PASS design by applying the design procedure previously described [37].
(c) As a preliminary experiment, make a linear array composed of identical strip elements (no slots) arranged along a feed line. Derive the frequency dependence of the average input admittance $\bar{Y}_{st}^{st}$ from its measured input reflection coefficient Γ_{in}, where the superscript st denotes a strip-element array. Then, by testing some arrays which have different coupling gaps, obtain the depen-

dence of resonant conductance $\bar{G}_{st}^{st}$ and resonant length a_{st}^{st} on the coupling gap δ_{st} at f_c. Fig. 13.20 shows measured values of typical dependences of $\bar{G}_{st}^{st}$ and a_{st}^{st} on δ_{st} at resonance.

Table 13.1 *Design data for five-set Chebÿchev-array CP–PASS*

m	1	2	3	4	5
A_{rm}	1·000	1·609	1·932	1·609	1·000
G_{stm} (mS)	0·420	1·116	1·641	1·148	0·445
R_{slm} (Ω)	2·10	5·58	8·21	5·74	2·23
δ_{stm} (mm)	1·67	1·32	1·21	1·31	1·64
δ_{slm} (mm)	2·80	1·62	1·28	1·59	2·72
a_{stm} (mm)	35·38	34·93	34·73	34·91	35·35
a_{slm} (mm)	38·39	38·02	37·84	38·01	38·37

Substrate: DICLAD 522 ($h = 0.8$ mm, $\varepsilon_r = 2.6$) $w = 2.0$ mm ($Z_C = 50$ Ω), $\alpha_g = 2$ dB/m, $\eta_l = 0.68$ $f_c = 3.0$ GHz, $D_S = D_P = D_R = 68.0$ mm, $Z_L = 0$ Ω, $b_{st} = b_{sl} = 2.0$, $a_{wd} = 55.0$, $b_{wd} = 20.0$, $e_{sw} = 0.3$ mm

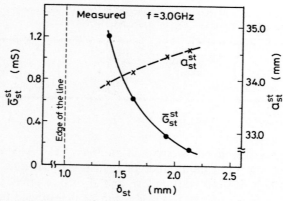

Fig. 13.20 *Measured typical dependences of average resonant conductance and resonant length of the coupling gap for strip-element arrays [38] (© IEE)*

(d) In a similar way, obtain the dependence of the average resonant resistance $\bar{R}_{sl}^{sl}$ and resonant length a_{sl}^{sl} on the coupling gap δ_{sl} for some slot arrays. Typical measured dependences are shown in Fig. 13.21.

(e) By combining the strip elements with the slots, construct a small array as shown in Fig. 13.9. To facilitate the procedure, the coupling gaps are chosen from Figs. 13.20 and 13.21 as the relationship for circular polarisation

$$\bar{R}_{sl}^{sl} = 2Z_c^2 \bar{G}_{st}^{st} \qquad (13.4)$$

is satisfied, where Z_c is the characteristic impedance of the line. Then, measure the axial ratio of the array in the broadside direction. If the axial ratio at f_c is greater than a specified value AR_{max}, the element lengths should be adjusted until the axial ratio becomes less than AR_{max}. Then obtain the frequency dependences of average input immittances $\bar{Y}_{st}^{cp}$ and $\bar{Z}_{sl}^{cp}$ of the elements, where the superscript cp denotes a circularly polarised array [38].

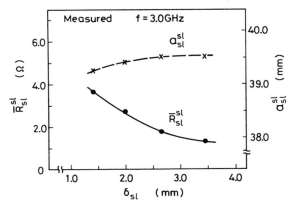

Fig. 13.21 *Measured typical dependences of average resonant resistance and resonant length on the coupling gap for slot arrays [38] (© IEE)*

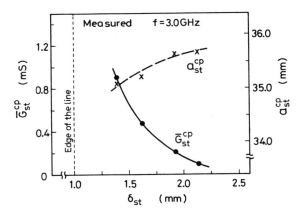

Fig. 13.22 *Measured typical dependences of average resonant conductance and resonant length on the coupling gap for circularly polarised arrays [38] (© IEE)*

(*f*) By testing some circularly polarised arrays with different coupling gaps, obtain the dependences of resonant conductance $\bar{G}_{st}^{cp}$ and length a_{st}^{cp} on δ_{st} and the dependences of resonant resistance $\bar{R}_{sl}^{cp}$ and length a_{sl}^{cp} on δ_{sl} at f_c. Some typical measured dependences are shown in Figs. 13.22 and 13.23, where AR_{max} was chosen to be 3 dB. It was found that the difference between, for example, Fig. 13.20 and Fig. 13.22 arose from mutual coupling between the strip elements and the slots.

(g) Determine all the element lengths and coupling gaps required for the design of CP–PASS from the experimental curves obtained in (f). There will be almost no need to correct the element dimensions because the curves will involve the inter-element coupling effects.

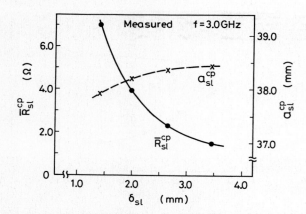

Fig. 13.23 *Measured typical dependences of average resonant resistance and resonant length on the coupling gap for circularly polarised arrays [38] (© IEE)*

In addition, when a reflector is placed under the ground plane, the same design procedure will also be available.

To demonstrate the validity of this design method, a Chebyshev-array CP–PASS consisting of five element sets was designed and measured at S-band [38]. The sidelobe level of -20 dB was specified. Using this design procedure and employing Figs. 13.22 and 13.23, design data for the array were obtained as shown in Table 13.1. A_{rm} is the amplitude ratio of each element set. For simplicity in the experiment, a reflector was not used.

Fig. 13.24 shows the front view of the array producing RHCP in the broadside direction. Slots and windows are indicated by broken lines.

Fig. 13.25 shows the measured radiation patterns in the yz-plane at f_c. The sidelobe levels of the co-polar radiation were less than -20 dB. The maximum cross-polar radiation was about -20 dB because circular polarisation was not achieved at f_c, as shown next.

Fig. 13.26 shows the measured axial ratio versus frequency. Circular polarisation was obtained at 2·95 GHz, which is 1·7% below f_c. The 3 dB bandwidth of the axial ratio was 9·3%.

Fig. 13.27 shows the measured input impedance of the array versus frequency. Good impedance matching was achieved as predicted, and the bandwidth for VSWR $\leqslant$ 2·0 was 8·5%.

For stricter designs of CP–PASS, e.g. a requirement for lower sidelobe levels, the cross-polarisation behaviour of the array has to be taken into account. In

Circularly polarised antenna arrays 775

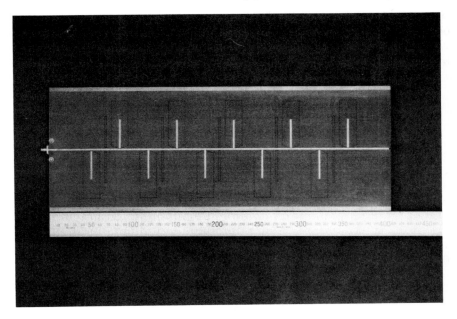

Fig. 13.24 *Front view of designed CP–PASS [38] (© IEE)*
Slots and windows are indicated by broken lines

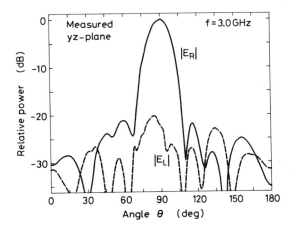

Fig. 13.25 *Measured radiation patterns in the yz-plane [38] (© IEE)*
$f = 3.0$ GHz (Fig. 13.24)

776   Circularly polarised antenna arrays

this case, the strip element will have to be represented instead by a T-type equivalent circuit in the design procedure [40].

For the case of a linear-array-type CP–PASS, the arrangement of the radiating elements shown in Fig. 13.9 seems to be the most suitable and efficient.

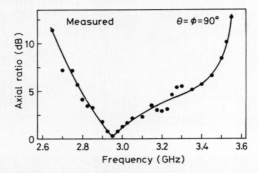

Fig. 13.26   *Frequency dependence of axial ratio measured at broadside [38] (© IEE)*

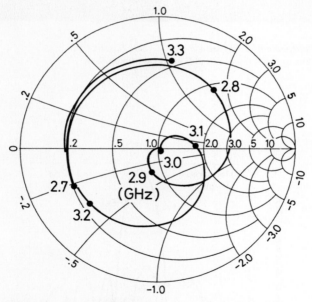

Fig. 13.27   *Measured input impedance versus frequency [38] (© IEE)*

However, if a planar array is formed using this arrangement, the spacing between elements on adjacent feed lines would be so small that the design would become more complicated because of strong inter-element coupling effects. To avoid this, radiating elements on one side of a microstrip line should be used for a planar array [39].

13.2.3 Design of travelling-wave arrays

Section 13.1.3 gave an outline of the four types of circularly polarised travelling-wave-array antennas. In this Section the crank-type microstrip-line antenna is chosen as representative and the method for its design is described.

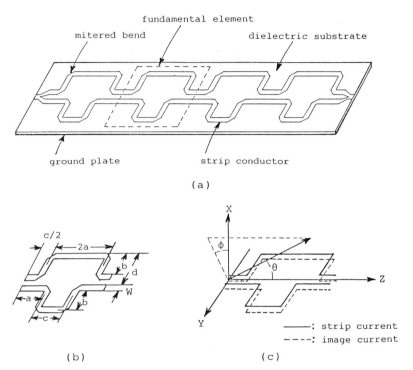

Fig. 13.28 *Crank-type microstrip-line antenna*
 a Antenna configuration (four elements)
 b Fundamental element
 c Idealised fundamental element

(a) Equations for the circularly polarised radiation: Fig. 13.28a shows a crank-type microstrip-line antenna with the fundamental element surrounded by a dotted line. Fig. 13.28b shows the shape of the fundamental element; however, we will study the idealised form shown in Fig. 13.28c. We make the following assumptions:

(i) The neighbourhood of the strip conductor is the medium with the effective relative permittivitiy.
(ii) There are two currents: the strip current which is concentrated along the central line of the strip conductor, and the image current formed by the ground plate.

778   Circularly polarised antenna arrays

(iii) The amplitude of the strip current is uniform, and the reflection of the electric current on the bent parts can be neglected. Thus, mutual coupling and substrate surface wave effects can also be neglected.

To obtain the conditions for the radiation of circularly polarised waves in $\theta = \theta_m$, $\phi = 0°$ direction, the following equation must be satisfied:

$$E_\phi = \pm jE_\theta \tag{13.5}$$

i.e. both components E_θ and E_ϕ of the electromagnetic wave radiated from the fundamental element must have the same absolute value, and if the phase difference is 90° the resulting wave will be circularly polarised.

When a linear-array antenna is constructed from these fundamental elements, the condition for the formation of the main beam in the θ_m direction, i.e. the condition that electromagnetic waves radiated in the θ_m direction from both end points of the fundamental element are in phase, is given by

$$kL\cos\theta_m - \beta(l - 4\delta) = 2n\pi \tag{13.6}$$

where $k = 2\pi/\lambda_0$, $\beta = 2\pi/\lambda_g$, $L = 2a + c$, $l = 2a + 2b + c$, λ_0 is the free space wavelength, λ_g is the guide wavelength of the microstrip line, L is the length of the fundamental element, l is the strip-conductor length of the fundamental element along the central line, a, b, c are the lengths of the crank segments, δ is the correction length for the crank segments and n is an integer. The physical meaning of δ is the difference between the physical and electrical lengths of the crank segments.

From the components E_θ and E_ϕ of the electromagnetic wave radiated from the fundamental element, and from eqns. 13.5 and 13.6, we can obtain a condition which ensures that the circularly polarised wave is radiated in the θ_m direction when $n = -2$. The condition is

$$a = \frac{1}{2U}\left\{\left(1 \mp \frac{1}{\pi}\tan^{-1}\Theta\right)\lambda_g - b + 2\delta\right\} \tag{13.7}$$

$$c = \frac{1}{U}\left\{\left(1 \pm \frac{1}{\pi}\tan^{-1}\Theta\right)\lambda_g - b + 2\delta\right\} \tag{13.8}$$

where

$$\Theta = \frac{(W-\delta)\sin\theta_m}{WU - \delta} \cdot \frac{\sin\frac{\beta}{2}(b-\delta) - \frac{\delta}{WU}\sin\frac{\beta}{2}(b + WU - 2\delta)}{\sin\frac{\beta}{2}(b-\delta) - \frac{\delta}{W}\sin\frac{\beta}{2}(b - W)} \tag{13.9}$$

$U = (1 - \zeta\cos\theta_m)$, $\zeta = \lambda_g/\lambda_0$ and W is the width of the crank segment. The upper sign represents left-handed circular polarisation and the lower sign represents right-handed circular polarisation. When the values for λ_g, ζ, θ_m and b

are given, the values for a and c can be obtained from eqns. 13.7 and 13.8. The choices of b is free; however, the optimal value for b is near $\lambda_g/2$.

Fig. 13.28b shows the distance d between the two strip conductors. The value of d should be chosen according to

$$d \leq \frac{\lambda_0}{2} \tag{13.10}$$

which is an experimentally derived inequality.

The conditions for circularly polarised radiation of the three remaining types of travelling-wave arrays are shown in Table 13.2.

(b) Details of the design: We will give the important factors in the construction of the four types of array antennas. However, a full explanation will be given only for the crank-type antenna. The length L of the fundamental element and the electrical length l' of the other three types can be obtained from Table 13.2.

(i) Correction of bent parts [41]
As shown in Fig. 13.29 the phase constant of the straight-line portion is β, and the phase constant of the bent parts is the effective phase constant β'. In this case, when the width of the straight line is W, the correction length for the crank segments will be

$$\delta = \left(1 - \frac{\beta}{\beta'}\right)W \tag{13.11}$$

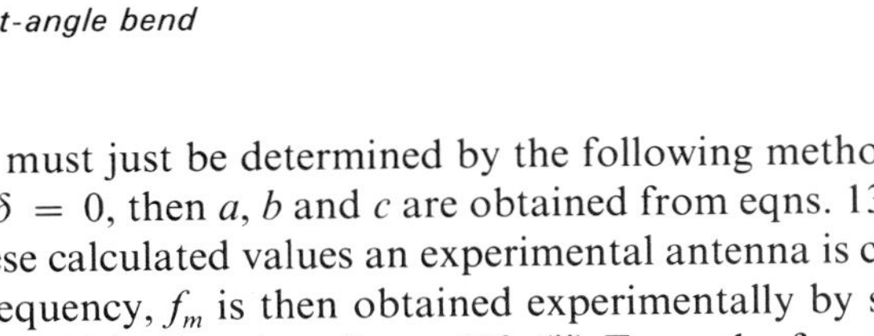

Fig. 13.29 *Right-angle bend*

The value of δ must just be determined by the following method: (i) By setting $\theta_m = 90°$ and $\delta = 0$, then a, b and c are obtained from eqns. 13.7 and 13.8. On the basis of these calculated values an experimental antenna is constructed. The value of the frequency, f_m is then obtained experimentally by setting the main beam in the broadside direction, $\theta_m = 90°$. (ii) From the frequency f_m and the design frequency f_d the value for δ is obtained from

$$\delta = \frac{\zeta V}{2}\left(\frac{1}{f_d} - \frac{1}{f_m}\right) \tag{13.12}$$

where V is the velocity of light. In this manner, a crank-type fundamental

Table 13.2 *Condition for circularly polarised radiation*

Type	Length of element L	Line length of element l	Electrical line length of element l'	Condition for circular polarisation ($\phi = 0°$)
(a) Rampart line	$2a + c$	$2a + 2b + c$	$l - 4\delta$	Eqns. 13.7 and 13.8
(b) Chain	$2\left(2a\sin\dfrac{\alpha}{2} + c\right)$	$2(2a + 2b + c)$	$l - 10\delta$*	$\tan\dfrac{\alpha}{2}\sin\theta_m = \tan\left(\dfrac{k_0(a-\delta)\,T}{2}\right)$, $T = \dfrac{\beta}{k} - \cos\theta_m \sin\dfrac{\alpha}{2}$, $\cos\theta_m = \dfrac{1}{L}\left(\dfrac{\beta l'}{k} - 2\lambda_0\right)$
(c) Square-loop	$2a + c$	$2a + 4b$	$l - 5\delta$*	when $\theta_m = 90°$, $a = \dfrac{1}{2}(\lambda_g + \delta)$, $b = \dfrac{\lambda_g}{4} + \delta$
(d) Crank	$2a + c$	$2a + 2b + c$	$l - 4\delta$	Eqns. 13.7 and 13.8

(a) Rampart line, (b) Chain, (c) Square-loop and (d) Crank
*Correction values for angles other than 90° are considered to be δ

Circularly polarised antenna arrays 781

element which radiates circular polarisation in the θ_m direction can be determined.

(ii) Main beam direction θ_m and frequency f
The crank-type microstrip-line antenna is a series-fed linear-array antenna, and when the frequency changes, the main beam direction will also change. The relationship between these two factors is given by

$$\cos\theta_m = \frac{1}{L}\left(\frac{l - 4\delta}{\zeta} + \frac{nV}{f}\right) \qquad (13.13)$$

where n represents a negative even number.

(iii) Length of the fundamental element
Normally we make $n = -2$ in eqn. 13.13 for the construction of a linear-array antenna. However, depending on the choice of the main beam direction θ_m, a grating lobe can appear simultaneously. Because of this, there are limitations on the values of θ_m when only the main beam, corresponding to $n = -2$, is present in the visible region. In other words,

$$\frac{L}{\lambda_0} < \frac{v}{\cos\theta_m + 1} \qquad (13.14)$$

$v = 1$ for rampart, loop

$v = 2$ for chain, crank

According to this equation, when the lengths of the crank segments a, b and c are calculated from eqns. 13.7 and 13.8 for a given value for θ_m, the value $L = 2a + c$ must satisfy eqn. 13.14. The direction θ_m is usually larger than 60°.

(iv) Return loss
A microstrip-line antenna has its microstrip lines periodically bent. If there are any reflected waves from the bent parts, such an antenna will exhibit a high return loss, no matter how small the reflection. High return loss is a consequence of the total sum of the reflected waves which come from every fundamental element, and are in phase when the electrical length of the fundamental element, $l' = l - 4\delta$, is equal to a multiple of $\lambda_g/2$. Therefore, the condition $l' = m\lambda_g/2$ (m is an integer) must be avoided.

When $\theta_m = 90°$, then $l' = 2\lambda_g$; therefore, there is always high return loss. In this case the following counter-measures can be applied: (i) The bent parts must be matched very carefully; and (ii) the length of each crank segment must be chosen carefully, so that the sum of the reflected waves from all the four bent parts in the fundamental element equals zero.

(v) Transmission loss and efficiency
In the case of a microstrip-line antenna, the strip current fed through the feeding end decreases exponentially on its way from the feeding end to the matched load, because the strip current is radiated from each fundamental element. Therefore, amplitude distribution, as shown in Fig. 13.30, can be assumed. In this case, the efficiency η of the microstrip-line antenna can be represented by

782 Circularly polarised antenna arrays

$$\eta = \eta_a \eta_b \eta_c$$
$$= \frac{2(1-R)}{(1+R)\ln(1/R)}(1-R^2)\eta_c \qquad (13.15)$$

where η_a is the aperture efficiency, η_b is the feeding efficiency $(1-R^2)$, η_c is the radiating efficiency and R^2 represents the power dissipated at the matched load. Here, η_c is determined by both the conductor loss and dielectric loss.

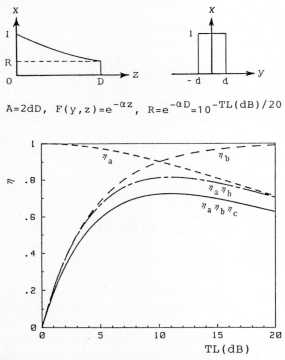

Fig. 13.30 *Efficiency of travelling-wave antenna ($\eta_c = 0.9$)*

Fig. 13.30 shows the results when $\eta_c = 0.9$. As can be seen, there is an optimal range for the transmission loss of between 10 and 12 dB, in which case the antenna efficiency is about 72%.

An example of the crank-type microstrip-line antenna is given in Chapter 19.

13.3 Practical design problems

13.3.1 Mutual coupling

Mutual coupling between radiating elements in microstrip arrays results in both distortion of the element radiation pattern and also errors in the element feed

Circularly polarised antenna arrays

voltages. In addition, in the case of circularly polarised arrays, it causes deterioration of the polarisation characteristics. Theoretical studies of mutual coupling in microstrip arrays have been presented by Pozar [42] and Malkomes [43], and experimental work has been done by Jedlicka *et al.* [44] and Haneishi *et al.* [45]. All these works, however, dealt with linear polarisation.

From a knowledge of mutual coupling in a linearly polarised array, the mutual coupling effects on polarisation characteristics of a circularly polarised array can be derived.

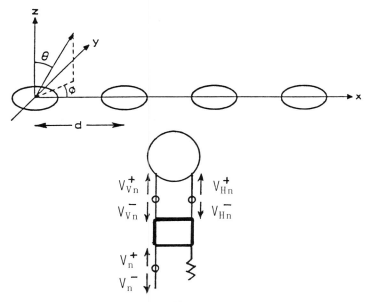

Fig. 13.31 *Geometry of circularly polarised linear array*

Let us consider a circularly polarised linear array composed of microstrip patches with feed hybrids as shown in Fig. 13.31. The array can be represented in terms of a scattering matrix [46], which expresses the complex coupling coefficients between the incident $^+$ and reflected $^-$ voltage at each feed part. Each element of the scattering matrix is given by E-plane and H-plane coupling coefficients for the linearly polarised array. Thus, for a circularly polarised array, if we assume the mutual coupling between orthogonal linearly polarised V- and H-components to be neglected, we have

$$\begin{pmatrix} V_{V1}^- \\ V_{V2}^- \\ \cdot \\ \cdot \\ \cdot \\ V_{VN}^- \end{pmatrix} = \begin{pmatrix} S_{V11} & S_{V12} & \cdots & S_{V1N} \\ S_{V21} & S_{V22} & \cdots & S_{V2N} \\ \cdot & & & \\ \cdot & & & \\ \cdot & & & \\ S_{VN1} & S_{VN2} & \cdots & S_{VNN} \end{pmatrix} \begin{pmatrix} V_{V1}^+ \\ V_{V2}^+ \\ \cdot \\ \cdot \\ \cdot \\ V_{VN}^+ \end{pmatrix} \qquad (13.16)$$

and

$$\begin{pmatrix} V_{H1}^- \\ V_{H2}^- \\ \cdot \\ \cdot \\ \cdot \\ V_{HN}^- \end{pmatrix} = \begin{pmatrix} S_{H11} & S_{H12} & \cdots & S_{H1N} \\ S_{H21} & S_{H22} & \cdots & S_{H2N} \\ \cdot \\ \cdot \\ \cdot \\ S_{HN1} & S_{HN2} & \cdots & S_{HNN} \end{pmatrix} \begin{pmatrix} V_{H1}^+ \\ V_{H2}^+ \\ \cdot \\ \cdot \\ \cdot \\ V_{HN}^+ \end{pmatrix} \quad (13.17)$$

The excitation voltages V_{Vn} and V_{Hn} at the V- and H-ports of the nth element are given by

$$V_{Vn} = (V_{Vn}^+ + V_{Vn}^-) = V_{Vn}^+ (1 + \Gamma_{Vn}) \quad (13.18)$$

where

$$\Gamma_{Vn} = \sum_{m=1}^{N} S_{Vnm} \frac{V_{Vm}^+}{V_{Vn}^+} \quad (13.19)$$

and

$$V_{Hn} = (V_{Hn}^+ + V_{Hn}^-) = V_{Hn}^+ (1 + \Gamma_{Hn}) \quad (13.20)$$

where

$$\Gamma_{Hn} = \sum_{m=1}^{N} S_{Hnm} \frac{V_{Hm}^+}{V_{Hn}^+} \quad (13.21)$$

If each hybrid is ideal, V_{Vn}^+ and V_{Hn}^+ can be expressed by the input voltage V_n^+ to the hybrid as

$$V_{Vn}^+ = \frac{1}{\sqrt{2}} V_n^+ \quad (13.22)$$

and

$$V_{Hn}^+ = \frac{\pm j}{\sqrt{2}} V_n^+ \quad (13.23)$$

The sign $\pm$ corresponds to the sense of polarisation rotation. Thus, the co- and cross-polarised components radiated from the nth element become

$$E_{nc} = V_n^+ \left(1 + \frac{\Gamma_{Vn} + \Gamma_{Hn}}{2}\right) \quad (13.24)$$

and

$$E_{nx} = V_n^+ \left(\frac{\Gamma_{Vn} - \Gamma_{Hn}}{2}\right) \quad (13.25)$$

If the main beam is scanned in the direction θ_0, where θ_0 is the angle from the normal to the array, as shown in Fig. 13.31, the co- and cross-polarised fields radiating from the array are expressed by

$$E_c(\theta) = \sum_{n=1}^{N} V_n^+ \left(1 + \frac{\Gamma_{Vn} + \Gamma_{Hn}}{2}\right) e^{jkd(n-1)(\sin\theta - \sin\theta_0)}, \qquad (13.26)$$

and

$$E_x(\theta) = \sum_{n=1}^{N} V_n^+ \left(\frac{\Gamma_{Vn} - \Gamma_{Hn}}{2}\right) e^{jkd(n-1)(\sin\theta - \sin\theta_0)} \qquad (13.27)$$

Γ_{Vn} and Γ_{Hn} depend on the scan angle θ_0, and thus the polarisation characteristics vary with the scan angle due to mutual coupling.

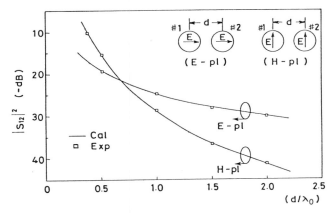

Fig. 13.32 *Measured and calculated mutual coupling $|S_{12}|^2$ for circular patches [47]*

Next, let us consider the element spacing of circularly polarised arrays. In order to suppress grating lobes and scan the beam widely, the spacing between adjacent element should be small. Small spacing, however, causes large mutual coupling, and deteriorates the radiation characteristics. Thus the spacing for a circularly polarised array should be chosen from practical trade-offs. The calculated and measured mutual coupling of a two-element circular-patch array versus element spacing for both the E-plane and H-plane are shown in Fig. 13.32 [47]. From this Figure, it can be seen that, as the spacing increases, the E-plane coupling becomes larger than the H-plane coupling owing to the stronger surface wave, and the magnitude of coupling for both planes is almost equal at $d = 0.68\lambda_0$. Fig. 13.33 shows the effect of orientation on coupling between circular patches for several element spacings. From this Figure, $d = 0.68\lambda_0$ may be suggested for circularly polarised arrays, because, at this spacing, the mutual couplings for the E- and the H-planes are nearly equal and less than -22 dB; furthermore, they are almost independent of the orientation angle of the array. This fact almost holds good for rectangular microstrip patches.

786 Circularly polarised antenna arrays

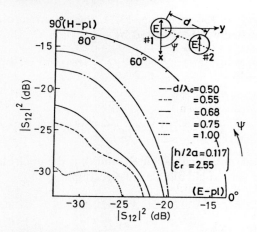

Fig. 13.33 *Mutual coupling between microstrip circular patches as a function of orientation* [47]

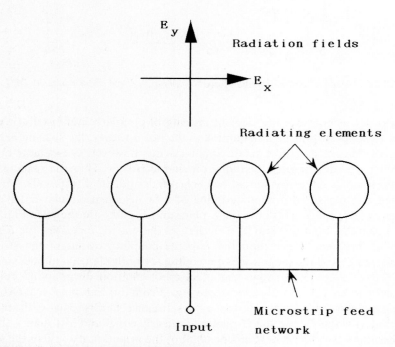

Fig. 13.34 *Concept of circularly polarised array*

13.3.2 Unwanted radiation

For practical design of microstrip-array antennas, and particularly for circularly polarised antennas, so-called 'unwanted radiation' should be taken into account. Unwanted radiation may result from the generation of higher-order modes in microstrip patches (see Chapter 4 for details), co-planar microstrip feed lines and discontinuities [18], microstrip feed transitions such as connectors [48], surface waves excited on the substrate [49], secondary current sources on the substrate edges [50], and so forth.

Unwanted radiation, in general, sometimes causes high cross-polarisation level, degradation of antenna gain, reduction of frequency bandwidth, and alteration to radiation patterns such as sidelobes and nulls.

Recently, cross-polarisation effects in linearly polarised microstrip antennas have been studied by, for example, Hall and James [51] and Hansen [52]. For circularly poalrised antennas, the axial ratio will readily be degraded by cross-polarisation. In the following, the relationship between axial ratio and cross-polarisation level will be discussed.

Fig. 13.34 shows a typical circularly polarised array composed of identical radiating elements with a normal microstrip feed network. The array will radiate nearly circular polarised waves in the broadside direction, which can be decomposed into the two orthogonal fields E_x and E_y. The circularly polarised radiation fields can be written as

$$|E_R| = |E_x + jE_y|/\sqrt{2} \qquad (13.28)$$

$$|E_L| = |E_x - jE_y|/\sqrt{2} \qquad (13.29)$$

where the subscripts R and L represent right-hand and left-hand circular polarisation, respectively. Then the axial ratio AR is given by

$$AR = (|E_R| - |E_L|)/(|E_R| + |E_L|) \qquad (13.30)$$

Suppose that the total radiation fields E_{xt} and E_{yt} are simply expressed as

$$E_{xt} = E_x + |E_{xc}|.\exp(jp) \qquad (13.31)$$

$$E_{yt} = E_y \qquad (13.32)$$

where $|E_{xc}| \exp(jp)$ represents the unwanted radiation field caused by, say, higher-order modes in the elements or feed network as mentioned above.

If right-hand circular polarisation is desired, the relationship $E_x = jE_y$ must be satisfied for the case of no unwanted radiation. However, we then obtain

$$E_x + |E_{xc}| \exp(jp) = jE_y \qquad (13.33)$$

Next UR is defined as

$$UR = |E_{xc}|/|E_x| \qquad (0 < UR < 1) \qquad (13.34)$$

in order to examine the influence of unwanted radiation on the axial ratio. Putting $E_x = |E_x|$ and substituting eqn. 13.34 into eqns. 13.28 and 13.29, we obtain

$$\frac{|E_L|}{|E_R|} = \frac{UR}{|2 + UR\exp(jp)|} \quad (13.35)$$

Then the axial ratio can be written as

$$AR = \frac{|2 + UR\exp(jp)| - UR}{|2 + UR\exp(jp)| + UR} \quad (13.36)$$

Fig. 13.35 shows the degradation of the axial ratio caused by the unwanted radiation defined in eqn. 13.34. The solid and the broken lines indicate the estimates for the cases $p = -\pi$ (the worst case) and $p = \pm\pi/2$, respectively.

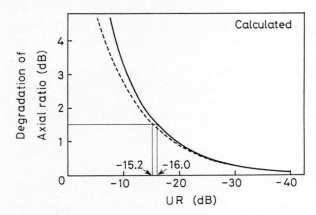

Fig. 13.35 *Influence of unwanted radiation on axial ratio*
——— $p = -\pi$ (worst case)
- - - - $p = \pm\pi/2$

The above discussion illustrates that the influence of unwanted radiation on the axial ratio can be quite large. Therefore, it is quite important to suppress such unwanted radiation in the design of circularly polarised antennas.

Various methods for suppressing unwanted radiation in circularly polarised microstrip antennas have been proposed. Chapter 4 elaborates on the suppression of higher-order modes in microstrip patches and describes an effective suppression method using 'paired elements'. Another effective method is to form a 'sequential array', which is a generalised method of paried elements. The sequential array is described in detail in Section 13.4.3.

13.3.3 Limitations and trade-offs

In a large array, the achievable gain is limited owing to the conduction loss and the dielectric loss in the microstrip feed line, and the radiation loss generated from discontinuities in hybrids, impedance transformers and right-angled corners of the microstrip line.

As the array size increases, the directional gain will increase proportionately. However, as array size increases, the feed-line length become longer, and the feeder loss will eventually increase faster than the directional gain; the power gain will therefore decrease.

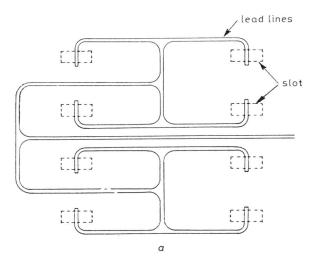

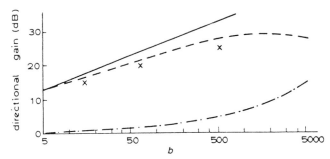

Fig. 13.36 *Microstrip ground-plane slot array losses [35]*
  a Array corporate feed arrangement
  b Power gain of array against number of elements
  ——— theoretical gain of array with no feeder losses
  - - - - theoretical gain of array with feeder losses
  –·–·– feeder attenuation
  × × × measured values of gain

Corporate feed arrays in particular, have longer feed lines and larger feeder loss than series-fed arrays. Reference 53 has quantified the limit for corporate feed arrays. The feed geometry is shown in Fig. 13.36a; each line feeds a ground-plane slot and there are $2L/\lambda_0$ and L/λ_0 elements in the E- and H-planes, respectively, where L^2 is the area of the square array. Although this array is designed for linear polarisation, circularly polarised array can be treated in a similar manner. For the geometry shown, the length of feeder line from the input point to any element is $3L/2$. thus the power gain of the array is given by

$$G = G_{el} + 10 \log \frac{2L^2}{\lambda_0^2} - \frac{3}{2} LF \tag{13.37}$$

where G_{el} is the element gain in decibels and F is the feeder loss in dB/unit-length [35].

The optimum value of L which gives the maximum power gain is

$$L = \frac{13 \cdot 3 \log e}{F} = \frac{5 \cdot 776}{F} \tag{13.38}$$

Fig. 13.36b shows the power gain versus number of elements for a 12 GHz array with a feeder loss $F = 0.075$ dB/cm. The maximum gain can be seen from the Figure to be about 30 dB.

An effective method of reducing feed losses in the corporately fed arrays is to replace part of the microstrip feed lines by a low-loss medium such as a coaxial line or waveguide. This approach is used in a synthetic-aperture radar antenna for the SEASAT satellite [54], and in a circularly polarised microstrip array for reception in a 12 GHz direct broadcasting satellite (DBS) [55]. The latter array consists of four 256-disc-element sub-arrays which are connected by a waveguide power combiner mounted on the rear side of the array. A measured gain of more than 33 dB was achieved for this 1024-element array.

In co-planar arrays, not only conductor loss and dielectric loss of feed lines, but also spurious radiations from power dividers, impedance transformers and corners become significant loss factors. These losses in microstrip lines represent the major limitation of microstrip antennas. Estimates of these losses are summarised as follows:

(a) *Dielectric loss* [56]:

$$\alpha_d = 27 \cdot 3 \left(\frac{q\varepsilon}{\varepsilon_e}\right) \frac{\tan \delta}{\lambda_g}$$

$$= 4 \cdot 34 \frac{q}{\sqrt{\varepsilon_e}} \sqrt{\frac{\mu_0}{\varepsilon_0}} \sigma \text{ (dB/cm)} \tag{13.39}$$

where ε is the dielectric constant of the substrate, $\varepsilon_e = 1 + q(\varepsilon - 1)$, q is the dielectric filling factor and σ is the conductivity of the substrate.

(b) *Conductor loss* [56]:

$$\alpha_c = \frac{8 \cdot 68 R_s}{WZ_m} \text{ (dB/cm)}, \tag{13.40}$$

$$R_s = \sqrt{\frac{\omega \mu_0}{2\sigma_c}} \tag{13.41}$$

where W is the width of the strip conductor, Z_m is the characteristic impedance, and σ_c is the conductivity of strip.

(c) Radiation loss [57]: All the expressions for radiated power due to discontinuities take the general form

$$P = 60(kt)^2 F(\varepsilon) \tag{13.42}$$

where t is the dielectrics thickness, k is the free-space wave number and $F(\varepsilon)$ depends only on the substrate permittivity, but is different for each microstrip configuration. This equation is for a unit incident current wave, the reflection at the open circuit being assumed complete. The $F(\varepsilon)$s for fundamental circuits were derived by Lewin [57].

Open circuit

$$F_1(\varepsilon) = \frac{\varepsilon + 1}{\varepsilon} - \frac{(\varepsilon - 1)^2}{2\varepsilon\sqrt{\varepsilon}} \ln \frac{\sqrt{\varepsilon + 1}}{\sqrt{\varepsilon - 1}} \tag{13.43}$$

Short circuit

$$F_2(\varepsilon) = 3 - \frac{1}{\varepsilon} - \left(3 + \frac{1}{\varepsilon}\right) \frac{\varepsilon - 1}{2\varepsilon\sqrt{\varepsilon}} \ln \frac{\sqrt{\varepsilon + 1}}{\sqrt{\varepsilon - 1}} \tag{13.44}$$

Matched termination

$$F_3(\varepsilon) = 1 - \frac{\varepsilon - 1}{2\sqrt{\varepsilon}} \ln \frac{\sqrt{\varepsilon + 1}}{\sqrt{\varepsilon - 1}} \tag{13.45}$$

Right-angle corner

$$F_4(\varepsilon) = \frac{\varepsilon + 1}{\sqrt{\varepsilon}} \ln \frac{\sqrt{\varepsilon + 1}}{\sqrt{\varepsilon - 1}} - \frac{2\varepsilon}{\sqrt{2\varepsilon - 1}} \ln \frac{\sqrt{2\varepsilon}}{\sqrt{2\varepsilon - 2}} \tag{13.46}$$

The dielectric loss is almost constant for the substrate thickness. Conductor loss and radiation loss, however, depend on the thickness, and furthermore the variations of losses with thickness are quite different. Thus, a study of the optimum thickness of the substrate is necessary.

Fig. 13.37 shows a calculated example of conductor losses and the radiation losses against thickness of the microstrip substrate [58]. The antenna consists of a 256-element circularly polarised planar array at 12 GHz. It was assumed that each feed line from the input port to each element has a length of 40 cm, eight 2-way dividers, and seven right-angle corners.

From the Figure, it can be seen that the conductor losses increase rapidly as the substrate thickness decreases; on the other hand, the radiation losses increase as the substrate becomes thicker. The losses also depend strongly on the characteristic impedance of the microstrip lines. Since an optimum thickness exists at which the sum of conductor loss and radiation loss becomes a minimum, we can determine the substrate with the optimum thickness.

13.3.4 Non-planar scanning arrays

User terminals in aeronautical-satellite and inter-satellite communication links require high-gain beams, capable of being steered to wide angles over a full hemisphere. Scanning losses in planar phased arrays increase rapidly beyond about 60°, necessitating employment of non-planar scanning arrays.

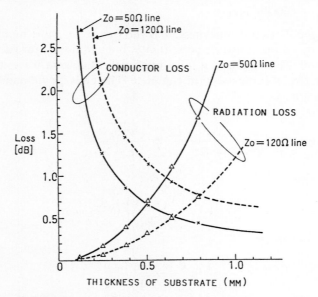

Fig. 13.37 *Conductor loss and radiation loss versus substrate thickness [58]*
f = 12 GHz, ε_r = 2·17

Electronically switched spherical arrays are simple wide-angle scanning arrays; an example using microstrip antennas has been described in a low-orbiting satellite pointing its beam at at geostationary data-relay satellite [59]. Another example is a dome-shaped switching array which consists of 120 circularly polarised microstrip disc elements, and which provides 14 dB of gain with good uniformity over almost 300° of total angle [60]. The array produces a beam by exciting a 12-element sub-array with non-phase compensation, and changes the beam-pointing direction by selecting another sub-array which may or may not contain some of the elements in the first sub-array. A disadvantage of such electronically switched spherical arrays is that they require a large number of elements to cover the hemisphere with high-gain beams; consequently, they are bulky and heavy.

A phase-compensated switched-element spherical array for mobile earth stations for satellite communication was proposed by Hori *et al.* [61]; thus enables one to achieve a high-gain beam with a small number of elements, by providing the same number of variable phase shifters as excited elements.

Let us consider a general spherical array, whose co-ordinate system is shown in Fig. 13.38. The array is composed of N elements which are located on the limited sphere tilted at an angle α from the vertical axis (z-axis). If M elements are excited and phase-shifted at a time, the radiation pattern is calculated from the following equation:

$$E(\theta, \phi) = \left| \sum_{n=1}^{M} g(\lambda, \mu) \exp[jk(e_r \cdot e_{pn}) + j\Psi_n] \right| \quad (13.47)$$

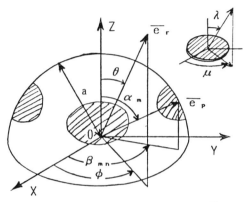

Fig. 13.38 *Co-ordinate system for spherical array [61] (© IEE)*

where, k is the wave number and Ψ_n is the phase of the nth element. e_r and e_{pn} are the vectors of the radiation and element position, respectively, and are given by

$$e_r = (\sin \theta \cos \phi, \sin \theta \sin \phi, \cos \theta) \quad (13.48)$$

$$e_{pn} = a(\sin \alpha \cos \beta_n, \sin \alpha \sin \beta_n, \cos \alpha) \quad (13.49)$$

where a is the radius of the sphere and $g(\lambda, \mu)$ is the radiation pattern of the array element.

The minimum coverage gain of a simple switched-element spherical array in which the elements are switched one at a time is represented by the crossover level of the radiation patterns of the adjacent array elements within the coverage area ($\theta_0 - \Delta\theta$ and $\theta_0 + \Delta\theta$ from the vertical axis). Since the minimum coverage gain of the antenna is limited, in order to improve it without altering the number of elements, the switched-element array proposed employs the same number of phase shifters as excited elements, as shown in Fig. 13.39. This antenna has both switched-element-array and phased-array functions. It uses the first function to achieve wide scanning and the seond to increase crossover level between adjacent beams.

The L-band array developed is shown in Fig. 13.40. It is 40 cm in diameter and 20 cm in height. The antenna is composed of the radiator section, the switching circuit and the controller. The radiator section consists of six microstrip discs with parasitic elements for broadening the bandwidth, which will be described in the next Section.

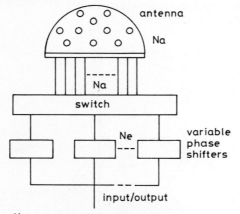

Fig. 13.39 *Switching spherical array with phase shifters [61] (© IEE)*

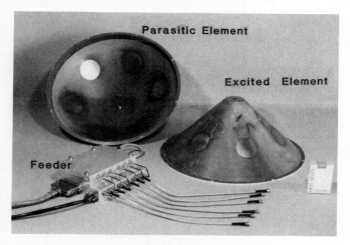

Fig. 13.40 *Inner construction of the spherical switching array (Courtesy: NTT, Japan)*

The two-element-excitation method is applied in this test antenna. To implement two-element excitation, only one phase shifter consisting of two bits, one of 45° and the other of 90°, is required. Consequently, two elements can generate seven beams, and a six-element array can radiate 42 beams.

The minimum coverage gain of a six-element array can be improved by 2·3 dB by using the proposed method, when the coverage area is within 20°–60° from the vertical axis.

In the design of conformal arrays, the mutual coupling between elements arranged on a curved or folded plane has to be investigated. Hori [62] studied experimentally the mutual coupling between microstrip discs with parasitic elements arranged on a roof-shaped plane as shown in Fig. 13.41. In the same

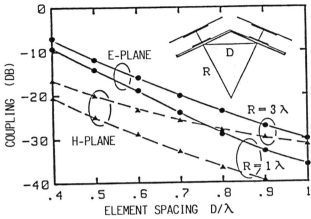

Fig. 13.41 *Measured mutual coupling of the non-planar two-element array with parasitic elements [62]*

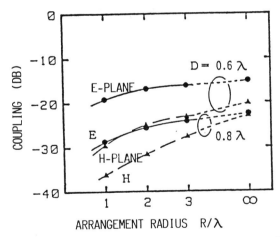

Fig. 13.42 *Measured mutual coupling of the non-planar array versus radius of the arrangement [62]*

figure, the mutual coupling versus element spacing D/λ (λ is the wavelength) is also shown. Fig. 13.42, however, shows the mutual coupling versus the arrangement radius R/λ. From the Figure, it can be seen that at $R > 3\lambda$, the mutual coupling becomes almost constant, and at $R < 2\lambda$, the mutual coupling in the non-planar arrangement is $5 \sim 10\,\text{dB}$ less than in planar arrangement.

796 Circularly polarised antenna arrays

13.4 Wideband circularly polarised arrays

In most communication systems, transmitting and receiving frequencies are separated by several percent (typically 7–10%). In general, a microstrip antenna has a narrow frequency bandwidth; therefore, in the practical design of circularly polarised arrays, techniques for achieving wideband polarisation characteristics, as well as wideband impedance characteristics, are important. There are the following techniques for achieving wideband circularly polarised arrays:

(i) Employment of wideband circularly polarised radiating elements
(ii) Stacked elements for dual-frequency resonance
(iii) Special configuration for wideband circularly polarised array.

Details of these techniques are described below.

13.4.1 Arrays of wideband elements

(a) Radiating element with substrate of low dielectric constant: Since the bandwidth of a microstrip antenna is given by [63]

$$BW = \frac{VSWR - 1}{Q\sqrt{VSWR}} \tag{13.50}$$

the bandwidth can be increased by reducing the Q-factor. It is well known that the Q-factor is proportional to dielectric constant of the substrate, and inversely proportional to its thickness. Therefore, in order to broaden the bandwidth, the utilisation of a thicker substrate with lower dielectric constant is effective.

One such substrate consists of two thin layers of PTFE (polytetrafluoroethylene) bonded on each side of honeycomb material. This method, however, frequently generates higher-order modes in a microstrip antenna. In circularly polarised antennas, the higher-order modes become one of the sources of cross-polarised waves, and therefore they must be suppressed. In a microstrip circular patch, the dominant mode is TM_{110} and the first higher-order mode is TM_{210}. One method of suppressing the undesired TM_{210} is to excite a microstrip radiator by four feeds with 0°, 90°, 180° and 270° phase differentials [3], as shown in Fig. 13.43. The TM_{210} mode is found by measuring the cross-coupling between orthogonal ports. Fig. 13.44 gives a measured example of the coupling between the orthogonal ports; the solid curve shows the case of the four-probe feed in Fig. 13.43, while the broken curve shows the case of a conventional two-probe feed. By using a thicker substrate with four-probe feeds, relativley wide impedance and axial-ratio bandwidths ($\geqslant 10\%$) can be achieved.

For a large array, a multiple-probe feed system would become complicated, more expensive, and more prone to RF loss.

Another technique for suppressing the effect of higher-order modes generated in a thick substrate of low dielectric constant is to cut two notches on the patches [64]. In a normal microstrip circularly polarised antenna using a thick substrate, higher-order modes are excited because of the asymmetrical feed structure, and they generate cross-polarisation. However, it is known that notches in circular patch provide arbitrary elliptical polarisation, and therefore properly designed notches can cancel the cross-polarisatin caused by the asymmetical feed structure.

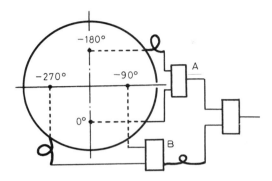

Fig. 13.43 *Four-probe feed for higher-order mode suppression [3] (© 1982 IEEE)*

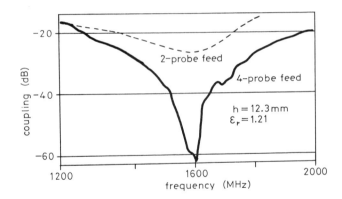

Fig. 13.44 *Comparison of coupling between feed ports [3] (© 1982 IEEE)*

Fig. 13.45 shows a seven-element array, in which each element has two notches and is excited from orthogonally located feed points with a phase differential of 90°. Fig. 13.46 shows the measured axial ratio of this array, and it can be seen that the effect of the notches on the suppression of higher-order modes is obvious.

(b) Application of parasitic elements: A two-layer microstrip antenna capable of broadband performance with excellent circular polarisation was proposed [65]. Such antennas are also referred to as electromagnetically coupled

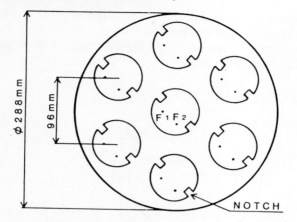

Fig. 13.45 *Wideband microstrip array composed of notched elements [64]*

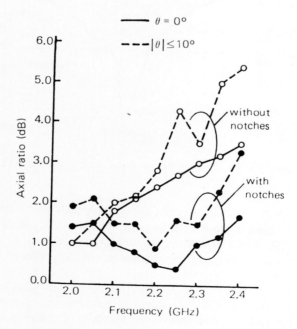

Fig. 13.46 *Improvement of axial ratio of the 7-element array by use of notches [64]*

patches (EMCP), which have been shown to be broadband radiators for linear polarisation. Fig. 13.47 illustrates the structure of the EMCP. The antenna element consists of two circular patches of diameter D_f and D_r separated by a

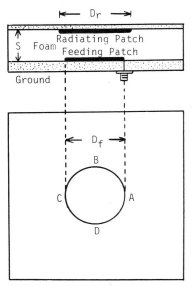

Fig. 13.47 *Electromagnetic coupled patch antenna [65] (© 1984 IEEE)*

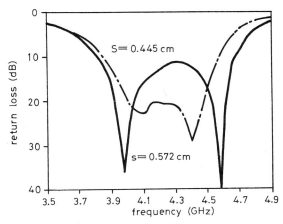

Fig. 13.48 *Measured return loss [65] (© 1984 IEEE)*

distance S. The top patch is excited by the bottom patch (the feeding patch) which in turn, is fed by a coaxial line from underneath, or by a microstrip line on the same plane as the feeding patch. The return loss of the EMCP, shown in Fig. 13.48, is characterised by two resonant frequencies which vary with separation. In general, the upper resonant frequency shifts downward and the lower shifts upward when the separation increases.

By using EM-coupled patches, a broadband circularly polarised array can be produced. However, when fed at two points (A and B in Fig. 13.47), the EMCP generates highly elliptical polarisation because of its asymmetrical feed structure. One technique of achieving a good axial ratio is shown in Fig. 13.49. This

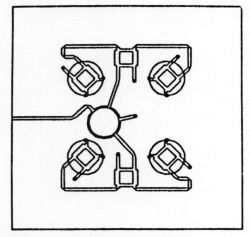

Fig. 13.49 *Circularly polarised EMCP array [65] (© 1984 IEEE)*

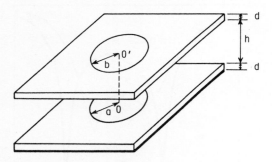

Fig. 13.50 *Microstrip disc antenna with parasitic element [67]*

array employs symmetrical deployment of the radiating elements, which are equally excited at two points. The configuration can cancel the radiation difference as a result of the symmetrical arrangement of array elements.

The EM-coupled patches are also capable of having only one resonant frequency, and they have broadband characteristics. The antenna was first described by Taga *et al.* [66], and more recently it has been applied to a circularly polarised shipborne antenna array for mobile satellite communication [67]. The structure of the broadband microstrip antenna with a parasitic element is illustrated in Fig. 13.50, where a parasitic element of radius b is mounted over a microstrip antenna of radius a at height h. The parameters a, d and the dielectric constant ε_r are determined from the substrate and the operating frequency, while b and h are related to the bandwidth. An example of the relative

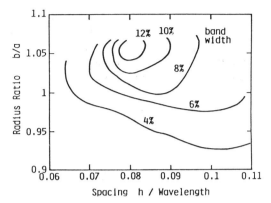

Fig. 13.51 *Relative bandwidth variation against element spacing and radius ratio of circular disc elements [67]*

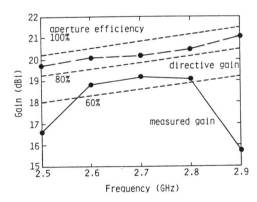

Fig. 13.52 *Frequency dependence of measured gain of the 4 × 4-element array with parasitic elements [67]*

bandwidth variation is shown in Fig. 13.51 when h/λ_0 (λ_0 is the wavelength) and b/a are varied; the relative bandwidth is defined as the ratio of the frequency bandwidth over which the VSWR remains below 1·5 to the centre frequency.

A wideband circularly polarised array can be obtained by combining a radiating and a parasitic array. Fig. 13.52 shows the frequency dependence of the gain of an S-band array. Both the radiating array and the parsitic array are composed of 4 × 4 elements. The radiating array has co-planar feeding circuits, and each element is excited by a branch-line coupler. The element spacing is 0·78 wavelength, and the height between the radiating array and the parasitic array is 2 cm. The measured gain shows that the aperture efficiency exceeds 62% over a frequency range of 2·6–2·8 GHz.

13.4.2 Arrays of dual-frequency stacked elements

For many uses, the increased bandwidth is actually needed at only two distinct frequencies, for transmission and reception, which may be too far apart for a single antenna to operate efficiently at both frequencies. The behaviour of the antenna characteristic at the range of intermediate frequencies may be of no concern. In such cases, an antenna operating in dual-frequency bands is useful.

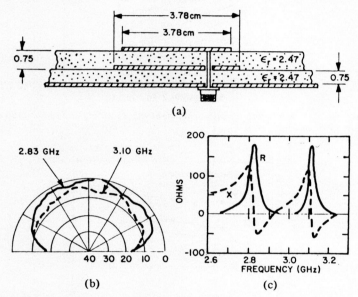

Fig. 13.53 *Dual-frequency stacked-patch antenna [63] (© 1981 IEEE)*
 a Cross section of typical stacked circular disc antenna
 b Measured E_θ patterns at 2·83 and 3·10 GHz
 c Measured input impedance showing resonance at 2·83 and 3·10 GHz

One technique is to stack one circular patch on top of another in a sandwich construction as shown in Fig. 13.53 [68, 63]. This antenna can also be applied as an element of an array, and one can obtain a dual-frequency circularly polarised array. An example is shown in Fig. 13.54. This antenna is nine-element airborne phased array for aeronautical satellite communication, the spacing of which is 94 mm (about half a wavelength at 1·6/1·5 GHz), and the dimensions are about 300 × 300 × 10 mm [69]. This phased array can scan its beam by ± 45° at least, for gain coverage greater than 12 dBi. The element is a newly developed stacked patch antenna as shown in Fig. 13.55. The upper and lower parts play a role in transmission and reception, respectively. And each layer of the antenna is individually fed at two points with 90° phase shift, in order to obtain circular polarisation. Each upper-layer element is a conventional microstrip antenna, while the lower one is a circular microstrip antenna with an

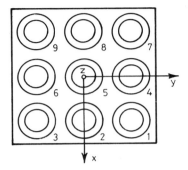

Fig. 13.54 *3 × 3 stacked-element array for dual-frequency operation [69] (© IEE)*

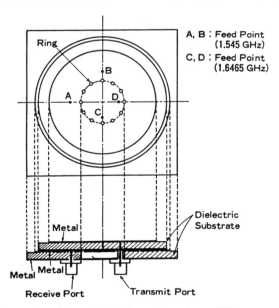

Fig. 13.55 *Configuration of the stacked elements [69] (© IEE)*

electrical shielding ring that provides enough space for the upper antenna to be easily fed [70]. Owing to the shielding ring, this antenna has good isolation characteristics between the transmit and receive ports. The measured coupling between two ports is less than −30 dB. Consequently, the antenna can be designed optimally for both transmission and reception frequency bands independently.

13.4.3 Wideband-array techniques

For a large array, the use of conventional dual-feed circularly polarised elements has the disadvantage of complicated structure, RF losses due to many feed cables, hybrids and power dividers, and high cost. The application of microstrip elements with single-point feeds is attractive for large circularly polarised arrays, but, in general, these antennas have narrow ellipticity bandwidth. Therefore, the techniques of configuring a circularly polarised arrays with elliptically or linearly polarised elements become important in practice.

Three techniques were outlined briefly in Section 13.1.1. In this Section more detailed explanations are provided.

(a) Array of microstrip-patch pairs: This method is to construct a circularly polarised array with pairs of microstrip patches [71]. In each pair, two elements are arranged with an angular orientation of 90° to each other and fed with 90° phase difference, as shown in Fig. 13.4. Using the pairs (or two-element subarrays), a circularly polarised array can be constructed.

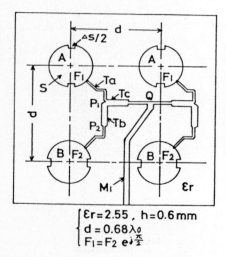

Fig. 13.56 *Four-element coplanar sub-array unit of CP pairs [71]*

Fig. 13.56 shows a four-element co-planar sub-array unit composed of the pairs. In the Figure, T_b is a quarter-wavelength impedance transformer which provides both 90° differential phase shift and impedance matching. Fig. 13.57 shows a 64-element circularly polarised array composed of the sub-array units. Typical measured data show that the 3 dB ellipticity bandwidth of the array with paired elements is about 6%, and it is greatly improved as compared with the conventional array.

(b) Array of 4-element subarrays: A circularly polarised array can also be formed by using 4-element sub-arrays, each of which is composed of 2 × 2 linearly polarised elements with unique angular and phase arrangements as shown in Fig. 13.5.

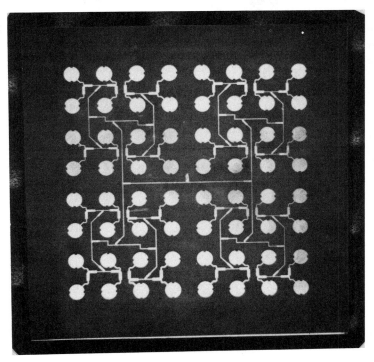

Fig. 13.57 *64-element array composed of the sub-array units [71]*

It is well known that circular polarisation can be achieved in the broadside direction of an array composed of two linearly polarised elements with angle and phase arranged in a 0°, 90° fashion. This circular polarisation, however, becomes very poor at angles greater than 5° off broadside owing to the spatial phase delay between the two elements, as shown in Fig. 13.58. This spatial phase delay no longer exists in the 2 × 2 subarray shown in Fig. 13.5, since the spatial phase delay in one row or column is opposed to that of the other row or column, and consequently they cancel each other out. The calculated radiation pattern shown in Fig. 13.59 indicates that a 2 × 2 sub-array has excellent quality of circular polarisation over a wide angular region of ± 40° from boresight [72].

(c) Sequential arrays: A more generalised circularly polarised array which consists of an arbitrary number of identical elements with arbitrary polarisation has been proposed by Teshirogi *et al.* [14]. In this array, the incremental angular

806 Circularly polarised antenna arrays

orientation and the excitation phase difference are provided sequentially to each element; therefore, the array is called a sequential array. The configuration of an N-element sequential array is illustrated in Fig. 13.6. The nth element is located at an arbitrary position on a plane, but with an orientation angle of

$$\phi_n = P(n-1)\pi/N \qquad (13.51)$$

where P is an integer and $1 \leqslant P \leqslant N - 1$, with respect to the first element, say element 1. The nth element is also fed with a differential phase shift of ϕ_n radians and the same orientation angle.

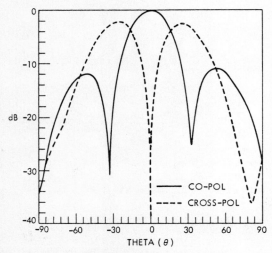

Fig. 13.58 *Calculated radiation pattern in principal plane of two-element CP array [72] (© 1985 IEEE)*
Element spacing = 0·9 wavelength

We assume that the polarisation of the radiated field from element 1 is elliptical in the boresight direction, and expressed by

$$\boldsymbol{E}_1 = a\boldsymbol{U}_1 + jb\boldsymbol{V}_1 \qquad (13.52)$$

where $\boldsymbol{U}_1$ and $\boldsymbol{V}_1$ are orthogonal unit vectors corresponding to the major and the minor axes, respectively, of polarisation ellipse, and a and b are the amplitudes of both components. The total radiated field $\boldsymbol{E}$ from the sequential array in the boresight direction can be derived as

$$\boldsymbol{E} = \sum_{n=1}^{N} \boldsymbol{E}_n = \frac{(a+b)}{2} N(\boldsymbol{U}_1 + j\boldsymbol{V}_1) \qquad (13.53)$$

This means that the sequential array radiates perfect circularly polarised waves in the boresight direction regardless of the polarisation of the element.

A sequential array has another advantage with regard to VSWR. According to the differential path length of each feed line, the reflected wave back to the input terminal from the nth element has a differential phase shift of $2\phi_n$.

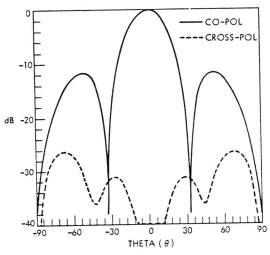

Fig. 13.59 *Calculated radiation pattern in principal plane of 2 × 2 array shown in Fig. 13.5 [72] (© 1985 IEEE)*
Element spacing = 0·9 wavelength.

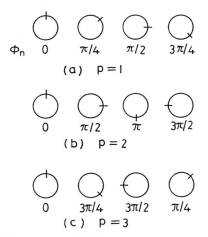

Fig. 13.60 *Possible 4-element sequential linear array*

Therefore, if the reflection coefficients of all the elements are the same, the sum of all the reflected waves V_r returning to the input terminal of the array becomes

$$V_r = V_0 \Gamma \sum_{n=1}^{N} \exp(j2\phi_n) = 0 \qquad (13.54)$$

Consequently, it can be seen that the sequential array provides not only perfect circular polarisation in the boresight, but also no reflection at the input terminal.

There are several configurations of the sequential array corresponding to P in eqn. 13.51. Fig. 13.60 shows three examples for a four-element linear sequential array. Since each phase difference ϕ_n is usually given by adjusting the feed-line length, these three arrays differ in ellipticity and VSWR bandwidth. Fig. 13.61 shows the improvement factor of cross-polarisation discrimination (XPD) and the ratio of the XPD of the sequential array to that of conventional array. From this Figure, it is clear that XPD is improved as N increases, and the case $P = 1$ is the best. It should be noticed that the examples in Figs. 13.5a and b can be interpreted in terms of a sequential array, as a two-pair array of a two-element sequential sub-array, and a four-element sequential array in the case of $P = 2$. Therefore, from the point of view of ellipticity bandwidth, the configuration in Fig. 13.60a is the widest of these three 4-element arrays.

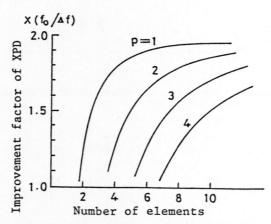

Fig. 13.61 *Improvement factor of cross-polarisation discrimination [14]*

Two-dimensional arrays can also be composed by combining linear sequential arrays. Fig. 13.62a shows an example of a square array. This is a recurrent type of sequential array, in which each row and each column array is a linear sequential array. For rectangular arrays, a more generalised configuration is available, as shown in Fig. 13.62b. In this method, the orientation angle and the differential phase shift to be applied to the (m, n)th element ϕ_{nm} is defined by

$$\phi_{nm} = {}_N\phi_n + {}_M\phi_m \qquad (13.55)$$

where

$$_N\phi_n = (n - 1)\pi/N, \quad {}_M\phi_m = (m - 1)\pi/M$$

Verification experiments have demonstrated wideband characteristics of the sequential array. The test array comprises two 2 × 4-element arrays. One is a

Circularly polarised antenna arrays 809

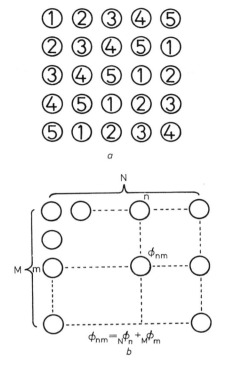

Fig. 13.62 *Arrangement of 2-dimensional sequential arrays*
  a Recurrence arrangment for a square array
  b Generalised arrangement for a rectangular array

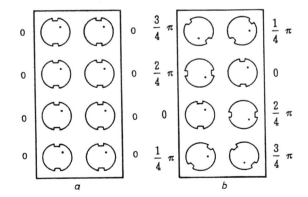

Fig. 13.63 *Two test arrays [14]*
  a Conventional array
  b Sequential array

conventional array and the other is a sequential array, as shown in Fig. 13.63. All the elements are identical microstrip patches which have small notches and are excited by single-point rear-side feeding. The material of the substrate is glass-cloth PTFE, the dielectric constant is 2·6 and the thickness is 4 mm.

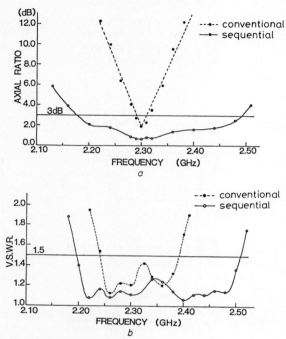

Fig. 13.64 *Measured bandwidth characteristics [14]*
  a Axial ratio
  b VSWR

Fig. 13.64*a* and *b* show the axial ratio and VSWR of these arrays. From the Figure, it is clear that the sequential array has much more wideband characteristics of circular polarisation and impedance than the conventional array. For example, 3 dB axial-ratio bandwidth of the sequential array exceeds 14%, which is about 15 times that of the conventional array; while 1·5 VSWR bandwidth is 13·7%, which is about twice that of the conventional array.

Several applications of sequential arrays have been presented for an airborne microstrip phased array in aeronautical satellite communication [73] and for a feed array of a contoured-beam reflector antenna [74].

13.5 References

1 HOWELL, J. Q.: 'Microstrip antennas', *IEEE Trans.*, 1975, **AP-23**, pp. 90–93

2 SANFORD, G.G.: 'Conformal microstrip phased array for aircraft tests with ATS-6', *IEEE Trans.*, 1978, **AP-26**, pp. 642–646
3 CHIBA, T., SUZUKI, Y., and MIYANO, N.: 'Suppression of higher modes and cross polarised component for microstrip antennas'. IEEE AP-S Int. Symposium Antennas and Propagat. Digest, 1982, pp. 285–288
4 KERR, J. L.: 'Microstrip polarisation techniques'. Proc. Antenna Applications Symposium, 1978
5 CARVER, K. R., and KOFFEY, E. L.: 'Theoretical investigation of the microstrip antenna'. Physic, and Sci. Lab., New Mexico State Univ., Technical Report PT-00929, 1979
6 RICHARDS, W. F., LO, Y. T., SIMON, P., and HARRISON, D. D.: 'Theory and applications for microstrip antennas'. Proc. Workshop on Printed Circuit Antenna Technology, New Mexico State Univ., 1979, pp. 8/1–23
7 SHEN, L. C.: 'The elliptical microstrip antenna with circular polarisation ', *IEEE Trans.*, 1981, **AP-29**, pp. 90–94
8 LONG, S. A., SHEN, L. C., SHAUBERT, D. H., and FARRAR, F. G.: 'An experimental study of the circular-polarised, elliptical, printed circuit antenna', *IEEE Trans.*, 1981, **AP-29**, pp. 95–99
9 WEINSCHEL, H. D.: 'A cylindrical array of circularly polarised microstrip antennas'. IEEE AP-S Int. Symposium Antennas and Propagat. Digest, 1975, pp. 177–180
10 HANEISHI, M., and YOSHIDA, S.: 'A design method of circularly polarised microstrip disk antenna by one-point feed', *Trans. IECE Japan*, 1981, **J64-B**, pp. 225–231 (in Japanese)
11 RODGERS, A.: 'Wideband squintless linear arrays', *Marconi Rev.*, 1972, **35**, pp. 221–243
12 HANEISHI, M., YOSHIDA, S., and GOTO, N.: 'A broadband microstrip array composed of single-feed type circularly polarised microstrip antennas'. IEEE AP-S Int. Symposium Antennas and Propagat. Digest, 1982, pp. 160–163
13 HUANG, J.: 'A technique for an array to generate circular polarisation with linearly polarised elements', *IEEE Trans.*, 1986, **AP-34**, pp. 1113–1124
14 TESHIROGI, T., TANAKA, M., and CHUJO, W.: 'Wideband circularly polarised array antenna with sequential rotations and phase shifts of elements'. Proc. Int. Symposium on Antennas and Propagat., Japan, 1985, pp. 117–120
15 ITOH, K., ARIGA, T., and SHINADA, H.: 'On some design examples of a circularly polarized wave array antenna using slot antenna combined with parasitically excited dipoles', *Trans. IECE Japan*, 1982, **J65-B**, pp. 1385–1392 (in Japanese)
16 ITO, K., AIZAWA, N., and GOTO, N.: 'Circularly polarised printed array antennas composed of strips and slots', *Electron. Lett.*, 1979, **15**, pp. 811–812
17 ITO, K.: 'Circularly polarized printed array antenna composed of end-fed strip dipoles and slots', *Electro. & Commun. Japan*, 1984, **67-B**, pp. 56–69
18 LEWIN, L.: 'Radiation from discontinuities in strip-line', *IEE Proc.*, 1960, **107C**, pp. 163–170
19 HALL, P. S.: 'Rampart microstrip line antennas'. European Patent Application 79301340.0, 1979
20 WOOD, C., HALL, P. S., and JAMES, J. R.: 'Design of wideband circularly polarised microstrip antennas and arrays'. 1st IEE Int. Conference on Antennas and Propagat., London, 1978, pp. 312–316
21 HENRIKSSON, J., MARKUS, K., and TIURI, M.: 'A circularly polarised travelling-wave chain antenna'. Proc. 9th European Microwave Conference, Brighton, 1979, pp. 174–178
22 MAKIMOTO, T., and NISHIMURA, S.: 'Circularly polarised microstrip line antenna'. US Patent 4 398 199, 1983
23 NISHIMURA, S., SUGIO, Y., and MAKIMOTO, T.: 'Crank-type circularly polarised microstrip line antenna'. IEEE AP-S Int. Symposium Antennas and Propagat. Digest, 1983, pp. 162–165
24 HALL, P. S.: 'Microstrip linear array with polarisation control', *IEE Proc.*, 1983, **130H**, pp. 215–224
25 WOLFF, E. A.: 'Antenna analysis' (John Wiley, NY 1966) Chap. 8

26 DOUVILLE, R. J. P., and JAMES, D. S.: 'Experimental study of symmetric microstrip bends and their compensation', *IEEE Trans.*, 1978, **MTT-26**, pp. 175–182
27 DUBOST, G.: 'Broadband circularly polarized flat antenna'. Proc. Int. Symposium on Antennas and Propagat., Japan, 1978, pp. 89–92
28 DUBOST, G., SAMSON, J., and FRIN, R.: 'Large-bandwidth flat cylindrical array with circular polarisation and omnidirectional radiation', *Electron. Lett.*, 1979, **15**, pp. 102–103
29 ITO, K., AIZAWA, N., and GOTO, N.: 'Circularly polarized microstrip antennas'. Report of Technical Group, IECE of Japan, AP78-90, 1978, pp. 21–26 (in Japanese)
30 NAKAOKA, K., ARAI, K., and ITOH, K.: 'Some problems on circularly polarised microstrip line slot arrays'. National Conv. Records on Optical and Wave Sect., IECE of Japan, 1982, S2-15 (in Japanese)
31 OLTMAN, H. G., and HUEBNER, D. A.,: 'Electromagnetically coupled microstrip dipoles', *IEEE Trans.*, 1981, **AP-29**, pp. 151–157
32 ITOH, K., BABA, H., OGAWA, Y., WATANABE, F., and YASUNAGA, M.: 'L-band airborne antenna using crossed slots'. Report of Technical Group, IECE of Japan, AP85-101, 1986, pp. 65–72 (in Japanese)
33 HENDERSON, A., and JAMES, J. R.: 'Low-cost flat-plate array with squinted beam for DBS reception', *IEE Proc.*, 1987, **134H**, pp. 509–514
34 JAMES, J. R., and HALL, P. S.: 'Microstrip antennas and arrays. Pt. 2: New array-design technique', *IEE J. MOA*, 1977, **1**, pp. 175–181
35 JAMES, J. R., HALL, P. S., and WOOD, C.: 'Microstrip antenna theory and design' (Peter Peregrinus, 1981)
36 COLLIN, R. E., and ZUCKER, F. J.: 'Antenna theory: Pt. 1' (McGraw-Hill, 1969) chap. 14
37 ITO, K., and KITAJIMA, H.: 'Design of series-fed circularly polarised printed array antenna'. 4th IEE Int. Conference on Antennas Propagat., Coventry, 1985, pp. 103–107
38 ITO, K., ITOH, K., and KOGO, H.: 'Improved design of series-fed circularly polarised printed linear arrays', *IEE Proc.*, 1986, **133H**, pp. 462–466
39 ITO, K., ITOH, K., OHTAKE, T., and KOGO, H.: 'Circularly polarized printed planar array composed of strip dipoles and slots'. IEEE AP-S Int. Symposium Antennas and Propagat. Digest, 1986, pp. 561–564
40 KOBAYASHI, A., ITO, K., and BAN, M.: 'A precise measurement of input immittances of elements coupled to a microstrip line (Continuation)'. Report of Technical Group, IECE of Japan, MW85-106, 1985, pp. 25–30 (in Japanese)
41 NISHIMURA, S., SUGIO, Y., and MAKIMOTO, T.: 'Side-looking circularly polarised microstrip line planar antenna'. Proc. Int. Symposium on Antennas and Propagat., Japan, 1985, pp. 129–132
42 POZER, D. M.: 'Input impedance and mutual coupling of rectangular microstrip antennas', *IEEE Trans.*, 1982, **AP-30**, pp. 1191–1196
43 MALKOMES, M.: 'Mutual coupling between microstrip patch antennas', *Electron. Lett.*, 1982, **18**, pp. 520–522
44 JEDLICKA, R. P., POE, M. T., and CARVER, K. R.: 'Measured mutual coupling between microstrip antennas', *IEEE Trans.*, 1981, **AP-29**, pp. 147–149
45 HANEISHI, M., YOSHIDA, S., and TABATA, M.: 'A design of back-feed type circularly polarized microstrip disk antennas having symmetrical perturbation element by one-point feed', *Trans. IECE Japan*, 1981, **J64-B**, pp. 612–618 (in Japanese)
46 BAILEY, M. C., and PARKS, F. G.: 'Design of microstrip disk antenna arrays'. NASA Technical Memorandum 78631, 1978
47 HANEISHI, M.: 'Studies on circularly polarised microstrip antennas'. Doctoral Thesis, Tokyo Inst. Tech., 1981 (in Japanese)
48 HENDERSON, A., and JAMES, J. R.: 'Design of microstrip antenna feeds. Pt. 1: Estimation of radiation loss and design implications', *IEE Proc*, 1981, **128H**, pp. 19–25
49 JAMES, J. R., and WILSON, G. J.: 'Microstrip antennas and arrays. Pt. 1: Fundamental action and limitations', *IEE J. MOA*, 1977, **1**, pp. 165–174

50 HUANG, J.: 'The finite ground plane effect on the microstrip antenna radiation patterns', *IEEE Trans.*, 1983, **AP-31**, pp. 649–653
51 HALL, P. S., and JAMES, J. R.: 'Crosspolarisation behaviour of series-fed microstrip linear arrays', *IEE Proc.*, 1984, **131H**, pp. 247–257
52 HANSEN, R. C.: 'Cross polarization of microstrip patch antennas', *IEEE Trans.*, 1987, **AP-35**, pp. 731–732
53 COLLIER, M.: 'Microstrip antenna array for 12 GHz TV', *Microwave J.*, 1977, **20**, pp. 67–71
54 MURPHY, L. R.: 'SEASAT and SIR–A microstrip antennas'. Proc. Workshop on Printed Antenna Technology, New Mexico State Univ., 1979, pp. 18/1–20
55 HANEISHI, M., HAKURA, Y., SAITO, S., and HASEGAWA, T.: 'A low-profile antenna for DBS reception'. IEEE AP-S Int. Symposium Antennas and Propagat. Digest, 1987, pp. 914–917
56 PUCEL, R. A., MASSE, D. J., and HARTWIG, C. P.: 'Losses in microstrip', *IEEE Trans.*, 1968, **MTT-16**, pp. 342–350
57 LEWIN, L.: 'Spurious radiation from microstrip', *IEE Proc.*, 1978, **125**, pp. 633–642
58 MURATA, T., and OHMARU, K.: 'Characteristics of circularly polarised printed antenna with two layer structure'. Report of Technical Group, IECE of Japan, AP86-101, 1986, pp. 83–87 (in Japanese)
59 STOCKTON, R., and HOCKENSMITH, R.: 'Application of spherical arrays — A simple approach'. IEEE AP-S Int. Symposium Antennas and Propagat. Digest, 1977, pp. 202–205
60 MAILLOUX, R. J., MCILEVENNA, J. F., and KERNWEIS, N. P.: 'Microstrip array technology', *IEEE Trans.*, 1981, **AP-29**, pp. 25–37
61 HORI, T., TERADA, N., and KAGOSHIMA, K.: 'Electronically steerable spherical array antenna for mobile earth station'. 5th IEE Int. Conference on Antennas and Propagat., York, 1987, pp. 55–58
62 HORI, T.: 'Mutual coupling between broadband microstrip antennas'. National Conv. Records, IECE of Japan, 1984, p. 714 (in Japanese)
63 CARVER, K. R., and MINK, J. W.: 'Microstrip antenna technology', *IEEE Trans.*, 1981, **AP-29**, pp. 2–24
64 TESHIROGI, T., and GOTO, N.: 'Recent phased array work in Japan'. ESA/COST 204 Phased-Array Antenna Workshop, 1983, pp. 37–44
65 CHEN, C. H., TULINTSEFF, A., and SORBELLO, R. M.: 'Broadband two layer microstrip antenna'. IEEE AP-S Int. Symposium Antennas and Propagat. Digest, 1984, pp. 251–254
66 TAGA, T., MISHIMA, H., and KANEHORI, T.: 'A broadband microstrip antenna at UHF band'. National Conv. Records, IECE of Japan, 1979, pp. 254–255 (in Japanese)
67 HORI, T., and NAKAJIMA, N.: 'Broadband circularly polarised microstrip array antenna with co-planar feed', *Trans. IECE Japan*, 1985, **J68-B**, pp. 515–522 (in Japanese)
68 LONG, S. A., and WALTON, M. D.: A dual-frequency stacked circular-disk antenna', *IEEE Trans.*, 1979, **AP-27**, pp. 270–273
69 YASUNAGA, M., WATANABE, F., SHIOKAWA, T. and YAMADA, M.: 'Phased array antennas for aeronautical satellite communications'. 5th IEE Int. Conference on Antennas and Propagat., York, 1987, pp. 47–50
70 GOTO, N., and KANETA, K.: 'Ring patch antennas for dual frequency use'. IEEE AP-S Int. Symposium Antennas and Propagat. Digest, 1987, pp. 944–947
71 HANEISHI, M., SAITO, S., YOSHIDA, S., and GOTO, N.: 'A circularly polarized planar arrays composed of the microstrip pairs element'. Report of Technical Group, IECE of Japan, AP 83-64, 1983, pp. 1–4 (in Japanese)
72 HUANG, J.: 'Circularly polarised microstrip array with wide axial ratio bandwidth and single feed L.P. elements'. IEEE AP-S Int. Symposium Antennas and Propagat. Digest, 1985, pp. 705–708
73 TESHIROGI, T., TANAKA, M., and OHMORI, S.: 'Airborne phased array antenna for mobile satellite communications'. IEEE AP-S Int. Symposium Antennas and Propagat. Digest, 1986, pp. 735–738
74 BALLING, P.: 'Design and analysis of contoured-beam reflector antenna feed arrays and contoured-beam array antennas'. JINA, 1986, Nice, pp. 315–329

Index

Abberation, 96
Acid, copper, 936
Activation energy, 875
Active angle, 1241
Active element pattern, 701, 739, 1249
Active elements in a cylindrical array, 1249
Active impedance, 732
Active patches, 34
Active reflection coefficient, 708, 739
Additives, brightener, 936
Adhesion, 936
Adjustable height patch, 30
Admittance, aperture, 727
Aerospace systems, 1057
Air gap, adjustable, 193
Air-filled waveguide, 858
Airborne antenna, 1139
Airborne phased arrays, 802, 810
Alumina substrates, 14, 1025
Aluminium, 931, 946
Aluminium foil, 1075
Amplitude distributions, 830
Analytical methods
    aperture radiation model, 583
    cavity model, 458, 580, 1058, 1228
    Cohns method, 611
    contour integral method, 472
    desegmentation, 456, 494
    direct form of network analogue, 462
    EMF, 249
    Filton Integration method, 285
    integral equations, 715
    method of moments, 364, 708
    moments method, 593
    multiport network approach, 455
    multiport network model, 462
    network analysis, 488
    plane wave spectrum method, 285
    real-space integration method, 286
    reciprocity method, 278
    Richmonds reaction equation, 597
    segmentation, 456, 488
    spectral domain approach, 590
    steepest descent method, 278
    surface current model, 1228
    synthesis, 361
    transmission line model, 456
    variational method, 235
    Wiener-Hopf method, 478
Angled slot array, 39
Anisotropy, 881, 909, 917
Annealed, 946
Annular ring, 25, 879, 1262
Annular slot, 111, 611, 1086
Annular slots
    attenuation coefficient, 614
    guide wavelength, 611
    impedance, 611
    reflector planes, 617
Antennas
    conical, 601
    DBS, 859, 861
    dielectric, 1287
    directly coupled three resonator, 507
    hybrid microstrip, 580
    integrated, 14
    gap coupled multiresonator, 507
    millimeter wave hybrid, 1285
    multi-terminal, 236
    open microstrip, 580
    window, 1283
    wrap-around, 85
Antenna location, received signal
    fluctuations, 1084
Antennas for portable equipment, 1092
Aperture admittance, 727
Aperture blockage, 96, 104
Aperture coupled patches, 330

Index

Aperture coupling, 35, 332, 823
Aperture distribution, 762, 821, 822, 825, 851
Aperture distribution Dolph-Chebyshev, 855
Aperture radiation analysis method, 583
Apertures, dichroic, 1282
Applications, 8
Array factor, 829
Array lattice
    rectangular, 706
    triangular, 706, 1243
Array structures, 36
Arrays, 7, 35
    angled slots, 39
    architecture, 745
    asymmetric step, 39
    bonding, 916
    brick wall, 848
    cascaded patch, 822
    chain, 37, 763, 816
    Chebyshev, 774
    circularly polarised, 755, 765
    co-planar, 758
    comb line, 37, 766, 822
    composite element, 759, 770
    conformal, 795, 1153, 1227
    conical, 1154
    constant-conductance, 846
    corner fed patches, 647
    corporate-fed, 789
    crank, 844
    crank-line, 777
    crank-type, 764
    cross fed, 36, 655, 843
    cross printed-dipole, 766
    cross slot, 766
    cylindrical, 34, 765, 1227
    DBS, 766, 790, 1288
    design of planar, 622
    discontinuity, 843
    dual frequency, circular polarised, 802
    dual polarised, 1073
    electronically switched, spherical, 792
    finite, 301, 731
    flat, circular polarised, 765
    four element, 1217
    Franklin, 816, 848
    Franklin line, 38
    herringbone, 835
    herringbone line, 38
    infinite, 698, 731
    lattice, 848
    linear, 345, 449
    linear centre fed, 839
    linearly polarised, 846
    log periodic, 39, 387
    microstrip, 443
    millimeter wave, 1282
    monopulse, 849, 1068, 1153
    non uniform, 630
    omnidirectional, 39, 382
    parasitic, 801
    parasitic patch, 825, 833
    parasitically coupled, 37, 345
    patch, 1263
    planar, 345, 776, 790
    rampart, 816, 844
    rampart line, 38, 223, 763, 1222
    recurrent sequential, 808
    resonant, 769, 816
    scanning, 792
    sector beam, 1081
    sequential, 759, 788, 805
    sequentially rotated, 12
    series, 816
    series fed patch, 515
    series fed patches, 624
    series-fed, 789, 818
    series-fed circular polarised, 770
    serpent, 816, 833, 843
    serpent line, 38
    sparse, 1273
    spherical, 34, 793, 1239
    square patches, 1073
    square-loop-type, 764
    squintless, 758, 841
    strip dipole and slot, 39
    sub arrays, 655
    synthesis, 662
    tapered, 650
    transposed, 821, 822
    travelling wave, 755, 762, 765, 769, 816, 1222
    triangle line, 38
    two dimensional, 808
    two sided, 746
    untransposed, 821
    wide bandwidth, 372
    wideband, 804
    wire grid, 848
Assemblies, 916
Asymmetric feeds, 797
Asymmetric step array, 39
Attachment mode, 433
Attenuation, 1016

Index

Attenuators, 1241
Axial polarisation, 1239
Axial ratio, 723, 787, 1185
Axial-ratio bandwidth, 758, 829, 859
Azimuthal modes, 53

Backward firing beams, 834
Bag-moulding, 930
Balanced stripline, 1003
Bandwidth, 5, 11, 111, 128, 219, 335, 338, 554, 743, 796, 895
Bandwidth, 9
    axial-ratio, 829, 859
    extension, 9
    gain, 823
    VSWR, 118, 823
Base-station antenna, 1083
Basis functions, 282
    entire domain, 429, 437
    Maxwell, 282
    Maxwell, modified, 282
    pulse, 282
    subdomain, 282, 437
    subsectional, 423, 440
Beam squint, 758
Beam steering, 1241, 1249, 1273
Beam-forming, adaptive, 862
Beams
    backward firing, 834
    forward firing, 834
Beamwidth, 104
Bend measurement, 976
Bending, 916
Bends, 1168
Beryllia substrates, 1025
Bessel function, 1058, 1204
Bidirectional communications, 1107
Bismaleimide, 951
Bismaleimide-triazine-epoxy, 878
Black copper oxide, 880
Blind spots, 336, 729
Blockage, aperture, 96
Boards, printed wiring, 916
Bolometer, 1194
Bonding
    lead, 952
    thermal-compression, 938
    thermosonic, 938
    wire, ultrasonic systems, 938
Boundary conditions, 49, 115, 395
Branch-line coupler, 854
Branching network, 825

Brass, 946
Brick wall arrays, 848
Broadband microstrip antenna, 1083, 1086, 1129, 1134
Broadband matching, 558
Built in antennas, 1092
Butler matrix, 855, 860, 1083

Cabin antenna, 1085
CAD, 14, 1031
    Acline, 1033
    Analop, 1036
    Autoart, 1037
    CADEC, 1033
    CiAO, 1035
    computer graphics, 1175
    Esope, 1033
    LINMIC, 1034
    Mama, 1036
    Micad, 1037
    Micpatch, 1035
    Microkop/Suspend, 1036
    Micros, 1038
    microstrip, 1001
    microwave software applications, 1036
    Midas, 1034
    Multimatch, 1036
    photoplots, 1175
    Planim, 1036
    S/Filsyn, 1037
    Supercompact, 1033
    techniques, 517
    Temcad, 1039
    Touchstone, 1032
    Transcad, 1037
    triplate, 1001
    fringing, 901
Capacitance, 901
Capacitive tuning, 329
Capacitors, 1040
Car telephones, 1079
Carbide, 922
Carbon fibre reinforced plastic, 1075
Cavity feeds, 859
Cavity model, 112, 458, 580, 884, 1058, 1228
Cavity-perturbation, 884
CCIR-TVRO conditions, 671
Central shorting pin, 85
Centred fed dipole, 287
Ceramic-PTFE, 953

Index

Chain array, 37, 763, 816, 940
Chain scission, 940
Chebyschev taper, 774, 1069
Chloride
    acid cupric, 949
    alkaline cupric, 949
    ferric, 949
Choke, peripheral, 104
Chokes, 97
Chromic-acid, 950
Circuit, etching, 1028
Circular array, 1127
Circular array feed, 1132
Circular array of slots, 1132
Circular array of strips, 1132
Circular microstrip discs, 1136
Circular patches, 45, 63, 111, 720, 1058, 1202
Circular polarisation, 5, 130, 219, 722, 744, 755, 787, 846, 848, 1127, 1132, 1261, 1273
Circular polarised radiation, 824
Circular polariser, 766
Circular-disc patches, 756
Circular-patch array, 1113
Circular-patch-slot array, 1119
Circularly polarised, singly fed patches, 221
Circularly polarised arrays, 755, 765
Circularly polarised circular patches, 232
Circularly polarised composite type patches, 222
Circularly polarised dipoles, 356
Circularly polarised elements, 12
Circularly polarised line antennas, 762
Circularly polarised patches, 27, 499, 821, 1218
Circulators, 1041
Circumferential polarisation, 1234
Co-axial feeds, 276
Co-planar arrays, 758
Co-planar coupling, 817
Co-planar feeds, 815
Co-polymers, 945
Coating, conformal, 950
Coaxial excitation, 432
Coaxial probe, 29, 433, 1194
Cohns method, 611
Collected volatile condensable materials (CVCM), 940
Colloidal, 881
Comb array, 37, 766, 822, 833
Combined feeds, 839
Comparator, monopulse, 859

Compensating hole transitions, 967
Components, 1039
Composite element, 755, 759, 770
Computational efficiency, 709
Computer graphics, 1175
Conductance
    edge, 476
    radiation, 476, 554, 822, 835, 859
    surface wave, 476
Conductivity, 879
Conductor, losses, 788, 790, 815, 1075
Conformal antenna, 1227
Conformal arrays, 795, 1153
Conformal mapping, 1008
Conical antennas, 601
Conical arrays, 1154
Conical beam, 1112, 1127, 1129, 1132, 1239
Conically depressed patch, 29
Connections, 332
Connector characterisation, 962
Connector test fixture, 963
Connectors, microstrip to coax, 1007
Constant-conductance arrays, 846
Contour integral method, 472
Coplanar line probes, 968
Coplanar stripline patch, 31
Copolymer substrates, 679
Corporate feeds, 301
Copper electroless, 935
Copper foil, 1075
Copper-Invar-copper, 946
Corner fed patches, 647
Corner reflector, 1082
Corporate feeds, 13, 35, 757, 789, 816, 850, 1069, 1288
Correction, end-fringing, 888
Correlation, 905
Corrugated ground plane, 28
Cosecant squared pattern, 671
Coupled triplate lines, 1169
Coupled-resonator, 340
Coupler
    branch-line, 854
    hybrid-ring, 854
Coupling, 817, 887, 895, 897
Coupling
    agents, 875
    aperture, 823
    co-planar, 817
    direct, 822
    electromagnetic, 824
    factor, 835

probe, 822
proximity, 818
Coupling gaps, 762
Coupling mechanisms, 816
Crank arrays, 764, 777, 844, 1115
Cross-printed dipole arrays, 766
Cross fed arrays, 36, 655, 843
Cross patch, 501
Cross polarisation, 68, 74, 76, 100, 104, 590, 1234
Cross slot arrays, 766, 1136
Crossed dipoles, 356
Crossed slot, 28
Crossover level, 793
Crystalline, 873, 878, 942
Current
    distributions, 112
    magnetic, 726
    sources, 112
Current sheet model, 731
Current-ribbon, 151
Cyanide, copper, 936
Cycling, wet/dry, 941
Cylindrical antenna, 1227
Cylindrical array, 1109
    active-element, 1249
    array feed, 1248
    design, 1240
    gain, 1248
    impedance, 1243
    mutual coupling, 1244
    pattern synthesis, 1251
    scanning, 1249
Cylindrical modes, 1230
Cylindrical near field scanning, 981
Cylindrical patches, 1227
Cylindrical wave propagation, 1072

Dantzig algorithm, 662
Data-relay satellite, 1146
DBS antennas, 766, 790, 859, 1112
De-smearing, 929
Decomposition, thermal, 919
Deformation, 924
Degenerate modes, 756
Dendrites, 879
Dents, 917
Desegmentation method, 456, 494
Design procedure, series-feed, 836
Device attachment, 936
Diagnostics, 1159, 1193. 1214
Diagnostics, liquid crystal, 984

Diagonal slot square patch, 499
Diagonally fed nearly square patch, 499
Dielectric
    antennas, 1287
    losses, 788, 815, 1016
Dielectric constant, 790, 897
Dielectric constant, effective, 481
Dielectric filling factor, 790
Dielectric image guide, 822, 858
Dielectric loss, 1075
Dielectric rod array, 39
Dielectric spheres, 31
Difference radiation pattern, 1068
Diffraction coefficient, 1069
Diffraction effects, 1069
Diodes, 1041
Dipole
    centre fed, 287
    flat folded, 765
    horizontal electric, 276, 403
    infinitesimal, 276
    printed, 765
    circularly polarised, 356
    crossed, 356
    efficiency, 367
    EMC, 295
    flat, 353
    multiple, 299
    mutual impedance, 359
    parasitic, 759
    polarisation, 361
    printed, 706, 732
    stacked, 299
    strip, 761
    synthesis, 361
    variable directivity, 372
Direct coupling, 822
Direct form of network analogue, 462
Direct-Broadcast-Satellite, 1288
Directional coupler
    hybrid-ring, 855
    rat-race, 855
Directivity, 127, 372, 554, 831
Directivity, linear arrays, 625
Directly coupled three resonator antenna
    disc, 25, 507, 843, 1258
Discontinuity arrays, 843
Discontinuity radiation, 791
Dispersion, microstrip, 1015
Dissipation, loss, 1285
Dissipation factor, 871, 878, 886, 895
Distribution
    amplitude, 830

Index

aperture, 825
  Dolph-Chebyshev, 830
  Taylor, 830
Dolph Chebyshev distributions, 830, 855
Dose rate, 940
Double tuning, 1064
Doubly diffracted field, 1069
Drills, 922
Dual aperture-fed patches, 823
Dual feeds, 219
Dual frequency circularly polarised arrays, 802
Dual polarised array, 1073
Dual-fed circularly polarised patches, 220
Dual-frequency patches, 30, 188, 197, 200, 312, 313, 796, 802
Dual-polarisation patches, 312, 318
Ductility, 879, 936
Dust protection shield, 1063
Dyadic Green's function, 284, 399

E-H antenna, 1083
Earth stations, 1112
Edge
  conductance, 476
  effects, 1069
  equivalent admittance network, 463
  ground plane, 12, 731
  non radiating, 481
Effective dielectric constant, 481
Effective loss tangent, 461
Effective permittivity two-layer medium, 196
Effective radius, 65, 137
Effective width, 115
Effects, environmental, 939
Efficiency, 5, 13, 117, 313, 346, 367, 413, 554, 781, 837, 1258
Efficiency
  computational, 709
  measurement, 991
  radiation, 9, 598, 734, 872
  spill-over, 104
Elastic, 924
Electromagnetically coupled patch, 31
Electric current analysis method, 583
Electric dipole, 403
Electric shielding ring, 803
Electric source, 1260
Electric walls, 113
Electric-current source method, 762, 765
Electric-field integral equation, 715, 726, 732

Electro-etch, 879
Electromagnetic coupling, 207, 797, 824
Electronically switched spherical arrays, 792
Electroplating, 935
Electrostatic charging, 1065
Element
  endfire, 749
  tapered-slot, 751
Element factor, 831
Element grid, 1241
Element pattern, 732
Elliptical patch, 25, 182, 235, 756, 1129
Elliptical polarisation, 186
EM-field, far-zone, 117
EMC dipole, 295
EMF method, 249
End fringing, 887
Endfire arrays, 1279
Endfire elements, 749
Energy, conformational, 881
Entire-domain basis functions, 429, 437
Environmental conditions, 875
Environmental effects, 679
Epoxy, 878, 951
Equitriangular patches, 111
Equivalence
  external, 47
  internal, 49
Equivalent sources, 112
Equivalent circuit, 237, 728
Equivalent circuit model, 769, 776
Equivalent current sources, 114
Equivalent edge admittance network, 463
Equivalent magnetic current, 116
Equivalent slot, 534
Equivalent surface currents, 49
Equivalent waveguide model, 817
Etch
  plasma, 929
  sodium, 929
Etchant, 949
Etching, 1028
Eulers constant, 474
Excitation, coaxial, 432
Excitation field, 396
Excitation voltage, 784
Expansion functions, 53
External equivalence, 47
External matching circuits, 28

Fabric, 922
  woven-glass, 935

Index

Far-field approximations, 408
Feed isolation, 320
Feed structures, 35
Feeder of a cylindrical array, 1248
Feeds, 32, 756, 767, 1001
   3 dB hybrid, 220
   architecture, 13
   asymmetric, 797
   cavity, 859
   co-axial, 276
   co-planer, 815
   coaxial probe, 29
   combined, 839
   corporate, 13, 35, 306, 757, 816, 850, 1069
   corporate, triplate, 1288
   dual, 219
   four point, 28
   four-probe, 756
   Lecher line, 353
   microstrip line, 29
   novel, 12
   overlaid, 1263
   parallel, 335, 757, 816, 825, 828
   perpendicular, 747
   phased array, 862, 1248
   radial waveguide, 859
   rear, 758
   reflector, 96
   resonant, 835, 1074
   sequentially rotated, 36
   series, 335, 758, 767, 816, 832
   series compensated, 36
   single, 219
   spurious-radiation, 332
   squintless, 832
   stripline, 353
   travelling-wave, 832
   two-dimensional, 839
   two-line, 276
Ferrimagnetic substrates, 1027
Ferrite superstrates, 1292
Fibres, glass, 922
Field, excitation, 396
Filler, ceramic, 922
Films, barrier, 930
Filton integration method, 285
Finite arrays, 301, 731
Finite ground plane, 12, 1069
Flat dipoles, 353
Flat folded dipole, 765
Floquet modes, 703
Flouborate, copper, 936

Flush mounted antennas, 1092
Foam, 953
Foils
   adhesion, 875
   aluminium, 1075
   copper, 1075
   electrodeposited, 879
   rolled, 879
   wrought, 879
Folded slots, 353
Forming rolls, 924
Forward firing beams, 834
Four element arrays, 1217
Four element sub-arrays, 805
Four point feeding, 28
Four-probe feeds, 756
Fourier Transform, 403, 406, 708, 716, 721, 726
Franklin array, 38, 816, 848, 1105
Free radical, 941
Frequency, resonant, 895
Frequency agility, 187
Frequency diversity, 1084
Fringing fields, 113
Full sheet resonance test method, 884, 897, 901
Fumes, 919
Functions
   testing, 53
   triangular, 54
Future prospects, 1

GaAs substrate, 332
GaAs superstrate, 281
GaAs transitions, 968
Gain, 127, 554, 831
   minimum coverage, 793
   bandwidth, 823
   factor, 104
Gain of a cylindrical array, 1248
Galerkin solution, 284
Gap coupled multiresonator antenna, 507
Gap-coupled patch, 817
Geometric optics field, 1069
Giotto spacecraft, 1061
Glass transition temperature, 875
Glossary, 24
Grain structure, 879
Grating lobes, 428, 826, 834, 1183
Green's function, 50, 116, 398, 411, 421, 426, 462, 695, 726
   dyadic, 399

Index

planar configurations, 519
  spectral domain, 327
Ground plane
  corrugated, 28
  slot, 32
  edge, 12, 731
GTD, 1183
Guide
  dielectric image, 822
  insular, 822
Gunn diode, 34

H-shaped patch, 26
Half-wave patch, 313, 817
Hand held message communication terminal, 1125
Handbook, 17
Hankel Transform, 406
Hankel function, 264, 473, 1230
Hard boundary diffraction coefficient, 1069
Herringbone array, 38, 835
Higher order modes, 356, 787, 796, 1058, 1129
Historical development, 1
History
  mechanical, 874
  thermal, 874
Holes, burr-free, 922
Honeycomb substrate, 796
Horizontal dipole, 403
Humidity, 875
Huygens sources, 1262
Hybrid coupler, 1064, 1081, 1164
Hybrid microstrip antenna, 25, 580
Hybrid sources, 353
Hybrid-ring coupler, 854, 855
Hydrocarbon, 951
Hydrolysis, 875
Hydrophobic, 941
Hyperthermia applicator, 1293

Image, 116
Impedance
  active, 732
  cylindrical array, 1243
  input, 9, 118, 439, 590, 708
  matching, 655, 1190
  matrix, 708, 716
  port matrix, 737, 739
  surface, 400
  transformer, 1064

Incoherent radar, 1073
Indoor communications antenna, 1083
Indoor receiving antenna, 1084
Inductance feed probe, 329
Inductors, 1041
Infinite arrays, 698, 731
Infinite phased arrays, 698
Infinitesimal dipole, 276
Infra-red, 875
Input
  impedance, 9, 118, 238, 432, 439, 708, 875
  admittance, 770
  conductance, 554
Input resistance of patches, 324
Inserted connector transitions, 967
Insular guide, 822
Integral equation, 47
Integral equation, electric-field, 715, 726, 732
Integrated antennas, 14
Integrated phased arrays, 742
Internal equivalence, 49
Iris, 906
Isolated power dividers, 852
Isolators, 1041

Junction effects, 1166

K connector transitions, 967
K′, thermal coefficient of, 947
Kevlar epoxy, 1075

Land mobile satellite communications, 1127
Lattice arrays, 848
Launchers, 1287
Layer, resistive, 947
Leaky cavity, 119
Lecher line feeds, 353
Light, ultra-violet, 929
Limiting oxygen index, 919
Line analysis
  dielectric Green's function, 1020
  Fourier transform, 1020
  integral equations, 1020
  TEM models, 1020
  variational techniques, 1020
Line antenna, circularly polarised, 762
Linear arrays, 345, 449

Index

Linear centre fed arrays, 839
Linear polarisation, 130
Linear slots, 606
Linearly polarised arrays, 846
Lines
    dielectric, 1285
    discontinuities, 762, 1017
    losses, 624
    open circuited, 1212
    parallel-coupled, 856
    parameters, 971
    parameter measurement, 970
    synthesis, 1015
    width, 943
Liquid crystal diagnostics, 984, 1058, 1159
Loading, reactive, 204, 826
Lobes, grating, 826, 834
Log periodic arrays, 39, 387, 834, 1273
Longitudinal polarisation, 822, 846
Lorentz's gauge, 398
Loss tangent, 117, 461
Losses, 174, 407
    conductor, 788, 815, 942, 1075
    dielectric, 1016, 1075
    dissipation, 1285
    line, 624
    ohmic, 1016
    radiation, 1016, 1075
    reflection, 816
    resistance, 887
    resistive, 895
    surface wave, 816, 1075
Low cost substrates, 674

Machining, 916
Magnetic
    current, 116, 726, 762
    materials, 1292
    source, 1260
    walls, 113
Main beam direction, 781
Mandrel, 931
Manpack radars, 1079
Manufacture, 5, 14
Marine radars, 1079
Maritime satellite communications, 1127
Matched terminations, 964
Matching, 5
Matching circuits
    external, 28
    gaps, 28
Mated connector test method, 963

Materials, magnetic, 1292
Mathematical modelling, 14
Matrix
    Butler, 861
    excitation, 58
    formulation, 50
    impedance, 708, 716
    Maxson-Blass, 861
Maxon Blass matrix, 860
Maxwell basis functions, 282
Maxwell, modified basis functions, 282
Measurements, 957, 1006
    bends, 976
    efficiency, 991
    line parameters, 970
    radiometric, 993
    resonant techniques, 976
    T-junctions, 977
Melt point, crystalline, 879
Melt viscosity, 928
Metal failure, 934
Metallic ring transitions, 967
Method of moments, 282, 295, 364, 401, 423, 708
Metrology, 1193
Microstrip
    circuit realisation, 1028
    impedance, 1013
    line, 1004
    materials and manufacture, 1023
Microstrip antenna
    hybrid, 25
    frequency variable, 1092, 1101
    post loaded, 1092
    quarter wavelength, 1092
    window attached, 1092
Microstrip dispersion, 1015
Microstrip field diagnostics, 1193
Microstrip line, 537
Microstrip line feeds, 29
Millimetre wave hybrid antennas, 1285
Minimum coverage gain, 793
Mismatch thermal, 953
Mixed potential integral equation, 400
Mobile communications base stations, 1081
Mobile communications antenna, 1083
Mobile satellite communications, 800, 1142
Mobile systems, 1079
Modal expansional method, 235
Mode ambiguous, 901
Modelling mathematical, 14
Modelling accuracy, 16
Modes

Index

azimuthal, 53
degenerate, 756
higher order, 356, 787, 796, 1058
orthogonal, 221
parallel-plate, 816
resonant, 900
suppressing pins, 1065
transverse electric field, 887
unwanted, 258
Modularity, 744
Modulus, 953
Moisture, 875, 941
Moments method, 593
Monopole probes, 1197
Monopulse arrays, 849, 1068, 1153
Monopulse comparator, 859
Moulding vacuum-bag, 926
Multi-terminal antenna, 236
Multibeam antenna, 1083, 1146
Multilayer substrates, 679, 944
Multipath fading, 1096
Multiple beam-forming networks, 817, 859
Multiple dipoles, 299
Multiple feed point patches, 262
Multiple frequency patches, 320
Multiple layer patches, 30
Multiple tuning, 341
Multiport network approach, 455
Multiresonator patch, 507
Mutual coupling, 249, 306, 337, 445, 561, 631, 702, 731, 739, 770, 782, 783, 1240
Mutual coupling in a cylindrical array, 1244
Mutual coupling network, 464, 482
Mutual impedance, 237, 291, 359, 378

Narrow pin transitions, 967
Near field mapping 989
Near-field probes, 981
Network analysis techniques, 488
Networks
  multiple-beam-forming, 817
  special-purpose, 859
Nitrogen, 930
Nodules, 879
Non radiating edge admittance network, 482
Non radiating edge characterisation, 481
Non uniform arrays, 630
Notched patch, 27, 796
Numerical analysis, 16
Numerical techniques, 417

Off centred pin transitions, 967
Offset fed patches, 221
Ohmic losses, 1016
Omnidirectional arrays, 382, 765
Open circuit end, 534
Open circuited lines, 1212
Open microstrip antennas, 580
Operational factors, 5
Optical modulator probe, 983
Optically tuned patches, 192
Orthogonal fields, 787
Orthogonal polarisation, 30
Overlaid feeds, 1263
Overlaid patches, 35, 1277

Pagers, 1092
Paired elements, 263, 270, 758, 788, 804
Parallel feeds, 335, 757, 816, 825, 828
Parallel plate resonator test method, 959
Parallel-coupled lines, 856
Parallel-plate modes, 816
Parallel-plate polariser, 766
Parallel-plate waveguide, 858
Parasitic arrays, 801
Parasitic patches, 29, 214, 264, 797, 825, 833, 1083, 1086, 1129
Parasitically coupled array, 37
Passivate, 950
Patch arrays, 822, 1263
Patches
  annular ring, 25, 111
  aperture-coupled, 330, 723
  bandwidth, 111, 554
  cavity model, 1228
  circular, 45, 63, 111, 232, 580, 720, 1058, 1202
  circular-disc, 756
  circularly polarised, 27, 499, 755, 821, 1218
  composite type, 222
  conically depressed, 29
  coplanar stripline, 31
  corner fed, 647
  cross, 501
  cylindrical, 1227
  design, 557
  diagonal slot square, 499
  diagonally fed nearly square, 499
  directivity, 554
  disc, 1258
  dual aperture-fed, 823
  dual band circularly polarised, 1061

Index

dual-fed, 220
dual-frequency, 30, 188, 197, 200, 312, 313, 796, 802
dual-polarisation, 312, 318
efficiency, 554
electromagnetically coupled, 31, 207, 797
elliptical, 25, 182, 235, 756
equitriangular, 111
gain, 554
Green's functions, 462
H shaped, 26
half-wavelength, 313
input conductance, 554
input resistance, 324
multiple feed points, 262
multiple frequency, 320
multiresonator, 507
mutual impedance, 378
notched, 27, 796
offset fed, 221
optically tuned, 192
overlaid, 1277
paired, 263, 270, 804
parasitic, 214, 797
pentagon, 756
pentagonal, 25, 505, 1218
piggy-back, 313
polarisation, 378
post-tuned, 315
probe-fed, 713, 720
quarter wavelength, 313, 1154
rectangular, 111, 224, 235, 436, 553, 580, 1215, 1234
rectangular, corner-fed, 756
rectangular ring, 26
resonant frequency, 324
short circuit, 374
short circuited, 353
short circuited ring, 346
shorted, 1281
singly fed, 221
slotted, 27
square, 25
square ring, 501
stacked, 29, 320
stacked circular-disc, 197
star, 26
stepped, 29
strip line, 111
surface-current model, 1232
thick, 253
tilted slot, 756
triangular, 25, 235, 1209
truncated corner, 27
truncated corner square, 499
two port, 511
wide, bandwidth, 320
wideband, 28, 796
Pattern, active element, 739
Pattern synthesis of cylindrical arrays, 1251
Peel strength, 937
Peel-test, 878
Pentagon patches, 25, 505, 756, 1218
Performance trade-offs, 7
Peripheral choke, 104
Permittivity
  complex, 872
  effective, 481
  relative, 871, 878, 881, 886, 895
  very-high, 1293
Perpendicular feeds, 747
Persulfate, 949
Perturbation segment, 224
Perturbation cavity, 914
Phase centre, 96, 106, 1161
Phase constant, 779
Phase shifters, 864, 1081, 1241
Phase shifters, Schiffmann, 848
Phased arrays, 378, 741, 802, 810, 862, 1241
Phased arrays
  infinite, 698
  integrated, 742
Photolithographic techniques, 1002
Photomask, 888
  adhesion, 918
Photoplots, 1175
Photoresist, 1029
Piggy-back patches, 313
Pinholes, 880
Pits, 917
Planar arrays, 345, 776, 790
Planar near field scanning, 981
Planar segments, characterisation by Z matrix, 467
Plane wave spectrum method, 285
Plastic substrates, 14, 1025
Platen-press, 929
Plating holes, 916
Point dipole approximation, 287
Poisson sum formula, 699
Polarisation, 5, 743, 783
Polarisation
  45 deg, 821
  axial, 1239

Index

circular, 5, 130, 219, 722, 744, 755, 846, 848, 1261, 1273
circumferential, 1234
cross, 68, 74, 76, 100, 104, 590, 1234
ellipse, 1187
elliptical, 186
ellipticity, 231
linear, 130
longitudinal, 822, 846
orthogonal, 30
tracking, 755
transverse, 821, 846
Poles, surface wave, 697
Polyethylene, 873
Polymer fume fever, 919
Polymer systems, thermoset, 878
Polymerisation, 881
Polymides, 951
Polypropylene substrates, 678
Polypropylene-ethylene substrates, 679
Polytetraflouroethylene, 873
Post-tuned patches, 315
Posts, 235
Potential, 411
    diffracted field, 397
    scalar, 399, 405
    vector, 50, 399, 403, 405
Power combiners, 816
Power dividers
    isolated, 852
    rat race, 1069
    split-tee, 852
    three-port, 852
    Wilkinson, 307, 852, 947, 1069
    in-line, 850
Precision, 891
Press, platen, 926
Pressure vessel, 929
Printed dipoles, 706, 732, 765
Probe coupling, 822
Probe-fed patches, 713
Probes
    coaxial, 433, 1194
    coplanar line, 968
    errors, 1200
    monopole, 1197
    near-field, 981
    optical modulator, 983
    scanning network, 1195
    short monopole, 982
    small loop, 982
    split coaxial balun, 982
    square law, 983

wafer, 968
Processing, 916
Proximity coupling, 818
PTFE, 873, 881, 919
    ceramic, 884
    glass fibre, 883
    woven glass, 884
Pulse basis functions, 282
Pyrophosphate, copper, 936

Q factor, 111, 119, 128, 174, 324, 590, 593, 796, 887, 895
Quarter wave resonance, 353
Quarter-wave patch, 313, 817
Quasi-log-periodic, 1279

Radar, 1105
Radar reflector, 1107
Radial waveguide feeds, 859
radiated, electric field, 411
Radiation
    circular polarised, 824
    conductance, 817
    cosmic, 940
    damage, 941
    dose, 941
    efficiency, 9, 126, 598, 734
    exposure, 939
    feeds, 12
    fields, 112
    high-energy, 940
    losses, 788, 791, 1016, 1075
    nuclear, 940
    patterns, 122, 335, 449
    resistance, 413
    slot, 222
    spurious, 353, 743, 816, 824
    ultra-violet, 941
    unwanted, 787
Radiometric measurement, 993
Radius, effective, 65
Radome, 14, 926, 945, 1273
Railway antennas, 1087
Rampart array, 38, 763, 816, 844, 1222
Rat race hybrid, 855, 1069
Reactance compensation, 761, 823
Real space integration method, 286
Rear feeds, 758
Reciprocity method, 278
Rectangular array lattice, 706
Rectangular patch, 7, 25, 111, 235, 436,

553, 1215, 1234
Rectangular ring patch, 26
Rectangular-slot array, 1118
Reflect array, 33
Reflection coefficient active, 708, 739
Reflection losses, 816
Reflector feeds, 96, 1258
Relaxation synthesis, 662
Residue, 278
Resin
    poly(tetraflouroethylene) (PTFE), 879
    polycyanate, 878
    polyetherimide, 879
    polyethersulfone, 879
    polyimide, 878
    polystyrene, 878
    polysulfone, 879
    triazine, 878
Resin laminates, 929
Resistance, radiation, 413
Resistive box terminations, 964
Resistive layer, 950
Resistivity, surface, 948
Resistors, 1040
Resonant arrays, 769, 816
Resonant cavity, 111
Resonant feed networks, 835, 1074
Resonant frequency, 115, 324, 593
Resonant peak, 887
Resonant ring test method, 961
Resonant-mode, 117
Resonator
    microstrip, 914
    stripline, 914
Resonator strip test method, 960
Richmonds reaction equation, 597
Rotation, sequential, 12, 263, 828, 859
Rotman lenses, 860
Routers, 922
Rutile substrates, 1025

S-matrix, 1240
Safety, 916, 919
Sapphire substrates, 1025
Satellite
    ERS-1, 1073
    ETS-V, 1112
    NAVSTAR (GPS), 1124
    antennas, 1146
    communications, 1136
    systems, 1079
Scalar potential, 399, 405

Scan angle, 785
Scan blindness, 700, 709, 719, 723, 731
Scan range, 743
Scanning a cylindrical array, 1249
Scanning arrays, 792
Scanning losses, 792
Scanning network probes, 1195
Scattering matrix, 702, 783
Scatterometer, 1073
Schartz-Christoffel transform, 1011
Schiffmann phase shifters, 848
Schotky barrier diode, 1194
Secondary surveillance radar, 1068
Sector beam array, 1081
Sector patterns, 664
Segmentation method, 456, 488
Semi infinite substrate, 731
Semiconductor substrates, 1027
Sequential arrays, 12, 36, 263, 788, 805,
    828, 859, 1142, 1265
Series arrays, 816
Series compensated feeds, 36
Series fed patch arrays, 515
Series feeds, 335, 758, 767, 816, 832, 789,
    818
Series-fed circularly polarised arrays, 770
Series-feed design procedure, 836
Serpent array, 38, 816, 833, 843
Shear, 937
Shipbourne antenna, 1136
Shock, thermal, 934
Short circuit patches, 25, 346, 353, 374,
    1087, 1281
Short monopole probe, 982
Shorting pin, 85
Sidelobe level, 5, 774
Simplex synthesis, 662
Single feeds, 219
Singly diffracted field, 1069
Skin effect, 943
Slot antenna, 1136
Slot combiner, 1084
Slotline transitions, 962
Slots
    annular, 611
    crossed, 28
    folded, 353
    linear, 606
Slotted line measurements, 975
Slotted patch, 27
Small loop probe, 982
Smear, 923, 934
Sodium-bisulphite, 950

Index

Solder reflow, 938
Solvents, 875
Sommerfeld equation, 276, 285, 407, 417
Sources, hybrid, 353
Space diversity, 939, 1084
Sparse arrays, 1273
Spatial phase delay, 805
Special-purpose networks, 859
Specimen, test, 909, 907
Spectral domain method, 265, 327, 404, 590
Spectral domain Green's function, 327
Spherical arrays, 34, 792, 793, 1134, 1239
Spherical dielectric overlays, 1267
Spherical near field scanning, 981
Spill-over efficiency, 104
Spiral, 30
Spiral slot, 33
Split coaxial balun probe, 982
Split-tee power dividers, 852
Splitters, Wilkinson, 356
Spurious radiation, 332, 353, 743, 816, 824
Square law probe, 983
Square patch, 25
Square patch array, 1073, 1115
Square ring patch, 501
Square-loop-type arrays, 764
Squintless arrays, 758, 826, 832, 841
Stacked antenna, 312
Stacked circular-disc patches, 197
Stacked dipoles, 299
Stacked patches, 29, 320
Stainless-steel, 946
Standing wave distribution, 887, 1199
Star patch, 26
Steepest descent method, 278, 408
Stepped patch, 29
Strain, internal, 920
Strain relief, 920
Stress riser, 934
Strip combiner, 1084
Strip conductors, 275
Strip dipole, 761, 1132
Strip line patches, 111
Strip slot, 1132
Stripline, balanced, 1003
Stripline
    suspended, 816, 823, 857
    thickness, 1011
    triplate, 816, 823
    feeds, 353
    transitions, 962
    resonator, 884
Structures
    array, 36
    feed, 35
Sub-arrays
    2★2, 759
    four element, 805
    two element, 804
Subdomain basis functions, 282, 437
Subsectional basis functions, 423, 440
Substrates, 15
    alumina, 14, 1025
    bending, 1177
    beryllia, 1025
    copolymer, 679
    effects, 279
    environmental effects, 679
    ferrimagnetic, 1027
    foam, 1288
    GaAs, 332
    honeycomb, 796
    low cost, low loss, 674
    materials, 871
    measurements, 958
    metal deposition, 1028
    metallisation, 1027
    multilayer, 679
    non-woven glass-PTFE, 892
    perpendicular, 337
    plastic, 14, 1025
    polypropylene, 678
    polypropylene-ethylene, 679
    rutile, 1025
    sapphire, 1025
    semi infinite, 731
    semiconductor, 1027
    technology, 14
    thick, 28, 113, 275, 356
    thickness, 74, 791
    thick metal backed, 678
    thin, 112
Sum radiation pattern, 1068
Summary, of chapters, 21
Summary, of topic areas, 18
Superstrate, 275, 597
    GaAs, 281
    teflon, 281
Superstrate effects, 281
Surface
    charge, 1196
    current density, 1196
    currents, 441
    fields, 67
    gradient, 52

Index

resistivity, 117
tangent, 52
wave, 406, 407, 410
Surface analytical techniques, 1194
Surface current model, 1228
Surface currents, equivalent, 49
Surface field metrology, 1193
Surface impedance, 400
Surface resistivity, 815
Surface treatment, 879, 929
Surface wave conductance, 476
Surface wave losses, 816
Surface wave poles, 593
Surface waves, 9, 12, 592, 600, 703, 728, 731, 734, 740, 751
Surface waves
   circle diagram, 706
   excitation, 116
   loss, 1075
   poles, 697
   excitation, 127
Surveillance radars, 1079
Susceptance, edge, 479
Suspended stripline, 816, 823, 857
Suspension, 881
Switched-element spherical array, 1132
Synthesis, 361
Synthesis, lines, 1015
Synthesis methods, 662
Synthetic aperture radar antenna, 790, 1146

T junction, 977, 850, 1166
Tab patches, 30
Tapered absorbing film terminations, 964
Tapered arrays, 650
Tapered-slot elements, 751
Taylor distribution, 830
Teflon superstrate, 281
Telemetry, 924
Temperature, 875, 941
   transition, 921
   variation of K' with, 942
Terminations
   matched, 964
   resistive box, 964
   tapered absorbing film, 964
   thin film pad, 964
Test method
   evaluation, 914
   fluid-displacement, 883
   full-sheet-resonance, 897

mated connector, 963
microstrip-resonator, 893
parallel plate resonator, 959
resonant ring, 961
resonant-cavity perturbation, 906
resonator strip, 960
stripline-resonator, 882
stripline resonator test, 886
Testing functions, 53, 427
Thermal paint, 1065
Thermal-expansion coefficient, 878, 936
Thermoplastic, 927
Thermoset, 927, 941
Thick film, 1031
Thick patches, 253
Thick substrate, 28, 275
Thin film pad terminations, 964
Three-faced array, 1111
Three-port power dividers, 852
Throwing power, 936
Tilt angle, 1188
Time-domain-reflectometry, 979
TM210-mode microstrip antenna, 1129
Total mass loss (TML), 940
Toxity, 919
Tracking slope, 1188
Train antenna, 1087
Transistors, 1041
Transition, 942
Transition, crystalline, 881
Transition phase, 875
Transitions, 1288
   compensating hole, 967
   GaAs, 967
   inserted connector, 967
   K connector, 967
   metallic ring, 967
   narrow pin, 967
   off centred pin, 967
   slotline, 962
   stripline, 962
   thermal, 881
   waveguide, 962
Transmission line analysis, 295, 527
Transmission line model, 112, 317, 456, 769
Transmission lines, 1001
Transmission-line matrix method, 1023
Transportable earth station, 1125
Transposed arrays, 821
Transverse polarisation, 821, 846
Travelling wave arrays, 13, 762, 765, 769, 816, 1222

Index

Travelling wave feeds, 832
Travelling-wave array design, 777
Triangle line array, 38
Triangular array lattice, 706, 1243
Triangular functions, 54
Triangular patch, 25, 235, 1209
Triazene, 951
Triplate, 857
Triplate, balanced, 1003
Triplate CAD, 1001
Truncated corner patch, 27
Truncated corner square patch, 499
Tuned circuit, 9
Two element sub-arrays, 804
Two port patch, 511
Two sided arrays, 746
Two-dimensional feeds, 839
Two-line feeds, 276
Two-sided configuration, 332

UHF pagers, 1079
Uniform lines, 1006
Untransposed arrays, 821
Unwanted modes, 258
Unwanted radiation, 787
Unwanted waves, 762
Urban mobile communications, 1086

Vacuum outgassing, 939
Vacuum-bag, 929
Variational method, 235
Vector potential, 50, 399
Vehicle antennas, 1086
Very-high permittivity, 1293
Vias, 945
    plated-through hole, 953
    reliability of, 953
Visco-elastic, 873, 920, 924
Voids, 935
Voltage maxima, 887
Voltage vector, 716
VSWR bandwidth, 118, 823, 829, 836

Wafer probing, 968
Walls
    electric, 113
    magnetic, 113
Waveguide, 906
    air-filled, 858
    dielectric, 944
    model, 817, 1023
    parallel-plate, 858, 897
    simulator, 711, 719
    transitions, 962
Waves
    cylindrical, 1072
    surface, 406, 407, 410, 592, 600, 703, 731, 734, 740, 751
    unwanted, 762
Weathering, 941
Welding, 937
    electron-beam, 938
    laser, 938
    parallel gap, 938
    percussive arc, 938
    resistance, 938
    ultrasonic, 938
Wettability, 935
Wheeled vehicles, 1085
Wicking, 875, 941
Wide bandwidth arrays, 372
Wide bandwidth patches, 28, 320, 796
Wide-bandwidth, 324, 1273
Wideband arrays, 804
Wideband baluns, 968
Wideband techniques, 253
Wiener-Hopf method, 478
Wilkinson power dividers, 307, 356, 852, 1069
Window antennas, 1283
Wire grid arrays, 848
Wrap around antenna, 34, 85

X-ray, 933

Z matrix
    arbitrary segments, 472
    circular segments, 471
    from Greens functions, 468
    planer segments, 467
    rectangular segments, 469